H. Baumgärtel · M. Wollenberg

**Mathematical Scattering Theory**

Mathematische Lehrbücher und Monographien
Herausgegeben von der Akademie der Wissenschaften der DDR
Institut für Mathematik

II. Abteilung
Mathematische Monographien
Band 59

Mathematical Scattering Theory

von H. Baumgärtel und M. Wollenberg

# Mathematical Scattering Theory

by Hellmut Baumgärtel and Manfred Wollenberg

Akademie-Verlag Berlin
1983

Verfasser:

Prof. Dr. habil. Hellmut Baumgärtel

Dr. sc. Manfred Wollenberg

Institut für Mathematik
der Akademie der Wissenschaften der DDR

Erschienen im Akademie-Verlag, DDR-1086 Berlin
Leipziger Str. 3—4
Lektor: Dipl.-Math. Gesine Reiher

Lizenznummer: 202 · 100/503/82
Einband und Schutzumschlag: Dietmar Kunz
Gesamtherstellung: VEB Druckhaus „Maxim Gorki“, 7400 Altenburg
Bestellnummer: 763 024 5 (6660) · LSV 1065
Printed in GDR
DDR 80,— M
ISSN 0076-5430

# Preface

The aim of this book is to give a systematic and self-contained presentation of the Mathematical Scattering Theory within the framework of operator theory in Hilbert space. The term Mathematical Scattering Theory denotes that theory which is on the one hand the common mathematical foundation of several physical scattering theories (scattering of quantum objects, of classical waves and particles) and on the other hand a branch of operator theory devoted to the study of the behavior of the continuous part of perturbed operators (some authors also use the term Abstract Scattering Theory).

Essential contributions to the development of this theory are due to K. Friedrichs, J. Cook, T. Kato, J. M. Jauch, S. T. Kuroda, M. Š. Birman, M. G. Kreĭn, L. D. Faddeev, R. Lavine, W. O. Amrein, B. Simon, D. Pearson, V. Enss, and others. It seems to the authors that the theory has now reached a sufficiently developed state that a self-contained presentation of the topic is justified.

The authors themselves in the seventies contributed to the theory on the one hand by solving open problems and on the other hand by giving new aspects for the structure of the topic. Therefore some parts of the book may be considered as a synoptical presentation of results of the author's research within this area in the last few years. It contains numerous results either not published or not to be found in any other book on scattering theory. Additional important features of this book are:

(I) The book uses the two-space formulation, several limit processes on the real line (mainly the so-called Absolute Abelian Limit) and the theory of asymptotic constants. It is devoted to the operator-theoretical aspect of the theory.

(II) Applications of the general results to special scattering systems, for instance derivation of the essential results of potential scattering, are treated mainly for the purpose of illustration (examples and counterexamples). Therefore in the Examples and Applications we prefer simplicity to generality,

(III) The connection with the physical source of the theory is not a direct topic of the book.

These features distinguish this book from others on scattering theory whose main emphasis is the concrete quantum mechanical (also classical) scattering, e.g. the investigation of Schrödinger operators (see Amrein/Jauch/Sinha [1], Reed/Simon [3]).

It is hoped that this book will be useful to advanced students in mathematics, mathematicians interested in this field of operator theory, and theoretical physicists paying attention to mathematically rigorous methods in scattering theory.

The mathematical foundations of the theory belong to Functional Analysis, in particular to the Theory of Operators in Hilbert Space. The reader is assumed to have a basic knowledge of these topics. As general references to them we shall use Kato [1], Reed/Simon [1, 2], Weidmann [1], and Yosida [1].

To establish a common language, the essential notions and facts of operator theory and measure theory (needed in the text) are described in Part I. There are five parts each prefaced by a summary. The chapters are divided into paragraphs, and paragraphs into sections. Theorems, propositions, corollaries, lemmas, examples, and remarks are numbered consecutively within each chapter, as are formulas. Reference to a theorem, lemma, etc. from another chapter is done by adding the number of the chapter to the number of the theorem, lemma, etc. For example Theorem 10.2 means Theorem 2 of Chapter 10 and formula (5.1) (or shortly (5.1)) means formula (1) of Chapter 5. The symbol ■ will denote the end of a proof.

At the end of each part there is an appendix entitled Notes and Remarks. It contains additional material, historical remarks, and references to the literature of this field. So far as it was possible we tried to include all essential papers on scattering theory on the abstract level of operator theory. Papers with more concrete methods as the theory of partial differential equations are included in the Notes and Remarks if they are concerned with the applications of the abstract theorems given in our examples.

There are a bibliography, a subject index, and a list of symbols.

It is a pleasure to thank here the people who in one way or another have contributed to the preparation of this book.

In particular we express our thanks to Professors M. Š. Birman and M. G. Kreĭn, and also to Professor B. Simon, whose suggestions greatly stimulated our work. We are especially grateful to Dr. W. O. Amrein and to Professors C. Chandler and A. G. Gibson who were so kind to go through some parts of the manuscript, correcting innumerable errors, mathematical as well as linguistic. We are also grateful to Professor R. S. Phillips who red Chapter 12. We are particularly indebted to our colleague Dr. H. Neidhardt for going through the whole manuscript and correcting many mathematical errors.

We wish to thank the Akademie-Verlag, especially R. Helle and G. Reiher for their careful work in preparing the print.

Berlin, Octobre 1982 The authors

# Contents

# List of Symbols

## General symbols

| | |
|---|---|
| a.e. | almost everywhere |
| arg | argument of a complex number |
| $\mathbb{C}$ | the complex numbers |
| clo | closure of sets (operators) |
| dis $(\cdot, \cdot)$ | distance |
| dom | domain |
| ess sup | essential supremum |
| $f(A)$ | image of $A$ |
| $f^{-1}(A)$ | inverse image of $A$ |
| iff | if and only if |
| Im | imaginary part |
| ima | image (range) |
| inf | infimum |
| ker | kernel (null space) |
| lim inf | limes inferior |
| lim sup | limes superior |
| $\mathbb{N}$ | the natural numbers 1, 2, ... |
| $\mathbb{R}$ | the real numbers |
| $\mathbb{R}_+$ | $[0, \infty)$ |
| $\mathbb{R}_-$ | $(-\infty, 0]$ |
| Re | real part |
| spa | linear span |
| sup | supremum |
| $\in$ | belongs to |
| $\notin$ | does not belong to |
| $\subset$ | proper inclusion |
| $\subseteq$ | inclusion |
| $\cup, \bigcup_\alpha$ | union |
| $\cap, \bigcap_\alpha$ | intersection |
| $\setminus$ | set difference |
| $\emptyset$ | empty set |
| $\equiv$ | identical, equivalent |
| $\perp$ | orthogonal, disjoint |
| $\cong$ | isomorphism |
| $\lvert\cdot\rvert$ | absolute value, Lebesgue measure, norm |
| $'$ | derivative |
| $\overline{\phantom{x}}$ | complex conjugate |
| $:=$ | the left side is defined by the right |
| $=:$ | the right side is defined by the left |
| $\times, \mathop{\times}_\alpha$ | Cartesian product |
| $\oplus, \bigoplus_\alpha$ | direct sum |
| $\ominus$ | direct difference |
| $\otimes, \bigotimes_\alpha$ | tensor product |
| $\delta_{jk}$ | Kronecker symbol |

## Symbols defined in the text

# Introduction

1. Mathematical Scattering Theory today denotes a branch of operator theory devoted to the study of the existence properties and the calculation of several kinds of limits of operators of the form

$$e^{itH} J e^{-itH_0}, \qquad t \in \mathbb{R}. \tag{1}$$

Here $H$, $H_0$ denote selfadjoint operators on Hilbert spaces $\mathscr{H}$, $\mathscr{H}_0$, and $J$ is a bounded operator from $\mathscr{H}_0$ into $\mathscr{H}$. These limits are considered on certain subspaces $\mathscr{U}_0$ of $\mathscr{H}_0$.

Historically physical theories have often been strongly motivated by the process of idealization of real phenomena, in order to put important features of nature in a very clear light. It frequently happens that such idealizations can be represented mathematically by some sort of limit processes, where the transition to the limit is the crucial step of the idealization. Actually this method is a primary source of mathematical physics. For example such an idealization makes it possible to speak mathematically of velocity (differentiation process of Newton). Other limit processes of this kind are encountered in statistical physics (limit of infinite degrees of freedom, of an infinite number of particles, of an infinite volume). With respect to scattering theories the crucial limit is the time-limit $t \to \pm\infty$, which is mathematically the center of every sort of scattering theory.

The simplest motivation for the importance of the special expression (1) can be given by the scattering problem of quantum mechanics in which a noninteracting (unperturbed) system $\{H_0, \mathscr{H}_0\}$ is perturbed by an interaction producing the new system $\{H, \mathscr{H}\}$. Further, in general, some rule is given which provides a connection between unperturbed and perturbed states at fixed time, say $t = 0$. Mathematically this rule is represented by an (identification) operator $J$ acting from $\mathscr{H}_0$ into $\mathscr{H}$ which is for simplicity assumed to be bounded; the simplest case is $\mathscr{H} = \mathscr{H}_0$ and $J = 1$. Then the quantum mechanical evolutions in $\mathscr{H}_0$ and $\mathscr{H}$ determine the identification operator $J(t)$ for an arbitrary time $t$,

$$J(t) = e^{itH} J e^{-itH_0}. \tag{2}$$

Now the above mentioned idealization leads one to investigate the time limits

$$W_\pm u := \operatorname*{Lim}_{t \to \pm\infty} e^{itH} J e^{-itH_0} u, \qquad u \in \mathscr{U}_0, \tag{3}$$

on certain subspaces $\mathscr{U}_0$. These operators $W_\pm$ are called *wave operators*. Further the so-called *scattering operator*

$$S := W_+^* W_- \tag{4}$$

is of special interest. If the wave operators exist on an a priori given subspace (in many cases the subspace of absolute continuity of $H_0$), the system $\{H, H_0, J\}$ is called a *scattering system*.

2. Historically Mathematical Scattering Theory has two sources, one of them is purely mathematical, the other is more physically oriented.

2.1. The mathematical source is the spectral theory of perturbed selfadjoint operators. This theory has been largely developed, in particular after the appearance of the famous spectral theorem of selfadjoint operators in its abstract formulation, the creation of the theory of unitary invariants of selfadjoint operators, and the classification of several unitary invariant parts of the spectrum.

At first a rather complete spectral theory of the perturbation of isolated eigenvalues (of finite multiplicity) was established (see Rellich [1]). Subsequently the behavior of the continuous spectrum upon perturbation was systematically studied. Pioneering work in this direction is due to Friedrichs [1938a], [1948a], see also Putnam [1954a] and Aronszjan [1957a].

In this connection it should be noted that the development of an abstract operator theoretical spectral theory (including perturbation) was influenced in an essential way by the construction and establishing of the abstract quantum mechanical formalism (see J. v. Neumann [1]).

It is a remarkable fact that already in the spectral theory of selfadjoint operators $H$, without any contact to physical questions, one can easily verify strongly different behavior of vectors belonging to different spectral parts of $H$ with respect to the "time" limit $t \to \pm\infty$, where $t$ is the parameter of the unitary group $e^{itH}$ generated by $H$. If $u$ is an eigenvector of $H$, then

$$|(u, e^{itH}u)| = \|u\|^2 \tag{5}$$

is independent of $t$, and hence $|(u, e^{itH}u)| \to 0$ as $t \to \pm\infty$ cannot be true. If, on the other hand, $u$ is from the absolutely continuous subspace of $H$, then

$$|(u, e^{itH}u)| \to 0, \qquad t \to \pm\infty. \tag{6}$$

Also the mean ergodic theorem

$$E_H(\{\lambda\}) = \underset{T\to\infty}{\text{s-lim}}\, T^{-1} \int_0^T dt\, e^{it(H-\lambda)}, \tag{7}$$

where $E_H(\Delta)$ denotes the spectral measure of $H$, expresses some internal connection between spectral theory and "time-like" limits. Therefore, in retrospect, it is not surprising that the methods used by Friedrichs for the investigations in his paper [1938a] are based on the construction of wave operators. Thus a definition of wave operators from purely spectral theoretic sources could have been given at this stage of development. But this did not happen.

2.2. In some contrast to pure spectral theoretical investigations the notions of wave and scattering operator were introduced by physicists (Møller [1945a], Heisenberg [1943a]). This is the physical source of the theory, the importance of the idealization $t \to \pm\infty$ in scattering theories mentioned above. The development which arose from these notions was purely heuristic for a long time, and in particular was not strongly connected with the mathematical foundations of quantum mechanics given by J. v. NEUMANN. For example there were no proofs of existence of the proposed limits of MØLLER.

The first step in the direction of a mathematically rigorous theory of the ideas of MØLLER and HEISENBERG was made by COOK [1957a] who gave the first exact proof of the existence of wave operators for a physically interesting example (two-body scattering with a potential $V \in L^2(\mathbb{R}^3, dx)$). After that JAUCH [1958a] formulated and developed the quantum mechanical scattering theory in a mathematically precise language on the level of abstract operator theory. The wave operators $W_\pm(H, H_0)$ (in the simple case of single channel scattering) were defined (with a slight modification) by

$$W_\pm(H, H_0) := \operatorname*{s-lim}_{t\to\pm\infty} e^{itH} e^{-itH_0} P^{ac}_{H_0}. \tag{8}$$

where $P^{ac}_{H_0}$ denotes the projection onto the absolutely continuous subspace of $H_0$ and the *S-operator* (or scattering operator) was defined by $S := W_+{}^*W_-$. The important notion of asymptotic completeness, $W_+W_+{}^* = W_-W_-{}^* = P^{ac}_H$, which implies the unitarity of the $S$-operator in $P^{ac}_{H_0}\mathscr{H}_0$ was introduced at that time. In this connection see also the important work of KURODA [1959a].

Since that time the problems of existence and (asymptotic) completeness of the wave operators have turned out to be the main topic of the mathematical scattering theory.

Also at that time the more mathematically oriented branch of the theory was greatly enriched by the famous result of ROSENBLUM [1957a] and KATO [1957a], [1957b] which says that the absolutely continuous parts of selfadjoint operators $H, H_0$ where $H - H_0$ is a trace operator are unitarily equivalent. The methods of the proof of this result, already influenced by the physical notions, were based on the existence of the corresponding wave operators.

Independently of this development towards a rigorous theory of wave and scattering operators there was a successful attack upon the inverse scattering problem for spherically symmetrical potentials in the fifties (for details see AGRANOVIČ/MARČENKO [1], CHADAN/SABATIER [1]). This problem was solved with the help of more concrete methods (solutions of certain integral equations, use of eigenfunctions for special Schrödinger operators).

2.3. At the end of the fifties the two directions, the mathematical (oriented towards spectral theory) and the physical one (oriented towards scattering theory), came together to form a unified topic of mathematical physics that has been a strong source for further development. This area is now called Mathematical Scattering Theory (on the operator theoretical level) and has the two aspects: (I) Spectral theory of perturbed selfadjoint operators; (II) Scattering theory of physical models (exact theory of wave operators, scattering operators, scattering cross sections, and other important physical notions). Thus, scattering theory arises in purely mathematical monographs devoted to spectral theory (of abstract operators or of perturbed partial differential operators), where it is used as a technical tool (see for example KATO [1], WEIDMANN [1]), and it is the object of works by mathematically oriented physicists (as an example see AMREIN/JAUCH/SINHA [1]).

3. Now we sketch the most important points in the further development of the theory. They include new methods for studying scattering systems, generalizations of the

basic notions of scattering theory, applications to concrete systems, and the solutions of certain central problems.

(I) The trace class theorem of ROSENBLUM and KATO was extended to the case where the difference of the resolvents of the unperturbed and perturbed operator is a trace operator (see e.g. BIRMAN/KREĬN [1962a]). These results were further developed and reached a culmination in the theorem that the wave operators

$$W_\pm(H, H_0; J) := \underset{t\to\pm\infty}{\text{s-lim}}\, e^{itH} J\, e^{-itH_0} P^{ac}_{H_0} \tag{9}$$

exist if $HJ - JH_0$ is a trace operator (see PEARSON [1978a]).

(II) Also under trace class conditions on $H$ and $H_0$ ($\mathscr{H} = \mathscr{H}_0$, $J = 1$) BIRMAN [1962b] showed that the wave operators $W_\pm(\Phi(H), \Phi(H_0))$ (see (8)) exist and are independent of $\Phi$ for a broad class of functions $\Phi$. This remarkable feature of the wave operators was called the invariance principle of the wave operators and was extended to a wider class of functions by KATO [1965a]. The conjecture that the existence of $W_\pm(H, H_0)$ always implies the existence of $W_\pm(\Phi(H), \Phi(H_0))$ (in sense of (8)) was shown to be false by WOLLENBERG [1975a]. However in the same paper it was proved that the existence of both $W_\pm(H, H_0)$ and $W_\pm(\Phi(H), \Phi(H_0))$ implies their equality.

(III) Scattering theory for many particle systems was already developed by JAUCH [1958b]. A precise existence proof of the corresponding wave operators was made by HACK [1959a] under the assumption that the potentials $q_{ij}(x_i - x_j)$ are square integrable in $x_i - x_j$. The first complete analysis of scattering for a many particle system was given by FADDEEV [1963a]. Following this work, scattering theory for many particle systems has been developed by many other authors (see e.g. GINIBRE/MOULIN [1974a], THOMAS [1975a], and SIGAL [1978a]).

(IV) DOLLARD [1964a] showed that the wave operators $W_\pm(H, H_0)$ do not exist in the physical important case of the scattering by a Coulomb potential; here $H_0 = -\Delta$, $H = -\Delta + \dfrac{c}{|x|}$, $\mathscr{H}_0 = L^2(\mathbb{R}^3, dx)$. However he was able to prove that generalized wave operators

$$W_{\pm,x}(H, H_0) := \underset{t\to\pm\infty}{\text{s-lim}}\, e^{itH}\, e^{-itH_0 - iX_t} \tag{10}$$

with $X_t = cH_0^{-1/2} \log |t|$ exist and have the same properties as the usual wave operators defined by (8). His results were extended to more general long range potentials (see e.g. HÖRMANDER [1976a]). Another way to overcome the difficulties of the scattering theory for long range potentials was proposed by LAVINE [1970a], AMREIN/MARTIN/MISRA [1970a], and others. The essential idea is to assume that for all elements of a certain $C^*$-algebra $\mathscr{A}$ of operators which commute with the unperturbed operator $H_0$ the limits

$$A_\pm := \underset{t\to\pm\infty}{\text{s-lim}}\, e^{itH} A e^{-itH} P^{ac}_H \tag{11}$$

exist and that the operators $A_\pm$ give the necessary information about the scattering process. AMREIN/MARTIN/MISRA [1970a] showed that this also implies the existence of generalized wave operators of the form (10) if $\mathscr{A}$ is large enough.

(V) There was further important progress in the so-called stationary theory of scattering in which the wave operators $W_\pm(H, H_0)$ are constructed via limits of operator spectral integrals

$$\int E_H(d\lambda)\left(1-(H-H_0)\,R_{H_0}(\lambda\pm i\varepsilon)\right) \tag{12}$$

or

$$\int \left(1-(H-H_0)\,R_H(\lambda\pm i\varepsilon)\right)^{-1} E_{H_0}(d\lambda). \tag{13}$$

These spectral integrals were interpreted directly (see Baumgärtel [1967a], Amrein/Georgescu/Jauch [1971a]) or with the help of auxiliary spaces $\mathscr{X} \subseteq \mathscr{H}_0$ (see especially Kato/Kuroda [1970a]). The abstract stationary theory on the basis of (12) was reformulated, refined and essentially finished by Baumgärtel/Obermann/Wollenberg [1979a].

(VI) The theory of smooth operators was created by Kato [1966a], [1968a]. It was further developed and applied to some scattering problems for Schrödinger operators by Lavine (see e.g. [1971a], [1972a]).

(VII) The development of the trace class, smoothness, and stationary methods made it possible to improve greatly the results on potential scattering and to obtain results on the spectrum of more general differential operators. For instance, it has been shown that the wave operators $W_\pm(H, H_0)$ (see (8)) for $H_0 = -\Delta$ and $H = -\Delta + q(x)$ on $\mathscr{H}_0 = L^2(\mathbb{R}^3, dx)$ exist and are complete if $q(x)$ is real and $|q(x)| \leqq c(1 + |x|)^{-\gamma}$, $\gamma > 1$ (see Kato [1970a]). This is in a certain sense optimal because of the result of Dollard for the Coulomb potential $q(x) = c|x|^{-1}$ (see (IV)). For more general differential operators see e.g. Birman [1969a], Schechter [1976a].

(VIII) The mathematical two-space scattering theory was initiated by Kato [1967a]. This was a very important step because it allowed a more natural formulation of certain classical scattering problems, for example scattering by electromagnetic and acoustic waves (see e.g. Wilcox [1966a] and Reed/Simon [1977a]). From our present point of view the theory is also motivated by the following additional arguments.

(a) Very simple formulations of quantum scattering problems are also possible in this language (e.g. multi-channel scattering, see Chandler/Gibson [1973a]).

(b) Technically it is often easier to prove the existence of wave operators with the help of some auxiliary operator $J$ which is asymptotically equivalent to the identity. For example, this is the essential idea of the proof of the Kupsch/Sandhas result [1966a] that local singularities are unessential for the existence of wave operators in potential scattering.

(c) In relativistic quantum scattering theories the interacting and the unperturbed states are best considered as elements of completely distinct spaces. Then the identification operator is naturally defined by equation (2) (see e.g. Coester [1965a]).

(d) The theory can also be applied to problems in quantum field theory, in particular the Haag-Ruelle scattering formalism can be usefully reformulated in terms of the two-space scattering theory (see Reed/Simon [3]).

(IX) A different approach to scattering theory was developed by Lax/Phillips [1]. Here the main objects of study are the so-called outgoing and incoming subspaces of the Hilbert space of the perturbed system. This theory applies to classical wave equations, for instance to scattering of acoustical waves by obstacles. The Lax/Phillips theory can be embedded into the two-space scattering theory.

(X) In connection with the abstract characterization of wave operators (see KREĬN [1964a]), BAUMGÄRTEL [1973a] undertook the first systematic investigation of the algebra of asymptotic constants, that is those operators $A$ for which the limits (11) exist. The theory of asymptotic constants was greatly refined, developed, and applied to some scattering problems by WOLLENBERG [1979a], [1979b].

(XI) New kinds of limit processes such as

$$\lim_{T\to\infty} T^{-1} \int_0^T dt\, \|e^{itH} J e^{-itH_0}u - W_+u\|^2 = 0 \tag{14}$$

instead of (3) were introduced in scattering theory by OBERMANN [1977a]. One can interpret this limit as an average-time-limit. These limits are related in concept to the well-known Cesaro limit

$$\lim_{T\to\infty} \|T^{-1} \int_0^T (e^{itH} J e^{-itH_0}u - W_+u)\, dt\| = 0$$

or Abelian limit

$$\lim_{\varepsilon\to+0} \|\varepsilon \int_0^\infty e^{-\varepsilon t}(e^{itH} J e^{-itH_0}u - W_+u)\, dt\| = 0$$

often used in physical literature (see e.g. GELL-MANN/GOLDBERGER [1953a]) but do not have its disadvantages. They are natural extensions of the strong limit. With the help of these limits the theory allows a much smoother formulation and leads to nice new results. For example, it was possible to solve the problem of the general validity of the invariance principle for wave operators affirmatively (see OBERMANN/WOLLENBERG [1978b]).

(XII) Last but not least there was a renaissance of Cook's method. ENSS [1978a] introduced a new approach to the completeness and spectral problems for Schrödinger operators. Roughly speaking, it proves the existence and completeness of the wave operators $W_\pm(H, H_0)$ under the assumption

$$\left| \int_0^{\pm\infty} dt\, \|(H - H_0)\, \varphi(H_0)\, e^{-itH_0}P_\pm\| \right| < \infty$$

where $\varphi$ are certain smooth functions, $P_+ + P_- = P^{ac}_{H_0}$, and $P_\pm$ satisfy some asymptotic relations. This approach was also applied to scattering problems for more general differential operators (see especially SIMON [1979a]).

# Part I.
# Selfadjoint Operators in Hilbert Spaces

The main objects of this book are selfadjoint operators in Hilbert spaces and related concepts. Also some special Banach spaces are of interest, for example Banach spaces consisting of operators in Hilbert spaces. Further the analysis of vector- and operator-valued functions (integration, differentiation etc.) defined on the real line or on the complex plane plays an essential role in this book. Therefore the reader is assumed to be familiar with Hilbert and Banach spaces (including the elementary theory of operators in such spaces) and with the analysis of vector-valued functions (including the fundamental concepts of measure theory, in particular on the $\mathbb{R}^n$). General references will be given in the Notes and Remarks to this part.

The purpose of this part is — for the convenience of the reader — to collect basic results concerning our central objects and results frequently used in what follows, to fix our notation and to call special attention to facts of particular importance for the topics to be presented later. No proofs are given if they can be found in the mentioned references. For convenience of the reader proofs will be given, in general, if the stated results are crucial but the proofs are available only in original papers.

# 1. Preliminaries

## 1.1. Fundamental concepts

### 1.1.1. Banach spaces and Hilbert spaces

*Complex linear manifolds* are denoted by $\mathscr{L}, \mathscr{M}, \mathscr{D}, \ldots$, and their elements, called *vectors*, by $f, g, \ldots, u, v, \ldots$ If $M$ is a set of vectors, the (linear) *submanifold generated* by $M$ is denoted by spa $M$. A *sesquilinear form* $s(f, g)$ on $\mathscr{L}$ is a complex-valued function which is linear in the second argument and anti-linear (conjugate linear) in the first. The corresponding *quadratic form* is denoted by $s(f)$. The relation

$$s(f, g) = (1/4) \big(s(f + g) - s(f - g) + is(f - ig) - is(f + ig)\big) \tag{1}$$

is called the *polarization identity*. A sesquilinear form $s$ is called *symmetric* if $\overline{s(f, g)} = s(g, f)$. A (symmetric) sesquilinear form is called *nonnegative* if $s(f) \geqq 0$ for all $f \in \mathscr{L}$ and *positive* if $s(f) = 0$ implies $f = 0$. A nonnegative sesquilinear form is called a *semiscalar product*, a positive sesquilinear form is called a *scalar product*. Recall Schwarz' inequality

$$|s(f, g)|^2 \leqq s(f)\, s(g), \qquad f, g \in \mathscr{L}, \tag{2}$$

for semiscalar products. Usually we denote scalar products and semiscalar products by $(\cdot, \cdot)$, sometimes with an index.

A *seminorm* $p(f)$ on $\mathscr{L}$ is a real-valued function which is nonnegative and satisfies $p(\alpha f) = |\alpha|\, p(f)$, $\alpha \in \mathbb{C}$, $f \in \mathscr{L}$, and the *triangle inequality*

$$p(f + g) \leqq p(f) + p(g). \tag{3}$$

A seminorm $p(\cdot)$ is called a *norm* if $p(f) = 0$ implies $f = 0$. Usually norms and seminorms are denoted by $\|\cdot\|$, sometimes with an index, but also $|||\cdot|||$ and $|\cdot|$ are used in this book. Note that $\|f\| := (f, f)^{1/2}$ defines a seminorm or a norm if $(\cdot, \cdot)$ denotes a semiscalar product or a scalar product, respectively.

Let $(\cdot, \cdot)$ be a semiscalar product. Then the set of all $f$ with $(f, f) = 0$ defines a submanifold $\mathscr{M}$ of $\mathscr{L}$, the so-called *kernel* of $(\cdot, \cdot)$, denoted by $\mathscr{M} = \ker (\cdot, \cdot)$ or $\mathscr{M} = \ker \|\cdot\|$. Recall that on the *factor space* $\mathscr{L}$ mod $\mathscr{M}$ (quotient space $\mathscr{L}/\mathscr{M}$) a scalar product can be defined uniquely by $(\tilde{f}, \tilde{g})_\sim := (f, g)$ where the equivalence class of $f \in \mathscr{L}$ is denoted by $\tilde{f} \in \mathscr{L}$ mod $\mathscr{M}$. In the following other notations for equivalence classes are also used; sometimes, if no confusion is possible, we omit a special indication for the classes.

A complex linear space $\mathscr{L}$ equipped with a scalar product is called a *pre-Hilbert space*. $\mathscr{L}$ equipped with a norm is called a *normed space*. In a normed space the norm defines a (locally convex) topology, the so-called *strong topology*. A sequence $f_n \in \mathscr{L}$ is called *strongly convergent* with *strong limit* $f \in \mathscr{L}$ if $\lim_{n\to\infty} \|f_n - f\| = 0$. The strong limit $f$ is denoted by $f = \text{s-}\lim_{n\to\infty} f_n$; sometimes we write $f_n \underset{s}{\to} f$. Recall the notion of a *Cauchy sequence* with respect to $\|\cdot\|$. A normed space is called *complete* if each Cauchy sequence is (strongly) convergent.

A complete normed space is called a *Banach space*. Banach spaces are denoted by $\mathscr{X}, \mathscr{Y}, \ldots$ A complete pre-Hilbert space is called a *Hilbert space*. Hilbert spaces are denoted by $\mathscr{H}, \mathscr{K}, \ldots$ Recall the canonical *completion* of a normed space (e.g. pre-Hilbert space) by Cauchy sequences.

A closed (linear) submanifold of a Banach space $\mathscr{X}$ is called a *subspace*. If $\mathscr{L}$ is a submanifold of $\mathscr{X}$, then the *closure* of $\mathscr{L}$ is denoted by clo $\mathscr{L}$ or $\text{clo}_{\|\cdot\|}\, \mathscr{L}$; clo $\mathscr{L}$ is called the subspace *generated by* $\mathscr{L}$. If $M$ is a set of vectors, then clo spa $M$ is called the subspace *generated by* $M$. In a Hilbert space $\mathscr{H}$ the *orthogonal complement* of a subspace $\mathscr{U}$ is denoted by $\mathscr{U}^\perp$.

The Banach space consisting of all *continuous linear functionals* $l$ of $\mathscr{X}$ is denoted by $\mathscr{X}^*$. Recall $\mathscr{H}^* = \mathscr{H}$ for Hilbert spaces (canonical identification by $l(f) = (l, f)$).

The system $p_f(u) = |(f, u)|$ of seminorms on $\mathscr{H}$ defines a (locally convex) topology, the so-called *weak topology*. A sequence $f_n \in \mathscr{H}$ is called *weakly convergent* with *weak limit* $f$ if $\lim_{n\to\infty} (h, f_n) = (h, f)$ for all $h \in \mathscr{H}$. The weak limit $f$ is denoted by $f = \text{w-}\lim_{n\to\infty} f_n$; sometimes we write $f_n \xrightarrow[\text{w}]{} f$.

A set $M$ of vectors from $\mathscr{H}$ is called *dense* (in $\mathscr{H}$) if clo $M = \mathscr{H}$. For example, if $\mathscr{H}_0$ is a pre-Hilbert space and if $\mathscr{H}$ denotes the (canonical) completion of $\mathscr{H}_0$ (such that $\mathscr{H}_0$ is canonically embedded within $\mathscr{H}$), then $\mathscr{H}_0$ is dense in $\mathscr{H}$. $M$ is called *generating* (with respect to $\mathscr{H}$) if clo spa $M = \mathscr{H}$.

A Hilbert space is called *separable* if there is a countable dense set or, equivalently, if there is a countable generating set. Note that in separable Hilbert spaces there exist *countable orthonormal bases* $B = \{e_1, e_2, \ldots\}$; note further that the *dimension* of a Hilbert space $\mathscr{H}$ is defined by $\dim \mathscr{H} := \text{card}\, B$ (card $M$ denotes the *cardinal number* (power) of a set $M$). Thus the dimension of a separable Hilbert space can take the values 1, 2, 3, ... and $\infty$ $(= \aleph_0)$. In this book the Hilbert spaces are assumed to be separable.

The *Cartesian product* of two Hilbert spaces $\mathscr{H}, \mathscr{K}$ is denoted by $\mathscr{H} \times \mathscr{K}$, the *direct sum* of two subspaces $\mathscr{K}_1, \mathscr{K}_2 \subset \mathscr{H}$ by $\mathscr{K}_1 \oplus \mathscr{K}_2$. The *tensor product* is denoted by $\mathscr{H} \otimes \mathscr{K}$. Sometimes we use the notations $\times_\alpha \mathscr{H}_\alpha$, $\oplus \mathscr{K}_\alpha$, $\otimes \mathscr{H}_\alpha$ for finite index sets.

**Examples.** The simplest example of a Hilbert space with finite dimension $n < \infty$ is $\mathbb{C}^n$ where $(f, g) := \sum_{j=1}^{n} \bar{f}_j g_j$ with $f = \{f_1, f_2, \ldots, f_n\}$, $g = \{g_1, g_2, \ldots, g_n\}$. An easy example for the case $\dim \mathscr{H} = \infty$ is given by the space $l^2_{\mathbb{C}}$, that is, the set of all sequences $f = \{f_1, f_2, \ldots\}$, $f_j \in \mathbb{C}$, with $\sum |f_j|^2 < \infty$; in this case $(f, g) := \sum \bar{f}_j g_j$.

### 1.1.2. Borel measures and integrals

Let $\Gamma$ be a set. A set $\mathscr{B}$ of subsets of $\Gamma$ is called a *Borel field* if: $\emptyset \in \mathscr{B}$, the union of any countably many members of $\mathscr{B}$ is a member of $\mathscr{B}$, and the complement of a member of $\mathscr{B}$ is a member of $\mathscr{B}$. The members of $\mathscr{B}$ are called *Borel sets*. A *Borel space* is a set $\Gamma$ equipped with a Borel field $\mathscr{B}$. A Borel field $\mathscr{B}$ is called *generated* by a set $\mathscr{M}$ of subsets of $\Gamma$ if it coincides with the smallest Borel field containing all members of $\mathscr{M}$.

A *measure* $\mu$ on a Borel space is a map $\mu : \mathscr{B} \to \mathbb{R}_+ \cup \{\infty\} = [0, \infty) \cup \{\infty\}$ with the properties: $\mu(\emptyset) = 0$ and $\mu\left(\bigcup_{j=1}^{\infty} A_j\right) = \sum \mu(A_j)$ for mutually disjoint Borel sets $A_j$, $A_i \cap A_s = \emptyset$ for all $i \neq s$. A measure $\mu$ is called *σ-finite* if there exists a sequence $\{A_j\}$

of Borel sets with the properties $\Gamma = \bigcup_{j=1}^{\infty} A_j$ and $\mu(A_j) < \infty$ for all $j$; it is called *finite* if $\mu(\Gamma) < \infty$.

Measures on topological spaces are of special interest. We restrict ourselves to topological spaces $\Gamma$ which are *metrizable, separable,* and *locally compact.* To $\Gamma$ there are associated the Borel fields generated by the open subsets of $\Gamma$, by the compact subsets of $\Gamma$, and by the compact $G_\delta$-sets of $\Gamma$ (a $G_\delta$-set is a countable intersection of open sets). Note that under our assumptions on $\Gamma$ all these Borel fields coincide. We call this Borel field the Borel field *generated by the topology* and denote it by $\mathcal{B}_\Gamma$ in the following.

Further under our assumption on $\Gamma$ the so-called Borel measures and Baire measures coincide (see the Notes and Remarks); thus in our framework *Borel measures* $\mu$ can be characterized as measures on $\mathcal{B}_\Gamma$ satisfying the condition:

(I) $\mu(C) < \infty$ for all compact sets $C$.

Then also

(II) $\mu(B) = \sup\{\mu(C);\ C \subset B,\ C \text{ compact}\} = \inf\{\mu(O);\ B \subset O,\ O \text{ open}\}$ is valid.

Note that a Borel measure $\mu$ is $\sigma$-finite and separable (that is, the metric space defined by $\operatorname{dis}(A, B) := \mu\big((A \setminus B) \cup (B \setminus A)\big)$, $A, B \in \mathcal{B}_\Gamma$, is separable). The support $\operatorname{supp}\mu$ of a Borel measure is defined by $\operatorname{supp}\mu := \{\bigcap \Delta;\ \Delta \text{ closed},\ \mu(\Gamma \setminus \Delta) = 0\}$.

The most important example for $\Gamma$ in the following is $\mathbb{R}^n$ with the usual topology defined by $\operatorname{dis}(x, y) = |x - y|$, $x, y \in \mathbb{R}^n$. The *Lebesgue measure* on $\mathbb{R}^n$ is a Borel measure. It is denoted by $|A|$, $A$ a Borel set.

In order to prepare some definitions and properties centered around the concept of vector-valued functions, we collect some facts on *nonnegative functions* $f\colon \Gamma \mapsto \mathbb{R}_+ = [0, \infty)$. $f$ is called a *Borel function,* if the inverse image $f^{-1}(B)$ is a Borel set for all Borel sets $B$ of $\mathbb{R}_+$. $f$ is called a *step function,* if $f(x) = \sum_{\varrho=1}^{m} \alpha_\varrho \chi_{\Delta_\varrho}(x)$ with $\alpha_\varrho \in \mathbb{R}_+$, $\Delta_\varrho \in \mathcal{B}_\Gamma$; $\chi_M(x)$ denotes the characteristic function of a set $M$, which is defined by 1 if $x \in M$, and by 0 elsewhere. Note that all step functions are Borel functions.

If a Borel measure $\mu$ is given on $\Gamma$, then two functions $f, g$ are called *$\mu$-equivalent,* $f \equiv g(\operatorname{mod}\mu)$, if $f(x) = g(x)$ a.e. with respect to $\mu$; we also use the notation $f(x) = g(x)\ (\operatorname{mod}\mu)$ in this case. This relation defines equivalence classes of functions. Note: if $f$ is a Borel function, then a function $g$ with $g \equiv f(\operatorname{mod}\mu)$ is not a Borel function in general.

A function $f$ is called *$\mu$-measurable,* if there is a sequence $t_n$ of step functions such that $f(x) = \lim_{n\to\infty} t_n(x)\ (\operatorname{mod}\mu)$. Note that together with $f$ all functions $g$ with $g \equiv f(\operatorname{mod}\mu)$ are $\mu$-measurable.

A function $f$ is $\mu$-measurable iff there is a Borel function $g$ with $f \equiv g(\operatorname{mod}\mu)$.

For nonnegative functions $f$ the *integral* with respect to a Borel measure $\mu$ can be defined as follows: For a step function $t(x) = \sum_{\varrho=1}^{m} \alpha_\varrho \chi_{\Delta_\varrho}(x)$ the integral is defined by $\int_\Gamma t(x)\,\mu(dx) := \sum_{\varrho=1}^{m} \alpha_\varrho \mu(\Delta_\varrho)$. For a $\mu$-measurable function $f(x) = \lim_{n\to\infty} t_n(x)\ (\operatorname{mod}\mu)$ the integral is defined by $\int_\Gamma f(x)\,\mu(dx) = \lim_{n\to\infty} \int_\Gamma t_n(x)\,\mu(dx)$. Note that this definition is independent of the special sequence $t_n$ and that $\int_\Gamma f(x)\,\mu(dx) = \int_\Gamma g(x)\,\mu(dx)$ if $f \equiv g(\operatorname{mod}\mu)$. $f$ is called *$\mu$-integrable* if $\int_\Gamma f(x)\,\mu(dx) < \infty$.

### 1.1.3. Lebesgue decomposition of Borel measures

Let $\mu_1, \mu_2$ be Borel measures on $\Gamma$. $\mu_1$ is called *absolutely continuous* with respect to $\mu_2$ or simply *$\mu_2$-continuous*, $\mu_1 \ll \mu_2$, if $\mu_2(\Delta) = 0$ implies $\mu_1(\Delta) = 0$ for all Borel sets $\Delta$, or, equivalently, (Radon-Nikodym-property), if there exists a nonnegative Borel function $\varrho$ such that $\mu_1(\Delta) = \int_\Delta \varrho(\lambda)\, \mu_2(\mathrm{d}\lambda)$ ($\varrho$ is uniquely determined mod $\mu_2$). If $\mu_1 \ll \mu_2$ and $\mu_2 \ll \mu_1$, then $\mu_1$ and $\mu_2$ are called *equivalent*, $\mu_1 \equiv \mu_2$. $\mu_1$ and $\mu_2$ are called *disjoint*, $\mu_1 \perp \mu_2$, if there is a Borel set $\Delta$ such that $\mu_1(\Delta) = \mu_2(\Gamma \setminus \Delta) = 0$.

If $\mu$, $\nu$ are Borel measures on $\Gamma$, then there exist two Borel measures $\mu_1, \mu_2$ with $\mu_1 \ll \nu$, $\mu_2 \perp \nu$ such that $\mu = \mu_1 + \mu_2$ (Lebesgue decomposition); $\mu_1, \mu_2$ are uniquely determined. A point $x \in \Gamma$ is called a *pure point* with respect to $\mu$ if $\mu(\{x\}) > 0$. $\mu$ is called *continuous* if there is no pure point. $\mu$ is called a *pure point measure* if $\mu(\Delta) = \sum_{x \in \Delta} \mu(\{x\})$. Note that there is a unique decomposition $\mu = \mu_c + \mu_p$ where $\mu_c$ is continuous and $\mu_p$ is a pure point measure.

In particular, if $\Gamma = \mathbb{R}$ and $\nu(\cdot) = |\cdot|$, then we obtain according to LEBESGUE's decomposition $\mu_c = \mu_{ac} + \mu_{sc}$, where $\mu_{ac}$ is absolutely continuous (with respect to the Lebesgue measure) and where $\mu_{sc}$ is continuous and disjoint to the Lebesgue measure.

## 1.2. Vector-valued functions and vector measures

### 1.2.1. Borel functions, measurable functions, and other types of functions

Let $\mathfrak{X}$ be a Banach space. A function $f: \Gamma \mapsto \mathfrak{X}$ is called a *Borel function* (or *Borel measurable*) if for each open set $B$ from $\mathfrak{X}$ the inverse image $f^{-1}(B)$ is a member of $\mathscr{B}_\Gamma$. $f$ is called a *step function* if $f(x) = \sum_{i=1}^{m} g_i \chi_{\Delta_i}(x)$ with $\Delta_i \in \mathscr{B}_\Gamma$, $g_i \in \mathfrak{X}$; note that all step functions are Borel functions, furthermore the set of all step functions is a linear manifold.

If a Borel measure $\mu$ is given on $\Gamma$, then two functions $f, g: \Gamma \mapsto \mathfrak{X}$ are called *$\mu$-equivalent*, $f \equiv g (\mathrm{mod}\, \mu)$, if $f(x) = g(x)$ (mod $\mu$). This relation defines equivalence classes of functions with respect to $\mu$. In the following (as usual in functional analysis) assertions are often to be interpreted in the sense that there is a representer of the equivalence class with the described property.

A function $f: \Gamma \mapsto \mathfrak{X}$ is called *$\mu$-measurable* (or *strongly $\mu$-measurable*) if there is a sequence of step functions $t_n$ such that $\operatorname{s-lim}_{n\to\infty} t_n(x) = f(x)$ (mod $\mu$). Note that if $f$ is $\mu$-measurable and $g \equiv f (\mathrm{mod}\, \mu)$, then also $g$ is $\mu$-measurable, that is, this property is compatible with the equivalence relation $f \equiv g (\mathrm{mod}\, \mu)$; whereas the property of a function to be a Borel function is not compatible with the equivalence relation.

A function $f: \Gamma \mapsto \mathfrak{X}$ is called *weakly $\mu$-measurable* if for each $l \in \mathfrak{X}^*$ the scalar function $l(f): \Gamma \mapsto \mathbb{C}$ is $\mu$-measurable.

Note that strong $\mu$-measurability of $f$ implies the existence of a Borel function $g$ with $g \equiv f (\mathrm{mod}\, \mu)$; in other words, if $f$ is strongly $\mu$-measurable, then there is a representer in the equivalence class of $f$ mod $\mu$ which is Borel measurable.

Furthermore, if $f$ is a Borel function, then $f$ is also weakly $\mu$-measurable. If $\mathfrak{X}$ is separable, then the weak $\mu$-measurability of $f$ implies the strong $\mu$-measurability of $f$. In other words, in this case the three definitions of measurability are equivalent mod $\mu$. This will be used very often in this book because $\mu$-null sets are unimportant in most cases. Measurable functions (all types) are stable (mod $\mu$) under linear operations, multiplication with scalar measurable functions, $\sup_{n\in\mathbb{N}}$, $\inf_{n\in\mathbb{N}}$ and pointwise (mod $\mu$) limits.

According to the strong (weak) topology in $\mathfrak{X}$ we have strongly (weakly) continuous functions $f: \Gamma \to \mathfrak{X}$. Strongly continuous functions are strongly $\mu$-measurable and weakly continuous functions are weakly $\mu$-measurable.

If $\Gamma = \mathbb{R}$ (or $\Gamma = \mathbb{C}$), then the strong derivative $\operatorname{s-lim}_{h\to 0} h^{-1}\big(f(x+h) - f(x)\big)$ of $f$ at the point $x$ is denoted by $f'(x)$. The strong differentiability at $x$ implies strong continuity at $x$. Sometimes also the weak derivative $\operatorname{w-lim}_{h\to 0} h^{-1}\big(f(x+h) - f(x)\big)$ is used. If $\Gamma = \mathbb{C}$, we have the concepts of strong or weak holomorphy for functions $f(z)$ defined on open sets in $\mathbb{C}$. Note that these notions coincide.

### 1.2.2. Bochner integral

A $\mu$-measurable function $f: \Gamma \to \mathfrak{X}$ is called *Bochner integrable* (with respect to $\mu$) if there is a sequence of step functions $t_n$ such that $\int_\Gamma \|f(x) - t_n(x)\| \, \mu(\mathrm{d}x) \to 0$ as $n \to \infty$. In this case $\int_\Gamma f(x)\, \mu(\mathrm{d}x) := \operatorname{s-lim}_{n\to\infty} \int_\Gamma t_n(x)\, \mu(\mathrm{d}x)$. A $\mu$-measurable function $f$ is Bochner integrable iff $\|f(x)\|$ is integrable; note that

$$\left\| \int_\Gamma f(x)\, \mu(\mathrm{d}x) \right\| \leqq \int_\Gamma \|f(x)\|\, \mu(\mathrm{d}x).$$

If $\Delta \in \mathfrak{B}_\Gamma$, we put $\int_\Delta f(x)\, \mu(\mathrm{d}x) := \int_\Gamma \chi_\Delta(x) f(x)\, \mu(\mathrm{d}x)$. Let the functions $f_n$ and $f$ be $\mu$-measurable. The sequence $f_n$ is called *measure-convergent* (with respect to $\mu$) with *measure limit* $f := \operatorname{m-lim}_{n\to\infty} f_n$ if $\lim_{n\to\infty} \mu(A_{n,\varepsilon}) = 0$ for all $\varepsilon > 0$ where $A_{n,\varepsilon} = \{x; \|f_n(x) - f(x)\| \geqq \varepsilon\}$. By Lebesgue's dominated convergence theorem, we have

$$\int_\Gamma \|f(x) - f_n(x)\|\, \mu(\mathrm{d}x) \to 0 \quad \text{as} \quad n \to \infty \quad \text{if} \quad f(x) = \operatorname{m-lim}_{n\to\infty} f_n(x)$$

and if $\|f_n(x)\| \leqq \alpha(x)$ (mod $\mu$) where $\alpha$ is integrable with respect to $\mu$.

Strongly continuous functions are Bochner integrable on compact sets. If $\Gamma$ is compact and $f$ is Bochner integrable, then there is a sequence of strongly continuous functions $g_n$ on $\Gamma$ such that $\lim_{n\to\infty} \int_\Gamma \|g_n(x) - f(x)\|\, \mu(\mathrm{d}x) = 0$. It is even possible to choose $g_n$ as polynomials, $g_n(x) = \sum_{l=1}^{m(n)} u_{n,l} x^l$ with $u_{n,l} \in \mathfrak{X}$.

### 1.2.3. Banach and Hilbert spaces of vector-valued functions

Let $\mathscr{H}$ be a separable Hilbert space and $p \geqq 1$. The Banach space of all (equivalent classes of) $\mu$-measurable vector-valued functions $f: \Gamma \to \mathscr{H}$ with $\int_\Gamma \|f(x)\|^p\, \mu(\mathrm{d}x) < \infty$ is denoted by $L^p(\Gamma, \mu; \mathscr{H})$ and its norm by $\|\cdot\|_{p,\mu}$ or $\|\cdot\|_{L_p}$. If $p = 2$, then it is a Hilbert space with the scalar product $(f, g) = \int_\Gamma (f(x), g(x))\, \mu(dx)$. The Banach space of all essentially bounded $\mu$-measurable vector functions is denoted by $L^\infty(\Gamma, \mu; \mathscr{H})$. Recall that here the different notions of measurability coincide in the sense of $\mu$-equivalence. The Banach space of all bounded strongly continuous vector functions is denoted by $C(\Gamma, \mathscr{H})$. The norms in $L^\infty(\Gamma, \mu; \mathscr{H})$ and $C(\Gamma, \mathscr{H})$ are denoted by $|\cdot|_{\infty,\mu}$ and $|\cdot|$. Occasionally, we shall write $f(\cdot)$ or $\{f(x)\}$ for elements $f$ from these spaces of vector functions. Note that for each $f \in L^\infty(\Gamma, \mu; \mathscr{H})$ there exists a sequence $f_n \in C(\Gamma, \mathscr{H})$ with $\sup_n |f_n| < \infty$ such that $\operatorname{s-lim}_{n\to\infty} f_n(x) = f(x) \pmod{\mu}$. If $\Gamma$ is locally compact, then the Banach space of all strongly continuous vector functions vanishing at infinity equipped with the norm $|\cdot|$ is denoted by $C_\infty(\Gamma, \mathscr{H})$; the normed space (norm $|\cdot|$) of all strongly continuous vector functions with compact support $\operatorname{supp} f = \operatorname{clo} \{x; f(x) \neq 0\}$ is denoted by $C_0(\Gamma, \mathscr{H})$. Note that $C_\infty(\Gamma, \mathscr{H}) = \operatorname{clo}_{|\cdot|} C_0(\Gamma, \mathscr{H})$. In case $\mathscr{H} = \mathbb{C}$ we drop this symbol in general.

If $\Gamma \subseteq \mathbb{R}^n$ and $\mu(\cdot) = |\cdot|$, we write $L^p(\Gamma, dx; \mathscr{H})$, $\|\cdot\|_{L_p}$ and $L^\infty(\Gamma, dx; \mathscr{H})$, $|\cdot|_\infty$ for the corresponding $L^p$- and $L^\infty$-spaces. The set of all infinitely strongly differentiable functions $f\colon \mathbb{R}^n \to \mathscr{H}$ for which

$$\|f\|_{l,p} := \sup_{x\in\mathbb{R}^n} \left\| |x|^l \frac{\partial^{p_1+\cdots+p_n}}{\partial x_1^{p_1} \ldots \partial x_n^{p_n}} f(x) \right\| < \infty \tag{4}$$

with $p = \{p_1, \ldots, p_n\}$, $p_i \in \mathbb{N}$, and $l \in \mathbb{N}$, is called the *Schwartz space* $\mathscr{S}(\mathbb{R}^n, \mathscr{H})$. Correspondingly, we define the spaces $C^\infty(\mathbb{R}^n, \mathscr{H})$ (infinitely strongly differentiable functions) and $C_0^\infty(\mathbb{R}^n, \mathscr{H}) \subset C^\infty(\mathbb{R}^n, \mathscr{H})$ (compact support).

In the Hilbert space $L^2(\mathbb{R}^n, dx; \mathscr{H})$ the *Fourier transformation* $F$ is defined by $(Ff)(k) := (2\pi)^{-n/2} \int e^{-i(k,x)} f(x)\, dx$ on the dense subset $\mathscr{S}(\mathbb{R}^n, \mathscr{H})$ where $(k, x) := \sum_{j=1}^n k_j x_j$. Note *Parseval's equality*

$$(Ff, Fg)_{L^2(\mathbb{R}^n, dk; \mathscr{H})} = (f, g)_{L^2(\mathbb{R}^n, dx; \mathscr{H})}.$$

The inverse $F^{-1}$ of the Fourier transformation is given by

$$(F^{-1}g)(k) = (2\pi)^{-n/2} \int_{\mathbb{R}^n} e^{i(x,k)} g(k)\, dk.$$

Sometimes we write $\hat{f}$ for the Fourier transform $Ff$ of $f$, $\hat{f}(x) = (Ff)(x)$, and $\check{f}$ for the inverse Fourier transform $F^{-1}f$ of $f$, $\check{f}(x) = (F^{-1}f)(x)$.

The *Hardy class* consisting of all functions $f\colon \mathbb{C}_\pm \to \mathscr{H}$ which are holomorphic in $\mathbb{C}_\pm$ (upper or lower halfplane without the real line) and satisfy $\sup_{\varepsilon>0} \int \|f(\lambda \pm i\varepsilon)\|^p\, d\lambda < \infty$ is denoted by $\mathscr{H}_\pm{}^p(\mathbb{R}, d\lambda; \mathscr{H})$, $p \geqq 1$. In particular, we are interested in the case $p = 2$. Note that the theorem of Paley-Wiener is valid: If $f \in \mathscr{H}_+{}^2(\mathbb{R}, d\lambda; \mathscr{H})$, then $\operatorname{s-lim}_{\varepsilon\to+0} f(\lambda + i\varepsilon) =: f(\lambda + i0)$ exists a.e. $\operatorname{mod} |\cdot|$, $f(\cdot + i0) \in L^2(\mathbb{R}, d\lambda; \mathscr{H})$, and $\lim_{\varepsilon\to+0} \int \|f(\lambda + i\varepsilon) - f(\lambda + i0)\|_{\mathscr{H}}^2\, d\lambda = 0$. Further $f(z)$, $z \in \mathbb{C}_+$, is uniquely determined by its boundary value function $f(\lambda + i0)$ (Cauchy's integral formula). Thus we may understand $\mathscr{H}_\pm{}^2(\mathbb{R}, d\lambda; \mathscr{H})$ as the space of the boundary value functions. Note that the Hardy classes $\mathscr{H}_\pm{}^2(\mathbb{R}, d\lambda; \mathscr{H})$ are Hilbert spaces; further note that the Fourier transformation $F$ maps $L^2(\mathbb{R}_+, dx; \mathscr{H})$ onto $\mathscr{H}_-^2(\mathbb{R}, d\lambda; \mathscr{H})$ and $L^2(\mathbb{R}_-, dx; \mathscr{H})$ onto $\mathscr{H}_+^2(\mathbb{R}, d\lambda; \mathscr{H})$, but $F^{-1}$ maps $L^2(\mathbb{R}_+, dx; \mathscr{H})$ onto $\mathscr{H}_+^2(\mathbb{R}, d\lambda; \mathscr{H})$ and $L^2(\mathbb{R}_-, dx; \mathscr{H})$ onto $\mathscr{H}_-^2(\mathbb{R}, d\lambda; \mathscr{H})$. ∎

### 1.2.4. Vector measures

A *vector measure* is a map $\Phi\colon \mathscr{B}_\Gamma \to \mathscr{X}$ with the properties $\Phi(\emptyset) = 0$ and $\Phi(\bigcup \Delta_j) = \sum \Phi(\Delta_j)$ (with respect to the strong topology of $\mathscr{X}$) where the Borel sets $\Delta_j$ are pairwise disjoint. The *variation* of $\Phi$ is defined by $|\Phi|(\Delta) = \sup\limits_{\pi} \sum\limits_{A \in \pi} \|\Phi(A)\|$ where the supremum is taken over all partitions $\pi$ of $\Delta$ into a finite number of disjoint members of $\mathscr{B}_\Gamma$. $|\Phi|$ is a nonnegative measure. If $|\Phi|(\Delta) < \infty$ for compact sets, then $\Phi$ is called a *measure of bounded variation.* $|\Phi|(\Gamma)$ is called the *total variation of* $\Phi$; if $|\Phi|(\Gamma) < \infty$, then $\Phi$ is called a *measure of finite total variation.* $\Phi$ is called *absolutely continuous* with respect to a nonnegative Borel measure $\mu$, or simply *$\mu$-continuous*, if $\mu(\Delta) = 0$ implies $\Phi(\Delta) = 0$ or, equivalently, if $\|\Phi(\Delta)\| \to 0$ as $\mu(\Delta) \to 0$. Then we write $\Phi \ll \mu$.

If a measure $\Phi$ is of bounded variation, then $|\Phi|$ is a (nonnegative) Borel measure with $\Phi \ll |\Phi|$; hence $\Phi$ itself is a *Borel vector measure* in the sense that for each Borel set $\Delta$ and $\varepsilon > 0$ there are a compact set $K$ and an open set $G$ with $K \subset \Delta \subset G$ such that $\|\Phi(\Delta_0)\| < \varepsilon$ if $\Delta_0 \subseteq G \setminus K$. This is in agreement with the notion of a Borel measure for nonnegative measures $\mu$ (see 1.1.2).

If a function $f\colon \Gamma \mapsto \mathscr{X}$ is Bochner integrable with respect to $\mu$ on compact sets, then $\Phi(\Delta) := \int\limits_\Delta f(x)\, \mu(dx)$ is a $\mu$-continuous vector measure of bounded variation where $|\Phi|(\Delta) = \int\limits_\Delta \|f(x)\|\, \mu(dx)$.

In particular, if $\Gamma = \mathbb{R}^n$ and $\mu(\cdot) = |\cdot|$, then the strong derivative property $\operatorname*{s-lim}\limits_{|\Delta| \to 0} |\Delta|^{-1} \int\limits_\Delta f(x)\, dx = f(\lambda) \pmod{|\cdot|}$ is valid where $\lambda \in \Delta$ and $\Delta$ is an $n$-dimensional interval.

Note that if $\Phi$ is $\mu$-continuous and of bounded variation, then a representation $\Phi(\Delta) = \int\limits_\Delta f(x)\, \mu(dx)$ (Radon-Nikodym property), with a function $f$ which is Bochner integrable on compact sets, is not valid in general. But there are many special Banach spaces where this property is valid. For example, if $\mathscr{X}$ is a separable dual ($\mathscr{X}$ separable and $\mathscr{X} = \mathscr{Y}^*$), then to each $\mu$-continuous vector measure $\Phi$ of bounded variation there exists a function $f\colon \Gamma \to \mathscr{X}$ which is Bochner integrable on compact sets such that $\Phi(\Delta) = \int\limits_\Delta f(x)\, \mu(dx)$.

Let $\Gamma = \mathbb{R}$, let $\mathscr{X}$ be a separable dual and $\Phi$ be a Borel vector measure of finite total variation. Then there exists a nonnegative Borel measure $\mu$ ($\mu = |\Phi|$) such that $\Phi \ll \mu$ and we have the representation $\Phi(\Delta) = \int\limits_\Delta f(x)\, \mu(dx)$ where $|\Phi|(\Gamma) = \int\limits_\Gamma \|f(x)\|\, \mu(dx)$ is valid. Moreover, $\Phi(\lambda) := \Phi\big((-\infty, \lambda)\big)$ is strongly differentiable a.e. with respect to $|\cdot|$; that is, $\operatorname*{s-lim}\limits_{h \to 0} h^{-1}\Phi\big([\lambda, \lambda + h)\big)$ exists a.e. and equals $f(\lambda)\, \mu'(\lambda)$ a.e. $\pmod{|\cdot|}$. ($\mu'(\lambda)$ denotes the derivative of $\mu(\lambda) := \mu\big((-\infty, \lambda)\big)$ which exists a.e. on $\mathbb{R} \pmod{|\cdot|}$).

If $\mathscr{X} = \mathscr{Y}^*$ and $\mathscr{Y}$ is separable, then a vector measure of bounded variation has a weak * differentiation property with respect to all nonnegative Borel measures $\mu$ on $\mathbb{R}$:

There is a vector function $g: \mathbb{R} \to \mathcal{X}$, defined a.e. mod $\mu$, such that for all $\lambda$, where $g(\lambda)$ is defined, the relation

$$\lim_{h\to 0} \mu([\lambda, \lambda + h))^{-1} \left(\Phi([\lambda, \lambda + h))\right)(f) = \left(g(\lambda)\right)(f), \qquad f \in \mathcal{Y}$$

is valid.

### 1.2.5. Vector-valued Cauchy-Stieltjes integrals

Let $\Phi: \mathcal{B}_{\mathbb{R}} \to \mathcal{X}$ be a vector measure of finite total variation. If $t: \mathbb{R} \mapsto \mathbb{C}$ is a step function, $t(\lambda) = \sum_{\varrho=1}^{m} c_\varrho \chi_{\Delta_\varrho}(\lambda)$, then $\int t(\lambda)\, \Phi(\mathrm{d}\lambda)$ is defined by

$$\int_{\mathbb{R}} t(\lambda)\, \Phi(\mathrm{d}\lambda) := \sum_{\varrho=1}^{m} c_\varrho \Phi(\Delta_\varrho).$$

If $f: \mathbb{R} \mapsto \mathbb{C}$ is $|\Phi|$-measurable and Bochner integrable with respect to $|\Phi|$, that is, if there is a sequence $t_n$ of step functions with $\lim_{n\to\infty} \int |f(\lambda) - t_n(\lambda)|\, |\Phi|\, (\mathrm{d}\lambda) = 0$, then $\int f(\lambda)\, \Phi(\mathrm{d}\lambda)$ is defined by $\int f(\lambda)\, \Phi(\mathrm{d}\lambda) := \operatorname{s-lim}_{n\to\infty} \int t_n(\lambda)\, \Phi(\mathrm{d}\lambda)$.

The *Cauchy-Stieltjes integral* with respect to $\Phi$ is defined by

$$f(z) := \int_{-\infty}^{\infty} \frac{1}{z - \lambda}\, \Phi(\mathrm{d}\lambda), \qquad \operatorname{Im} z > 0. \tag{5}$$

The function $f: \mathbb{C}_+ \to \mathcal{X}$ is holomorphic in $\mathbb{C}_+$. We put

$$\gamma_\varepsilon(x) = x(x^2 + \varepsilon^2)^{-1}, \qquad \delta_\varepsilon(x) = \varepsilon\pi^{-1}(x^2 + \varepsilon^2)^{-1}. \tag{6}$$

Then $(x + \mathrm{i}\varepsilon)^{-1} = \gamma_\varepsilon(x) - \mathrm{i}\pi\delta_\varepsilon(x)$. Further we put

$$g_\varepsilon(x) = \int_{-\infty}^{\infty} \gamma_\varepsilon(x - \lambda)\, \Phi(\mathrm{d}\lambda), \qquad h_\varepsilon(x) = \int_{-\infty}^{\infty} \delta_\varepsilon(x - \lambda)\, \Phi(\mathrm{d}\lambda). \tag{7}$$

Then $f_\varepsilon(x) := f(x + \mathrm{i}\varepsilon) = g_\varepsilon(x) - \mathrm{i}\pi h_\varepsilon(x)$. Let $\mathcal{X}$ be a separable dual. Then, according to 1.2.2, there are a nonnegative Borel measure $\mu$ and a function $\varphi \in L^1(\mathbb{R}, \mu; \mathcal{X})$ such that

$$\Phi(\Delta) = \int_{\Delta} \varphi(x)\, \mu(\mathrm{d}x),$$

hence $f(z) = \int (z - x)^{-1}\, \varphi(x)\, \mu(\mathrm{d}x)$. We are interested in the behavior of $f_\varepsilon(x)$, $g_\varepsilon(x)$, $h_\varepsilon(x)$ as $\varepsilon \to +0$.

**Proposition 1.** $\operatorname{s-lim}_{\varepsilon\to+0} h_\varepsilon(x)$ *exists a.e.* mod $|\cdot|$ *and equals* $\varphi(x)\, \mu'(x)$ *a.e.* mod $|\cdot|$.

**Proposition 2.** *If* $\mathcal{X}$ *is a Hilbert space, then* $\operatorname{s-lim}_{\varepsilon\to+0} f_\varepsilon(x)$ *exists a.e.* mod $|\cdot|$.

If $\Phi(\cdot)$ is absolutely continuous with respect to $|\cdot|$, but not necessarily of finite total variation, we have $f(z) = \int (z - x)^{-1}\, \varphi(x)\, \mathrm{d}x$. Further let $\mathcal{X} = \mathcal{H}$ be a Hilbert space.

**Proposition 3.** *Let* $\varphi \in L^p(\mathbb{R}, dx; \mathscr{H})$, $p \geqq 1$. *If* $\Phi'(x) = \varphi(x)$ *at the point* $x$, *then* $\operatorname*{s-lim}_{\varepsilon\to+0} g_\varepsilon(x) =: g_+(x)$ *and* $\operatorname*{s-lim}_{\varepsilon\to+0} h_\varepsilon(x) =: h_+(x)$ *exist and we have*

$$g_+(x) = \int_{-\infty}^{\infty} (x-\lambda)^{-1}\varphi(\lambda)\,d\lambda := \operatorname*{s-lim}_{\delta\to+0} \int_{|x-\lambda|\geqq\delta} (x-\lambda)^{-1}\varphi(\lambda)\,d\lambda, \quad h_+(x) = \varphi(x). \tag{8}$$

For $\mu = |\cdot|$, Prop. 2 is a special case ($p = 1$) of Prop. 3. Concerning $L^p$-properties of $f_\varepsilon(x)$ and $f_+(x)$ we note the following results where it is convenient to treat $g_\varepsilon(x)$ and $h_\varepsilon(x)$ separately.

**Proposition 4.** *Let* $\varphi \in L^p(\mathbb{R}, dx; \mathscr{H})$, $p \geqq 1$. *Then* $h_\varepsilon \in L^p(\mathbb{R}, dx; \mathscr{H})$ *for all* $\varepsilon > 0$, $\|h_\varepsilon\|_p \leqq \|\varphi\|_p$, *and* $\|h_\varepsilon - \varphi\|_p \to 0$ *as* $\varepsilon \to +0$.

**Proposition 5.** *Let* $\varphi \in L^p(\mathbb{R}, dx; \mathscr{H})$, $p > 1$. *Then* $g_\varepsilon \in L^p(\mathbb{R}, dx; \mathscr{H})$ *for all* $\varepsilon > 0$, $g_+ \in L^p(\mathbb{R}, dx; \mathscr{H})$, *and* $\|g_+\|_p \leqq M_p \|\varphi\|_p$ ($M_p$ independent of $\varphi$).

If $p = 1$, then Prop. 5 is not true in general. For example, for $\varphi(x) = (x^2 + 1)^{-1}$ one obtains $f(z) = \pi/(z + i)$, hence $g_\varepsilon(x) = \pi x/\left(x^2 + (\varepsilon + 1)^2\right)$ which is not a member of $L^1$ for $\varepsilon \geqq 0$. The results concerning $\|g_\varepsilon - g_+\|_p \to 0$ as $\varepsilon \to +0$ seem to be meager for arbitrary $p \geqq 1$. But in case $p = 2$ we have stronger results. According to Prop. 4 and Prop. 5, we have $f_\varepsilon \in L^2(\mathbb{R}, dx; \mathscr{H})$ for all $\varepsilon > 0$. Let $\hat\varphi = F\varphi$ and $\hat f_\varepsilon = Ff_\varepsilon$.

**Proposition 6.** $\hat f_\varepsilon$ *is given by*

$$\hat f_\varepsilon(t) = -2i\pi\chi_{[0,\infty)}(t)\,e^{-\varepsilon t}\hat\varphi(t), \qquad \varepsilon \geqq 0, \tag{9}$$

*and*

$$\|f_\varepsilon\|_2 \leqq 2\pi\,\|\varphi\|_2, \qquad \varepsilon \geqq 0, \tag{10}$$

*is valid.*

**Proof.** According to the convolution theorem the Fourier transform of $f_\varepsilon(x)$ is given by $(2\pi)^{1/2} F\left(\frac{1}{\cdot + i\varepsilon}\right)\hat\varphi$. But

$$F\left(\frac{1}{\cdot + i\varepsilon}\right)(t) = -i(2\pi)^{1/2}\chi_{[0,\infty)}(t)\,e^{-\varepsilon t}.$$

By Parseval's equation,

$$\|f_\varepsilon\|_2 = 2\pi\,\|\chi_{[0,\infty)}(\cdot)\,\hat\varphi(\cdot)\,e^{-\varepsilon\cdot}\|_2 \leqq 2\pi\,\|\hat\varphi(\cdot)\,e^{-\varepsilon\cdot}\|_2 \leqq 2\pi\,\|\hat\varphi\|_2 = 2\pi\,\|\varphi\|_2$$

is obtained. ∎

From (10) it follows that $f(x + i\varepsilon)$ is a member of the Hardy class $\mathscr{H}_+^2(\mathbb{R}, dx; \mathscr{H})$; (9) gives the explicit form of its Fourier transform as a member of $L^2(\mathbb{R}_+, dt; \mathscr{H})$. According to the general properties of the Hardy class (see 1.2.3), we have $\|f_\varepsilon - f_+\| \to 0$ as $\varepsilon \to +0$.

## 1.3. Linear operators

### 1.3.1. Elementary properties

Linear operators from $\mathscr{H}$ to $\mathscr{K}$ are denoted by $A, B, \ldots, T, \ldots$ etc. The *domain* of $A$ is denoted by dom $A$, the *image* (range) by ima $A$, the *kernel* by ker $A$. We use the notation supp $A = (\ker A)^{\perp}$. An operator $A$ is called *bounded* if dom $A = \mathscr{H}$ and $\|A\| := \sup_{f \neq 0} (\|Af\|/\|f\|) < \infty$; $\|A\|$ is called the *norm* of $A$. The set of all bounded operators from $\mathscr{H}$ to $\mathscr{K}$ is denoted by $\mathscr{L}(\mathscr{H}, \mathscr{K})$. With respect to the operator norm $\mathscr{L}(\mathscr{H}, \mathscr{K})$ is a Banach space. The topology given by the norm is called the *norm topology*. We write $\text{n-lim}_{j\to\infty} A_j = A$ or $A_j \xrightarrow[n]{} A$, if $\|A_j - A\| \to 0$ as $j \to \infty$ (*norm convergence*). The system $\varrho_f(A) := \|Af\|$, $f \in \mathscr{H}$, of seminorms defines a locally convex topology, the so-called *strong topology*. We write $A = \text{s-lim}_{n\to\infty} A_n$ or $A_n \xrightarrow[s]{} A$ if $\text{s-lim}_{n\to\infty} A_n f = Af$ for all $f \in \mathscr{H}$ (*strong convergence*). The system $\varrho_{fg}(A) = |(g, Af)|$ of seminorms defines a locally convex topology, the so-called *weak topology*. We write $A = \text{w-lim}_{n\to\infty} A_n$ or $A_n \xrightarrow[w]{} A$, if $\text{w-lim}_{n\to\infty} A_n f = Af$ for all $f \in \mathscr{H}$ (*weak convergence*). If $\mathscr{H} = \mathscr{K}$ we write $\mathscr{L}(\mathscr{H})$ for $\mathscr{L}(\mathscr{H}, \mathscr{H})$. $\mathscr{L}(\mathscr{H})$ is a Banach algebra, $\|AB\| \leqq \|A\| \, \|B\|$. Note that the *principle of uniform boundedness* is valid: if $A_n$ is weakly convergent, then $\{\|A_n\|\}$ is bounded.

The set of all *closed operators* is denoted by $\mathscr{C}(\mathscr{H}, \mathscr{K})$. If $B$ is an *extension* of $A$, then we write $A \subseteqq B$. The *closure* of an operator $A$, if it exists, is denoted by clo $A$. If clo $B = A$, then dom $B$ is called a *core* of $A$. Note that $A \in \mathscr{C}(\mathscr{H}, \mathscr{K})$ and dom $A = \mathscr{H}$ imply $A \in \mathscr{L}(\mathscr{H}, \mathscr{K})$ (*closed graph theorem*). If dom $A$ is dense in $\mathscr{H}$, the *adjoint* of $A$ (from $\mathscr{H}$ to $\mathscr{K}$) is denoted by $A^*$; one has $(A^*f, g) = (f, Ag)$ for all $g \in \text{dom}\, A$.

If $T_j \in \mathscr{L}(\mathscr{H}_j)$, $j = 1, 2$, an operator $A \in \mathscr{L}(\mathscr{H}_1, \mathscr{H}_2)$ is called an *intertwining operator* (with respect to $T_1, T_2$) if $AT_1 = T_2A$. An operator $A$ is called *relatively bounded* with respect to $B$ or simply *B-bounded*, if dom $B \subseteqq$ dom $A$ and $\|Af\| \leqq a \|f\| + b \|Bf\|$ for all $f \in \text{dom}\, B$. The greatest lower bound $b_0$ of all $b$ in this inequality is called the *B-bound* of $A$.

### 1.3.2. Operator-valued functions

A function $A: \Gamma \to \mathscr{L}(\mathscr{H}, \mathscr{K})$ is called *norm μ-measurable* if it is $\mu$-measurable with respect to the Banach space $\mathscr{L}(\mathscr{H}, \mathscr{K})$ and the Borel measure $\mu$. It is called *strongly μ-measurable* if $A(\cdot)\, f$ is $\mu$-measurable for all $f \in \mathscr{H}$ with respect to $\mathscr{K}$. It is called *weakly μ-measurable* if $A(\cdot)\, f$ is weakly $\mu$-measurable for all $f$ with respect to $\mathscr{K}$. Note that strong and weak measurability coincide. Similarly we define norm continuity, strong continuity, and weak continuity and, if $\Gamma = \mathbb{R}$ or $\Gamma = \mathbb{C}$, norm, strong, and weak differentiability for operator functions. Note that the three types of holomorphic operator functions coincide.

### 1.3.3. Selfadjoint, unitary, and other special operators

A densely defined operator $A$ from $\mathscr{H}$ to $\mathscr{H}$ is called *selfadjoint* if $A^* = A$. If $A \in \mathscr{C}(\mathscr{H}, \mathscr{K})$ and dom $A$ is dense, then $A^*A$ is selfadjoint. $A$ is called *normal* if $A^*A = AA^*$. A selfadjoint operator $A$ is called *nonnegative*, $A \geqq 0$, if $(f, Af) \geqq 0$ for all $f \in \text{dom}\, A$. The square root of a bounded nonnegative operator $A \geqq 0$ is denoted by $A^{1/2}(\geqq 0)$. $A$ is called *accretive* if $\mathrm{Re}\,(f, Af) \geqq 0$ for all $f \in \mathrm{dom}\, A$. We use the notation $|A| := (A^*A)^{1/2}$. A *projection* $P$ is always an orthoprojection ($P^2 = P$, $P^* = P$) in this book. A *unitary operator* is always an operator from $\mathscr{L}(\mathscr{H})$ with $U^*U = UU^* = 1$. An *isometric operator* $V$ from $\mathscr{H}$ onto $\mathscr{K}$ is defined by $V^*V = 1_{\mathscr{H}}$, $VV^* = 1_{\mathscr{K}}$. An operator $V \in \mathscr{L}(\mathscr{H}, \mathscr{K})$ is called *partially isometric* if $V^*V$ is a projection (then also $VV^*$ is a projection). $P := V^*V$ is called the *initial projection* (ima $P$ the *initial subspace*) and $Q := VV^*$ is called the *final projection* (ima $Q$ the *final subspace*) of $V$.

The *polar decomposition* of $T \in \mathscr{L}(\mathscr{H}, \mathscr{K})$ is denoted by $T = \mathrm{sgn}\, T\, |T|$ where $|T| \in \mathscr{L}(\mathscr{H})$, $\mathrm{sgn}\, T \in \mathscr{L}(\mathscr{H}, \mathscr{K})$. $\mathrm{sgn}\, T$ is partially isometric; $\ker(\mathrm{sgn}\, T) = \ker T$, that is, the initial subspace of $\mathrm{sgn}\, T$ is given by $\mathrm{supp}\, T = \mathrm{clo}\,(\mathrm{ima}\, |T|)$; $\mathrm{ima}\,(\mathrm{sgn}\, T) = \mathrm{clo}\,(\mathrm{ima}\, T)$, that is, the final subspace of $\mathrm{sgn}\, T$ is given by $\mathrm{clo}\,(\mathrm{ima}\, T)$.

Projections $P$, $Q$ with $PQ = QP = 0$ are called *orthogonal*, we write $P \perp Q$. Further we use the notation $P^\perp := 1 - P$.

If $T \in \mathscr{C}(\mathscr{H})$, then a subspace $\mathscr{M}$ with $T\,(\mathscr{M} \cap \mathrm{dom}\, T) \subseteqq \mathscr{M}$ is called an *invariant subspace*. If $P$ denotes the projection onto $\mathscr{M}$, this is equivalent to $TP = PTP$. If $\mathscr{M}$ and $\mathscr{M}^\perp$ are invariant subspaces, $\mathscr{M}$ is called a *reducing subspace* and $P$ a *reducing projection*, $PT \subseteqq TP$. If $\mathscr{M}$ is reducing, then the *part* of $T$ in $\mathscr{M}$ is denoted by $T \restriction \mathscr{M}$.

Two operators $T_1 \in \mathscr{L}(\mathscr{H}_1)$, $T_2 \in \mathscr{L}(\mathscr{H}_2)$ are called *unitarily equivalent* if there is an isometric operator $V$ from $\mathscr{H}_1$ onto $\mathscr{H}_2$ which is intertwining for $T_1$, $T_2$: $VT_1 = T_2V$, or, equivalently, $T_2 = VT_1V^*$.

If $H_1$ and $H_2$ are selfadjoint on $\mathscr{H}_1$ and $\mathscr{H}_2$, respectively, an operator $A \in \mathscr{L}(\mathscr{H}_1, \mathscr{H}_2)$ is called *intertwining* with respect to $H_1$, $H_2$ if $H_2A \supseteq AH_1$. In case $H_1 = H_2$, $A$ is called *commuting*. Note that if $A$ is intertwining with respect to $H_1$, $H_2$, then $\mathrm{sgn}\, A$ is also intertwining; further the initial projection $P_1$ of $\mathrm{sgn}\, A$ and $|A|$ commute with $H_1$. On the other hand, $|A^*|$ and $P_2$ (the final projection of $\mathrm{sgn}\, A$) commute with $H_2$. Finally, the parts $H_1 \restriction P_1\mathscr{H}_1$ and $H_2 \restriction P_2\mathscr{H}_2$ are unitarily equivalent (via $\mathrm{sgn}\, A$).

### 1.3.4. Compact operators

The set of all *compact operators* of $\mathscr{L}(\mathscr{H}, \mathscr{K})$ is denoted by $\mathscr{L}_\infty(\mathscr{H}, \mathscr{K})$. If $A \in \mathscr{L}(\mathscr{H}', \mathscr{H})$, $B \in \mathscr{L}(\mathscr{K}, \mathscr{K}')$, $K \in \mathscr{L}_\infty(\mathscr{H}, \mathscr{K})$, then $BKA \in \mathscr{L}_\infty(\mathscr{H}', \mathscr{K}')$. If $A \in \mathscr{L}_\infty(\mathscr{H}, \mathscr{K})$, then $A^* \in \mathscr{L}_\infty(\mathscr{K}, \mathscr{H})$. In particular, $\mathscr{L}_\infty(\mathscr{H}) := \mathscr{L}_\infty(\mathscr{H}, \mathscr{H})$ is a *two-sided $*$-ideal* in $\mathscr{L}(\mathscr{H})$. The set of all *finite-dimensional operators* of $\mathscr{L}(\mathscr{H}, \mathscr{K})$ is denoted by $\mathscr{L}_{\mathrm{fin}}(\mathscr{H}, \mathscr{K})$. $\mathscr{L}_\infty$ is closed with respect to the norm-topology and $\mathrm{clo}_{\|\cdot\|}\, \mathscr{L}_{\mathrm{fin}}\,(\mathscr{H}, \mathscr{K}) = \mathscr{L}_\infty(\mathscr{H}, \mathscr{K})$. A compact operator $K$ has a *Schmidt representation* $K = \sum \alpha_j(e_j, \cdot)_{\mathscr{H}}\, f_j$ where $\alpha_1 \geqq \alpha_2 \geqq \cdots \geqq 0$, $\alpha_j \to 0$ as $j \to \infty$, and where the systems $\{e_j\}$, $\{f_j\}$ are orthonormal in $\mathscr{H}$, $\mathscr{K}$, respectively. The numbers $\alpha_j$ are the eigenvalues of $(K^*K)^{1/2}$, the so-called *singular values* of $K$.

The set of all *Hilbert-Schmidt operators* is denoted by $\mathscr{L}_2(\mathscr{H}, \mathscr{K})$, the *Hilbert-Schmidt norm* is denoted by $\|A\|_2 := (\sum \|Ae_j\|^2)^{1/2}$ (which is independent of the choice of the orthonormal basis $\{e_j\}$). Note that $\|A_2\| = \|A^*\|_2$ and $\|BAC\|_2 \leqq \|B\|\, \|A\|_2\, \|C\|$ where $B$, $C$ are bounded. Further, $\mathscr{L}_2 \subset \mathscr{L}_\infty$ and $\|A\| \leqq \|A\|_2$ are valid. $\mathscr{L}_2(\mathscr{H}, \mathscr{K})$ is the closure of $\mathscr{L}_{\mathrm{fin}}(\mathscr{H}, \mathscr{K})$ with respect to $\|\cdot\|_2$. In particular, $\mathscr{L}_2(\mathscr{H}) := \mathscr{L}_2(\mathscr{H}, \mathscr{H})$ is a two-sided $*$-ideal in $\mathscr{L}(\mathscr{H})$. $\mathscr{L}_2(\mathscr{H}, \mathscr{K})$ is a Hilbert space with respect to the scalar product $(A, B) := \sum (Ae_j, Be_j)$. An operator $K \in \mathscr{L}_\infty$ is a Hilbert-Schmidt operator iff $\sum \alpha_j^2 < \infty$; in this case $\|K\|_2 = (\sum \alpha_j^2)^{1/2}$.

The set of all *trace operators* is denoted by $\mathscr{L}_1(\mathscr{H}, \mathscr{K})$, the *trace norm* is denoted by $\|A\|_1 := \| |A|^{1/2}\|_2^2$. Note that $\|A\|_1 = \|A^*\|_1$ and $\|BAC\|_1 \leqq \|B\| \|A\|_1 \|C\|$ if $B, C$ bounded; further, $\mathscr{L}_1(\mathscr{H}, \mathscr{K}) \subset \mathscr{L}_2(\mathscr{H}, \mathscr{K})$ and $\|A\|_2 \leqq \|A\|_1$. $\mathscr{L}_1$ is a Banach space with respect to $\|\cdot\|_1$ and $\mathscr{L}_1 = \operatorname{clo}_{\|\cdot\|_1} \mathscr{L}_{\text{fin}}$ is valid. In particular, $\mathscr{L}_1(\mathscr{H}) := \mathscr{L}_1(\mathscr{H}, \mathscr{H})$ is a two-sided $*$-ideal in $\mathscr{L}(\mathscr{H})$. Note also that $\|AB\|_1 \leqq \|A\|_2 \|B\|_2$ (if $A, B \in \mathscr{L}_2$, then $AB \in \mathscr{L}_1$ and the inequality is valid). Conversely, any operator $A \in \mathscr{L}_1$ may be written as the product of two $\mathscr{L}_2$-operators. An operator $K \in \mathscr{L}_\infty$ is a trace operator iff $\sum \alpha_j < \infty$; in this case $\|K\|_1 = \sum \alpha_j$.

Further we denote by $\mathscr{L}_p(\mathscr{H}, \mathscr{K})$ the set of all compact operators $K$ with $\sum \alpha_j^p < \infty$ and put $\|K\|_p := (\sum \alpha_j^p)^{1/p}$, $1 \leqq p < \infty$. $\mathscr{L}_p(\mathscr{H}, \mathscr{K})$ is a Banach space with respect to $\|\cdot\|_p$; $\mathscr{L}_p(\mathscr{H}) := \mathscr{L}_p(\mathscr{H}, \mathscr{H})$ is a two-sided $*$-ideal. Note that $\|AB\|_1 \leqq \|A\|_p \|B\|_q$, where $p^{-1} + q^{-1} = 1$ (if $A \in \mathscr{L}_p$, $B \in \mathscr{L}_q$, then $AB \in \mathscr{L}_1$ and the inequality is valid). If $A \in \mathscr{L}_p$, $p \geqq 1$, then there exist compact operators $K, K'$ and operators $B, C \in \mathscr{L}_p$ such that $A = BK = K'C$.

The *trace* of a trace operator $A$ is denoted by $\operatorname{tr} A := \sum (e_j, Ae_j)$ (which is independent of the choice of the orthonormal basis $\{e_j\}$). Note that $\operatorname{tr} AB = (A^*, B)_{\mathscr{L}_2}$ for $A, B \in \mathscr{L}_2$ and $\operatorname{tr} AB = \operatorname{tr} BA$ if $A \in \mathscr{L}_\infty$, $B \in \mathscr{L}$ such that $AB, BA \in \mathscr{L}_1$; in particular this is valid if $A \in \mathscr{L}_p$, $B \in \mathscr{L}_q$, $p^{-1} + q^{-1} = 1$. Further $\operatorname{tr} A^* = \overline{\operatorname{tr} A}$, $\operatorname{tr} A$ is linear with respect to $A$ and invariant with respect to $A \to S^{-1}AS$ where $S, S^{-1} \in \mathscr{L}$, that is, $\operatorname{tr}(S^{-1}AS) = \operatorname{tr} A$. If $A, B \geqq 0$, then $\operatorname{tr} AB \geqq 0$. The trace is continuous with respect to $\|\cdot\|_1$, $|\operatorname{tr} A| \leqq \|A\|_1$; if $A \geqq 0$, then $\operatorname{tr} A = \|A\|_1$.

Note that $\mathscr{L}_1 = (\mathscr{L}_\infty)^*$ in the sense that $A \to \operatorname{tr}(A\,\cdot)$ establishes an isometric isomorphism from $\mathscr{L}_1$ onto $\mathscr{L}_\infty{}^*$. Further $\mathscr{L} = \mathscr{L}_1{}^*$ in the sense that $B \to \operatorname{tr}(B\,\cdot)$ establishes an isometric isomorphism from $\mathscr{L}$ onto $\mathscr{L}_1{}^*$.

The trace can also be understood as *spectral trace*, namely we have $\operatorname{tr} A = \sum \lambda_j(A)$, where $\lambda_1(A), \lambda_2(A), \ldots$ denote the eigenvalues of $A$. Because of $\sum |\lambda_j(A)| \leqq \|A\|_1$, for trace operators the *determinant* $\det(1 - A)$ can be defined by $\det(1 - A) := \prod_j (1 - \lambda_j(A))$. Note that $\det(1 - AB) = \det(1 - BA)$ if $A \in \mathscr{L}_\infty$, $B \in \mathscr{L}$ such that $AB, BA \in \mathscr{L}_1$. $\det(1 - A)$ is continuous with respect to $\|\cdot\|_1$, $\det(1 - A^*) = \overline{\det(1 - A)}$ and

$$\det\{(1 - A)(1 - B)\} = \det(1 - A)\det(1 - B).$$

Further $1 - A$ is bounded invertible iff $\det(1 - A) \neq 0$.

*Relative compactness* of an operator $A$ with respect to a closed operator $H$ is defined by $A \in \mathscr{L}_\infty(\mathscr{X}_H, \mathscr{K})$ where $\operatorname{dom} H \subseteqq \operatorname{dom} A$ and $\mathscr{X}_H$ denotes the Banach space $\operatorname{dom} H$ with respect to the norm $[f] := \|f\| + \|Hf\|$; $\mathscr{K}$ is the image space of $A$.

For later application we need the following lemma.

**Lemma 7.** *Let* $T_n \in \mathscr{L}_2$, $T \in \mathscr{L}$, $T_n \xrightarrow[w]{} T$, *and*

$$\alpha = \sup_n \|T_n\|_2 < \infty. \tag{11}$$

*Then* $T \in \mathscr{L}_2$ *and* $\lim_{n\to\infty} \|B(T_n - T)\,C\|_2 = 0$ *for all* $B, C \in \mathscr{L}_\infty$.

Proof. At first we prove $T \in \mathscr{L}_2$. Let $A = (f, \cdot)\,g$, then we have

$$(T, A)_{\mathscr{L}_2} = \operatorname{tr} T^*A = \operatorname{tr} AT^* = (f, T^*g) = (Tf, g).$$

Correspondingly we obtain $(T_n, A)_{\mathscr{L}_2} = (T_n f, g)$. Thus $(T_n, C)_{\mathscr{L}_2}$ converges for all $C \in \mathscr{L}_{\text{fin}}$ which is dense in $\mathscr{L}_2$. Because of (11) we obtain: $T_n$ is $\mathscr{L}_2$-weakly convergent, but the limit is $T$, hence $T \in \mathscr{L}_2$ and $\|T\|_2 \leqq \alpha$ follows. Now we prove the second assertion. We define the operator $\mathscr{T}_n \in \mathscr{L}(\mathscr{L}_\infty(\mathscr{H}), \mathscr{L}_2(\mathscr{H}))$ by $\mathscr{T}_n(C) := B(T_n - T)\, C$. Then we have $\|\mathscr{T}_n\| \leqq \|B\|\, \|T_n - T\|_2 \leqq 2\alpha\, \|B\|$, that is, $\|\mathscr{T}_n\|$ is uniformly bounded. Further, for $C = (f, \cdot)\, g$, we obtain $\|\mathscr{T}_n(C)\|_2 = \|(f, \cdot)\, B(T_n - T)\, g\|_2 = \|f\|\, \|B(T_n - T)\, g\| \to 0$ as $n \to \infty$, because of $B \in \mathscr{L}_\infty$. By the uniform boundedness of $\mathscr{T}_n$, this can be extended to all compact operators $C$. ∎

### 1.3.5. Spectra and resolvents

If $T \in \mathscr{C}(\mathscr{H})$, the *resolvent set* of $T$ is denoted by

$$\operatorname{res} T := \{z \in \mathbb{C};\ (z1 - T)^{-1} \in \mathscr{L}(\mathscr{H})\}.$$

The *spectrum* of $T$ is denoted by $\operatorname{spec} T := \mathbb{C} \setminus \operatorname{res} T$. The *resolvent* is denoted by $R_T(z) := (z1 - T)^{-1}$. $R_T(z)$ is holomorphic on $\operatorname{res} T$ which is an open set and $\operatorname{ima} R_T(z) = \operatorname{dom} T$. The identity

$$R_T(z) - R_T(z') = (z' - z)\, R_T(z)\, R_T(z') \tag{12}$$

s called the *first resolvent identity*, the identity

$$R_T(z)\, B - B R_{T_0}(z) = R_T(z)\, (TB - BT_0)\, R_{T_0}(z) \tag{13}$$

is called the *second resolvent identity* where $T \in \mathscr{C}(\mathscr{H})$, $T_0 \in \mathscr{C}(\mathscr{H}_0)$, $B \in \mathscr{L}(\mathscr{H}_0, \mathscr{H})$, and $B \operatorname{dom} T_0 \subseteqq \operatorname{dom} T$. $\operatorname{spec} T$ is closed and

$$\|R_T(z)\| \geqq (\operatorname{dis}\{z, \operatorname{spec} T\})^{-1}. \tag{14}$$

Further $\operatorname{spec} T^* = \overline{\operatorname{spec} T}$ and $R_{T^*}(z) = R_T(\bar{z})^*$. If $T \in \mathscr{L}(\mathscr{H})$ then $\operatorname{spec} T$ is nonempty and compact. The series $R_T(z) = \sum_{n=0}^{\infty} T^n z^{n+1}$, which is norm convergent for $|z| > \operatorname{spr} T := \lim_{n\to\infty} \|T^n\|^{1/n} = \sup_{z \in \operatorname{spec} T} |z|$, is called the *Neumann series*, $\operatorname{spr} T$ is called the *spectral radius*; Note that $\operatorname{spr} T \leqq \|T\|$.

If $A$ is normal, then $\operatorname{spr} A = \|A\|$. If $A = A^*$, then $\operatorname{spec} A \subseteqq \mathbb{R}$ and $\|R_A(z)\| = (\operatorname{dis}\{z, \operatorname{spec} A\})^{-1} \leqq |\operatorname{Im} z|^{-1}$. If $U$ is unitary, then $\operatorname{spec} U \subseteqq \{z;\ |z| = 1\} =: S_1$, the unit sphere in $\mathbb{C}$.

### 1.3.6. Dunford's functional calculus

If $T \in \mathscr{L}(\mathscr{H})$, then polynomials of $T$ can be defined elementarily, namely by

$$p(T) := \alpha_0 1 + \alpha_1 T + \cdots + \alpha_n T^n$$

if $p(\lambda) = \sum_{j=1}^{n} \alpha_j \lambda^j$, $\alpha_j \in \mathbb{C}$. If $f(z)$ is complex-valued and holomorphic in some $\varepsilon$-neighborhood $N_\varepsilon$ of $\operatorname{spec} T$, then $f(T) := (2i\pi)^{-1} \int_L f(z)\, (z - T)^{-1}\, dz$, where $L$ lies in $N_\varepsilon \setminus \operatorname{spec} T$ and consists of finitely many positively oriented closed curves, is a bounded operator (the integral may be considered as a Bochner integral with respect to $\mathscr{L}(\mathscr{H})$). The operator $f(T)$ satisfies the following properties (Dunford's calculus):

(I) $(\alpha f + \beta g)(T) = \alpha f(T) + \beta g(T)$, $\alpha, \beta \in \mathbb{C}$,

(II) $(f \cdot g)(T) = f(T)\, g(T)$,

(III) if $f(z) = \sum \alpha_s z^s$ for $|z| > \operatorname{spr} T + \varepsilon$, $\varepsilon > 0$, then $f(T) = \sum \alpha_s T^s$ with respect to the norm convergence,

(IV) $\operatorname{spec} \{f(T)\} = f(\operatorname{spec} T)$.

Thus this calculus is compatible with the elementary formation of polynomials and also with the formation of $A^{1/2}$ for selfadjoint operators $A \geqq \alpha 1$, $\alpha > 0$.

**Example.** Let $A \in \mathscr{L}$ and $1 \in \operatorname{res} A$. Take a positively oriented simple closed integration path $L \subset \operatorname{res} A$ such that spec $A$ lies in the interior of $L$ and 1 in the exterior of $L$ (see the figure):

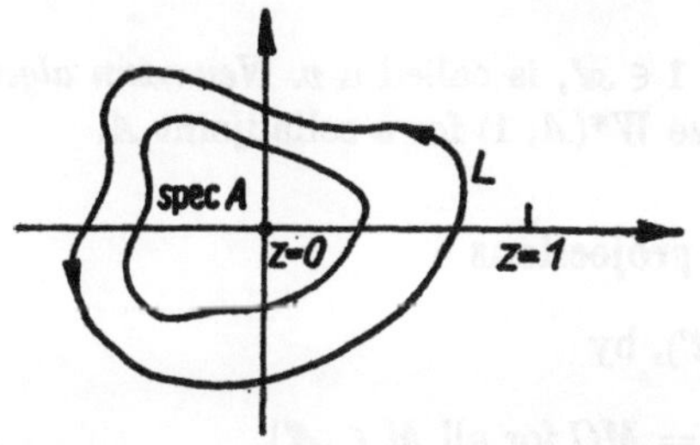


Take $f(z) := \log(1 - z)$ and use that branch of $f(z)$ which is defined by $f(0) = 0$. Then $\log(1 - A) := (2i\pi)^{-1} \int_L \log(1 - z)(z - A)^{-1}\, dz$ is well-defined according to Dunford's calculus.

**Proposition 8.** *If* $A \in \mathscr{L}_1(\mathscr{H})$ *and* $\det(1 - A) \neq 0$, *then* $\log(1 - A) \in \mathscr{L}_1$ *and* $\operatorname{tr} \log(1 - A) = \log \det(1 - A)$.

Proof. Because of $f(0) = 0$ one has $f(A) = (2i\pi)^{-1} \int_L z^{-1} f(z)\, A(z - A)^{-1}\, dz$. But $A(z - A)^{-1} \in \mathscr{L}_1$ and this operator function is continuous with respect to $\|\cdot\|_1$. Hence Dunford's integral can be interpreted as a Bochner integral with respect to $\mathscr{L}_1(\mathscr{H})$, thus $f(A) \in \mathscr{L}_1(\mathscr{H})$. Further, according to (IV),

$$\operatorname{spec} \log(1 - A) = \{\log(1 - \lambda);\ \lambda \in \operatorname{spec} A\}.$$

spec $A$ consists of eigenvalues $\lambda_1(A), \lambda_2(A), \ldots$ only. Hence $\operatorname{tr} \log(1 - A) = \sum \log(1 - \lambda_j(A))$. On the other hand $\log \det(1 - A) = \log \prod (1 - \lambda_j(A)) = \sum \log(1 - \lambda_j(A))$. ■

## 1.4. C*- and W*-algebras

### 1.4.1. *-algebras

A set $\mathscr{A} \subseteq \mathscr{L}(\mathscr{H})$ which is closed under addition, multiplication, and multiplication by complex numbers is called an *algebra* (of operators). $\mathscr{A}$ is called an *$*$-algebra* if $A \in \mathscr{A}$ implies $A^* \in \mathscr{A}$. Two $*$-algebras $\mathscr{A}_1 \subseteq \mathscr{L}(\mathscr{H}_1)$, $\mathscr{A}_2 \subseteq \mathscr{L}(\mathscr{H})_2$ are called *spatial isomorphic*, $\mathscr{A}_1 \cong \mathscr{A}_2$, if there is an isometric operator $V$ from $\mathscr{H}_1$ onto $\mathscr{H}_2$ such that $A_2 = VA_1V^*$ is a bijection from $\mathscr{A}_1$ onto $\mathscr{A}_2$.

A $*$-algebra which is closed with respect to the norm topology is called a *C*-algebra*. The intersection of $C^*$-algebras is a $C^*$-algebra. If $\mathscr{M}$ denotes a set of operators, by

$C^*(\mathcal{M}) := \bigcap_{\mathcal{A} \supseteq \mathcal{M}} \mathcal{A}$ ($\mathcal{A}$ $C^*$-algebras) we denote the $C^*$-algebra which is generated by $\mathcal{M}$. If $A$ is normal, then the famous theorem of Gelfand-Neumark (on Abelian $C^*$-algebras) says that $C^*(A, 1)$ is isometrically isomorphic to $C(\operatorname{spec} A)$, $C(\operatorname{spec} A) \ni f \to f(A) \in C^*(A, 1)$ (Gelfand's isomorphism), where $f(\lambda) \equiv \lambda$ corresponds to $A$, $f(\lambda) \equiv \bar{\lambda}$ to $A^*$, $f(\lambda) \equiv 1$ to $1$, and where $|f| = \|f(A)\|$. If $A$ is selfadjoint, $f \geqq 0$ implies $f(A) \geqq 0$. If $f(z)$ is holomorphic in some $\varepsilon$-neighborhood of $\operatorname{spec} A$, then $f(A)$ coincides with Dunford's $f(A)$.

A $*$-algebra $\mathcal{A}$ which is closed with respect to the weak topology is called a *$W^*$-algebra*. The intersection of $W^*$-algebras is a $W^*$-algebra. If $\mathcal{M}$ denotes a set of operators, by $W^*(\mathcal{M}) := \bigcap_{\mathcal{A} \supseteq \mathcal{M}} \mathcal{A}$ ($\mathcal{A}$ $W^*$-algebras) we denote the $W^*$-algebra generated by $\mathcal{M}$. Obviously $W^*(\mathcal{M}) \supseteqq C^*(\mathcal{M})$.

A $W^*$-algebra $\mathcal{A}$ with unit operator $1$, $1 \in \mathcal{A}$, is called a *v. Neumann algebra*. One aim of the spectral theory is to characterize $W^*(A, 1)$ for a selfadjoint $A$.

### 1.4.2. Commutant and equivalence of projections

If $\mathcal{M}$ denotes a set of operators from $\mathcal{L}(\mathcal{H})$, by

$$\operatorname{com} \mathcal{M} := \{C; C \in \mathcal{L}(\mathcal{H}), \quad CM = MC \text{ for all } M \in \mathcal{M}\}$$

we denote the *commutant* of $\mathcal{M}$. $\operatorname{com} \mathcal{M}$ is a v. Neumann algebra. If $\mathcal{M}_1 \subseteqq \mathcal{M}_2$ then $\operatorname{com} \mathcal{M}_2 \subseteqq \operatorname{com} \mathcal{M}_1$.

If $\mathcal{A}$ is a $*$-algebra with 1, then obviously $\mathcal{A} \subseteqq \operatorname{clo}_s \mathcal{A} \subseteqq \operatorname{clo}_w \mathcal{A} \subseteqq \operatorname{com}(\operatorname{com} \mathcal{A})$. Note that in this case $\operatorname{com}(\operatorname{com} \mathcal{A}) \subseteqq \operatorname{clo}_s \mathcal{A}$, hence $\mathcal{A} \subseteqq \operatorname{clo}_s \mathcal{A} = \operatorname{clo}_w \mathcal{A} = \operatorname{com}(\operatorname{com} \mathcal{A})$. Thus for $*$-algebras with 1 weak and strong closure coincide and the v. Neumann algebras can be characterized by the equation $\mathcal{A} = \operatorname{com}(\operatorname{com} \mathcal{A})$ (*double commutant theorem*). Note that an operator $T \in \operatorname{clo}_s \mathcal{A}$, where $\mathcal{A}$ is an $*$-algebra with 1, can be "reached" by *sequences* of $\mathcal{A}$, that is, there is a sequence $A_n \in \mathcal{A}$ such that $\operatorname{s-lim}_n A_n = T$ (Kaplansky's lemma).

If $\mathcal{A}_1$, $\mathcal{A}_2$ are v. Neumann algebras, then $\operatorname{com} W^*(\mathcal{A}_1, \mathcal{A}_2) = \operatorname{com} \mathcal{A}_1 \cap \operatorname{com} \mathcal{A}_2$. The *center* $\mathcal{A} \cap \operatorname{com} \mathcal{A}$ of a v. Neumann algebra $\mathcal{A}$ is denoted by $\operatorname{cen} \mathcal{A}$; it is an *Abelian* v. Neumann algebra ($AB = BA$ for all $A, B \in \operatorname{cen} \mathcal{A}$) and $\operatorname{cen} \mathcal{A} = \operatorname{cen} \operatorname{com} \mathcal{A}$. A v. Neumann algebra $\mathcal{A}$ is Abelian iff $\mathcal{A} \subseteqq \operatorname{com} \mathcal{A}$ or, equivalently, if $\mathcal{A} = \operatorname{cen} \mathcal{A}$, then also $\mathcal{A} = \operatorname{cen} \operatorname{com} \mathcal{A}$. $\mathcal{A}$ is called *maximal Abelian* if $\mathcal{A} = \operatorname{com} \mathcal{A}$.

The *equivalence of projections* $P$, $Q$ of a v. Neumann algebra $\mathcal{A}$ is denoted by $P \sim Q \pmod{\mathcal{A}}$; $P \sim Q$ means that there exists a partial isometry $U \in \mathcal{A}$ such that $U^*U = P$, $UU^* = Q$. Note that in an Abelian v. Neumann algebra there are no different equivalent projections, that is, $P \sim Q$ implies $P = Q$ in this case.

If $P \in \operatorname{com} \mathcal{A}$ is a projection, then we denote the *part of* $\mathcal{A}$ with respect to the Hilbert space $P\mathcal{H}$ by $\mathcal{A} \mid P\mathcal{H}$, that is, $\mathcal{A} \mid P\mathcal{H}$ is the v. Neumann algebra consisting of all operators $A \mid P\mathcal{H}$, where $A \in \mathcal{A}$.

If $\mathcal{A}_1$ and $\mathcal{A}_2$ are v. Neumann algebras on $\mathcal{H}_1$ and $\mathcal{H}_2$, respectively, then the *Cartesian product* of $\mathcal{A}_1$ and $\mathcal{A}_2$ on $\mathcal{H}_1 \times \mathcal{H}_2$ is denoted by $\mathcal{A}_1 \times \mathcal{A}_2$ and the tensor product $W^*(A_1 \otimes A_2, A_1 \in \mathcal{A}_1, A_2 \in \mathcal{A}_2)$ on $\mathcal{H}_1 \otimes \mathcal{H}_2$ by $\mathcal{A}_1 \otimes \mathcal{A}_2$.

# 2. Multiplicity Theorie

In this chapter we collect the necessary facts centered around the concept of *multiplicity* which is connected with the simple process of "doubling" $A \times A \in \mathscr{L}(\mathscr{H} \times \mathscr{H})$ of an operator $A \in \mathscr{L}(\mathscr{H})$. We are mainly interested in selfadjoint operators but it is quite natural to deal with v. Neumann algebras at first. Multiplicity theory is one of the decisive foundations for an understanding of the unitary invariants of selfadjoint operators. The other foundation is the spectral theory to be presented in the next chapter. Multiplicity theory is decivise for the transition from the spectral theorem (see Chapter 3) to the spectral representation of a selfadjoint operator, that is, to a representation where the unitary invariants of the operator become evident. In principle multiplicity theory is independent of spectral theory (but usually technical tools of spectral theory appear in certain proofs of multiplicity theory).

## 2.1. Simple algebras and algebras with multiplicity

### 2.1.1. Simple algebras, simple and disjoint projections

A v. Neumann algebra $\mathscr{A}$ is called *simple* if there are no projections $P, Q \in \operatorname{com} \mathscr{A}$ with $P, Q \neq 0$, $PQ = QP = 0$, such that $P \sim Q \pmod{\operatorname{com} \mathscr{A}}$. This definition is a "geometrical" one, but a simple algebra can be characterized algebraically by the property "com $\mathscr{A}$ is Abelian". Thus if $\mathscr{A}$ is "large" (hence com $\mathscr{A}$ is „small"), then there is a chance that $\mathscr{A}$ is simple. For example, $\mathscr{L}(\mathscr{H})$ is simple.

Correspondingly, a projection $P \in \operatorname{com} \mathscr{A}$ is called *simple* if $\mathscr{A} \mid P\mathscr{H}$ is simple. Equivalently, $P$ is simple if there is no resolution $P = Q_1 + Q_2$ with $Q_1, Q_2 \neq 0$, $Q_1Q_2 = Q_2Q_1 = 0$, $Q_1, Q_2 \in \operatorname{com} \mathscr{A}$ such that $Q_1 \sim Q_2 \pmod{\operatorname{com} \mathscr{A}}$. A crucial point in connection with this definition is the question whether there are (enough) simple projections in a given v. Neumann algebra $\mathscr{A}$. In the case of our interest (Abelian algebras) the question can be answered affirmatively.

If projections $P, Q \in \operatorname{com} \mathscr{A}$ are given, it may be that certain parts $\mathscr{A} \mid P'\mathscr{H}$, $\mathscr{A} \mid Q'\mathscr{H}$, where $P' \leqq P$, $Q' \leqq Q$, are spatial isomorphic. If there are no such parts, then the projections $P, Q$ are called *disjoint*, $P \mathbin{\delta} Q$. In other words, $P \mathbin{\delta} Q$ if there are no projections $P'$, $Q' \in \operatorname{com} \mathscr{A}$, $P' \leqq P$, $Q' \leqq Q$, such that $P' \sim Q' \pmod{\operatorname{com} \mathscr{A}}$. Again this definition is a "geometrical" one, but there is also an "algebraic" characterization, namely $P \mathbin{\delta} Q$ iff $PTQ \in \operatorname{com} \mathscr{A}$ implies $PTQ = 0$ for each $T \in \mathscr{L}(\mathscr{H})$. In other words, $P \mathbin{\delta} Q$ is valid iff there is no nontrivial intertwining operator between the algebras $\mathscr{A} \mid P\mathscr{H}$ and $\mathscr{A} \mid Q\mathscr{H}$ (with respect to single operators see 1.3.3).

Two v. Neumann algebras are called *disjoint* if there are no (nontrivial) parts being spatial isomorphic.

### 2.1.2. Central covers

The projections of cen $\mathscr{A}$ are called *central projections*. The *central cover* of a projection $P \in \text{com}\,\mathscr{A}$ is denoted and defined by $\text{cov}\,P := \inf_{E \geqq P, E \in \text{cen}\,\mathscr{A}} E$, that is, cov $P$ is the minimal central projection $E$ with the property $E \geqq P$. This algebraic definition can be transformed in a geometrical language: cov $P$ is just the projection onto the subspace $\mathscr{K} = \text{clo spa}\,\{Cf; C \in \text{com}\,\mathscr{A}, f = Pf\}$. Note the following simple properties of cov $P$:

1. $P \sim Q \pmod{\text{com}\,\mathscr{A}}$ implies $\text{cov}\,P = \text{cov}\,Q$ and
2. $P \mathbin{\delta} Q$ iff $\text{cov}\,P \perp \text{cov}\,Q$.

Further, note a crucial property giving a link between the concepts of simple projection and central cover: If $P, Q \in \text{com}\,\mathscr{A}$ are projections with $\text{cov}\,Q \leqq \text{cov}\,P$ and if $Q > 0$ is simple, then there is a projection $P' \in \text{com}\,\mathscr{A}$, $0 < P' < P$, such that $Q \sim P' \pmod{\text{com}\,\mathscr{A}}$.

### 2.1.3. Algebras and projections with multiplicity

A v. Neumann algebra $\mathscr{A}$ is said to have *multiplicity n*, $n \geqq 2$, if there is a system $P_1, P_2, \ldots, P_n \in \text{com}\,\mathscr{A}$ of $n$ simple and pairwise orthogonal projections with $1 = P_1 + P_2 + \cdots + P_n$ such that $P_i \sim P_j \pmod{\text{com}\,\mathscr{A}}$ for $i \neq j$. The algebraic counterpart to this geometrical definition is the proposition "$\mathscr{A}$ has multiplicity $n$ iff $\mathscr{A} \cong \mathscr{A}_1 \otimes 1_n$ where $\mathscr{H} \cong \mathscr{H}_1 \otimes \mathscr{K}_n$, $\dim \mathscr{K}_n = n$ and $\mathscr{A}_1$ is simple". Note that this concept of multiplicity $n$ is justified by the fact that $\mathscr{A}_1 \otimes 1_n \cong \mathscr{A}_1' \otimes 1_m$ with simple algebras $\mathscr{A}_1, \mathscr{A}_1'$ implies $n = m$.

Correspondingly, a projection $P \in \text{com}\,\mathscr{A}$ is said to have *multiplicity n* if $\mathscr{A} \restriction P\mathscr{H}$ has multiplicity $n$ or, equivalently, if there is a resolution $P = Q_1 + Q_2 + \cdots + Q_n$ with $n$ simple and pairwise orthogonal projections such that $Q_i \sim Q_j \pmod{\text{com}\,\mathscr{A}}$ for $i \neq j$. The multiplicity of a projection $P$ is denoted by mul $P$. The terms simple and multiplicity 1 are used synonymously. We collect some properties of "multiple" projections:

1. Let $P \in \text{com}\,\mathscr{A}$ be a projection with $\text{mul}\,P = n$ and let $E$ be a central projection with $0 < E < P$. Then $\text{mul}\,E = n$.
2. Let $E_1, \ldots, E_n$ be a system of pairwise orthogonal central projections and let $P_1, \ldots, P_n \in \text{com}\,\mathscr{A}$ be a system of projections with $0 < P_j < E_j$ and $\text{mul}\,P_j = 1$. Then also $\text{mul}\,(\sum P_j) = 1$.
3. Let $E_1, \ldots, E_m$ be a system of pairwise orthogonal central projections with $\text{mul}\,E_j = n$. Then also $\text{mul}\,(\sum E_j) = n$.
4. Let $E_1 \leqq E_2 \leqq \cdots$ be a sequence of central projections with $\sup E_j = E$ and $\text{mul}\,E_j = n$. Then also $\text{mul}\,E = n$.

## 2.2. Algebras of type I, decomposition theorem

### 2.2.1. Algebras of type I

According to the material collected in 2.1 the crucial question connected with the concept of *multiplicity* (of a projection) is whether there exist a number $n$ and a central projection $E$ such that $\text{mul}\, E = n$. The answer is affirmatively if there are enough simple projections.

A v. Neumann algebra $\mathscr{A}$ is called of *type I* if every nonvanishing central projection $E$ contains a nonvanishing simple projection $P$, $P \leqq E$ and $P \in \text{com}\, \mathscr{A}$. With respect to algebras of type I we have the following deep result.

**Theorem 1.** *If $\mathscr{A}$ is of type I, then there are a number $n \geqq 1$ and a central projection $E \neq 0$ such that* $\text{mul}\, E = n$.

With the help of this theorem the resolution of $\mathscr{A}$ according to different multiplicities easily follows. If $n = \infty, 1, 2, \ldots$ is given, denote by $\mathscr{E}_n$ the system of all central projections $E$ with $\text{mul}\, E = n$ and put $E_n = \sup_{E \in \mathscr{E}_n} E$; $E_n$ is the *maximal central projection* $E$ with $\text{mul}\, E = n$. The system $E_\infty, E_1, E_2, \ldots$ is pairwise orthogonal and $\sum E_n = 1$ is true.

### 2.2.2. Abelian v. Neumann algebras

We introduce a special class of projections of $\text{com}\, \mathscr{A}$. If $f \in \mathscr{H}$, $f \neq 0$, then the projection $Q = Q_f$ onto $\text{clo spa}\, \{Af; A \in \mathscr{A}\}$ belongs to $\text{com}\, \mathscr{A}$. A projection $Q$ of this type is called a *cyclic projection*. The corresponding subspace $\text{clo spa}\, \{Af, A \in \mathscr{A}\}$ is called a *cyclic subspace*.

A set $\mathscr{M}$ of vectors is called *generating with respect to $\mathscr{H}$ and $\mathscr{A}$* if $\text{clo spa}\, \{Af; f \in \mathscr{M}, A \in \mathscr{A}\} = \mathscr{H}$. Thus, if $Q = Q_f$ is cyclic, then $f$ is generating with respect to $Q\mathscr{H}$ and $\mathscr{A} \mid Q\mathscr{H}$. Note that this terminology is in agreement with the term generating set with respect to a Hilbert space $\mathscr{H}$ (see 1.1.1). In the latter case the corresponding algebra is $\mathscr{A} = \{\lambda 1\}$.

A v. Neumann algebra $\mathscr{A}$ is called *cyclic* if there is a single generating vector $f \neq 0$ (or equivalently, if $1_{\mathscr{H}}$ is a cyclic projection, that is, if $\mathscr{H}$ itself is cyclic).

The set of all cyclic projections is a subset of $\text{com}\, \mathscr{A}$. A simple projection is automatically cyclic, but the conversion is not true in general. But if $\mathscr{A}$ is Abelian, the conversion is true.

**Proposition 2.** *If $\mathscr{A}$ is Abelian, then a projection $Q \in$* $\text{com}\, \mathscr{A}$ *is simple iff $Q$ is cyclic.*

Thus for Abelian algebras the concepts of simple algebra and cyclic algebra coincide. Obviously, Prop. 2 implies

**Proposition 3.** *Abelian v. Neumann algebras are of type* I.

Let $\mathscr{A}$ be an Abelian v. Neumann algebra. Then, according to 2.2.1, $\mathscr{A}$ can be represented as a direct sum, which is uniquely determined by $\mathscr{A}$, of the algebras $\mathscr{A} \mid E_n\mathscr{H}$ which have multiplicity $n$. The algebras $\mathscr{A} \mid E_n\mathscr{H}$ are called the *parts of homogeneous multiplicity* or simply the *homogeneous parts* of $\mathscr{A}$. Note that two Abelian v. Neumann algebras $\mathscr{A}$ on $\mathscr{H}$ and $\mathscr{A}'$ on $\mathscr{H}'$ with corresponding maximal central projections $E_n$,

$E_n'$ of multiplicity $n$ are spatial isomorphic, $\mathscr{A} \cong \mathscr{A}'$, iff $\mathscr{A} \mid E_n\mathscr{H} \cong \mathscr{A}' \mid E_n'\mathscr{H}'$ for $n = \infty$ 1, 2, ... In particular, if $\mathscr{A} = W^*(A, 1)$, $\mathscr{A}' = W^*(A', 1')$ where $A$, $A'$ are selfadjoint in $\mathscr{H}$, $\mathscr{H}'$, we have: $A$ is unitarily equivalent to $A'$ iff $A \mid E_n\mathscr{H}$ is unitarily equivalent to $A' \mid E_n'\mathscr{H}'$ for all $n = \infty, 1, 2, \ldots$ Thus, in order to investigate the unitary invariants of an operator $A$, it is sufficient to investigate the unitary invariants of its homogeneous parts.

# 3. Spectral Theory

## 3.1. Spectral measures, scalar spectral integrals, and the spectral theorem

### 3.1.1. Spectral measures on Borel spaces and the Banach space $L^\infty(\Gamma, E)$

Let $\mathscr{H}$ be separable and $\Gamma$ be a Borel space with a Borel field $\mathscr{B}$. A spectral measure $E: \mathscr{B} \ni \Delta \to E(\Delta) \in \mathscr{L}(\mathscr{H})$ is a projection-valued function on $\mathscr{B}$ with the following properties:

(I) $E\left(\bigcup_{j=1}^{\infty} \Delta_j\right) = \sum_{j=1}^{\infty} E(\Delta_j)$ where the Borel sets $\Delta_j$ are pairwise disjoint and $\sum_{j=1}^{\infty}$ means strong convergence,

(II) $E(\emptyset) = 0$,

(III) $E(\Gamma \setminus \Delta) = E(\Delta)^\perp$,

(IV) $E(\Delta_1 \cap \Delta_2) = E(\Delta_1)\, E(\Delta_2)$.

Then also $E(\Delta_1 \cup \Delta_2) = E(\Delta_1) + E(\Delta_2) - E(\Delta_1)\, E(\Delta_2)$ and "$\Delta_1 \subseteq \Delta_2$ implies $E(\Delta_1) \leqq E(\Delta_2)$" are valid.

Simple examples of spectral measures are given by $\Gamma = \{1, 2, \ldots\}$, $\mathscr{B}$: all subsets, $E(\{n\}) := P_n$ where the projections $P_n$ are pairwise orthogonal and $\sum P_n = 1$; or by $\Gamma = \mathbb{R}$, $\mathscr{B} = \mathscr{B}_{\mathbb{R}}$, $\mathscr{H} = L^2(\mathbb{R}, \mathrm{d}x)$, $E(\Delta) := \chi_\Delta$ (the multiplication operator by the characteristic function of $\Delta$).

A Borel set $\Delta \in \mathscr{B}$ is said to have $E$-measure zero if $E(\Delta) = 0$. The Banach space of all $E$-essentially bounded Borel functions $\alpha: \Gamma \to \mathbb{C}$ is denoted by $L^\infty(\Gamma, E)$; it is equipped with the norm $|\alpha|_E = \operatorname*{ess\,sup}_{\lambda \in \Gamma} |\alpha(\lambda)| := \inf_{\Delta, E(\Delta) = 1} \sup_{\lambda \in \Delta} |\alpha(\lambda)|$. Note that all step functions belong to $L^\infty(\Gamma, E)$ for all $E(\cdot)$. Moreover the linear manifold of all step functions is dense in $L^\infty(\Gamma, E)$ for all $E(\cdot)$. If $\Gamma$ is compact, then the Banach space $C(\Gamma)$ is a proper subspace of $L^\infty(\Gamma, E)$.

### 3.1.2. The spectral integral and the v. Neumann algebra $T(L^\infty)$

The complex-valued measures $\mu_{f,g}(\Delta) := (f, E(\Delta)\, g)$, $f, g \in \mathscr{H}$, are finite, $|\mu_{f,g}(\Delta)| \leqq \|f\|\, \|g\|$, and they have a finite total variation, $|\mu_{f,g}|\, (\Gamma) \leqq 4 \sup_{\Delta \in \mathscr{B}_\Gamma} |\mu_{f,g}(\Delta)|$.

The *spectral integral* for step functions $\tau$ is defined and denoted by

$$\int_\Gamma \tau(x)\, E(\mathrm{d}x) := \sum_{\varrho=1}^{m} c_\varrho E(\Delta_\varrho), \quad \tau(x) = \sum_{\varrho=1}^{m} c_\varrho \chi_{\Delta_\varrho}(x), \quad c_\varrho \in \mathbb{C}.$$

Because of $\left\| \int_\Gamma \tau(x)\, E(\mathrm{d}x) \right\| \leqq 4\, |\tau|_E$ the spectral integral can be defined for all $\alpha \in L^\infty(\Gamma, E)$ by

$$\mathscr{L}(\mathscr{H}) \ni T(\alpha) := \int_\Gamma \alpha(x)\, E(\mathrm{d}x) := \operatorname*{n-lim}_{q \to \infty} \int_\Gamma \tau_q(x)\, E(\mathrm{d}x), \ |\alpha - \tau_q|_E \to 0 \text{ as } q \to \infty.$$

If $\Delta \in \mathscr{B}_\Gamma$, then we define $\int\limits_\Delta \alpha(x)\, E(\mathrm{d}x) := \int\limits_\Gamma \chi_\Delta(x)\, \alpha(x)\, E(\mathrm{d}x)$. The spectral integral is countably additive (with respect to strong convergence), it is linear with respect to $\alpha$ and depends only on the equivalence class of $\alpha$ (with respect to $E$). Further we collect the properties:

(I) $T(\alpha\beta) = T(\alpha)\, T(\beta)$,
(II) $T(\alpha)^* = T(\bar{\alpha})$, that is, all $T(\alpha)$ are normal,
(III) $\left(u, \int\limits_\Gamma \alpha(x)\, E(\mathrm{d}x)\, v\right) = \int\limits_\Gamma \alpha(x)\, \mu_{u,v}(\mathrm{d}x)$ and

$$\left(\int\limits_\Gamma \alpha(x)\, E(\mathrm{d}x)\, u, \int\limits_\Gamma \beta(x)\, E(\mathrm{d}x)\, v\right) = \int\limits_\Gamma \overline{\alpha(x)}\, \beta(x)\, \mu_{u,v}(\mathrm{d}x),$$

(IV) $\|T(\alpha)\| = |\alpha|_E$,
(V) $T(\alpha)$ has a bounded inverse iff $1/\alpha$ is essentially bounded on $\Gamma$,
(VI) $\operatorname{spec} T(\alpha) = \bigcap\limits_{\Delta, E(\Delta)=1} \operatorname{clo} \operatorname{ima} (\alpha \mid \Delta)$.

Obviously the set of all $T(\alpha)$, $\alpha \in L^\infty(\Gamma, E)$, is an Abelian $C^*$-algebra $\mathscr{A}$ with 1. Note that $\mathscr{A}$ is a v. Neumann algebra and $\mathscr{A} = W^*(E(\Delta); \Delta \in \mathscr{B})$.

If $\Gamma$ is a topological space (metrizable, separable, locally compact), then all measures $\mu_{f,f}(\cdot)$ are nonnegative Borel measures and all $\mu_{f,g}(\cdot)$ are complex-valued Borel measures. As usual we define the *support* of $E(\cdot)$ by

$$\operatorname{supp} E := \{\bigcap \Delta; E(\Delta) = 1, \Delta \text{ closed}\},$$

that is, supp $E$ is the smallest closed set $\Delta_0$ such that $E(\Delta_0) = 1$. Obviously, $E(\cdot)$ may be considered only on supp $E$ in every case, that is, we can choose $\Gamma = \operatorname{supp} E$. There are two essentially different cases:

(I) supp $E$ is compact,
(II) supp $E$ is noncompact.

If supp $E$ is compact, then $C(\operatorname{supp} E) \subset L^\infty(\operatorname{supp} E, E) = \mathscr{A}$. Then the $C^*$-algebra $\mathscr{C}$ of all $T(\alpha)$, $\alpha \in C(\operatorname{supp} E)$, is a subalgebra of $\mathscr{A}$ and $\mathscr{A} = \operatorname{clo}_s \mathscr{C} = \operatorname{clo}_w \mathscr{C}$.

If $\Gamma = \mathbb{C}$ and supp $E$ is compact, then the bounded normal operator $A = \int \lambda E(\mathrm{d}\lambda)$ generates an Abelian $C^*$-algebra $C^*(A, 1)$, consisting of all $T(\alpha) = \int \alpha(x)\, E(\mathrm{d}x)$, $\alpha \in C(\operatorname{supp} E)$, such that the spectral integral realizes the Gelfand isomorphism (see 1.4.1). The spectral theorem to be presented in the next section states that the Gelfand isomorphism can be always implemented by a spectral integral.

### 3.1.3. The spectral theorem for bounded normal operators

The spectral theorem states a one-to-one correspondence between bounded normal operators $A$ and spectral measures $E(\cdot)$ on $\mathscr{B}_\mathbb{C}$ with supp $E$ compact such that supp $E$ = spec $A$ and

$$C(\operatorname{spec} A) \ni \alpha \to \alpha(A) = \int\limits_\mathbb{C} \alpha(x)\, E(\mathrm{d}x) \in C^*(A, 1) \tag{1}$$

is a realization of the Gelfand isomorphism between $C(\operatorname{spec} A)$ and $C^*(A, 1)$. Note that the spectral theorem is an easy consequence of Riesz' theorem on the characterization of $(C(\operatorname{spec} A))^*$ by Borel measures on spec $A$. Via the spectral theorem we are able to define the function $\alpha(A)$ of a bounded normal operator $A = \int \lambda E(d\lambda)$ for all essentially bounded Borel functions $\alpha \in L^\infty(\operatorname{spec} A, E)$ by

$$\alpha(A) := \int \alpha(x)\, E(dx).$$

Note that $W^*(A, 1)$ coincides with $W^*(E(\Delta), \Delta \in \mathcal{B}_{\mathbb{C}})$. Thus $W^*(A, 1)$ consists of all essentially bounded Borel functions of $A$, whereas $C^*(A, 1)$ consists of all continuous functions of $A$. Note that the *spectral projections* $E(\Delta)$ of the spectral measure coincide with the *central projections* of $W^*(A, 1)$.

If $\alpha \in L^\infty(\operatorname{spec} A, E)$, then the spectral measure $F(\cdot)$ of $\alpha(A)$ is given by $F(\Delta) = E(\alpha^{-1}(\Delta))$, $\Delta \in \mathcal{B}_{\mathbb{C}}$.

### 3.1.4. The spectral theorem and the functional calculus for unbounded selfadjoint operators

The spectral theorem for an unbounded selfadjoint operator $A$ states a one-to-one correspondence between these operators $A$ and spectral measures $E(\cdot)$ in $\mathcal{B}_{\mathbb{R}}$ with noncompact support supp $E$ such that $\operatorname{spec} A = \operatorname{supp} E$ and

(I) $\operatorname{dom} A = \{f; \int \lambda^2 (f, E(d\lambda) f) < \infty\}$,

(II) $Af = \int_{-\infty}^{\infty} \lambda E(d\lambda) f := \operatorname{s-lim}_{n\to\infty} \int_{-n}^{n} \lambda E(d\lambda) f$

are valid. Note that this theorem can be proved with the help of the correspondence between $A$ and $R_A(i) = (i - A)^{-1}$, which is bounded and normal, where $\lambda \in \operatorname{spec} A \cup \{\infty\}$ corresponds to $\mu \in \operatorname{spec} R_A(i)$ via $\mu = (i - \lambda)^{-1}$ and $\lambda = \infty$ corresponds to $\mu = 0$ which always belongs to $\operatorname{spec} R_A(i)$ because $A$ is unbounded.

The spectral theorem implies a general functional calculus for (bounded and unbounded) selfadjoint operators. Let $A$ be selfadjoint, $E(\cdot)$ be its spectral measure and let $\alpha$ denote a complex-valued Borel function on $R$. The function $\alpha(A)$ can be defined if $\int |\alpha(\lambda)|^2 (f, E(d\lambda) f) < \infty$ for a dense set of vectors $f \in \mathcal{H}$ (note that this implies that $\alpha(\cdot)$ is a.e. finite with respect to $E$). The set of all such vectors $f$ is $\operatorname{dom} \alpha(A)$. On its domain $\alpha(A)$ is defined by

$$\alpha(A) f = \int_{-\infty}^{\infty} \alpha(\lambda)\, E(d\lambda) f := \operatorname{s-lim}_{n\to\infty} \int_{-n}^{n} \alpha(\lambda)\, E(d\lambda) f.$$

We collect the following properties of the *closed* operator $\alpha(A)$:

(I) $(a\alpha + b\beta)(A) = a\alpha(A) + b\beta(A)$, $\quad a, b \in \mathbb{C}$.

(II) $(\alpha\beta)(A) = \alpha(A)\, \beta(A)$.

(III) $(\bar{\alpha})(A) = \alpha(A)^*$ in particular, if $\alpha$ is real-valued, then $\alpha(A)$ is selfadjoint.

(IV) If $\alpha_n(\lambda) \to \alpha(\lambda)$ a.e. with respect to $E$ and $\sup_n |\alpha_n|_E < \infty$ (hence also $\alpha \in L^\infty(\mathbb{R}, E)$ and $\alpha_n(A)$, $\alpha(A)$ are bounded), then $\operatorname{s-lim}_{n\to\infty} \alpha_n(A)$ exists and equals $\alpha(A)$.

(V) $\alpha(A)$ is bounded iff $\alpha \in L^\infty(\mathbb{R}, E)$.

(VI) If $B$ is bounded and $BA \subseteq AB$, then $B\alpha(A) \subseteq \alpha(A)\, B$.

In the following $E_H(\Delta)$ will denote the spectral measure of a selfadjoint operator $H$.

The concepts of generating set, simple (cyclic), multiplicity $n$, introduced in Chapter 2 for v. Neumann algebras, are used in the following also for a single selfadjoint (bounded or unbounded) operator $H$.

A set $\mathscr{M}$ of vectors is called *generating with respect to $\mathscr{H}$ and $H$* if $\mathscr{M}$ is generating with respect to $\mathscr{H}$ and $W^*(E_H(\Delta);\ \Delta \in \mathscr{B}_{\mathbb{R}})$. Sometimes $\mathscr{M}$ is called generating with respect to $\mathscr{H}$ and $E_H(\cdot)$ in this case.

$H$ is called *simple* if $W^*(E_H(\Delta);\ \Delta \in \mathscr{B}_{\mathbb{R}})$ is simple (note that the concepts of simple and cyclic coincide with respect to $H$ because $W^*(E_H(\Delta),\ \Delta \in \mathscr{B}_{\mathbb{R}})$ is Abelian).

$H$ is said to have *multiplicity* $n$ if $W^*(E_H(\Delta),\ \Delta \in \mathscr{B}_{\mathbb{R}})$ has multiplicity $n$. Sometimes this property is expressed by saying "the spectrum of $H$ has multiplicity $n$".

### 3.1.5. Spectral integrals and Bochner integrals

For applications it is useful to transform spectral integrals into Bochner integrals. In this direction we have the following result.

**Proposition 1.** *Let $A$ be selfadjoint with spectral measure $E_A(\cdot)$, let $\Gamma \subset \mathbb{R}$ be compact and let $\alpha(x, y)$ be a complex-valued Borel function: $\mathbb{R} \times \Gamma \ni \{x, y\} \to \alpha(x, y) \in \mathbb{C}$. Assume the existence of a dense linear set $\mathscr{D}$ of vectors $f \in \mathscr{H}$ such that for almost all $y \in \Gamma (\mathrm{mod}|\cdot|)$ and all $f \in \mathscr{D}$, $\int |\alpha(x, y)|^2\, \sigma_f(\mathrm{d}x) < \infty$ and*

$$\int_{\mathbb{R}\times\Gamma} |\alpha(x, y)|^2\, \sigma_f(\mathrm{d}x)\, \mathrm{d}y < \infty \tag{2}$$

*where $\sigma_f(\Delta) = (f, E_A(\Delta) f)$. Then:*

(i) *$\alpha(A, y)$ exists a.e. on $\Gamma$* (mod $|\cdot|$) *where* dom $\alpha(A, y) \supseteq \mathscr{D}$,

(ii) *$\beta(x) = \int_\Gamma \alpha(x, y)\, \mathrm{d}y$ is a.e. finite (with respect to $E_A$) and $\beta(A)$ exists with* dom $\beta(A) \supseteq \mathscr{D}$,

(iii) *$\alpha(A, y) f$, $f \in \mathscr{D}$, is strongly measurable and the Bochner integral $\int_\Gamma \alpha(A, y) f\, \mathrm{d}y$ exists,*

(iv) *$\int_\Gamma \alpha(A, y) f\, \mathrm{d}y = \beta(A) f$, $f \in \mathscr{D}$.*

**Proof.** (i) is obvious.

(ii) We have $\int_\Gamma |\alpha(x, y)|\, \mathrm{d}y \leq |\Gamma|^{1/2} \left(\int_\Gamma |\alpha(x, y)|^2\, \mathrm{d}y\right)^{1/2}$. Thus from (2) follows that $\int_\Gamma |\alpha(x, y)|\, \mathrm{d}y$ is finite a.e. with respect to all $\sigma_f(\cdot)$, $f \in \mathscr{D}$, and we have

$$\int_{\mathbb{R}} \sigma_f(\mathrm{d}x) \left(\int_\Gamma |\alpha(x, y)|\, \mathrm{d}y\right)^2 \leq |\Gamma| \int_{\mathbb{R}} \sigma_f|(\mathrm{d}x) \int_\Gamma |\alpha(x, y)|^2\, \mathrm{d}y < \infty.$$

This implies (ii).

(iii) The scalar function $(f, \alpha(A, y) f) = \int \alpha(x, y)\, \sigma_f|(\mathrm{d}x)$, $f \in \mathscr{D}$, is Borel measurable because of

$$\int_{\mathbb{R}\times\Gamma} |\alpha(x, y)|\, \sigma_f(\mathrm{d}x)\, \mathrm{d}y \leq |\Gamma|^{1/2}\, \sigma_f(\mathbb{R}) \left(\int_{\mathbb{R}\times\Gamma} |\alpha(x, y)|^2\, \sigma_f(\mathrm{d}x)\, \mathrm{d}y\right)^{1/2} < \infty. \tag{3}$$

Namely according to Fubini's theorem, (3) implies that $\int |\alpha(x, y)|\, \sigma_f(\mathrm{d}x)$ is Borel measurable and a.e. finite mod $|\cdot|$. According to the polarization identity we obtain that $(g, \alpha(A, y) f)$ is Borel measurable for all $f. g \in \mathscr{D}$. But $\mathscr{D}$ is dense in $\mathscr{H}$, hence

$\alpha(A, y) f$ is weakly measurable for $f \in \mathcal{D}$, hence strongly measurable. Further we obtain

$$\int_\Gamma \|\alpha(A, y) f\| \, dy = \int_\Gamma \left( \int_{\mathbb{R}} |\alpha(x, y)|^2 \sigma_f(dx) \right)^{1/2} dy \leq |\Gamma|^{1/2} \left( \int_{\mathbb{R} \times \Gamma} |\alpha(x, y)|^2 \sigma_f(dx) \, dy \right)^{1/2} < \infty,$$

hence $\alpha(A, y) f$ is Bochner integrable on $\Gamma$.

(iv) It is sufficient to prove $\left(h, \int_\Gamma \alpha(A, y) f \, dy\right) = (h, \beta(A) f)$ for a dense set of vectors $h$, for example, for $h \in \mathcal{D}$. Because of the polarization identity it is sufficient to prove $\left(f, \int_\Gamma \alpha(A, y) f \, dy\right) = (f, \beta(A) f)$ for all $f \in \mathcal{D}$. But the latter equation explicitely means

$$\int_\Gamma \left( \int_{\mathbb{R}} \alpha(x, y) \, \sigma_f(dx) \right) dy = \int_{\mathbb{R}} \left( \int_\Gamma \alpha(x, y) \, dy \right) \sigma_f(dx),$$

and this is true according to Fubini's theorem which is applicable because of (3). ∎

### 3.1.6. Spectral measures and resolvents

As an application of the functional calculus and the transformation of spectral integrals into Bochner integrals we consider the calculation of the spectral measure $E_A(\cdot)$ from the resolvent $R_A(z)$ for a selfadjoint operator $A$.

**Proposition 2.** *Let $A$ be selfadjoint and let $(a, b)$ be a bounded open interval. Then*

$$E_A\big((a, b)\big) = \operatorname*{s-lim}_{\delta \to +0} \operatorname*{s-lim}_{\varepsilon \to +0} (2i\pi)^{-1} \int_{a+\delta}^{b-\delta} \big(R_A(x - i\varepsilon) - R_A(x + i\varepsilon)\big) \, dx. \tag{4}$$

Proof. We consider

$$\alpha(x; \delta, \varepsilon) := (2i\pi)^{-1} \int_{a+\delta}^{b-\delta} \left( \frac{1}{y - i\varepsilon - x} - \frac{1}{y + i\varepsilon - x} \right) dy$$
$$= \pi^{-1} \left( \arctan \frac{b - \delta - x}{\varepsilon} - \arctan \frac{a + \delta - x}{\varepsilon} \right)$$

and obtain $|\alpha(x; \delta, \varepsilon)| \leq 1$,

$$\lim_{\varepsilon \to +0} \alpha(x; \delta, \varepsilon) = \chi_{(a+\delta, b-\delta)}(x), \qquad x \neq a + \delta, b - \delta,$$

$$\lim_{\delta \to +0} \lim_{\varepsilon \to +0} \alpha(x; \delta, \varepsilon) = \chi_{(a,b)}(x);$$

that is, we have pointwise convergence and uniform boundedness. Hence according to 3.1.4 (IV), we obtain $\operatorname*{s-lim}_{\delta \to +0} \operatorname*{s-lim}_{\varepsilon \to +0} \alpha(A; \delta, \varepsilon) = E_A\big((a, b)\big)$. The remaining assertion is

$$\alpha(A; \delta, \varepsilon) = (2i\pi)^{-1} \int_{a+\delta}^{b-\delta} \big(R_A(x - i\varepsilon) - R_A(x + i\varepsilon)\big) \, dx. \tag{5}$$

The function $\beta_\varepsilon(x, y) := (y - i\varepsilon - x)^{-1} - (y + i\varepsilon - x)^{-1}$ is holomorphic and uniformly bounded for $(x, y) \in \mathbb{R} \times [a + \delta, b - \delta]$. Hence Prop. 1 is applicable and (4) is proved. ∎

## 3.2. Unitary evolution groups

### 3.2.1. Stone's theorem

The unitary evolution group generated by the selfadjoint operator $H$ is always denoted by

$$U(t) := e^{-itH} = \int_{-\infty}^{\infty} e^{-it\lambda} E_H(d\lambda), \qquad t \in \mathbb{R}. \tag{6}$$

Note that $U(t)$ is a strongly continuous one-parameter unitary group. $U(t)$ is norm-continuous iff $H$ is bounded. dom $H$ can be characterized by

$$\operatorname{dom} H = \{f;\ \operatorname*{s-lim}_{t\to 0} t^{-1}\big(U(t) - 1\big) f \text{ exists}\}$$

and in this case $\operatorname*{s-lim}_{t\to 0} t^{-1}\big(U(t) f - f\big) = -iHf$ is valid.

The famous Stone theorem states that to each strongly continuous one-parameter unitary group $U(t)$ there exists a selfadjoint operator (the generator of the group) such that $U(t) = e^{-itH}$. Note that weak measurability of a group $U(t)$ already implies its strong continuity.

### 3.2.2. Operator Fourier transformations

The representation (6) of a strongly continuous unitary group by a spectral integral with respect to the spectral measure of its generator can be used to represent functions of $H$ by operator Fourier integrals. As an example we state

**Proposition 3.** *Let $\varphi \in L^1(\mathbb{R}, dx) \cap L^2(\mathbb{R}, dx)$ and let $H$ be selfadjoint. Then*

$$\phi(H) = (2\pi)^{-1/2} \int_{-\infty}^{\infty} \varphi(x)\, e^{-ixH}\, dx \tag{7}$$

*where $\phi$ denotes the Fourier transform of $\varphi$ and the integral on the right-hand side is defined as strong limit of Bochner integrals on compact intervals.*

Proof. Let $\Gamma \subset \mathbb{R}$ be compact and $\phi_\Gamma(\lambda) = (2\pi)^{-1/2} \int_\Gamma \varphi(x)\, e^{-i\lambda x}\, dx$. Then according to Prop. 1 we have $\phi_\Gamma(H) = (2\pi)^{-1/2} \int_\Gamma \varphi(x)\, e^{-ixH}\, dx$. This Bochner integral is strongly (even with respect to the norm) convergent if $\Gamma \to \mathbb{R}$. On the other hand, we have $\phi(H) = \operatorname*{s-lim}_{\Gamma\to\mathbb{R}} \phi_\Gamma(H)$ because $\phi_\Gamma(\lambda) \to \phi(\lambda)$ pointwise on $\mathbb{R}$, $|\phi_\Gamma(\lambda)|$ is uniformly bounded and hence 3.1.4 (IV) is applicable. ∎

As an example we consider the special case $\varphi(x) = \chi_{(-\infty,0]}(x)\, e^{\varepsilon x}$, $\varepsilon > 0$. In this case we have

$$\phi(\lambda) = (2\pi)^{-1/2} \int_{-\infty}^{0} e^{-i\lambda x}\, e^{\varepsilon x}\, dx = (2\pi)^{-1/2}\, i(\lambda + i\varepsilon)^{-1},$$

hence from (7) we get $i(2\pi)^{-1/2}\,(H + i\varepsilon)^{-1} = (2\pi)^{-1/2} \int_{-\infty}^{0} e^{\varepsilon x}\, e^{-ixH}\, dx$ or, if we substitute $H$ by $\lambda - H$,

$$i(2\pi)^{-1/2}\, R_H(\lambda + i\varepsilon) = (2\pi)^{-1/2} \int_{-\infty}^{0} e^{-i\lambda x}\, e^{\varepsilon x}\, e^{ixH}\, dx, \quad \varepsilon > 0. \tag{8}$$

Formula (8) can also be considered as the calculation of the vector-valued Fourier transform of the vector-valued function $f_\varepsilon(x) = \chi_{(-\infty,0]}(x)\, e^{\varepsilon x}\, e^{ixH}f$, $f \in \mathscr{H}$. This point of view can also be used with respect to the general formula (7) if we replace $H$ by $\lambda - H$. Then we obtain

$$\phi(\lambda - H) = (2\pi)^{-1/2} \int e^{-i\lambda x} \varphi(x)\, e^{ixH}\, dx, \tag{9}$$

that is, the Fourier transform of the vector-valued function $g(x) = \varphi(x)\, e^{ixH} g$ is given by $\phi(\lambda - H)\, g$. Since $\varphi \in L^2(\mathbb{R}, dx)$, we have Parseval's equation

$$\int_{-\infty}^{\infty} \|\phi(\lambda - H)\, g\|_{\mathscr{H}}^2\, d\lambda = \|g\|^2 \int |\varphi(x)|^2\, dx.$$

For (8) this gives

$$(2\pi)^{-1} \int_{-\infty}^{\infty} \|R_H(\lambda + i\varepsilon)\, g\|^2\, d\lambda = (2\varepsilon)^{-1}, \quad \|g\| = 1, \quad \varepsilon > 0. \tag{10}$$

If (8) is multiplied from the left by an operator $X$, then we get

$$i(2\pi)^{-1/2}\, X R_H(\lambda + i\varepsilon) = (2\pi)^{-1/2} \int_{-\infty}^{0} e^{-i\lambda x}\, e^{\varepsilon x}\, X\, e^{ixH}\, dx.$$

Applying now Parseval's equation, we obtain

$$(2\pi)^{-1} \int_{-\infty}^{\infty} \|X R_H(\lambda + i\varepsilon)\, g\|^2\, d\lambda = \int_{0}^{\infty} e^{-2\varepsilon t}\, \|X\, e^{-itH} g\|^2\, dt. \tag{11}$$

### 3.2.3. The v. Neumann algebra generated by $e^{-itH}$

The algebra $W^*(e^{-itH};\, t \in \mathbb{R})$ is the v. Neumann algebra generated by $e^{-itH}$. Obviously, one has $W^*(e^{-itH};\, t \in \mathbb{R}) \subseteq W^*\big(E_H(\Delta);\, \Delta \subseteq \mathscr{B}_{\mathbb{R}}\big)$ since $e^{-itH}$ is a spectral integral. But (8) shows that all resolvents $R_H(z)$, $\operatorname{Im} z \neq 0$, are members of $W^*(e^{-itH};\, t \in \mathbb{R})$, hence according to Prop. 2 all $E_H\big((a, b)\big)$ are members of $W^*(e^{-itH};\, t \in \mathbb{R})$ where $(a, b)$ is an open interval. Therefore $W^*(e^{-itH}; t \in \mathbb{R}) = W^*(E_H(\Delta); \Delta \subseteq \mathscr{B}_{\mathbb{R}})$ is valid. Note that in particular this implies

$$\text{clo spa}\,\{E_H(\Delta)\,f; \Delta \in \mathscr{B}_{\mathbb{R}}\} = \text{clo spa}\,\{e^{-itH}f; t \in \mathbb{R}\}, \qquad f \in \mathscr{H}.$$

Thus there is always a sequence of trigonometrical polynomials $\tau_n(\lambda)$ such that $\operatorname{s-lim}_{n\to\infty} \tau_n(H)\,f = E_H(\Delta)\,f$, $\Delta$ compact. Note that if two selfadjoint operators $H_1$, $H_2$ on $\mathscr{H}_1$, $\mathscr{H}_2$ and two vectors $f_1$, $f_2$ are given, then it is possible to choose *one* sequence $\tau_n(\lambda)$ to approximate $E_{H_1}(\Delta)\,f_1$ and $E_{H_2}(\Delta)\,f_2$. If $H$ is bounded, then there is a positive number $a$ such that even

$$\text{clo spa}\,\{e^{-itH}f; t \in \mathbb{R}\} = \text{clo spa}\,\{e^{-imaH}f; m = 0, \pm 1, \pm 2, \ldots\}$$

is valid for all $f \in \mathscr{H}$, for $e^{-itH}f = \int e^{-it\lambda}\,E(d\lambda)\,f$ may be approximated by $\int g_n(\lambda)\,E_H(d\lambda)\,f$ with $g_n$ continuous functions of compact support, and $g_n$ may be approximated by trigonometrical polynomials $\sum c_m\,e^{-ima\lambda}$ where $a$ depends on spec $H$ (for example $a\,\|H\| \leqq 2\pi$ is possible).

## 3.3. Classification of selfadjoint operators and their spectra by properties of their spectral measures

### 3.3.1. The spectrally discrete subspace

If $f \in \text{dom}\,H$ and $Hf = \lambda f$, then $f$ is called an *eigenvector* of $H$ with the corresponding *eigenvalue* $\lambda$. Every eigenvalue belongs to spec $H$. The set of all eigenvalues is at most countable and is called the *eigenvalue spectrum* denoted by eig $H$. $\lambda \in \mathbb{R}$ is eigenvalue iff $E_H(\{\lambda\}) \neq 0$; in this case $E_H(\{\lambda\})\,\mathscr{H}$ is the subspace of all eigenvectors belonging to $\lambda$ iand dim $E_H(\{\lambda\})\,\mathscr{H}$ is called the *multiplicity* of $\lambda$. The subspace

$$\mathscr{H}_p(H) := \text{clo spa}\,\{f; f \text{ is eigenvector of } H\} = \text{clo spa}\,\{E_H(\{\lambda\})\,\mathscr{H}; \lambda \in \text{eig}\,H\}$$

s called the *spectrally discrete subspace* of $H$. The projection onto $\mathscr{H}_p(H)$, the *spectrally discrete projection*, is denoted by $P_H^p$. A vector $f$ belongs to $\mathscr{H}_p(H)$ iff there is a countable set $\Delta_f$ such that $(f, E_H(\mathbb{R} \setminus \Delta_f)\,f) = 0$. Actually, $\Delta_f$ can be chosen as eig $H$ uniformly for all $f \in \mathscr{H}_p\,(H)$. The projection $P_H^p$ is given by $P_H^p = \sum_{\lambda \in \text{eig}\,H} E_H(\{\lambda\}) = E_H(\text{eig}\,H)$, thus $P_H^p \in \text{com}\,(\text{com}\,H) = W^*(H, 1)$ is a spectral projection. Further we use the notation $\text{spec}_p\,H := \text{spec}\,(H \upharpoonright P_H^p\mathscr{H}) = \text{clo eig}\,H$. Note that eig $H$ itself is not a spectrum, in general. The operator $H \upharpoonright P_H^p\mathscr{H}$ is called the *spectrally discrete part* of $H$.

### 3.3.2. The spectrally continuous subspace

The subspace $\mathscr{H}_c(H) := \mathscr{H}_p(H)^\perp$ is called the *spectrally continuous subspace* and the projection $P_H^c := (P_H^p)^\perp$, the *spectrally continuous projection*. Note that $f \in \mathscr{H}_c(H)$ iff $(f, E_H(\{\lambda\})\,f) = 0$ for all $\lambda \in \mathbb{R}$, that is, iff $\|E_H((-\infty, \lambda))\,f\|^2$ is continuous on $\mathbb{R}$ or, equivalently, iff $E_H((-\infty, \lambda))\,f$ is strongly continuous on $\mathbb{R}$. Obviously $P_H^c \in \text{com}\,(\text{com}\,H,)$ that is, $P_H^c$ is a spectral projection, namely $P_H^c = E_H\,(\text{spec}\,H \setminus \text{eig}\,H)$. The operator $H \upharpoonright P_H^c\mathscr{H}$ is called the *spectrally continuous part* of $H$ and its spectrum $\text{spec}_c\,H = \text{spec}\,(H \upharpoonright P_H^c\mathscr{H})$ is called the *continuous spectrum* of $H$.

### 3.3.3. The singularly continuous subspace

The *singularly continuous subspace* $\mathscr{H}_{sc}(H)$ is defined by the set of all vectors $f \in \mathscr{H}_c(H)$ such that there is a Borel set $\Delta_f$ with $|\Delta_f| = 0$ and $E_H(\Delta_f) f = f$ ($f \in \mathscr{H}_{sc}(H)$ iff $\mu_f(\cdot) \perp |\cdot|$). Obviously $\mathscr{H}_{sc}(H)$ is a subspace of $\mathscr{H}_c(H)$. The projection $P_H^{sc}$ onto $\mathscr{H}_{sc}(H)$ is called the *singular continuous projection*.

**Proposition 4.** *$P_H^{sc}$ is a spectral projection, that is,* $P_H^{sc} \in$ com com $H$.

**Proof.** If $C \in$ com $H$, then $E_H(\Delta) C = C E_H(\Delta)$ for all $\Delta \in \mathscr{B}_{\mathbb{R}}$. Let $f \in \mathscr{H}_{sc}(H)$, that is, $E_H(\Delta_f) f = f$ for some $\Delta_f$ with $|\Delta_f| = 0$. Then $Cf = C E_H(\Delta_f) f = E_H(\Delta_f) Cf$, hence $Cf \in P_H^{sc}\mathscr{H}$. Thus we obtain $P_H^{sc} C = C P_H^{sc}$. ■

By $P_H^{sc} \in$ com com $H$, there is a Borel set $\Delta^{sc}$ such that $P_H^{sc} = E_H(\Delta^{sc})$. Note that $\Delta^{sc}$ is not unique. For example $\Delta^{sc} \cup \left\{\bigcup_{i=1}^{\infty} \lambda_i\right\}$ with $\lambda_i \notin$ eig $H$ is again such a set.

**Proposition 5.** *There is a set $\Delta^{sc}$ such that* eig $H \cap \Delta^{sc} = \emptyset$ *and* $|\Delta^{sc}| = 0$.

**Proof.** The first assertion is obvious. Because of $\mathscr{H}_{sc}(H) \subseteq \mathscr{H}$, the subspace $\mathscr{H}_{sc}(H)$ is separable. Let $f_1, f_2, \dots$ denote a dense sequence and $\Delta_1, \Delta_2, \dots$ the corresponding sequence of null sets with $E_H(\Delta_i) f_i = f_i$. Let $P = E_H\left(\bigcup_{i=1}^{\infty} \Delta_i\right)$ and $\Delta^{sc} = \bigcup_{i=1}^{\infty} \Delta_i$. Then we have $Pf_i = f_i$ and therefore $Pf = f$ for all $f \in \mathscr{H}_{sc}(H)$. Thus $P_H^{sc} \leq P$. On the other hand $Pf = f$ implies $E_H(\Delta^{sc}) f = f$ with $|\Delta^{sc}| = 0$ by construction. Thus $P \leq P_H^{sc}$. ■

A consequence of this statement is that the null set $\Delta_f$ appearing in the definition of $\mathscr{H}_{sc}(H)$ can be chosen independently of $f \in \mathscr{H}_{sc}(H)$, namely $\Delta_f := \Delta^{sc}$. The operator $H \restriction P_H^{sc}\mathscr{H}$ is called the *singularly continuous part* of $H$ and its spectrum spec$_{sc}$ $H$ := spec($H \restriction P_H^{sc}\mathscr{H}$) is called the *singularly continuous spectrum* (of $H$). Note that spec$_{sc}$ $H \subseteq$ clo $\Delta^{sc}$. In general $|\text{spec}_{sc}\, H| > 0$ despite of $|\Delta^{sc}| = 0$. The operators $H$ with $|\text{spec}_{sc}\, H| = 0$ are of special interest.

### 3.3.4. The absolutely continuous subspace

The *absolutely continuous subspace* $\mathscr{H}_{ac}(H)$ is defined by the set of all vectors $f \in \mathscr{H}$ such that $\mu_f(\Delta) = (f, E_H(\Delta) f) = 0$ for all Borel sets $\Delta$ with $|\Delta| = 0$. In other words, $f \in \mathscr{H}_{ac}(H)$ iff $\mu_f(\cdot)$ is absolutely continuous (with respect to $|\cdot|$), $\mu_f \ll |\cdot|$. $\mathscr{H}_{ac}(H)$ is a subspace and $\mathscr{H}_{ac}(H) \subseteq \mathscr{H}_c(H)$. The projection $P_H^{ac}$ onto $\mathscr{H}_{ac}(H)$ is called the *absolutely continuous projection*.

**Proposition 6.** *$P_H^{ac}$ is a spectral projection,* $P_H^{ac} \in$ com com $H$, *and the relations* $\mathscr{H}_{ac}(H) \perp \mathscr{H}_{sc}(H)$ *and* $\mathscr{H}_{ac}(H) \oplus \mathscr{H}_{sc}(H) = \mathscr{H}_c(H)$ *are valid.*

**Proof.** At first we prove $\mathscr{H}_{sc} \perp \mathscr{H}_{ac}$: let $f \in \mathscr{H}_{ac}$, $g \in \mathscr{H}_{sc}$, that is $g = E_H(\Delta^{sc}) g$, where $|\Delta^{sc}| = 0$. Therefore we obtain $(f, g) = (f, E_H(\Delta^{sc}) g) = (E_H(\Delta^{sc}) f, g) = 0$. Since $P_H^{ac} \leq P_H^c$ we obtain $P_H^{ac} \leq P_H^c - P_H^{sc}$. Now we prove $P_H^c - P_H^{sc} \leq P_H^{ac}$. Let $f \in (P_H^c - P_H^{sc}) \mathscr{H}$, then $f \in \mathscr{H}_c$ and $P_H^{sc} f = 0$, hence

$$P_H^{sc} E_H(\Delta) f = E_H(\Delta) P_H^{sc} f = 0, \qquad \Delta \text{ Borel set.}$$

Therefore $\|E_H(\Delta) f\|^2 = (f, E_H(\Delta) f) = 0$ if $|\Delta| = 0$, because in the opposite case $E_H(\Delta) f \neq 0$ (for some null set $\Delta$) we obtain a contradiction to the above relation. That is, we have $f \in P_H^{ac}\mathscr{H}$ and finally $P_H^{ac} = P_H^{c} - P_H^{sc}$, hence $P_H^{ac}$ is a spectral projection and the last assertion is valid too. ∎

The operator $H \restriction P_H^{ac}\mathscr{H}$ is called the *absolutely continuous part* of $H$ and its spectrum $\operatorname{spec}_{ac} H := \operatorname{spec} (H \restriction P_H^{ac}\mathscr{H})$ is called the *absolutely continuous spectrum* of $H$. According to Prop. 6 there is a Borel set $\Delta^{ac}$ with $P_H^{ac} = E_H(\Delta^{ac})$. $\Delta^{ac}$ is not unique and $\operatorname{spec}_{ac} H \subseteq \operatorname{clo} \Delta^{ac}$. Furthermore, because of $P_H^c = P_H^{ac} + P_H^{sc}$, $P_H^{ac}P_H^{sc} = P_H^{sc}P_H^{ac} = 0$, we obtain $\Delta^{sc} \cap \Delta^{ac}$ is an $E_H$-null set and $\operatorname{spec} \mathscr{H} \setminus \operatorname{eig} H = \Delta^{sc} \cup \Delta^{ac} (\operatorname{mod} E_H)$. An operator $H$ is called *absolutely continuous* if $P_H^{ac} = 1$.

### 3.3.5. The essential spectrum

The set of all isolated eigenvalues of $H$ with finite multiplicity is denoted by $\operatorname{dis} H$. The closed set $\operatorname{spec}_e H = \operatorname{spec} H \setminus \operatorname{dis} H$ is called the *essential spectrum* of $H$. Note that $\lambda \in \operatorname{spec}_e H$ iff $\dim E_H((\lambda - \varepsilon, \lambda + \varepsilon)) = \infty$ for all $\varepsilon > 0$. $\operatorname{spec}_e H$ consists of all accumulation points of $\operatorname{spec} H$ and of all eigenvalues of infinite multiplicity. The projection $P_H^e := 1 - \sum_{\lambda \in \operatorname{dis} H} E_H(\{\lambda\})$ is called the *essential projection* and the operator $H \restriction P_H^e\mathscr{H}$ is called the *essential part* of $H$. Note that $\operatorname{spec}_e H = \operatorname{spec} (H \restriction P_H^e\mathscr{H})$ and $\operatorname{spec}_c H \subseteq \operatorname{spec}_e H$.

### 3.3.6. The spectral core

If $H$ is selfadjoint, we call a Borel set $\Delta_0 \subseteq \operatorname{spec} H$ a *spectral core* of $H$ if $E_H(\Delta_0) = 1$ and $|\Delta_0|$ is minimal. Let $\inf_{\Delta, E(\Delta)=1} |\Delta| = m \geq 0$. Note that there is always a Borel set $\Delta_0$ with $E_H(\Delta_0) = 1$ and $|\Delta_0| = m$ (take a sequence $\Delta_\varrho$ with $E_H(\Delta_\varrho) = 1$ and $|\Delta_\varrho| \to m$ and put $\Delta_0 = \bigcap_{\varrho=1}^{\infty} \Delta_\varrho$). Note further that the condition "$|\Delta_0|$ is minimal" may be replaced by the condition "if $\Delta' \subseteq \Delta_0$ and $E_H(\Delta') = 1$, then $|\Delta_0 \setminus \Delta'| = 0$". Obviously a spectral core of $H$ is not unique. Despite this fact we use the notation $\operatorname{spec}^c H$ for a spectral core of $H$. Note that $\operatorname{clo} (\operatorname{spec}^c H) = \operatorname{spec} H$.

If $P_H^{sc} = E_H(\Delta^{sc})$ with $|\Delta^{sc}| = 0$, $\Delta^{sc} \subseteq \operatorname{spec}_{sc} H$, and $P_H^{ac} = E_H(\Delta^{ac})$ where $\Delta^{ac} \cap \Delta^{sc} = \emptyset$ and if $\Delta^{ac}$ is a spectral core of $H \restriction P_H^{ac}\mathscr{H}$ then $\Delta_0 = \operatorname{eig} H \cup \Delta^{sc} \cup \Delta^{ac}$ is a spectral core of $H$. Thus the concept of a spectral core is essentially related to the absolutely continuous part of $H$.

In order to illustrate the difference between spectrum and spectral core, take a Borel set $\Delta \subseteq [0, 1]$, $|\Delta| > 0$ with $\operatorname{clo} \Delta = [0, 1]$, but $|\operatorname{clo} \Delta \setminus \Delta| > 0$. Now consider the multiplication operator $H \sim \lambda$ on the space $L^2(\Delta, d\lambda)$. Then $\Delta$ is a spectral core of $H$, but $\operatorname{spec} H$ is not a spectral core. Note that $H$ is absolutely continuous.

**Example.** 7. An example for a Borel set $\Delta$ mentioned above is

$$\Delta = \bigcup_{n=1}^{\infty} (r_n - \varepsilon^n, r_n + \varepsilon^n) \cap [0, 1], \varepsilon = 1/4,$$

where $\{r_1, r_2, r_3, \ldots\}$ denotes the set of all rational numbers from $(0, 1)$. Obviously, $|\Delta| > 0$ and $|\Delta| < 2 \sum_n \varepsilon^n = 2/3$ hence $|\operatorname{clo} \Delta \setminus \Delta| > 0$.

We call an absolutely continuous operator $H$ *absolutely continuous on its spectrum* if spec $H$ is a spectral core of $H$.

**Example 8.** Here we consider an example where $H$ is absolutely continuous and spec $H$ = spec$^c$ $H$). Let $\mathscr{H} = L^2(R^n, dx)$ and let $H$ denote the natural selfadjoint extension of $-\Delta = \sum_{j=1}^{n} \left(-\frac{\partial^2}{\partial x_j^2}\right)$ ($\Delta$ the Laplacian) defined on $C_0^\infty(\mathbb{R}^n)$. $H$ restriced on $C_0^\infty(\mathbb{R})$ is essentially selfadjoint, that is, $C_0^\infty(\mathbb{R}^n)$ is a core of $H$. Further, we have that the Fourier transform $FHF^{-1}$ of $H$ is multiplication by $k^2$ on $L^2(\mathbb{R}^n, dk)$. This leads to the well-known fact that $H$ is absolutely continuous and its spectral core is $[0, \infty)$ which coincides with its spectrum.

## 3.4. Spectral properties of evolution groups and resolvents, and "sandwiched" resolvents

### 3.4.1. Characterization of the spectrally discrete and the spectrally continuous subspaces

**Proposition 9** (Mean ergodic theorem). *For any real* $\lambda$

$$E_H(\{\lambda\}) = \operatorname*{s-lim}_{|b-a|\to\infty} \frac{1}{b-a} \int_a^b e^{-it(\lambda-H)}\, dt. \tag{12}$$

Proof. Because $(b-a)^{-1} \int_a^b e^{-it(\lambda-H)}\, dt$ is uniformly bounded, it is sufficient to prove the limit relation for a dense set. Because of $\mathscr{H} = E_H(\{\lambda\})\, \mathscr{H} \oplus$ clo ima $(\lambda - H)$ one can choose $E_H(\{\lambda\})\, \mathscr{H} \oplus$ ima $(\lambda - H)$. For $f = E_H(\{\lambda\})\, f$, (12) is obvious. If $f = (\lambda - H) g$, then $e^{-it(\lambda-H)} f = i \frac{d}{dt} (e^{-it(\lambda-H)} g)$, hence $(b-a)^{-1} \int_a^b e^{-it(\lambda-H)} f\, dt = i(b-a)^{-1} (e^{-ib(\lambda-H)} g - e^{-ia(\lambda-H)} g)$ which tends to $0$ for $|b-a| \to \infty$. On the other hand, $E(\{\lambda\})\, f = 0$ in this case. ∎

**Proposition 10.** *A vector* $f \in \mathscr{H}$ *belongs to* $P_H^c \mathscr{H}$ *iff*

$$\lim_{T\to\infty} (2T)^{-1} \int_{-T}^{T} |(f, e^{-itH} f)|^2\, dt = 0. \tag{13}$$

Proof. We have $(f, e^{-itH} f) = \int e^{-it\lambda} \mu_f(d\lambda)$. According to Wiener's theorem,

$$\lim_{T\to\infty} (2T)^{-1} \int_{-T}^{T} |(f, e^{-itH} f)|^2\, dt = \sum_{\lambda\in\mathbb{R}} \mu_f(\{\lambda\})$$

is valid. But $f \in P_H^c \mathscr{H}$ iff $\mu_f(\{\lambda\}) = 0$ for all $\lambda \in \mathbb{R}$. ∎

If $f \in P_H^c \mathscr{H}$ and $g \in \mathscr{H}$, then also

$$\lim_{T\to\infty} (2T)^{-1} \int_{-T}^{T} |(g, e^{-itH} f)|^2\, dt = 0 \tag{14}$$

is valid. This follows easily from the polarization identity and from $g = P_H^c g + (P_H^c)^\perp g$.

### 3.4.2. Convergence properties of the resolvent

At first note that if one drops the outer limit $\operatorname{s-lim}_{\delta \to +0}$ in the general formula (4), one obtains the formula

$$\frac{1}{2}\left(E_H([a,b]) + E_H((a,b))\right) = \operatorname*{s-lim}_{\varepsilon \to +0} (2i\pi)^{-1} \int_a^b \left(R_H(x - i\varepsilon) - R_H(x + i\varepsilon)\right) dx. \tag{15}$$

The reason for this is that (see the proof of (4))

$$\lim_{\varepsilon \to +0} \alpha(x; 0, \varepsilon) = \begin{cases} 1 & x \in (a,b), \\ 1/2 & x = a, x = b, \\ 0 & x \notin [a,b]. \end{cases}$$

Thus, if neither $a$ nor $b$ is an eigenvalue of $H$, then from (15) we obtain

$$E_H\left((a,b)\right) = \operatorname*{s-lim}_{\varepsilon \to +0} (2i\pi)^{-1} \int_a^b \left(R_H(x - i\varepsilon) - R_H(x + i\varepsilon)\right) dx. \tag{16}$$

Note further that the formula

$$(2i\pi)^{-1} \left(R_H(\lambda - i\varepsilon) - R_H(\lambda + i\varepsilon)\right) = \int_{-\infty}^{\infty} \delta_\varepsilon(\lambda - x)\, E_H(dx) \tag{17}$$

is valid. (17) follows easily from the functional calculus.

Formulas such as (17), (16) or (15) express connections between resolvents and spectral measures where there is an averaging (integration) with respect to the spectral parameter $x$. Of course the question arises if there are connections between resolvents and spectral measures in the pointwise sense (a.e. mod $|\cdot|$ with respect to $x$). In the following we present propositions in this direction which are useful in the other parts of this book.

### 3.4.3. The "sandwiched" spectral measure

**Lemma 11.** *Let $E(\cdot)$ be an arbitrary spectral measure on $\mathbb{R}$ and let $A, B \in \mathscr{L}_2$. Then $\Phi(\Delta) := AE(\Delta)B$ is a $\mathscr{L}_1$-valued Borel measure with finite total variation.*

Proof. According to 1.2.4 it is sufficient to show that $AE(\Delta)B$ is an $\mathscr{L}_1$-valued measure with finite total variation. Concerning the property of $AE(\Delta)B$ to be an $\mathscr{L}_1$-valued measure we use factorizations $A = C_1A_1$ and $B = C_2B_1$ with $A_1, B_1 \in \mathscr{L}_2$ and $C_1, C_2 \in \mathscr{L}_\infty$. Then $C_1\left(A_1E(\Delta)\right)C_2$ is an $\mathscr{L}_2$-valued measure (according to Lemma 1.7) and $\Phi(\Delta)$ is an $\mathscr{L}_1$-valued measure. Further, we obtain

$$\begin{aligned}\sum_\varrho \|AE(\Delta_\varrho)B\|_1 &\leq \sum_\varrho \|AE(\Delta_\varrho)\|_2 \|E(\Delta_\varrho)B\|_2 \\ &\leq \left(\sum \|AE(\Delta_\varrho)\|_2^2\right)^{1/2} \left(\sum_\varrho \|E(\Delta_\varrho)B\|_2^2\right)^{1/2},\end{aligned}$$

but $\sum_\varrho \|AE(\Delta_\varrho)\|_2^2 = \sum_\varrho \|E(\Delta_\varrho) A\|_2^2 = \|A\|_2^2$. Thus we get $\sum_\varrho \|AE(\Delta_\varrho) B\|_1 \leq \|A_2\| \|B\|_2$. ■

According to the fact that $\mathscr{L}_1 = \mathscr{L}_\infty{}^*$ is a separable dual and according to 1.2.2, Lemma 11 yields

**Proposition 12.** *$\Phi(\Delta)$ is absolutely continuous with respect to a certain nonnegative Borel measure $\mu(\cdot)$, that is, there is an $\mathscr{L}_1$-valued Bochner integrable function $K \in L^1(\mathbb{R}, \mu; \mathscr{L}_1)$ such that*

$$AE(\Delta) B = \int_\Delta K(\lambda) \mu(d\lambda). \tag{18}$$

From Prop. 12 and again 1.2.2 we easily get

**Proposition 13.** *The $\mathscr{L}_1$-valued measure $AE(\Delta) B$ is $\mathscr{L}_1$-differentiable a.e.* (mod $|\cdot|$) *on $\mathbb{R}$ where the derivative is given by $K(\lambda) \mu'(\lambda)$ a.e.* (mod $|\cdot|$), *that is, we have*

$$\lim_{h \to 0} \left\| \frac{AE([\lambda, \lambda + h)) B}{h} - K(\lambda) \mu'(\lambda) \right\|_1 = 0 \tag{19}$$

*for almost all $\lambda \in \mathbb{R}$* (mod $|\cdot|$) (if $h < 0$, we put $E([\lambda, \lambda + h)) = -E((\lambda + h, \lambda])$).

Note that the $\mathscr{L}_1$-differentiability of $AE(\Delta) B$ a.e. (mod $|\cdot|$) can be proved also independently of formula (18) using the facts that $\mathscr{L}_\infty$ is separable, $\mathscr{L}_1 = \mathscr{L}_\infty{}^*$ and that there exists a weak derivative (see 1.2.2). The transition from weak derivatives to $\mathscr{L}_1$-derivatives can be performed with the help of the principle of uniform boundedness and of Lemma 1.7. The $\mathscr{L}_1$-derivative of $AE(\Delta) B$ is denoted by $\big(AE(d\lambda) B\big)/d\lambda$.

### 3.4.4. The "sandwiched" resolvent

According to (18) we have

$$AR_H(z) B = \int_{-\infty}^{\infty} (z - \lambda)^{-1} K(\lambda) \mu(d\lambda) \tag{20}$$

where $K \in L^1(R, \mu; \mathscr{L}_1)$. This formula leads to

**Proposition 14.** *The limits of $AR_H(\lambda \pm i\varepsilon) B$ exist as $\varepsilon \to +0$ with respect to $\|\cdot\|_2$ a.e.* mod $|\cdot|$. *They are denoted by $AR_H(\lambda \pm i0) B$. The limit of $AR_H(\lambda - i\varepsilon) B - AR_H(\lambda + i\varepsilon) B$ exists as $\varepsilon \to +0$ even with respect to $\|\cdot\|_1$ a.e.* mod $|\cdot|$. *The equation*

$$AR_H(\lambda - i0) B - AR_H(\lambda + i0) B = 2i\pi \frac{AE_H(d\lambda) B}{d\lambda} \tag{21}$$

*is valid a.e.* mod $|\cdot|$.

Proof. Because of $\|K\|_2 \leq \|K\|_1$ we have $K \in L^1(\mathbb{R}, \mu; \mathscr{L}_2)$, but $\mathscr{L}_2$ is a Hilbert space. Hence the first result follows from Prop. 1.2. From (17) we get

$$(2i\pi)^{-1} \big(AR_H(\lambda - i\varepsilon) B - AR_H(\lambda + i\varepsilon) B\big) = \int_{-\infty}^{\infty} \delta_\varepsilon(\lambda - x) K(x) \mu(dx).$$

Hence the second result follows from Prop. 1.1. ■

Now we consider the expressions $AR_H(\lambda \pm i\varepsilon) f$ and $AE_H(\Delta) f$ for $f \in \mathscr{H}$, $A \in \mathscr{L}_2$ and obtain

**Proposition 15.** *For all $f \in \mathscr{H}$ the strong limits* $\underset{\varepsilon \to +0}{\text{s-lim}}\, AR_H(\lambda \pm i\varepsilon) f$ *exist a.e.* mod $|\cdot|$. *Further, the function $AE_H((-\infty, \lambda)) f$ is strongly differentiable a.e.* mod $|\cdot|$, *that is,* $(AE_H(d\lambda) f)/d\lambda := \underset{h \to 0}{\text{s-lim}}\, h^{-1} AE_H([\lambda, \lambda + h)) f$ *exists a.e.* mod $|\cdot|$ *and the equation*

$$AR_H(\lambda - i0) f - AR_H(\lambda + i0) f = 2i\pi (AE_H(d\lambda) f)/d\lambda \tag{22}$$

*is valid a.e.* mod $|\cdot|$.

**Proof.** The assertion follows easily from Prop. 13 and Prop. 14 if $B$ is chosen such that $Bf = f$. ∎

If we choose $A = (f, \cdot) f$, that is, if we consider $(f, R_H(\lambda \pm i\varepsilon) f)$ and $(f, E_H(\Delta) f)$, we have $(2i\pi)^{-1} \big((f, R_H(\lambda - i\varepsilon) f) - (f, R_H(\lambda + i\varepsilon) f)\big) = \int_{-\infty}^{\infty} \delta_\varepsilon(\lambda - x)\, (f, E_H(dx) f)$ and we obtain

**Proposition 16.** *For any $f \in \mathscr{H}$ and for each point $\lambda$ where $(f, E_H(d\lambda) f)/d\lambda$ exists, we have:* $\lim_{\varepsilon \to +0} \int \delta_\varepsilon(\lambda - x)\, (f, E_H(dx) f)$ *exists and equals* $(f, E_H(d\lambda) f)/d\lambda$.

**Proof.** It follows again from Prop. 1.1. In this case we have $\varphi(\lambda) \equiv 1$ and we use additionally the fact that the limit exists at that points where the derivative exists. ∎

## 3.5. Properties of the absolutely continuous subspace

### 3.5.1. $L^1$-properties and implications

**Proposition 17.** *Let $f \in P_H^{ac}\mathscr{H}$, $g \in \mathscr{H}$. Then*

(i) *$(g, E_H(\Delta) f)$ is absolutely continuous and*

$$\left| \frac{(g, E_H(d\lambda) f)}{d\lambda} \right|^2 \leq \frac{(f, E_H(d\lambda) f)}{d\lambda} \frac{(g_0, E_H(d\lambda) g_0)}{d\lambda} \tag{23}$$

*is valid a.e.* mod $|\cdot|$; $g_0 = P_H^{ac} g$,

(ii) *$(g, E_H(d\lambda) f)/d\lambda \in L^1(\mathbb{R}, d\lambda)$ and*

$$\int_{-\infty}^{\infty} \left| \frac{(g, E_H(d\lambda) f)}{d\lambda} \right| d\lambda \leq \|f\|\, \|P_H^{ac} g\|; \quad \int_{-\infty}^{\infty} \frac{(g, E_H(d\lambda) f)}{d\lambda}\, d\lambda = (P_H^{ac} g, f). \tag{24}$$

**Proof.** (23) follows from $|(g, E_H(\Delta) f)|^2 \leq (f, E_H(\Delta) f)\, (g, E_H(\Delta) g)$ (SCHWARZ' inequality for the sesquilinear form $(g, E_H(\Delta) f)$); (24) follows from (23) and

$\int ((f, E_H(d\lambda) f)/d\lambda)^{1/2} ((g_0, E_H(d\lambda) g_0)/d\lambda)^{1/2}\, d\lambda \leq \left(\int ((f, E_H(d\lambda) f)/d\lambda\right)^{1/2}$
$\left(\int ((g_0, E_H(d\lambda) g_0)/d\lambda)\, d\lambda\right)^{1/2}$. ∎

In particular $\{(f, E_H(d\lambda) f)/d\lambda\} \in L^1(\mathbb{R}, d\lambda)$ for all $f \in P_H^{ac}\mathscr{H}$. Thus Prop. 16 may be supplied by the following

**Proposition 18.** *If $f \in P_H^{ac}\mathscr{H}$, then*

$$\lim_{\varepsilon \to +0} \int_{-\infty}^{\infty} \left| \int_{-\infty}^{\infty} \delta_\varepsilon(\lambda - x) \left(f, E_H(\mathrm{d}x) f\right) - \left((f, E_H(\mathrm{d}\lambda) f)/\mathrm{d}\lambda\right) \right| \mathrm{d}\lambda = 0, \tag{25}$$

$$\int_{-\infty}^{\infty} \left| \int_{-\infty}^{\infty} \delta_\varepsilon(\lambda - x) \left(f, E_H(\mathrm{d}x) f\right) \right| \mathrm{d}\lambda \leqq \|f\|^2, \qquad \varepsilon > 0; \tag{26}$$

*in other words, the convergence of $\int \delta_\varepsilon(\lambda - x) \left(f, E_H(\mathrm{d}x) f\right)$ to $\left((f, E_H(\mathrm{d}\lambda) f)/\mathrm{d}\lambda\right)$ is valid also in the $L^1(\mathbb{R}, \mathrm{d}\lambda)$-sense.*

Proof. It follows immediately from Prop. 1.4 for $p = 1$ because of $\left(f, E_H(\mathrm{d}\lambda) f\right)/\mathrm{d}\lambda \in L^1(\mathbb{R}, \mathrm{d}\lambda)$ and $\int \delta_\varepsilon(\lambda - x) \left(f, E_H(\mathrm{d}x) f\right) = \int \delta_\varepsilon(\lambda - x) \left((f, E_H(\mathrm{d}\lambda) f)/\mathrm{d}\lambda\right) \mathrm{d}\lambda$. ■

### 3.5.2. Some $L^\infty$- and $L^p$-properties ($p > 1$)

We introduce a linear submanifold $\mathscr{M}_\infty$ of $P_H^{ac}\mathscr{H}$ which is very useful in the following. We put

$$|||f|||_\infty := \left(\operatorname*{ess\,sup}_\lambda \left((f, E_H(\mathrm{d}\lambda) f\lambda)/\mathrm{d}\lambda\right)\right)^{1/2} \tag{27}$$

and define the subset $\mathscr{M}_\infty$ of $P_H^{ac}\mathscr{H}$ by $\mathscr{M}_\infty = \{f;\ |||f|||_\infty < \infty\}$; if necessary, we write $\mathscr{M}_\infty(H)$ and $|||\cdot|||_{\infty,H}$. Then we have

**Proposition 19.** *$\mathscr{M}_\infty$ is a linar submanifold of $P_H^{ac}\mathscr{H}$ which is dense in $P_H^{ac}\mathscr{H}$. $|||\cdot|||_\infty$ is a norm on $\mathscr{M}_\infty$.*

Proof. If $f, g \in P_H^{ac}\mathscr{H}$, then we immediately obtain from (23)

$$\left| \frac{(f + g, E_H(\mathrm{d}\lambda)(f + g))}{\mathrm{d}\lambda} \right| \leqq \frac{(f, E_H(\mathrm{d}\lambda) f)}{\mathrm{d}\lambda} + \frac{(g, E_H(\mathrm{d}\lambda) g)}{\mathrm{d}\lambda} + 2 \left(\frac{(f, E_H(\mathrm{d}\lambda) f)}{\mathrm{d}\lambda}\right)^{1/2} \left(\frac{(g, E_H(\mathrm{d}\lambda) g)}{\mathrm{d}\lambda}\right)^{1/2}$$

a.e. mod $|\cdot|$, hence $|||f + g|||_\infty \leqq |||f|||_\infty + |||g|||_\infty$. Thus $\mathscr{M}_\infty$ is linear and $|||\cdot|||_\infty$ is a seminorm. But from $|||f|||_\infty = 0$ also $f = 0$ follows, that is, $|||\cdot|||_\infty$ is a norm. Now we show that $\mathscr{M}_\infty$ is dense in $P_H^{ac}\mathscr{H}$. Let $f \in P_H^{ac}\mathscr{H}$. We put $\alpha(\lambda) := \left(f, E_H(\mathrm{d}\lambda) f\right)/\mathrm{d}\lambda$. Because of $\int \alpha(\lambda)\,\mathrm{d}\lambda < \infty$ the function $\alpha$ is finite a.e. mod $|\cdot|$. Let $\Delta_n$ be the Borel set defined by the inequality $\alpha(\lambda) > n$. Then $\Delta_1 \supseteqq \Delta_2 \supseteq \cdots$. Put $\Delta_\infty = \bigcap_{n=1}^{\infty} \Delta_n$. Then $\lim_{n\to\infty} |\Delta_n| = |\Delta_\infty| = 0$. Put further $u_n := E_H(\mathbb{R} \setminus \Delta_n) f = f - E_H(\Delta_n) f$. Then we obtain $\|E_H(\Delta_n) f\|^2 = \left(f, E_H(\Delta_n) f\right) \to 0$ as $n \to \infty$ since $f \in P_H^{ac}\mathscr{H}$. Moreover,

$$\left(u_n, E_H(\Delta) u_n\right) = \int_\Delta \left(1 - \chi_{\Delta_n}(\lambda)\right) \alpha(\lambda)\,\mathrm{d}\lambda,$$

hence $\left(u_n, E(\mathrm{d}\lambda) u_n\right)/\mathrm{d}\lambda = (1 - \chi_{\Delta_n}(\lambda)\, \alpha(\lambda) \leqq n$ a.e. (mod $|\cdot|$), that is, $u_n \in \mathscr{M}_\infty$ and $u_n \xrightarrow[s]{} f$. ■

Similarly we may define the linear submanifold $\mathscr{M}_p$, $p > 1$, of $P_H^{ac}\mathscr{H}$ by

$$\mathscr{M}_p = \left\{ f;\ |||f|||_p := \left( \int_{-\infty}^{\infty} \left( (f, E_H(\mathrm{d}\lambda)\, f)/\mathrm{d}\lambda \right)^p \,\mathrm{d}\lambda \right)^{1/2p} < \infty \right\}. \tag{28}$$

Note that $|||\cdot|||_p$ is a norm on $\mathscr{M}_p$. Note further that for $p = 1$ one obtains $|||f|||_1 = \|f\|$ and $\mathscr{M}_1 = P_H^{ac}\mathscr{H}$. If $p > 1$, we have the inclusion $\mathscr{M}_\infty \subset \mathscr{M}_p$ which is obvious because of the inequality

$$|||f|||_p^{2p} \leqq |||f|||_\infty^{2(p-1)}\, |||f|||_1^2. \tag{29}$$

In particular, $\mathscr{M}_p$ is dense in $P_H^{ac}\mathscr{H}$ for $p > 1$. Note that the set of all $f \in \mathscr{M}_\infty$ with the property $f = E_H(\Delta)\, f$ where $\Delta$ is some compact Borel set is dense in $\mathscr{M}_\infty$.

For convenience we mention the following identity which is often useful: If $u \in \mathscr{H}$, then

$$(2\mathrm{i}\pi)^{-1} \left(u, (R_H(\lambda - \mathrm{i}\varepsilon) - R_H(\lambda + \mathrm{i}\varepsilon))\, u\right) = \frac{\varepsilon}{\pi} \|R_H(\lambda + \mathrm{i}\varepsilon)\, u\|^2$$

$$= -\pi^{-1} \operatorname{Im} \left(u, R_H(\lambda + \mathrm{i}\varepsilon)\, u\right). \tag{30}$$

We introduce the following subset of $\mathscr{H}$:

$$\mathscr{N}_\infty := \left\{ f;\ \sup_{\lambda \in \mathbb{R}, \varepsilon > 0} (\varepsilon/\pi)\, \|R_H(\lambda + \mathrm{i}\varepsilon)\, f\|^2 < \infty \right\}. \tag{31}$$

Note that $\mathscr{N}_\infty$ is a linear submanifold of $\mathscr{H}$. Comparing $\mathscr{N}_\infty$ with $\mathscr{M}_\infty$ and $P_H^{ac}\mathscr{H}$, we obtain

**Proposition 20.** *We have $\mathscr{M}_\infty \subset \mathscr{N}_\infty \subset P_H^{ac}\mathscr{H}$, in particular $\mathscr{N}_\infty$ is dense in $P_H^{ac}\mathscr{H}$.*

**Proof.** Let $(a, b)$ be an open and bounded interval. Then (see (15))

$$\left(u, E_H((a, b))\, u\right) \leqq \lim_{\varepsilon \to +0} \int_a^b (\varepsilon/\pi)\, \|R_H(\lambda + \mathrm{i}\varepsilon)\, u\|^2 \,\mathrm{d}\lambda, \qquad u \in \mathscr{H}.$$

If $f \in \mathscr{N}_\infty$, we obtain

$$\left(f, E_H((a, b))\, f\right) \leqq \sup_{\mathbb{R} \ni \lambda, \varepsilon > 0} (\varepsilon/\pi)\, \|R_H(\lambda + \mathrm{i}\varepsilon)\|^2\, (b - a). \tag{32}$$

From (32) immediately the absolute continuity of $\left(f, E_H(\Delta)\, f\right)$ follows. Thus $f \in P_H^{ac}\mathscr{H}$. Now we prove the first inclusion: We have $R_H(\lambda + \mathrm{i}\varepsilon)\, u = \int (\lambda + \mathrm{i}\varepsilon - x)^{-1}\, E_H(\mathrm{d}x)\, u$, hence for $u = f \in P_H^{ac}\mathscr{H}$

$$\|R_H(\lambda + \mathrm{i}\varepsilon)\, f\|^2 = \int_{-\infty}^{\infty} \left((\lambda - x)^2 + \varepsilon^2\right)^{-1} \left((f, E_H(\mathrm{d}x)\, f)/\mathrm{d}x\right) \mathrm{d}x \tag{33}$$

follows. If $f \in \mathscr{M}_\infty$, then we obtain

$$(\varepsilon/\pi)\, \|R_H(\lambda + \mathrm{i}\varepsilon)\, f\|^2 \leqq |||f|||_\infty^2 \int_{-\infty}^{\infty} \delta_\varepsilon(\lambda - x)\, \mathrm{d}x = |||f|||_\infty^2. \ \blacksquare$$

Similarly we can assert a $L^p$-analogue of Prop. 20. For this reason we introduce the following subset of $\mathscr{H}$:

$$\mathscr{N}_p = \left\{f;\ \sup_{\varepsilon>0} \int_{-\infty}^{\infty} \left((\varepsilon/\pi)\ \|R_H(\lambda + i\varepsilon) f\|^2\right)^p \mathrm{d}\lambda < \infty\right\}, \qquad p \geqq 1. \tag{34}$$

Note that for $p = 1$ we have $\mathscr{N}_1 = \mathscr{H}$ because of (10). The subset $\mathscr{N}_p$ has the following property:

**Proposition 21.** *For $p > 1$ the inclusion $\mathscr{M}_p \subset \mathscr{N}_p \subset P_H^{\mathrm{ac}}\mathscr{H}$ is valid, in particular $\mathscr{N}_p$ is dense in $P_H^{\mathrm{ac}}\mathscr{H}$.*

**Proof.** For abbreviation we put $\alpha_\varepsilon(\lambda) = \varepsilon\pi^{-1}\ \|R_H(\lambda + i\varepsilon) f\|^2$. Let $\Delta = \bigcup_{j=1}^{n} (a_j, b_j)$ be a finite union of pairwise disjoint and open intervals. Then we obtain, using HÖLDER's inequality $\|fg\|_1 \leqq \|f\|_p\ \|g\|_q$,

$$\left(f, E_H(\Delta) f\right) \leqq \lim_{\varepsilon\to+0} \int_\Delta \alpha_\varepsilon(\lambda)\ \mathrm{d}\lambda \leqq \sup_{\varepsilon>0} \left(\int_\Delta \alpha_\varepsilon(\lambda)^p\ \mathrm{d}\lambda\right)^{1/p} |\Delta|^{1/q}$$

where $p^{-1} + q^{-1} = 1$. If $f \in \mathscr{N}_p$, then from this inequality again the absolute conti nuity of $\left(f, E_H(\Delta) f\right)$ follows, that is, we have $f \in P_H^{\mathrm{ac}}\mathscr{H}$. Now we prove the first inclusion: According to (33), we have $\alpha_\varepsilon(\lambda) = \int \delta_\varepsilon(\lambda - x) \left(\left(f, E(\mathrm{d}x) f\right)/\mathrm{d}x\right) \mathrm{d}x$. Let $f \in \mathscr{M}_p$ and put $\varphi(x) = \left(f, E(\mathrm{d}x) f\right)/\mathrm{d}x$. Then, using again HÖLDER's inequality, we get

$$\alpha_\varepsilon(\lambda)^p \leqq \left(\int_{-\infty}^{\infty} \delta_\varepsilon(\lambda - x)\ \mathrm{d}x\right)^{p/q} \left(\int_{-\infty}^{\infty} \delta_\varepsilon(\lambda - x)\ \varphi(x)^p\ \mathrm{d}x\right),$$

hence

$$\int_{-\infty}^{\infty} \alpha_\varepsilon(\lambda)^p\ \mathrm{d}\lambda \leqq \int_{-\infty}^{\infty} \mathrm{d}\lambda \left(\int_{-\infty}^{\infty} \delta_\varepsilon(\lambda - x)\ \varphi(x)^p\ \mathrm{d}x\right) = \int_{-\infty}^{\infty} \varphi(x)^p\ \mathrm{d}x = |||f|||_p^{\,2p}$$

follows. Thus $f \in \mathscr{N}_p$. The last assertion is obvious because of $\mathscr{M}_\infty \subset \mathscr{M}_p \subset \mathscr{N}_p$. ■

### 3.5.3. $L^2$-properties

**Proposition 22.** *Let $f \in \mathscr{M}_\infty$, $g \in \mathscr{H}$. Then*

(i) $\left\{(g, E_H(\mathrm{d}\lambda) f)/\mathrm{d}\lambda\right\} \in L^2(\mathbb{R}, d\lambda)$ *and* $\int \left|(g, E_H(\mathrm{d}\lambda) f)/\mathrm{d}\lambda\right|^2 \mathrm{d}\lambda \leqq |||f|||_\infty{}^2\ \|P_H^{\mathrm{ac}} g\|^2$,
(ii) $\{(g, \mathrm{e}^{-itH} f)\} \in L^2(\mathbb{R}, \mathrm{d}t)$ *and*

$$\int_{-\infty}^{\infty} |(g, \mathrm{e}^{-itH} f)|^2\ \mathrm{d}t = 2\pi \int_{-\infty}^{\infty} \left|\left((g, E_H(\mathrm{d}\lambda) f)/\mathrm{d}\lambda\right)\right|^2 \mathrm{d}\lambda \tag{35}$$

*are valid.*

**Proof.** (i) follows from (23). Further, we have

$$(g, \mathrm{e}^{-itH} f) = \int_{-\infty}^{\infty} \mathrm{e}^{-it\lambda} \left((g, E_H(\mathrm{d}\lambda) f)/\mathrm{d}\lambda\right) \mathrm{d}\lambda.$$

Hence (ii) follows; (35) is just Parseval's equation for $(2\pi)^{1/2}\big((g, E_H(\mathrm{d}\lambda) f)/\mathrm{d}\lambda\big)$ and its Fourier transform $(g, \mathrm{e}^{-\mathrm{i}tH}f)$. ■

The next proposition plays an important role in the abstract scattering theory.

**Proposition 23.** *The relation* $\operatorname*{w-lim}_{t\to\pm\infty} \mathrm{e}^{-\mathrm{i}tH}P_H^{\mathrm{ac}} = 0$ *is valid.*

Proof. Because of $\|\mathrm{e}^{-\mathrm{i}tH}P_H^{\mathrm{ac}}\| = 1$ it is sufficient to prove the assertion for a dense set of vectors $f$ from $P_H^{\mathrm{ac}}\mathscr{H}$. By Prop. 19, we can choose $\mathscr{M}_\infty$. Then because of Prop. 17 and Prop. 22, we obtain $\{(g, E_H(\mathrm{d}\lambda) f)/\mathrm{d}\lambda\} \in L^2(\mathbb{R}, \mathrm{d}\lambda) \cap L^1(\mathbb{R}, \mathrm{d}\lambda)$. Therefore $(g, \mathrm{e}^{-\mathrm{i}tH}f)$ is from $C_\infty(\mathbb{R})$. ■

Also the following proposition is decisive for some proofs (see for example Part IV, Chapter 18). Note that according to Prop. 15 the vector-valued function $\big(TE(\mathrm{d}\lambda) f\big)/\mathrm{d}\lambda$ is defined a.e. mod $|\cdot|$ if $T \in \mathscr{L}_2$.

**Proposition 24.** *Let* $T \in \mathscr{L}_2(\mathscr{H})$ *and* $f \in \mathscr{M}_\infty$. *Then*

$$\int_{-\infty}^{\infty} \|T\mathrm{e}^{-\mathrm{i}tH}f\|^2 \,\mathrm{d}t \leq 2\pi \,\|T\|_2^2 \,|||f|||_\infty^2, \tag{36}$$

*that is, in particular we have* $\{T\,\mathrm{e}^{-\mathrm{i}tH}f\} \in L^2(\mathbb{R}, \mathrm{d}t; \mathscr{H})$. *Further,* $\{\big(TE(\mathrm{d}\lambda) f\big)/\mathrm{d}\lambda\} \in L^2(\mathbb{R}, \mathrm{d}\lambda; \mathscr{H})$ *and*

$$2\pi \int_{-\infty}^{\infty} \|\big(TE(\mathrm{d}\lambda) f\big)/\mathrm{d}\lambda\|^2 \,\mathrm{d}\lambda = \int_{-\infty}^{\infty} \|T\,\mathrm{e}^{-\mathrm{i}tH}f\|^2 \,\mathrm{d}t \tag{37}$$

*is valid.*

Proof. We have $\|Tg\|^2 = \|\,|T|\,g\|^2$ where $|T| \geq 0$, $|T| \in \mathscr{L}_2(\mathscr{H})$. Thus $|T| = \sum c_\varrho(g_\varrho, \cdot)g_\varrho$ where $\sum c_\varrho^2 < \infty$ and $\{g_1, g_2, \ldots\}$ denotes a certain orthonormal system. Then

$$\|Tg\|^2 = (g, |T|^2 g) = \sum_{\varrho=1}^{\infty} c_\varrho^2 \,|(g_\varrho, g)|^2. \tag{38}$$

Further, setting $g = \mathrm{e}^{-\mathrm{i}tH}f$ and integrating over the whole real line, we obtain, using (23) and the definition of $|||\cdot|||_\infty$,

$$\int_{-\infty}^{\infty} \|T\,\mathrm{e}^{-\mathrm{i}tH}f\|^2 \,\mathrm{d}t = \sum_{\varrho=1}^{\infty} c_\varrho^2 \int_{-\infty}^{\infty} |(g_\varrho, \mathrm{e}^{-\mathrm{i}tH}f)|^2 \,\mathrm{d}t = 2\pi \sum_{\varrho=1}^{\infty} c_\varrho^2 \int_{-\infty}^{\infty} \left|\frac{(g_\varrho, E_H(\mathrm{d}\lambda) f)}{\mathrm{d}\lambda}\right|^2 \mathrm{d}\lambda$$

$$\leq 2\pi \,|||f|||_\infty^2 \sum_{\varrho=1}^{\infty} c_\varrho^2.$$

Further, from (38) it follows that $TE_H(\Delta) f$ is an absolutely continuous vector measure with respect to $|\cdot|$, hence $TE_H(\Delta) f = \int_\Delta \big((TE_H(\mathrm{d}\lambda) f)/\mathrm{d}\lambda\big) \,\mathrm{d}\lambda$. Therefore we obtain

$$T\,\mathrm{e}^{-\mathrm{i}tH}f = \int_{-\infty}^{\infty} \mathrm{e}^{-\mathrm{i}t\lambda}\big(TE_H(\mathrm{d}\lambda) f/\mathrm{d}\lambda\big) \,\mathrm{d}\lambda,$$

that is, $(2\pi)^{1/2} \big(TE_H(\mathrm{d}\lambda) f\big)/\mathrm{d}\lambda$ is the inverse Fourier transform of $T\,\mathrm{e}^{-\mathrm{i}tH}f$, hence $\{\big(TE_H(\mathrm{d}\lambda) f\big)/\mathrm{d}\lambda\} \in L^2(\mathbb{R}, \mathrm{d}\lambda; \mathscr{H})$ follows. (37) is just Parseval's equation for $(2\pi)^{1/2} \big(TE_H(\mathrm{d}\lambda) f\big)/\mathrm{d}\lambda$ and $T\,\mathrm{e}^{-\mathrm{i}tH}f$. ■

Now we consider $TR_H(z) f = \int (z - x)^{-1} \left((TE_H(\mathrm{d}x) f)/\mathrm{d}x\right) \mathrm{d}x$ for $T \in \mathscr{L}_2$ and $f \in \mathscr{M}_\infty$. According to (8) and (10) we have

$$\mathrm{i}(2\pi)^{-1/2}\, TR_H(\lambda + \mathrm{i}\varepsilon) = (2\pi)^{-1/2} \int_{-\infty}^{\infty} \mathrm{e}^{-\mathrm{i}\lambda x} \chi_{(-\infty,0]}(x)\, \mathrm{e}^{\varepsilon x} T\, \mathrm{e}^{\mathrm{i}xH}\, \mathrm{d}x, \quad \varepsilon > 0,$$

$$\mathrm{i}(2\pi)^{-1/2}\, TR_H(\lambda - \mathrm{i}\varepsilon) = (2\pi)^{-1/2} \int_{-\infty}^{\infty} \mathrm{e}^{-\mathrm{i}\lambda x} \chi_{[0,\infty)}(x)\, \mathrm{e}^{-\varepsilon x} T\, \mathrm{e}^{\mathrm{i}xH}\, \mathrm{d}x, \quad \varepsilon > 0,$$

and, correspondingly,

$$\int_{-\infty}^{\infty} \|TR_H(\lambda + \mathrm{i}\varepsilon) f\|^2\, \mathrm{d}\lambda = 2\pi \int_0^{\infty} \mathrm{e}^{-2\varepsilon t}\, \|T\, \mathrm{e}^{-\mathrm{i}tH} f\|^2\, \mathrm{d}t, \quad \varepsilon > 0, \tag{39}$$

$$\int_{-\infty}^{\infty} \|TR_H(\lambda - \mathrm{i}\varepsilon) f\|^2\, \mathrm{d}\lambda = 2\pi \int_{-\infty}^{0} \mathrm{e}^{2\varepsilon t}\, \|T\, \mathrm{e}^{-\mathrm{i}Ht} f\|^2\, \mathrm{d}t, \quad \varepsilon > 0. \tag{40}$$

Therefore we can assert the following Hardy-class properties of $TR_H(\lambda \pm \mathrm{i}\varepsilon) f$.

**Proposition 25.** *If $T \in \mathscr{L}_2$ and $f \in \mathscr{M}_\infty$, then $\{TR_H(\lambda + \mathrm{i}\varepsilon) f\} \in \mathscr{H}_+{}^2(\mathbb{R}, \mathrm{d}\lambda; \mathscr{H})$ and*

$$\sup_{\varepsilon > 0} \int_{-\infty}^{\infty} \|TR_H(\lambda + \mathrm{i}\varepsilon) f\|^2\, \mathrm{d}\lambda \leqq 2\pi \int_0^{\infty} \|T\, \mathrm{e}^{-\mathrm{i}tH} f\|^2\, \mathrm{d}t. \tag{41}$$

*Further, $\{TR_H(\lambda - \mathrm{i}\varepsilon) f\} \in \mathscr{H}_-{}^2(\mathbb{R}, \mathrm{d}\lambda; \mathscr{H})$ and*

$$\sup_{\varepsilon > 0} \int_{-\infty}^{\infty} \|TR_H(\lambda - \mathrm{i}\varepsilon) f\|^2\, \mathrm{d}\lambda \leqq 2\pi \int_{-\infty}^{0} \|T\, \mathrm{e}^{-\mathrm{i}tH} f\|^2\, \mathrm{d}t. \tag{42}$$

**Proof.** It follows immediately from (39) and (40). ∎

**Remark 26.** From Prop. 25 it easily follows — according to the Hardy-class theory — that the limits $TR_H(\lambda \pm \mathrm{i}0) f$ whose existence was already proved in Prop. 15, are functions from $L^2(\mathbb{R}, \mathrm{d}\lambda; \mathscr{H})$ and that the convergence of $TR(\lambda \pm \mathrm{i}\varepsilon) f$ to the limits is simultaneously a convergence with respect to $\|\cdot\|_{L_2(\mathbb{R}, \mathrm{d}\lambda, \mathscr{H})}$. Further, note that the equation $\left(TE_H(\mathrm{d}\lambda) f\right)/\mathrm{d}\lambda = (2\mathrm{i}\pi)^{-1} \left(TR_H(\lambda - \mathrm{i}0) f - TR_H(\lambda + \mathrm{i}0) f\right)$ can be considered as the resolution of $\left(TE_H(\mathrm{d}\lambda) f\right)/\mathrm{d}\lambda \in L^2(\mathbb{R}, \mathrm{d}\lambda; \mathscr{H})$ into the sum of mutual orthogonal components with respect to the direct sum resolution

$$L^2(\mathbb{R}, \mathrm{d}\lambda; \mathscr{H}) = \mathscr{H}_+{}^2(\mathbb{R}, \mathrm{d}\lambda; \mathscr{H}) \oplus \mathscr{H}_-{}^2(\mathbb{R}, \mathrm{d}\lambda; \mathscr{H}).$$

**Proposition 27.** *Let $H$ be absolutely continuous and let $\alpha \in L^2(\mathbb{R}, \mathrm{d}t)$. Then $\check{\alpha}(H)$ is densely defined, $\mathscr{M}_\infty \subset \operatorname{dom} \check{\alpha}(H)$, and $\mathscr{M}_\infty$ is a core of $\check{\alpha}(H)$. Further*

$$\check{\alpha}(H) f = (2\pi)^{-1/2} \int_{-\infty}^{\infty} \alpha(t)\, \mathrm{e}^{-\mathrm{i}tH} f\, \mathrm{d}t, \qquad f \in \mathscr{M}_\infty,$$

*where $\int_{-\infty}^{\infty} \mathrm{d}t$ means* $\operatorname*{s-lim}_{n\to\infty} \int_{-n}^{n} \mathrm{d}t$.

Proof. Let $f \in \mathcal{M}_\infty$. Then $\int |\hat{\alpha}(\lambda)|^2 \mu_f(d\lambda) \leq |||f|||_\infty^2 \int |\hat{\alpha}(\lambda)|^2 d\lambda < \infty$ and $\|\hat{\alpha}(H) E_H(\Delta) f\|^2 \leq \int_\Delta |\hat{\alpha}(\lambda)|^2 d\lambda \, |||f|||_\infty^2$. Hence $\mathcal{M}_\infty \subseteq \operatorname{dom} \hat{\alpha}(H)$. Now let $f \in \operatorname{dom} \hat{\alpha}(H)$. Then there exists a sequence $\Delta_n \subset \mathbb{R}$ such that $f_n := E_H(\Delta_n) f \in \mathcal{M}_\infty$ and $\operatorname{s-lim}_{n\to\infty} f_n = f$. Since $\hat{\alpha}(H) f_n = \hat{\alpha}(H) E_H(\Delta_n) f = E_H(\Delta_n) \hat{\alpha}(H) f$, $\hat{\alpha}(H) f_n$ converges strongly to $\hat{\alpha}(H) f$, that is, $\hat{\alpha}(H)$ restricted on $\mathcal{M}_\infty$ has the closure $\hat{\alpha}(H)$ and $\mathcal{M}_\infty$ is a core of $\hat{\alpha}(H)$. Further, because of $\alpha\chi_{(-n,n)} \in L^1(\mathbb{R}, d\lambda)$ and Prop. 3 we have $\hat{\alpha}_n(H) = (2\pi)^{-1/2} \int_{-n}^{n} \alpha(t) e^{-itH} dt$ where $\hat{\alpha}_n(\lambda) = (2\pi)^{-1/2} \int_{-n}^{n} \alpha(t) e^{-it\lambda} dt$. But for $f \in \mathcal{M}_\infty$ we have $\|\hat{\alpha}(H) f - \hat{\alpha}_n(H) f\|^2 = \int |\hat{\alpha}(\lambda) - \hat{\alpha}_n(\lambda)|^2 \big((f, E_H(d\lambda) f)/d\lambda\big) d\lambda \leq |||f|||_\infty^2 \int |\hat{\alpha}(\lambda) - \hat{\alpha}_n(\lambda)|^2 d\lambda \to 0$ as $n \to \infty$. ■

If $T \in \mathcal{L}_2(\mathcal{H})$ and $E_H(\Delta)$ is an arbitrary spectral measure, then $\|TE_H(\Delta)\|_2^2$ is a nonnegative Borel measure (see the proof of Lemma 11). Concerning the absolutely continuous part $E_H(\Delta) P_H^{ac}$, we obtain

**Proposition 28.** *If $T \in \mathcal{L}_2$ and $E_H(\Delta)$ denotes a spectral measure, then the nonnegative Borel measure $\mu(\Delta) = \|TE_H(\Delta) P_H^{ac}\|_2^2$ is absolutely continuous with respect to the Lebesgue measure.*

Proof. Recall the measure property of $\mu$ which comes from the relation

$$\sum \|TE_H(\Delta_\varrho) P_H^{ac}\|_2^2 = \|TP_H^{ac} E_H(\Delta)\|_2^2$$

where $\Delta = \cup \Delta_\varrho$ and $\{\Delta_\varrho\}$ denotes a system of pairwise disjoint Borel sets. Now assuming $|\Delta| = 0$, $E_H(\Delta) P_H^{ac} = 0$ and the assertion follows. ■

### 3.5.4. The absolutely continuous subspace for functions $\alpha(H)$

Let $\alpha(\cdot)$ be real-valued on $\mathbb{R}$ and let $\alpha(H)$ be selfadjoint and defined according to the functional calculus. We formulate some properties which relate the absolutely continuous subspaces of $H$ and $\alpha(H)$. For this reason note that

$$\alpha^{-1}\big(\alpha(\Delta)\big) \supseteq \Delta \tag{43}$$

and

$$E_{\alpha(H)}(\Delta) = E_H\big(\alpha^{-1}(\Delta)\big), \tag{44}$$

hence

$$E_H(\Delta) \leq E_{\alpha(H)}\big(\alpha(\Delta)\big) \tag{45}$$

is valid.

**Proposition 29.** *If $|\Delta| = 0$ implies $|\alpha(\Delta)| = 0$, then $P_{\alpha(H)}^{ac} \leq P_H^{ac}$.*

Proof. Let $f \in P_{\alpha(H)}^{ac}\mathcal{H}$ and $\Delta$ such that $|\Delta| = 0$. Then, according to (45), $\big(f, E_H(\Delta) f\big) \leq \big(f, E_{\alpha(H)}(\alpha(\Delta)) f\big) = 0$, hence $f \in P_H^{ac}\mathcal{H}$. ■

**Proposition 30.** *If $|\Delta| = 0$ implies $|\alpha(\Delta)| = 0$ and further $|\Delta| = 0$ implies $|\alpha^{-1}(\Delta) \cap \operatorname{spec}^c(H \mid P_H^{ac}\mathcal{H})| = 0$, then $P_{\alpha(H)}^{ac} = P_H^{ac}$.*

**Proof.** Let $f \in P_H^{ac}\mathscr{H}$ and $|\Delta| = 0$. Then according to (44) we have

$$\left(f, E_{\alpha(H)}(\Delta) f\right) = \left(f, E_H(\alpha^{-1}(\Delta)) f\right) = \left(f, E_H(\alpha^{-1}(\Delta) \cap \operatorname{spec}^c(H \upharpoonright P_H^{ac}\mathscr{H})) f\right).$$

Hence $f \in P_{\alpha(H)}^{ac}\mathscr{H}$ follows. The second assertion is proved in Prop. 29. ∎

**Remark 31.** If $\alpha(\lambda)$ is continuously differentiable on spec $H$, then the assumption of Prop. 29 is fulfilled. If, additionally, $\alpha'(\lambda) > 0$ on spec $H$, then also the assumptions of Prop. 30 are satisfied and thus $P_H^{ac} = P_{\alpha(H)}^{ac}$.

### 3.5.5. Spectral manifolds

A linear manifold $\mathscr{K} \subset P_H^{ac}\mathscr{H}$ is called *spectral* if there is a spectral core $\operatorname{spec}^c(H \upharpoonright P_{ac}^H \mathscr{H}) = \Gamma(\operatorname{clo} \Gamma = \operatorname{spec}_{ac} H)$ such that $\left(f, E_H(d\lambda) f\right)/d\lambda$ exists for all $f \in \mathscr{K}$ and for all $\lambda \in \Gamma$. Note that a finite-dimensional manifold $\mathscr{K}$ is always spectral: If $e_1, e_2, \dots, e_m$ is an orthonormal basis of $\mathscr{K}$, then there is a Borel set $\Delta_0$ with $|\Delta_0| = 0$ such that $\left(e_i, E(d\lambda) e_j\right)/d\lambda$ exists for all $i, j = 1, 2, \dots, m$ and for all $\lambda \in \operatorname{spec}^c(H \upharpoonright P_H^{ac}\mathscr{H}) \setminus \Delta_0 \left(E(\cdot) = E_H(\cdot)\right)$. But this set is also a spectral core. By a similar argument it is easy to show that the following is true:

**Proposition 32.** *There exist spectral manifolds which are generating with respect to $P_H^{ac}\mathscr{H}$ and $H$.*

**Proof.** Take a basis $e_1, e_2, \dots$ in $P_H^{ac}\mathscr{H}$ and put $\mathscr{K} = \operatorname{spa}\{e_1, e_2, \dots\}$. ∎

Let $\mathscr{K}$ be a spectral manifold with a corresponding spectral core $\Gamma$. Then, according to Prop. 17,

$$(k, k')_{E,\lambda} := \left(k, E(d\lambda) k'\right)/d\lambda, \quad \lambda \in \Gamma, \quad k, k' \in \mathscr{K},$$

is a semiscalar product in $\mathscr{K}$. Thus we can form the Hilbert spaces $\hat{\mathscr{K}}_\lambda := \operatorname{clo}_{|\cdot|_{E,\lambda}} \mathscr{K} \bmod (\ker |\cdot|_{E,\lambda})$ where $|k|_{E,\lambda} = (k, k)_{E,\lambda}^{1/2}$ for all $\lambda \in \Gamma$. For later application we introduce the following linear manifolds. Let $\lambda$ be an arbitrary point of $\mathbb{R}$. Then we introduce the notation

$$[f]_{E,\lambda}^2 := \limsup_{h \to 0} \left(f, E([\lambda, \lambda + h)) f\right)/h. \tag{46}$$

Thus $[f]_{E,\lambda}^2 < \infty$ means that the set of all derivative numbers of the function $\left(f, E((-\infty, \lambda)) f\right)$ is bounded at the point $\lambda$. We denote the set of all vectors $f \in P_H^{ac}\mathscr{H}$ with $[f]_{E,\lambda} < \infty$ by $\mathscr{D}_\lambda$. Note that an arbitrary $f \in P_H^{ac}\mathscr{H}$ necessarily belongs to $\mathscr{D}_\lambda$ for almost all $\lambda$ with respect to $|\cdot|$, because $\left(f, E(d\lambda) f\right)/d\lambda$ exists a.e. mod $|\cdot|$.

**Proposition 33.** *$\mathscr{D}_\lambda$ is a linear manifold and $[\cdot]_{E,\lambda}$ is a seminorm on $\mathscr{D}_\lambda$.*

**Proof.** We have $\|E(\Delta)(f + g)\| \leqq \|E(\Delta) f\| + \|E(\Delta) g\|$; hence

$$|\Delta|^{-1/2} \|E(\Delta)(f + g)\| \leqq |\Delta|^{-1/2} \|E(\Delta) f\| + |\Delta|^{-1/2} \|E(\Delta) g\|$$

and

$$\limsup_{h \to 0} h^{-1/2} \left\|E([\lambda, \lambda + h))(f + g)\right\| \leqq \limsup_{h \to 0} h^{-1/2} \left\|E([\lambda, \lambda + h)) f\right\| + \limsup_{h \to 0} h^{-1/2} \left\|E([\lambda, \lambda + h)) g\right\|$$

for $h > 0$ (and correspondingly for $h < 0$). Hence $[f + g]_{E,\lambda} \leqq [f]_{E,\lambda} + [g]_{E,\lambda}$. $[cf]_{E,\lambda} = |c| \, [f]_{E,\lambda}$, $c \in \mathbb{C}$, is obvious. The linearity of $\mathscr{D}_\lambda$ then follows. ■

According to Prop. 33, $\mathscr{D}_\lambda \bmod (\ker [\cdot]_{E,\lambda})$ is a linear normed space and we can form the Banach space $\hat{\mathscr{D}}_\lambda = \mathrm{clo}_{[\cdot]_{E,\lambda}} \mathscr{D}_\lambda \bmod (\ker [\cdot]_{E,\lambda})$.

If $\mathscr{X} \subset P_H^{\mathrm{ac}} \mathscr{H}$ is a spectral manifold with spectral core $\Gamma$, then $\mathscr{X} \subset \mathscr{D}_\lambda$ for all $\lambda \in \Gamma$ and

$$|f|_{E,\lambda} = [f]_{E,\lambda}, \qquad f \in \mathscr{X}. \tag{47}$$

Obviously $\ker |\cdot|_{E,\lambda} = \ker [\cdot]_{E,\lambda} \cap \mathscr{X}$. Hence for vectors $f \in \mathscr{X}$ the equivalence classes with respect to $\mathscr{D}_\lambda$ are larger than those with respect to $\mathscr{X}$, but there is a canonical embedding

$$\mathscr{X} \bmod (\ker |\cdot|_{E,\lambda}) \subseteqq \mathscr{D}_\lambda \bmod (\ker [\cdot]_{E,\lambda}) \tag{48}$$

because from $[f_1 - f_2]_{E,\lambda} = 0$, $f_1, f_2 \in \mathscr{X}$, according to $f_1 - f_2 \in \mathscr{X}$ and (47), immediately $|f_1 - f_2|_{E,\lambda} = 0$ follows. (48) implies the following diagram

$$\begin{array}{ccccc} & \subseteqq & \mathscr{D}_\lambda \bmod (\ker [\cdot]_{E,\lambda}) & \subseteqq & \\ \mathscr{X} \bmod (\ker |\cdot|_{E,\lambda}) & & & & \hat{\mathscr{D}}_\lambda . \\ & \subseteqq & \hat{\mathscr{X}}_\lambda & \subseteqq & \end{array} \tag{49}$$

The operator of the canonical embedding (identification operator) of $\mathscr{D}_\lambda \subset \mathscr{H}$ in $\mathscr{D}_\lambda \bmod (\ker [\cdot]_{E,\lambda})$ is denoted by $D_\lambda$.

# 4. Direct Integrals and Spectral Representations

## 4.1. Direct and maximal integrals

### 4.1.1. Preliminaries

Let $\Gamma$ be a topological space (metrizable, separable, locally compact) as before and let $\mu$ be a nonnegative Borel measure on $\Gamma$. Further, let $\mathcal{H}_\lambda$ be a family of separable Hilbert spaces indexed by $\lambda \in \Gamma$. We call a subsystem $\mathcal{S}$ of the Cartesian product $\mathop{\times}\limits_{\lambda\in\Gamma} \mathcal{H}_\lambda$ *admissible* if the following properties are satisfied:

(I) $\mathcal{S}$ is linear and closed with respect to multiplication by functions from $L^\infty(\Gamma, \mu)$.

(II) If $f \in \mathcal{S}$, then $\|f(\lambda)\|_{\mathcal{H}_\lambda}$ is Borel measurable and $\int\limits_\Gamma \|f(\lambda)\|^2_{\mathcal{H}_\lambda} \mu(d\lambda) < \infty$.

(III) $\text{clo}_{\|\cdot\|_{\mathcal{H}_\lambda}} \text{spa} \{f(\lambda); f \in \mathcal{S}\} = \mathcal{H}_\lambda \pmod{\mu}$.

(IV) If $\int\limits_\Delta \|f(\lambda)\|^2_{\mathcal{H}_\lambda} \mu(d\lambda) = 0$ for all $f \in \mathcal{S}$, then $\mu(\Delta) = 0$.

The functions $f$ from $\mathcal{S}$ are called *admissible functions*. Often we write $\|\cdot\|_{\mathcal{H}_\lambda} =: \|\cdot\|_\lambda$.

**Example 1.** We put $\mathcal{H}_\lambda = \mathcal{H}$ and $\mathcal{S} := \left\{\sum\limits_{\varrho=1}^m \alpha_\varrho(\lambda)\, \chi_{\Delta_\varrho}(\lambda)\, f_\varrho; \Delta_\varrho \text{ compact}, \alpha_\varrho \in L^\infty(\Gamma, \mu), f_\varrho \in \mathcal{H}\right\}$. Then obviously the conditions (I)–(IV) are satisfied.

**Example 2.** Let $\mathcal{K}$ be an arbitrary linear manifold and let $s(\lambda; \cdot, \cdot)$ be a family of semiscalar products indexed by $\lambda \in \Gamma$. Assume that for all $f \in \mathcal{K}$ the following conditions are satisfied:

(i) $s(\lambda; f) := s(\lambda; f, f)$ is Borel measurable and $\int\limits_\Gamma s(\lambda; f)\, \mu(d\lambda) < \infty$.

(ii) $\int\limits_\Delta s(\lambda; f)\, \mu(d\lambda) = 0$ for all $f \in \mathcal{K}$ implies $\mu(\Delta) = 0$.

Put $(k, k')_\lambda := s(\lambda; k, k')$ and $\|k\|_\lambda^2 := s(\lambda; k)$. Now form the Hilbert spaces $\hat{\mathcal{K}}_\lambda := \text{clo}_{\|\cdot\|_\lambda}$ (spa $\mathcal{K}$ mod ker $\|\cdot\|_\lambda$). We denote the canonical embedding of $\mathcal{K}$ into $\hat{\mathcal{K}}_\lambda$ by $J_\lambda: \mathcal{K} \ni k \to J_\lambda k \in \hat{\mathcal{K}}_\lambda$. In other words, $J_\lambda k$ denotes the equivalence class of $k$ with respect to $\|\cdot\|_\lambda$. Now we define $\mathcal{S}$ by

$$\mathcal{S} := \left\{\sum_{\varrho=1}^m \alpha_\varrho(\lambda)\, J_\lambda f_\varrho; f_\varrho \in \mathcal{K}, \alpha_\varrho \in L^\infty(\Gamma, \mu)\right\}. \tag{1}$$

Obviously, (I) is satisfied. Concerning (II) we can restrict ourselves to $f(\lambda) = J_\lambda k$. But $\|J_\lambda k\|_\lambda^2 = s(\lambda; k)$, hence (II) is satisfied. (III) is obvious, because $\{J_\lambda k\}$ is dense in $\hat{\mathcal{K}}_\lambda$. (IV) is satisfied because of (ii). Note that condition (ii) expresses a certain minimality of $\Gamma$ in the following sense: If there is a $\Delta$ with $\mu(\Delta) > 0$ but $\int\limits_\Delta s(\lambda; k)\, \mu(d\lambda) = 0$ for all $k \in \mathcal{K}$, then $\Gamma$ can be replaced by $\Gamma' = \Gamma \setminus \Delta$ because of $\int\limits_{\Gamma\setminus\Delta} s(\lambda, k)\, \mu(d\lambda) = \int\limits_\Gamma s(\lambda; k)\, \mu(d\lambda)$.

### 4.1.2. Direct integrals

Let $\mathcal{S}$ be admissible. We call a function $f \in \mathop{\times}\limits_{\lambda\in\Gamma} \mathcal{H}_\lambda$ *strongly measurable with respect to* $\mathcal{S}$ if there exists a sequence $t_n \in \mathcal{S}$ such that $\lim\limits_{n\to\infty} \|f(\lambda) - t_n(\lambda)\|_\lambda = 0 \pmod{\mu}$ is valid. Then the *direct integral* of the family $\mathcal{H}_\lambda$ with respect to $\Gamma$, $\mu$ and $\mathcal{S}$ is defined by the set of all strongly measurable functions $f$ such that $\int_\Gamma \|f(\lambda)\|_\lambda^2 \,\mu(\mathrm{d}\lambda) < \infty$. In the direct integral we introduce a semiscalar product by

$$(f, g) := \int_\Gamma \big(f(\lambda), g(\lambda)\big)_\lambda \,\mu(\mathrm{d}\lambda). \tag{2}$$

This definition is justified because $\big(f(\lambda), g(\lambda)\big)_\lambda$ is Borel measurable according to the fact that $\big(f(\lambda), g(\lambda)\big)_\lambda = \lim_{n\to\infty} \big(t_n(\lambda), s_n(\lambda)\big)$ where

$$\lim_{n\to\infty} \|f(\lambda) - t_n(\lambda)\|_\lambda = 0 \quad \text{and} \quad \lim_{n\to\infty} \|g(\lambda) - s_n(\lambda)\|_\lambda = 0 \ (\mathrm{mod}\ \mu).$$

Then as usual the norm is defined by $\|f\|^2 := (f, f)$ and one obtains $\|f\| = 0$ iff $f(\lambda) = 0$ (mod $\mu$). As usual we take the corresponding factor space, then $(f, g)$ is a scalar product and the direct integral is obviously a pre-Hilbert space. Moreover

**Proposition 3.** *The direct integral is a Hilbert space.*

**Proof.** It is standard and omitted here. ■

The direct integral is denoted by $L^2(\Gamma, \mu; \mathscr{H}_\lambda; \mathscr{S})$. If $\mathscr{S}$ is chosen as in Example 1, then the direct integral is nothing else than $L^2(\Gamma, \mu; \mathscr{H})$ because in this case the strong measurability with respect to $\mathscr{S}$ coincides with the usual $\mu$-measurability of a function with values in $\mathscr{H}$. If $\mathscr{S}$ is chosen as in Example 2, then we drop the symbol $\mathscr{S}$ in the notation and write $L^2(\Gamma, \mu; \hat{\mathscr{K}}_\lambda)$ for the direct integral in this case. These direct integrals are important in the following. They are generalizations of Example 1, namely if $s(\lambda; k, k') \equiv (k, k')$ is independent of $\lambda$ and $\hat{\mathscr{K}} = \mathrm{clo}\, \mathscr{K} \bmod \|\cdot\|$, then we obtain again $L^2(\Gamma, \mu; \hat{\mathscr{K}})$.

### 4.1.3. Maximal integrals

Let $\mathscr{S}$ be admissible. Then according to Zorn's lemma, which is applicable in this situation, there are maximal admissible subsystems $\mathscr{S}_{\max}$ of $\mathsf{X}_{\lambda\in\Gamma} \mathscr{H}_\lambda$ containing $\mathscr{S}$. $\mathscr{S}_{\max}$ has the property: If $\mathscr{S}'$ is admissible and $\mathscr{S}' \supseteqq \mathscr{S}_{\max}$, then $\mathscr{S}' = \mathscr{S}_{\max}$. We call a maximal system $\mathscr{S}_{\max} \supseteqq \mathscr{S}$ a *maximal integral* with respect to $\mathscr{S}$. In general, $\mathscr{S}_{\max}$ is not uniquely determined by $\Gamma$, $\mu$, $\mathscr{H}_\lambda$, $\mathscr{S}$. In $\mathscr{S}_{\max}$ a semiscalar product is defined as before by $(f, g) = \int_\Gamma \big(f(\lambda), g(\lambda)\big)_\lambda \,\mu(\mathrm{d}\lambda)$, and $\|f\| = 0$ iff $f(\lambda) = 0 \bmod \mu$. As before we take the factor space and obtain

**Proposition 4.** *A maximal integral $\mathscr{S}_{\max} \supseteqq \mathscr{S}$ is a Hilbert space.*

**Proof.** It is standard and omitted here. ■

If $\mathscr{S}$ is chosen according to Example 1, then $L^2(\Gamma, \mu; \mathscr{H})$ is maximal. Namely let $\mathscr{S}'$ be admissible and $\mathscr{S}' \supseteqq L^2(\Gamma, \mu; \mathscr{H})$. Let $s \in \mathscr{S}'$. Then $\|s(\lambda)\|_\lambda$ is measurable and $\int_\Gamma \|s(\lambda)\|_\lambda^2 \,\mu(\mathrm{d}\lambda) < \infty$. Further $s(\lambda) + c\chi_\Delta(\lambda)\, h$, $h \in \mathscr{H}$, are functions from $\mathscr{S}'$. From the polarization identity we obtain that $\big(s(\lambda), \chi_\Delta(\lambda)\, h\big)_\lambda$ is Borel measurable for all compact sets $\Delta$, hence $(s(\lambda), h)$ is Borel measurable for all $h \in \mathscr{H}$, and thus $s(\lambda)$ is strongly $\mu$-measurable, that is, $s \in L^2(\Gamma, \mu; \mathscr{H})$. Moreover in this case there is only one maximal system containing $\mathscr{S}$: If $\mathscr{S}_{\max} \supseteqq \mathscr{S}$, then $\mathscr{S}_{\max} = L^2(\Gamma, \mu; \mathscr{H})$ for if $s \in \mathscr{S}_{\max}$, then as before we obtain that $(s(\lambda), h)$ is Borel measurable, hence $s(\lambda)$ is $\mu$-measurable, that is, $s \in L^2(\Gamma, \mu; \mathscr{H})$. In general we have

**Proposition 5.** *If $\mathscr{S}$ is admissible and $\mathscr{S}_{\max} \supseteqq \mathscr{S}$, then $L^2(\Gamma, \mu; \mathscr{H}_\lambda; \mathscr{S}) \subseteqq \mathscr{S}_{\max}$.*

**Proof.** Let $f \in L^2$. Then there is a sequence $s_n \in \mathscr{S}$ with $\|f - s_n\| \to 0$ as $n \to \infty$; but $\mathscr{S}_{\max}$ is a Hilbert space, hence $f \in \mathscr{S}_{\max}$. ∎

**Remark 6.** It seems to be difficult to give explicit counterexamples which show that $\mathscr{S}_{\max} \supset L^2(\Gamma, \mu; \mathscr{H}_\lambda; \mathscr{S})$ is possible. In 4.3 we present a large class of admissible systems where direct and maximal integrals coincide.

## 4.2. Direct sums and tensor products

The following properties are valid for direct and also for maximal integrals; the formulation is for direct integrals. The proofs are obvious and will be omitted.

If $\mathscr{S}$ is admissible and $\Gamma = \Gamma_1 \cup \Gamma_2$, $\Gamma_1 \cap \Gamma_2 = \emptyset$, then also $\mathscr{S} \mid \Gamma_1$ and $\mathscr{S} \mid \Gamma_2$ are admissible.

**Proposition 7.** *Let $\mathscr{S}$ and $\Gamma$ as above. Then*

$$L^2(\Gamma, \mu; \mathscr{H}_\lambda; \mathscr{S}) \cong L^2(\Gamma_1, \mu; \mathscr{H}_\lambda; \mathscr{S} \mid \Gamma_1) \times L^2(\Gamma_2, \mu; \mathscr{H}_\lambda; \mathscr{S} \mid \Gamma_2), \tag{3}$$

*where $\times$ denotes the Cartesian product and the isomorphism is given by $f \leftrightarrow \{f \mid \Gamma_1, f \mid \Gamma_2\}$.*

The second possibility for direct sums is related to subspaces $\mathscr{K}_\lambda$ if $\mathscr{H}_\lambda$. We denote the projection onto $\mathscr{K}_\lambda$ by $P_\lambda$. We call $\mathscr{S}$ closed with respect to $P_\lambda$ if $f \in \mathscr{S}$ implies $g \in \mathscr{S}$ where $g(\lambda) = P_\lambda f(\lambda)$.

**Proposition 8.** *If $\mathscr{S}$ is closed with respect to $P_\lambda$, then*

$$L^2(\Gamma, \mu; \mathscr{H}_\lambda; \mathscr{S}) \cong L^2(\Gamma, \mu; \mathscr{K}_\lambda; P_\lambda \mathscr{S}) \times L^2(\Gamma, \mu; \mathscr{K}_\lambda^\perp; P_\lambda^\perp \mathscr{S}) \tag{4}$$

*where the isomorphism is given by $f \leftrightarrow \{P_\lambda f(\lambda), P_\lambda^\perp f(\lambda)\}$.*

If $\mathscr{S}$ is admissible and $\mathscr{K}$ is a Hilbert space, we define the following admissible system $\mathscr{S} \otimes \mathscr{K}$ with respect to the family $\mathscr{H}_\lambda \otimes \mathscr{K}$: $\mathscr{S} \otimes \mathscr{K} = \text{spa}\,\{f(\cdot) \otimes k; f \in \mathscr{S}, k \in \mathscr{K}\}$. It is easy to verify the properties (I)—(IV) for $\mathscr{S} \otimes \mathscr{K}$: (II) Let $f_1, f_2 \in \mathscr{S}$, $k_1, k_2 \in \mathscr{K}$, then

$$\big(f_1(\lambda) \otimes k_1, f_2(\lambda) \otimes k_2\big)_{\mathscr{H}_\lambda \otimes \mathscr{K}} = \big(f_1(\lambda), f_2(\lambda)\big)_\lambda (k_1, k_2)_{\mathscr{K}},$$

hence this function is Borel measurable and

$$\int_\Gamma \big(f_1(\lambda) \otimes k_1, f_2(\lambda) \otimes k_2\big)_{\mathscr{H}_\lambda \otimes \mathscr{K}} \mu(\mathrm{d}\lambda) = (f_1, f_2)\,(k_1, k_2)_{\mathscr{K}};$$

thus (II) follows. The other properties are obvious.

**Proposition 9.** *We have*

$$L^2(\Gamma, \mu; \mathscr{H}_\lambda \otimes \mathscr{K}; \mathscr{S} \otimes \mathscr{K}) \cong L^2(\Gamma, \mu; \mathscr{H}_\lambda; \mathscr{S}) \otimes \mathscr{K} \tag{5}$$

*where the isomorphism is defined by $f(\lambda) \otimes k \leftrightarrow f \otimes k$ and $f(\lambda) \otimes k$ means the vector-valued function $\Gamma \ni \lambda \to f(\lambda) \otimes k \in \mathscr{H}_\lambda \otimes \mathscr{K}$.*

**Example 10.** We consider a special case, namely $\mathscr{H}_\lambda = \mathscr{H}$ and choose $\mathscr{S}$ as in Example 1. Then we obtain

$$L^2(\Gamma, \mu; \mathscr{H} \otimes \mathscr{K}) \cong L^2(\Gamma, \mu; \mathscr{H}) \otimes \mathscr{K}, \tag{6}$$

where the isomorphism is given by $f(\lambda) \otimes k \leftrightarrow f \otimes k$. In particular, if $\mathscr{H} = \mathbb{C}$, then we obtain

$$L^2(\Gamma, \mu; \mathscr{K}) \cong L^2(\Gamma, \mu) \otimes \mathscr{K} \tag{7}$$

where $f(\lambda)\, k \leftrightarrow f \otimes k$.

## 4.3. Direct integrals for separable admissible systems

### 4.3.1. Generating sequences of functions

Let $\mathscr{S}$ be admissible. A sequence $f_1, f_2, \ldots$ of functions from the direct integral $L^2(\Gamma, \mu; \mathscr{H}_\lambda; \mathscr{S})$ (not necessarily from $\mathscr{S}$ itself) is called *generating* if the relation $\mathscr{H}_\lambda = \text{clo spa}\,\{f_1(\lambda), f_2(\lambda), \ldots\}$ (mod $\mu$) is valid or, equivalently, if the sequence $f_1(\lambda)$, $f_2(\lambda), \ldots$ is generating in $\mathscr{H}_\lambda$ for almost all $\lambda$ (mod $\mu$). An admissible system $\mathscr{S}$ is called *separable* if the corresponding direct integral contains a generating sequence.

Concerning Example 1, let $\{h_1, h_2, \ldots\}$ be a generating sequence for $\mathscr{H}$. Further, let $\alpha \in L^2(\Gamma, \mu)$ and $\alpha(\lambda) \neq 0$ a.e. mod $\mu$. Then $\alpha(\lambda)\, h_1, \alpha(\lambda)\, h_2, \ldots$ is an admissible sequence and generating. Thus the admissible system $\mathscr{S}$ of Example 1 is separable. Recall that in this case we have the equality $L^2(\Gamma, \mu; \mathscr{H}) = \mathscr{S}_{\max}$. The next proposition shows that this equality is a consequence of the separability of $\mathscr{S}$.

**Proposition 11.** *If $\mathscr{S}$ is separable, then $L^2(\Gamma, \mu; \mathscr{H}_\lambda; \mathscr{S})$ is maximal.*

Proof. Let $\mathscr{S}_{\max} \supseteq L^2(\Gamma, \mu; \mathscr{H}_\lambda; \mathscr{S})$ be maximal. Take a function $f$ from $\mathscr{S}_{\max} \ominus L^2$, that is, we have $f \perp s$ for all $s \in L^2$ or, equivalently,

$$\int_\Gamma (f(\lambda), s(\lambda))_\lambda\, \mu(\mathrm{d}\lambda) = 0, \qquad s \in L^2.$$

Together with $s$ also $\chi_\Delta s$ belongs to $L^2$, hence $\int_\Delta (f(\lambda), s(\lambda))_\lambda\, \mu(\mathrm{d}\lambda) = 0$ follows for all Borel sets $\Delta$. Thus $(f(\lambda), s(\lambda))_\lambda = 0$ (mod $\mu$). Now put $s(\lambda) = e_\varrho(\lambda)$ where $e_1, e_2, \ldots$ is a generating sequence. Hence we obtain

$$(f(\lambda), e_\varrho(\lambda))_\lambda = 0 (\text{mod}\, \mu), \qquad \varrho = 1, 2, \ldots \tag{8}$$

Let $\Delta_\varrho$ be the Borel set where (8) is not valid. Then $\Delta_\varrho$ is a $\mu$-null set and so is $\bigcup \Delta_\varrho$.

Thus we have $(f(\lambda), e_\varrho(\lambda))_\lambda = 0$, $\varrho = 1, 2, \ldots$, for all $\lambda \in \Gamma (\text{mod}\ \mu)$. Because of $\text{clo spa}\,\{e_\varrho(\lambda)\} = \mathscr{H}_\lambda (\text{mod}\, \mu)$ we obtain $f(\lambda) = 0 (\text{mod}\, \mu)$. ∎

### 4.3.2. Some properties of direct integrals with separable admissible systems

In this section we collect some properties for $L^2(\Gamma, \mu; \mathscr{H}_\lambda; \mathscr{S})$ where $\mathscr{S}$ is separable. Some proofs are omitted (see the Notes and Remarks). At first note that $m(\lambda) := \dim \mathscr{H}_\lambda$ is Borel measurable if $\mathscr{S}$ is separable.

**Proposition 12.** *Let $\mathscr{S}$ be separable. Then there is a sequence $e_1(\lambda), e_2(\lambda), \ldots$ of vector-valued functions from $\Gamma$ to $\mathscr{H}_\lambda$ with $\chi_\Delta e_\varrho \in L^2(\Gamma, \mu; \mathscr{H}_\lambda; \mathscr{S})$ for compact $\Delta$ such that $e_1(\lambda), \ldots, e_{m(\lambda)}(\lambda)$ is an orthonormal basis in $\mathscr{H}_\lambda$ a.e.* mod $\mu$ *and $e_\varrho(\lambda) = 0$ for $\varrho > m(\lambda)$.*

**Proposition 13.** *Let $\mathscr{S}$ be separable and $e_1(\lambda), e_2(\lambda), \ldots$ a pointwise orthonormal basis* mod $\mu$ *according to Prop. 12. Then $f \in L^2(\Gamma, \mu; \mathscr{H}_\lambda; \mathscr{S})$ iff $|(e_\varrho(\lambda), f(\lambda))_\lambda|^2$ is integrable with respect to $\mu$ and $\sum \int_\Gamma |(e_\varrho(\lambda), f(\lambda))_\lambda|^2 \mu(d\lambda) < \infty$. Moreover*

$$f(\lambda) = \operatorname*{s-lim}_{n\to\infty} \sum_{\varrho=1}^{n} \big(e_\varrho(\lambda), f(\lambda)\big)_\lambda e_\varrho(\lambda)$$

*and*

$$(g, f) = \sum_{\varrho=1}^{\infty} \int_\Gamma \overline{\big(e_\varrho(\lambda), f(\lambda)\big)_\lambda} \big(e_\varrho(\lambda), g(\lambda)\big)_\lambda \mu(d\lambda).$$

**Proposition 14.** *Let $\mathscr{S}$ be separable. Then $L^2(\Gamma, \mu; \mathscr{H}_\lambda; \mathscr{S})$ is separable.*

**Proof.** Let $e_1(\lambda), e_2(\lambda), \ldots$ be a pointwise orthonormal basis mod $\mu$. Further $L^2(\Gamma, \mu)$ is separable. Let $\alpha_1, \alpha_2, \ldots$ denote a dense sequence in $L^2(\Gamma, \mu)$. Then the functions $\alpha_{m_1}(\lambda)\, e_1(\lambda) + \cdots + \alpha_{m_r}(\lambda)\, e_r(\lambda)$ are dense in $L^2(\Gamma, \mu; \mathscr{H}_\lambda; \mathscr{S})$. ■

There is a basic isomorphism with respect to direct integrals with separable admissible systems $\mathscr{S}$. Put $\Lambda_\varrho = \{\lambda; m(\lambda) = \varrho\}$, $\varrho = \infty, 1, 2, \ldots$ Further, let $\mathscr{K}_\varrho$ be a fixed Hilbert space with $\dim \mathscr{K}_\varrho = \varrho$. Then

**Proposition 15.** *$L^2(\Lambda_\varrho, \mu; \mathscr{H}_\lambda; \mathscr{S} \mid \Lambda_\varrho)$ is isomorphic to $L^2(\Lambda_\varrho, \mu; \mathscr{K}_\varrho)$ such that the isomorphism is compatible with the multiplication by functions $\alpha \in L^\infty(\Lambda_\varrho, \mu)$. More precisely: There is an isometry $V$ from $L^2(\Lambda_\varrho, \mu; \mathscr{H}_\lambda; \mathscr{S} \mid \Lambda_\varrho)$ onto $L^2(\Lambda_\varrho, \mu; \mathscr{K}_\varrho)$ such that $V(\alpha f) = \alpha V f$ for all $f \in L^2(\Lambda_\varrho, \mu; \mathscr{H}_\lambda; \mathscr{S} \mid \Lambda_\varrho)$ and all $\alpha \in L^\infty(\Lambda_\varrho, \mu)$.*

**Proof.** Let $e_1(\lambda), e_2(\lambda), \ldots$ be a pointwise orthonormal basis in $\mathscr{H}_\lambda$ mod $\mu$. Further, let $k_1, k_2, \ldots$ be an orthonormal basis in $\mathscr{K}_\varrho$. Now we define $V_\lambda: \mathscr{H}_\lambda \ni e_j(\lambda) \to k_j \in \mathscr{K}_\varrho$, $\lambda \in \Lambda_\varrho$, $j = 1, 2, \ldots, \varrho$. Hence $V_\lambda$ is isometric from $\mathscr{H}_\lambda$ onto $\mathscr{K}_\varrho$. Now let $f \in L^2(\Lambda_\varrho, \mu; \mathscr{H}_\lambda; \mathscr{S})$. Then $\big(e_j(\lambda), f(\lambda)\big)_\lambda$ is Borel measurable, hence

$$\big(V_\lambda^* k_j, f(\lambda)\big)_\lambda = \big(k_j, V_\lambda f(\lambda)\big)_{\mathscr{K}_\varrho}$$

is Borel measurable, that is, $V_\lambda f(\lambda) \in L^2(\Lambda_\varrho, \mu; \mathscr{K}_\varrho)$. Conversely, if $k(\lambda) \in L^2(\Lambda_\varrho, \mu; \mathscr{K}_\varrho)$, then $V_\lambda^* k(\lambda) \in L^2(\Lambda_\varrho, \mu; \mathscr{H}_\lambda; \mathscr{S})$ because $\big(e_j(\lambda), V_\lambda^* k(\lambda)\big)_\lambda = \big(k_j, k(\lambda)\big)_{\mathscr{K}_\varrho}$ is Borel measurable and

$$\sum \int_{\Lambda_\varrho} |(e_j(\lambda), V_\lambda^* k(\lambda))_\lambda|^2 \mu(d\lambda) = \sum \int_{\Lambda_\varrho} |(k_j, k(\lambda))_{\mathscr{K}_\varrho}|^2 \mu(d\lambda)$$
$$= \int_{\Lambda_\varrho} \|k(\lambda)\|^2_{\mathscr{K}_\varrho} \mu(d\lambda) < \infty.$$

Hence $V: L^2(\Lambda_\varrho, \mu; \mathscr{H}_\lambda; \mathscr{S} \mid \Lambda_\varrho) \ni f \to \{(Vf)(\lambda)\} = \{V_\lambda f(\lambda)\} \in L^2(\Lambda_\varrho, \mu; \mathscr{K}_\varrho)$ is an isometry from $L^2(\Lambda_\varrho, \mu; \mathscr{H}_\lambda; \mathscr{S} \mid \Lambda_\varrho)$ onto $L^2(\Lambda_\varrho, \mu; \mathscr{K}_\varrho)$. The property $V(\alpha f) = \alpha V f$ is obvious. ■

## 4.4. The algebra of scalar operators and its commutant

**Throughout this paragraph $\mathscr{S}$ is separable.**

### 4.4.1. Scalar operators

A scalar operator $L_\alpha$ with respect to a direct integral $L^2(\Gamma, \mu; \mathscr{H}_\lambda; \mathscr{S})$ is defined by a scalar function $\alpha \in L^\infty(\Gamma, \mu)$ and the equation

$$(L_\alpha f)(\lambda) = \alpha(\lambda) f(\lambda), \qquad f \in L^2. \tag{9}$$

**Proposition 16.** *$L_\alpha \in \mathscr{L}(L^2)$ and $\|L_\alpha\| = |\alpha|_\infty$.*

Proof. Only the last assertion has to be proved. We have

$$\|L_\alpha f\|^2 = \int_\Gamma \|\alpha(\lambda) f(\lambda)\|_\lambda^2 \, \mu(\mathrm{d}\lambda) \leqq \operatorname*{ess\,sup}_\lambda |\alpha(\lambda)|^2 \int_\Gamma \|f(\lambda)\|_\lambda^2 \, \mu(\mathrm{d}\lambda) = |\alpha|_\infty \|f\|^2,$$

that is, $\|L_\alpha\| \leqq |\alpha|_\infty$ follows. Assume $\|L_\alpha\| < |\alpha|_\infty$. Let $c$ be defined by $\|L_\alpha\| < c < |\alpha|_\infty$. Then there is a Borel set $\Lambda'$, $\mu(\Lambda') > 0$, such that $|\alpha(\lambda)| > c$ for $\lambda \in \Lambda'$. Hence

$$\int_{\Lambda'} |\alpha(\lambda)|^2 \|f(\lambda)\|_\lambda^2 \, \mu(\mathrm{d}\lambda) > c^2 \int_{\Lambda'} \|f(\lambda)\|_\lambda^2 \, \mu(\mathrm{d}\lambda)$$

follows for all $f \in L^2$, but this is a contradiction to $\|L_\alpha\| < c$. ■

The set of all $L_\alpha$ is denoted by $\mathscr{M}$.

**Proposition 17.** *$\mathscr{M}$ is a v. Neumann algebra and $L_\alpha{}^* = L_{\bar\alpha}$.*

Proof. Only the assertion that $\mathscr{M}$ is weakly closed has to be proved. It is sufficient (see 1.4.2) to prove that $\mathscr{M}$ is strongly closed. Let $L_{\alpha_n}$ be strongly convergent, $L_{\alpha_n} \xrightarrow[s]{} A \in \mathscr{L}(L^2)$. Then $\|L_{\alpha_n}\| < C$, that is $|\alpha_n|_\infty < C$. Further,

$$\int_\Gamma |\alpha_n(\lambda) - \alpha_m(\lambda)|^2 \|f(\lambda)\|_\lambda^2 \, \mu(\mathrm{d}\lambda) \to 0$$

as $n, m \to \infty$. If $\Delta$ is a compact set, then there is a function $f \in L^2$ such that $\|f(\lambda)\|_\lambda = \chi_\Delta(\lambda) \pmod{\mu}$. Hence $\int_\Delta |\alpha_n(\lambda) - \alpha_m(\lambda)|^2 \, \mu(\mathrm{d}\lambda) \to 0$ as $n, m \to \infty$. Thus $\alpha_n$ is locally $L^2(\Gamma, \mu)$-convergent. Hence there is a subsequence $\alpha_{n'}$ which is convergent a.e. mod $\mu$ on $\Gamma$. We denote the limit by $\alpha_0(\lambda)$. Because of $|\alpha_0(\lambda)| \leqq |\alpha_{n'}(\lambda)| + |\alpha_0(\lambda) - \alpha_{n'}(\lambda)|$ for all $n'$, we obtain ess sup $|\alpha_0(\lambda)| < \infty$, hence $L_{\alpha_0} \in \mathscr{M}$ and

$$\|L_{\alpha_{n'}} f - L_{\alpha_0} f\|^2 = \int_\Gamma |\alpha_{n'}(\lambda) - \alpha_0(\lambda)|^2 \|f(\lambda)\|_\lambda^2 \, \mu(\mathrm{d}\lambda) \to 0$$

as $n' \to \infty$ because $|\alpha_{n'}(\lambda) - \alpha_0(\lambda)|^2$ is uniformly bounded in $n'$ and $\lambda$. Therefore $L_{\alpha_0} = A$. ■

### 4.4.2. The commutant of the scalar operators

An operator function $A(\lambda): \Gamma \ni \lambda \to A(\lambda) \in \mathscr{L}(\mathscr{H}_\lambda)$ is called *admissible* if for every $f \in L^2$ also $A(\lambda) f(\lambda)$ is from $L^2$. For admissible operator functions an operator $A$ from $L^2$ into $L^2$ can be defined by

$$(Af)(\lambda) := A(\lambda) f(\lambda), \qquad f \in L^2. \tag{10}$$

If $A(\lambda)$ is admissible, then $\|A(\lambda)\|_\lambda$ is Borel measurable. This is an easy consequence of the separability of $\mathscr{S}$. Namely take a generating sequence $f_1(\lambda), f_2(\lambda), \dots$ and put $g(\lambda) = \sum_{\varrho=1}^{m} \alpha_{n_\varrho}(\lambda) f_\varrho(\lambda)$. The set of all $g$ is countable (see the proof of Prop. 14). Then

$$\|A(\lambda)\|_\lambda := \sup_{g \neq 0} \frac{\|A(\lambda) g(\lambda)\|_\lambda}{\|g(\lambda)\|_\lambda},$$

hence $\|A(\lambda)\|_\lambda$ is Borel measurable. The operator $A$, corresponding to an admissible operator function $A(\lambda)$, belongs to $\mathscr{L}(L^2)$ iff $\operatorname{ess\,sup}_\lambda \|A(\lambda)\|_\lambda < \infty$ (the proof corresponds to the proof of Prop. 16 and is omitted). We denote the set of all $A$ such that $A$ is defined by (10) and $A \in \mathscr{L}(L^2)$ by $\mathscr{A}$.

**Proposition 18.** *$\mathscr{A}$ is a v. Neumann algebra and $\mathscr{A} = \operatorname{com} \mathscr{M}$. If $A$ corresponds to $A(\lambda)$, then $A^*$ corresponds to $A(\lambda)^*$*

Proof. $\mathscr{A} \subseteq \operatorname{com} \mathscr{M}$ is obvious. Only $\operatorname{com} \mathscr{M} \subseteq \mathscr{A}$ has to be proved. Let $C \in \operatorname{com} \mathscr{M}$. The aim is to construct a corresponding family $C(\lambda) \in \mathscr{L}(\mathscr{H}_\lambda)$ with $\operatorname{ess\,sup}_\lambda \|C(\lambda)\|_\lambda < \infty$. We have

$$CL_\alpha = L_\alpha C, \qquad \alpha \in L^\infty(\Gamma, \mu). \tag{11}$$

For convenience let us assume that $\mu(\Gamma) < \infty$. Take a pointwise orthonormal basis (mod $\mu$) in $\mathscr{H}_\lambda$, $e_1(\lambda), e_2(\lambda), \dots$ Then $e_1, e_2, \dots \in L^2$. We put $(Ce_\varrho)(\lambda) = f_\varrho(\lambda)$, $\varrho = 1, 2, \dots$ Because of (11) we obtain with $\alpha_1, \alpha_2, \dots, \alpha_m \in L_2(\Gamma, \mu)$

$$\left(C\left(\sum_{\varrho=1}^{m} \alpha_\varrho e_\varrho\right)\right)(\lambda) = \sum_{\varrho=1}^{m} \alpha_\varrho(\lambda) f_\varrho(\lambda).$$

If $\lambda$ is fixed, then $\left\{\sum_{\varrho=1}^{m} \alpha_\varrho(\lambda) e_\varrho(\lambda)\right\}$ is dense in $\mathscr{H}_\lambda$ a.e. mod $\mu$. Now we define an operator $C(\lambda)$ for fixed $\lambda$, by

$$C(\lambda)\left(\sum_{\varrho=1}^{m} c_\varrho e_\varrho(\lambda)\right) := \sum_{\varrho=1}^{m} c_\varrho f_\varrho(\lambda), \qquad c_\varrho \in \mathbb{C}.$$

$C(\lambda)$ is for almost all $\lambda$ mod $\mu$ defined on a dense set in $\mathscr{H}_\lambda$. Then for all $g(\lambda) = \sum_{\varrho=1}^{m} \alpha_\varrho(\lambda) e_\varrho(\lambda)$, which are dense in $L^2$, already the equation

$$(Cg)(\lambda) = C(\lambda) g(\lambda) \tag{12}$$

is obtained. Now we assert: $C(\lambda)$ is bounded on $\mathscr{H}_\lambda$ a.e. mod $\mu$ and $\operatorname{ess\,sup}_\lambda \|C(\lambda)\|_\lambda < \infty$. Because of $CL_{\chi_\Delta} = L_{\chi_\Delta}C$ we have

$$\int_\Delta \|(Cg)(\lambda)\|_\lambda^2 \, \mu(\mathrm{d}\lambda) \leq \|C\|^2 \int_\Delta \|g(\lambda)\|_\lambda^2 \, \mu(\mathrm{d}\lambda)$$

for all Borel sets $\Delta$. Thus we obtain

$$\|C(\lambda)\, g(\lambda)\|_\lambda \leq \|C\| \, \|g(\lambda)\|_\lambda$$

a.e. mod $\mu$ for $g$ fixed. Now we choose a sequence $g_1(\lambda), g_2(\lambda), \ldots$ such that for fixed $\lambda$ the sequence is dense in $\mathscr{H}_\lambda$(mod $\mu$). Then we obtain

$$\|C(\lambda)\, g_j(\lambda)\|_\lambda / \|g_j(\lambda)\|_\lambda \leq \|C\|, \qquad j = 1, 2, \ldots,$$

hence $\|C(\lambda)\|_\lambda \leq \|C\|$ for almost all $\lambda$, that is, $\operatorname{ess\,sup}_\lambda \|C(\lambda)\|_\lambda \leq \|C\|$. Thus the equation (12) can be extended to all $f \in L^2$. ■

Note that the conclusions of this proof can also be used to prove the following

**Corollary 19.** *Let $L^2(\Gamma, \mu; \mathscr{H}_\lambda; \mathscr{S})$ and $L^2(\Gamma, \mu; \mathscr{K}_\lambda; \mathscr{S}')$ denote two direct integrals, where $\mathscr{S}, \mathscr{S}'$ are separable. The scalar operators corresponding to $\alpha$ are denoted by $L_\alpha, L_\alpha'$. If $C$ is an intertwining operator for $L_\alpha, L_\alpha'$, that is, $L_\alpha' C = CL_\alpha$, $\alpha \in L^\infty(\Gamma, \mu)$, $C \in \mathscr{L}(L^2(\Gamma, \mu; \mathscr{H}_\lambda; \mathscr{S}), L^2(\Gamma, \mu; \mathscr{K}_\lambda; \mathscr{S}'))$, then there is a family $C(\lambda) \in \mathscr{L}(\mathscr{H}_\lambda, \mathscr{K}_\lambda)$ with* ess sup $\|C(\lambda)\|_\lambda < \infty$ *such that $(Cf)(\lambda) = C(\lambda) f(\lambda)$ a.e.* mod $\mu$ *for all $f \in L^2(\Gamma, \mu; \mathscr{H}_\lambda, \mathscr{S})^\lambda$.*

### 4.4.3. A special case

Let dim $\mathscr{H}_\lambda = 1$, that is $\mathscr{H}_\lambda = \mathbb{C}$. Then from Prop. 18 we obtain in this case $\mathscr{M} = \operatorname{com} \mathscr{M}$, that is, $\mathscr{M}$ is maximal Abelian (hence $\mathscr{M}$ is cyclic). A case of particular interest is: $\Gamma \subset \mathbb{R}$, $\Gamma$ compact, $\mu$ a finite Borel measure on $\Gamma$. Then $\mathscr{M} = \mathscr{M}_\mu$ is cyclic. A special operator in $\mathscr{M}_\mu$ is $(L_\lambda f)(\lambda) = \lambda f(\lambda)$; $\mathscr{M}_\mu = W^*(L_\lambda, 1)$. A special generating vector for $L^2(\Gamma, \mu)$ with respect to $\mathscr{M}_\mu$ is $f(\lambda) \equiv 1$.

A very important question is the following: Let $L^2(\Gamma, \mu)$, $L^2(\Gamma', \mu')$ be given. One asks for conditions on $\mu, \mu'$ such that $\mathscr{M}_\mu, \mathscr{M}_{\mu'}$ are spatial isomorphic and in particular $L_\lambda$ in $\mathscr{M}_\mu$ and $L_\lambda'$ in $\mathscr{M}_{\mu'}$ are unitarily equivalent.

**Lemma 20.** *The correspondence $\mathscr{M}_\mu \ni L_\alpha \leftrightarrow L_\alpha' \in \mathscr{M}_{\mu'}$ is a spatial isomorphism* (that is, $L_\alpha$ and $L_\alpha'$ are unitarily equivalent) *iff $\mu$ and $\mu'$ are equivalent measures, $\mu \equiv \mu'$. $\mathscr{M}_\mu$ and $\mathscr{M}_{\mu'}$ are disjoint iff $\mu$ and $\mu'$ are disjoint measures, $\mu \perp \mu'$.*

**Proof.** Let $L_\alpha$ and $L_\alpha'$ be unitarily equivalent, that is, we have an isometry from $L^2(\Gamma, \mu)$ onto $L^2(\Gamma', \mu')$, $L^2(\Gamma, \mu) \ni f \to g \in L^2(\Gamma', \mu')$ such that $\alpha(\lambda) f(\lambda) \to \alpha(\lambda) g(\lambda)$. In particular $\chi_\Delta(\lambda) f(\lambda) \to \chi_\Delta(\lambda) g(\lambda)$ is valid. Let $\Delta$ be compact and take $f(\lambda) = \chi_\Delta(\lambda)$, then $\chi_\Delta(\lambda) \to \chi_\Delta(\lambda) g(\lambda)$, hence $\mu(\Delta) = \int_\Delta |g(\lambda)|^2 \mu'(\mathrm{d}\lambda)$, that is $\mu \ll \mu'$. Correspondingly $\mu' \ll \mu$, hence $\mu \equiv \mu'$ follows. If $\mu \equiv \mu'$, then $\mu(\Delta) = \int_\Delta \varrho(\lambda)\, \mu(\mathrm{d}\lambda)$ and $\mu'(\Delta) = \int_\Delta \varrho(\lambda)^{-1} \mu(\mathrm{d}\lambda)$. If $f \in L^2(\Gamma, \mu)$, then $\int_\Gamma |f(\lambda)|^2 \mu(\mathrm{d}\lambda) = \int_\Gamma |f(\lambda)|^2 \varrho(\lambda)\, \mu'(\mathrm{d}\lambda)$ that is, $L^2(\Gamma, \mu) \ni f(\cdot)$

$\to f(\cdot)\,\varrho(\cdot)^{1/2} \in L^2(\Gamma', \mu')$ is an isometry from $L^2(\Gamma, \mu)$ onto $L^2(\Gamma', \mu')$. Obviously $\alpha(\lambda)\, f(\lambda) \to \alpha(\lambda)\, f(\lambda)\, \varrho(\lambda)^{1/2}$ is valid, that is, $L_\alpha$ and $L_\alpha'$ are unitarily equivalent. The proof for the second part of the lemma is omitted. ∎

## 4.5. Unitary invariants and spectral representations

### 4.5.1. Spectral forms, the corresponding direct integrals, and spectral representations

In this section we start with a special kind of Example 2. Let $\mu$ be a Borel measure on $\Gamma$, let $\mathscr{H}$ be a Hilbert space and $E(\cdot)$ be a spectral measure defined on $\mathscr{B}_\Gamma$. Assume that $E(\Delta) = 0$ implies $\mu(\Delta) = 0$. Further, let $\mathscr{K}$ be a generating linear manifold for $\mathscr{H}$ with respect to $E(\cdot)$.

A family of semiscalar products $s(\lambda; k, k')$, $\lambda \in \Gamma$; $k, k' \in \mathscr{K}$ is called a *spectral form* with respect to $E(\cdot)$ if

$$\int_\Delta s(\lambda; k, k)\, \mu(\mathrm{d}\lambda) = \big(k, E(\Delta)\, k'\big), \qquad k, k' \in \mathscr{K}, \quad \Delta \in \mathscr{B}_\Gamma \tag{13}$$

is valid. Then the condition (i) of Example 2 is satisfied. If

$$\int_\Delta s(\lambda; k, k')\, \mu(\mathrm{d}\lambda) = \|E(\Delta)\, k\|^2 = 0$$

for all $k \in \mathscr{K}$, then $E(\Delta)\, f = 0$ for all $f \in \mathscr{H}$ because $\mathscr{K}$ is generating. Hence $E(\Delta) = 0$ and $\mu(\Delta) = 0$, that is, (ii) is satisfied. In this case we can form the direct integral $L^2(\Gamma, \mu; \hat{\mathscr{K}}_\lambda)$ where $\mathscr{S}$ is chosen according to Example 2. Then it turns out that there is an isometry from $\mathscr{H}$ onto $L^2(\Gamma, \mu; \hat{\mathscr{K}}_\lambda)$ such that $E(\Delta)$ corresponds to the characteristic function $\chi_\Delta(\lambda)$ or, equivalently, that $W^*\big(E(\Delta)\big)$ corresponds to $\mathscr{M}$. More precisely, we have

**Proposition 21.** *Let $s(\lambda; \cdot, \cdot)$, $\lambda \in \Gamma$, be a spectral form with respect to $E(\cdot)$ defined on $\mathscr{K}$ where $\mathscr{K}$ is generating with respect to $\mathscr{H}$ and $E(\cdot)$. Then there is an isometric operator $V$ from $L^2(\Gamma, \mu; \hat{\mathscr{K}}_\lambda)$ onto $\mathscr{H}$ such that*

$$V(\{\chi_\Delta(\lambda)\, J_\lambda k\}) = E(\Delta)\, k, \qquad k \in \mathscr{K},\ \Delta \in \mathscr{B}_\Gamma,$$
$$V(\alpha f) = \int_\Gamma \alpha(\lambda)\, E(\mathrm{d}\lambda)\, Vf, \qquad \alpha \in L^\infty(\Gamma, \mu),\ f \in L^2(\Gamma, \mu; \hat{\mathscr{K}}_\lambda).$$

**Proof.** First we define $V$ for the functions $\alpha(\lambda)\, J_\lambda k$ from $\mathscr{S}$:

$$\{\alpha(\lambda)\, J_\lambda k\} \to V\big(\{\alpha(\lambda)\, J_\lambda k\}\big) := \int_\Gamma \alpha(\lambda)\, E(\mathrm{d}\lambda)\, k, \qquad k \in \mathscr{K},\ \alpha \in L^\infty(\Gamma, \mu).$$

Now we calculate the corresponding scalar products. On the one hand we have, $f = \{\alpha(\lambda)\, J_\lambda k\}$ and $g = \{\beta(\lambda)\, J_\lambda h\}$,

$$(f, g) = \int_\Gamma \big(\alpha(\lambda)\, J_\lambda k, \beta(\lambda)\, J_\lambda h\big)_\lambda\, \mu(\mathrm{d}\lambda)$$
$$= \int_\Gamma \overline{\alpha(\lambda)}\, \beta(\lambda)\, (J_\lambda k, J_\lambda h)_\lambda\, \mu(\mathrm{d}\lambda) = \int_\Gamma \overline{\alpha(\lambda)}\, \beta(\lambda)\, s(\lambda; k, h)\, \mu(\mathrm{d}\lambda).$$

On the other hand we obtain (see 3.1.2)

$$\left(\int_\Gamma \alpha(\lambda)\, E(\mathrm{d}\lambda)\, k, \int_\Gamma \beta(\lambda)\, E(\mathrm{d}\lambda)\, h\right) = \int_\Gamma \overline{\alpha(\lambda)}\, \beta(\lambda)\, \big(k, E(\mathrm{d}\lambda)\, h\big).$$

But from (13) it immediately follows that $\big(k, E(\Delta)\, h\big)$ is absolutely continuous with respect to $\mu$ and that the two integrals coincide. Hence

$$(Vf, Vg) = (f, g), \tag{14}$$

that is, the scalar products are invariant under $V$. Thus by linearity we may extend $V$ to the whole $\mathscr{S} = \left\{\sum_{\varrho=1}^{m} \alpha_\varrho(\lambda)\, J_\lambda k_\varrho\right\}$ as an isometric operator. Further, by continuity this operator may be extended to the whole Hilbert space $L^2(\Gamma, \mu; \hat{\mathscr{K}}_\lambda)$ as an isometric operator. But ima $V = \mathscr{H}$ because $\mathscr{K}$ is generating. The relation $V(\alpha f) = \int_\Gamma \alpha(\lambda)\, E(\mathrm{d}\lambda)\, Vf$ follows immediately from $V(\{J_\lambda, k\}) = k$ and $V\big(\{\alpha(\lambda)\, J_\lambda k\}\big) = \int_\Gamma \alpha(\lambda)\, E(\mathrm{d}\lambda)\, V(\{J_\lambda k\})$. ∎

**Remark 22.** A representation of a spectral measure of the form presented in Prop. 21, namely by the characteristic function with respect to a direct integral, is called a *spectral representation* of the spectral measure.

Note that if $\mathscr{H}$ and $E(\cdot)$ are given, then it is always possible to choose a linear manifold $\mathscr{K}$, a Borel measure $\mu$ and a family of semiscalar products such that this family is a spectral form. It is even possible to choose $\mathscr{K}$ in such a way that it is the linear span of a countable set of vectors (that is, the so-called algebraic dimension of $\mathscr{K}$ is countable): Let $e_1, e_2, \ldots$ be an orthonormal basis in $\mathscr{H}$. Put $\mu_j(\Delta) = \big(e_j, E(\Delta)\, e_j\big)$ and $\mu(\Delta) = \sum 2^{-n}\mu_n(\Delta)$. Then the Borel measures $\mu_1, \mu_2, \ldots$ are absolutely continuous with respect to $\mu$. Put $\mathscr{K} = \operatorname{spa}\{e_1, e_2, \ldots\}$. Then $\big(e_i, E(\Delta)\, e_j\big) = \int_\Gamma \varrho_{ij}(\lambda)\, \mu(\mathrm{d}\lambda)$ and one can choose a Borel set $\Gamma_0 \subset \Gamma$ with $\mu(\Gamma_0) = 0$ such that $\varrho_{ij}$ is defined on $\Gamma \setminus \Gamma_0$ for all $i, j$. Put $\varrho_{ij}(\lambda) = 0$ on $\Gamma_0$. Now let $f = \sum_{i=1}^{m} c_i e_i$, $g = \sum_{i=1}^{n} a_i e_i$ and define $s(\lambda; f, g)$ by $s(\lambda; f, g) = \sum \bar{c}_i a_j \varrho_{ij}(\lambda)$. Then $s(\lambda; \cdot, \cdot)$ is a spectral form of $\mathscr{K}$. If $E(\Delta) = 0$, that is, $\big(f, E(\Delta)\, f\big) = 0$ for all $f \in \mathscr{H}$, than $\mu_j(\Delta) = 0$, $j = 1, 2, \ldots$, hence $\mu(\Delta) = 0$.

In other words, the method described in Prop. 21 leads to a spectral representation for every spectral measure. Thus, in particular, we can associate to every bounded selfadjoint operator $A$ a spectral representation with respect to its spectral measure where $\Gamma = \operatorname{supp} E \subset \mathbb{R}$ is compact and $A$ itself is represented by $L_\lambda$.

### 4.5.2. Simple operators

If $A$ is simple, then the construction of 4.5.1 is very easy. In this case one can choose $\mathscr{K} = \{ce; c \in \mathbb{C}\}$ where $e$ is generating with respect to $A$, $\Gamma = \operatorname{spec} A$, $\mu(\Delta) = \big(e, E_A(\Delta)e\big)$ and $s(\lambda; ae, be) = \bar{a}b$. Hence in this case $L^2(\Gamma, \mu; \hat{\mathscr{K}}_\lambda)$ is nothing else than $L^2(\operatorname{spec} A, \mu)$ and the isometric operator $V$ of Prop. 21 is defined by

$$L^\infty(\operatorname{spec} A, \mu) \ni \alpha \to V\alpha = \alpha(A)\, e \in \mathscr{H}.$$

Hence we obtain

**Corollary 23.** *Let $A$ be simple in $\mathscr{H}$. Then there is a finite Borel measure $\mu$ defined on* spec $A$ *such that $A$ is unitarily equivalent to $L_\lambda$ in $L^2(\text{spec } A, \mu)$. The measure class $\{\mu\}$ is a complete system of unitary invariants of $A$. The invariant* spec $A$ *is given by* spec $A = \text{supp } \mu$.

Proof. The first part is obvious. The second part follows from Lemma 20. ∎

### 4.5.3. Unitary invariants of an arbitrary bounded selfadjoint operator and the corresponding spectral representations

First we reduce the problem of the construction of a complete set of unitary invariants of a selfadjoint bounded operator to the case of a simple operator. According to 2.2.1 and 2.2.2, there is a resolution of the identity $1 = \sum E_n$ corresponding to a given operator $A$ (where the $E_n$ are uniquely determined) such that $A \restriction E_n\mathscr{H}$ is of multiplicity $n$. Moreover two operators $A$, $A'$ are unitarily equivalent iff $A \restriction E_n\mathscr{H}$, $A' \restriction E_n'\mathscr{H}'$ are unitarily equivalent for all $n = \infty, 1, 2, \ldots$ Note that the $E_n$ are spectral projections with respect to $A$, hence $E_n = E(\Delta_n)$ with Borel sets $\Delta_n$, $\Delta_n \cap \Delta_m = \emptyset \pmod{\mu}$, $\bigcup_n \Delta_n = \text{spec } A \pmod{\mu}$, $\Delta_n$ are unique mod $E$. If $A$ is of homogeneous multiplicity $n$, then, according to 2.1.3, we have: $A$ is unitarily equivalent to $A_1 \otimes 1_{\mathscr{K}}$ in $\mathscr{H}_1 \otimes \mathscr{K}$ where $A_1$ is simple and $\dim \mathscr{K} = n$. Thus according to 4.5.2 we obtain the following result:

**Proposition 24.** *Let $A$ be bounded selfadjoint in $\mathscr{H}$, let $A_n$ denote its homogeneous parts of multiplicity $n$, let $A_n$ be unitarily equivalent to $B_n \otimes 1_n$ where $B_n$ is simple and let $\{\mu_n\}$ be the corresponding measure class. Then the sequence $\{\mu_\infty\}, \{\mu_1\}, \ldots$ of pairwise disjoint measure classes (of finite Borel measures on $\mathbb{R}$) is a complete system of unitary invariants of $A$.*

The invariant spec $A$ can be derived from this system by

$$\text{spec } A = \text{clo } \{\text{supp } \mu_\infty \cup \text{supp } \mu_1 \cup \text{supp } \mu_2 \cup \cdots\}.$$

With the help of a special spectral representation of $A$, which we will derive from Prop. 24, the complete system of unitary invariants can be transformed into a slightly changed form:

First we can renormalize each measure $\mu_j$ such that $\sum \mu_j(\mathbb{R}) < \infty$. Then $\mu(\Delta) := \mu_\infty(\Delta) + \mu_1(\Delta) + \cdots$ defines a finite Borel measure on $\mathbb{R}$. We choose a sequence of pairwise disjoint Borel sets $\Delta_\infty, \Delta_1, \ldots$ with $\bigcup_j \Delta_j = \mathbb{R} \pmod{\mu}$ such that $\mu(\Delta \cap \Delta_j) = \mu_j(\Delta)$, $j = \infty, 1, 2, \ldots$ where the $\Delta_j$ are unique mod $\mu$. Now we define a family of Hilbert spaces $\mathscr{H}_\lambda$ by $\mathscr{H}_\lambda := \mathscr{K}_n$ for $\lambda \in \Delta_n$ where $\mathscr{K}_n$ is a fixed Hilbert space of dimension $n$, and put

$$\mathscr{S} := \left\{\sum_{n,\varrho} \alpha_\varrho(\lambda)\, \chi_{\Sigma_\varrho \cap \Delta_\varrho}(\lambda)\, k_{n,\varrho};\ \alpha_\varrho \in L^\infty(\mathbb{R}, \mu),\ k_{n,\varrho} \in \mathscr{K}_n\right\}.$$

Then we can form the direct integral $L^2(\mathbb{R}, \mu; \mathscr{H}_\lambda; \mathscr{S})$. According to 4.2 we have

$$L^2(\mathbb{R}, \mu; \mathscr{H}_\lambda; \mathscr{S}) \cong \bigtimes_{n=\infty,1}^{\infty} L^2(\Delta_n, \mu_n; \mathscr{K}_n).$$

We obtain

**Proposition 25.** (i) *There is an isometric operator $V$ from $\mathscr{H}$ onto $L^2(\mathbb{R}, \mu; \mathscr{H}_\lambda; \mathscr{S})$ such that $V(E_A(\Delta) f)(\lambda) = \chi_\Delta(\lambda) (Vf)(\lambda)$, that is, $L^2(\mathbb{R}, \mu; \mathscr{H}_\lambda; \mathscr{S})$ is the direct integral of a spectral representation of $E_A(\cdot)$; the relations $VAV^* = L_\lambda$ and $VE_A(\Delta) V^* = L_{\chi_\Delta}$ are valid.*

(ii) *The direct integral $L^2(\mathbb{R}, \mu; \mathscr{H}_\lambda; \mathscr{S})$ (where $\mathscr{S}$ is separable) is uniquely determined by this property up to equivalence of the measure $\mu$ and equality of* $\dim \mathscr{H}_\lambda$ *a.e.* mod $\mu$. *That is: if $L^2(\mathbb{R}, \mu'; \mathscr{H}_\lambda'; \mathscr{S}')$ denotes a second direct integral ($\mathscr{S}'$ separable), then $L_\lambda$ and $L_\lambda'$ are unitarily equivalent iff $\mu \equiv \mu'$ and* $\dim \mathscr{H}_\lambda = \dim \mathscr{H}_\lambda'$ *a.e.* mod $\mu$.

**Proof.** (i) We have $\mathscr{H} \cong \bigoplus_n E_n\mathscr{H} \cong \bigtimes_n \mathscr{H}_n \otimes \mathscr{K}_n \cong \bigtimes_n L^2(\Delta_n, \mu_n) \otimes \mathscr{K}_n \cong \bigtimes L^2(\Delta_n, \mu_n; \mathscr{K}_n) \cong L^2(\mathbb{R}, \mu; \mathscr{H}_\lambda; \mathscr{S})$. By the unitary equivalence of $\mathscr{H}$ and $L^2(\mathbb{R}, \mu; \mathscr{H}_\lambda; \mathscr{S})$, the operator $A$ on $\mathscr{H}$ is transformed into $L_\lambda$ on $L^2(\mathbb{R}, \mu; \mathscr{H}_\lambda; \mathscr{S})$.
(ii) If $L^2(\mathbb{R}, \mu; \mathscr{H}_\lambda; \mathscr{S})$ is given where $\mathscr{S}$ is separable, then according to 4.3.2 (basic isomorphism) we have $L^2(\mathbb{R}, \mu; \mathscr{H}_\lambda; \mathscr{S}) \cong L^2 \bigtimes(\Lambda_\varrho, \mu_\varrho; \mathscr{K}_\varrho)$ where the Borel sets $\Lambda_\varrho$ are pairwise disjoint and $\mu_\varrho(\Delta) = \mu(\Delta \cap \Delta_\varrho)$. This direct sum resolution corresponds simultaneously to the resolution of the identity, which is due to the multiplicity theory, with respect to $L_\lambda$. Thus, if a second direct integral $L^2(\mathbb{R}, \mu'; \mathscr{H}_\lambda'; \mathscr{S}') \cong \bigtimes L^2(\Lambda_\varrho', \mu_\varrho'; \mathscr{K}_\varrho')$ is given, then $L_\alpha$ and $L_\alpha'$ are unitarily equivalent iff $L_\lambda \restriction L^2(\Lambda_\varrho, \mu_\varrho; \mathscr{K}_\varrho)$ is unitarily equivalent to $L_\lambda' \restriction L^2(\Lambda_\varrho', \mu_\varrho'; \mathscr{K}_\varrho')$ and this is the case, according to Lemma 20, iff $\mu_\varrho \equiv \mu_\varrho'$ hence also $\Lambda_\varrho = \Lambda_\varrho' (\mathrm{mod}\ \mu_\varrho)$. ∎

**Remark 26.** According to Prop. 25, a complete system of unitary invariants of a bounded selfadjoint operator $A$, in terms of $L^2(\mathbb{R}, \mu; \mathscr{H}_\lambda; \mathscr{S})$, is given by the *measure class* $\{\mu\}$ and the so-called *multiplicity function* $m(\lambda) := \dim \mathscr{H}_\lambda$.

The following technical lemmas are often useful.

**Lemma 27.** *Let $H$ be selfadjoint and absolutely continuous on $\mathscr{H}$. Then there are a Hilbert space $\mathscr{H}_1$, a selfadjoint absolutely continuous operator $H_0$ on $\mathscr{H}_0 := \mathscr{H} \times \mathscr{H}_1$ with $H_0 \restriction \mathscr{H} = H$ and a set $\mathscr{D}_0 \subset \mathscr{M}_\infty(H_0)$, generating with respect to $\mathscr{H}_0$ and $H_0$, such that for each $u \in \mathscr{D}_0$ there is a positive number $c$ with the property that $\{e^{imcH_0}u\}$, $m = 0, \pm 1, \pm 2, \ldots$, is an orthonormal system which is generating in* clo spa $\{e^{itH_0}u; t \in \mathbb{R}\}$. *Furthermore, the set $\mathscr{D} = P_{\mathscr{H}}\mathscr{D}_0$, where $P_{\mathscr{H}}$ is the projection onto $\mathscr{H}$, is contained in $\mathscr{M}_\infty(H)$ and is generating with respect to $\mathscr{H}$ and $H$. If $\mathscr{H}_0$ runs through all possible extensions of $\mathscr{H}$, with the properties above, then the union of all corresponding sets $\mathscr{D}$ is generating in $\mathscr{M}_\infty(H)$ (with respect to $|||\cdot|||_{H,\infty}$).*

**Proof.** If suffices to take a part of $H$ which is bounded and has multiplicity 1. Thus we can assume that $H$ is multiplication by $\lambda$ on $\mathscr{H} = L^2(\Delta, d\lambda)$, $|\Delta| < \infty$. Let $\Delta = \Delta_1 \cup \Delta_2$ with disjoint Borel sets $\Delta_1, \Delta_2$. Assume $[c_1, d_1] \supseteq \Delta_1$ and $[c_2, d_2] \supseteq \Delta_2$. Then we put

$$\mathscr{H}_1 = L^2([c_1, d_1] \setminus \Delta_1, d\lambda) \times L^2([c_2, d_2] \setminus \Delta_2, d\lambda)$$

and can identify $\mathscr{H}_0 = \mathscr{H} \times \mathscr{H}_1$ with $L^2([c_1, d_1], d\lambda) \times L^2([c_2, d_2], d\lambda)$. Set $\mathscr{D}_0 = \{u_1, u_2\}$ with $u_1(\cdot) = |c_1 - d_1|^{-1/2} \chi_{[c_1,d_1]}(\cdot) \in L^2([c_1, d_1], d\lambda)$ and $u_2(\cdot) = |c_2 - d_2|^{-1/2}\chi_{[c_2,d_2]}(\cdot) \in L^2([c_2, d_2], d\lambda)$. It is a straight-forward calculation to show that $\mathscr{D}_0$ has the desired properties. Further, we have that $\mathscr{D} = \{v_1, v_2\}$ with $v_1(\cdot) = |c_1 - d_1|^{-1/2}\chi_{\Delta_1}(\cdot)$, $v_2(\cdot) = |c_2 - d_2|^{-1/2}\chi_{\Delta_2}(\cdot)$. Clearly, $\mathscr{D} \subset \mathscr{M}_\infty(H)$ and is generating with respect to $\mathscr{H}$ and $H$. By varying the set $\Delta_1$ we obtain all functions $\chi_{\Delta_0}(\cdot)\, a_0$ with $\Delta_0 \subseteq \Delta$ arbitrary and $a_0$ some positive number. This set of functions is evidently generating in $\mathscr{M}_\infty(H)$ with respect to the norm $|||\cdot|||_{H,\infty}$. ■

**Lemma 28.** *Let $H$ be simple and let $Q_1, Q_2 \in \operatorname{com} H$ be projections. Then $Q_1 = Q_2$ iff* $\operatorname{spec^c} \{H \restriction Q_1\mathscr{H}\} = \operatorname{spec^c} \{H \restriction Q_2\mathscr{H}\}$.

**Proof.** The "only if"-part is obvious. Let $\operatorname{spec^c} \{H \restriction Q_1\mathscr{H}\} = \operatorname{spec^c} \{H \restriction Q_2\mathscr{H}\}$. We denote this Borel set by $\Delta_0$. Because of $W^*(E_H(\Delta); \Delta \text{ Borel set}) = \operatorname{com} H$ (note that $W^*(E_H(\Delta); \Delta \text{ Borel set})$ is maximal Abelian, see 4.4.3) we have $Q_1 = E_H(\Delta_1)$, $Q_2 = E_H(\Delta_2)$, where $\Delta_1, \Delta_2$ can be taken as spectral cores of $H \restriction Q_1\mathscr{H}$, $H \restriction Q_2\mathscr{H}$. Thus $Q_1 = E_H(\Delta_0) = Q_2$. ■

At the end of this section let us make some remarks to the spectral representations of a selfadjoint operator given by Prop. 21 and Prop. 25. In both cases $A$ is represented by $L_\lambda$ with respect to a direct integral. The direct integral of Prop. 21 is constructed by a spectral form on a certain linear manifold $\mathscr{K}$, and the direct integral of Prop. 25, by a separable $\mathscr{S}$. It seems to be difficult (and we could not decide this question) to prove that the special admissible system $\mathscr{S}$ of Prop. 21 is separable, in general. This gap has the consequence that we cannot apply Prop. 25 to this special case; thus we do not know, in general, if the measure $\mu$ of Prop. 21 is from the measure class $\{\mu_A\}$ and if $\dim \hat{\mathscr{K}}_\lambda = m(\lambda)$ a.e. mod $\mu$ where $\{\mu_A\}$, $m(\lambda)$ are the unitary invariants of $A$.

If $\mathscr{K}$ has a countably algebraic dimension ($\mathscr{K} = \operatorname{spa} \{k_1, k_2, \ldots\}$), then $\mathscr{S}$ is separable since $\hat{\mathscr{K}}_\lambda = \operatorname{clo} \operatorname{spa} \{J_\lambda k_1, J_\lambda k_2, \ldots\}$ and the sequence $J_\lambda k_1, J_\lambda k_2, \ldots$ is generating. Thus, in this case we can apply Prop. 25.

Note that when $\mathscr{K}$ is given where $\mathscr{K}$ is generating with respect to $E(\cdot)$, then there is always a submanifold $\mathscr{K}_0 \subset \mathscr{K}$ with a countably algebraic dimension such that $\mathscr{K}_0$ is already generating. Hence we may restrict a spectral form, given an $\mathscr{K}$, to $\mathscr{K}_0$. Then $\mathscr{S} = \left\{\sum_\varrho \alpha_\varrho(\lambda)\, J_\lambda k_\varrho; \alpha_\varrho \in L^\infty, k_\varrho \in \mathscr{K}_0\right\}$ is separable and Prop. 25 is applicable.

The advantage of the spectral representation in Prop. 21 is its explicit construction. This representation is especially very useful in the stationary scattering theory. The advantage of the spectral representation in Prop. 25, which was constructed in a more abstract way via the multiplicity theory, is to contain the unitary invariants explicitely and also that the direct integral contains a generating set $\{f_j\}$. Both representations will be needed in this book. For convenience in the following parts we use sometimes for $L^2(\Delta, \mu; \mathscr{H}_\lambda; \mathscr{S})$ the shorter notation $L^2(\Delta, \mu; \mathscr{H}_\lambda)$ without explicit referring to the admissible system $\mathscr{S}$. Further it should be noted that in most cases we use separable admissible systems $\mathscr{S}$ (the only exception is Chapter 14).

# 5. Operator Spectral Integrals

## 5.1. Operator spectral integrals defined by Riemann-Stieltjes sums

### 5.1.1. Definitions and connections to scalar spectral integrals

Let $E(\cdot)$ be a spectral measure with a compact support $\operatorname{supp} E \subset [a, b)$, $-\infty < a < b < \infty$. Let $A(\lambda)$ be an operator-valued function defined on $[a, b)$. We consider partitions $\mathfrak{Z}$ of the interval $[a, b)$ of the form $[a, b) = [\lambda_0, \lambda_1) \cup [\lambda_1, \lambda_2) \cup \cdots \cup [\lambda_{n-1}, \lambda_n)$ where $\lambda_0 = a$, $\lambda_n = b$, and put $\Delta_m := [\lambda_{m-1}, \lambda_m)$, $m = 1, \ldots, n$. Thus $[a, b) = \bigcup_{\varrho=1}^{n} \Delta_\varrho$ and the $\Delta_\varrho$ are pairwise disjoint. Put $|\mathfrak{Z}| = \max_\varrho |\Delta_\varrho|$. We define an operator $F_\mathfrak{Z}(A)$ by

$$F_\mathfrak{Z}(A) := \sum_{\varrho=1}^{n} A(x_\varrho)\, E(\Delta_\varrho), \qquad x_\varrho \in \Delta_\varrho. \tag{1}$$

(1) is a so-called Riemann-Stieltjes sum with respect to $A(\cdot)$ and $E(\cdot)$. If there is an operator $F_0 \in \mathscr{L}(\mathscr{H})$ such that $\lim_{|\mathfrak{Z}| \to 0} \|F_\mathfrak{Z}(A) - F_0\| = 0$ independent of $\mathfrak{Z}$ and $\{x_\varrho\}$, then $F_0$ is called the *operator spectral integral* of $A(\cdot)$ with respect to $E(\cdot)$ and is denoted by

$$F_0 =: \int_a^b A(\lambda)\, E(\mathrm{d}\lambda).$$

Obviously the operator spectral integral is linear with respect to $A(\cdot)$ and we have $B \int_a^b A(\lambda)\, E(\mathrm{d}\lambda) = \int_a^b BA(\lambda)\, E(\mathrm{d}\lambda)$ for $B \in \mathscr{L}(\mathscr{H})$.

**Proposition 1.** *If* $A(\lambda) = \alpha(\lambda)\, 1$ *where* $\alpha \in C([a, b])$, *then* $\int_a^b A(\lambda)\, E(\mathrm{d}\lambda)$ *exists and coincides with the scalar spectral integral* $\int_a^b \alpha(\lambda)\, E(\mathrm{d}\lambda)$.

**Proof.** We have $F_\mathfrak{Z}(A) = \sum_{\varrho=1}^{n} \alpha(x_\varrho)\, E(\Delta_\varrho) = \int t_\mathfrak{Z}(\lambda)\, E(\mathrm{d}\lambda)$ where $t_\mathfrak{Z}(\lambda) = \sum_{\varrho=1}^{n} \alpha(x_\varrho)\, \chi_{\Delta_\varrho}(\lambda)$. Because of $|t_\mathfrak{Z}(\lambda)| \leq |\alpha|$ and $|t_\mathfrak{Z}(\lambda) - \alpha(\lambda)| \to 0$ a.e. mod $|\cdot|$ we obtain the existence of $n\text{-}\lim_{|\mathfrak{Z}| \to 0} F_\mathfrak{Z}(A)$ independent of $\mathfrak{Z}$ and $\{x_\varrho\}$ and its equality with $\int_a^b \alpha(\lambda)\, E(\mathrm{d}\lambda)$. ■

In particular Prop. 1 ensures the existence of the operator spectral integral for operator polynomials $P(\lambda) = \sum_{\varrho=1}^{m} A_\varrho \lambda^\varrho$, $A_\varrho \in \mathscr{L}(\mathscr{H})$. If we put $H = \int_a^b \lambda E(\mathrm{d}\lambda)$, we obtain $\int_a^b P(\lambda)\, E(\mathrm{d}\lambda) = \sum_{\varrho=1}^{m} A_\varrho H^\varrho$.

**Proposition 2.** *If* $\int_a^b A(\lambda)\,E(d\lambda)$ *exists and* $\alpha \in C([a, b])$, *then also* $\int_a^b \alpha(\lambda)\,A(\lambda)\,E(d\lambda)$ *exists and we have*

$$\int_a^b \alpha(\lambda)\,A(\lambda)\,E(d\lambda) = \int_a^b A(\lambda)\,E(d\lambda) \int_a^b \alpha(\lambda)\,E(d\lambda).$$

Proof. Let $F_3 = \sum \alpha(x_\varrho)\,A(x_\varrho)\,E(\Delta_\varrho)$. Then $F_3 = \sum A(x_\varrho)\,E(\Delta_\varrho) \sum \alpha(x_\sigma)\,E(\Delta_\sigma)$, but $\sum \alpha(x_\sigma)\,E(\Delta_\sigma)$ is norm-convergent to $\int_a^b \alpha(\lambda)\,E(d\lambda)$ and $\sum A(x_\varrho)\,E(\Delta_\varrho)$ is norm-convergent to $\int_a^b A(\lambda)\,E(d\lambda)$. ■

**Remark 3.** Similarly the operator spectral integral $\int_a^b E(d\lambda)\,B(\lambda)$ can be defined for operator-valued functions $B(\lambda)$ as a norm-limit of Riemann-Stieltjes sums of the form $\sum E(\Delta_\varrho)\,B(x_\varrho)$.

If $\Delta$ is a Borel set, we define $\int_\Delta A(\lambda)\,E(d\lambda)$ or $\int_\Delta E(d\lambda)\,B(\lambda)$ by $\int_a^b \chi_\Delta(\lambda)\,A(\lambda)\,E(d\lambda)$ or $\int_a^b E(d\lambda)\,\chi_\Delta(\lambda)\,B(\lambda)$. If supp $E$ is noncompact, we define $\int A(\lambda)\,E(d\lambda)$ or $\int E(d\lambda)\,B(\lambda)$ by $\operatorname{s-lim}_{n\to\infty} \int_{-n}^{n} A(\lambda)\,E(d\lambda)$ or $\operatorname{s-lim}_{n\to\infty} \int_{-n}^{n} E(d\lambda)\,B(\lambda)$. The operator spectral integral is countably additive with respect to the strong convergence.

If $f(\lambda)$ denotes a vector-valued function, then the vector spectral integral $\int_a^b E(d\lambda)\,f(\lambda)$ can be defined similarly. This integral will be considered in 5.2 from another point of view.

Also the operator spectral integrals $\int A(\lambda)\,E(d\lambda)$ and $\int E(d\lambda)\,B(\lambda)$ can be defined with respect to strong convergence with the help of Riemann-Stieltjes sums $\sum_\varrho A(x_\varrho)\,E(\Delta_\varrho)\,f$ and $\sum_\varrho E(\Delta_\varrho)\,A(x_\varrho)\,f$. Then $\left(\int A(\lambda)\,E(d\lambda)\right) f = \int A(\lambda)\,E(d\lambda)\,f$ and $\left(\int E(d\lambda)\,B(\lambda)\right) f = \int E(d\lambda)\,B(\lambda)\,f$. The latter type is a special case of a vector spectral integral.

### 5.1.2. Existence of operator spectral integrals

Let again $E(\cdot)$ be a spectral measure with compact support supp $E \subset [a, b]$. There is a large class of operator functions $A(\lambda)$ such that $\int_a^b A(\lambda)\,E(d\lambda)$ exists.

**Proposition 4.** *Let* $A(\lambda)$ *be defined on* $[a, b)$ *and assume the existence of the derivative* $A'(\lambda)$ *with respect to the operator norm a.e.* mod $|\cdot|$ *on* $[a, b)$. *Further, let* $A'(\lambda)$ *be Bochner integrable on* $[a, b)$ *and assume that* $A(\lambda) = A(a) + \int_a^\lambda A'(x)\,dx$. *Then* $\int_a^b A(\lambda)\,E(d\lambda)$ *exists and the estimate*

$$\left\| \int_a^b A(\lambda)\,E(d\lambda) \right\| \le \|A(a)\| + \int_a^b \|A'(\lambda)\|\,d\lambda \tag{2}$$

*is valid.*

Proof. First we derive an estimate for $\|F_{\mathfrak{Z}}(A)\|$. We put $E(\lambda) := E((-\infty, \lambda))$. Then we obtain, putting $x_n = b$,

$$F_{\mathfrak{Z}}(A) = \sum_{\varrho=1}^{n} A(x_\varrho)\, E([\lambda_{\varrho-1}, \lambda_\varrho)) = \sum_{\varrho=1}^{n} A(x_\varrho)\, \big(E(\lambda_\varrho) - E(\lambda_{\varrho-1})\big)$$

$$= A(x_n)\, E(\lambda_n) - A(x_1)\, E(\lambda_0) - \sum_{\varrho=2}^{n} \big(A(x_\varrho) - A(x_{\varrho-1})\big)\, E(\lambda_\varrho)$$

$$= A(b) - \sum_{\varrho=2}^{n} \big(A(x_\varrho) - A(x_{\varrho-1})\big)\, E(\lambda_\varrho).$$

Hence

$$\|F_{\mathfrak{Z}}(A)\| \leqq \|A(b)\| + \sum_{\varrho=1}^{n} \|A(x_\varrho) - A(x_{\varrho-1})\| \leqq \|A(b)\| + \int_a^b \|A'(\lambda)\|\, \mathrm{d}\lambda \tag{3}$$

follows. Now we choose a sequence of operator polynomials $Q_n(\lambda)$ such that

$$\int_a^b \|A'(\lambda) - Q_n(\lambda)\|\, \mathrm{d}\lambda \to 0 \quad \text{as} \quad n \to \infty$$

(see 1.2.2) and put $P_n(\lambda) := \int_b^\lambda Q_n(x)\, \mathrm{d}x + A(b)$. Then $P_n'(\lambda) = Q_n(\lambda)$ and $P_n(b) = A(b)$. Applying (3) to $F_{\mathfrak{Z}}(A - P_n)$, we obtain $\|F_{\mathfrak{Z}}(A - P_n)\| \leqq \int_a^b \|A'(\lambda) - Q_n(\lambda)\|\, \mathrm{d}\lambda$. Note that $F_{\mathfrak{Z}}(A)$ is linear in $A$. Finally we choose partitions $\mathfrak{Z}$ and $\mathfrak{Z}'$ of $[a, b)$ and obtain

$$\|F_{\mathfrak{Z}}(A) - F_{\mathfrak{Z}'}(A)\| \leqq \|F_{\mathfrak{Z}}(A - P_n)\| + \|F_{\mathfrak{Z}}(P_n) - F_{\mathfrak{Z}'}(P_n)\| + \|F_{\mathfrak{Z}'}(P_n - A)\|.$$

Let $\varepsilon > 0$ and fix $n$ such that $\|F_{\mathfrak{Z}}(P_n - A)\| < \varepsilon/3$ uniformly with respect to $\mathfrak{Z}$. Then because $\int_a^b P_n(\lambda)\, E(\mathrm{d}\lambda)$ exists, the second term $\|F_{\mathfrak{Z}}(P_n) - F_{\mathfrak{Z}'}(P_n)\| < \varepsilon/3$ if $|\mathfrak{Z}|, |\mathfrak{Z}'| < \delta(\varepsilon)$. Hence $\operatorname*{n-lim}_{|\mathfrak{Z}| \to 0} F_{\mathfrak{Z}}(A)$ exists independently of $\mathfrak{Z}$ and moreover (2) is valid. ■

**Remark 5.** A similar existence theorem can be proved also for the other types of spectral integrals. As a consequence we note that (for example) the vector spectral integral exists if $f(\lambda)$ is strongly continuous on $[a, b]$, strongly differentiable on $[a, b]$ and if $f'(\lambda)$ is also strongly continuous. Further, the operator and vector spectral integrals exist if the integrands $A(\lambda)$, $B(\lambda)$, $f(\lambda)$ are holomorphic in $[a, b]$.

It is easy to derive from (2) a convergence property of the operator spectral integral for sequences $A_n(\lambda)$.

### 5.1.3. Integration by parts

In this section we consider $\int_a^b A(\lambda)\, E(\mathrm{d}\lambda)\, f$ as an operator spectral integral with respect to strong convergence and assume that the conditions of Prop. 4 are satisfied (with respect to the operator norm). Note that $E(\lambda)\, f$ is Borel measurable for all $f \in \mathscr{H}$, hence $A'(\lambda)\, E(\lambda)\, f$ is also measurable. Further,

$$\int_a^b \|A'(\lambda)\, E(\lambda)\, f\|\, d\lambda \leqq \|f\| \int_a^b \|A'(\lambda)\|\, d\lambda < \infty,$$

hence $A'(\lambda)\, E(\lambda)\, f$ is Bochner integrable in the vector sense. The operator spectral integral $\int_a^b A(\lambda)\, E(d\lambda)\, f$ can be transformed into the Bochner integral $\int_a^b A'(\lambda)\, E(\lambda)\, f\, d\lambda$ by the formula of integration by parts.

**Proposition 6.** *Suppose the assumptions of Prop. 4 are satisfied. Let* $[c, d] \subseteq [a, b]$. *Then*

$$\int_c^d A(\lambda)\, E(d\lambda)\, f + \int_c^d A'(\lambda)\, E(\lambda)\, f\, d\lambda = A(d)\, E(d)\, f - A(c)\, E(c)\, f. \tag{4}$$

**Proof.** It is sufficient to prove

$$\left(g, \int_c^d A(\lambda)\, E(d\lambda)\, f\right) + \left(g, \int_c^d A'(\lambda)\, E(\lambda)\, f\, d\lambda\right) = \left(g, A(d)\, E(d)\, f - A(c)\, E(c)\, f\right)$$

for all $g \in \mathscr{H}$. Further it is sufficient to prove the assertion for polynomials $P(\lambda)$ (recall the convergence property). Thus, it is sufficient to prove the assertion for the case $A(\lambda) = \alpha(\lambda)\, 1$ where $\alpha(\lambda)$, $\alpha'(\lambda)$ are continuous on $[a, b]$. In this case we have to prove

$$\int_c^d \alpha(\lambda)\, \big(g, E(d\lambda)\, f\big) + \int_c^d \alpha'(\lambda)\, \big(g, E(\lambda) f\big)\, d\lambda = \big(g, \alpha(d)\, E(d)\, f - \alpha(c)\, E(c)\, f\big).$$

But this is nothing else than the formula of integration by parts for scalar Stieltjes integrals which can be applied in this case. ■

**Remark 7.** Formula (4) is valid also for the other integrals under the conditions of Prop. 4 (or the corresponding ones for vector functions). In the case of a vector spectral integral we have

$$\int_c^d E(d\lambda)\, f(\lambda) + \int_c^d E(\lambda)\, f'(\lambda)\, d\lambda = E(d)\, f(d) - E(c)\, f(c). \tag{5}$$

### 5.1.4. An intertwining formula

In this section we derive another transformation of certain operator spectral integrals into Bochner integrals via an intertwining property. This formula will be given for vector spectral integrals.

**Proposition 8.** *Let* $[a, b]$, $(c, d)$ *be intervals where* $[a, b]$ *is compact. Let the vector-valued function* $g(t)$ *be strongly continuous on* $(c, d)$ *and assume* $\int_c^d \|g(t)\|\, dt < \infty$. *Let* $\alpha(\lambda, t)$ *be a scalar function defined for* $\{\lambda, t\} \in [a, b] \times (c, d)$, *assume* $\alpha(\lambda, t)$ *to be continuous and* *uniformly bounded on* $[a, b] \times (c, d)$ *and let* $\frac{\partial}{\partial\lambda}\alpha(\lambda, t)$ *be continuous and*

$$\int_c^d \left\| \frac{\partial}{\partial\lambda} \alpha(\lambda, t)\, g(t) \right\| dt < \infty.$$

*We put* $\alpha(H, t) = \int_a^b \alpha(\lambda, t)\, E(d\lambda)$ *where* $H = \int_a^b \lambda E(d\lambda)$ *and* $f(\lambda) := \int_c^d \alpha(\lambda, t)\, g(t)\, dt$. *Then the vector spectral integral* $\int_a^b E(d\lambda)\, f(\lambda)$ *and the Bochner integral* $\int_c^d \alpha(H, t)\, g(t)\, dt$ *exist and coincide, that is,*

$$\int_a^b E(d\lambda) \int_c^d \alpha(\lambda, t)\, g(t)\, dt = \int_c^d dt \int_a^b \alpha(\lambda, t)\, E(d\lambda)\, g(t).$$

**Proof.** Recall the formula $\|\alpha(H, t)\| = |\alpha|(\cdot, t)$. Hence we have $\|\alpha(H, t_1) - \alpha(H, t_2)\| = |\alpha(\cdot, t_1) - \alpha(\cdot, t_2)| = \sup_{\lambda \in [a,b]} |\alpha(\lambda, t_1) - \alpha(\lambda, t_2)|$, that is, $\alpha(H, t)\, g(t)$ is strongly continuous on $(c, d)$ and

$$\int_c^d \|\alpha(H, t)\, g(t)\|\, dt \le \sup_{t \in (c,d)} \|\alpha(H, t)\| \int_c^d \|g(t)\|\, dt \le \sup_{\lambda, t} |\alpha(\lambda, t)| \int_c^d \|g(t)\|\, dt < \infty,$$

hence $\int_c^d \alpha(H, t)\, g(t)\, dt$ exists. Further, $f'(\lambda) = \int_c^d \frac{\partial}{\partial\lambda} \alpha(\lambda, t)\, g(t)\, dt$ is strongly continuous on $[a, b]$ and according to Prop. 4 this implies the existence of $\int_a^b E(d\lambda)\, f(\lambda)$. The coincidence of these two integrals can be proved by integration by parts. We have

$$\int_a^b E(d\lambda)\, f(\lambda) = [E(\lambda)\, f(\lambda)]_b^a - \int_a^b E(\lambda)\, f'(\lambda)\, d\lambda,$$

further

$$\int_a^b E(\lambda)\, f'(\lambda)\, d\lambda = \int_a^b E(\lambda) \left( \int_c^d \frac{\partial}{\partial\lambda} \alpha(\lambda, t)\, g(t)\, dt \right) d\lambda.$$

The vector-valued function $E(\lambda) \frac{\partial}{\partial\lambda} \alpha(\lambda, t)\, g(t)$ is strongly measurable on $[a, b] \times (c, d)$ and

$$\int_a^b \int_c^d \left\| E(\lambda) \frac{\partial}{\partial\lambda} \alpha(\lambda, t)\, g(t) \right\| dt\, d\lambda \le \int_a^b \int_c^d \left\| \frac{\partial}{\partial\lambda} \alpha(\lambda, t)\, g(t) \right\| dt\, d\lambda < \infty.$$

Therefore we can apply Fubini's theorem and obtain

$$\int_a^d E(\lambda)\, f'(\lambda)\, d\lambda = \int_c^d dt \left( \int_a^b \frac{\partial}{\partial\lambda} \alpha(\lambda, t)\, E(\lambda)\, g(t)\, d\lambda \right).$$

Applying integration by parts to the Bochner integral $\int_a^b \left[\frac{\partial}{\partial\lambda} \alpha(\lambda, t)\right] E(\lambda)\, g(t)\, d\lambda$, we obtain

$$\int_a^b \left[\frac{\partial}{\partial\lambda} \alpha(\lambda, t)\right] E(\lambda)\, g(t)\, d\lambda = [\alpha(\lambda, \cdot)\, E(\lambda)\, g(t)]_a^b - \int_a^b \alpha(\lambda, t)\, E(d\lambda)\, g(t)$$

hence

$$\int_a^b E(\lambda)\, f'(\lambda)\, d\lambda = \int_c^d [\alpha(\lambda, t)\, E(\lambda)\, g(t)]_a^b\, dt - \int_c^d \alpha(H, t)\, g(t)\, dt.$$

But $\int_c^d [\alpha(\lambda, t)\, E(\lambda)\, g(t)]_a^b\, dt = E(b) \int_c^d \alpha(b, t)\, g(t)\, dt - E(a) \int_c^d \alpha(a, t)\, g(t)\, dt = E(b)\, f(b)$ $- E(a)\, f(a)$, thus finally we obtain $\int_a^b E(d\lambda)\, f(\lambda) = \int_c^d \alpha(H, t)\, g(t)\, dt.$ ■

## 5.2. Vector spectral integrals for absolutely continuous spectral measures

In the following $E(\cdot)$ denotes an absolutely continuous spectral measure with compact support $\operatorname{supp} E \subseteq [a, b]$. This paragraph is devoted to a definition of the vector spectral integral which is more closely related to the definition of the Bochner (Lebesgue) integral.

### 5.2.1. Step functions

Let $\Delta_1, \Delta_2, \ldots, \Delta_m$ denote a partition of $[a, b]$ into a finite union of pairwise disjoint Borel sets. For step functions $t(\lambda) = \sum_{\varrho=1}^{m} \chi_{\Delta_\varrho}(\lambda)\, t_\varrho$ the vector spectral integral is defined by $\int_a^b E(d\lambda)\, t(\lambda) := \sum_{\varrho=1}^{m} E(\Delta_\varrho)\, t_\varrho = g$. Note that $\big(t_\varrho, E(d\lambda)\, t_\varrho\big)/d\lambda$ exists a.e. mod $|\cdot|$ for $\varrho = 1, 2, \ldots, m$, hence $\big(t(\lambda), E(dx)\, t(\lambda)\big)/dx\,|_{x=\lambda}$ exists a.e. mod $|\cdot|$ and we obtain

$$\|g\|^2 = \sum_{\varrho=1}^{m} \big(t_\varrho, E(\Delta_\varrho)\, t_\varrho\big) = \sum_{\varrho=1}^{m} \int_{\Delta_\varrho} \frac{\big(t_\varrho, E(dx)\, t_\varrho\big)}{dx}\, dx = \int_a^b \frac{\big(t(\lambda), E(dx)\, t(\lambda)\big)}{dx}\bigg|_{x=\lambda} d\lambda,$$

further

$$\|E(\Delta)\, g\|^2 = \int_\Delta \frac{\big(t(\lambda), E(dx)\, t(\lambda)\big)}{dx}\bigg|_{x=\lambda} d\lambda = \int_\Delta \frac{\big(g, E(d\lambda)\, g\big)}{d\lambda}\, d\lambda,$$

hence

$$\frac{\big(g, E(d\lambda)\, g\big)}{d\lambda} = \frac{\big(t(\lambda), E(dx)\, t(\lambda)\big)}{dx}\bigg|_{x=\lambda} \quad (\text{mod } |\cdot|).$$

We put $|g|_\lambda^2 := (g, E(\mathrm{d}\lambda)\, g)/\mathrm{d}\lambda$, then $|g|_\lambda = |t(\lambda)|_\lambda$ (mod $|\cdot|$) and $\left\| \int\limits_b^a E(\mathrm{d}\lambda)\, t(\lambda) \right\|^2 = \int\limits_a^b |t(\lambda)|_\lambda^2 \,\mathrm{d}\lambda$.

### 5.2.2. A general class $\mathscr{F}$ of functions

The vector spectral integral may be extended to a larger class of vector-valued functions. We consider a class $\mathscr{F}$ of functions $f(\lambda)$ such that the following conditions are satisfied:

(I) $\mathscr{F}$ is linear,

(II) all step functions belong to $\mathscr{F}$,

(III) for all $f \in \mathscr{F}$ the function $|f(\lambda)|_\lambda$ is defined a.e. mod $|\cdot|$ and this function is Borel measurable on $[a, b]$.

For example, we can choose $\mathscr{F}$ such that all absolutely continuous functions (on $[a, b]$) belong to $\mathscr{F}$. For this purpose we have to show that an absolutely continuous function $f$ satisfies condition (III) and, further, that $\left(g, E(\mathrm{d}x)\, f(\lambda)\right)/\mathrm{d}x|_{x=\lambda}$ exists a.e. mod $|\cdot|$ and is Borel measurable for all $g \in \mathscr{H}$.

**Proposition 9.** *If $f(\cdot)$ is absolutely continuous on $[a, b]$, then $|f(\lambda)|_\lambda$ and $\left(g, E(\mathrm{d}x)\, f(\lambda)\right)/\mathrm{d}x|_{x=\lambda}$ exist a.e.* mod $|\cdot|$ *(the latter for all $g \in \mathscr{H}$) and these functions are Borel measurable.*

**Proof.** First we show that $\|E(\lambda)\, f(\lambda)\|^2$ is a function of finite total variation on $[a, b]$. Let $[a, b] = \bigcup\limits_{\varrho=1}^{n} [\lambda_{\varrho-1}, \lambda_\varrho]$ and put $E_\varrho = E(\lambda_\varrho)$, $f_\varrho = f(\lambda_\varrho)$, $\Delta_\varrho = [\lambda_{\varrho-1}, \lambda_\varrho]$. Then

$$\sum_{\varrho=1}^{n} \left| \|E_\varrho f_\varrho\|^2 - \|E_{\varrho-1} f_{\varrho-1}\|^2 \right| \leq \sum_{\varrho=1}^{n} |(f_\varrho, E(\Delta_\varrho)\, f_\varrho)| + 2 \sup_\lambda \|f(\lambda)\| \sum_{\varrho=1}^{n} \|f_\varrho - f_{\varrho-1}\|$$

and (see the proof of Prop. 4)

$$\sum_{\varrho=1}^{n} |(f_\varrho, E(\Delta_\varrho)\, f_\varrho| = \left\| \sum_{\varrho=1}^{n} E(\Delta_\varrho)\, f_\varrho \right\|^2 \leq \left( \|f(b)\| + \sum_{\varrho=2}^{n} \|f_\varrho - f_{\varrho-1}\| \right)^2.$$

Hence

$$\sum_{\varrho=1}^{n} \left| \|E_\varrho f_\varrho\|^2 - \|E_{\varrho-1} f_{\varrho-1}\|^2 \right| \leq \left( \|f(b)\| + \int_a^b \|f'(\lambda)\| \,\mathrm{d}\lambda \right)^2 + 2 \sup_\lambda \|f(\lambda)\| \int_a^b \|f'(\lambda)\| \,\mathrm{d}\lambda$$

follows, that is the total variation of $\|E(\lambda)\, f(\lambda)\|^2$ is finite,

$$\operatorname*{var}_{a \leq \lambda \leq b} \|E(\lambda)\, f(\lambda)\|^2 \leq \left( \|f(b)\| + \int_a^b \|f'(\lambda)\| \,\mathrm{d}\lambda \right)^2 + 2 \sup_\lambda \|f(\lambda)\| \int_a^b \|f'(\lambda)\| \,\mathrm{d}\lambda. \tag{6}$$

Thus $\dfrac{\mathrm{d}}{\mathrm{d}\lambda} \left(f(\lambda), E(\lambda)\, f(\lambda)\right)$ exists a.e. mod $|\cdot|$ on $[a, b]$. Further, we have

$$\frac{\mathrm{d}}{\mathrm{d}\lambda} \left(f(\lambda), E(\lambda)\, f(\lambda)\right) = \left(f(\lambda), E(\mathrm{d}x)\, f(\lambda)\right)/\mathrm{d}x|_{x=\lambda} + \left(f'(\lambda), E(\lambda)\, f(\lambda)\right) + \left(f(\lambda), E(\lambda)\, f'(\lambda)\right). \tag{7}$$

$(f'(\lambda), E(\lambda) f(\lambda))$ exists a.e. mod $|\cdot|$. Hence from (7) we obtain the existence of $|f(\lambda)|_\lambda$ a.e. mod $|\cdot|$ and the Borel measurability of this function. In order to prove the second assertion we consider $(g, E(\lambda) f(\lambda))$ and we show that also this function has a finite total variation. We have

$$\sum_{\varrho=1}^{n} |(g, E_\varrho f_\varrho) - (g, E_{\varrho-1} f_{\varrho-1})| = \sum_{\varrho=1}^{n} |(g, E_\varrho f_\varrho - E_{\varrho-1} f_{\varrho-1})|$$
$$= \sum_{\varrho=1}^{n} |(g, (E_\varrho - E_{\varrho-1}) f_\varrho) + (g, E_{\varrho-1}(f_\varrho - f_{\varrho-1}))|$$
$$\leq \sum_{\varrho=1}^{n} |(E(\Delta_\varrho) g, E(\Delta_\varrho) f_\varrho)| + \|g\| \sum_{\varrho=1}^{n} \|f_\varrho - f_{\varrho-1}\|$$
$$\leq \left(\sum_{\varrho=1}^{n} \|E(\Delta_\varrho) g\|^2\right)^{1/2} \left(\sum_{\varrho=1}^{n} \|E(\Delta_\varrho) f_\varrho\|^2\right)^{1/2} + \|g\| \int_a^b \|f'(\lambda)\| \, d\lambda$$
$$\leq \|g\| \left\| \sum_{\varrho=1}^{n} E(\Delta_\varrho) f_\varrho \right\| + \|g\| \int_a^b \|f'(\lambda)\| \, d\lambda \leq \|g\| \left( \|f(b)\| + 2 \int_a^b \|f'(\lambda)\| \, d\lambda \right),$$

that is,

$$\operatorname*{var}_{a \leq \lambda \leq b} (g, E(\lambda) f(\lambda)) \leq \|g\| \left( \|f(b)\| + 2 \int_a^b f'(\lambda)\| \, d\lambda \right).$$

Hence $\frac{d}{d\lambda} (g, E(\lambda) f(\lambda))$ exists a.e. mod $|\cdot|$ on $[a, b]$ and is Borel measurable. Further we have

$$\frac{d}{d\lambda} (g, E(\lambda) f(\lambda)) = (g, E(dx) f(\lambda))/dx \,|_{x=\lambda} + (g, E(\lambda) f'(\lambda)). \tag{8}$$

From (8) we obtain the existence of $(g, E(d\lambda) f(x))/d\lambda|_{\lambda=x}$ a.e. mod $|\cdot|$ and its Borel measurability. ■

In the following we assume that $\mathscr{F}$ contains all absolutely continuous vector-valued functions. Further we assume $\mathscr{F}$ to be closed with respect to multiplication by scalar functions $\alpha \in L^\infty([a, b], d\lambda)$.

### 5.2.3. Vector spectral integral for functions from $\mathscr{F}$

A function $f \in \mathscr{F}$ is called *integrable* with respect to the spectral measure $E(\cdot)$ if there is a sequence of step functions $t_n(\cdot)$ such that

$$\int_a^b |f(\lambda) - t_n(\lambda)|_\lambda^2 \, d\lambda \to 0, \qquad n \to \infty. \tag{9}$$

The sequence $t_n(\cdot)$ may be chosen such that $|f(\lambda) - t_n(\lambda)|_\lambda \to 0$ as $n \to \infty$ a.e. mod $|\cdot|$. Namely, (9) implies the measure convergence of $|f(\lambda) - t_n(\lambda)|_\lambda$ to 0, hence there is a subsequence which converges a.e. mod $|\cdot|$. If (9) is satisfied, we define

$$\int_a^b E(\mathrm{d}\lambda)\, f(\lambda) := \operatorname*{s-lim}_{n\to\infty} \int_a^b E(\mathrm{d}\lambda)\, t_n(\lambda). \tag{10}$$

This strong limit exists because of

$$\left\| \int_a^b E(\mathrm{d}\lambda)\, t_n(\lambda) - \int_a^b E(\mathrm{d}\lambda)\, t_m(\lambda) \right\|^2 = \int_a^b |t_n(\lambda) - t_m(\lambda)|_\lambda^2 \,\mathrm{d}\lambda \to 0$$

as $n, m \to \infty$. Obviously, the limit is independent of the special sequence $t_n(\cdot)$. Further. we define

$$\int_\Delta E(\mathrm{d}\lambda)\, f(\lambda) := \int_a^b E(\mathrm{d}\lambda)\, \chi_\Delta(\lambda)\, f(\lambda). \tag{11}$$

The set of integrable functions is linear. We have

**Proposition 10.** *If $f \in \mathcal{F}$ is integrable, then*

$$\left\| \int_a^b E(\mathrm{d}\lambda)\, f(\lambda) \right\| = \left( \int_a^b |f(\lambda)|_\lambda^2 \,\mathrm{d}\lambda \right)^{1/2}, \tag{12}$$

*in particular* $\int_a^b |f(\lambda)|_\lambda^2 \,\mathrm{d}\lambda < \infty$.

Proof. We put for abbreviation $[f] := \left( \int_a^b |f(\lambda)|_\lambda^2 \,\mathrm{d}\lambda \right)^{1/2}$. If $f$ is integrable, then there is a sequence $t_n$ with $[f - t_n] \to 0$ as $n \to \infty$. Now $[t_n] < \infty$, hence $|[f] - [t_n]| \leq [f - t_n] \to 0$ that is, $[f] < \infty$ and moreover $[f] = \lim_{n\to\infty} [t_n]$. Now we have $\left\| \int_a^b E(\mathrm{d}\lambda) t_n(\lambda) \right\| = [t_n]$, hence $\left\| \int_a^b E(\mathrm{d}\lambda)\, f(\lambda) \right\| = \lim_{n\to\infty} \left\| \int_a^b E(\mathrm{d}\lambda)\, t_n(\lambda) \right\| = [f]$. ■

**Proposition 11.** *Let $f \in \mathcal{F}$ be integrable and let $\alpha \in L^\infty([a, b], \mathrm{d}\lambda)$. Then $\alpha f$ is also integrable and*

$$\int_a^b E(\mathrm{d}\lambda)\, \alpha(\lambda)\, f(\lambda) = \alpha(H) \int_a^b E(\mathrm{d}\lambda)\, f(\lambda) \tag{13}$$

*is valid where $H = \int \lambda E(\mathrm{d}\lambda)$.*

Proof. Let $\tau_n$ be a sequence of scalar step functions with $\operatorname*{ess\,sup}_\lambda |\alpha(\lambda) - \tau_n(\lambda)| \to 0$ as $n \to \infty$. Further, let $t_n(\lambda)$ be a sequence of (vector-valued) step functions with $[f - t_n] \to 0$ as $n \to \infty$. Then we obtain

$$|\alpha(\lambda)\, f(\lambda) - \tau_n(\lambda)\, t_n(\lambda)|_\lambda \leq |\alpha(\lambda) - \tau_n(\lambda)|\, |f(\lambda)|_\lambda + |\tau_n(\lambda)|\, |f(\lambda) - t_n(\lambda)|_\lambda.$$

Hence

$$\left(\int_a^b |\alpha(\lambda) f(\lambda) - \tau_n(\lambda) t_n(\lambda)|_\lambda^2 \,d\lambda\right)^{1/2} \leq \operatorname*{ess\,sup}_\lambda |\alpha(\lambda) - \tau_n(\lambda)| \left(\int_a^b |f(\lambda)|_\lambda^2 \,d\lambda\right)^{1/2}$$

$$+ \sup_n \left\{\operatorname*{ess\,sup}_\lambda |\tau_n(\lambda)| \left(\int_a^b f|(\lambda) - t_n(\lambda)|_\lambda^2 \,d\lambda\right)^{1/2}\right\} \to 0 \text{ as } n \to \infty,$$

that is, $\alpha(\cdot) f(\cdot)$ is integrable and we obtain

$$\int_a^b E(d\lambda)\, \alpha(\lambda) f(\lambda) = \operatorname*{s-lim}_{n\to\infty} \int_a^b E(d\lambda)\, \tau_n(\lambda) t_n(\lambda) = \operatorname*{s-lim}_{n\to\infty} \int_a^b \tau_n(\lambda) E(d\lambda) \int_a^b E(d\lambda)\, t_n(\lambda)$$

$$= \operatorname*{s-lim}_{n\to\infty} \int_a^b \tau_n(\lambda) E(d\lambda) \operatorname*{s-lim}_{n\to\infty} \int_a^b E(d\lambda) f(\lambda) = \alpha(H) \int_a^b E(d\lambda) f(\lambda). \blacksquare$$

According to Prop. 11 and (11), we have

$$\int_\Delta E(d\lambda) f(\lambda) = E(\Delta) \int_a^b E(d\lambda) f(\lambda). \tag{14}$$

Thus we further obtain:

(I) The vector spectral integral is absolutely continuous, that is, $\left\|\int_\Delta E(d\lambda) f(\lambda)\right\| \to 0$ as $|\Delta| \to 0$.

(II) It is $\sigma$-additive, that is $\int_\Delta E(d\lambda) f(\lambda) = \sum_{\varrho=1}^\infty \int_{\Delta_\varrho} E(d\lambda) f(\lambda)$, $\Delta = \bigcup_\varrho \Delta_\varrho$, where $\{\Delta_\varrho\}$ is a family of pairwise disjoint Borel sets.

From Prop. 11 it also follows that vector polynomials are integrable. From Prop. 10 and Prop. 11 we obtain

**Corollary 12.** *Let $f \in \mathcal{F}$ be integrable and put $g = \int_a^b E(d\lambda) f(\lambda)$. Then $(g, E(d\lambda) g)/d\lambda = |f(\lambda)|_\lambda^2 \pmod{|\cdot|}$.*

**Proof.** We have $(g, E(\Delta) g) = \int_\Delta ((g, E(d\lambda) g)/d\lambda)\, d\lambda = \int_\Delta |f(\lambda)|_\lambda^2 \,d\lambda$ for all Borel sets, hence the assertion follows. $\blacksquare$

**Proposition 13.** *If $f \in \mathcal{F}$ and if there is a sequence $t_n$ of step functions with $|f(\lambda) - t_n(\lambda)|_\lambda \to 0$ as $n \to \infty$ a.e. mod $|\cdot|$, then $f$ is integrable iff $[f] := \left(\int_a^b |f(\lambda)|_\lambda^2 \,d\lambda\right)^{1/2} < \infty$.*

**Proof.** Only one part has to be proved. Let $[f] < \infty$. Further let

$$s_n(\lambda) = \begin{cases} t_n(\lambda) & \text{if } \; |t_n(\lambda)|_\lambda \leq |f(\lambda)|_\lambda (1 + 1/n), \\ 0 & \text{otherwise.} \end{cases}$$

Then $s_n(\cdot)$ is a sequence of step functions and $|f(\lambda) - s_n(\lambda)|_\lambda \leq 2\, |f(\lambda)|_\lambda (1 + 1/n)$. Hence the integrability of $f$ follows via Lebesgue's dominated convergence theorem. $\blacksquare$

Finally we show that absolutely continuous functions on $[a, b]$ are integrable.

**Proposition 14.** *Let $f$ be absolutely continuous on $[a, b]$. Then $f$ is integrable.*

**Proof.** First from (7)

$$\int_a^b |f(\lambda)|_\lambda^2 \,\mathrm{d}\lambda \leqq \int_a^b \frac{\mathrm{d}}{\mathrm{d}\lambda} \big(f(\lambda), E(\lambda) f(\lambda)\big) \,\mathrm{d}\lambda + 2 \sup_\lambda \|f(\lambda)\| \int_a^b \|f'(\lambda)\| \,\mathrm{d}\lambda$$

follows. If $T(\lambda)$ denotes the total variation of $\big(f(\lambda), E(\lambda) f(\lambda)\big)$ with respect to $[a, \lambda]$ and if we put for abbreviation $\varphi(\lambda) = \big(f(\lambda), E(\lambda) f(\lambda)\big)$, then $\varphi(\lambda) = T(\lambda) - \big(T(\lambda) - \varphi(\lambda)\big)$ where $T(\lambda)$ and $T(\lambda) - \varphi(\lambda)$ are monotone functions. Hence

$$\left| \int_a^b \varphi'(\lambda) \,\mathrm{d}\lambda \right| = \left| \int_a^b T'(\lambda) \,\mathrm{d}\lambda - \int_a^b \big(T(\lambda) - \varphi(\lambda)\big)' \,\mathrm{d}\lambda \right|$$
$$\leqq |T(b) - T(a)| + \big|T(b) - \varphi(b) - \big(T(a) - \varphi(a)\big)\big| \leqq 2T(b) + \varphi(b)$$

follows, that is, we obtain with the help of (6)

$$\int_a^b |f(\lambda)|_\lambda^2 \,\mathrm{d}\lambda \leqq 2 \left( \left( \|f(b)\| + \int_a^b \|f'(\lambda)\| \,\mathrm{d}\lambda \right)^2 + 2 \sup_\lambda \|f(\lambda)\| \int_a^b \|f'(\lambda)\| \,\mathrm{d}\lambda \right)$$
$$+ \|f(b)\|^2 + 2 \sup_\lambda \|f(\lambda)\| \int_a^b \|f'(\lambda)\| \,\mathrm{d}\lambda. \tag{15}$$

Now we choose a sequence of polynomials $q_n(\cdot)$ such that $\int_a^b \|f'(\lambda) - q_n(\lambda)\| \,\mathrm{d}\lambda \to 0$ and put $p_n(\lambda) = \int_b^\lambda q_n(x) \,\mathrm{d}x + f(b)$. Then

$$\sup_\lambda \|f(\lambda) - p_n(\lambda)\| \leqq \int_a^b \|f'(\lambda) - q_n(\lambda)\| \,\mathrm{d}\lambda \to 0 \quad \text{as} \quad n \to \infty,$$

and from (15) we obtain $\int_a^b |f(\lambda) - p_n(\lambda)|_\lambda^2 \,\mathrm{d}\lambda \to 0$, $n \to \infty$. Because $p_n$ is integrable, also $f$ is integrable and $\int_a^b E(\mathrm{d}\lambda) f(\lambda) = \operatorname{s-lim}_{n\to\infty} \int_a^b E(\mathrm{d}\lambda) p_n(\lambda)$ follows. ■

Note that for (piecewise) absolutely continuous functions $f(\lambda)$ on $[a, b]$ the spectral integral $\int_a^b E(\mathrm{d}\lambda) f(\lambda)$ (defined according to (10)) coincides with the spectral integral defined according to 5.1.1. (the existence is ensured by Prop. 4).

### 5.2.4. A spectral representation by vector spectral integrals

The vector spectral integral $g = \int E(\mathrm{d}\lambda) f(\lambda)$ for piecewise absolutely continuous functions can be used for the construction of a special representation of $E(\Delta)$ which turns out to be a spectral representation if a spectral manifold is given. In the following supp $E$ is not necessarily compact, but $E(\cdot)$ is an absolutely continuous spectral measure, as before.

The class of all piecewise absolutely continuous vector-valued functions such that $\int E(\mathrm{d}\lambda)\, f(\lambda)$ exists is denoted by $\mathscr{F}_0$. If $f \in \mathscr{F}_0$, then $f(\lambda) \in \mathscr{D}_\lambda$ a.e. mod $|\cdot|$ and $\left\|\int E(\mathrm{d}\lambda)\, f(\lambda)\right\| = \int |f(\lambda)|_\lambda^2\, \mathrm{d}\lambda$ (for the definition of the linear manifold $\mathscr{D}_\lambda$ see 3.5.5). Note that $\int E(\mathrm{d}\lambda)\, g = g$, that is, each vector $g \in \mathscr{H}$ can be represented as a vector spectral integral by "itself", $f(\lambda) \equiv g$. We define a map $\Phi$ from $\mathscr{F}_0$ into $\mathscr{H}$ by

$$\mathscr{F}_0 \ni f \to \Phi f = \int E(\mathrm{d}\lambda)\, f(\lambda) \in \mathscr{H}.$$

We have ima $\Phi = \mathscr{H}$ and ker $\Phi = \{f;\, |f(\lambda)|_\lambda = 0 \text{ a.e. mod } |\cdot|\}$ Recall the identification operator $D_\lambda$ from $\mathscr{D}_\lambda$ into $\mathscr{D}_\lambda$ mod (ker $|\cdot|_\lambda$).

Now we consider the map $\hat{\Phi}$ from the set $\hat{\mathscr{F}}_0$ of all functions $\{D_\lambda f(\lambda)\}$, $f \in \mathscr{F}_0$, into $\mathscr{H}$ given by

$$\hat{\mathscr{F}}_0 \ni \{D_\lambda f(\lambda)\} \to \hat{\Phi}\big(\{D_\lambda f(\lambda)\}\big) = \int_{-\infty}^{\infty} E(\mathrm{d}\lambda)\, f(\lambda) \in \mathscr{H}. \tag{16}$$

The map $\hat{\Phi}$ is bijective from $\hat{\mathscr{F}}_0$ onto $\mathscr{H}$. Further, the representation of $\mathscr{H}$ given by $\hat{\Phi}$ is "almost" a spectral representation because

$$\hat{\Phi}\big(\{\chi_\Delta(\lambda)\, D_\lambda f(\lambda)\}\big) = E(\Delta)\, \hat{\Phi}\big(\{D_\lambda f(\lambda)\}\big),$$

but the value-spaces of the functions $D_\lambda f(\lambda)$ are Banach spaces $\hat{\mathscr{D}}_\lambda$.

Now let $\mathscr{K}$ be a linear manifold in $\mathscr{H}$ which is spectral and generating with respect to $E(\Delta)$. Then $\hat{\mathscr{K}}_\lambda \subset \hat{\mathscr{D}}_\lambda$ (see 3.5.5). Note that this implies $D_\lambda k \in \hat{\mathscr{K}}_\lambda$, $k \in \mathscr{K}$. Furthermore, according to 4.5.1, by $s(\lambda; k, k') := \big(k, E(\mathrm{d}\lambda)\, k'\big)/\mathrm{d}\lambda$, $\lambda \in \Gamma$, ($\Gamma$ spectral core of $H = \int \lambda E(\mathrm{d}\lambda)$), $k, k' \in \mathscr{K}$, is defined a spectral form with respect to $E(\cdot)$ and $|\cdot|$ (note that $E(\Delta) = 0$ for $\Delta \subset \Gamma$ implies $|\Delta| = 0$ because $\Gamma$ is a spectral core), that is we have

$$\int_\Delta s(\lambda; k, k')\, \mathrm{d}\lambda = \big(k, E(\Delta)\, k'\big), \qquad k, k' \in \mathscr{K}, \qquad \Delta \in \mathscr{B}_\Gamma.$$

Hence (according to Example 4.2) we can form the direct integral $L^2(\Gamma, \mathrm{d}\lambda; \hat{\mathscr{K}}_\lambda)$. Then we obtain

**Proposition 15.** *Let $D_\lambda f(\lambda)$ be the representer of $g \in \mathscr{H}$ with respect to* (16) *and let $\tilde{g}(\lambda)$ be its representer with respect to $L^2(\Gamma, \mathrm{d}\lambda; \hat{\mathscr{K}}_\lambda)$. Then we have*

$$\tilde{g}(\lambda) = D_\lambda f(\lambda) \ (\text{mod } |\cdot|). \tag{17}$$

Proof. It is sufficient to prove (17) for the representer $D_\lambda g$, $g \in \mathscr{H}$. Then we have to prove $|D_\lambda g - \tilde{g}(\lambda)|_\lambda = 0$ a.e. mod $|\cdot|$ or $\int |D_\lambda g - \tilde{g}(\lambda)|_\lambda^2\, \mathrm{d}\lambda = 0$. Now we choose $g_n \in \text{spa}\,\{E(\Delta)\, k,\, k \in \mathscr{K}\}$ such that $\|g_n - g\| \to 0$ as $n \to \infty$. Then we obtain

$$\left(\int_{-\infty}^{\infty} |D_\lambda g - \tilde{g}(\lambda)|_\lambda^2\, \mathrm{d}\lambda\right)^{1/2} \leqq 2\,\|g - g_n\| + \left(\int_{-\infty}^{\infty} |D_\lambda g_n - \tilde{g}_n(\lambda)|_\lambda^2\, \mathrm{d}\lambda\right)^{1/2}$$

that is, it is sufficient to prove (17) for $g = k \in \mathscr{K}$. But in this case $D_\lambda k = \hat{k}(\lambda) = J_\lambda k$ is obvious (see 3.5.5., where the "enlarging" of equivalence classes with respect to $\mathscr{K}$ mod (ker $|\cdot|_\lambda$) and $\mathscr{D}_\lambda$ mod (ker $|\cdot|_\lambda$) was pointed out). ∎

### 5.2.5. Vector spectral integrals and Hilbert-Schmidt operators

Let $A \in \mathscr{L}_2(\mathscr{H})$ and let $E(\cdot)$ be an absolutely continuous spectral measure. We consider vector-valued functions $f(\lambda)$ of the form $f(\lambda) = Ag(\lambda)$. The corresponding vector spectral integral is given by $\int E(\mathrm{d}\lambda)\, Ag(\lambda)$, which maps the vector-valued function $g(\lambda)$ into a vector from $\mathscr{H}$. The question arises if there is a simple condition on $A$ such that $\int E(\mathrm{d}\lambda)\, A\{\cdot\}$ is a bounded operator from $L^2(\mathbb{R}, \mathrm{d}\lambda; \mathscr{H})$ into $\mathscr{H}$. Recall that $\big(A^*E(\mathrm{d}\lambda)\, A\big)/\mathrm{d}\lambda \in \mathscr{L}_1(\mathscr{H})$ a.e. mod $|\cdot|$ (see 3.4.3).

**Proposition 16.** *If*

$$\operatorname*{ess\,sup}_{\lambda} \left\| \frac{A^*E(\mathrm{d}\lambda)\, A}{\mathrm{d}\lambda} \right\| < \infty, \tag{18}$$

*then* $\int E(\mathrm{d}\lambda)\, A\{\cdot\} \in \mathscr{L}\big(L^2(\mathbb{R}, \mathrm{d}\lambda; \mathscr{H}), \mathscr{H}\big)$.

**Proof.** We obtain for piecewise absolutely continuous functions $f \in L^2(\mathbb{R}, \mathrm{d}\lambda; \mathscr{H})$ that

$$\left\| \int_{-\infty}^{\infty} E(\mathrm{d}\lambda)\, Af(\lambda) \right\|^2 = \int_{-\infty}^{\infty} \frac{\big(Af(\lambda), E(\mathrm{d}x)\, Af(\lambda)\big)}{\mathrm{d}x}\bigg|_{x=\lambda} \mathrm{d}\lambda$$

$$= \int_{-\infty}^{\infty} \frac{\big(f(\lambda), A^*E(\mathrm{d}x)\, Af(\lambda)\big)}{\mathrm{d}x}\bigg|_{x=\lambda} \mathrm{d}\lambda = \int_{-\infty}^{\infty} \left(f(\lambda), \frac{A^*E(\mathrm{d}x)\, A}{\mathrm{d}x} f(\lambda)\right)_{x=\lambda} \mathrm{d}\lambda.$$

Recall that $\big(A^*E(\mathrm{d}x)\, A\big)/\mathrm{d}x$ exists a.e. mod $|\cdot|$ as an $\mathscr{L}_1$-derivative. Further, we get

$$\int_{-\infty}^{\infty} \left(f(\lambda), \frac{A^*E(\mathrm{d}\lambda)\, A}{\mathrm{d}\lambda} f(\lambda)\right) \mathrm{d}\lambda \le \operatorname*{ess\,sup}_{\lambda} \left\| \frac{A^*E(\mathrm{d}\lambda)\, A}{\mathrm{d}\lambda} \right\| \int_{-\infty}^{\infty} \big(f(\lambda), f(\lambda)\big)_{\mathscr{H}}\, \mathrm{d}\lambda.$$

Finally note that the piecewise absolutely continuous functions $f$ with $f \in L^2(\mathbb{R}\, \mathrm{d}\lambda; \mathscr{H})$ are dense in $L^2(\mathbb{R}, \mathrm{d}\lambda; \mathscr{H})$, that is, the vector spectral integral $\int E(\mathrm{d}\lambda)\, Af(\lambda)$ can be extended continuously to a (bounded) operator defined on the whole space $L^2(\mathbb{R}, \mathrm{d}\lambda; \mathscr{H})$. ∎

## 5.3. Approximate operator spectral integrals

For applications in the stationary scattering theory we introduce spectral integrals for approximate spectral measures. If $E(\cdot)$ denotes a spectral measure, we put

$$E_\varepsilon(\Delta) := (2i\pi)^{-1} \int_\Delta \big(R_H(\lambda - i\varepsilon) - R_H(\lambda + i\varepsilon)\big)\, \mathrm{d}\lambda = \int_\Delta \delta_\varepsilon(\lambda - H)\, \mathrm{d}\lambda \tag{19}$$

where $\delta_\varepsilon(\lambda - H) = \int \delta_\varepsilon(\lambda - x)\, E_H(\mathrm{d}x)$. Recall that $\operatorname*{s-lim}_{\varepsilon \to +0} E_\varepsilon\big((a, b)\big) = E_H\big((a, b)\big)$ if neither $a$ nor $b$ are eigenvalues of $H$. According to (19), an operator spectral integral with respect to the approximate spectral measure $E_\varepsilon(\cdot)$ can be defined by

$$\int_a^b E_\varepsilon(\mathrm{d}\lambda)\, f(\lambda) := \int_a^b \delta_\varepsilon(\lambda - H)\, f(\lambda)\, \mathrm{d}\lambda, \quad \varepsilon > 0, \tag{20}$$

where the right-hand side is a usual Bochner integral. The following proposition states a convergence property of an approximate spectral integral to a usual spectral integral when the spectral measure $E_H(\cdot)$ is absolutely continuous.

**Proposition 17.** *If $E_H(\cdot)$ is an absolutely continuous spectral measure and if $f$ is an absolutely continuous vector-valued function on $[a, b]$, then the convergence relation*

$$\operatorname*{s-lim}_{\varepsilon\to+0} \int_a^b E_\varepsilon(\mathrm{d}\lambda)\, f(\lambda) = \int_a^b E_H(\mathrm{d}\lambda)\, f(\lambda) \tag{21}$$

*is valid.*

**Proof.** Integration by parts leads to

$$\int_a^b E_H(\mathrm{d}\lambda)\, f(\lambda) = [E_H(\lambda)\, f(\lambda)]_a^b - \int_a^b E_H(\lambda)\, f'(\lambda)\, \mathrm{d}\lambda$$

and

$$\int_a^b E_\varepsilon(\mathrm{d}\lambda)\, f(\lambda) = [E_\varepsilon(\lambda)\, f(\lambda)]_a^b - \int_a^b E_\varepsilon(\lambda)\, f'(\lambda)\, \mathrm{d}\lambda,$$

where $E_H(\lambda) = E_H\big((-\infty, \lambda)\big)$ and $E_\varepsilon(\lambda) = E_\varepsilon\big((-\infty, \lambda)\big)$. Then $\operatorname*{s-lim}_{\varepsilon\to+0} E_\varepsilon(b)\, f(b) = E_H(b)\, f(b)$ and $\operatorname*{s-lim}_{\varepsilon\to+0} E_\varepsilon(a)\, f(a) = E_H(a)\, f(a)$. Further we have $\sup_{\varepsilon>0,\lambda} \|E_\varepsilon(\lambda)\| < \infty$, hence $\|E_\varepsilon(\lambda)\, f'(\lambda) - E_H(\lambda)\, f'(\lambda)\| \leq \big(\sup_{\varepsilon>0,\lambda} \|E_\varepsilon(\lambda)\| + 1\big)\, \|f'(\lambda)\|$ follows. Therefore we obtain, because of $\operatorname*{s-lim}_{\varepsilon\to+0} E_\varepsilon(\lambda)\, f'(\lambda) = E_H(\lambda)\, f'(\lambda)$ for almost all $\lambda \bmod |\cdot|$,

$$\lim_{\varepsilon\to+0} \left\| \int_a^b E_\varepsilon(\lambda)\, f'(\lambda)\, \mathrm{d}\lambda - \int_a^b E_H(\lambda)\, f'(\lambda)\, \mathrm{d}\lambda \right\| \leq \lim_{\varepsilon\to+0} \int_a^b \|E_\varepsilon(\lambda)\, f'(\lambda) - E_H(\lambda) f'(\lambda)\|\, \mathrm{d}\lambda = 0$$

according to Lebesgue's dominated convergence theorem. ■

# Notes and Remarks to Part I

For functional analysis, in particular the theory of Hilbert spaces and their selfadjoint operators, we refer to Achieser/Glasman [1], Birman/Solomjak [1], Dunford/Schwartz [1, 2], Kato [1], Reed/Simon [1, 2], Triebel [1], Weidmann [1], and Yosida [1]. See also Amrein/Jauch/Sinha [1]. For real-valued functions, see Natanson [1]. For standard measure theory, the reader may consult Halmos [1]. For vector measures, we recommend Diestel/Uhl [1].

*1.1.2.* Concerning our terminology Borel field, Borel set, Borel space, we refer to Takesaki [1]. For the various types of Borel fields, Baire sets, Borel measures, Baire measures and their relations, see Halmos [1].

*1.2.1.* PETTIS' measurability theorem, the relation between strong and weak $\mu$-measurability, can be found in DIESTEL/UHL [1, p. 42]. For the connection between Borel and $\mu$-measurable functions, see REED/SIMON [1, p. 116].

*1.2.3.* The theory of the Hardy classes can be found in STEIN/WEISS [1, Ch. III]; for $\mathscr{H}^2_\pm$ and the Paley-Wiener theorem, see also YOSIDA [1, p. 161].

*1.2.4.* The Radon-Nikodym property is extensively discussed in DIESTEL/UHL [1]. The representation $\Phi(\Delta) = \int_\Delta f(\lambda) |\Phi| (d\lambda)$ with the properties described in this section is proved in DIESTEL/UHL [1, p. 79]. The weak differentiation property for a vector measure $\Phi$ of bounded variation is proved in GELFAND/SCHILOW [3, p. 154].

*1.2.5.* The proofs of Prop. 1 and 2 are given in ASANO [1967a]. For Prop. 3, 4, and 5, the reader may consult STEIN/WEISS [1, Ch. II and VI]. The convolution theorem may be found in YOSIDA [1, p. 160] or REED/SIMON [2, p. 6]. It says $\hat{f}\hat{g} = \hat{f} * \hat{g}$ with $(f * g)(y) = (2\pi)^{-n/2} \int f(y - x) g(x) dx$.

*1.3.2.* The equality of the various types of holomorphis operator functions is proved in YOSIDA [1, p. 218].

*1.3.4.* For trace operators, trace, and determinant, see GOHBERG/KREĬN [1]; see also DUNFORD/SCHWARTZ [2], Lemma 7 is due to BIRMAN/ÈNTINA [1967a].

*1.3.6.* For DUNFORD's calculus, see DUNFORD/SCHWARTZ [1].

*1.4.1.* For the theorem of GELFAND-NEĬMARK (= NEUMARK), GELFAND's isomorphism, see NEUMARK [1, p. 228]; see also REED/SIMON [1].

*1.4.2.* KAPLANSKY's lemma, the double commutant theorem, and related facts are proved in ARVESON [1].

*Chapter 2.* A general reference is ARVESON [1]. In particular the famous Theorem 1 is proved there. The material presented in Chapter 2 is organized according to this reference.

*3.1.3.* There are many references for the famous spectral theorem, see e.g. KATO [1, p. 360] and DUNFORD/SCHWARTZ [2, p. 48].

*3.2.1.* A proof that weak measurability of a group implies strong continuity can be found in REED/SIMON [1, p. 268].

*3.3.6.* The concept of spectral core (not the term itself) is due to KATO [1968a].

*3.4.1.* WIENER's theorem is proved in REED/SIMON [3, p. 340].

*3.5.2.* The relation between $\mathscr{N}_\infty(H)$ and $P_H^{ac}\mathscr{H}$ has been noticed first by GUSTAFSON/JOHNSON [1974a].

*3.5.3.* The inequality (36) is due to ROSENBLUM (ROSENBLUM's lemma), see for example ACHIESER/GLASMAN [1, p. 300].

*3.5.4.* Remark 31 is an easy consequence of the fact that absolutely continuous functions map null sets into null sets (see NATANSON [1, p. 277]), Prop. 29, 30 and that $\alpha^{-1}(x)$ exists and is absolutely continuous on $\alpha(\text{spec}\, H)$ when $\alpha'(x) > 0$ on spec $H$ (see NATANSON [1, p. 308]). See also KATO [1, p. 547]. Note that even continuous functions, in general, do not map null sets into null sets and therefore $P^{ac}_{\alpha(H)} \leq P^{ac}_H$ is not true for arbitrary continuous functions $\alpha$. For example, let $\alpha(x) = x + \alpha_c(x)$ on $[0, 1]$ where $\alpha_c$ is the Cantor function (see NATANSON [1, p. 239] or REED/SIMON [1, p. 21]). $\alpha(x)$ is strictly monotone increasing on $[0, 1]$ and continuous. Thus $\alpha^{-1}(x)$ exists, is continuous and monotone on $[\alpha(0), \alpha(1)]$. Put $\mathscr{H} = L^2([0, 1], dx)$, $\mathscr{H}' = L^2([\alpha(0), \alpha(1)], dx)$, $H = x$ on $\mathscr{H}$ and $U: \mathscr{H} \to \mathscr{H}'$ the unitary map defined by $(Uf)(x) = f(\alpha^{-1}(x))$. Then we have that $U$ transforms $\alpha(H)$ on $\mathscr{H}$ into $H' = x$ on $\mathscr{H}'$. Thus $\alpha(H)$ is absolutely continuous, but $H$ has a singular continuous part since $\alpha(x)$ has the singular continuous part $\alpha_c(x)$. Therefore $1 = P^{ac}_{\alpha H} > P^{ac}_H$.

*Chapter 4.* General references for direct integrals are DIXMIER [1] and NEĬMARK/FOMIN [1955a]. The proofs of Prop. 12 and 13 can be found in DIXMIER [1]. For 4.4, see e.g. NEĬMARK/FOMIN [1955a]. Lemma 20 is given in ARVESON [1].

Spectral forms were introduced by KATO and KURODA (see KATO/KURODA [1970a], [1971a]). Also Prop. 21 is essentially due to KATO/KURODA [1970a]. The arguments in the remark after Prop. 21 can be found for example in GELFAND/WILENKIN [1, p. 125].

*Chapter 5.* Operator spectral integrals were introduced, in the form presented here, by DALECKĬ KREĬN [1965a]. The first application of such operator spectral integrals to scattering theory is due to BAUMGÄRTEL [1967a]. Other definitions of operator spectral integrals, applications, and related problems are to find in GALINDO-TIXAIRE [1959a], BIRMAN/SOLOMJAK [1966a], [1967a], [1973a], PRUGOVECKI [1969a], [1971b], PEARSON [1971a], AMREIN/GEORGESCU/JAUCH [1971a], and THOMAS [1975a].

# Part II. Algebras of Asymptotic Constants

This part is devoted to the study of asymptotic constants, that is, those operators $X$ for which $X_H(t) = e^{itH} X e^{-itH} P_H^{ac}$ tends to a limit $X_+$ as $t \to \infty$ ($H$ denotes a fixed selfadjoint operator on a Hilbert space $\mathscr{H}$). The restriction to the absolutely continuous subspace $P_H^{ac}\mathscr{H}$ of $H$ is suitable in many applications and makes the theory sufficiently rich in useful mathematical results. With each of both limiting processes described in the introduction of this book, s-lim and absolute Abelian limit, the set of asymptotic constants forms an algebra, also called wave algebra (for $H$).

Sometimes one is also interested in the asymptotic behavior of $X_H(t)$ for large negative values of the parameter $t$. But since $X_H(t) = X_{-H}(-t)$, this is equivalent to the asymptotic behavior of $X_{-H}(t)$ for $t \to \infty$. Thus, to ask whether $X_H(t)$ tends to a limit as $t \to -\infty$ means to ask whether $X$ is an asymptotic constant for $-H$.

Here the main motivation for the investigation of these algebras is the fact that the wave operators — or some other objects connected with them — are asymptotic constants. Thus we study such properties above all which are important in the theory of the wave operator. For instance, projections and partial isometries are considered in detail. Further, we deal with topological properties of wave algebras. It is remarkable that by introducing a suitable topology one gets that every asymptotic constant $X$ is a limit of a sequence $X_n = A_n + C_n$ where $A_n \in \text{com}\, H$ and $C_n$ are finite-dimensional operators. In particular, we also discuss the problem to determine the set of all selfadjoint operators $H'$ which generate (in the sense of the asymptotics of $X_{H'}(t)$) the same algebra of asymptotic constants as $H$. This problem is completely solved in the special case of operators $H'$ which commute with $H$.

Another motivation to study asymptotic constants comes from the so-called algebraic approach to scattering problems. Physically one can interpret $H$ as the Hamiltonian of a quantum mechanical system, $P_H^{ac}\mathscr{H}$ as the set of scattering states (at least in many physically interesting cases) and $X_H(t)$ as the time evolution of an observable $X$ on the scattering states in the Heisenberg picture. If $H$ represents the Hamiltonian of an interacting system and $H_0$ that of the corresponding free system, then the usual scattering theory compares the evolutions $e^{-itH}$ and $e^{-itH_0}$ asymptotically. From a physical point of view it is more natural, in quantum mechanical scattering situations, to compare the corresponding time evolutions of certain observables (characteristical for scattering experiments), i.e. $X_H(t)$ and $X_{H_0}(t)$. In particular, if one only considers observables characterizing the free motion (e.g. functions of momenta and angular momenta), then one expects that these observables are asymptotic constants for $H$ and for $-H$, i.e. the time evolutions $X_H(t)$ of these observables $X$ tend to limits $X_+$ and $X_-$ when $t \to \infty$ and $t \to -\infty$, and that the limits $X_+$ and $X_-$ provide information on the scattering process. This is the physical background for the algebraic scattering theory, partly studied in Chapter 10 (Part III).

In this part $\mathscr{H}$ denotes a separable Hilbert space and $H$ is a selfadjoint operator acting on it. To avoid trivial cases, we assume $P_H^{ac} > 0$ throughout this part. Further, we shall often use the notations $P^{ac} := P_H^{ac}$, $R(z) := R_H(z)$ and $E(\cdot) := E_H(\cdot)$. When $\mathscr{A}$ is an algebra of operators and $Q \in \mathscr{A}$ is a projection, then we sometimes write $\mathscr{A} \mid Q$ for the subalgebra $\{A \in \mathscr{A};\, A = AQ = QA\}$.

# 6. General Theory of Asymptotic Constants

## 6.1. Special limiting processes for vector- and operator-valued functions

As a preliminary to the study of limits of operator-valued functions $e^{itH'}Xe^{-itH}$ we here collect some facts about different limiting processes for functions from $L^\infty(\mathbb{R}_+, dt; \mathscr{H})$.

### 6.1.1. A class of equivalent limiting processes

Let $u \in L^\infty(\mathbb{R}_+, dt; \mathscr{H})$. Then besides the strong limit, $\operatorname*{s-lim}_{t\to\infty} u(t)$, we consider other limiting methods which are defined by means of certain nonnegative functions $g_\varepsilon(t)$, $\varepsilon > 0$. First we define a special class of such functions.

**Definition 1.** The function $\omega_\varepsilon \in L^\infty(\mathbb{R}_+, dt)$, $\varepsilon > 0$, is called an *averaging function* if it satisfies the following conditions:

(I) $\omega_\varepsilon(t) \geqq 0$,

(II) $\omega_\varepsilon(t) := \varepsilon\psi(\varepsilon t)$,

(III) $\int k(t)\, dt = 1$ with $k(t) := e^{-t}\psi(e^{-t})$; i.e. $\omega_\varepsilon \in L^1(\mathbb{R}_+, dt)$,

(IV) $\hat{k}(p) := (2\pi)^{-1/2} \int k(t)\, e^{-ipt}\, dt \neq 0$ for all $p \in \mathbb{R}$.

Now we have the following important statement.

**Lemma 2.** *Let* $u \in L^\infty(\mathbb{R}_+, dt; \mathscr{H})$. *Assume that the strong limit* $u_+ := \operatorname*{s-lim}_{\varepsilon\to+0} \int_0^\infty \omega_{\varepsilon,1}(t)\, u(t) dx$ *exists for some averaging function* $\omega_{\varepsilon,1}$. *Then for any other averaging function* $\omega_{\varepsilon,2}$ *the strong limit* $\operatorname*{s-lim}_{\varepsilon\to+0} \int_0^\infty \omega_{\varepsilon,2}(t)\, u(t)\, dt$ *also exists and is equal to* $u_+$ (the integrals are Bochner integrals with respect to $\mathscr{H}$).

**Proof.** First we use the special structure of the averaging functions $\omega_{\varepsilon,j}$ to obtain

$$\int_0^\infty \omega_{\varepsilon,j}(t)\, u(t)\, dt = \int_0^\infty \varepsilon\psi_j(\varepsilon t)\, u(t)\, dt = \int_0^\infty e^{-\xi}\psi_j(e^{-\xi}t)\, u(t)\, dt$$

$$= \int_{-\infty}^\infty e^{-\xi}\psi_j(e^{\eta-\xi})\, u(e^\eta)\, e^\eta\, d\eta = \int_{-\infty}^\infty k_j(\xi - \eta)\, u(e^\eta)\, d\eta,$$

where we have made the transformations $\varepsilon = e^{-\xi}$ and $t = e^\eta$. Thus the limit $\varepsilon \to +0$ is equivalent to the limit $\xi \to \infty$. That means that the limits in the lemma correspond to

$$u_+ = \operatorname*{s-lim}_{\xi\to\infty} \int_{-\infty}^\infty k_j(\xi - \eta)\, u(e^\eta)\, d\eta. \tag{1}$$

Now we use the following theorem (see Hille/Phillips [1, p. 156]): Let $k_1(\cdot) \in L^1(\mathbb{R}, dt)$ and let its Fourier transform be different from zero for all real arguments. Let $k_2(\cdot) \in L^1(\mathbb{R}, dt)$ and let $v(x)$ be a uniformly bounded and $|\cdot|$-measurable function from $\mathbb{R}$ into a Banach space $\mathscr{X}$. Assume $\operatorname{s-lim}_{y\to\infty} \int k_1(y-x)\, v(x)\, dx = v_+ \int k_1(x)\, dx$ exists with $v_+ \in \mathscr{X}$. Then $\operatorname{s-lim}_{y\to\infty} \int k_2(y-x)\, v(x)\, dx$ exists and equals $v_+ \int k_2(x)\, dx$.

Applying this statement to our situation and using the properties of $k_j$, we get that (1) for $j = 1$ implies (1) for $j = 2$, and this proves the lemma. ■

**Remark 3.** This lemma shows that all limiting processes generated by averaging functions are equivalent. Thus it is sufficient to consider one special averaging function and the corresponding limiting process.

Our next goal is to deduce that the class of averaging functions is wide enough to include many different means.

**Example 4** (Abel mean). The Abelian limit is defined by $\operatorname{s-lim}_{\varepsilon\to+0} \varepsilon \int_0^\infty e^{-\varepsilon t} u(t)\, dt$. Thus $\omega_\varepsilon(t) = \varepsilon\, e^{-\varepsilon t}$ and $\varphi(t) = e^{-t}$ in this case. Further, $k(t) = e^{-t}\, e^{-(e^{-t})}$. Then we get $\int k(t)\, dt = \int e^{-t} \varphi(e^{-t})\, dt = \int_0^\infty \varphi(x) \times dx = \int_0^\infty e^{-x}\, dx = 1$; i.e. condition (III) of Definition 1 is satisfied. The Fourier transform of $k(t)$ is given by $\hat{k}(p) = (2\pi)^{-1/2}\, \Gamma(1 + ip)$ (see Gradštein/Ryžik [1, p. 322]) and therefore different from zero for all real $p$. Thus all assumptions of Definition 1 are fulfilled, that is, $\varepsilon\, e^{-\varepsilon t}$ is an averaging function.

**Example 5** (Cesaro mean). The Cesaro limit is defined by $\operatorname{s-lim}_{T\to\infty} T^{-1} \int_0^T u(t)\, dt$. Setting $T = \varepsilon^{-1}$, we get the standard form $\operatorname{s-lim}_{\varepsilon\to+0} \varepsilon \int_0^\infty \chi_{[0,1]}(\varepsilon t)\, u(t)\, dt$. Thus $\omega_\varepsilon(t) = \varepsilon \chi_{[0,1]}(\varepsilon t)$ and $\varphi(t) = \chi_{[0,1]}(t)$ in this case. Further, $k(t) = e^{-t} \chi_{[0,\infty)}(t)$ and $\int_{-\infty}^\infty k(t)\, dt = \int_0^\infty e^{-t}\, dt = 1$. The Fourier transform of $k(t)$ can be easily calculated to $\hat{k}(p) = (2\pi)^{-1/2} (1 + ip)^{-1} \neq 0$ for all real $p$. All together we have that the function $\varepsilon \chi_{[0,1]}(\varepsilon t)$ representing the Cesaro limit is an averaging function.

**Example 6** (second Hölder mean). The second Hölder limit is defined by

$$\operatorname{s-lim}_{T\to\infty} T^{-1} \int_0^T dt\, t^{-1} \int_0^t u(t')\, dt'.$$

This can be transformed into

$$\operatorname{s-lim}_{T\to\infty} T^{-1} \int_0^T u(t')\, dt' \int_0^T t^{-1} \chi_{[0,\infty)}(t - t')\, dt = \operatorname{s-lim}_{T\to\infty} T^{-1} \int_0^T f_T(t)\, u(t)\, dt$$

with $f_T(t) = \log (T/t)$. Setting $T = \varepsilon^{-1}$ we get the standard form

$$\operatorname{s-lim}_{\varepsilon\to+0} \varepsilon \int_0^\infty \chi_{[0,1]}(\varepsilon t) \log (1/\varepsilon t)\, u(t)\, dt.$$

Thus $\omega_\varepsilon(t) = \varepsilon \chi_{[0,1]}(\varepsilon t) \log (1/\varepsilon t)$ and $\varphi(t) = \chi_{[0,1]}(t) \log (t^{-1})$. Further, $k(t) = \chi_{[0,\infty]}(t)\, e^{-t} t$, $\int k(t)\, dt = 1$. Then (see Gradštein/Ryžik [1, p. 324]) $\hat{k}(p) = (2\pi)^{-1/2} (1 + ip)^{-2}$, and this proves that (III) and (IV) of Definition 1 are fulfilled. Hence the function $\varepsilon \chi_{[0,1]}(\varepsilon t) \log (1/\varepsilon t)$ representing the second Hölder mean is an averaging function.

From these examples of averaging functions we see that one can take the Abel mean as a representer for all limiting processes with averaging functions. Thus we are led to

**Definition 7.** Let $u \in L^\infty(\mathbb{R}, dt; \mathscr{H})$.

(a) The vector $u_+ \in \mathscr{H}$ is called *Abelian limit* of $u(t)$, denoted by $u_+ = \text{A-}\lim_{t\to\infty} u(t)$, if for some (and thus all) averaging function $\omega_\varepsilon$, $\text{s-}\lim_{\varepsilon\to+0} \int_0^\infty \omega_\varepsilon(t)\,[u(t) - u_+]\,dt = 0$ exists.

(b) The vector $u_+ \in \mathscr{H}$ is called *absolute Abelian limit* of $u(t)$, denoted by $u_+ = |\text{A}|\text{-}\lim_{t\to\infty} u(t)$, if for some (and thus all) averaging function $\omega_\varepsilon$, $\text{s-}\lim_{\varepsilon\to+0} \int_0^\infty \omega_\varepsilon(t)\,\|u(t) - u_+\|^2\,dt = 0$ exists.

Later we shall see that for abstract considerations the Abelian averaging function $\omega_\varepsilon(t) = \varepsilon\,e^{-\varepsilon t}$ is very useful whereas for concrete calculation the Cesaro averaging function $\omega_\varepsilon(t) = \varepsilon\chi_{[0,1]}(\varepsilon t)$ is more important. An essential property of the limiting processes which we consider is that they are invariant with respect to translations, this means that they are invariant means.

**Lemma 8.** *The strong, Abelian, and absolute Abelian limit are invariant with respect to translations, that is, when* $u \in L^\infty(\mathbb{R}, dt; \mathscr{H})$ *and* $u_a(x) := u(x + a)$, $a \in \mathbb{R}$, *then* $\text{s-}\lim_{t\to\infty} u(t) = u_+$, $\text{A-}\lim_{t\to\infty} u(t) = u_+$ *or* $|\text{A}|\text{-}\lim_{t\to\infty} u(t) = u_+$ *implies* $\text{s-}\lim_{t\to\infty} u_a(t) = u_+$, $\text{A-}\lim_{t\to\infty} u_a(t) = u_+$ *or* $|\text{A}|\text{-}\lim_{t\to\infty} u_a(t) = u_+$, *respectively.*

**Proof.** For the strong limit this is trivial. A straightforward calculation shows that the Cesaro mean is invariant with respect to translations, and so are the Abelian mean and absolute Abelian mean because of Lemma 2 and Definition 7. ∎

Next we prove that under additional assumptions on $u(t)$, $|\text{A}|\text{-}\lim_{t\to\infty} u(t) = 0$ implies that a certain discrete mean of $u(t)$ vanishes, similar to the trivial fact that $\text{s-}\lim_{t\to\infty} u(t) = 0$ implies $\text{s-}\lim_{n\to\infty} u(t_n) = 0$ for each sequence $t_n \to \infty$ as $n \to \infty$.

**Lemma 9.** *Assume that* $u \in L^\infty(\mathbb{R}_+, dt; \mathscr{H})$, $u(\cdot)$ *is uniformly continuous in* $t \in \mathbb{R}_+$ *and that* $|\text{A}|\text{-}\lim_{t\to\infty} u(t) = 0$ *exists. Then for each* $a \in \mathbb{R}_+$, $\lim_{N\to\infty} N^{-1} \sum_{n=1}^{N} \|u(a(n-1))\|^2 = 0$.

**Proof.** First we show that $u_N(t) := N^{-1} \sum_{n=1}^{N} \|u(a(n-1) + t)\|^2$, $0 \leqq t \leqq a$, is equicontinuous with respect to $N$. We have

$$|u_N(t_1) - u_N(t_2)| \leqq N^{-1} \sum_{n=1}^{N} \Big|\|u(a(n-1) + t_1)\|^2 - \|u(a(n-1) + t_2)\|^2\Big|$$

$$\leqq \left(N^{-1} \sum_{n=1}^{N} \|u(a(n-1) + t_1) - u(a(n-1) + t_2)\|^2\right)^{1/2} 2\,|u(\cdot)|_\infty .$$

By the uniform continuity of $u(t)$, we get that each summand tends uniformly in $n$ to zero as $|t_1 - t_2| \to 0$. Thus the left side vanishes uniformly in $N$ as $|t_1 - t_2| \to 0$.

Now we have $\lim_{T\to\infty} T^{-1}\int_0^T \|u(t)\|^2\,dt = 0$. Set $T = aN$. Then

$$(aN)^{-1}\int_0^{aN} \|u(t)\|^2\,dt = (aN)^{-1}\sum_{n=1}^{N}\int_{a(n-1)}^{an} \|u(t)\|^2\,dt$$

$$= a^{-1}\int_0^a N^{-1}\sum_{n=1}^{N} \left\|u\left(a(n-1)+t'\right)\right\|^2\,dt'$$

$$= a^{-1}\int_0^a u_N(t)\,dt \to 0 \quad \text{as} \quad N\to\infty.$$

By the equicontinuity of $u_N(t)$, this implies $u_N(t)\to 0$ as $N\to\infty$ for each $t\in[0,a]$ and proves the lemma. ∎

Occasionally, we shall also need

**Lemma 10.** *Let* $f\in L^\infty(\mathbb{R}_+, dt)$ *and* $f(t)\geqq 0$ *a.e.* (mod $|\cdot|$). *Then* $\text{A-}\lim_{t\to\infty} f(t) = 0$ *is equivalent to* $\text{A-}\lim_{t\to\infty} f(t)^2 = 0$.

**Proof.** The proof is a consequence of the estimates

$$T^{-1}\int_0^T f(t)^2\,dt \leqq |f(\cdot)|_\infty\, T^{-1}\int_0^T f(t)\,dt,$$

$$T^{-1}\int_0^T f(t)\,dt = \int_0^T (T^{-1/2})\left(T^{-1/2}f(t)\right)dt$$

$$\leqq \left(\int_0^T T^{-1}\,dt\right)^{1/2}\left(\int_0^T T^{-1}f(t)^2\,dt\right)^{1/2} = \left(T^{-1}\int_0^T f(t)^2\,dt\right)^{1/2}. \ ∎$$

Sometimes it is useful to formulate the absolute Abelian limit with the help of a family of semiscalar products and seminorms on $L^\infty(\mathbb{R}_+, dt; \mathscr{H})$. Let $u, v\in L^\infty(\mathbb{R}_+, dt; \mathscr{H})$. Then we define a semiscalar product $[\cdot,\cdot]_\varepsilon$ by ($\omega_\varepsilon$ averaging function)

$$[u, v]_\varepsilon := \int_0^\infty \omega_\varepsilon(t)\left(u(t), v(t)\right)dt, \qquad \varepsilon > 0. \tag{2}$$

The corresponding seminorm $[\cdot]_\varepsilon$ is given by $[u]_\varepsilon := [u, u]_\varepsilon^{1/2}$. Some elementary properties and the relation to |A|-lim are collected in

**Lemma 11.** *Let* $u, v\in L^\infty(\mathbb{R}_+, dt; \mathscr{H})$. *Then:*

(i) $|[u, v]_\varepsilon| \leqq [u]_\varepsilon\,[v]_\varepsilon$,

(ii) $[u]_\varepsilon \leqq |u|_\infty$,

(iii) $[u, v]_\varepsilon = (u_0, v_0)$ *for functions* $u(t) := u_0$, $v(t) := v_0$ *for all* $t\in\mathbb{R}_+$.

(iv) $|\text{A}|\text{-}\lim_{t\to\infty} u(t) = u_0$ *is equivalent to* $\lim_{\varepsilon\to+0} [u - \tilde{u}_0]_\varepsilon = 0$, *where* $\tilde{u}_0$ *denotes the vector function* $\tilde{u}_0(t) := u_0$ *for all* $t\in\mathbb{R}_+$.

**Proof.** The proof is elementary and is omitted. ■

By means of this definition of the absolute Abelian limit, we prove

**Lemma 12.** *Let* $u, v \in L^\infty(\mathbb{R}_+, \mathrm{d}t; \mathscr{H})$. *Suppose that* $u_+ = |\mathrm{A}|\text{-}\lim_{t\to\infty} u(t)$ *and* $v_+ = |\mathrm{A}|\text{-}\lim_{t\to\infty} v(t)$ *exist. Set* $w(t) := \big(u(t), v(t)\big)$ *and* $w_+ := (u_+, v_+)$. *Then* $\mathrm{A}\text{-}\lim_{t\to\infty} w(t) = w_+$ *exists.*

**Proof.** By Lemma 11, we have

$$|[u, v]_\varepsilon - (u_+, v_+)| = |[u, v]_\varepsilon - [\tilde{u}_+, \tilde{v}_+]_\varepsilon| = |[u - \tilde{u}_+, v]_\varepsilon - [\tilde{u}_+, v - \tilde{v}_+]_\varepsilon|$$

$$\leqq [u - \tilde{u}_+]_\varepsilon\, [v]_\varepsilon + [v - \tilde{v}_+]_\varepsilon\, [u]_\varepsilon \leqq [u - \tilde{u}_+]_\varepsilon\, |v|_\infty + [v - \tilde{v}_+]_\varepsilon\, |u|_\infty$$

with $\tilde{u}_+(t) = u_+$ and $\tilde{v}_+(t) = v_+$ for all $t \in \mathbb{R}_+$. This estimate proves the lemma. ■

### 6.1.2. Strong, Abelian and absolute Abelian limits

In this section we establish the relations between the three different limits: strong, Abelian and absolute Abelian limits. The next lemma is basic for this.

**Lemma 13.** *Let* $u \in L^\infty(\mathbb{R}_+, \mathrm{d}t; \mathscr{H})$. *Then* "$|\mathrm{A}|\text{-}\lim_{t\to\infty} u(t) = u_+$ *exists" is equivalent to the existence of a Borel set* $\Delta \subseteq \mathbb{R}_+$ *such that*

(i) $\lim_{t\to\infty} t^{-1} \int_0^t \chi_\Delta(t')\, \mathrm{d}t' = 0,$

(ii) $\operatorname{s-lim}_{t\in\mathbb{R}_+\setminus\Delta, t\to\infty} u(t) = u_+$, *that is, for each* $\varepsilon > 0$ *there is* $T(\varepsilon) > 0$ *such that* $\|u(t) - u_+\| \leqq \varepsilon$ *for* $t \in \mathbb{R}_+ \setminus \Delta$ *and* $t \geqq T(\varepsilon)$.

**Proof.** 1. First we prove that (i) and (ii) imply $\lim_{T\to\infty} T^{-1} \int_0^T \|u(t) - u_+\|^2\, \mathrm{d}t = 0$ i.e. $|\mathrm{A}|\text{-}\lim_{t\to\infty} u(t) = u_+$. By (ii), for each $\varepsilon > 0$ there is $T(\varepsilon) > 0$ such that $\|u(t) - u_+\| \leqq \varepsilon$ for all $t \geqq T(\varepsilon)$ and $t \in \mathbb{R}_+ \setminus \Delta$. Thus, for $T > T(\varepsilon)$,

$$T^{-1} \int_0^T \|u(t) - u_+\|^2\, \mathrm{d}t = T^{-1} \int_0^T \chi_\Delta(t)\, \|u(t) - u_+\|^2\, \mathrm{d}t$$

$$+ T^{-1} \int_0^{T(\varepsilon)} \big(1 - \chi_\Delta(t)\big)\, \|u(t) - u_+\|^2\, \mathrm{d}t + T^{-1} \int_{T(\varepsilon)}^T \big(1 - \chi_\Delta(t)\big)\, \|u(t) - u_+\|^2\, \mathrm{d}t$$

$$\leqq |u - u_+|_\infty^{\,2} \left\{ T^{-1} \int_0^T \chi_\Delta(t)\, \mathrm{d}t + T(\varepsilon)/T \right\} + \varepsilon^2 T^{-1}\big(T - T(\varepsilon)\big).$$

Then (i) and the arbitrariness of $\varepsilon > 0$ prove that the left side vanishes as $T \to \infty$, and, therefore, $|\mathrm{A}|\text{-}\lim_{t\to\infty} u(t) = u_+$ exists.

2. Now we assume $\lim_{T\to\infty} T^{-1}\int_0^T \|u(t)-u_+\|^2\,dt = 0$. Then there is a zero sequence $\{\varepsilon_n\}$ such that $\varepsilon_n \geqq n^{-1}\int_0^n \|u(t)-u_+\|^2\,dt$ and $\varepsilon_n \geqq \varepsilon_{n+1}$. Now let $\mathfrak{N}_n = \{t \in \mathbb{R}_+;\, t \in [0, n),\, \|u(t)-u_+\|^2 \geqq \sqrt{\varepsilon_n}\}$, $\mathfrak{M}_n = \mathfrak{N}_n \cap [n-1, n)$ and $\mathfrak{M} = \bigcup_{n=1}^{\infty} \mathfrak{M}_n$. Clearly $\mathfrak{M} \cap [0, n) \subseteqq \mathfrak{N}_n$. Thus we get, using the definitions of $\varepsilon_n$ and $\mathfrak{N}_n$,

$$T^{-1}\int_0^T \chi_{\mathfrak{M}}(t)\,dt \leqq T^{-1}\int_0^N \chi_{\mathfrak{N}_N}(t)\,dt + (T-N)/T$$

$$\leqq (T-N)/T + N^{-1}\int_0^N \chi_{\mathfrak{N}_N}(t)\,\|u(t)-u_+\|^2\,\varepsilon_N^{-1/2}\,dt$$

$$\leqq (T-N)/T + \left(\sqrt{\varepsilon_N}\,N\right)^{-1}\int_0^N \|u(t)-u_+\|^2\,dt$$

$$\leqq (T-N)/T + \sqrt{\varepsilon_N} \to 0 \quad \text{as} \quad T\to\infty \ (\text{thus } N\to\infty)$$

where $N$ is the greatest natural number contained in $(0, T]$. Setting $\Delta = \mathfrak{M}$, this proves (i). By construction, it is immediately clear that on the set $\mathbb{R}_+ \setminus \mathfrak{M}$ the function $\|u(t)-u_+\|$ tends to zero as $t\to\infty$. This shows that (ii) is also satisfied. ■

From this statement if follows easily that the absolute Abelian limit is a regular extension of the usual strong limit.

**Corollary 14.** *Let $u \in L^\infty(\mathbb{R}_+, dt; \mathscr{H})$. Assume $u_+ = \text{s-}\lim_{t\to\infty} u(t)$ exists. Then $|\mathrm{A}|\text{-}\lim_{t\to\infty} u(t)$ also exists and is equal to $u_+$.*

Now we describe the relation between the Abelian and the absolute Abelian limits.

**Lemma 15.** *Let $u \in L^\infty(\mathbb{R}_+, dt; \mathscr{H})$. Then the following two conditions are equivalent:*

(i) $u_+ = |\mathrm{A}|\text{-}\lim_{t\to\infty} u(t)$ *exists,*

(ii) $u_+ = \mathrm{A}\text{-}\lim_{t\to\infty} u(t)$ *and* $\|u_+\|^2 = \mathrm{A}\text{-}\lim_{t\to\infty} \|u(t)\|^2$ *exist.*

**Proof.** 1. First we show that (i) implies (ii). The first assertion of (ii) follows from the estimate

$$\int_0^\infty \omega_\varepsilon(t)\,\|u(t)-u_+\|\,dt \geqq \left\|\int_0^\infty \omega_\varepsilon(t)\,\left(u(t)-u_+\right)dt\right\|$$

and Lemma 10. The second assertion of (ii) is a consequence of Lemma 12.

2. Now we prove the other direction. We have

$$\int_0^\infty \omega_\varepsilon(t)\,\|u(t)-u_+\|^2\,dt = \int_0^\infty \omega_\varepsilon(t)\,\left(\|u(t)\|^2 + \|u_+\|^2\right)dt$$

$$- \int_0^\infty \omega_\varepsilon(t)\left[\left(u(t), u_+\right) + \left(u_+, u(t)\right)\right]dt.$$

$\operatorname{A-lim}_{t\to\infty} u(t) = u_+$ implies that the second term on the right side tends to $-2\,\|u_+\|^2$. By the second condition in (ii), the first term tends to $2\,\|u_+\|^2$. Thus the right side of the equation above vanishes as $\varepsilon \to +0$ and this proves that $|\mathrm{A}|\text{-}\lim_{t\to\infty} u(t) = u_+$ exists. ■

From this statement we obtain that the Abelian limit is a regular extension of the absolute Abelian limit.

Now we consider these limiting methods for operator-valued functions.

**Definition 16.** Let $X(t)$ be a uniformly bounded, strongly $|\cdot|$-measurable function from $\mathbb{R}_+$ into $\mathscr{L}(\mathscr{H}, \mathscr{H}')$ ($\mathscr{H}'$ a separable Hilbert space).

(I) If there is a bounded operator $X_0$ such that for all $u \in \mathscr{H}$, $|\mathrm{A}|\text{-}\lim_{t\to\infty} X(t)\, u = X_0 u$ exists, then we write $|\mathrm{A}|\text{-}\lim_{t\to\infty} X(t) = X_0$ and say $X_0$ is the *absolute Abelian limit* of $X(t)$.

(II) If there is a bounded operator $X_0$ such that for all $u \in \mathscr{H}$, $\operatorname{A-lim}_{t\to\infty} X(t)\, u = X_0 u$ exists, then we write $\operatorname{A-lim}_{t\to\infty} X(t) = X_0$ and say $X_0$ is the *Abelian limit* of $X(t)$.

We use the same notations as in the case of vector-valued functions because it is convenient and cannot give any confusion. Obviously, the limit $X_0$ is uniquely determined by both limiting processes.

**Lemma 17.** *Let $X(t)$ be a uniformly bounded, strongly $|\cdot|$-measurable function from $\mathbb{R}_+$ into $\mathscr{L}(\mathscr{H}, \mathscr{H}')$ ($\mathscr{H}'$ a separable Hilbert space). Then $|\mathrm{A}|\text{-}\lim_{t\to\infty} X(t) = X_0$ is equivalent to the existence of a Borel set $\Delta \subset \mathbb{R}_+$ such that*

(i) $\lim_{t\to\infty} t^{-1} \int_0^t \chi_\Delta(t')\, dt' = 0$,

(ii) $\operatorname{s-lim}_{t\in\mathbb{R}_+\setminus\Delta,\, t\to\infty} X(t) = X_0$.

**Proof. 1.** Assume that (i) and (ii) are satisfied for $X(t)$ and $X_0$. Then $X_0$ is bounded, and an application of Lemma 13 proves $|\mathrm{A}|\text{-}\lim_{t\to\infty} X(t) = X_0$.

**2.** Now assume $|\mathrm{A}|\text{-}\lim_{t\to\infty} X(t) = X_0$ exists. Let $\{u_n\}_1^\infty$ be dense in the unit sphere $\|u\| = 1$ of $\mathscr{H}$ and define $s(t) = \sum 2^{-n}\, \|(X(t) - X_0)\, u_n\|^2$. Then $\|(X(t) - X_0)\, u_n\|^2 \leq 2^n s(t)$. Further,

$$T^{-1} \int_0^T s(t)\, dt = \sum_{n=1}^\infty 2^{-n} T^{-1} \int_0^T \left\|\left(X(t) - X_0\right) u_n\right\|^2 dt \leq \sum_{n=1}^\infty 2^{-n} \sup_{t\in\mathbb{R}_+} \|X(t) - X_0\|^2.$$

Since each term under the sum (in the middle) tends to zero as $T \to \infty$ and since the series is uniformly convergent in $T \in \mathbb{R}_+$, we get that the left side tends to zero as $T \to \infty$. Thus we can apply Lemma 13 to $s(t)$ ($\mathscr{H} = \mathbb{C}$ in this case) and get that there is a set $\Delta \subset \mathbb{R}_+$ such that $\lim_{t\to\infty} t^{-1} \int_0^t \chi_\Delta(t')\, dt' = 0$ and $\lim_{t\in\mathbb{R}_+\setminus\Delta,\, t\to\infty} s(t) \to 0$. Because of $\|(X(t) - X_0)\, u_n\|^2 \leq 2^n s(t)$ this implies $\operatorname{s-lim}_{t\in\mathbb{R}_+\setminus\Delta,\, t\to\infty} X(t)\, u_n = X_0\, u_n$ for all $n$. Since $X(t) - X_0$ is uniformly bounded, this can be extended to all vectors $u \in \mathscr{H}$ which proves (i) and (ii). ■

**Remark 18.** Lemma 17 shows that the absolute Abelian limit does not differ very much from the strong limit and that they have common properties. For instance, when $X_1(t): \mathbb{R}_+ \to \mathscr{L}(\mathscr{H}, \mathscr{H}')$ and $X_2(t): \mathbb{R}_+ \to \mathscr{L}(\mathscr{H}', \mathscr{H}'')$ are strongly continuous, are uniformly bounded and have the limits $X_1 = |\mathrm{A}|\text{-}\lim_{t\to\infty} X_1(t)$ and $X_2 = |\mathrm{A}|\text{-}\lim_{t\to\infty} X_2(t)$, then the limit $|\mathrm{A}|\text{-}\lim_{t\to\infty} X_2(t)\, X_1(t) = X_2 X_1$ also exists.

Since we shall often get the same results for both s-lim and |A|-lim, we introduce the common symbol Lim if the statements are valid for both limiting methods.

## 6.2. Definitions and general properties

### 6.2.1. The definition of asymptotic constants

Before turning to the definition of our main objects, the asymptotic constants, we present some simple features of operator functions $X(t) = \mathrm{e}^{\mathrm{i}tH'} X\, \mathrm{e}^{-\mathrm{i}tH}$ where $H, H'$ are selfadjoint on the separable Hilbert spaces $\mathscr{H}$, $\mathscr{H}'$ and $X \in \mathscr{L}(\mathscr{H}, \mathscr{H}')$. Since $\mathrm{e}^{-\mathrm{i}tH}$, $\mathrm{e}^{\mathrm{i}tH'}$ are strongly continuous and $X$ is bounded, $X(t)$ is also strongly continuous and uniformly bounded. The following statement shows that the domain of the limits of $X(t)$, as $t \to \infty$, is given by a subspace $\mathscr{X}$ invariant under $H$.

**Proposition 19.** *Let* $X(t): \mathbb{R} \to \mathscr{L}(\mathscr{H}, \mathscr{H}')$ *be as above. Then the set of all* $f \in \mathscr{H}$ *such that* $\operatorname{Lim}_{t\to\infty} X(t) f = f_+$ *exists is a subspace* $\mathscr{X}$ *invariant under* $H$*; that is, the projection* $P$ *onto* $\mathscr{X}$ *commutes with* $H$.

Proof. 1. Obviously, $\mathscr{X}$ is linear. Let $f_n \in \mathscr{X}$ and $f = \text{s-}\lim_{n\to\infty} f_n$. Then $f \in \mathscr{X}$ follows from the uniform boundedness of $X(t)$ and the estimate $\|(X(t) - X(t'))\, f\| \leq \|(X(t) - X(t'))\, f_n\| + \|X(t) - X(t')\|\, \|f - f_n\|$.

2. To prove that $\mathscr{X}$ is invariant under $H$, by clo spa $\{E(\Delta) f;\ \Delta$ Borel set$\}$ $=$ clo spa $\{\mathrm{e}^{-\mathrm{i}tH} f;\ t \in \mathbb{R}\}$ (see Section 3.2.3), it is sufficient to verify that $f \in \mathscr{X}$ implies $\mathrm{e}^{-\mathrm{i}tH} f \in \mathscr{X}$ for all $t \in \mathbb{R}$. But this follows from $X(t)\, \mathrm{e}^{-\mathrm{i}sH} f = \mathrm{e}^{-\mathrm{i}sH'} X(t+s)\, f$, the fact that $\operatorname{Lim}_{t\to\infty} X(t) f = f_+$ implies the existence of $\operatorname{Lim}_{t\to\infty} X(t+s) f = f_+$ (Lemma 8), and $\operatorname{Lim}_{t\to\infty} BX(t) f = B \operatorname{Lim}_{t\to\infty} X(t) f$ for $B \in \mathscr{L}(\mathscr{H}', \mathscr{H}'')$. ■

From the proof of this proposition we immediately obtain

**Corollary 20.** *Let* $X(t)$ *and* $\mathscr{X}$ *be as in Prop.* 19. *Let* $\mathscr{M}$ *be a subset of* $\mathscr{X}$. *Then* clo spa $\{E(\Delta) f;\ f \in \mathscr{M},\ \Delta$ *Borel set*$\} \subseteq \mathscr{X}$.

This technical corollary reduces existence considerations of limits of $X(t)$ to the corresponding ones on certain subsets. Recall the definition of a generating set with respect to a subspace $Q\mathscr{H}$ and $H$ (see Section 3.1.4). Then this notion and Corollary 20 leads immediately to

**Corollary 21.** *Let* $X(t)$, $P$ *and* $\mathscr{X}$ *be as in Prop.* 19. *Let* $\mathscr{M}$ *be a generating set with respect to* $Q\mathscr{H}$ *and* $H$ ($Q$ a projection) *and assume* $\mathscr{M} \subseteq \mathscr{X}$. *Then* $Q \leq P$.

After these preliminaries we can now define the main objects of this part. Recall that always $P^{\mathrm{ac}} > 0$ is assumed.

**Definition 22.** The symbol $\mu^H$ denotes the map from the set of all $X \in \mathscr{L}(\mathscr{H})$ for which $X_0 := \operatorname{Lim}_{t\to\infty} e^{itH} X e^{-itH} P^{ac}$ exists into $\mathscr{L}(\mathscr{H})$ defined by $\mu^H: X \to X_0$ shortly written $X_0 = \mu^H(X)$. The domain, image, and kernel of the map $\mu^H$ are denoted by dom $\mu^H$, ima $\mu^H$, and ker $\mu^H$, respectively. The map $\mu^H$ is called the *wave morphism*. In case Lim = |A|-lim we write $\mu_a{}^H$ and in case Lim = s-lim we write $\mu_s{}^H$.

Sometimes we shall say $X$ is an *asymptotic constant* when $X$ belongs to dom $\mu_a{}^H$. The next statement asserts that $\mu_a{}^H$ is a regular extension of the wave morphism $\mu_s{}^H$. Thus we can also use the notation asymptotic constant for operators $X \in$ dom $\mu_s{}^H$. Occasionally, we shall also say *strong asymptotic constant* in this case to emphasize that $e^{itH} X e^{-itH} P^{ac}$ even tends strongly to its limit.

**Theorem 23.** *If $X \in$ dom $\mu_s{}^H$, then $X \in$ dom $\mu_a{}^H$ and $\mu_a{}^H(X) = \mu_s{}^H(X)$.*

Proof. The proof is a simple consequence of Corollary 14. ■

An easy consequence of Corollary 21 is the following useful result.

**Proposition 24.** *Let $X \in \mathscr{L}(\mathscr{H})$ and let $\mathscr{M}$ be a generating set with respect to $P^{ac}\mathscr{H}$ and $H$. Then the existence of $\operatorname{Lim}_{t\to\infty} e^{itH} X e^{-itH} P^{ac} u$ for all $u \in \mathscr{M}$ implies $X \in$ dom $\mu^H$.*

### 6.2.2. Simple properties of the wave morphism

First we collect some basic facts on the wave morphism $\mu^H$.

**Theorem 25.** *The map $\mu^H$ has the following properties:*

(i) *If $X \in$ dom $\mu^H$, then $\mu^H(X) \in \mathscr{L}(\mathscr{H})$ and $\|\mu^H(X)\| \leq \|X\|$.*

(ii) *dom $\mu^H$ is a linear space and $\mu^H(a_1X_1 + a_2X_2) = a_1\mu^H(X_1) + a_2\mu^H(X_2)$, $a_1, a_2 \in \mathbb{C}$ and $X_1, X_2 \in$ dom $\mu^H$.*

(iii) ima $\mu^H = \{X; X \in$ com $H$, $X = P^{ac}X = XP^{ac}\} = ($com $H) \restriction P^{ac}$.

(iv) *dom $\mu^H$ = ima $\mu^H \oplus$ ker $\mu^H$ and $\mu^H(X) = X$ iff $X \in$ ima $\mu^H$.*

(v) *If $X \in$ ker $\mu^H$, then $BXC \in$ ker $\mu^H$ for all $B \in \mathscr{L}(\mathscr{H})$ and $C \in$ com $H$.*

(vi) *dom $\mu^H$ is an algebra and $\mu^H(X_1X_2) = \mu^H(X_1)\,\mu^H(X_2)$ for all $X_1, X_2 \in$ dom $\mu^H$.*

(vii) *If $X \in$ dom $\mu^H$ and $X$ is selfadjoint, then $\mu^H(X)$ is also selfadjoint.*

(viii) *If $X \in$ dom $\mu^H$ and $X$ is a projection, then $\mu^H(X)$ is also a projection.*

Proof. (i). This follows easily from the fact that $\mu^H(X)$ is a strong limit of $e^{it_nH} X e^{-it_nH} P^{ac}$ for a certain sequence $t_n \to \infty$ as $n \to \infty$ (see Lemma 17) and $\|e^{it_nH} X e^{-it_nH} P^{ac}\| \leq \|X\|$.

(ii) The linear properties are obvious for the strong limit, that is $\mu_s{}^H$. For $\mu_a{}^H$ this follows from the fact that we can define |A|-lim via the seminorms $[\cdot]_\varepsilon$ (see Lemma 11) and have

$$\begin{aligned} &[e^{itH}(a_1X_1 + a_2X_2)\, e^{-itH} P^{ac}u - a_1\mu^H(X_1)\, u - a_2\mu^H(X_2)\, u]_\varepsilon \\ &\leq [e^{itH}(a_1X_1)\, e^{-itH} P^{ac}u - \mu^H(a_1X_1)\, u]_\varepsilon \\ &\quad + [e^{itH}(a_2X_2)\, e^{-itH} P^{ac}u - \mu^H(a_2X_2)\, u]_\varepsilon . \end{aligned}$$

(iii) The relation ima $\mu^H \supsetneqq \{X; X \in \text{com}\, H, X = XP^{ac} = P^{ac}X\}$ follows directly from the definition of ima $\mu^H$. It remains to show ima $\mu^H \subsetneqq \{X; X \in \text{com}\, H, X = XP^{ac} = P^{ac}X\}$. Let $X_0 \in \text{dom}\, \mu^H$ and $\mu^H(X_0) = X$. Let $u \in \mathscr{H}$. Then $\lim_{t\to\infty} e^{itH} X_0\, e^{-itH} P^{ac} u = Xu$ and $\lim_{t\to\infty} e^{i(t+s)H} X_0\, e^{-i(t+s)H} P^{ac} u = Xu$ for each $s \in \mathbb{R}$, by Lemma 8. Thus $e^{isH} X\, e^{-isH} u = Xu$ for all $u \in \mathscr{H}$ and $s \in \mathbb{R}$ and this implies $X \in \text{com}\, H$ (see Section 3.2.3). The relation $X = P^{ac}X = XP^{ac}$ follows from the definition of $X = \mu^H(X_0)$ and the fact that $P^{ac}$ is a spectral projection and thus $P^{ac}B = BP^{ac}$ for all $B \in \text{com}\, H$.

(iv) The proof is simple and we omit it.

(v) The proof is an easy consequence of the definition of ker $\mu^H$ and the estimate $\|e^{itH} BXC\, e^{-itH} P^{ac} u\| \leq \|B\|\, \|X\, e^{-itH} C P^{ac} u\| = \|B\|\, \|e^{itH} X\, e^{-itH} P^{ac} u'\|$ with $u' = Cu$.

(vi) Let $X_1, X_2 \in \text{dom}\, \mu^H$ and $X_i = X_{i,0} + C_i$ the corresponding decompositions with respect to dom $\mu^H$ = ima $\mu^H \oplus$ ker $\mu^H$, i.e. $C_i \in \ker \mu^H$ and $X_{i,0} \in \text{ima}\, \mu^H$. Then $X_1X_2 = X_{1,0}X_{2,0} + X_{1,0}C_2 + C_1X_{1,0} + C_1C_2$. The first term on the right side belongs to ima $\mu^H$ (see (iii)) and the other terms belong to ker $\mu^H$ because of (v). By (iii) and (iv), we then obtain $\mu^H(X_1X_2)$ exists and equals $\mu^H(X_1)\, \mu^H(X_2)$.

(vii) Assume $X = X^* \in \text{dom}\, \mu^H$. Then, by (vi) and (iii), $\mu^H(X) = \mu^H(P^{ac}X)$. Now $\mu^H(P^{ac}X)$ is a strong limit of a sequence of selfadjoint operators $e^{it_nH} P^{ac} X\, e^{-it_nH} P^{ac} = e^{it_nH} P^{ac} X P^{ac}\, e^{-it_nH}$ (see Lemma 17), thus $\mu^H(P^{ac}X)$ is also selfadjoint.

(viii) We must prove that $X \in \text{dom}\, \mu^H$ with $X = X^* = X^2$ implies $\mu^H(X) = (\mu^H(X))^* = \mu^H(X)\, \mu^H(X)$. By (vi), $\mu^H(XX) = \mu^H(X) = \mu^H(X)\, \mu^H(X)$ when $X^2 = X$. According to (vii), $X^* = X$ implies $(\mu^H(X))^* = \mu^H(X)$. Thus $\mu^H(X)$ is a projection. ∎

**Remark 26.** This theorem states among other things that dom $\mu^H$ is an algebra, that ker $\mu^H$ is a left ideal with respect to $\mathscr{L}(\mathscr{H})$ and that ker $\mu^H$ is an ideal with respect to dom $\mu^H$. The algebra dom $\mu^H$ is called *wave algebra* and the ideal ker $\mu^H$ is called *wave ideal*. Sometimes we shall also use the notation *algebra of asymptotic constants* for dom $\mu^H$.

**Remark 27.** In the language of operator algebras this theorem asserts that $\mu^H$ is a homomorphism from dom $\mu^H$ onto ima $\mu^H \cong \text{com}\, H \restriction P^{ac}\mathscr{H}$. Moreover, (i) says that $\mu^H$ is continuous in norm and dom $\mu^H$ is a Banach algebra.

**Theorem 28.** *Let $A_1, A_2, \ldots, A_n$ be selfadjoint operators belonging to* dom $\mu^H$. *Further, suppose that the operators $A_i$ pairwise commute. Then for every complex-valued function $f(x_1, \ldots, x_n)$ which is continuous on the set $\Delta = \text{spec}\, A_1 \times \cdots \times \text{spec}\, A_n$ the normal operator $f(A_1, \ldots, A_n)$ belongs to* dom $\mu^H$ *and*

$$f(\mu^H(A_1), \ldots, \mu^H(A_n))\, P^{ac} = \mu^H(f(A_1, \ldots, A_n)). \tag{3}$$

Proof. Since $\mu^H$ is continuous in norm, $X_s \in \text{dom}\, \mu^H$ and $X_s \xrightarrow[n]{} X$ imply $X \in \text{dom}\, \mu^H$ and $\mu^H(X_s) \xrightarrow[n]{} \mu^H(X)$. $f(A_1, \ldots, A_n)$ is the norm limit of a sequence of polynomials $p_m(A_1, \ldots, A_n) = \sum a^{(m)}_{\nu_1\nu_2\cdots\nu_n} A_1^{\nu_1} A_2^{\nu_2} \cdots A_n^{\nu_n}$. From Theorem 25 it follows that $p_m(A_1, \ldots, A_n) \in \text{dom}\, \mu^H$ and $\mu^H(p_m(A_1, \ldots, A_n)) = p_m(\mu^H(A_1), \ldots, \mu^H(A_n))\, P^{ac}$. This together with the remarks above leads to the desired result. ∎

If $X \in \text{dom}\, \mu^H$ is arbitrary, then all polynomials in $X$ are elements of dom $\mu^H$. The behavior of the resolvent $(z - X)^{-1}$ is investigated in the next theorem.

**Theorem 29.** *Let $X \in \operatorname{dom} \mu^H$. Let $\operatorname{res}_\infty X$ denote the maximal connected subregion of $\operatorname{res} X$ which contains the set $\{z; |z| > \operatorname{spr} X\}$ (i.e. a neighborhood of $\infty$). Then:*

(i) *For all $z \in \operatorname{res}_\infty X$ the relation $(z - X)^{-1} = R_X(z) \in \operatorname{dom} \mu^H$ is valid, furthermore $\operatorname{res}_\infty X \setminus \{0\} \subseteq \operatorname{res} \mu^H(X)$, in particular $\operatorname{spr} \mu^H(X) \leq \operatorname{spr} X$, and*

$$\mu^H((z - X)^{-1}) = (z - \mu^H(X))^{-1} P^{ac}, \quad z \in \operatorname{res}_\infty X \setminus \{0\}. \tag{4}$$

(ii) $\operatorname{res}_\infty X \subseteq \operatorname{res} \{\mu^H(X) \restriction P^{ac}\mathscr{H}\}$ *and*

$$(z1_{P^{ac}\mathscr{H}} - \mu^H(X) \restriction P^{ac}\mathscr{H})^{-1} = \mu^H((z - X)^{-1}) \restriction P^{ac}\mathscr{H}, \quad z \in \operatorname{res}_\infty X.$$

**Proof.** (i) First, let $|z| > \operatorname{spr} X$. Then (see Section 1.3.5)

$$(z - X)^{-1} = \sum_{l=0}^{\infty} z^{-(l+1)} X^l$$

in the sense of operator norm convergence. Because of $X^l \in \operatorname{dom} \mu^H$ for $l = 2, 3, \ldots$ and because $\operatorname{dom} \mu^H$ is a Banach algebra we obtain $(z - X)^{-1} \in \operatorname{dom} \mu^H$ and

$$\mu^H((z - X)^{-1}) = z^{-1}\mu^H(1) + \sum_{l=1}^{\infty} z^{-(l+1)} \mu^H(X^l)$$

$$= z^{-1}P^{ac} + \sum_{l=1}^{\infty} z^{-(l+1)} (\mu^H(X))^l P^{ac} = (z - \mu^H(X))^{-1} P^{ac}.$$

This implies $\operatorname{spr} \mu^H(X) \leq \operatorname{spr} X$, $z \in \operatorname{res} \mu^H(X)$ and

$$(z - \mu^H(X))^{-1} = z^{-1}(P^{ac})^\perp + \mu^H((z - X)^{-1}).$$

Now let $z \in \operatorname{res}_\infty X$. Then there is a point $\zeta$ with $|\zeta| > \operatorname{spr} X$ such that $R_X(\cdot)$ can be holomorphically continued (by repeated Taylor expansion) from $\zeta$ to $z$ along a smooth curve $\mathscr{K} \subset \operatorname{res}_\infty X$ (recall that $R_X(\cdot)$ is holomorphic in $\operatorname{res} X$). But from

$$R_X(\eta) = \sum_{n=0}^{\infty} (n!)^{-1} R_X^{(n)}(\zeta) (\eta - \zeta)^n, \quad R_X^{(n)}(\zeta) = (-1)^n n! R_X(\zeta)^{n+1}$$

we immediately obtain that $R_X(\zeta) \in \operatorname{dom} \mu^H$ implies $R_X(\eta) \in \operatorname{dom} \mu^H$ for all $\eta$ such that the series is norm convergent. Repeating this procedure, we get $R_X(z) \in \operatorname{dom} \mu^H$. If additionally $z \neq 0$, then we obtain

$$(z - \mu^H(X)) (z^{-1}(P^{ac})^\perp + (\mu^H(z - X)^{-1}))$$

$$= (z(P^{ac})^\perp + \mu^H(z - X)) (z^{-1}(P^{ac})^\perp + \mu^H((z - X)^{-1}))$$

$$= (P^{ac})^\perp + \mu^H(1) = 1,$$

hence $z \in \operatorname{res} \mu^H(X)$ and

$$(z - \mu^H(X))^{-1} = z^{-1}(P^{ac})^\perp + \mu^H((z - X)^{-1}),$$

that is, (4) follows and (i) is proved.

(ii) If $z \in \operatorname{res}_\infty X$, then by (i) we have $(z - X)^{-1} \in \operatorname{dom} \mu^H$, hence

$$\left(z1_{P^{ac}\mathscr{H}} - \mu^H(X) \mid P^{ac}\mathscr{H}\right) \left(\mu^H((z - X)^{-1} \mid P^{ac}\mathscr{H}\right)$$
$$= \left(\mu^H(z - X) \mid P^{ac}\mathscr{H}\right) \left(\mu^H((z - X)^{-1} \mid P^{ac}\mathscr{H}\right) = 1 \mid P^{ac}\mathscr{H},$$

that is $z \in \operatorname{res} \{\mu^H(X) \mid P^{ac}\mathscr{H}\}$ and (ii) is proved. ∎

Using Dunford's functional calculus, we get from this statement

**Corollary 30.** *Let* $X \in \operatorname{dom} \mu^H$ *and let* $f(z)$ *be holomorphic in a neighborhood* $\mathscr{N}$ *of* $\mathbb{C} \setminus \operatorname{res}_\infty X$. *Then* $f(X) \in \operatorname{dom} \mu^H$ *and* $f(\mu^H(X)) P^{ac} = \mu^H(f(X))$.

**Proof.** Let $\Gamma \subset \operatorname{res}_\infty X \cap \mathscr{N}$. Then $f(X) = (2i\pi)^{-1} \int_\Gamma f(\lambda) (\lambda - X)^{-1} d\lambda$ exists as a Bochner integral ($\Gamma$ is a closed curve). By Theorem 29, the integrand belongs to dom $\mu^H$, and so does $f(X)$ as a norm limit of finite sums $\sum_{i=1}^N f(\lambda_i) (\lambda_i - X)^{-1} s_i$, $s_i \in \mathbb{C}$. The last assertion follows from the corresponding result for polynomials. ∎

Examples for functions satisfying the assumption of Corollary 30 are $e^{a\lambda}$, $\sin a\lambda$, $\cos a\lambda$ etc. with $a \in \mathbb{R}$.

Another immediate consequence of Theorem 29 is the following interesting assertion on the relation between the spectra of $X$ and $\mu^H(X)$.

**Corollary 31.** *Let* $X$ *be selfadjoint and* $X \in \operatorname{dom} \mu^H$. *Then* $\operatorname{res} X \setminus \{0\} \subseteq \operatorname{res} \mu^H(X)$, *that is,*

$$\operatorname{spec} \mu^H(X) \subseteq \operatorname{spec} X \cup \{0\} \quad \text{and} \quad \operatorname{spec} \left(\mu^H(X) \mid P^{ac}\mathscr{H}\right) \subseteq \operatorname{spec} X. \tag{5}$$

Proof. For selfadjoint operators $\operatorname{res}_\infty X = \operatorname{res} X$, and this together with Theorem 29 implies (5). ∎

Relation (5) suggests the investigation of points $\lambda$ with the property "$\lambda \in \operatorname{spec} X$ but $\lambda \notin \operatorname{spec} \mu^H(X) \mid P^{ac}\mathscr{H}$". In other words, (5) suggests the possibility of the disappearance of parts of spec $X$ if the wave morphism acts on $X$. Later (Chapter 7) we shall see that with the help of such disappearing parts of the spectrum it becomes possible to construct special projections belonging to ker $\mu^H$.

### 6.2.3. Some properties of wave ideals

The wave ideal $\ker \mu^H$ is defined to be the set of all $X \in \mathscr{L}(\mathscr{H})$ with $\operatorname{Lim}_{t\to\infty} e^{itH} X e^{-itH} P^{ac} = 0$. In case $\mu^H = \mu_a{}^H$ it is possible to give a characterization of elements of ker $\mu_a{}^H$ in terms of the resolvent of $H$. This is the content of the next statements. A corresponding simple characterization of elements of ker $\mu_s{}^H$ does not seem to exist.

**Proposition 32.** *A bounded operator* $X$ *belongs to* ker $\mu_a{}^H$ *iff*

$$\lim_{\varepsilon\to+0} \varepsilon \int_{-\infty}^{\infty} \|XR(\lambda + i\varepsilon) P^{ac}u\|^2 \, d\lambda = 0 \quad \textit{for all} \quad u \in \mathscr{H}. \tag{6}$$

**Proof.** According to (3.11), $(2\pi)\int_0^\infty e^{-2\varepsilon t}\,\|X\,e^{-itH}P^{ac}u\|^2\,dt = \int_{-\infty}^\infty \|XR(\lambda+i\varepsilon)\,P^{ac}u\|^2\,d\lambda$.

This and the fact that $X \in \ker \mu_a{}^H$ iff $\lim_{\varepsilon\to+0}\int_0^\infty \omega_\varepsilon(t)\,\|X\,e^{-itH}P^{ac}u\|^2\,dt = 0$ with $\omega_\varepsilon(t) = \varepsilon\,e^{-\varepsilon t}$ imply the desired result. ∎

**Proposition 33.** *The wave ideal* $\ker \mu_a{}^H$ *consists of all bounded operators* $X$ *satisfying the following condition: For each* $u \in \mathscr{H}$, *the function* $\sqrt{\varepsilon}\,XR(\lambda+i\varepsilon)\,P^{ac}u$ *is (strongly) measure-convergent to* 0 *as* $\varepsilon \to +0$.

**Proof.** 1. Assume $X \in \ker \mu_a{}^H$. Then, by (6), the function $f_\varepsilon(\lambda) = \sqrt{\varepsilon}\,XR(\lambda+i\varepsilon)\,P^{ac}u$ converges in $L^2$-sense to zero as $\varepsilon \to +0$. Now $L^2$-convergence implies measure convergence. This proves the condition in Prop. 33.

2. Conversely assume that $\sqrt{\varepsilon}\,XR(\lambda+i\varepsilon)\,P^{ac}u$ is strongly measure convergent to zero as $\varepsilon \to +0$ for all $u \in \mathscr{H}$. From the results of Section 3.5.2 we get that there exists a dense set $\mathscr{D}$ in $P^{ac}\mathscr{H}$ such that for all $\varepsilon > 0$, $u \in \mathscr{D}$,

$$\varepsilon\,\|XR(\lambda+i\varepsilon)\,u\|^2 \leqq \varepsilon\,\|X\|^2\,\|R(\lambda+i\varepsilon)\,u\|^2 \leqq \|X\|^2\,\Phi_u(\lambda), \qquad \Phi_u \in L_1(\mathbb{R}, d\lambda).$$

According to a well-known theorem of Lebesgue, measure convergence of a sequence $f_\varepsilon(\lambda)$ to $f(\lambda)$ and $|f_\varepsilon(\lambda)| \leqq \Phi(\lambda)$, $\Phi \in L^1(\mathbb{R}, d\lambda)$, imply $f \in L^1(\mathbb{R}, d\lambda)$ and $\lim_{\varepsilon\to+0}\int f_\varepsilon(\lambda)\,d\lambda = \int f(\lambda)\,d\lambda$. Applying this to our situation, we get (6) for all $u \in \mathscr{D}$. Because of

$$2\varepsilon\int_{-\infty}^\infty \|XR(\lambda+i\varepsilon)\,P^{ac}u\|^2 = 4\varepsilon\pi\int_0^\infty e^{-2\varepsilon t}\,\|X\,e^{-itH}P^{ac}u\|^2\,dt \leqq 2\pi\,\|X\|^2\,\|u\|^2,$$

this can be extended to all $u \in \mathscr{H}$ proving $X \in \ker \mu_a{}^H$. ∎

**Remark 34.** Note that measure convergence in $\varepsilon$ implies the existence of a sequence $\varepsilon_n \to +0$ such that $\sqrt{\varepsilon_n}\,XR(\lambda+i\varepsilon_n)\,P^{ac}u$ is a.e. (mod $|\cdot|$) strongly convergent to zero as $n \to \infty$. The sequence $\varepsilon_n$ may depend on $u$.

Obviously, wave ideals contain all bounded operators $X$ with $XP^{ac} = 0$. Our next aim is to show that also the class of all compact operators belongs to every wave ideal.

**Theorem 35.** *The compact operators are contained in every wave ideal, i.e.* $\mathscr{L}_\infty(\mathscr{H}) \subseteq \ker \mu^H$.

**Proof.** From $\operatorname{dom}\mu_s{}^H \subseteq \operatorname{dom}\mu_a{}^H$ and $\operatorname{ima}\mu_a{}^H = \operatorname{ima}\mu_s{}^H$ (see Theorems 23 and 25) it follows $\ker \mu_s{}^H \subseteq \ker \mu_a{}^H$. Thus it is sufficient to prove $\mathscr{L}_\infty(\mathscr{H}) \subseteq \ker \mu_s{}^H$. Further, since $\mu_s{}^H$ is linear and continuous in norm and since every compact operator is a norm limit of finite-dimensional operators of the form $\sum_{j=1}^n f_j(g_j, \cdot)$ with $f_j, g_j \in \mathscr{H}$ it suffices to prove that for arbitrary $f, g \in \mathscr{H}$, $\lim_{t\to\infty}\|e^{itH}\,f(g, e^{-itH}P^{ac}u)\| = \lim_{t\to\infty}\|f\|\,|(g, e^{-itH}P^{ac}u)| = 0$ for all $u \in \mathscr{H}$. But this follows from Prop. 3.23. ∎

We have already seen that wave ideals are left ideals with respect to $\mathscr{L}(\mathscr{H})$. The next statement extends this to the assertion that they are the greatest left ideals contained in the corresponding wave algebras.

**Proposition 36.** *Let $X \in \mathscr{L}(\mathscr{H})$. Then $BX \in \operatorname{dom} \mu^H$ for all $B \in \mathscr{L}(\mathscr{H})$ if and only if $X \in \ker \mu^H$.*

**Proof.** That $X \in \ker \mu^H$ implies $BX \in \ker \mu^H$ was already proved (see Theorem 25). Now we prove the converse direction. For this aim we first show that $\operatorname{dom} \mu^H$ is a proper subset of $\mathscr{L}(\mathscr{H})$.

There is a Borel set $\Delta \subset \operatorname{spec}^c H$ such that $E(\Delta)$ and $E(\mathbb{R} \setminus \Delta)$ are different from zero and $\dim E(\Delta)\,\mathscr{H} = \dim E(\mathbb{R} \setminus \Delta)\,\mathscr{H} = \infty$. Therefore there is a unitary operator $U$ on $\mathscr{H}$ such that $UE(\Delta) = E(\mathbb{R} \setminus \Delta)\,U$. Now assume $U \in \operatorname{dom} \mu^H$. Then we obtain $\mu^H(U)\,E(\Delta) = \mu^H\big(UE(\Delta)\big) = \mu^H\big(E(\mathbb{R} \setminus \Delta)\,UE(\Delta)\big) = E(\mathbb{R} \setminus \Delta)\,\mu^H(U)\,E(\Delta) = \mu^H(U)\,E(\mathbb{R} \setminus \Delta)\,E(\Delta) = 0$ since $\mu^H(U) \in \operatorname{com} H$, and similarly $\mu^H(U)\,E(\mathbb{R} \setminus \Delta) = 0$; that is $\mu^H(U) = 0$. On the other hand we have $\|e^{itH} U\, e^{-itH} P^{ac} f\| = \|P^{ac} f\|$ and this implies $\mu^H(U) \neq 0$ leading to a contradiction. Thus $U \notin \operatorname{dom} \mu^H$.

Now we assume $BX \in \operatorname{dom} \mu^H$ for each $B \in \mathscr{L}(\mathscr{H})$. First, set $B = 1$, then it follows $X \in \operatorname{dom} \mu^H$. Set $X_0 = \mu^H(X)$ and let $E$ be the projection onto $\operatorname{clo\,spa} \{E(\Delta)\,u;\ u \in X_0\mathscr{H},\ \Delta \text{ Borel set}\}$. Then $E$ commutes with $H$ and the set $X_0\mathscr{H}$ is generating with respect to $E\mathscr{H}$ and $H$. Further, we have $E \leq P^{ac}$. By $BX \in \operatorname{dom} \mu^H$ and $X - X_0 \in \ker \mu^H$, we get $BX_0 \in \operatorname{dom} \mu^H$ for all $B \in \mathscr{L}(\mathscr{H})$, and according to Corollary 21, we obtain that $\operatorname{Lim}_{t\to\infty} e^{itH} B\, e^{-itH} E$ exists for all $B \in \mathscr{L}(\mathscr{H})$.

Set $H' = H \restriction E\mathscr{H}$. Then this implies $\operatorname{Lim}_{t\to\infty} e^{itH'} B\, e^{-itH'}$ exists for all $B \in \mathscr{L}(E\mathscr{H})$, i.e. $\operatorname{dom} \mu^{H'} = \mathscr{L}(E\mathscr{H})$. However, we have proved above that $\operatorname{dom} \mu^{H'} \subset \mathscr{L}(E\mathscr{H})$ when $P^{ac}_H > 0$; in our case $P^{ac}_{H'} = 1_{E\mathscr{H}}$. Thus $E = 0$ is only possible and this gives $X_0 = 0$. Therefore $X \in \ker \mu^H$ proving the assertion. ∎

**Remark 37.** In the course of the proof we have verified the additional assertion that $\operatorname{dom} \mu^H$ is always a proper subset of $\mathscr{L}(\mathscr{H})$.

The next point is to prove a spectral property of elements of wave ideals (for non-selfadjoint operators $X$ the essential spectrum of $X$ is defined to be $\operatorname{spec} X \setminus \operatorname{dis} X$ where $\operatorname{dis} X$ is defined as in Section 3.3.1).

**Proposition 38.** *Let $X \in \ker \mu^H$. Then the point zero belongs to the essential spectrum of $X$.*

**Proof.** Assume $X \in \ker \mu^H$ and $X$ has zero at most as isolated eigenvalue of finite multiplicity. Let $Q$ be the Riesz spectral projection belonging to the eigenvalue 0. Then $QX = XQ$ and $0 \in \operatorname{res}\big(X \restriction (1 - Q)\,\mathscr{H}\big)$, that is, $X \restriction (1 - Q)\,\mathscr{H}$ is bounded invertible. Then $Y = X(1 - Q) + Q$ is bounded invertible in $\mathscr{H}$ and we have $Y \in \operatorname{dom} \mu^H$ and $\mu^H(Y) = \mu^H(X) = 0$. On the other hand, $Y \in \ker \mu^H$ implies $Y^{-1}Y = 1 \in \ker \mu^H$ since $\ker \mu^H$ is a left ideal, but this is obviously false. Hence, when $X \in \ker \mu^H$, zero must belong to the essential spectrum. ∎

**Remark 39.** This spectral property of elements from wave ideals also expresses the fact that wave ideals are "disjoint" to the class of Fredholm operators because "$\operatorname{spec} X \setminus \operatorname{dis} X \not\ni 0$" is one of the assumptions of the definition of Fredholm operators $X$.

## 6.3. Relations between the algebras $\mathcal{L}_\infty(\mathcal{H})$, $\mathcal{L}(\mathcal{H})$, dom $\mu_s^H$ and dom $\mu_a^H$

### 6.3.1. Some technical criteria

The aim of this section is to establish some statements which are useful in proving $X \in \operatorname{dom} \mu^H$. We begin with a statement saying that certain linear transformations of $H$ do not change the wave morphism $\mu^H$.

**Proposition 40.** *Let $a > 0$ and $b \in \mathbb{R}$. Then $\mu^{aH+b} = \mu^H$, that means* dom $\mu^{aH+b}$ = dom $\mu^H$ *and $\mu^H(X) = \mu^{aH+b}(X)$ for $X \in \operatorname{dom} \mu^H$.*

**Proof.** The assertion follows easily from the definition of $\mu^{aH+b}$ and the fact that the transformation $H \to aH + b$ corresponds to the one $t \to at$ in the expression $e^{itH} X e^{-itH} P_{ac}$. ■

Occasionally it is easier to work with some discrete versions of the strong and the absolute Abelian limit.

**Lemma 41.** *Let $H, X \in \mathcal{L}(\mathcal{H})$ and let $u \in \mathcal{H}$. Further, let $a \in (0, \infty)$ such that* clo spa $\{e^{-inaH}u;\ n = 0, \pm 1, \ldots\}$ = clo spa $\{e^{-itH}u;\ t \in \mathbb{R}\}$, *e.g. if $0 < a\,\|H\| \leq 2\pi$ (see 3.2.3). Then*

(i) $\operatorname{s-lim}_{t\to\infty} e^{itH} X e^{-itH} u = 0$ *iff* $\lim_{n\to\infty} \|X e^{-inaH} u\| = 0,\quad n \in \mathbb{N}$,

(ii) $|\mathrm{A}|\text{-}\lim_{t\to\infty} e^{itH} X e^{-itH} u = 0$ *iff* $\lim_{N\to\infty} N^{-1} \sum_{n=1}^{N} \|X e^{-i(n-1)aH} u\|^2 = 0,\quad N \in \mathbb{N}$.

**Proof.** (i) The nontrivial part is to show that $\lim_{n\to\infty} \|X e^{-inaH} u\| = 0$ implies $\lim_{t\to\infty} \|X e^{-itH} u\| = 0$. Obviously, we obtain $\lim_{n\to\infty} \|X e^{-inaH} e^{-iNaH} u\| = 0$ for all $N = 0, \pm 1, \ldots$ Hence $\lim_{n\to\infty} \|X e^{-inaH} f\| = 0$ is valid for all $f \in$ clo spa $\{e^{-itH}u;\ t \in \mathbb{R}\}$. Therefore $\sup_{0\leq\tau\leq 1} \|X e^{-inaH} e^{-i\tau aH} u\|$ also tends to 0 as $n \to \infty$ because $[0, 1]$ is compact. Now $t = [t] + \tau$ ($[t]$ natural number, $\tau \in [0, 1)$). Hence we have

$$\|X e^{-itaH} u\| = \|X e^{-i[t]aH} e^{-i\tau aH} u\| \leq \sup_{0\leq\tau\leq 1} \|X e^{-i[t]aH} e^{-i\tau aH} u\|,$$

and this implies the assertion.

(ii) Since $e^{itH} X e^{-itH} u$ is uniformly continuous in $t$, we can apply Lemma 9 and get that $|\mathrm{A}|\text{-}\lim_{t\to\infty} e^{itH} X e^{-itH} u = 0$ implies

$$\lim_{N\to\infty} N^{-1} \sum_{n=1}^{N} \|X e^{-i(n-1)aH} u\|^2 = 0, \qquad N \in \mathbb{N}.$$

Now we assume that $\lim_{N\to\infty} N^{-1} \sum_{n=1}^{N} \|X e^{-i(n-1)aH} u\|^2 = 0$. Then it is an easy calculation to show that also

$$\lim_{N\to\infty} N^{-1} \sum_{n=1}^{N} \|X e^{-i(n-1)aH} e^{imaH} u\|^2 = 0$$

for each integer $m$. Since $N^{-1} \sum_{n=1}^{N} \|X\, e^{-i(n-1)aH} f\|^2 \leq \|X\|^2\, \|f\|^2$, we thus obtain $\lim_{N\to\infty} N^{-1} \sum_{n=1}^{N} \|X\, e^{-i(n-1)aH} f\|^2 = 0$ for all $f \in$ clo spa $\{e^{-imaH}u;\, m = 0, \pm 1, \ldots\}$ = clo spa $\{e^{-itH}u;\, t \in \mathbb{R}\}$.

Thus $a_N(\tau) := N^{-1} \sum_{n=1}^{N} \|X\, e^{-i(n-1)aH}\, e^{-i\tau H}u\|^2 \to 0$ as $N \to \infty$ for all $\tau \in [0, a]$. Further, we have

$$N^{-1} \int_0^{aN} \|X\, e^{-itH}u\|^2\, dt = N^{-1} \sum_{n=1}^{N} \int_{a(n-1)}^{an} \|X\, e^{-itH}u\|^2\, dt$$

$$= \int_0^a \left( N^{-1} \sum_{n=1}^{N} \|X\, e^{-i(n-1)aH}\, e^{-i\tau H}u\|^2 \right) d\tau = \int_0^a a_N(\tau)\, d\tau.$$

That this expression tends to zero as $N \to \infty$ follows immediately from $\|a_N(\tau)\| \leq \|X\|^2 \|u\|^2$ and Lebesgue's dominated convergence theorem. The extension to the assertion $\lim_{T\to\infty} T^{-1} \int_0^T \|X\, e^{-itH}u\|^2\, dt = 0$ is obvious. ∎

**Corollary 42.** *Let $H$ and $a$ be as in Lemma 41. Let $\mathcal{M}$ be a generating set with respect to $P^{ac}\mathcal{H}$ and $H$. Then:*

(i) $X \in \ker \mu_s^H$ *iff* $X \in \mathcal{L}(\mathcal{H})$ *and* $\lim_{n\to\infty} \|X\, e^{-inaH}u\| = 0$ *for all* $u \in \mathcal{M}$,

(ii) $X \in \ker \mu_a^H$ *iff* $X \in \mathcal{L}(\mathcal{H})$ *and* $\lim_{N\to\infty} N^{-1} \sum_{n=1}^{N} \|X\, e^{-i(n-1)aH}u\|^2 = 0$ *for all* $u \in \mathcal{M}$.

Proof. The proof follows immediately from Lemma 41 and Prop. 24. ∎

At the end of this section we come to two simple, but often very useful, statements.

**Proposition 43.** *Let $X_1$ and $X_2$ be two bounded nonnegative operators on $\mathcal{H}$ such that $X_1 \leq X_2$. Then $X_2 \in \ker \mu^H$ implies $X_1 \in \ker \mu^H$.*

Proof. From $X_1 \leq X_2$ immediately $\|X_1^{1/2}u\|^2 \leq \|X_2u\|\, \|u\|$ follows for all $u \in \mathcal{H}$. Hence we get $\|X_1^{1/2}\, e^{-itH}P^{ac}u\|^2 \leq \|u\|\, \|X_2\, e^{-itH}P^{ac}u\|$ for all $t \in \mathbb{R}$, $u \in \mathcal{H}$. Using $X_2 \in \ker \mu^H$, we obtain $X_1^{1/2} \in \ker \mu^H$ and therefore $X_1 = X_1^{1/2}X_1^{1/2} \in \ker \mu^H$ because $\ker \mu^H$ is a left ideal. ∎

**Proposition 44.** *Let $\{Q_n\}_{n=1}^N$ be a set of mutually disjoint projections ($N \leq \infty$) such that $Q_n \in$ com $H$, $n \in \{1, \ldots, N\}$ and $\sum_{n=1}^{N} Q_n \geq P^{ac}$. Then a bounded operator $X$ belongs to* dom $\mu^H$ *iff* $XQ_n \in$ dom $\mu^H$ *for each* $n \in \{1, 2, \ldots, N\}$.

Proof. $XQ_n \in$ dom $\mu^H$ for each $n \in \{1, 2, \ldots, N\}$ implies that $\operatorname{Lim}_{t\to\infty} e^{itH}X\, e^{-itH}P^{ac}u$ exists for all $u \in \mathcal{D} = \bigcup_{n=1}^{N} P^{ac}Q_n\mathcal{H}$. By $\sum_{n=1}^{N} Q_n \geq P^{ac}$, $\mathcal{D}$ is a generating set in $P^{ac}\mathcal{H}$ and thus, by Prop. 24, $X \in$ dom $\mu^H$. The other direction is trivial. ∎

### 6.3.2. Wave ideals and compact operators

We have already seen that the set of compact operators belongs to each wave ideal. Now it is a natural question to ask whether the inclusion $\mathscr{L}_\infty(\mathscr{H}) \subseteq \ker \mu^H$ is always proper or not. This is answered in

**Theorem 45.** $\mathscr{L}_\infty(\mathscr{H}) \subset \ker \mu^H$. *Moreover, there is an operator* $X \in \ker \mu^H$ *with* $X = P^{ac}X = XP^{ac}$ *and* $X \notin \mathscr{L}_\infty(\mathscr{H})$.

**Proof.** By $P^{ac} > 0$, $|\mathrm{spec}^c (H \restriction P^{ac}\mathscr{H})| > 0$. Thus there is a sequence of Borel sets $\Delta_n$ such that $\Delta_n \cap \Delta_{n'} = \emptyset$ for $n \neq n'$, $\Delta_n \subset \mathrm{spec}^c(H \restriction P^{ac}\mathscr{H})$, and $|\Delta_n| > 0$. Then the projections $Q_n = E(\Delta_n) P^{ac} > 0$ for all $n \in \mathbb{N}$ and they are mutually disjoint. Further, let $P_n$ be a sequence of projections with $P_n < Q_n$ and $\dim P_n\mathscr{H} = 1$. Thus $X := \sum P_n$ is a projection and $XQ_n = P_n$ is compact and $XQ_n \in \mathscr{L}_\infty(\mathscr{H})$ for each $n$. Now we can apply Prop. 44 and we get $X \in \mathrm{dom}\, \mu^H$. Clearly, $\mu^H(X)\, Q_n = \mu^H(XQ_n) = 0$ and therefore $X \in \ker \mu^H$. By construction, $X \notin \mathscr{L}_\infty(\mathscr{H})$ and $X = XP^{ac} = P^{ac}X$. ■

**Remark 46.** Notice that the operator $X$ constructed in the proof even satisfies a stronger statement, namely that it belongs to the intersection of all wave ideals $\ker \mu^{\alpha(H)}$ where $\alpha$ runs over all real-valued Borel functions such that $\alpha(H)$ is a well-defined operator and $P^{ac}_{\alpha(H)} \geq P^{ac}_H$.

**Remark 47.** Since the wave ideals are closed in norm and contain properly the set of compact operators, they cannot be (two-sided) ideals with respect to $\mathscr{L}(\mathscr{H})$ because $\mathscr{L}_\infty(\mathscr{H})$ is the largest norm closed ideal in $\mathscr{L}(\mathscr{H})$.

Whereas Theorem 45 states that every wave ideal, and also the intersection of certain wave ideals, is wider than the set of compact operators, the next assertion says that the intersection of all wave ideals coincides with $\mathscr{L}_\infty(\mathscr{H})$.

**Proposition 48.** *Let* $\mathscr{A}$ *denote the set of all selfadjoint operators* $H'$ *on* $\mathscr{H}$ *which are unitarily equivalent to* $H$. *Then* $\bigcap_{H' \in \mathscr{A}} \ker \mu^{H'} = \mathscr{L}_\infty(\mathscr{H})$.

**Proof.** Suppose on the contrary there is an operator $X$ with $X \notin \mathscr{L}_\infty(\mathscr{H})$ and $X \in \bigcap_{H' \in \mathscr{A}} \ker \mu^{H'}$. Since $\{\ker \mu^{H'}\}$ are left ideals, $|X|$ has the same properties as $X$. By $|X| \notin \mathscr{L}_\infty(\mathscr{H})$, there is a number $\varepsilon > 0$ such that $Q := E_{|X|}([\varepsilon, \infty))$ is also not compact. Further, because of $\varepsilon Q \leq |X|$, we get $Q \in \bigcap_{H' \in \mathscr{A}} \ker \mu^{H'}$ from Prop. 43. By construction, $\dim Q\mathscr{H} = \infty$ and Prop. 38 gives $\dim (1 - Q)\, \mathscr{H} = \infty$. Now let $P$ be a projection from com (com $H$) with the properties $P < P^{ac}$, $\dim P\mathscr{H} = \dim (1 - P)\, \mathscr{H} = \infty$. Clearly such a projection exists. Then there is a unitary operator $U$ on $\mathscr{H}$ with $UP = QU$ and thus $U(1 - P) = (1 - Q)\, U$. Now we define $H'' = UHU^*$. Then $H''Q = UHU^*Q = UHPU^* = QH''$, and, since $P < P^{ac}$, $Q < P^{ac}_{H''}$. This gives

$$\|Q\, e^{-itH''} P^{ac}_{H''} u\| = \|e^{-itH''} Qu\| = \|Qu\|.$$

Thus $Q \notin \ker \mu^{H''}$. This contradicts $Q \in \bigcap_{H' \in \mathscr{A}} \ker \mu^{H'}$ because $H'' \in \mathscr{A}$. Therefore the assumptions $X \notin \mathscr{L}_\infty(\mathscr{H})$ and $X \in \bigcap_{H' \in \mathscr{A}} \ker \mu^{H'}$ lead to a contradiction. ■

To illustrate the class of operators which are not compact but belong to the wave ideal $\ker \mu^H$, we consider a simple example.

**Example 49.** Let $H$ be the natural selfadjoint extension of $-\mathrm{i}d/dx$ on the Hilbert space $\mathscr{H} = L^2(\mathbb{R}, dx)$. $H$ is absolutely continuous, i.e. $P^{ac} = 1$. The action of $e^{-itH}$ on vectors $u = u(\cdot) \in \mathscr{H}$ is given by $(e^{-itH}u)(x) = u(x-t)$. Further, let $F$ be the multiplication operator by the characteristic function $\chi_\Delta(\cdot)$ where $\Delta \subset [-M, M], M < \infty$. We want to show that $F \in \ker \mu_s^H \cap \ker \mu_s^{-H}$. Let $u(\cdot) \in L^2(\mathbb{R}, dx)$ with a finite support contained in $[a, b]$. Then

$$\|e^{itH} F e^{-itH} u\|^2 = \int dx\, \chi_\Delta(x)\, |u(x-t)|^2 \leq \int_{-M}^{M} dx\, |u(x-t)|^2$$

$$= \int_{-M-t}^{M-t} dx\, |u(x)|^2.$$

The right side vanishes if $M - t < a$ and if $-M - t > b$. The first case is fulfilled for all sufficiently large $t$ and the second, for all sufficiently large $-t$. Thus $\|F e^{-itH}u\| \to 0$ as $t \to \pm\infty$ for such vectors $u$ and thus for all $u \in \mathscr{H}$ since these vectors are dense in $\mathscr{H}$. This proves $F \in \ker \mu_s^H \cap \ker \mu_s^{-H}$. Clearly, $F$ is not compact, $\mathrm{spec}_e\, F = \{0, 1\}$.

### 6.3.3. The difference between $\ker \mu_s^H$ and $\ker \mu_a^H$

It was already remarked that $\mu_a^H$ is a regular extension of $\mu_s^H$ and hence $\ker \mu_s^H \subseteq \ker \mu_a^H$. Now we want to verify that the inclusion is proper.

**Theorem 50.** $\ker \mu_s^H \subset \ker \mu_a^H$, *that is, there is an operator $X$ such that $X \in \ker \mu_a^H$ but $X \notin \ker \mu_s^H$.*

**Proof. 1.** For each $H$ on $\mathscr{H}$ there is a decomposition $\mathscr{H} = \mathscr{H}_1 \oplus \mathscr{H}_2$ such that $\mathscr{H}_1$ reduces $H$, $H \upharpoonright \mathscr{H}_1$ is absolutely continuous, and $H \upharpoonright \mathscr{H}_1$ is simple. We shall construct the operator $X$ in the form $X = X_1 \oplus 0$ corresponding to the decomposition $\mathscr{H} = \mathscr{H}_1 \oplus \mathscr{H}_2$. Now $H \upharpoonright \mathscr{H}_1$ is unitarily equivalent to the multiplication operator by $x$ on a Hilbert space $L^2(\Delta, dx)$, $|\Delta| > 0$. Furthermore, this operator is unitarily equivalent to the operator $H_0 \upharpoonright E_{H_0}(\Delta)\, \mathscr{H}_0$ where $H_0 = -\mathrm{i}\, d/dx$ on $\mathscr{H}_0 = L^2(\mathbb{R}, dx)$ (see Example 49, and note that the Fourier transformation transforms $H_0$, $\mathscr{H}_0$ into $x$, $L^2(\mathbb{R}, dx)$). Thus, to prove this theorem, it is enough to find an operator $X_0$ on $\mathscr{H}_0$ with the following properties:

$$X_0 E_{H_0}(\Delta) = E_{H_0}(\Delta)\, X_0 = X_0,\ \ X_0 \in \ker \mu_a^{H_0},\ X_0 \notin \ker \mu_s^{H_0}.$$

The operator $X_1$ (from $X = X_1 \oplus 0$) can then be formed from $X_0$ via the described unitary equivalences.

**2.** The next step is to construct an operator $F$ with $F \in \ker \mu_a^{H_0}$ and $F \notin \ker \mu_s^{H_0}$. We define $F$ as multiplication operator by $\chi_\Delta(\cdot)$ where $\Delta = \bigcup_{n=1}^{\infty} [n^2, n^2 + 1]$ and show that $F$ has the desired properties. It is easy to verify that

$$\lim_{T\to\infty} T^{-1} \int_0^T \chi_\Delta(t)\, dt = 0. \tag{7}$$

Now let $u \in \mathscr{H}$ with supp $u(\cdot) \subset [c, d]$ (finite interval). Then

$$T^{-1} \int_0^T \|F e^{-itH} u\|^2 dt = T^{-1} \int_0^T dt \int_c^d \chi_\Delta(x+t) |u(x)|^2 dx$$

$$\leq \|u\|^2 \sup_{x \in [0,d]} T^{-1} \int_0^T \chi_\Delta(x+t) dt$$

$$\leq \|u\|^2 T^{-1} \int_{-|c|}^{T+|d|} \chi_\Delta(t) dt \to 0 \text{ as } T \to \infty$$

because of (7). Since the set of such vectors is dense in $\mathscr{H}_0$ this proves $F \in \ker \mu_a^{H_0}$. Now let $t_n = n^2$ and $u \in \mathscr{H}_0$ with supp $u(\cdot) \subset [0, 1]$. Then

$$\|F e^{-it_n H_0} u\|^2 = \int \chi_\Delta(x) |u(x - t_n)|^2 dx = \int_{n^2}^{n^2+1} |u(x - n^2)|^2 dx = \|u\|^2$$

because of the fact that the support of $u(x - n^2)$ lies in $[n^2, n^2 + 1]$. Thus $F \notin \ker \mu_s^{H_0}$.

**3.** It remains to construct from $F$ the operator $X_0$ with the properties described above. Set $A = WFE_{H_0}(\Delta)$ where $W$ is a partial isometry with $W^*W = 1$ and $WW^* = E_{H_0}(\Delta)$. Because of Theorem 25 (v), $A \in \ker \mu_a^{H_0}$. Now let $u \in \mathscr{H}_0$ such that $f := E_{H_0}(\Delta) u \neq 0$. Then there is an integer $m$ such that $\int_m^{m+1} |f(x)|^2 dx = a > 0$. Take the sequence $t_n = n^2 - m$. Then

$$\|A e^{-it_n H_0} u\|^2 = \|F e^{-it_n H_0} f\|^2 = \int \chi_\Delta(x) |f(x - t_n)|^2 dx$$

$$\geq \int_{n^2}^{n^2+1} |f(x - n^2 + m)|^2 dx = \int_m^{m+1} |f(x)|^2 dx = a > 0,$$

and therefore $A \notin \ker \mu_s^{H_0}$. By construction, $A = AE_{H_0}(\Delta) = E_{H_0}(\Delta) A$ and thus $X_0 = A$ has all desired properties. This concludes the proof. ■

**Remark 51.** The operator $X$ in Theorem 50 can have the additional property to be a projection. Namely take an operator $X$ with $X \in \ker \mu_a^H$ and $X \notin \ker \mu_s^H$. Clearly, $|X|$ has the same properties and so has $|X| E_{|X|}([\varepsilon, \infty))$ for some sufficiently small $\varepsilon > 0$ because of $|X| E_{|X|}(\Delta) \leq |X|$ (use Prop. 43) and $|X| E_{|X|}([\varepsilon, \infty)) \to |X|$, $\varepsilon \to +0$, in norm ($\mu^H$ is continuous in norm). Then the projection $P = E_{|X|}([\varepsilon, \infty))$ has also the properties $P \in \ker \mu_a^H$, $P \notin \ker \mu_s^H$. This follows from the estimates $\varepsilon P \leq |X| P$, $\|X\| P \geq |X| P$, and Prop. 43.

An example for an operator $X$ which belongs to $\ker \mu_a^H$ but not to $\ker \mu_s^H$ was constructed in the course of the proof of Theorem 50: $\mathscr{H} = L^2(\mathbb{R}, dx)$, $H = -i\, d/dx$ and $X = \chi_\Delta(\cdot)$ with $\Delta = \bigcup_{n=1}^{\infty} [n^2, n^2 + 1]$.

### 6.3.4. Some relations between $\mathscr{L}(\mathscr{H})$ and dom $\mu^H$

First we want to illustrate the fact that wave algebras are properly included in $\mathscr{L}(\mathscr{H})$ with the help of the following simple example.

**Example 52** Let $H$ and $\mathscr{H}$ be defined as in Example 49. Further, let $F$ denote multiplication by $\chi_\Delta(\cdot)$ where $\Delta = \bigcup_{n=1}^{\infty} (2n, 2n+1)$. Put $\Delta_1 = \bigcup_{n=1}^{\infty} (2n+1, 2n+3/2)$ and $\Delta_2 = \bigcup_{n=1}^{\infty} (2n, 2n+1/2)$. Let $u \in \mathscr{H}$ with $\operatorname{supp} u(\cdot) \subset [0, 1/2]$. Then we see that

$$(e^{itH} F e^{-itH} u)(x) = \chi_\Delta(x+t)\, u(x) = (e^{itH} \chi_\Delta(\cdot)\, u(\cdot - t))(x)$$

equals 0 for $t \in \Delta_1$ and equals $u(x)$ for $t \in \Delta_2$. Since neither $\Delta_1$ nor $\Delta_2$ satisfies $\lim_{T\to\infty} T^{-1} \int_0^T \chi_{\Delta t}(t)\, dt = 0$, we get from Lemma 13 that $|\mathrm{A}|\text{-}\lim_{t\to\infty} e^{itH} F e^{-itH} u$ cannot exist.

On the other hand, the following assertion says that already the sum of the wave ideals for $H$ and $-H$ coincides with the set of all bounded operators.

**Proposition 53.** $\ker \mu^H + \ker \mu^{-H} = \mathscr{L}(\mathscr{H})$, *i.e. for each* $X \in \mathscr{L}(\mathscr{H})$ *there is a decomposition* $X = X_+ + X_-$ *with* $X_+ \in \ker \mu^H$ *and* $X_- \in \ker \mu^{-H}$.

**Proof.** 1. For the proof we deduce that there exists a decomposition of the identity $1 = P_+ + P_-$ such that $P_+ \in \ker \mu^H$ and $P_- \in \ker \mu^{-H}$. Assume for a moment that we have such a decomposition. Then $X = XP_+ + XP_-$ for each $X \in \mathscr{L}(\mathscr{H})$. Since the wave ideals are left ideals, the relations for $P_+$ and $P_-$ imply $X_+ := XP_+ \in \ker \mu^H$ and $X_- := XP_- \in \ker \mu^{-H}$. Thus every bounded operator has the desired decomposition.

2. Note that it is enough to prove the existence of the operators $P_+$, $P_-$ in the special case: $H$ is bounded, is absolutely continuous, and is simple. The generalization to arbitrary $H$ is standard. Because of Prop. 40, we can even assume (without loss of generality) that $\operatorname{spec} H$ is contained in $[0, 2\pi]$. All together this implies that we can identify $H$, $\mathscr{H}$ with $X$, $L^2(\Delta, dx)$ where $\Delta \subseteq [0, 2\pi]$.

3. Now $L^2(\Delta, dx)$ may be regarded as a subspace of $L^2([0, 2\pi], dx)$. Denote by $H'$ the multiplication by $x$ on $\mathscr{H}' = L^2([0, 2\pi], dx)$ and denote by $P$ the projection onto $\mathscr{H} = L^2(\Delta, dx)$. Let $u_n$, $n = 0, \pm 1, \dots$, be defined by $u_n(x) = (2\pi)^{-1/2} e^{-inx}$. These vectors form an orthonormal basis in $\mathscr{H}'$. Set $\tilde{P}_+ = \sum_{n=0}^{-\infty} (u_n, \cdot)\, u_n$ and $\tilde{P}_- = \sum_{n=1}^{\infty} (u_n, \cdot)\, u_n = 1 - \tilde{P}_+$. By $e^{-inH'} u_0 = u_n$, we obtain

$$\|\tilde{P}_+ e^{-inH'} u_0\|^2 = \sum_{s=0}^{-\infty} |(u_s, u_n)|^2 = 0 \quad \text{for} \quad n \geqq 1,$$

$$\|\tilde{P}_- e^{inH'} u_0\|^2 = \sum_{s=1}^{\infty} |(u_s, u_{-n})|^2 = 0 \quad \text{for} \quad n \geqq 0.$$

Since $\mathscr{H}' = \operatorname{clo\,spa} \{e^{-inH'} u_0;\, n = 0, \pm 1, \dots\}$, Corollary 42 leads to $\tilde{P}_+ \in \ker \mu_s^{H'}$ and $\tilde{P}_- \in \ker \mu_s^{-H'}$. Now we define $P_+' = P\tilde{P}_+P$ and $P_-' = P\tilde{P}_-P$. This gives $P_+' + P_-' = P$, $P_+' \in \ker \mu_s^{H'}$ and $P_-' \in \ker \mu_s^{-H'}$ if we use that $P$ commutes with $H'$. Now we can identify $\ker \mu^{H'} \restriction P\mathscr{H}'$, $\ker \mu^{-H'} \restriction P\mathscr{H}'$ with $\ker \mu^H$, $\ker \mu^{-H}$. Thus the operators $P_+'$ and $P_-'$ correspond to operators $P_+$ and $P_-$ with the desired properties. ■

## 6.4. Topological characterizations of wave algebras

### 6.4.1. Definitions of special topologies

Wave algebras are Banach algebras with respect to the operator norm. However the topology generated by the operator norm does not reflect the special properties of these algebras. Now we introduce other topologies whose definitions are more connected with the structure of wave algebras.

Let $X \in \mathscr{L}(\mathscr{H})$ and $u \in \mathscr{H}$. Define

$$p_s(X; H, u) := \sup_{t \geq 0} \|X \,\mathrm{e}^{-\mathrm{i}tH} P^{\mathrm{ac}} u\| + \|X(1 - P^{\mathrm{ac}})\, u\|, \tag{8}$$

$$p_a(X; H, u) := \sup_{\varepsilon > 0} \varepsilon \int_0^\infty \mathrm{e}^{-\varepsilon t} \|X \,\mathrm{e}^{-\mathrm{i}tH} P^{\mathrm{ac}} u\| \,\mathrm{d}t + \|X(1 - P^{\mathrm{ac}})\, u\|, \tag{9}$$

and let $p(X; H, u)$ denote the common symbol for both $p_s(X; H, u)$ and $p_a(X; H, u)$ which will be always used when the statements are true for both.

**Proposition 54.** *The functions* $p(\cdot; H, u)\colon \mathscr{L}(\mathscr{H}) \to \mathbb{R}_+$, $u \in \mathscr{H}$, *satisfy the following relations*:

$$p(A + B; H, u) \leq p(A; H, u) + p(B; H, u), \qquad A, B \in \mathscr{L}(\mathscr{H}), \tag{10}$$

$$p(AB; H, u) \leq \|A\|\, p(B; H, u), \qquad A, B \in \mathscr{L}(\mathscr{H}), \tag{11}$$

$$p(cA; H, u) = |c|\, p(A; H, u), \qquad A \in \mathscr{L}(\mathscr{H}),\ c \in \mathbb{C}, \tag{12}$$

$$p(A; H, u) = 0 \text{ for all } u \in \mathscr{H} \text{ implies } A = 0. \tag{13}$$

Proof. The only part of the proof which is not quite trivial is (13) for $p_a(A; H, u)$. (13) implies $\|A(1 - P^{\mathrm{ac}})\, u\| = 0$ for all $u \in \mathscr{H}$ and $\sup_{\varepsilon>0} \varepsilon \int_0^\infty \mathrm{e}^{-\varepsilon t} \|A \,\mathrm{e}^{-\mathrm{i}tH} P^{\mathrm{ac}} u\| \,\mathrm{d}t = 0$ for all $u \in \mathscr{H}$. The first relation implies $A(1 - P^{\mathrm{ac}}) = 0$ and the second implies that the integrand is zero (a.e. mod $|\cdot|$) for each $\varepsilon > 0$, i.e. $\|A \,\mathrm{e}^{-\mathrm{i}tH} P^{\mathrm{ac}} u\| = 0$ for all $u \in \mathscr{H}$ and almost all $t$. This gives $AP^{\mathrm{ac}} = 0$. ■

This statement says that the functions $p(\cdot; H, u)$ with the index set $\mathscr{H} \ni u$ are seminorms separating points.

**Definition 55.** The topology generated by the family of seminorms $p(\cdot; H, u)$ with the index set $\mathscr{H} \ni u$ is called the *$\mu^H$-topology*. The sequence $C_n$, $C_n \in \mathscr{L}(\mathscr{H})$, is called *$\mu^H$-convergent* with the limit $C$, $C \in \mathscr{L}(\mathscr{H})$, if $p(C_n - C; H, u) \to 0$ as $n \to \infty$ for each $u \in \mathscr{H}$ with $\|u\| = 1$.

Note that the limit $C$ is uniquely determined by the $\mu^H$-convergence of the sequence $C_n$. An easy consequence of this definition is

**Proposition 56.** *The $\mu^H$-topology is not weaker than the strong operator topology on $\mathscr{L}(\mathscr{H})$.*

Proof. In case $\mu^H = \mu_s{}^H$ this follows from the estimate $p_s(C; H, u) \geq \|CP^{\mathrm{ac}} u\| + \|C(1 - P^{\mathrm{ac}})\, u\| \geq \|Cu\|$ (see (8)). In case $\mu^H = \mu_a{}^H$ we can get the same estimate from (9). Namely we have

$$\sup_{\varepsilon>0} \varepsilon \int_0^\infty \mathrm{e}^{-\varepsilon t} \|C \,\mathrm{e}^{-\mathrm{i}tH} P^{\mathrm{ac}} u\| \,\mathrm{d}t \geq \sup_{\varepsilon>0} \left| \varepsilon \int_0^\infty \mathrm{e}^{-\varepsilon t} \|C(\mathrm{e}^{-\mathrm{i}tH} - 1)\, P^{\mathrm{ac}} u\| \,\mathrm{d}t - \|CP^{\mathrm{ac}} u\| \right|$$

$$= \sup_{\varepsilon>0} \left| \|CP^{\mathrm{ac}} u\| - \int_0^\infty \mathrm{e}^{-t} \|C(\mathrm{e}^{-\mathrm{i}(t/\varepsilon)H} - 1)\, P^{\mathrm{ac}} u\| \,\mathrm{d}t \right| \geq \|CP^{\mathrm{ac}} u\|.$$

By (9), this implies $p_a(C; H, u) \geq \|Cu\|$. ■

A first insight into the connection between the new topology and the set of asymptotic constants is given by

**Theorem 57.** *The wave algebra* dom $\mu^H$ *is closed with respect to the* $\mu^H$*-convergence, that is, if* $X_n \in$ dom $\mu^H$ *and if* $X_n$ *is* $\mu^H$*-convergent to* $X$*, then* $X \in$ dom $\mu^H$ *and* $\operatorname{s-lim}_{n\to\infty} \mu^H(X_n) = \mu^H(X)$.

**Proof.** We prove this theorem only for $\mu_s^H$. For $\mu_a^H$ the proof is similar. Let $u \in \mathscr{H}$ with $\|u\| = 1$. For $s > 0$, $t > 0$ we have

$$\begin{aligned} \|e^{itH} X e^{-itH} P_{ac} u - e^{isH} X e^{-isH} P_{ac} u\| &\leq \|e^{itH}(X - X_n) e^{-itH} P_{ac} u\| \\ &+ \|e^{isH}(X - X_n) e^{-isH} P_{ac} u\| + \|e^{itH} X_n e^{-itH} P_{ac} u - e^{isH} X_n e^{-isH} P_{ac} u\| \\ &\leq 2p_s(X - X_n; H, u) + \|e^{itH} X_n e^{-itH} P_{ac} u - e^{isH} X_n e^{-isH} P_{ac} u\|. \end{aligned}$$

The right side tends to $2p_s(X - X_n; H, u)$ as $s, t \to \infty$ for each $n$. Since $n \in \mathbf{N}$ is arbitrary and $p_s(X - X_n; H, u) \to 0$ as $n \to \infty$, we get that the strong limit of $e^{itH} X e^{-itH} P_{ac} u$ exists. Thus $X \in$ dom $\mu_s^H$.

Further, for $t > 0$, we have that

$$\|e^{itH}(X - X_n) e^{-itH} P_{ac} u\| \leq \sup_{t \geq 0} \|(X - X_n) e^{-itH} P_{ac} u\| \leq p_s(X - X_n; H, u).$$

By our assumptions on $X - X_n$, this implies

$$\lim_{n\to\infty} \lim_{t\to\infty} \|e^{itH}(X - X_n) e^{-itH} P_{ac} u\| = 0$$

and therefore $\operatorname{s-lim}_{n\to\infty} \mu_s^H(X_n) = \mu_s^H(X)$. ■

Since dom $\mu^H$ contains the set of all compact operators and dom $\mu^H$ is a proper subset of $\mathscr{L}(\mathscr{H})$, it cannot be closed with respect to the strong convergence (the strong closure of the set of compact operators is $\mathscr{L}(\mathscr{H})$). Thus this theorem together with Prop. 56 says that the $\mu^H$-topology is properly stronger than the strong operator topology.

An easy consequence of Theorem 57 is

**Corollary 58.** *The wave ideal* ker $\mu^H$ *is closed with respect to the* $\mu^H$*-convergence; that is if* $X_n \in$ ker $\mu^H$ *and* $X_n$ *is* $\mu^H$*-convergent to* $X$*, then* $X \in$ ker $\mu^H$.

### 6.4.2. A dense subset in dom $\mu^H$

It is often desirable to have a subset of an algebra $\mathscr{A}$ which is dense in $\mathscr{A}$ and has a simpler structure. Then one can extend certain properties of elements of this subset to their limits. Such a subset is given by the next statement.

**Theorem 59.** *The wave ideal* ker $\mu^H$ *is the closure of the set of finite-dimensional operators with respect to the* $\mu^H$*-convergence. This means that for every operator* $X \in$ ker $\mu^H$ *there is a sequence of finite-dimensional operators* $X_n$ *such that* $X_n$ *is* $\mu^H$*-convergent to* $X$.

**Proof. 1.** First we note that it is sufficient to consider the convergence of the sequence $X_n$ on the subspace $P^{ac}\mathscr{H}$ because the statement is well-known on $(1 - P^{ac})\mathscr{H}$ ($\mu^H$-convergence coincides with the strong convergence on this subspace).

Now let $P_n$ be a sequence of projections with $\operatorname{s-lim}_{n\to\infty} P_n = 1$ and $\dim P_n\mathscr{H} = n$. Let $u \in P^{ac}\mathscr{H}$. Then for each finite-dimensional operator $Q$, we have $\lim_{n\to\infty} \|P_nQ - Q\| = 0$. Now $e^{-itH}u$ is strongly continuous in $t \in \mathbb{R}$, and thus, for each $\varepsilon > 0$ and finite interval $[0, T]$, we can find a finite-dimensional projection $Q_{T,\varepsilon}$ such that

$$\|Q_{T,\varepsilon}\, e^{-itH}u - e^{-itH}u\| \leq \varepsilon \quad \text{for all} \quad t \in [0, T].$$

Therefore

$$\begin{aligned}\|(1 - P_n)\, X\, e^{-itH}u\| &\leq \|(1 - P_n)\, XQ_{T,\varepsilon}\, e^{-itH}u\| + \|(1 - P_n)\, X(1 - Q_{T,\varepsilon})\, e^{-itH}u\| \\ &\leq \|(1 - P_n)\, XQ_{T,\varepsilon}\|\, \|u\| + \varepsilon\|X\| \quad \text{for all} \quad t \in [0, T],\end{aligned}$$

where the first term on the right side tends to zero as $n \to \infty$ since $XQ_{T,\varepsilon} \in \mathscr{L}_{\text{fin}}(\mathscr{H})$. This implies that, for a given $\eta > 0$,

$$\|(1 - P_n)\, X\, e^{-itH}u\| \leq \eta \quad \text{for all} \quad t \in [0, T] \quad \text{and} \quad n > N(\eta). \tag{14}$$

2. Next we consider the case $X \in \ker \mu_s^H$. Then there is for each $\varepsilon > 0$ a number $T > 0$ such that $\|X\, e^{-itH}u\| \leq \varepsilon$ for all $t > T$. This together with (14) leads to

$$\sup_{t\geq 0} \|(1 - P_n)\, X\, e^{-itH}u\| \leq \varepsilon \quad \text{for} \quad n > N(\varepsilon),$$

and this means $p_s\big((1 - P_n)\, X; H, u\big) \to 0$ as $n \to \infty$. Hence $X_n = P_nX$ is such a sequence of finite-dimensional operators which converges to $X$ because $P_n$ does not depend on the vector $u$.

3. Let $X \in \ker \mu_a^H$. Then there is for each $\eta > 0$ a number $\varepsilon_0 > 0$ such that $\varepsilon \int_0^\infty e^{-\varepsilon t}\, \|X\, e^{-itH}u\|\, dt \leq \eta$ for $\varepsilon \in (0, \varepsilon_0)$. Using this together with (14), we can easily calculate that

$$\varepsilon \int_0^\infty e^{-\varepsilon t}\, \|(1 - P_n)\, X\, e^{-itH}u\|\, dt \leq \eta \quad \text{for all} \quad \varepsilon > 0 \quad \text{and} \quad n > N(\eta).$$

This implies $p_a\big((1 - P_n)\, X; H, u\big) \to 0$ as $n \to \infty$ and thus the approximation of $X$ by $P_nX$ in $\mu_a^H$-convergence. ■

**Remark 60.** Since $\operatorname{dom} \mu^H = \operatorname{ima} \mu^H \oplus \ker \mu^H$ and $\operatorname{ima} \mu^H = (\operatorname{com} H) \mid P^{ac}$, this theorem leads to a characterization of the wave algebra as the closure (with respect to the $\mu^H$-convergence) of the set $\mathscr{A} = \mathscr{A}^H \oplus \mathscr{L}_{\text{fin}}(\mathscr{H})$ where $\mathscr{A}^H$ is strongly dense in $\operatorname{ima} \mu^H$.

### 6.4.3. Compatibility of the algebraic properties with the $\mu^H$-topology

By definition the $\mu^H$-topology makes $\operatorname{dom} \mu^H$ to a locally convex space. Now the question is whether or not this topological structure is compatible with the algebraic structure of $\operatorname{dom} \mu^H$, in particular whether the product $XY$ is continuous with respect to the $\mu^H$-topology. Clearly, $XY$ is continuous in $Y$ for fixed $X$, this follows from (11).

Unfortunately, $XY$ is not continuous in $X$ for each $Y$ from dom $\mu^H$, and thus dom $\mu^H$ is not a topological algebra with respect to the $\mu^H$-topology. This is the content of

**Proposition 61.** *There exists an operator $Y \in$ dom $\mu^H$ such that the map $T_Y$: dom $\mu^H \to$ dom $\mu^H$ given by $T_Y(x) = XY$ is not continuous with respect to the $\mu^H$-topology.*

Proof. 1. To prove this statement, we use that $T_Y$ (as a map from a locally convex space $\mathfrak{X}$ into $\mathfrak{X}$) is continuous iff for each seminorm $p(\cdot\,; H, u)$ there is a finite set of seminorms $p(\cdot\,; H, v_1), p(\cdot\,; H, v_2) \ldots, p(\cdot\,; H, v_s)$ and a number $c > 0$ such that

$$p\big(T_Y(X); H, u\big) \leqq c \sum_{j=1}^{s} p(X; H, v_j) \quad \text{for all} \quad X \in \text{dom}\, \mu^H. \tag{15}$$

We shall construct a special operator $Y \in$ dom $\mu^H$ such that (15) is not true. For simplicity we restrict ourselves to the special case in which $\mathscr{H} = L^2([0, 2\pi], dx)$, and $H$ is multiplication by $x$. The extension to arbitrary $H$ on $\mathscr{H}$ can be done in a similar way as in the proofs of Theorem 50 and Prop. 53.

2. First we give the proof under the assumption that for some $u_0 \in \mathscr{H}$ there exists a sequence of finite-dimensional projections $Q_n$ such that $Q_nQ_{n'} = Q_{n'}Q_n = 0$ for $n \neq n'$ and $p(Q_n; H, u_0) \geqq c > 0$ where $c$ is independent of $n$. This assumption will be proved later. Now we define a compact operator $C$ by

$$C = \sum_{n=1}^{\infty} \lambda_n Q_n, \qquad |\lambda_n| \to 0 \text{ as } n \to \infty \text{ and } \sqrt{n}\,|\lambda_n| \to \infty \text{ as } n \to \infty.$$

Further, let $P_n$ be a sequence of projections satisfying: $P_nP_{n'} = P_{n'}P_n = 0$ for $n \neq n'$, $P_n \in$ com $H$, and $P_n \neq 0$. Clearly, such a sequence exists. Further, let $W$ be a partial isometry with $WQ_n = P_nW$ for all $n$. Then we define $Y = WC$. By definition we get

$$p(P_nY; H, u_0) = p(\lambda_nQ_n; H, u_0) \geqq |\lambda_n|\, c. \tag{16}$$

On the other hand, when $\{v_j\}_{j=1}^s$ is any finite set of vectors from $\mathscr{H}$ with $\|v_j\| = 1$, then $\sum_{n=1}^{\infty} \|P_nv_j\|^2 \leqq \|v_j\|^2 = 1$ and thus $\|P_{n_p}v_j\| \leqq c_j\, |n_p|^{-1/2}$ (for a subsequence $\{n_p\}_1^{\infty}$) with a number $c_j$ depending only on $v_j$. Hence we obtain the estimate (set $a := \sum_{j=1}^{s} c_j$)

$$\sum_{j=1}^{s} p(P_{n_p}; H, v_j) = \sum_{j=1}^{s} \|P_{n_p}v_j\| \leqq a\, |n_p|^{-1/2}.$$

This inequality and (16 )show that (15) cannot be true for all $X = P_{n_p}$ and any fixed set $\{v_j\}_{j=1}^s$ because $\sqrt{n}\,|\lambda_n| \to \infty$ as $n \to \infty$. Therefore the special map $T_Y$ is not continuous.

3. It remains to verify the existence of such projections $Q_n$. First, consider the case $\mu^H = \mu_s^H$. Set $Q_n := u_n(u_n, \cdot)$ with $u_n(x) = (2\pi)^{-1/2}\, e^{-inx}$. Then, by $e^{-inH}u_0 = u_n$, we obtain

$$p_s(Q_n; H, u_0) = \sup_{t \geqq 0} \|Q_n\, e^{-itH}u_0\| \geqq \|Q_n\, e^{-inH}u_0\| = \|Q_nu_n\| = 1.$$

Thus this sequence of projections $Q_n$ satisfies the required assumption for the vector $u_0$.

Now consider the case $\mu^H = \mu_a{}^H$. Define $Q_n = \sum_{s=2^{n-1}+1}^{2^n} u_s(u_s, \cdot)$. Then $2^{-n} \sum_{s=1}^{2^n} \|Q_n u_s\| = 1/2$. Set $\varepsilon_n = 2^{-n}$. Then

$$p_a(Q_n; H, u_0) = \sup_{\varepsilon>0} \varepsilon \int_0^\infty e^{-\varepsilon t} \|Q_n e^{-itH} u_0\| dt$$

$$\geq \varepsilon_n \int_0^{\varepsilon_n^{-1}} e^{-\varepsilon_n t} \|Q_n e^{-itH} u_0\| dt \geq e^{-1} \varepsilon_n \int_0^{\varepsilon_n^{-1}} \|Q_n e^{-itH} u_0\| dt$$

$$= e^{-1} 2^{-n} \sum_{s=1}^{2^n} \int_0^1 \|Q_n e^{-isH} u_t\| dt$$

with $u_t = e^{itH} u_0$. Using this and the elementary estimate $\|u_t - u_0\| \leq ct$, we get for all sufficiently small $\eta > 0$,

$$p_a(Q_n; H, u_0) \geq e^{-1} 2^{-n} \sum_{s=1}^{2^n} \int_0^\eta dt \, \big| \|Q_n e^{-isH} u_0\| - \|Q_n(e^{-isH}(u_t - u_0)\| \big|$$

$$\geq \left| e^{-1} 2^{-n} \sum_{s=1}^{2^n} \left\{ \|Q_n u_s\| \eta - \int_0^\eta \|Q_n e^{-isH}(u_t - u_0)\| dt \right\} \right|$$

$$= e^{-1} \left| 2^{-1}\eta - 2^{-n} \sum_{s=1}^{2^n} \int_0^\eta \|Q_n e^{-isH}(u_t - u_0)\| dt \right| \geq a\eta > 0$$

with some number $a > 0$ independent of $n$. Here it was used that

$$\int_0^\eta \|Q_n e^{-isH}(u_t - u_0)\| dt \leq \int_0^\eta \|u_t - u_0\| dt \leq \int_0^\eta ct \, dt = (c/2) \eta^2.$$

This concludes the proof of the existence of such projections $Q_n$. ∎

Instead of having the continuity of $XY$ in the left factor, one has a weaker statement.

**Proposition 62.** *The product $XY$ is sequentielly continuous in $X$ for each fixed $Y$ from* dom $\mu^H$ *with respect to the $\mu^H$-topology; that is, if $X_n \in$ dom $\mu^H$ is $\mu^H$-convergent to $X$, then $X_n Y$ is $\mu^H$-convergent to $XY$ for each $Y \in$ dom $\mu^H$.*

Proof. Suppose $X_n$ is $\mu^H$-convergent to $X$ and $Y \in$ dom $\mu^H$. Then $Y = Y_1 + Y_2$ with $Y_1 \in$ ima $\mu^H$ and $Y_2 \in$ ker $\mu^H$. Thus we get

$$p\big((X - X_n) Y; H, u\big) \leq p\big((X - X_n) Y_1; H, u\big) + p\big((X - X_n) Y_2; H; u\big)$$

$$= p\big((X - X_n); H, Y_1 u\big) + p\big((X - X_n) Y_2; H, u\big).$$

The first term on the right side tends to zero as $n \to \infty$ by assumption. Therefore we have to consider only the second term. By Theorem 59, $Y_2$ can be approximated by finite-dimensional operators $Y_{2,m}$. Then

$$\begin{aligned} &p((X - X_n) Y_2; H, u) \\ &\leqq p((X - X_n)(Y_2 - Y_{2,m}); H, u) + p((X - X_n) Y_{2,m}; H, u) \\ &\leqq \|X - X_n\| \, p(Y_2 - Y_{2,m}; H, u) + p((X - X_n) Y_{2,m}; H, u). \end{aligned} \tag{17}$$

According to Prop. 56, $\mu^H$-convergence implies strong convergence, and therefore, by the principle of uniform boundedness, $\|X - X_n\| \leqq c < \infty$ with $c$ independent of $n$. Thus the first term on the right side of (17) tends to zero as $m \to \infty$ uniformly in $n$. The second term obeys the estimate $p((X - X_n) Y_{2,m}; H, u) \leqq \|(X - X_n) Y_{2,m}\| \times p(1; H, u)$, and it goes to zero as $n \to \infty$ for each $m$ because $X_n \xrightarrow[s]{} X$ and $Y_{2,m}$ is finite-dimensional. These considerations show that the right side of (17) is smaller than every $\varepsilon > 0$ for all $n > N(\varepsilon)$ and a sufficiently large $m$. This proves $p((X - X_n) Y_2; H, u) \to 0$ as $n \to \infty$. ∎

From the proof of this statement we see that the product $XY$ is continuous on uniformly bounded subsets of dom $\mu^H$.

## 6.5. Generalizations

In the preceding paragraphs the study of the asymptotic limits of $e^{itH} X e^{-itH}$ was restricted to the absolutely continuous subspace of $H$. Now we want to make some remarks to this choice and possible generalizations.

### 6.5.1. Universal spectral projections

Let us fix some property which determines a map $\mathcal{M}$ from the class of selfadjoint operators $H$ into the class of projections $P_H$ with the following properties:

(I) $P_H$ is a spectral projection, that is, $P_H \in$ com com $H$, thus $P_H = E_H(\Delta_H)$ with some Borel set $\Delta_H$.

(II) If $U$ is isometric from $\mathcal{H}$ onto another Hilbert space $\mathcal{H}'$ and $H' = UHU^*$, then $P_{H'} = UP_HU^*$.

(III) If $H' = H \mid Q\mathcal{H}$ where $Q$ is a reducing projection for $H$, then $P_{H'} = P_H \mid Q\mathcal{H}$.

The spectral property which determines the map $H \xrightarrow{\mathcal{M}} P_H$ is called *universal spectral property* and the projections $P_H$ given by the fixed map $\mathcal{M}$ are called *universal spectral projections*. Note that in general $\Delta_H$ depends on $H$ as we will see later.

In what follows we shall often use the phrase "let $P_A$ be a fixed universal (spectral) projection" instead of saying that we fix a map from the set of selfadjoint operators $A$ into the set of projections $P_A$ determined by a universal spectral property.

The properties (II) and (III) are compatible. We have

**Lemma 63.** *If $H$ is unitarily equivalent to $H \mid Q\mathcal{H}$, that is, $HQ = UHU^*$ with $U^*U = 1$, $UU^* = Q$ and $U \in$ com $H$, then $P_{H \mid Q\mathcal{H}} = P_H \mid Q\mathcal{H} = UP_HU^* \mid Q\mathcal{H}$.*

Proof. Because of $U \in$ com $H$ we obtain $P_HU = UP_H$. Then we also have $P_HUU^* = UP_HU^*$, hence $P_HQ = QP_H = UP_HU^*$. This, (II) and (III) imply the statement. ∎

The next result expresses a certain feature of the projection $P_H$ which is immediately connected with the unitary invariance of the corresponding spectral property.

**Proposition 64.** *Let $H_1$ be selfadjoint on $\mathscr{H}_1$. Further, let $Q_1$ be a reducing projection for $H_1$ and let $Q_0$ be a reducing projection for $H$. Assume $Q_0 \leqq P_H$ and $H_1 \mid Q_1\mathscr{H}_1$ is unitarily equivalent to $H \mid Q_0\mathscr{H}$. Then $Q_1 \leqq P_{H_1}$.*

**Proof.** We have $P_{H|Q_0\mathscr{H}} = P_H \mid Q_0\mathscr{H} = 1_{Q_0\mathscr{H}}$. Further, there is a partial isometry $U: \mathscr{H} \to \mathscr{H}_1$ with $U^*U = Q_0$, $UU^* = Q_1$ and $H_1 \mid Q_1\mathscr{H}_1 = U(H \mid Q_0\mathscr{H})\, U^* \mid Q_1\mathscr{H}_1$. Hence we obtain

$$P_{H_1|Q_1\mathscr{H}_1} = (UP_{H|Q_0\mathscr{H}}U^*) \mid Q_1\mathscr{H}_1 = U1_{Q_0\mathscr{H}}U^* \mid Q_1\mathscr{H}_1$$
$$= UU^* \mid Q_1\mathscr{H}_1 = Q_1 \mid Q_1\mathscr{H}_1 = 1_{Q_1\mathscr{H}_1}.$$

But $P_{H_1|Q_1\mathscr{H}_1} = P_{H_1} \mid Q_1\mathscr{H}_1$, hence $P_{H_1} \mid Q_1\mathscr{H}_1 = Q_1 \mid Q_1\mathscr{H}_1$. In every case we have $P_{H_1}Q_1 = Q_1P_{H_1} \leqq Q_1$. Thus we obtain $P_{H_1}Q_1 = Q_1P_{H_1} = Q_1$ and $Q_1 \leqq P_{H_1}$. ∎

Simple examples of universal spectral projections are $P_H^{ac}$, $P_H^{c}$, $P_H^{p}$, $E_H(\Delta)$ (where $\Delta$ is fixed), and $E_{H,1} + E_{H,2} + \cdots + E_{H,m}$ where $E_{H,m}$ denotes the spectral projection of (homogeneous) multiplicity $m$. To obtain other examples, we introduce for each $H$ a system $\mathscr{A}_H$ of bounded operators with the properties:

(I) Each operator $A \in \mathscr{A}_H$ is bounded, $A \in \mathscr{L}(\mathscr{H}, \mathscr{H}_A)$ where the Hilbert space $\mathscr{H}_A$ may depend on $A$.

(II) $\mathscr{A}_H$ has the left ideal property, i.e. $A \in \mathscr{A}_H$ and $B \in \mathscr{L}(\mathscr{H}_A, \mathscr{H}_B)$ imply $BA \in \mathscr{A}_H$

(III) If $A \in \mathscr{A}_H$ and $K \in \operatorname{com} H$, then $AK \in \mathscr{A}_H$.

(IV) If $H' = UHU^*$, $U: \mathscr{H} \to \mathscr{H}'$ is isometric, then $\mathscr{A}_{H'} = \mathscr{A}_H U^*$

(V) If $H' = H \mid Q\mathscr{H}$, $Q$ a reducing projection for $H$, then $\mathscr{A}_{H'} = \mathscr{A}_H \mid Q\mathscr{H}$.

The properties (IV) and (V) are compatible: Let $U \in \operatorname{com} H$ and $U^*U = 1$, $UU^* = Q$, $A \in \mathscr{A}_H$. Then $AQ = AUU^*$ and $AU \in \mathscr{A}_H$. On the other hand, $AU^* = AU^*Q$.

The simplest example is obtained when we take $\mathscr{A}_H = \bigcup_{\mathscr{H}'} \mathscr{L}_\infty(\mathscr{H}, \mathscr{H}')$ i.e. $\mathscr{A}_H$ is chosen as the set of all compact operators with arbitrary image space. The properties (I)—(V) can be easily verified.

**Definition 65.** Let $\mathscr{A}_H$ be given. Then we denote by $P_{\mathscr{A}_H,s}$ and $P_{\mathscr{A}_H,a}$ the projections onto the subspaces of all $u \in \mathscr{H}$ with

$$\lim_{t\to\infty} \|A\, e^{-itH}u\| = 0 \quad \text{for all } A \in \mathscr{A}_H \tag{18}$$

and

$$\lim_{T\to\infty} T^{-1} \int_0^T \|A\, e^{-itH}\, u\|^2\, dt = 0 \quad \text{for all } A \in \mathscr{A}_H, \tag{19}$$

respectively.

**Proposition 66.** *The projections $P_{\mathscr{A}_H,s}$ and $P_{\mathscr{A}_H,a}$ are universal spectral projections.*

**Proof.** (I) For convenience we write $P_{\mathscr{A}_H}$ for $P_{\mathscr{A}_H,s}$ or $P_{\mathscr{A}_H,a}$. It is sufficient to prove that $P_{\mathscr{A}_H}$ commute with all reducing projections for $H$. Let $Q$ be a reducing projection for $H$ and $f = P_{\mathscr{A}_H}f$. Then $\|A\, e^{-itH}\, Qf\| = \|AQ\, e^{-itH}f\|$. Hence, by (III) of $\mathscr{A}_H$, $P_{\mathscr{A}_H}Qf = Qf$

and $P_{\mathscr{A}_H} QP_{\mathscr{A}_H} = QP_{\mathscr{A}_H} = P_{\mathscr{A}_H} Q$ follow. (II) Let $A' \in \mathscr{A}_{H'}$. Then, by (IV) of $\mathscr{A}_H$ we have $A' = AU^*$ with $A \in \mathscr{A}_H$. Hence we obtain

$$\|A' \, e^{-itH'} f'\| = \|AU^*U \, e^{-itH} U^* f'\| = \|A \, e^{-itH} U^* f'\|,$$

that is, $f' \in P_{\mathscr{A}_{H'}} \mathscr{H}'$ iff $U^*f' \in P_{\mathscr{A}_H} \mathscr{H}$ or $P_{\mathscr{A}_{H'}} f' = f'$ iff $P_{\mathscr{A}_H} U^* f' = U^* f'$ or $P_{\mathscr{A}_{H'}} f' = f'$ iff $UP_{\mathscr{A}_H} U^* f' = f'$. Hence, $P_{\mathscr{A}_{H'}} = UP_{\mathscr{A}_H} U^*$. (III) We have $\|A' \, e^{-itH'} f'\|_{Q\mathscr{H}} = \|AQ \, e^{-itH} f'\|$. Thus $f' = P_{\mathscr{A}_{H'}} f'$ iff $f' = P_{\mathscr{A}_H} Qf$, that is, $P_{\mathscr{A}_{H'}} = P_{\mathscr{A}_H} \restriction Q\mathscr{H}$. ■

**Example 67.** Let $\mathscr{A}_H = \bigcup_{\mathscr{H}'} \mathscr{L}_\infty(\mathscr{H}, \mathscr{H}')$ and $P_H = P_{\mathscr{A}_H, a}$. In this case we obtain that the universal spectral projection $P_H = P_{\mathscr{A}_H, a}$ is equal to $P_H^c$. Namely, if $f = P_H^c f$, then $A \in \mathscr{L}_\infty(\mathscr{H}, \mathscr{H}')$ implies $\lim_{T \to \infty} T^{-1} \int_0^T \|A \, e^{-itH} f\|^2 \, dt = 0$. This follows easily from Prop. 3.10 by similar arguments as n the proof of Theorem 35. However, when $f \in P_H^p \mathscr{H} = (1 - P_H^c) \mathscr{H}$, i.e. $f = \sum c_\varrho u_\varrho$ where $u_\varrho$ are eigenvectors of $H$, then take the compact operator $A = u_s(u_{s}, \cdot)$ with $c_s \neq 0$. Thus $\|A \, e^{-itH} f\| = |c_s| > 0$ as a simple calculation shows.

**Example 68.** $\mathscr{A}_H = \bigcup_{\mathscr{H}'} \mathscr{L}_\infty(\mathscr{H}, \mathscr{H}')$ and $P_H = P_{\mathscr{A}_H, s}$. In this case we get $P_H \geq P_H^{ac}$ since $\lim_{t \to \infty} \|A \, e^{-itH} f\| = 0$ for all $f \in P_H^{ac} \mathscr{H}$ and all compact operators $A$ (see Theorem 35). On the other hand, the same reasoning as in Example 67 proves that $P_H \leq P_H^c$. Thus, $P_H^{ac} \leq P_H = P_{\mathscr{A}_H, s} \leq P_H^c$. Often $P_H^{ac} = P_H^c$ such that $P_H = P_H^{ac} = P_H^c$. But there exist operators $H$ such that $P_H^{ac} < P_H^c$ and $P_H$ is different from both projections. For instance, there are operators $H$ (even Schrödinger operators, see the Notes and Remarks) such that $P_H^{sc} \neq 0$ and there is a vector $u \in P_H^{sc} \mathscr{H}$ with $\lim_{t \to \infty} (u, e^{-itH} u) = 0$. This implies by a simple calculation that for the projection $P_u$ onto clo spa $\{e^{-itH} u; t \in \mathbb{R}\}$, we have $\text{w-lim}_{t \to \infty} e^{-itH} P_u = 0$. Thus we get $\text{s-lim}_{t \to \infty} C \, e^{-itH} P_u = 0$ for all $C \in \mathscr{A}_H$ (the same proof as for Theorem 35), and this implies $P_H = P_{\mathscr{A}_H, s} \geq P_H^{ac} + P_u > P_H^{ac}$ for these special operators $H$. On the other hand, there are operators $H$ with $P_H^{sc} \neq 0$ and a vector $u \in P_H^{sc} \mathscr{H}$ such that $\lim_{t \to \infty} (u, e^{-itH} u)$ does not exist. Thus, this implies by similar arguments as above that $P_H = P_{\mathscr{A}_H, s} < P_H^c$ for these operators $H$. From these two examples of operators $H$ we see that the universal spectral projections determined by (18) via the set $\bigcup_{\mathscr{H}'} \mathscr{L}_\infty(\mathscr{H}, \mathscr{H}')$ may considerably vary, in dependence of $H$, between $P_H^c$ and $P_H^{ac}$. Thus $P_{\mathscr{A}_H, s}$ leads to universal spectral projections different from the ones given after Prop. 64.

### 6.5.2. Wave algebras with universal spectral projections

Checking the proofs of Theorem 23 and Theorem 25, on finds that they remain true if one replaces the projection $P_H^{ac}$ in Definition 22 by a projection $E_H \in \text{com com } H$, i.e. $E_H = E_H(\Delta_H)$ with some Borel set $\Delta_H$. Further, if $H'$ is unitarily equivalent to $H$, i.e. $H' = UHU^*$ with $U^*U = 1$ and $UU^* = 1$, then it is reasonable to require that the corresponding wave algebras are also unitarily equivalent, i.e. dom $\mu^{H'} = U(\text{dom } \mu^H) \, U^*$. Thus, if one wants to generalize, the notion of wave algebra with the help of another projection $E_H$ determined by any universal property, one comes necessarily to the concept of the universal spectral projection (the third property of $P_H$ can also be motivated in the framework of wave algebras). This means that a reasonable generalization of Definition 22 is given if we replace $P_H^{ac}$ by a universal spectral projection $P_H$. Clearly, the important fact that the set of compact operators belongs to the wave ideal (see Theorem 35) and the important characterization of $\ker \mu^H$ by compact (even finite-dimensional) operators and by the corresponding topology (see Theorem 59) remain true only under additional assumptions on $P_H$. For instance, in case of the |A|-lim this remains true when $P_H \leqq P_H^c$ (see Example 67). In case of the s-lim this remains true when $P_H \leqq P_H^{ac}$ (for $P_H > P_H^{ac}$ there are counter-examples, see Example 68). In particular we note that the topological characterization of the wave algebra $\operatorname{dom} \mu_a^H$ is also true when $P_H^{ac}$ is replaced by some universal spectral projection $P_H \leqq P_H^c$.

However it should be noticed that in almost all practical interesting cases $P_H^c = P_H^{ac}$ and generalizations of the type $P_H^{ac} \to P_H \leqq P_H^c$ are trivial. Moreover, most of the results in the following chapters use essentially the fact that we restrict ourselves to vectors from the absolutely continuous subspace of $H$.

In the remainder of this section we introduce some useful notations.

**Definition 69.** Let $P_H$ be a universal spectral projection. Then $\mathscr{L}_{H,P_H}(\mathscr{H}, \mathscr{H}')$ denotes the set of all bounded operators $X$ from $\mathscr{H}$ to the Hilbert space $\mathscr{H}'$ which satisfy $\operatorname{Lim}_{t\to\infty} X \mathrm{e}^{-itH} P_H = 0$. We use the notations $\mathscr{L}_{H,P_H}(\mathscr{H}, \mathscr{H}')_s$ and $\mathscr{L}_{H,P_H}(\mathscr{H}, \mathscr{H}')_a$ if Lim = s-lim and Lim = |A|-lim, respectively.

**Proposition 70.** (i) *$\mathscr{L}_{H,P_H}(\mathscr{H}, \mathscr{H}')$ is a linear space closed in operator norm.*

(ii) *If $X \in \mathscr{L}_{H,P_H}(\mathscr{H}, \mathscr{H}')$ and $A \in \mathscr{L}(\mathscr{H}', \mathscr{H}'')$, then $AX \in \mathscr{L}_{H,P_H}(\mathscr{H}, \mathscr{H}'')$.*

(iii) *If $P_H \leqq P_H^c$, then $\mathscr{L}_\infty(\mathscr{H}, \mathscr{H}') \subset \mathscr{L}_{H,P_H}(\mathscr{H}, \mathscr{H}')_a$*

(iv) *If $P_H \leqq P_H^{ac}$, then $\mathscr{L}_\infty(\mathscr{H}, \mathscr{H}') \subset \mathscr{L}_{H,P_H}(\mathscr{H}, \mathscr{H}')_s$.*

Proof. (i) and (ii) are easy consequences of Definition 69. (iii) follows from Example 67 and the definition of |A|-lim. (iv) was proved in Example 68. ■

Because of $\mathscr{L}_{H,P_H^{ac}}(\mathscr{H}, \mathscr{H}) = \ker \mu^H$ and (ii) of Prop. 70, we see that in case $P_H = P_H^{ac}$ the study of $\mathscr{L}_{H,P_H^{ac}}(\mathscr{H}, \mathscr{H}')$ reduces to the study of the wave ideal $\ker \mu^H$.

Special cases of interest are when an operator is asymptotically isometric or equivalent to another operator.

**Definition 71.** Let $X \in \mathscr{L}(\mathscr{H}, \mathscr{H}')$. Suppose there is a partial isometry $W$ such that $X - W \in \mathscr{L}_{H,P_H}(\mathscr{H}, \mathscr{H}')$. Then $X$ is called *asymptotically partially isometric* with respect to $HP_H$ (in case $\mathscr{L}_{H,P_H}(\mathscr{H}, \mathscr{H}')_s$ we say *strongly asymptotically partially isometric*).

**Definition 72.** Let $X, X' \in \mathscr{L}(\mathscr{H}, \mathscr{H}')$. Assume $X - X' \in \mathscr{L}_{H,P_H}(\mathscr{H}, \mathscr{H}')$. Then we say $X$ is *asymptotically equivalent to* $X'$ with respect to $HP_H$. (In case $\mathscr{L}_{H,P_H}(\mathscr{H}, \mathscr{H}')_s$ we say *strongly asymptotically equivalent*). Sometimes we use the short notation $X \underset{H}{\approx} X'$ or $X - X' \in \mathscr{L}_{H,P_H}(\mathscr{H}, \mathscr{H}')$ if $P_H$ is fixed.

# 7. Special Classes of Asymptotic Constants

## 7.1. Projections

### 7.1.1. Spectral projections of selfadjoint asymptotic constants

Let $A \in \operatorname{dom} \mu^H$ be selfadjoint. Then it is a natural question to ask which spectral projections of $A$ belong to $\operatorname{dom} \mu^H$. The next statement gives a partial answer.

**Proposition 1.** *Let $A \in \operatorname{dom} \mu^H$ be selfadjoint and set $A_0 = \mu^H(A)$. Suppose $\Delta$ is an open neighborhood of a Borel set $\Delta_0$ and suppose* $\operatorname{dis}\left(\operatorname{clo} \Delta, (\operatorname{spec} A_0) \setminus \Delta_0\right) > 0$. *Then $E_A(\operatorname{clo} \Delta) \in \operatorname{dom} \mu^H$ and $\mu^H\left(E_A(\operatorname{clo} \Delta)\right) = E_{A_0}(\operatorname{clo} \Delta) P^{ac}$.*

**Proof.** Let $f \in C(\mathbb{R})$ with $f(x) \geqq 0$, $f(x) = 1$ on $\operatorname{clo} \Delta$ and $f(x) = 0$ on $\operatorname{spec} A_0 \setminus \Delta_0$. Then (see Theorem 6.28)

$$\mu^H\left(f(A)\right) = f(A_0) P^{ac} = E_{A_0}(\operatorname{clo} \Delta) P^{ac}.$$

Further, let $g \in C(\mathbb{R})$ with $g(x) \geqq 0$, $\operatorname{supp} g \cap \operatorname{spec} A_0 = \emptyset$ and $0 \leqq f(x) - \chi_{\operatorname{clo} \Delta}(x) \leqq g(x)$. Clearly, such a function $g$ exists. Then $\mu^H\left(g(A)\right) = g(A_0) P^{ac} = 0$, i.e. $g(A) \in \ker \mu^H$. Thus, by $f(A) - E_A(\operatorname{clo} \Delta) \leqq g(A)$, $f(A) - E_A(\operatorname{clo} \Delta) \in \ker \mu^H$. Since $f(A) \in \operatorname{dom} \mu^H$, this implies $E_A(\operatorname{clo} \Delta) \in \operatorname{dom} \mu^H$ and

$$\mu^H\left(E_A(\operatorname{clo} \Delta)\right) = \mu^H\left(f(A)\right) = E_{A_0}(\operatorname{clo} \Delta) P^{ac}. \blacksquare$$

For example, if $\Delta_0$ is an isolated closed part of $\operatorname{spec} \mu^H(A)$, then there is always an open neighborhood $\Delta$ of $\Delta_0$ such that the assumptions of Prop. 1 are satisfied. Note that in this case $\Delta_0$ is not necessarily also an isolated part of $\operatorname{spec} A$ (see (6.5)), that is, some parts of $\operatorname{spec} A$ may disappear under $\mu^H$. If this happens, then certain spectral projections of $A$ necessarily belong to $\ker \mu^H$, as is shown in the following corollary.

**Corollary 2.** *Let $A \in \operatorname{dom} \mu^H$ be selfadjoint. Assume $\lambda \in \operatorname{spec} A$ but $\lambda \notin \operatorname{spec} \mu^H(A)$. Further, assume $\lambda \in [a, b] \subset \operatorname{res} \mu^H(A)$. Then $E_A([a, b]) \in \ker \mu^H$.*

**Proof.** Set $\Delta = (a, b)$ and $\Delta_0 = \{\lambda\}$. In this case $\lambda \notin \operatorname{spec} \mu^H(A)$ and

$$\operatorname{dis} (\operatorname{clo} \Delta, \operatorname{spec} \mu^H(A) \setminus \{\lambda\}) = \operatorname{dis} \left([a, b], \operatorname{spec} \mu^H(A)\right) > 0$$

since $[a, b] \subset \operatorname{res} \mu^H(A)$. Applying Prop. 1, we get $E_A([a, b]) \in \operatorname{dom} \mu^H$ and $\mu^H\left(E_A([a, b])\right) = E_{\mu^H(A)}([a, b]) P^{ac} = 0$. $\blacksquare$

In particular, if $A$ is selfadjoint and $A \in \ker \mu^H$, then Corollary 2 implies that $E_A(\Delta) \in \ker \mu^H$ for each closed set $\Delta$ with $\Delta \cap \{0\} = \emptyset$.

Later we shall see that there are asymptotic constants $X$ (selfadjoint) and Borel sets $\Delta$ such that $E_X(\Delta) \notin \operatorname{dom} \mu^H$.

### 7.1.2. Projections $P$ with $\mu^H(P) \neq \mu^{-H}(P)$

In scattering theory one is interested in operators $X$ for which $X(t) = e^{itH} X e^{-itH} P^{ac}$ tends to different limits when $t \to \infty$ and $t \to -\infty$. Since the asymptotics for large negative values of $t$ corresponds to the wave morphism $\mu^{-H}$, this means that we have to look for operators $X \in \operatorname{dom} \mu^H \cap \operatorname{dom} \mu^{-H}$ with $\mu^H(X) \neq \mu^{-H}(X)$. In this section we consider projections $X$ with these and other properties.

First we need a technical lemma.

**Lemma 3.** *Let $P \in \operatorname{dom} \mu^H \cap \operatorname{dom} \mu^{-H}$ be a projection with $\mu^H(P) = P^{ac}$ and $\mu^{-H}(P) = 0$. Let $Q \in \operatorname{com} H$ be a projection with $Q \leqq P^{ac}$. Then there is a projection $F \leqq Q$ with $F \in \operatorname{dom} \mu^H \cap \operatorname{dom} \mu^{-H}$, $\mu^H(F) = Q$ and $\mu^{-H}(F) = 0$.*

**Proof.** Consider $A = QPQ = \int_0^1 \lambda E_A(d\lambda)$. Then $\mu^H(A) = \mu^H(QPQ) = Q$ and $\mu^{-H}(A) = 0$. Hence, according to Prop. 1, one can choose $F = E_A([\varepsilon, 1])$ with $\varepsilon \in (0, 1)$. ∎

**Theorem 4.** *There is a projection $P_+ \in \operatorname{dom} \mu^H \cap \operatorname{dom} \mu^{-H}$ with $\mu^H(P_+) = P^{ac}$, $\mu^{-H}(P_+) = 0$ and $P_+ < P^{ac}$.*

**Proof.** We can decompose $P^{ac}\mathscr{H}$ into a direct sum of subspaces $\mathscr{H}_n$ with the properties: Each subspace $\mathscr{H}_n$ reduces $H$, $H \restriction \mathscr{H}_n$ is simple and absolutely continuous. Since it suffices to construct the projection $P_+$ in each subspace $\mathscr{H}_n$ separately, we can restrict ourselves to the special case that $P^{ac} = 1$ and $H$ is simple. By the spectral representation theorem we can further assume $\mathscr{H} = L^2(\Delta, d\lambda)$ and $H$ is multiplication by $\lambda$. According to Lemma 3, it is sufficient to assume $\Delta = \mathbb{R}$ ($L^2(\Delta, d\lambda)$ can be regarded as a subspace of $L^2(\mathbb{R}, d\lambda)$ when $\Delta \subset \mathbb{R}$). Since $H = \lambda$ on $L^2(\mathbb{R}, d\lambda)$ is unitarily equivalent to $\hat{H} = -i d/dx$ on $L^2(\mathbb{R}, dx)$ (via Fourier transformation), it is sufficient to find a projection $P_+$ on $L^2(\mathbb{R}, dx)$ with $P_+ \in \operatorname{dom} \mu^{\hat{H}} \cap \operatorname{dom} \mu^{-\hat{H}}$, $\mu^{\hat{H}}(P_+) = 1$ and $\mu^{-\hat{H}}(P_+) = 0$. We claim that $P_+ = \chi_{[0,\infty)}$, where $\chi_{[0,\infty)}$ is multiplication by $\chi_{[0,\infty)}(x)$, has these properties. Namely (see also Example 6.49),

$$(e^{it\hat{H}} P_+ e^{-it\hat{H}} u)(x) = \chi_{[0,\infty)}(x + t)\, u(x) = \chi_{[-t,\infty)}(x)\, u(x)$$

and $\chi_{[-t,\infty)}$ tends strongly to 1 as $t \to \infty$ and to 0 as $t \to -\infty$. ∎

An example for such a projection was already constructed in the course of the proof. Another example is given by: $\mathscr{H} = L^2([0, 2\pi], dx)$, $H = x$ and $P_+ = \sum_{n \geq 1} u_n(u_n, \cdot)$ with $u_n(x) = (2\pi)^{-1/2} e^{-inx}$ (see the proof of Prop. 6.53 where $P_+ = \tilde{P}_-$).

With the help of such a projection $P_+$ one can easily construct operators $P$ from $\operatorname{dom} \mu^H \cap \operatorname{dom} \mu^{-H}$ which have arbitrary operators $Q_+$, $Q_-$ from $\operatorname{ima} \mu^H$ as limits: Set $P = P_+Q_+ + (1 - P_+)\, Q_-$ then $P \in \operatorname{dom} \mu^H \cap \operatorname{dom} \mu^{-H}$ and $\mu^H(P) = Q_+$, $\mu^{-H}(P) = Q_-$ as a simple calculation shows.

### 7.1.3. Equivalence relations between projections

In the set of projections belonging to a fixed wave algebra one can introduce a special equivalence relation.

**Definition 5.** Let $P, Q \in \operatorname{dom} \mu^H$ be projections. $P$ is called *equivalent to* $Q$, denoted by $P \approx Q$, if $P - Q \in \ker \mu^H$.

It is easy to verify that $\approx$ is an equivalence relation. First we show that there is a lot of special equivalent projections.

**Proposition 6.** *Let $Q \in \operatorname{dom} \mu^H$ be a projection with $\dim Q\mathscr{H} = \infty$. Then there is a projection $P$ such that: $P < Q$, $\dim (Q - P)\mathscr{H} = \infty$, $P \in \operatorname{dom} \mu^H$ and $\mu^H(P) = \mu^H(Q)$.*

**Proof.** 1. Assume $\mu^H(Q) = 0$. Then every projection $P < Q$ belongs also to $\operatorname{dom} \mu^H$ and $\mu^H(P) = 0$. This follows from Prop. 6.43. Now $Q = \sum u_n(u_n, \cdot)$ where $\{u_n\}$ is some orthonormal system in $\mathscr{H}$. Set $P := \sum u_{2n}(u_{2n}, \cdot)$. Clearly $P < Q$ and $\dim (Q - P)\mathscr{H} = \infty$. The other properties of $P$ follow from the remarks above.

2. Assume $Q_0 = \mu^H(Q) > 0$. Then $Q_0 \leqq P^{ac}$, $Q_0 \in \operatorname{com} H$ and $\dim Q_0\mathscr{H} = \infty$. The projection $Q_0$ satisfies the relation $\operatorname{Lim}_{t\to\infty} (Q\,e^{-itH}P^{ac}f - e^{-itH}Q_0f) = 0$ for all $f \in \mathscr{H}$. By Lemma 6.17, there is a sequence $t_n \to \infty$ such that

$$\lim_{n\to\infty} \|Q\,e^{-it_nH}P^{ac}f - e^{-it_nH}Q_0f\| = 0, \qquad f \in \mathscr{H}. \tag{1}$$

Now we introduce a sequence of disjoint projections $F_s$ with

$$Q_0 = \sum_{s=1}^{\infty} F_s, \qquad F_s \in \operatorname{com} H$$

(for example $F_s = E(\Delta_s)\,Q_0E(\Delta_s)$ where $\{\Delta_s\}$ is a suitable resolution of $\mathbb{R}$). Further, let $D_s < F_s$ be finite-dimensional projections. Then from (1) we get $\lim_{n\to\infty} \|Q\,e^{-it_nH}D_s - e^{-it_nH}D_s\| = 0$ for all $s \in \mathbb{N}$. Setting $D_s(n) := e^{-it_nH}D_s\,e^{it_nH}$, we obtain

$$\lim_{n\to\infty} \|QD_s(n) - D_s(n)\| = 0, \qquad s \in \mathbb{N}.$$

Note that $D_s(n) < F_s$ for all $n$ and $\dim D_s(n)\,\mathscr{H} = \dim D_s\mathscr{H}$. Now we choose a subsequence $n_s = n_s(\varepsilon)$ such that

$$\|QD_s(n_s) - D_s(n_s)\| < \varepsilon^s, \qquad s \in \mathbb{N}, \tag{2}$$

for some $\varepsilon > 0$. Obviously, $\tilde{P} := \sum D_s(n_s)$ is a projection and $\mu^H(\tilde{P}) = 0$ (use Prop. 6.44). Further from (2) we get $K = \sum \big(QD_s(n_s) - D_s(n_s)\big)$ is compact. Hence $T := \tilde{P} + K = \sum QD_s(n_s) = Q\tilde{P}$ is noncompact and we have $T = QT$, $T \in \ker \mu^H$. From $T^* = \tilde{P} + K^*$ it follows that $T^* \in \ker \mu^H$ and thus $\mu^H(|T^*|) = 0$. $|T^*|$ is noncompact and $|T^*| = Q\,|T^*|\,Q$. Consequently, there is $\varepsilon > 0$ such that $E := E_{|T^*|}\big([\varepsilon, \infty)\big)$ is infinite-dimensional. From Prop. 1 it follows that $E \in \ker \mu^H$. By construction $E = QEQ$ and $\dim E\mathscr{H} = \infty$. This ensures that $P := Q - E$ has the desired properties. ■

**Remark 7.** From the construction of the projection $P$ one obtains a little more, namely that $E = Q - P$ is even an element of $\ker \mu^{\alpha(H)}$ for each Borel function $\alpha$ such that $P^{ac}_{\alpha(H)} = P^{ac}$ and $\alpha(H)$ is well-defined.

In the theory of $W^*$-algebras two projections $P, Q$ are called equivalent, $P \sim Q$, if there exists a partial isometry $U$ from this $W^*$-algebra such that $U^*U = P$ and $UU^* = Q$. The next statement shows that the equivalence relation $\approx$ implies the other $\sim$.

**Theorem 8.** *Let $P, Q \in \operatorname{dom} \mu^H$ be projections. Assume $P \approx Q$, i.e. $P - Q \in \ker \mu^H$. Then there is a partial isometry $W$ such that $W^*W = P$, $WW^* = Q$, $W, W^* \in \operatorname{dom} \mu^H$ and $\mu^H(W) = \mu^H(W^*) = \mu^H(P)$.*

**Proof.** By Prop. 6, there are projections $P'$, $Q'$ such that: $P' < P$ and $Q' < Q$, $\dim (P - P')\,\mathscr{H} = \dim (Q - Q')\,\mathscr{H} = \infty$ and $P - P', Q - Q' \in \ker \mu^H$. Set $A = Q'P'$

and let $A = V|A|$ be the polar decomposition of $A$. Set $E := \mu^H(Q) = \mu^H(P) = \mu^H(A) = \mu^H(A^*)$. Then

$$\mu^H(|A|^2) = \mu^H(A^*A) = \mu^H(A^*)\,\mu^H(A) = E.$$

From this and Theorem 6.28 (for $\alpha(x) = \sqrt{x}$ on $\mathbb{R}_+$ and $\alpha(x) = 0$ on $\mathbb{R}_-$) we obtain $|A| \in \operatorname{dom}\mu^H$ and $\mu^H(|A|) = E$. By $VP' = V$, $\mu^H(P') = E$ and $\mu^H(|A| - E) = 0$, we have

$$\begin{aligned} E = \mu^H(A) = \mu^H(V|A|) &= \mu^H\big(V|A| + V(E - |A|)\big) = \mu^H(VE) \\ &= \mu^H\big(VP' + V(E - P')\big) = \mu^H(VP') = \mu^H(V), \end{aligned}$$

i.e. $V \in \operatorname{dom}\mu^H$ and $\mu^H(V) = E$. In the same manner we can prove $V^* \in \operatorname{dom}\mu^H$ and $\mu^H(V^*) = E$. From the definition of $V$ we see that $V^*V = P_1 \leqq P' < P$ and $VV^* = Q_1 \leqq Q' < Q$. This implies $Q_1, P_1 \in \operatorname{dom}\mu^H$ and $\mu^H(Q_1) = \mu^H(P_1) = E$. Since $\dim (Q - Q_1)\mathscr{H} = \dim (P - P_1)\mathscr{H} = \infty$, there is a partial isometry $U$ with $U^*U = P - P_1$ and $UU^* = Q - Q_1$. By $U = U(P - P_1)$ and $U^* = U^*(Q - Q_1)$, we obtain $U, U^* \in \ker\mu^H$. Now define $W := V + U$. It is easily verified that $W$ is a partial isometry with $W^*W = P$ and $WW^* = Q$. Furthermore, from the properties of $V$ and $U$ it follows that $W, W^* \in \operatorname{dom}\mu^H$ and $\mu^H(W) = \mu^H(W^*) = E = \mu^H(P)$. ■

**Remark 9.** In the same manner one can prove that when the projections $P, Q \in \operatorname{dom}\mu^H \cap \operatorname{dom}\mu^{-H}$ with $\mu^H(P) = \mu^H(Q)$ and $\mu^{-H}(P) = \mu^{-H}(Q)$, then the partial isometry $W$ constructed in the proof of this theorem has the additional properties: $W, W^* \in \operatorname{dom}\mu^{-H}$ and $\mu^{-H}(W) = \mu^{-H}(W^*) = \mu^{-H}(P)$.

The other direction, that the relation $P \sim Q$ implies $P \approx Q$, is not true, in general. To show this, we consider a simple example.

**Example 10.** Let $\mathscr{H} = L^2(\Delta, dx; \mathbb{C}^2)$ and let $H$ be multiplication by $x$. The operators on this space can be represented as $2 \times 2$-matrices. Let the projection $P$ be defined by the matrix $\begin{pmatrix} \chi_\Delta(x) & 0 \\ 0 & 0 \end{pmatrix}$ and let the unitary operator $W$ be given by $\begin{pmatrix} 0 & 1 \\ 1 & 0 \end{pmatrix}$. Then $P$ and $W$ commute with $H$ and we get $Q = WPW^*$ is given by the matrix $\begin{pmatrix} 0 & 0 \\ 0 & \chi_\Delta(x) \end{pmatrix}$. Thus $W, W^*, P$, and $Q$ belong to $\operatorname{com} H = \operatorname{ima}\mu^H$ in this case. Hence $P \sim Q$. Further $P - Q \in \operatorname{ima}\mu^H$ and $P - Q \neq 0$. This proves $P - Q \notin \ker\mu^H$ and thus $P \not\approx Q$.

If $H$ has not a simple absolutely continuous spectrum, then there are always projections $P, Q \in \operatorname{dom}\mu^H$ such that $P \sim Q$ is valid but $P \not\approx Q$. However, it is not hard to show that $P \sim Q$ implies $P \approx Q$ if $H$ has a simple absolutely continuous spectrum (use that $\operatorname{ima}\mu^H$ is Abelian and thus $\mu^H(P)$, $\mu^H(Q)$, $\mu^H(W)$, $\mu^H(W^*)$ commute with each other).

## 7.2. Adjoints and partial isometries

### 7.2.1. Adjoints

First we note a simple but essential feature of the adjoint of an asymptotic constant.

**Theorem 11.** *Suppose that $X, X^* \in \operatorname{dom}\mu^H$. Then $\mu^H(X^*) = (\mu^H(X))^*$; this means that the map $\mu^H$ is interchangeable with the map $^*\colon X \to X^*$ on every $C^*$-subalgebra of $\operatorname{dom}\mu^H$.*

**Proof.** Let $X, X^* \in \operatorname{dom} \mu^H$. Then from Theorem 6.25 we obtain

$$\mu^H(X) = \operatorname*{Lim}_{t\to\infty} P^{ac} e^{itH} X e^{-itH} P^{ac}, \quad \mu^H(X^*) = \operatorname*{Lim}_{t\to\infty} P^{ac} e^{itH} X^* e^{-itH} P^{ac}.$$

Therefore, for all $f, g \in \mathscr{H}$ we have $\big(f, \mu^H(X)\, g\big) = \operatorname*{Lim}_{t\to\infty} (f, P^{ac} e^{itH} X\, e^{-itH} P^{ac} g)$ $= \operatorname*{Lim}_{t\to\infty} (P^{ac} e^{itH} X^* e^{-itH} P^{ac} f, g) = \big(\mu^H(X^*)\, f, g\big)$, and this implies the assertion. ■

Occasionally, we shall use the next elementary result.

**Proposition 12.** (i) *Let $U$ be a partial isometry with $U^*U = P$. Then $U \in \ker \mu^H$ is equivalent to $P \in \ker \mu^H$.*

(ii) *Let $A$ be a normal operator. Then $A \in \ker \mu^H$ implies $A^* \in \ker \mu^H$.*

**Proof.** (i) $U \in \ker \mu^H$ implies $P = U^*U \in \ker \mu^H$ because $\ker \mu^H$ is a left ideal. Conversely, $P \in \ker \mu^H$ implies $U = UP \in \ker \mu^H$ because of the same argument. (ii) Let $A = V\,|A|$ be the polar decomposition of $A$. Since $A \in \ker \mu^H$, $|A| = V^*A \in \ker \mu^H$, and also $V^*\,|A| \in \ker \mu^H$ because $\ker \mu^H$ is a left ideal. Since $A$ is normal, we have $AA^* = A^*A$, and this implies $V\,|A| = |A|\,V$ and also $V^*\,|A| = |A|\,V^*$. Hence $A^* = |A|\,V^* = V^*\,|A| \in \ker \mu^H$. ■

**Theorem 13.** *The wave algebra is not a $*$-algebra. That is, there is an operator $X \in \operatorname{dom} \mu^H$ such that $X^* \notin \operatorname{dom} \mu^H$.*

**Proof.** Let $E \in \ker \mu^H$ be a projection such that $\dim E\mathscr{H} = \infty$. Such a projection exists by Prop. 6 (set $Q = P^{ac}$ and take $E = P^{ac} - P$). Then take a partial isometry $W$ with $W^*W = E$ and $WW^* = P^{ac}$. Since $\dim E\mathscr{H} = \dim P^{ac}\mathscr{H} = \infty$, such $W$ exists. Then $W \in \ker \mu^H$, by Prop. 12, but $W^* \notin \ker \mu^H$ because $P^{ac} \notin \ker \mu^H$. Thus, by Theorem 11, $W^*$ is not an element of $\operatorname{dom} \mu^H$, i.e. $X = W$ is the desired operator. ■

### 7.2.2. Properties of partial isometries from $\operatorname{dom} \mu^H$

The wave morphism $\mu^H$ maps selfadjoint operators and projections into selfadjoint operators and projections, respectively. Now the question is whether the images of partial isometries under $\mu^H$ are again partial isometries. First we give some sufficient criteria.

**Theorem 14.** *Let $U \in \operatorname{dom} \mu^H$ be a partial isometry with $U^*U = P$ and $UU^* = Q$. Set $U_+ = \mu^H(U)$. Then:*

(i) *If $P \in \operatorname{dom} \mu^H$, then $U_+$ is a partial isometry with $U_+^*U_+ = \mu^H(P)$.*

(ii) *If $U^* \in \operatorname{dom} \mu^H$, then $U_+$ is a partial isometry, $P$ and $Q$ are elements of $\operatorname{dom} \mu^H$, and $\mu^H(P) = U_+^*U_+$ and $\mu^H(Q) = U_+U_+^*$.*

(iii) *If $P, Q \in \operatorname{dom} \mu^H$ and $U_+U_+^* = \mu^H(Q)$, then $U^* \in \operatorname{dom} \mu^H$.*

**Proof.** (i) We have $P = U^*U = U^*(U - U_+) + U^*U_+$. Since $U - U_+ \in \ker \mu^H$ and $P \in \operatorname{dom} \mu^H$, we get $U^*U_+ \in \operatorname{dom} \mu^H$ and $\mu^H(U^*U_+) = \mu^H(P)$. On the other hand, $\mu^H(U_+^*U) = \big(\mu^H(U^*U_+)\big)^*$ because of Theorem 11 and this implies $U_+^*U_+ = \mu^H(P)$. Since $\mu^H(P)$ is a projection as an image of a projection, this shows that $U_+$ is a partial isometry. (ii) This is a consequence of $P = U^*U$, $Q = UU^*$ and (i). (iii) Set $Q_+ = \mu^H(Q)$. Then $U^* = U^*Q = U^*(Q - Q_+) + U^*Q_+ = U^*(Q - Q_+) + U^*U_+U_+^* = U^*(Q - Q_+)$

$+ U^*(U_+ - U) U_+^* + U^*UU_+ = U^*(Q - Q_+) + U^*(U_+ - U) U_+^* + PU_+^*$. By $Q - Q_+$, $U - U_+ \in \ker \mu^H$ and $P \in \operatorname{dom} \mu^H$, we get that the right side of this equality belongs to $\operatorname{dom} \mu^H$ and therefore $U^* \in \operatorname{dom} \mu^H$. ■

The next statement shows that the image of a partial isometry under $\mu^H$ is not always a partial isometry. Further it makes some assertions on the initial and final projections of a partial isometry $U$ from $\operatorname{dom} \mu^H$.

**Proposition 15.** (i) *There exists a partial isometry $U \in \operatorname{dom} \mu^H$ such that $\mu^H(U)$ is not partially isometric.*

(ii) *There exists a partial isometry $U \in \operatorname{dom} \mu^H$ such that $P = U^*U \notin \operatorname{dom} \mu^H$.*
(iii) *There exists a partial isometry $U \in \operatorname{dom} \mu^H$ such that $Q = UU^* \notin \operatorname{dom} \mu^H$.*

**Proof.** (i) By Prop. 6, there is a projection $Q \leq P^{ac}$ such that $Q \in \operatorname{dom} \mu^H$, $\mu^H(Q) = P^{ac}$, $Q_- := P^{ac} - Q \in \ker \mu^H$ and $\dim Q\mathscr{H} = \dim Q_-\mathscr{H} = \infty$. Then there is a basis $\{u_n\}_{-\infty}^{\infty}$ in $P^{ac}\mathscr{H}$ such that

$$Q = \sum_{n=1}^{\infty} u_n(u_n, \cdot) \text{ and } Q_- = \sum_{n=0}^{-\infty} u_n(u_n, \cdot).$$

Set $w_n := \left(1/\sqrt{2}\right)(u_n + u_{-(n-1)})$ for $n = 1, 2, \ldots$ By definition $\{w_n\}_1^{\infty}$ is an orthonormal system and $P = \sum w_n(w_n, \cdot)$ is a projection. Now we define $U = \sum u_n(w_n, \cdot)$. Clearly, $U$ is a partial isometry with $U^*U = P$. Further, $U = U_1 + U_2$ with $U_1 = \sum_{n\geq 1} \left(1/\sqrt{2}\right) u_n(u_n, \cdot) = \left(1/\sqrt{2}\right) Q$ and $U_2 = \sum \left(1/\sqrt{2}\right) u_n(u_{-(n-1)}, \cdot) = U_2Q_-$. Since $Q_- \in \ker \mu^H$, $U_2 = U_2Q_- \in \ker \mu^H$. Further, by $\mu^H(Q) = P^{ac}$, we get $\mu^H(U_1) = \mu^H\left(\left(1/\sqrt{2}\right) Q\right) = \left(1/\sqrt{2}\right) P^{ac}$. These two facts imply $U \in \operatorname{dom} \mu^H$ and $\mu^H(U) = \left(1/\sqrt{2}\right) P^{ac}$. Thus $\mu^H(U)$ is not partially isometric. (ii) Take $U$ from (i). Then $P = U^*U \notin \operatorname{dom} \mu^H$ because of Theorem 14 (i). (iii) Choose a projection $E \in \ker \mu^H$ with $\dim E\mathscr{H} = \infty$ (such $E$ exists by Prop. 6). Further, choose a projection $Q \notin \operatorname{dom} \mu^H$ (such $Q$ exists by (ii)). By $\dim E\mathscr{H} = \dim Q\mathscr{H} = \infty$, there is a partial isometry $U$ with $U^*U = E$ and $UU^* = Q$. According to Prop. 12, $U \in \ker \mu^H$, but by construction $UU^* = Q \notin \operatorname{dom} \mu^H$. ■

The following two special statements will be needed in Part III.

**Lemma 16.** *Let $V \in \operatorname{dom} \mu^H$ be partially isometric with $V^*V = Q$ and $VV^* = F$. Assume $Q \in \operatorname{com} H$ and $\mu^H(V) = Q$. Then $FQ \in \operatorname{dom} \mu^H$ and $\mu^H(FQ) = Q$.*

**Proof.** We have $FQ = VV^*Q = VV^*(Q - V) + VV^*V = F(Q - V) + V$. By $V - Q \in \ker \mu^H$, this implies $FQ \in \operatorname{dom} \mu^H$ and $\mu^H(FQ) = \mu^H(V) = Q$. ■

**Lemma 17.** *Let $F$ and $Q$ be projections. Assume $Q \in \operatorname{com} H$, $FQ \in \operatorname{dom} \mu^H$ and $\mu^H(FQ) = Q$. Then there is a partial isometry $V$ with $V^*V = Q$ and $VV^* = F$ such that $V$, $V^*Q \in \operatorname{dom} \mu^H$ and $\mu^H(V) = \mu^H(V^*Q) = Q$. If in addition $FQ \in \operatorname{dom} \mu^{-H}$ with $\mu^{-H}(FQ) = Q$, then $V$ can be chosen so that $V$, $V^*Q \in \operatorname{dom} \mu^{-H}$ and $\mu^{-H}(V) = \mu^{-H}(V^*Q) = Q$, too.*

**Proof.** The proof follows the same ideas as the proof of Theorem 8. Thus we only sketch some points of the proof. Let $A = FQ$ and $A = W|A|$ be its polar decomposition. Then we get, as in the proof of Theorem 8, $W \in \operatorname{dom} \mu^H$ and $\mu^H(W) = Q$, and similarly as in that proof we can construct from $W$ a partial isometry $V$ with $V \in \operatorname{dom} \mu^H$, $\mu^H(V) = Q$, $V^*V = Q$ and $VV^* = F$. Thus, by $V^*Q = V^*(Q - V) + V^*V$

$= V^*(Q - V) + Q$, we obtain $V^*Q \in \operatorname{dom} \mu^H$ and $\mu^H(V^*Q) = Q$. The other assertion of this lemma can be proved in the same manner. ■

### 7.2.3. Unitary operators

According to Theorem 14 the image of a unitary operator from dom $\mu^H$ under the map $\mu^H$ is a partial isometry with the initial subspace $P^{ac}\mathscr{H}$. Now we ask whether this partial isometry is even unitary on $P^{ac}\mathscr{H}$. A partial answer gives

**Proposition 18.** *Let $U \in \operatorname{dom} \mu^H$ be unitary on $\mathscr{H}$. Then:* (i) $\mu^H(U) \restriction P^{ac}\mathscr{H}$ *is unitary if $H$ has no part with infinite multiplicity. Moreover, then $U^* \in \operatorname{dom} \mu^H$.*

(ii) *If there is a number $\xi \in \operatorname{res} U$ with $|\xi| = 1$, then $\mu^H(U) \restriction P^{ac}\mathscr{H}$ is unitary and $U^* \in \operatorname{dom} \mu^H$.*

Proof. (i) By Theorem 14, we have $U_+^*U_+ = P^{ac}$ with $U_+ = \mu^H(U)$. Now $P^{ac}\mathscr{H}$, $H \restriction P^{ac}\mathscr{H}$ can be identified with $L^2(\Delta, \mathrm{d}x; \mathscr{H}_x)$, $x$ because of the spectral representation theorem (see 4.5.1). Since $U_+ \in \operatorname{com} H$ and $U_+ = P^{ac}U_+ = U_+P^{ac}$, $U_+$ corresponds to a multiplication operator by $U_+(x)$ on $L^2(\Delta, \mathrm{d}x; \mathscr{H}_x)$ which is an isometry a.e. on $\Delta$ (mod $|\cdot|$), i.e. $U_+^*(x)\, U_+(x) = 1_{\mathscr{H}_x}$ a.e. on $\Delta$. Since $\dim \mathscr{H}_x < \infty$ for all $x \in \Delta$, $U_+(x)$ is unitary a.e. on $\Delta$ and therefore $U_+ \restriction P^{ac}\mathscr{H}$ is unitary. It remains to show that $U^* \in \operatorname{dom} \mu^H$. But this follows from Theorem 14 (iii). (ii) Again we have $U_+^*U_+ = P^{ac}$ with $U_+ = \mu^H(U)$. Since there is $\xi \in \operatorname{res} U$ with $|\xi| = 1$ and $U$ is unitary, we get that $0 \in \operatorname{res}_\infty U$ (see Theorem 6.29) and therefore $U^{-1} = U^* \in \operatorname{dom} \mu^H$. This implies, by Theorem 14 (ii), that $U_+U_+^* = P^{ac}$. ■

The following example proves that $\mu^H(U) \restriction P^{ac}\mathscr{H}$ is not always unitary when $U$ is unitary on $\mathscr{H}$. Naturally, this example uses an operator $H$ with infinitely dimensional multiplicity of its spectrum.

**Example 19.** Let $H = -\mathrm{id}/\mathrm{d}x$ on $\mathscr{H} = L^2(\mathbb{R}, \mathrm{d}x; \mathscr{H}_0)$ with $\dim \mathscr{H}_0 = \infty$. Further, let $\{e_s\}_1^\infty$ denote an orthonormal basis in $\mathscr{H}_0$. The set $\mathscr{D}$ of vectors $u \in \mathscr{H}$ having the form $u(x) = \alpha(x)\, e_s$, $\alpha \in C_0(\mathbb{R})$, is generating in $\mathscr{H}$. Let $u \in \mathscr{D}$, $u(x) = \alpha(x)\, e_s$. Then we define

$$(Uu)(x) = \chi_{(-\infty,0]}(x)\, \alpha(x)\, e_s + \sum_{n=1}^{\infty} \chi_{(n-1,n]}(x)\, \alpha(x)\, B_n e_s$$

where $B_n$ are unitary operators on $\mathscr{H}_0$ given by

$$B_n e_j = \begin{cases} e_{j+1} & \text{for } 1 \leq j \leq n-1 \quad (\text{if } n > 1), \\ e_1 & \text{for } j = n, \\ e_j & \text{for } j > n. \end{cases}$$

It is easy to verify that $U$ is unitary on $\mathscr{H}$. From this definition we get $(U\,\mathrm{e}^{-\mathrm{i}tH}u)(x) = \alpha(x-t)\, e_{s+1}$, $u(x) = \alpha(x)\, e_s$, for all sufficiently large $t$. This follows from the fact that the support of $\alpha(x-t) = \tilde{\alpha}_t(x)$ lies in the interval $[n, \infty)$ with $n - 1 > s$ for large $t$. Thus $\operatorname{s-lim}_{t\to\infty} \mathrm{e}^{\mathrm{i}tH}U\,\mathrm{e}^{-\mathrm{i}tH}u = \{\alpha(x)\, e_{s+1}\}$. This implies $U \in \operatorname{dom} \mu_s^H$ and $\mu_s^H(U) = U_+$ with $U_+$ defined on $\mathscr{D}$ by $(U_+u)(x) = \alpha(x)\, e_{s+1}$ for $u(x) = \alpha(x)\, e_s$. Obviously, $U_+$ is not a unitary operator because the image of $U_+$ does not contain the set of vectors $\alpha(x)\, e_1$. Hence $U$ is the desired counterexample.

## 7.3. $C^*$- and $W^*$-subalgebras of the wave algebra

### 7.3.1. $\mu^H$ as $*$-homomorphism on $*$-subalgebras

In this section we study some properties of the wave morphism $\mu^H$ when we restrict it to $C^*$- or $W^*$-subalgebras of dom $\mu^H$. Recall that $\mu^H$ restricted to $*$-subalgebras is a $*$-homomorphism (see Theorem 11 and Remark 6.27).

**Proposition 20.** *Let $\mathscr{A}$ be a $C^*$-subalgebra of* dom $\mu^H$. *Then $\mu^H$ restricted to $\mathscr{A}$ is a $*$-homomorphism from $\mathscr{A}$ into* ima $\mu^H$ *and $\mu^H(\mathscr{A})$ is again a $C^*$-subalgebra of* ima $\mu^H$. *If in addition $\mu^H$ is injective on $\mathscr{A}$, i.e. $0 \neq A \in \mathscr{A}$ implies $\mu^H(A) \neq 0$, then $\mu^H$ is isometric on $\mathscr{A}$, that is $\|\mu^H(A)\| = \|A\|$ for all $A \in \mathscr{A}$.*

**Proof.** Only the last assertion has to be proved. Let $\mu^H$ be injective on $\mathscr{A}$. It is sufficient to prove $\|\mu^H(A)\| = \|A\|$ for all $A \in \mathscr{A}$ with $\|A\| = 1$. Obviously we have $\|\mu^H(A)\| \leqq \|A\|$. Assume the existence of an $A$ with $\|\mu^H(A)\| < 1$. Now $\|A^*A\| = \|A\|^2$, hence $\|A^*A\| = 1$. Further, we have $\mu^H(A^*) = (\mu^H(A))^*$. Hence we get $\mu^H(A^*A) = (\mu^H(A))^* \mu^H(A)$, that is, $\|\mu^H(A^*A)\| = \|\mu^H(A)\|^2$. We put $B := A^*A \geqq 0$, then $\|B\| = 1$ and $\|\mu^H(B)\| < 1$. Now let $f \in C(\mathbb{R}_+)$ and $f(\lambda) \geqq 0$, further $f(1) = 1$ and $f(\lambda) = 0$ on $[0, \|\mu^H(B)\| + \varepsilon]$, $\varepsilon > 0$, $\varepsilon + \|\mu^H(B)\| < 1$. Then $f(B) \in$ dom $\mu^H$, $f(B) \neq 0$ (because of $1 \in$ spec $B$) and $f(B) \geqq 0$. Further, $\mu^H(f(B)) = f(\mu^H(B))\, P^{ac} = 0$ because of supp $f$ $\cap$ spec $\mu^H(B) = \emptyset$, hence $f(B) \in$ ker $\mu^H$ and $f(B) \in \mathscr{A}$, which is a contradiction to $f(B) \neq 0$. ∎

An important special case is given when the $C^*$-algebra $\mathscr{A}$ is even a $W^*$-algebra. Then the question arises whether $\mu^H(\mathscr{A})$ is also a $W^*$-algebra. This is connected with the question of whether or not $\mu^H \restriction \mathscr{A}$ is a so-called $W^*$-homomorphism, i.e. if $\mu^H \restriction \mathscr{A}$ is continuous with respect to the weak $*$-topologies on $\mathscr{A}$ and $\mathscr{L}(\mathscr{H})$.

**Proposition 21.** *Let $\mathscr{A} \in$ dom $\mu^H$ be a $W^*$-algebra. Then $\mu^H \restriction \mathscr{A}$ is a $W^*$-homomorphism and in particular $\mu^H(\mathscr{A})$ is a $W^*$-algebra.*

**Proof.** Since $\mathscr{A}$ is a $W^*$-algebra, there is a unique Banach space $\mathscr{A}_*$ such that $\mathscr{A} = (\mathscr{A}_*)^*$ (see Sakai [1, p. 1]). By $\mathscr{A} \subset \mathscr{L}(\mathscr{H})$, we get $\mathscr{A}_* \supseteq (\mathscr{L}(\mathscr{H}))_* = \mathscr{L}_1(\mathscr{H})$. The weak $*$-topology in $\mathscr{A}$ is defined by the family of seminorms $|a_*(A)|$, $a_* \in \mathscr{A}_*$; the weak $*$-topology in $\mathscr{L}(\mathscr{H})$ is defined by the seminorms $|\mathrm{tr}\,(PX)|$, $P \in \mathscr{L}_1(\mathscr{H})$ and $X \in \mathscr{L}(\mathscr{H})$; $\mathscr{L}_1(\mathscr{H})$ is the indexing set.

First we note that $\lim\limits_{t\to\infty} \mathrm{tr}\, PB_t = 0$ if $P \in \mathscr{L}_1(\mathscr{H})$ and $\operatorname*{s-lim}\limits_{t\to\infty} B_t = 0$ where $B_t \in \mathscr{L}(\mathscr{H})$. Namely, let $P^* = \sum c_s g_s(f_s, \cdot)$ be the Schmidt representation of $P^*$ with $\sum |c_s| < \infty$ and choose an orthonormal basis $\{e_j\}$ which contains the system $\{f_s\}$. Then $\mathrm{tr}\, PB_t = \sum \bar{c}_s (g_s, B_t e_s)$ and the assertion follows easily. Secondly, for $A \in$ dom $\mu^H$ we have $\mu^H(A) = \operatorname*{s-lim}\limits_{T\to\infty} T^{-1} \int\limits_0^T e^{itH} A\, e^{-itH} P^{ac}\, dt$. Thus, applying the result above on the operator function $B_T = T^{-1} \int\limits_0^T e^{itH} A\, e^{-itH} P^{ac}\, dt - \mu^H(A)$, we get

$$\mathrm{tr}\,\{P\mu^H(A)\} = \lim_{T\to\infty} \mathrm{tr} \left\{ P(T^{-1} \int\limits_0^T e^{itH} A\, e^{-itH} P^{ac}\, dt) \right\}. \tag{3}$$

Now, because of $\operatorname{tr} CD = \operatorname{tr} DC$, we obtain

$$\operatorname{tr}\left\{P\left(T^{-1}\int_0^T e^{itH}A\, e^{-itH}P^{ac}\,dt\right)\right\} = \operatorname{tr}\left\{T^{-1}\int_0^T P^{ac}\, e^{-itH}P\, e^{itH}A\,dt\right\} = \operatorname{tr}\{P_T A\}$$

where $P_T = T^{-1}\int_0^T P^{ac}\, e^{-itH}P\, e^{itH}\,dt \in \mathscr{L}_1(\mathscr{H})$ for all $T > 0$. From this and (3) it follows that

$$\operatorname{tr}\{P\mu^H(A)\} = \lim_{T\to\infty} \operatorname{tr}\{P_T A\}, \qquad A \in \mathscr{A}. \tag{4}$$

Next we use that $\mathscr{A}_*$ is sequentielly complete with respect to the weak topology (see Sakai [1, p. 21]), i.e. there is $p_\infty^{(P)} \in \mathscr{A}_*$ such that $p_\infty^{(P)}(A) = \lim_{T\to\infty} \operatorname{tr}\{P_T A\}$, $A \in \mathscr{A}$. This relation and (4) prove $p_\infty^{(P)}(A) = \operatorname{tr}\{P\mu^H(A)\}$ for all $A \in \mathscr{A}$ and $P \in \mathscr{L}_1(\mathscr{H})$, and this means that $\mu^H(A)$ is continuous with respect to the corresponding weak $*$-topologies in $\mathscr{A}$ and $\mathscr{L}(\mathscr{H})$, i.e. $\mu^H \mid \mathscr{A}$ is a $W^*$-homomorphism. Since $W^*$-homomorphisms map $W^*$-algebras onto $W^*$-algebras (see Sakai [1, p. 40]), $\mu^H(\mathscr{A})$ is a $W^*$-algebra. ∎

### 7.3.2. $\mu^H$ as spatial isomorphism on $W^*$-subalgebras

In the preceding section it was shown that for a $W^*$-algebra $\mathscr{A}$, $\mu^H \mid \mathscr{A}$ is a $W^*$-homomorphism and $\mu^H(\mathscr{A})$ is a $W^*$-algebra. Now it is known that one can find a maximal projection $E \in \operatorname{cen} \mathscr{A}$ such that $\mu^H$ restricted to $\mathscr{A} \mid E := \{A \in \mathscr{A};\, A = AE = EA\} \cong A \mid E\mathscr{H}$ is injective, i.e. $\mu^H(A) \neq 0$ if $0 \neq A \in \mathscr{A} \mid E$. Note that if $\mathscr{A}$ is a v. Neumann algebra, then also $\mathscr{A} \mid E\,\mathscr{H}$ and $\mu^H(\mathscr{A}) \mid \mu^H(E)\mathscr{H}$ are v. Neumann algebras, and $\mu^H$ is a $W^*$-isomorphism of $\mathscr{A} \mid E$ onto $\mu^H(\mathscr{A} \mid E)$. The projection $P := \mu^H(E)$ belongs to the center of $\mu^H(\mathscr{A} \mid E)$ and we have $\mu^H(\mathscr{A} \mid E) = \mu^H(\mathscr{A} \mid E) \mid P = \mu^H(\mathscr{A}) \mid P$.

**Theorem 22.** *Let $\mathscr{A} \subset \operatorname{dom} \mu^H$ be a v. Neumann algebra. Assume there is a projection $E \in \operatorname{cen} \mathscr{A}$ such that $\mu^H$ restricted to $\mathscr{A} \mid E$ is injective and $(\operatorname{com} \mathscr{A}) \mid E$ is Abelian. Further, assume that $(\operatorname{com} \mu^H(\mathscr{A} \mid E)) \mid P$ with $P := \mu^H(E)$ is also Abelian. Then $\mu^H$ is a spatial isomorphism of $\mathscr{A} \mid E$ onto $\mu^H(\mathscr{A} \mid E)$, that is, there is a partial isometry $W$ such that $W^*W = E$, $WW^* = P$ and $\mu^H(A) = WAW^*$ for all $A \in \mathscr{A} \mid E$.*

**Proof.** This is a simple consequence of the fact that a $W^*$-isomorphism of a v. Neumann algebra $\mathscr{N}_1$ onto a v. Neumann algebra $\mathscr{N}_2$ is spatial if $\operatorname{com} \mathscr{N}_1$ and $\operatorname{com} \mathscr{N}_2$ are Abelian (see Sakai [1, p. 120]). ∎

If $\mathscr{A}$ satisfies the assumptions of Theorem 22, then we can define a strongly continuous unitary group $e^{-itK}$ on $\mathscr{H}$ by

$$e^{-itK} = W^*\, e^{-itH}W + (1 - E). \tag{5}$$

By definition, $K$ is selfadjoint and $K(1 - E) = (1 - E)\,K = 0$. Since $H$ commutes with all operators from $\mu^H(\mathscr{A})$, we get, by $\mu^H(A) = WAW^*$, that $K$ commutes with all operators from $\mathscr{A}$; that is $\mathscr{A} \subseteq \operatorname{com} K$.

The next statement shows that the partial isometry $W$ in Theorem 22 can be obtained as limit of an operator-valued function $e^{itH}T_t$ where $T_t$ is asymptotically isometric and is connected with $\mathscr{A}$, and thus with $K$.

**Theorem 23.** *Let $\mathscr{A}$, $E$, $P$, and $W$ be as in Theorem 22. Then there exist a function $T_t: \mathbb{R} \to \mathscr{C}(\mathscr{H})$ and a dense set $\mathscr{D}$ in $\mathscr{H}$ such that:*

(i) $\mathscr{D} \subseteq \operatorname{dom} T_t$ *for all* $t \in \mathbb{R}$, $A\mathscr{D} \subseteq \mathscr{D}$ *for all* $A \in \mathscr{A}$.
(ii) $T_tAu = AT_tu$ *for all* $t \in \mathbb{R}$, $u \in \mathscr{D}$ *and* $A \in \mathscr{A}$.
(iii) $\lim_{t\to\infty} \|T_tu\| = \|u\|$ *for* $u \in \mathscr{D}$ *in case* $\mu^H = \mu_s{}^H$ *and* A-$\lim_{t\to\infty} \|T_tu\| = \|u\|$ *for* $u \in \mathscr{D}$ *in case* $\mu^H = \mu_a{}^H$.
(iv) *The functions $T_tu$ are strongly continuous for each $u \in \mathscr{D}$, uniformly in $t \in \mathbb{R}$.*
(v) $\operatorname{Lim}_{t\to\infty} e^{itH}T_tEu$ *exists for each* $u \in \mathscr{D}$ *and equals* $Wu$.

**Proof.** **1.** (com $\mathscr{A}$) $\restriction E$ is Abelian. Thus there exists a vector $u_0 \in E\mathscr{H}$ such that $\{Au_0; A \in \mathscr{A} \restriction E\}$ is dense in $E\mathscr{H}$. We can assume $\|u_0\| = 1$ and set $\mathscr{D}_0 := \{Au_0; A \in \mathscr{A} \restriction E\}$ and $\mathscr{D} = \mathscr{D}_0 + (1 - E)\mathscr{H}$. Since $u_0$ is generating with respect to $E\mathscr{H}$ and $\mathscr{A} \restriction E$, $u_0$ is separating for (com $\mathscr{A} \restriction E$) $\restriction E$ = (com $\mathscr{A}$) $\restriction E$, i.e. $Bu_0 = 0$ and $B \in$ (com $\mathscr{A}$) $\restriction E$ imply $B = 0$ (use $ABu_0 = BAu_0$, $A \in \mathscr{A}$, for the proof). Now let $E_0$ be the projection onto clo $\{Bu_0; B \in \text{com } \mathscr{A}\}$. Then $E_0 \in$ com (com $\mathscr{A}$) $\restriction E = \mathscr{A} \restriction E$ as a simple calculation shows. Thus $E_0u_0 = u_0$. Further, if $Au_0 = 0$ for some $A \in \mathscr{A}$, then $BAu_0 = ABu_0 = 0$ for all $B \in$ com $\mathscr{A}$. This implies $Au = 0$ for all $u \in E_0\mathscr{H}$ and thus $AE_0 = 0$.

**2.** Now we define an operator-valued function $\tilde{T}_t$ on $\mathscr{D}$ by

$$\tilde{T}_t(1 - E) = 1 - E \quad \text{and} \quad \tilde{T}_tAu_0 = AE_0\, e^{-itH}Wu_0, \qquad A \in \mathscr{A}. \tag{6}$$

This definition is senseful because $A_1u_0 = A_2u_0$ with $A_1, A_2 \in \mathscr{A}$ implies $A_1E_0 = A_2E^0$ (see 1.) and therefore $A_1E_0\, e^{-itH}Wu_0 = A_2E_0\, e^{-itH}Wu_0$. $\tilde{T}_tAu_0$ is strongly continuous, uniformly in $t \in \mathbb{R}$, since $e^{-itH}Wu_0$ has this property. This proves (iv) if we can show that $\tilde{T}_t$ is closable and we denote its closure by $T_t$. By definition $\tilde{T}_t$ is densely defined and commutes with all operators $A \in \mathscr{A}$.

**3.** Next we prove (v). If $u = Au_0$, then

$$\begin{aligned}\|(W - e^{itH}\tilde{T}_t)\, u\| &= \|(WAE_0 - e^{itH}AE_0\, e^{-itH}W)\, u_0\| \\ &= \|(WAE_0W^* - e^{itH}AE_0\, e^{-itH}P^{ac})\, Wu_0\|\end{aligned}$$

where we have used that $W^*Wu_0 = E_0u_0 = u_0$ and $W = PW$ with $P \leqq P^{ac}$. Since $AE_0 \in \mathscr{A} \restriction E$ and $\mathscr{A} \restriction E \subset \operatorname{dom} \mu^H$, we get that the right side of the equality above converges to zero as $t \to \infty$ (use $\mu^H(B) = WBW^*$). This means that (v) is satisfied because of $T_t = \tilde{T}_t$ on $\mathscr{D}$ and $Au_0 \in E\mathscr{H}$ for all $A \in \mathscr{A}$. (iii) follows from (v), according to Lemma 6.12 (in case $\mu^H = \mu_a{}^H$), and the fact that s-$\lim_{t\to\infty} u_t = u_0$ implies $\lim_{t\to\infty} \|u_t\| = \|u_0\|$ following from $\|u_t - u_0\| \geqq |\|u_t\| - \|u_0\||$.

**4.** Now we verify that $\tilde{T}_t$ is closable. It is only necessary to do this on the subspace $E\mathscr{H}$. Since (com $\mathscr{A}$) $\restriction E$ is Abelian, we get for $C \in$ (com $\mathscr{A}$) $\restriction E$ that

$$\|C^*u_0\|^2 = (C^*u_0, C^*u_0) = (u_0, CC^*u_0) = (u_0, C^*Cu_0) = \|Cu_0\|^2.$$

By this and the fact that the set $\mathscr{D}_1 := \{Cu_0; C \in (\text{com } \mathscr{A}) \restriction E\}$ is dense in $E_0\mathscr{H}$, for any $A \in \mathscr{A}$ we have

$$\begin{aligned}\|E_0A^*u_0\| &= \sup_{u\in\mathscr{D}_1, u\neq 0} |(u, E_0A^*u_0)|/\|u\| = \sup_{Cu_0\neq 0, C\in(\text{com}\mathscr{A})\restriction E} \frac{|(Cu_0, E_0A^*u_0)|}{\|Cu_0\|} \\ &= \sup_{Cu_0\neq 0, C\in(\text{com}\mathscr{A})\restriction E} \frac{|(Au_0, C^*u_0)|}{\|C^*u_0\|} \leqq \|Au_0\|.\end{aligned} \tag{7}$$

Suppose now that $u_n \in \mathscr{D}_0$ and $u_n$ converges to zero as $n \to \infty$. Further, suppose that $\operatorname{Lim}_{n\to\infty} \tilde{T}_t u_n = v$ exists. By definition $u_n = A_n u_0$ with $A_n \in \mathscr{A}$. Then for any $B \in \mathscr{A}$, $(Bu_0, v) = \operatorname{Lim}_{n\to\infty} (Bu_0, A_n E_0 \,\mathrm{e}^{-\mathrm{i}tH} W u_0)$. By (7), this gives

$$|(Bu_0, A_n E_0 \,\mathrm{e}^{-\mathrm{i}tH} W u_0)| \leq \|E_0 A_n{}^* B u_0\| \, \|\mathrm{e}^{-\mathrm{i}tH} W u_0\|$$
$$\leq \|B^* A_n u_0\| \leq \|B^*\| \, \|A_n u_0\| \to 0 \quad \text{as} \quad n \to \infty.$$

Thus $(v, Bu_0) = 0$ for all $B \in \mathscr{A}$ and this implies $Ev = 0$. Since $\tilde{T}_t \mathscr{D}_0 \subseteq E\mathscr{H}$, we get $v = \lim_{n\to\infty} \tilde{T}_t u_n \in E\mathscr{H}$. Thus $v = 0$. This proves that $\tilde{T}_t$ is closable.

5. The last step is to show that (ii) is true. For $u \in \mathscr{D}$, this follows directly from the definition of $T_t$. Now let $u \in \operatorname{dom} T_t$. By definition of the closure there is a sequence $A_n \in \mathscr{A}$ such that $\lim_{n\to\infty} A_n u_0 = u$ and $\lim_{n\to\infty} T_t A_n u_0 = T_t u$. Let $C \in \mathscr{A}$. Then $Cu = \lim_{n\to\infty} CA_n u_0$ and $CT_t u = \lim_{n\to\infty} CT_t A_n u_0 = \lim_{n\to\infty} T_t CA_n u_0$. Thus $Cu \in \operatorname{dom} T_t$ for each $C \in \mathscr{A}$ and $CT_t u = T_t Cu$. ∎

It is possible to prove a little more about the operator-valued function $T_t$. Recall the definition (5) of the unitary group $\mathrm{e}^{-\mathrm{i}tK}$ and introduce a new operator-valued function by

$$G_t := \mathrm{e}^{\mathrm{i}tK} T_t. \tag{8}$$

Clearly, $G_t$ is defined on the same set of vectors as $T_t$. The next assertion says that $G_t$ is "feebly oscillating" and thus its behavior is very different from the one of groups $\mathrm{e}^{-\mathrm{i}tH}$.

**Proposition 24.** *Let $s \in \mathbb{R}$ and $u \in \mathscr{D}$* (see Theorem 23). *Then*

$$\operatorname{Lim}_{t\to\infty} (G_{t+s} - G_t)\, u = 0. \tag{9}$$

Proof. We have already noticed that $\mathrm{e}^{-\mathrm{i}tK} \in \operatorname{com} \mathscr{A}$. Since $\mathscr{A} \mid E\mathscr{H}$ is a v. Neumann algebra and $(\operatorname{com} \mathscr{A}) \mid E$ is Abelian, we have $(\operatorname{com} \mathscr{A}) \mid E \subseteq \mathscr{A}$. By $\mathrm{e}^{-\mathrm{i}tK} E \in (\operatorname{com} \mathscr{A}) \mid E$ and $\mathrm{e}^{-\mathrm{i}tK}(1 - E) = 1 - E$, then $\mathrm{e}^{-\mathrm{i}tK} \in \mathscr{A} \subset \operatorname{dom} \mu^H$. Thus we obtain $\mu^H(\mathrm{e}^{-\mathrm{i}tK} E) = W \mathrm{e}^{-\mathrm{i}tK} W^* = WW^* \mathrm{e}^{-\mathrm{i}tH} WW^* = \mathrm{e}^{-\mathrm{i}tH} P$. Now let $u = Au_0 \in \mathscr{D}_0$, $A \in \mathscr{A}$. From (8) and (6) we get

$$(G_{t+s} - G_t)\, u = \mathrm{e}^{\mathrm{i}(t+s)K} AE_0 \,\mathrm{e}^{-\mathrm{i}(t+s)H} W u_0 - \mathrm{e}^{\mathrm{i}tK} AE_0 \,\mathrm{e}^{-\mathrm{i}tH} W u_0$$
$$= \mathrm{e}^{\mathrm{i}(t+s)K} AE_0 \,\mathrm{e}^{-\mathrm{i}(t+s)K} [W u_0 - \mathrm{e}^{\mathrm{i}sH} \mathrm{e}^{\mathrm{i}tH} \mathrm{e}^{-\mathrm{i}sK} E \,\mathrm{e}^{-\mathrm{i}tH} W u_0]$$

where we have used $AE_0 = EAE_0 = AE_0 E$ $(E \geq E_0)$. By $\mu^H(\mathrm{e}^{-\mathrm{i}sK} E) = \mathrm{e}^{-\mathrm{i}sH} P$, we get from this equality the desired relation (9) on $\mathscr{D}_0$. The extension to $\mathscr{D}$ is trivial. ∎

A simple example for an operator-valued function $G_t$ satisfying (5) is given by $G_t = \mathrm{e}^{\mathrm{i}(\log t)H}$ as an easy calculation shows.

We conclude this section with a statement which is a converse to Theorem 23 in a certain sense.

**Proposition 25.** *Let $\mathscr{A}$ be a v. Neumann algebra and let $E \in \operatorname{cen} \mathscr{A}$ be a projection. Assume there is a strongly continuous function $T_t \colon \mathbb{R}_+ \to \mathscr{L}(\mathscr{H})$ such that:*

(I) *$T_t E$ is unitary on $E\mathscr{H}$ for all $t \in \mathbb{R}_+$,*

(II) *$T_t \in \operatorname{com} \mathscr{A}$ for all $t \in \mathbb{R}_+$,*

(III) *$V := \operatorname{Lim}_{t\to\infty} \mathrm{e}^{\mathrm{i}tH} T_t E$ exists. Then $\operatorname{Lim}_{t\to\infty} \mathrm{e}^{\mathrm{i}tH} A \,\mathrm{e}^{-\mathrm{i}tH} P = VAV^*$ exists for all $A \in \mathscr{A} \mid E$ where $P = VV^*$.*

**Proof.** From (III) and (I) it follows that $V$ is a partial isometry with $V^*V = E$ and $VV^* = P$. Then the equality

$$\begin{aligned}\|(ET_t^* \, e^{-itH}P - V^*) \, u\|^2 &= \|ET_t^* \, e^{-itH}Pu\|^2 + \|V^*u\|^2 \\ &\quad - (ET_t^* \, e^{-itH}Pu, V^*u) - (EV^*u, T_t^* \, e^{-itH}Pu) \\ &\leq 2 \, \|Pu\|^2 - 2 \, \mathrm{Re} \, \{(ET_t^* \, e^{-itH}Pu, V^*u)\} \\ &\leq 2 \, \|Pu\|^2 - 2 \, \mathrm{Re} \, \{(Pu, e^{itH}T_tEV^*u)\}\end{aligned}$$

together with (III) proves that also $\operatorname{Lim}_{t\to\infty} ET_t^* \, e^{-itH}P = V^*$ exists. Then the assertion follows from

$$e^{itH}AE \, e^{-itH}P = e^{itH}T_tEAT_t^* \, e^{-itH}P$$

and $\operatorname{Lim}_{t\to\infty} A(t) \, B(t) = \left(\operatorname{Lim}_{t\to\infty} A(t)\right)\left(\operatorname{Lim}_{t\to\infty} B(t)\right)$ if the limits of the right side exist. ∎

**Remark 26.** If $VV^* = P = P^{ac}$, then $\mathscr{A} \restriction E \subset \operatorname{dom} \mu^H$ and $\mu^H(A) = VAV^*$ for all $A \in \mathscr{A} \restriction E$. Furthermore, it is easy to see that assumption (I) can be replaced by an asymptotic condition.

### 7.3.3. Functions of asymptotic constants

Theorem 6.28 states, in particular, that the $C^*$-algebra $C^*(A, 1)$, generated by the continuous functions of a selfadjoint operator $A \in \operatorname{dom} \mu^H$, is a subset of $\operatorname{dom} \mu^H$. Sometimes one is interested in operators which are bounded functions of an unbounded selfadjoint operator. In such a situation the next result is useful.

**Proposition 27.** *Let $A$ be a selfadjoint operator such that*

*(I) $R_A(\mathrm{i})$ and $R_A(-\mathrm{i})$ belong to $\operatorname{dom} \mu^H$,*

*(II) $\mu^H(R_A(\mathrm{i})) = R_B(\mathrm{i}) \, P^{ac}$ and $\mu^H(R_A(-\mathrm{i})) = R_B(-\mathrm{i}) \, P^{ac}$ where $B$ is another selfadjoint operator.*

*Then for each $\alpha \in C(\mathbb{R})$, $\alpha(A) \in \operatorname{dom} \mu^H$ and $\mu^H(\alpha(A)) = \alpha(B) \, P^{ac}$.*

**Proof.** 1. First we prove that $\alpha(A) \in \operatorname{dom} \mu^H$ if $\alpha \in C_\infty(\mathbb{R})$. For this we use that each $\alpha \in C_\infty(\mathbb{R})$ is the uniform limit of a sequence of polynomials in $(\mathrm{i} - \lambda)^{-1}$ and $(-\mathrm{i} - \lambda)^{-1}$ by the complex Stone-Weierstrass theorem (applied to the one-point compactification of $\mathbb{R}$); let $\mathscr{A}$ be the closure in $|\cdot|$ of all polynomials in $(\mathrm{i} - \lambda)^{-1}$ and $(-\mathrm{i} - \lambda)^{-1}$ and note that 1. $\alpha \in \mathscr{A}$ implies $\bar{\alpha} \in \mathscr{A}$, 2. $\mathscr{A}$ separates points and 3. for each $\lambda \in \mathbb{R}$ there is $\alpha$ with $\alpha(\lambda) \neq 0$ (see Reed/Simon [1, p. 102, 121]).

Thus $\alpha(A)$ is a limit in norm of polynomials in $(\mathrm{i} - A)^{-1}$ and $(-\mathrm{i} - A)^{-1}$ when $\alpha \in C_\infty(\mathbb{R})$, and therefore $\alpha(A) \in \operatorname{dom} \mu^H$. This and (II) imply $\mu^H(\alpha(A)) = \alpha(B) \, P^{ac}$, $\alpha \in C_\infty(\mathbb{R})$.

2. Let $\alpha \in C(\mathbb{R})$. Then $\beta(x) = \alpha(x) \, (\mathrm{i} - x)^{-1}$ belongs to $C_\infty(\mathbb{R})$. Thus $\beta(A) \in \operatorname{dom} \mu^H$ and $\mu^H(\beta(A)) = \alpha(B) \, (\mathrm{i} - B)^{-1} P^{ac}$. This gives, by $\alpha(A) \, (\mathrm{i} - B)^{-1} = \alpha(A) \, (\mathrm{i} - A)^{-1} + \alpha(A) \, [(\mathrm{i} - B)^{-1} - (\mathrm{i} - A)^{-1}]$ and (II), that $\alpha(A) \, (\mathrm{i} - B)^{-1} \in \operatorname{dom} \mu^H$ and $\mu^H(\alpha(A) \, (\mathrm{i} - B)^{-1}) = \alpha(B) \, (\mathrm{i} - B)^{-1} P^{ac}$. But this means that

$$\operatorname{Lim}_{t\to\infty} e^{itH}\alpha(A) \, e^{-itH}P^{ac}u = \alpha(B) \, P^{ac}u, \qquad u \in (\mathrm{i} - B)^{-1}\mathscr{H}.$$

This set of vectors $u$ is dense in $\mathscr{H}$ and thus the assertion follows. ■

To illustrate the use of this statement we consider an easy example.

**Example 28.** Set $H' = H + V$ where $V$ is a selfadjoint compact operator. Then $R_{H'}(z) - R_H(z) = R_{H'}(z)\, V R_H(z)$ (see (13) in Chapter 1) is compact for each nonreal $z$, and thus $R_{H'}(z) - R_H(z) \in \ker \mu_s^H$. By Prop. 27, this proves that $\alpha(H') \in \operatorname{dom} \mu_s^H$ and $\mu^H(\alpha(H')) = \alpha(H)\, P^{ac}$ for all $\alpha \in C(\mathbb{R})$. In particular, we have $e^{-itH'} \in \operatorname{dom} \mu_s^H$ and $\mu_s^H(e^{-itH'}) = e^{-itH} P^{ac}$, $t \in \mathbb{R}$. Obviously, this function is an example where we cannot apply Theorem 6.28 when $H$ is unbounded.

The next example shows that condition (II) of Prop. 27 is not dispensable for the assertion.

**Example 29.** Set $H = -\mathrm{i}d/dx$ and $\mathscr{H} = L^2(\mathbb{R}, dx)$. Further, let $A$ denote the multiplication operator by $x$. Then

$$\|(z - A)^{-1} e^{-itH} u\|^2 = \int_{-\infty}^{\infty} dx\, |(z - x)^{-1}|^2\, |u(x - t)|^2 \to 0 \quad \text{as} \quad t \to \infty$$

as an easy calculation with vectors $u \in C_0(\mathbb{R})$ and a density argument show. Thus we have $(z - A)^{-1} \in \ker \mu_s^H$ for all nonreal $z$, and assumption (I) of Prop. 27 is fulfilled. However,

$$(e^{itH} e^{isA} e^{-itH} u)(x) = e^{is(x+t)} u(x) = e^{ist} e^{isx} u(x),$$

which shows that $e^{isA} \notin \operatorname{dom} \mu^H$. That is, not all operators $\alpha(A)$ with $\alpha \in C(\mathbb{R})$ belong to $\operatorname{dom} \mu^H$.

Another question is whether the $W^*$-algebra $W^*(A, 1)$, generated by the essentially bounded Borel functions of a selfadjoint operator $A \in \operatorname{dom} \mu^H$, is a subset of $\operatorname{dom} \mu^H$. Since $W^*(A, 1)$ is the closure in operator norm of sequences $\sum_{s=1}^{n} c_s E_A(\Delta_s)$, $c_s \in \mathbb{C}$ and $\Delta_s$ Borel sets, this is equivalent to the question of whether all spectral projections of $A$ belong to $\operatorname{dom} \mu^H$.

A simple class of asymptotic constants for which $W^*(A, 1) \subset \operatorname{dom} \mu^H$ is given by

**Proposition 30.** *Let $A \in \ker \mu^H$ be selfadjoint. Assume that 0 is an isolated eigenvalue of $A$. Then for each $\alpha \in L^\infty(\operatorname{spec} A, E_A)$, $\alpha(A) \in \operatorname{dom} \mu^H$ and $\mu^H(\alpha(A)) = \alpha(0)\, P^{ac}$.*

Proof. According to Corollary 2, we have $E_A([a, b]) \in \ker \mu^H$ if $[a, b] \cap \{0\} = \emptyset$. Hence $E_A(\{0\}) = E_A((-\delta, \delta)) \in \operatorname{dom} \mu^H (\delta > 0$ sufficiently small) and $\mu^H(E_A(\{0\})) = P^{ac}$, or $E_A(\{0\})^\perp \in \ker \mu^H$. Thus, by $\alpha(A) = \alpha(0)\, E_A(\{0\}) + \alpha(A)\, E_A(\{0\})^\perp$, we get the assertion. ■

The following example proves that, in general, not all spectral projections, and thus not all essentially bounded Borel functions, of an asymptotic constant again belong to the wave algebra.

**Example 31.** By Prop. 15, there is a projection $Q \notin \operatorname{dom} \mu^H$. We can write $Q = \sum u_n(u_n, \cdot)$ where $\{u_n\}_1^\infty$ is an orthonormal system. Now we define a compact operator $A$ by

$$A = \sum_{n=1}^{\infty} \lambda_n u_n(u_n, \cdot), \quad \lambda_n > 0, \quad \lambda_n \to 0 \quad \text{as} \quad n \to \infty.$$

$A$ is nonnegative and $E_A((0, \infty)) = Q$. That is, there is a spectral projection of $A$ which does not belong to $\operatorname{dom} \mu^H$ whereas $A$, by definition, is an operator from $\ker \mu^H$.

To prove that the $W^*$-algebra $W^*(A, 1)$ is a subset of $\operatorname{dom} \mu^H$, one can use the theory of the wave operator treated in Part III and Part IV. Namely, assume that the limit $W_+(H', H) := \operatorname*{Lim}_{t\to\infty} e^{itH'} e^{-itH} P^{ac}$ exists and $W_+(H', H)\, W_+(H', H)^* = P^{ac}_{H'}$. Then one can show, in a similar way as in the proof of Prop. 25 (replace $e^{-itH}$ and $T_t$ by $e^{-itH}$

and $e^{-itH}$), that com $H' \subset$ dom $\mu^H$; in particular, $A \in$ com $H'$ implies $W^*(A, 1) \subset$ dom $\mu^H$. Thus, using the results of Part IV, one can find many classes of operators $X$ such that $W^*(X, 1) \subset$ dom $\mu^H$.

## 7.4. Classes of asymptotic constants for special generators

In applications the generator $H$ of the wave algebra dom $\mu^H$ is usually a differential operator acting on a Hilbert space $\mathscr{H} = L^2(\mathbb{R}^n, dx)$ or a multiplication operator by $x$ on a Hilbert space $\mathscr{H} = L^2(\Delta, dx; \mathscr{H}_0)$. Correspondingly, the operators for which we have to study the question whether or not they are asymptotic constants are often multiplication operators in the first case or integral operators in the second. These operators are studied in Sections 2, 3, and 4. As a preliminary Section 1 deals with a special map $\Gamma_H$ from a subset of $\mathscr{L}(\mathscr{H})$ into ker $\mu_s^H$.

### 7.4.1. The map $\Gamma_H$ and the wave ideal ker $\mu_s^H$

First we introduce a special operation $\Gamma_H$ which is in a certain sense the inverse operation to the commutator $[\cdot, H]$ as we shall see later

**Definition 32.** Let $A \in \mathscr{L}(\mathscr{H})$. Assume that

$$A_0 = \operatorname{s-lim}_{t\to\infty} i \int_0^t e^{i\tau H} A\, e^{-i\tau H} P^{ac}\, d\tau \tag{10}$$

exists, then we write $\Gamma_H(A) = A_0$. The domain of the map $\Gamma_H$ is denoted by dom $\Gamma_H$ (the set of all $A \in \mathscr{L}(\mathscr{H})$ so that (10) exists), the image of $\Gamma_H$ by ima $\Gamma_H$, and the kernel (null space) of $\Gamma_H$ by ker $\Gamma_H$.

The next theorem collects some properties of this map and shows its tight relation to the wave morphism $\mu_s^H$.

**Theorem 33.** *The map $\Gamma_H$ has the following properties:*

(i) *dom $\Gamma_H$ is a linear manifold.*

(ii) *Let $A \in$ com $H$. Then $X \in$ dom $\Gamma_H$ implies $AX, XA \in$ dom $\Gamma_H$, $\Gamma_H(AX) = A\Gamma_H(X)$ and $\Gamma_H(XA) = \Gamma_H(X)\, A$.*

(iii) *If $B \in$ ima $\Gamma_H$, then $B = BP^{ac}$, $B(\text{dom } H) \subseteq$ dom $H$, $B \in$ ker $\mu_s^H$, $T := BH - HB \in$ dom $\Gamma_H$, and $\Gamma_H(T) = B$.*

(iv) *Assume that $B = BP^{ac}$, $B(\text{dom } H) \subseteq$ dom $H$, $B \in$ ker $\mu_s^H$, and $T := BH - HB$ is bounded, then $B \in$ ima $\Gamma_H$.*

(v) *ker $\Gamma_H = \{X; XP^{ac} = 0\}$.*

Proof. (i) and (ii) follow immediately from the definition of $\Gamma_H$.

(iii) Let $B \in$ ima $\Gamma_H$ with $B = \Gamma_H(X)$. Then we have

$$X(t) := e^{itH} B\, e^{-itH} = i \int_t^\infty e^{isH} X\, e^{-isH} P^{ac}\, ds = B - i \int_0^t e^{isH} X\, e^{-isH} P^{ac}\, ds$$

which follows from (10) by multiplying it by $e^{itH}$ from the left and by $e^{-itH}$ from the right. Clearly, $X(t)$ is strongly differentiable and $d/dt\, X(t) = -i\, e^{itH} X\, e^{-itH} P^{ac}$. From the definition of $X(t)$ we also get

$$e^{itH}Bu = X(t)\, e^{itH}u. \tag{11}$$

Assume $u \in \operatorname{dom} H$ and differentiate the right side of (11). This is possible since $X(t)$ is strongly differentiable and $e^{itH}u$ is strongly differentiable for $u \in \operatorname{dom} H$. Thus $e^{itH}Bu$ is also strongly differentiable and, since $e^{itH}v$ is only strongly differentiable for $v \in \operatorname{dom} H$, this implies $Bu \in \operatorname{dom} H$ and therefore $B(\operatorname{dom} H) \subseteq \operatorname{dom} H$. Differentiating (11) and setting $t = 0$, we get

$$iHBu = -iXP^{ac}u + iBHu \quad \text{or} \quad XP^{ac} = T := BH - HB.$$

Furthermore, from $X(t) = i\int_t^\infty e^{isH}X\, e^{-isH}P^{ac}\, ds$ we see that $X(t)$ tends strongly to zero as $t \to \infty$, $X \in \operatorname{dom} \Gamma_H$, and therefore $e^{itH}B\, e^{-itH} \xrightarrow[s]{} 0$, i.e. $B \in \ker \mu_s^H$. The relation $B = BP^{ac}$ is obvious. This concludes the proof of (iii).

(iv) Now we assume that $B$ satisfies the assumptions in (iv). Then

$$d/dt(e^{itH}B\, e^{-itH}u) = i\, e^{itH}(HB - BH)\, e^{-itH}u$$

exists for all $u \in \operatorname{dom} H$ because $B\, e^{-itH}(\operatorname{dom} H) \subseteq \operatorname{dom} H$. Setting $T := -(HB - BH)$ and using $B = BP^{ac}$, we get from the equality above that

$$e^{itH}B\, e^{-itH}P^{ac}u = Bu - i\int_0^t e^{isH}T\, e^{-isH}P^{ac}u\, ds.$$

By $B \in \ker \mu_s^H$, the left side tends strongly to zero as $t \to \infty$, and thus $T \in \operatorname{dom} \Gamma_H$ with $\Gamma_H(T) = B$.

(v) In the course of the proof of (iii) we have shown that $B \in \operatorname{ima} \Gamma_H$ with $\Gamma_H(X) = B$ implies $BH - HB = XP^{ac}$ on $\operatorname{dom} H$. Thus $B = 0$ implies $XP^{ac} = 0$. ∎

**Remark 34.** If $H$ is bounded, then we get the simpler relation $\operatorname{ima} \Gamma_H = \{X;\ X \in \ker \mu_s^H,\ X = XP^{ac}\}$. For unbounded operators $H$ the assumptions $X \in \ker \mu_s^H$ and $X = XP^{ac}$ imply neither $T := XH - HX \in \mathscr{L}(\mathscr{H})$ nor $X(\operatorname{dom} H) \subseteq \operatorname{dom} H$. It is easy to construct examples for these facts.

From the definition of $\Gamma_H$ it easily follows that $X \in \operatorname{dom} \mu_s^H$ when we have $X \in \mathscr{L}(\mathscr{H})$ and $XH - HX \in \operatorname{dom} \Gamma_H$. The following statement extends this criterion.

**Proposition 35.** *Let $X \in \mathscr{L}(\mathscr{H})$. Then the expression $T(u, v) := (XHu, v) - (Xu, Hv)$ is well-defined for all $u, v \in \operatorname{dom} H$. Assume there are two sets $\mathscr{D}_1$ and $\mathscr{D}_2$ such that:*

(I) *$\mathscr{D}_1 \subseteq \operatorname{dom} H$ and $\mathscr{D}_1$ is dense in $\mathscr{H}$,*
(II) *$\mathscr{D}_2$ is a generating set with respect to $H$ and $P^{ac}\mathscr{H}$, $\mathscr{D}_2 \subseteq \operatorname{dom} H \cap P^{ac}\mathscr{H}$,*
(III) $\lim\limits_{t>\tau\to\infty} \sup\limits_{v\in\mathscr{D}_1, \|v\|=1} \int_\tau^t |T(e^{-isH}u, e^{-isH}v)|\, ds = 0$ *for all $u \in \mathscr{D}_2$.*

*Then $X \in \operatorname{dom} \mu_s^H$ and $XP^{ac} = X_1 + X_2$ where $X_1 \in \operatorname{ima} \mu^H$ and $X_2$ is given by*

$(X_2u, v) = -\lim\limits_{t\to\infty} i\int_0^t T(e^{-isH}u, e^{-isH}v)\, ds$ *for all $v \in \mathscr{D}_1$ and $u \in$* spa $\{e^{itH}f; f \in \mathscr{D}_2, t \in \mathbf{R}\}$.

**Proof.** Let $v \in \mathscr{D}_1 \subseteq \operatorname{dom} H$ and $u \in \mathscr{D}_2 \subseteq \operatorname{dom} H \cap P^{ac}\mathscr{H}$. Then

$$\begin{aligned} d/ds(e^{isH} X e^{-isH} u, v) &= d/ds(X e^{-isH} u, e^{-isH} v) \\ &= i(XH e^{-isH} u, e^{-isH} v) - i(X e^{-isH} u, H e^{-isH} v) \\ &= iT(e^{-isH} u, e^{-isH} v) \end{aligned}$$

exists since $e^{-isH} u$ is strongly differentiable for $u \in \operatorname{dom} H$. This relation implies

$$(e^{isH} X e^{-isH} u, v) - (e^{i\tau H} X e^{-i\tau H} u, v) = i \int_\tau^s T(e^{-itH} u, e^{-itH} v)\, dt \tag{12}$$

and thus

$$\begin{aligned} \|e^{isH} X e^{-isH} u - e^{i\tau H} X e^{-i\tau H} u\| &= \sup_{v \in \mathscr{D}_1, \|v\|=1} |(e^{isH} X e^{-isH} u - e^{i\tau H} X e^{-i\tau H} u, v)| \\ &\leq \sup_{v \in \mathscr{D}_1, \|v\|=1} \int_\tau^s |T(e^{-itH} u, e^{-itH} v)|\, dt. \end{aligned}$$

By (III), this implies that $\operatorname{s-lim}_{t\to\infty} e^{itH} X e^{-itH} u =: X_1 u$ exists for all $u \in \mathscr{D}_2$ and thus, by Prop. 6.24, that $X \in \operatorname{ima} \mu_s^H$. This and (12) for $\tau = 0$ give the last assertion, about $(X_2 u, v)$, for $u \in \mathscr{D}_2$. A simple calculation extends this to all $u \in \operatorname{spa}\{e^{itH} f;\, f \in \mathscr{D}_2, t \in \mathbb{R}\}$. ∎

In case that $c(XH - HX) \geq 0$ for some $c \in \mathbb{C}$ we obtain that $X \in \operatorname{dom} \mu_s^H$ without other assumptions.

**Proposition 36.** *Let $X \in \mathscr{L}(\mathscr{H})$. Assume there are a set $\mathscr{D}$ dense in $\operatorname{dom} H \cap P^{ac}\mathscr{H}$ with $e^{-itH}\mathscr{D} \subseteq \mathscr{D}$, $t \in \mathbb{R}$, a number $c \neq 0$ and a nonnegative operator $T = TP^{ac} \in \mathscr{L}(\mathscr{H})$ such that $c[(XHu, v) - (Xu, Hv)] = (Tu, v)$ for all $u \in \mathscr{D}$ and $v \in \operatorname{dom} H$. Then $X \in \operatorname{dom} \mu_s^H$ and $XP^{ac} = X_1 + X_2$ where $X_1 \in \operatorname{com} H$ and*

$$c(X_2 u, v) = -\lim_{t\to\infty} i \int_0^t (e^{isH} T e^{-isH} u, v)\, ds \quad \text{for } u \in \mathscr{D}$$

*and $v \in \operatorname{dom} H$.*

**Proof.** Again as in the proof of Prop. 35 we find, for $u \in \mathscr{D}$ and $v \in \operatorname{dom} H$,

$$\begin{aligned} &c(e^{isH} X e^{-isH} u, v) - c(e^{itH} X e^{-itH} u, v) \\ &= i c \int_t^s (e^{i\tau H}[XH - HX] e^{-i\tau H} u, v)\, d\tau = i \int_t^s (e^{i\tau H} T e^{-i\tau H} u, v)\, d\tau. \end{aligned} \tag{13}$$

Set $u = v$ in this equality and use $T \geq 0$, then we obtain

$$|c| \int_t^s \|T^{1/2} e^{-i\tau H} u\|^2\, d\tau \leq 2\, \|X\|\, \|u\|^2$$

for a dense set of vectors $u$ in $P^{ac}\mathscr{H}$ and, by $T = TP^{ac}$, also for a dense set of vectors $u$ in $\mathscr{H}$. Using this and (13), we get

$$\int_t^s |(e^{i\tau H} T e^{-i\tau H} u, v)|\, d\tau \leq (2\, |c|^{-1}\, \|X\|)^{1/2}\, \|u\| \left( \int_t^s \|T^{1/2} e^{i\tau H} v\|^2\, d\tau \right)^{1/2} \to 0 \quad \text{as } s > t \to \infty.$$

Thus the assumptions of Prop. 35 are satisfied and this leads to the desired results. ∎

**Example 37.** Let $H = -\mathrm{id}/\mathrm{d}x$ on $\mathscr{H} = L^2(\mathbb{R}, \mathrm{d}x)$. Let $X$ be the multiplication operator by $\alpha(x)$ where $\alpha(x) = -cx^{-a}$ for $x \geqq 1$ and $\alpha(x) = 0$ for $x < 1$, $a > 0$ and $c > 0$. Then the operator $i[XH - HX]$ is well-defined on the dense set of vectors $u$ where $u(x)$ is continuously differentiable and $u(x) = 0$ in a neighborhood of $x = 1$. It is a multiplication operator by the function $t(x)$ where $t(x) = acx^{-(1+a)}$ for $x \geqq 1$ and $t(x) = 0$ for $x < 1$. Thus $\mathrm{i}[XH - HX]$ is a bounded non-negative operator and we can apply Prop. 36. This yields $X \in \operatorname{dom} \mu_s^H$.

### 7.4.2. Multiplication operators

In this section $H$ denotes the natural selfadjoint extension of

$$-\Delta = -\sum_{j=1}^{n} (\partial^2/\partial x_j^2) \quad \text{on} \quad \mathscr{H} = L^2(\mathbb{R}^n, \mathrm{d}x), \qquad x = \{x_1, x_2, \ldots, x_n\}$$

(see Example 3.8). Recall that $P^{ac} = 1$.

Further, let $\mathfrak{M}$ denote the class of multiplication operators by $\alpha(x)$ with $\alpha \in L^\infty(\mathbb{R}, \mathrm{d}x)$. We want to study those operators from $\mathfrak{M}$ which belong to $\ker \mu^H$. Before doing this, we note some facts. First, note that the set $\mathscr{D}$ of vectors $u_a \in \mathscr{H}$ with $u_a(x) = \mathrm{e}^{-|x-a|^2/2}$, $a \in \mathbb{R}^n$, is generating in $\mathscr{H}$ (see Kato [1, p. 496]). Secondly, we have that $\mathrm{e}^{-\mathrm{i}tH}$ acts on such vectors in a simple manner, namely

$$u_a(t, x) := (\mathrm{e}^{-\mathrm{i}tH} u_a)(x) = (1 + 2\mathrm{i}t)^{-n/2}\, \mathrm{e}^{-|x-a|^2/(2+4\mathrm{i}t)}. \tag{14}$$

This can be proved by a straightforward calculation which shows that $\mathrm{d}/\mathrm{d}t(u_a(t, x))$ $= \mathrm{i} \sum_{j=1}^{n} (\partial^2/\partial x_j^2)\, u_a(t, x)$.

**Theorem 38.** *Let $A \in \mathfrak{M}$ and let $\alpha(x)$ be the corresponding function for $A$. Then the following statements are equivalent:*

(i) $\lim_{R\to\infty} R^{-n} \int_{|x|\leq R} |\alpha(x)|^2 \,\mathrm{d}x = 0$,
(ii) $A \in \ker \mu_s^H$,
(iii) $A \in \ker \mu_s^{-H}$,
(iv) $A \in \ker \mu_a^H$,
(v) $A \in \ker \mu_a^{-H}$.

Proof. Let $u_a \in \mathscr{D}$. By (14), we then get

$$\|A\, \mathrm{e}^{-\mathrm{i}tH} u_a\|^2 = \int |\alpha(x)|^2\, |1 + 4t^2|^{-n/2}\, |\mathrm{e}^{-|x-a|^2/(2+4\mathrm{i}t)}|^2 \,\mathrm{d}x$$
$$= \int |\alpha(x)|^2\, |1 + 4t^2|^{-n/2}\, \mathrm{e}^{-|x-a|^2/(1+4t^2)} \,\mathrm{d}x. \tag{15}$$

Thus $\|A\, \mathrm{e}^{-\mathrm{i}tH} u_a\|^2 = \|A\, \mathrm{e}^{\mathrm{i}tH} u_a\|^2$ and this shows that (ii) is equivalent to (iii) and (iv) is equivalent to (v). From this reason it is sufficient to prove that (i) implies (ii) and that (iv) implies (i); that (ii) implies (iv) is known (see Theorem 6.23).

First we show that (i) implies (ii). By (15) we have

$$\|A\, \mathrm{e}^{-\mathrm{i}tH} u_a\|^2 \leq (2\,|t|)^{-n} \Big\{ \int_{|x+a|\leq N|t|} |\alpha(x + a)|^2 \,\mathrm{d}x^2$$
$$+ |\alpha(\cdot)|_\infty^2 \int_{|x+a|>N|t|} \mathrm{e}^{-|x|^2/(1+4t^2)} \,\mathrm{d}x \Big\}. \tag{16}$$

By (i) the first term on the right side of (16) tends to zero as $t \to \infty$ for arbitrary fixed $N > 0$. The second term converges to zero as $N \to \infty$, uniformly in $t \geqq 1$; this shows an easy estimate. Thus the left side of (16) vanishes as $t \to \infty$. Since the set of such vectors $u_a$ is generating in $\mathscr{H}$, this provides $A \in \ker \mu_s^H$.

Now it remains to verify that (iv) implies (i). From (15) we get

$$T^{-1} \int_0^T \|A\, e^{-itH} u_a\|^2 \, dt \geqq c_1 T^{-1} \int_1^T (1 + 4t^2)^{-n/2} \, dt \int_{|x| \leqq t} |\alpha(x)|^2 \, dx$$

$$\geqq c_1 c_2 T^{-1} \int_1^T t^{-n} \, dt \int_{r \leqq t} r^{n-1} \, dr \int_\Omega |\alpha(r, \omega)|^2 \, d\omega \tag{17}$$

where $\Omega$ is the surface of the unit sphere in $\mathbb{R}^n$, $r = |x|$, $\int_\Omega d\omega$ denotes the integration over $\Omega$, $\alpha(r, \omega) = \alpha(x)$ and $c_1, c_2$ are positive numbers determined by $c_1 = \min_{t \geqq 1, |x| \leqq t} e^{-|x-a|^2/(1+4t^2)}$ and by the requirement $c_2 t^{-n} \leqq (1 + 4t^2)^{-n/2}$ for $t \geqq 1$. Set $q(r) := \int_\Omega |\alpha(r, \omega)|^2 \, d\omega$. Clearly, $q \in L^\infty(\mathbb{R}_+, dr)$. From (17) we obtain

$$T^{-1} \int_0^T \|A\, e^{-itH} u_a\|^2 \, dt \geqq c_1 c_2 T^{-1} \int_1^T t^{-n} \, dt \int_0^t r^{n-1} q(r) \, dr$$

$$\geqq c_1 c_2 \varepsilon^{n-1} T^{-1} \int_1^T t^{-1} \, dt \int_{\varepsilon t}^t q(r) \, dr$$

for arbitrary $\varepsilon \in (0, 1)$. By $A \in \ker \mu_a^H$, $T^{-1} \int_1^T t^{-1} \, dt \int_{\varepsilon t}^t q(r) \, dr$ tends to zero as $T \to \infty$ for each $\varepsilon \in (0, 1)$ and thus also for $\varepsilon = 0$ since $q \in L^\infty(\mathbb{R}_+, dr)$. But this means that the second Hölder mean of $q$ is zero and thus, by Lemma 6.2, we get $\lim_{T \to \infty} T^{-1} \int_0^T q(r) \, dr = 0$. This together with the estimate

$$T^{-n} \int_{|x| \leqq T} |\alpha(x)|^2 \, dx = T^{-n} \int_0^T r^{n-1} \, dr \int_\Omega |\alpha(r, \omega)|^2 \, d\omega$$

$$= T^{-n} \int_0^T r^{n-1} q(r) \, dr \leqq T^{-1} \int_0^T q(r) \, dr$$

leads to (i). ■

It is an interesting fact that for these multiplication operators the two limiting methods, s-lim and |A|-lim, give the same results, and that also the two directions of the parameter $t$, $t \to \infty$ and $t \to -\infty$, represented by $\mu^H$ and $\mu^{-H}$ asymptotically, coincide on this set.

### 7.4.3. Integral operators

In this section $H$ is multiplication by $x$ on $\mathscr{H} = L^2(\Delta, dx; \mathscr{H}_0)$ where $\Delta$ is a Borel set of the real line and $\mathscr{H}_0$ is a separable Hilbert space. Here we are interested in integral operators $K$ of the form

$$(Ku)(x) = \int_\Delta k(x, x')\, u(x') \, dx' \tag{18}$$

where $k(x, x')$ is an operator-valued function from $\Delta \times \Delta$ into $\mathscr{L}(\mathscr{H}_0)$. Now we present a sufficient condition that $K$ belongs to $\ker \mu_s^H \cap \ker \mu_s^{-H}$.

**Proposition 39.** *Let $K$ be a bounded integral operator of the form* (18). *Assume $\{e_j\}_1^N$ is a generating set in $\mathscr{H}_0$ such that for each $j \in \{1, 2, \ldots, N\}$:*

(I) *The operators $V_j^\sigma$ on $L^2(\Delta, \mathrm{d}x)$ defined by $(V_j^\sigma u)(x) = \int\limits_\sigma \|k(x, x')\, e_j\|_{\mathscr{H}_0} u(x')\, \mathrm{d}x'$ are bounded where $\sigma$ runs through all Borel sets with $\sigma \subset \Delta$ and $|\sigma| < \infty$.*

(II) *$\|k(x, \cdot)\, e_j\|_{\mathscr{H}_0} \in L^1(\sigma, \mathrm{d}x')$ for almost all $x \in \Delta(\mathrm{mod}\, |\cdot|)$ and all $\sigma$ described in* (I). *Then $K \in \ker \mu_s^H \cap \ker \mu_s^{-H}$.*

**Proof.** It suffices to prove $\|K\, \mathrm{e}^{-\mathrm{i}tH} u\| \to 0$ as $t \to \infty$ for a generating set of vectors $u$ in $\mathscr{H}$. As such set $\mathscr{D}$ we choose $\mathscr{D} = \bigcup\limits_{j=1}^N \mathscr{D}_j$ where $u \in \mathscr{D}_j$ if $u(x) = \chi_\sigma(x)\, e_j$ with $\sigma \subset \Delta$ and $|\sigma| < \infty$.

Now let $u \in \mathscr{D}$ with $u(x) = \chi_\sigma(x)\, e_j$. Then

$$\|(K\, \mathrm{e}^{-\mathrm{i}tH} u)(x)\|_{\mathscr{H}_0} = \left\| \int k(x, x')\, \mathrm{e}^{-\mathrm{i}tx'} \chi_\sigma(x')\, e_j\, \mathrm{d}x' \right\|_{\mathscr{H}_0}$$
$$\leqq \int\limits_\sigma \|k(x, x')\, e_j\|_{\mathscr{H}_0} \chi_\sigma(x')\, \mathrm{d}x' = (V_j^\sigma \chi_\sigma)(x) = f(x) \tag{19}$$

where $f \in L^2(\Delta, \mathrm{d}x)$ because of (I). Thus, for the proof of $\|k\, \mathrm{e}^{-\mathrm{i}tH} u\| \to 0$ it is enough to verify that $\|(K\, \mathrm{e}^{-\mathrm{i}tH} u)(x)\|_{\mathscr{H}_0} \to 0$ as $|t| \to \infty$ for almost all $x(\mathrm{mod}\, |\cdot|)$. But this follows from (II); then let $v(x') := k(x, x')\, e_j \chi_\sigma(x')$ for a fixed $x$ such that $\|k(x, \cdot)\, e_j\|_{\mathscr{H}_0} \in L^1(\sigma, \mathrm{d}x')$, thus $\left\| \int v(x')\, \mathrm{e}^{-\mathrm{i}tx'}\, \mathrm{d}x' \right\|_{\mathscr{H}_0} \to 0$ as $|t| \to \infty$ since the Fourier transform of a norm integrable function, here $v(x')$, tends to zero as $|t| \to \infty$, and now apply this to (19). ■

One can easily get special examples where the assumptions of this statement are satisfied. For instance, consider integral operators on $L^2(\Delta, \mathrm{d}x)$ where the kernel $k(x, x') \geqq 0$ is locally integrable with respect to $x'$ for almost all $x(\mathrm{mod}\, |\cdot|)$. A concrete example is given by: $\mathscr{H} = L^2(\mathbb{R}_+, \mathrm{d}x)$ and $K$ defined by $(Ku)(x) = x^{-1} \int\limits_0^x u(y)\, \mathrm{d}y$, i.e. $k(x, y) = 0$ for $x \leqq y$ and $k(x, y) = x^{-1}$ for $x > y$. It is an easy exercise to verify the assumptions of Prop. 39 in this case.

The following statement considers operators of convolution type.

**Proposition 40.** *Let $K$ be an integral operator of the form* (18) *with $\Delta = \mathbb{R}_+$ and $k(x, x') = k(x - x')$. Assume that $k(x) = k_1(x) + k_2(x)$ with $\|k_1(\cdot)\| \in L^1(\mathbb{R}, \mathrm{d}x)$ and $\|k_2(\cdot)\| \in L^2(\mathbb{R}, \mathrm{d}x)$. Further, assume that the Fourier transform of $k_2(\cdot)$ is a uniformly bounded operator-valued function into $\mathscr{L}(\mathscr{H}_0)$. Then $K \in \ker \mu_s^H \cap \ker \mu_s^{-H}$.*

**Proof. 1.** First we claim that $K$ is bounded. Let $F$ denote the Fourier transformation in $L^2(\mathbb{R}, \mathrm{d}x; \mathscr{H}_0)$. Then

$$\|Ku\|^2 = \int\limits_{\mathbb{R}_+} \mathrm{d}x \left\| \int\limits_{\mathbb{R}_+} k(x - x')\, u(x')\, \mathrm{d}x' \right\|^2 \leqq \int\limits_{-\infty}^{\infty} \mathrm{d}x \left\| \int\limits_{\mathbb{R}_+} k(x - x')\, u(x')\, \mathrm{d}x' \right\|^2$$
$$= \int\limits_{-\infty}^{\infty} \mathrm{d}x \left\| F\left( \int\limits_{\mathbb{R}_+} k(x - x')\, u(x')\, \mathrm{d}x' \right) \right\|^2 = 2\pi \int\limits_{-\infty}^{\infty} \mathrm{d}s\, \|(Fk)(s)\, (Fu)(s)\|^2$$

where we have used the convolution theorem and the identification of $u \in L^2(\mathbb{R}_+, dx)$ with $u \in L^2(\mathbb{R}, dx)$ such that $u(x) = 0$ on $\mathbb{R}_-$. By our assumption on $k_1$ and $k_2$ we have $\|(Fk)(s)\| \leqq c < \infty$ with $c$ independent of $s$. Thus the inequality above provides $\|Ku\|^2 \leqq c^2 \|u\|^2$, i.e $K$ is bounded.

2. Now we show that $K$ satisfies the other assumptions of Prop. 39. (II) is satisfied because $k(x, x')$ is locally integrable in $\mathscr{L}(\mathscr{H}_0)$-norm. This follows from $\|k_1(\cdot)\| \in L^1(\mathbb{R}, dx)$ and $\|k_2(\cdot)\| \in L^2(\mathbb{R}, dx)$, since a square integrable function is locally absolutely integrable. It remains to verify that $\chi_\sigma(x') \|k(x - x') e_j\|$ is a kernel of a bounded integral operator on $L^2(\mathbb{R}, dx)$ for each $\sigma \subset \mathbb{R}$, $|\sigma| < \infty$, and for each $e_j$ from a generating set in $\mathscr{H}_0$. Now we have

$$\chi_\sigma(x') \|k(x - x') e_j\| \leqq \chi_\sigma(x') \|k_1(x - x')\| + \chi_\sigma(x') \|k_2(x - x')\|, \qquad \|e_j\| = 1.$$

Clearly, the first term on the right side is the kernel of a bounded integral operator because of the same arguments as in 1. The second term on the right side allows the estimate

$$\int dx \left| \int_\sigma dx' \|k_2(x - x')\| \|u(x')\| \right|^2 \leqq \int dx \int_\sigma \|k_2(x - x')\|^2 dx' \int \|u(x')\|^2 dx'$$
$$\leqq |\sigma| \|u\|^2 \int \|k_2(x - x')\|^2 dx < \infty,$$

showing that $\chi_\sigma(x') \|k_2(x - x')\|$ is the kernel of a bounded integral operator. Thus $\chi_\sigma(x') \|k_2(x - x') e_j\|$ is also a kernel of a bounded integral operator. This concludes the proof. ∎

**Remark 41.** One can extend the assertions of Prop. 40 to integral operators acting like $(Ku)(x) = \int_0^\infty u(x') \, dk(x - x')$ where $k(x)$ is a continuous function of bounded variation on $\mathbb{R}$. However, then $K$ is, in general, only an element of $\ker \mu_a{}^H \cap \ker \mu_a{}^{-H}$ because the Fourier-Stieltjes transform of $dk(x)$ vanishes at infinity only in Cesaro limit.

To illustrate Prop. 40, we consider the following easy example.

**Example 42.** Let $K$ be an integral operator on $L^2(\mathbb{R}_+, dx)$ with the kernel $k(x, x') = k_1(x - x') + k_2(x - x')$ where $k_1(x) = c_1(1 + |x|)^{-1} |x|^{-\alpha}$, $c_1 > 0$ and $\alpha \in (1/2, 1)$, and $k_2(x) = c_2 x^{-1}(e^{ix} - 1)$. It is obvious that $k_1 \in L^1(\mathbb{R}, dx)$ but $k_1 \notin L^2(\mathbb{R}, dx)$. Furthermore, $k_2 \in L^2(\mathbb{R}, dx)$ but $k_2 \notin L^1(\mathbb{R}, dx)$. The Fourier transform of $k_2$ is given by $i\sqrt{2\pi}\, c_2 \chi_{[0,1]}(x)$. Therefore $K$ satisfies the assumptions of Prop. 40, and hence $K \in \ker \mu_s{}^H \cap \ker \mu_s{}^{-H}$.

### 7.4.4. Integral operators with singularities of Cauchy type

As in the preceding section $H$ is multiplication by $x$ on the Hilbert space $\mathscr{H} = L^2(\Delta, dx; \mathscr{H}_0)$. For simplicity we shall assume that $\Delta = \mathbb{R}$. The extension of the results to arbitrary Borel sets $\Delta \subseteqq \mathbb{R}$ is not difficult.

Now let $K$ be an integral operator of the form

$$(Ku)(x) = \int_{-\infty}^{\infty} \frac{k(x, x')}{x - x'} u(x') \, dx' := \underset{\delta \to +0}{\text{s-lim}} \int_{|x - x'| > \delta} \frac{k(x, x')}{x - x'} u(x') \, dx' \tag{20}$$

where $k(x, x')$ is an operator-valued function from $\mathbb{R} \times \mathbb{R}$ into $\mathscr{L}(\mathscr{H}_0)$ satisfying some smoothness conditions.

**Proposition 43.** *Let $K$ be defined as in (20). Assume that $k(x, x')$ is the Fourier transform of an operator-valued function $h(s, s')$ from $\mathbb{R} \times \mathbb{R}$ into $\mathscr{L}(\mathscr{H}_0)$ with $\int ds \int \|h(s,s')\| ds' < \infty$. Then $K \in \operatorname{dom} \mu_s^H \cap \operatorname{dom} \mu_s^{-H}$ and $\mu_s^H(K) = \mu_s^{-H}(-K) = i\pi K_0$ where $K_0$ is the multiplication operator by $k(x) := k(x, x)$.*

**Proof. 1.** First we note that $k(x, x')$ is a continuous operator-valued function and tends to zero as $|x| \to \infty$ or $|x'| \to \infty$ since $k(x, x')$ is the Fourier transform of $h(s, s')$ with $\int ds \int ds' \|h(s, s')\| < \infty$ (the proof is a simple consequence of the corresponding statements for functions with values in $\mathbb{C}$). Thus $k(x) = k(x, x)$ is well-defined and defines a bounded operator $K_0$. Further, we use the relation between (20) and $(K_\varepsilon u)(x) := \int \frac{k(x, x')}{x - x' + i\varepsilon} u(x') dx'$ given by Prop. 1.3 in the special case $w(x')$ in place of $k(x, x') u(x')$. Namely assume $\operatorname{s-lim}_{\varepsilon \to +0} (K_\varepsilon u)(x)$ exists at a point $x$. Then, setting for a moment $w(x') = k(x, x') u(x')$, we get that $(Ku)(x)$ exists and $(Ku)(x) = \operatorname{s-lim}_{\varepsilon \to +0} (K_\varepsilon u)(x) + i\pi k(x) u(x)$. From this we find that the existence of the limit $\operatorname{s-lim}_{\varepsilon \to +0} K_\varepsilon$, implies the boundedness of the operator $K$.

2. Let us consider the expression

$$(e^{itH}(KH - HK) e^{-itH}u, v) = \int_{-\infty}^{\infty} dx \left( \int_{-\infty}^{\infty} k(x, x') e^{-itx'} u(x') dx', e^{-itx} v(x) \right)_{\mathscr{H}_0}.$$

Because of the properties of $k(x, x')$, this expression is uniformly bounded in $t$ for vectors $u$, $v$ such that $u(x)$, $v(x)$ have compact support. With the help of the Fourier transformation we can write

$$\begin{aligned} K_T(u, v) &:= -i \int_0^T dt\, (e^{itH}(KH - HK) e^{-itH}u, v) \\ &= -i \int_0^T dt \int_{-\infty}^{\infty} ds \left( \int_{-\infty}^{\infty} ds' h(s, -s')\, \hat{u}(s' - t), \hat{v}(s - t) \right)_{\mathscr{H}_0} \\ &= -i \int_{-\infty}^{\infty} ds \int_{-\infty}^{\infty} ds' \left( \left[ \int_0^T dt\, h(s + t, -s' - t) \right] \hat{u}(s'), \hat{v}(s) \right)_{\mathscr{H}_0}. \end{aligned}$$

Now we show that this expression defines a bounded integral operator for all $T > 0$ with a bound independent of $T$. Set $k_T(s, s') = i \int_0^T h(s + t, -s' - t)\, dt$. Then we have

$$\begin{aligned} \int_{\infty}^{\infty} ds \left\| \int_{-\infty}^{\infty} k_T(s, s')\, u(s')\, ds' \right\|^2 &\leq \int_{-\infty}^{\infty} ds \left( \int_{-\infty}^{\infty} \|k_T(s, s')\|^{1/2} \|k_T(s, s')\|^{1/2} \|u(s')\|\, ds' \right)^2 \\ &\leq \int_{-\infty}^{\infty} ds \int_{-\infty}^{\infty} \|k_T(s, s')\|\, ds' \int_{-\infty}^{\infty} \|k_T(s, s'')\|\, \|u(s'')\|^2\, ds'' \\ &= \int_{-\infty}^{\infty} ds''\, \|u(s'')\|^2 \int_{-\infty}^{\infty} ds\, \|k_T(s, s'')\| \int_{-\infty}^{\infty} \|k_T(s, s')\|\, ds'. \end{aligned} \tag{21}$$

Using that

$$\int_{-\infty}^{\infty} \|k_T(s, s')\| \, ds \leq \int_{-\infty}^{\infty} ds \int_0^T \|h(-s - t, s' - t)\| \, dt$$

$$\leq \int_{-\infty}^{\infty} ds \int_{-\infty}^{\infty} \|h(s, s')\| \, ds' \leq c < \infty$$

and similarly $\int \|k_T(s, s')\| \, ds' \leq c < \infty$, we obtain from (21) that the integral operator $K_T$ with the kernel $k_T(s, s')$ is bounded with a bound independent of $T$. Moreover, this shows that the strong limit of $K_T$ exists as $T \to \infty$ and is bounded. Now it is easy to prove that also the operator $\tilde{K}_\varepsilon$ defined by

$$(\tilde{K}_\varepsilon u, v) := -\mathrm{i} \int_0^{\infty} \mathrm{e}^{-\varepsilon t} (\mathrm{e}^{\mathrm{i}tH}(KH - HK) \, \mathrm{e}^{-\mathrm{i}tH} u, v) \, dt$$

is bounded for all $\varepsilon > 0$, and $\operatorname{s-lim}_{\varepsilon \to +0} \tilde{K}_\varepsilon$ exists and equals $\operatorname{s-lim}_{T \to \infty} K_T$. From

$$(\tilde{K}_\varepsilon u)(x) = \mathrm{i} \int_0^{\infty} dt \, \mathrm{e}^{-\varepsilon t} \int_{-\infty}^{\infty} \mathrm{e}^{\mathrm{i}t(x - x')} k(x, x') \, u(x') \, dx'$$

we easily get, by changing the order of integration

$$(\tilde{K}_\varepsilon u)(x) = -\int_{-\infty}^{\infty} \frac{k(x, x')}{x - x' + \mathrm{i}\varepsilon} \, u(x') \, dx' .$$

Thus $\tilde{K}_\varepsilon = -K_\varepsilon$, and since the strong limit of $\tilde{K}_\varepsilon$ exists we obtain that $K$ is bounded and $(Ku, v) = \lim_{T \to \infty} (-K_T u, v) + (\mathrm{i}\pi K_0 u, v)$. Now we can apply Prop. 35 to $K$ and get the desired result for $\mu_s{}^H(K)$. The other case, $\mu_s{}^{-H}(K)$ can be treated similarly. ∎

Now we come to another class of singular integral operators of the form (20). First we introduce a special norm for operator-valued functions from $\mathbb{R} \times \mathbb{R}$ into $\mathcal{L}(\mathcal{H}_0)$.

**Definition 44.** Let $p$ and $q$ be two real numbers with $p > 0$ and $q \in (0, 1)$. Let $k(\cdot, \cdot)$ be an $\mathcal{L}(\mathcal{H}_0)$-valued function of two real variables. Set $(L_h{}^1 k)(x, x') = k(x + h, x')$, $(L_h{}^2 k)(x, x') = k(x, x' + h)$ for $h \in \mathbb{R}$, set $A_h{}^1 k = h^{-q}(L_h{}^1 k - k)$, $A_h{}^2 k = h^{-q}(L_h{}^2 k - k)$ for $h \in (0, \infty)$ and set $(A_\infty^1 k)(x, x') = (1 + |x|^p) \, k(x, x')$, $(A_\infty^2 k)(x, x') = (1 + |x'|^p) \times k(x, x')$. Put

$$\|k\|_{p,q} := \sup_{\infty \geq h_1, h_2 > 0} \; \sup_{\infty > x, x' > -\infty} \|(A_{h_1}^1 A_{h_2}^2 k)(x, x')\|_{\mathcal{H}_0} .$$

Then $\mathcal{G}_{p,q}(\mathcal{H}_0)$ denotes the set of all $k(\cdot, \cdot)$ such that $\|k\|_{p,q} < \infty$.

We need the following important result (see Dunford/Schwartz [3, p. 2424]).

**Lemma 45.** *If $k \in \mathcal{G}_{p,q}(\mathcal{H}_0)$, then the improper integral*

$$(Ku)(x) := \int_{-\infty}^{\infty} \frac{k(x, x')}{x - x'} \, u(x') \, dx' \tag{22}$$

*exists for all $u \in C_0^\infty(\mathbb{R}, \mathscr{H}_0)$. The operator $K$ defined by (22) is bounded and $\|Ku\| \leqq c_{p,q} \|u\|_{L^2}$ where $c_{p,q}$ depends only on $k$, $p$ and $q$.*

Clearly, if $k \in \mathscr{G}_{p,q}(\mathscr{H}_0)$, then $k(x, x')$ is continuous in both variables in $\mathscr{L}(\mathscr{H}_0)$-norm and tends to zero like $|x|^{-p}$ or $|x'|^{-p}$ when $|x| \to \infty$ or $|x'| \to \infty$, respectively. As a simple example consider $k(x, x') = u(x) \big(v(x'), \cdot\big)$ where $u$ and $v$ are Hölder continuous $\mathscr{H}_0$-valued functions which vanish at inifinity like $|x|^{-p}$.

**Proposition 46.** *Let $K$ be defined as in (20). Assume $k \in \mathscr{G}_{p,q}(\mathscr{H}_0)$ for some numbers $p > 0$ and $q \in (0, 1)$. Then $K \in \operatorname{dom} \mu_s^H \cap \operatorname{dom} \mu_s^{-H}$ and $\mu_s^H(K) = \mu_s^{-H}(-K) = i\pi K_0$ where $K_0$ denotes the multiplication operator by $k(x) = k(x, x)$.*

**Proof.** It suffices to consider the convergence of $e^{itH} K e^{-itH}$ on the dense set $\mathscr{D} = C_0^\infty(\mathbb{R}, \mathscr{H}_0)$ in $\mathscr{H}$ since $K$ is bounded by Lemma 45. Let $u \in \mathscr{D}$ and let $\operatorname{supp} u \subseteqq [-a, a]$. Then

$$v_t(x) := (e^{itH} K e^{-itH} u)(x) = \int_{-a}^{a} (x - x')^{-1} k(x, x')\, e^{it(x-x')} u(x')\, dx'. \tag{23}$$

For $|x| \geqq a' > a$ we get the estimate $\|v_t(x)\|_{\mathscr{H}_0} < c\,|x|^{-1}$ where $c$ is independent of $t$. Thus, for each $\varepsilon > 0$, there is a number $b > a$ such that

$$\int_{\mathbb{R} \setminus [-b,b]} \|v_t(x)\|^2_{\mathscr{H}_0}\, dx < \varepsilon. \tag{24}$$

For $x \in (-b, b)$ we split up the integral over $x'$ in (23) into two terms, taking some number $b' > b$,

$$\int_{-b'}^{b'} (x - x')^{-1} k(x, x')\, e^{it(x-x')} u(x')\, dx' = \int_{-b'}^{b'} (x - x')^{-1} [k(x, x')\, u(x') - k(x, x)\, u(x)]\, e^{it(x-x')}\, dx'$$
$$+ k(x, x)\, u(x) \int_{-b'}^{b'} (x - x')^{-1} e^{it(x-x')}\, dx'. \tag{25}$$

The integrand of the first term on the right side is from $L^1(\mathbb{R}, dx'; \mathscr{H}_0)$ because of the Hölder continuity of $k(x, x')\, u(x')$. Thus, by the Riemann-Lebesgue Lemma, the first term on the right side tends to zero as $|t| \to \infty$ for each $x$. Let $w_t(x)$ denote this term and note that $\|w_t(x)\|_{\mathscr{H}_0} \leqq c$ for all $x \in [-b, b]$ (as a simple calculation shows), then the Lebesgue dominated convergence theorem yields $\lim_{t \to \pm\infty} \int_{-b}^{b} \|w_t(x)\|^2_{\mathscr{H}_0}\, dx = 0$. An elementary calculation shows that the second term tends strongly to $\pm i\pi k(x, x)\, u(x)$ as $t \to \pm\infty$ uniformly in $x \in [-b, b]$. Thus, for all $|t| > T(\varepsilon)$ we get, using (24) and the results on the terms in (25), that

$$\int_{-\infty}^{\infty} \|v_t(x) \mp i\pi k(x, x)\, u(x)\|^2_{\mathscr{H}_0}\, dx \leqq 2\varepsilon.$$

This gives the desired result. ∎

# 8. The Invariance of Wave Morphisms and Wave Algebras

If we consider the objects $\mu^H$, dom $\mu^H$ and ker $\mu^H$, then it is a natural question to ask under which transformations $H \to H'$ they remain invariant. Clearly, $\mu^H = \mu^{H'}$ implies dom $\mu^H =$ dom $\mu^{H'}$ and ker $\mu^H =$ ker $\mu^{H'}$, and we shall also see that dom $\mu^{H'} =$ dom $\mu^H$ implies ker $\mu^{H'} =$ ker $\mu^H$. However, the converse is not true.

We have already seen (see Prop. 6.40) that the transformations $H \to H' = aH + b$ with $a > 0$ and $b \in \mathbb{R}$ leave $\mu^H$ invariant, i.e. $\mu^H = \mu^{H'}$. On the other hand, we have ker $\mu^H \neq$ ker $\mu^{-H}$ (thus $\mu^H \neq \mu^{-H}$): Take the operator $X \notin$ dom $\mu^H$ from Prop. 6.36 and apply Prop. 6.53 which says $X = X_+ + X_-$ with $X_+ \in$ ker $\mu^H$ and $X_- \in$ ker $\mu^{-H}$, thus $X_- \notin$ ker $\mu^H$. This shows that not all functions of $H$ generate the same wave ideal.

Knowledge about the invariance of the wave morphism, wave algebra, and wave ideal is useful for two reasons. First, it gives us insight into the internal structure of these objects and into the problem how far these objects depend on the special operator $H$, which originally defined them, and second, it helps us to solve the question of whether or not an operator $X$ belongs to a wave algebra or wave ideal. For instance, when we have a bounded operator $X$ and we want to know if $X \in$ dom $\mu^H$, then it is sometimes easier to solve the question of whether $X \in$ dom $\mu^{H'}$ (with some $H'$) and to get $X \in$ dom $\mu^H$ from dom $\mu^H =$ dom $\mu^{H'}$.

In this chapter we call $H$ *generator of a wave algebra* (wave ideal) $\mathscr{A}$ if $\mathscr{A} =$ dom $\mu^H$ ($\mathscr{A} =$ ker $\mu^H$).

## 8.1. The class of generators of a wave algebra or wave ideal

### 8.1.1. Some classes of generators

The next statement provides a large and important class of generators for a given wave algebra.

**Proposition 1.** *Let $H'$ be a selfadjoint operator on $\mathscr{H}$ such that $H'P^{ac}_{H'} = UHU^*$ with $U^*U = P^{ac}$ and $UU^* = P^{ac}_{H'}$. Suppose that $U, U^* \in$ dom $\mu^H$ and $\mu^H(P^{ac}_{H'}) = P^{ac}$. Then* dom $\mu^H =$ dom $\mu^{H'}$.

**Proof.** Let $A \in$ dom $\mu^H$. Then, from the equality

$$\begin{aligned} e^{itH'} A\, e^{-itH' P^{ac}_{H'}} &= e^{itH' P^{ac}_{H'}} A\, e^{-itH' P^{ac}_{H'}} + e^{itH'}(1 - P^{ac}_{H'})\, A\, e^{-itH' P^{ac}_{H'}} \\ &= U\, e^{itH} U^* A U\, e^{-itH P^{ac}} U^* + e^{itH'}(1 - P^{ac}_{H'})\, AU\, e^{-itH P^{ac}} U^*, \end{aligned}$$

we obtain that $A \in \operatorname{dom} \mu^{H'}$ because of $(1 - P_{H'}^{ac}) \in \ker \mu^H$ and $U, U^* \in \operatorname{dom} \mu^H$. Thus $\operatorname{dom} \mu^H \subseteqq \operatorname{dom} \mu^{H'}$. Further, the above equality implies $\ker \mu^{H'} \supseteqq \ker \mu^H$.

Now assume $A \in \operatorname{dom} \mu^{H'}$. Then, from the equality

$$\begin{aligned} e^{itH} A\, e^{-itH} P^{ac} &= e^{itH} P^{ac} A\, e^{-itH} P^{ac} + e^{itH}(1 - P^{ac})\, A\, e^{-itH} P^{ac} \\ &= U^* e^{itH'} U A U^* e^{-itH'} P_{H'}^{ac} U + e^{itH}(1 - P^{ac})\, A U^* e^{-itH'} P_{H'}^{ac} U, \end{aligned}$$

$U, U^* \in \operatorname{dom} \mu^H \subseteqq \operatorname{dom} \mu^{H'}$ and $(1 - P^{ac}) \in \ker \mu^H \subseteqq \ker \mu^{H'}$, we obtain $A \in \operatorname{dom} \mu^H$. Thus $\operatorname{dom} \mu^H = \operatorname{dom} \mu^{H'}$. ∎

**Remark 2.** In the course of the proof of Prop. 1 we have verified that $\ker \mu^{H'} = \ker \mu^H$. That $\ker \mu^{H'} \supseteqq \ker \mu^H$ follows from $\operatorname{dom} \mu^{H'} \supseteqq \operatorname{dom} \mu^H$ is a general fact. Namely, let $A \in \ker \mu^H$, thus $DA \in \ker \mu^H$ for all $D \in \mathscr{L}(\mathscr{H})$ and this implies $DA \in \operatorname{dom} \mu^{H'}$ for all $D \in \mathscr{L}(\mathscr{H})$. By Prop. 6.36, this implies $A \in \ker \mu^{H'}$. Thus $\ker \mu^H = \ker \mu^{H'}$ is a necessary condition for the equality of the wave algebras $\operatorname{dom} \mu^H = \operatorname{dom} \mu^{H'}$.

**Remark 3.** In Part III we shall see that operators $H$, $H'$, which posses complete wave operators $U := W_+(H, H')$, satisfy the assumptions of Prop. 1 (i.e. $U$ and $U^*$ satisfy these assumptions). Thus the results of Part IV on existence and completeness of wave operators can be applied to the questions considered here.

An easy consequence of Prop. 1 is

**Corollary 4.** *Let $\mathscr{A}$ be a wave algebra, i.e. $\mathscr{A} = \operatorname{dom} \mu^H$ for some $H$. Then there is a selfadjoint operator $H'$ such that $\mathscr{A} = \operatorname{dom} \mu^{H'}$ and $H'$ is absolutely continuous.*

Proof. Assume $P^{ac} < 1$. Clearly, by definition, $\mu^H(1 - P^{ac}) = 0$. Thus, according to Theorem 7.8, there is a partial isometry $W$ such that $W, W^* \in \operatorname{dom} \mu^H$, $W^*W = P^{ac}$ and $WW^* = 1$. Now define $H' := WHW^*$. $H'$ is unitarily equivalent to the absolutely continuous part of $H$ and therefore $H'$ is absolutely continuous. The assumptions of Prop. 1 are satisfied. Thus $\operatorname{dom} \mu^H = \operatorname{dom} \mu^{H'}$. ∎

In the following statement the transformations $H \to aH + b$, $a > 0$ and $b \in \mathbb{R}$, which leave invariant $\mu^H$, are slightly generalized.

**Proposition 5.** *Let $\alpha$ be defined by:*

(I) $\alpha(x) = \sum_{n=1}^{N} \chi_{\Delta_n}(x)\,[a_n x + b_n]$ *with* $N \leqq \infty$, $a_n > 0$, $b_n \in \mathbb{R}$, *and* $\Delta_n$ *disjoint Borel sets,*

(II) $\bigcup_{n=1}^{N} \Delta_n \supseteqq \operatorname{spec} H$ *and* $\alpha(\Delta_n) \cap \alpha(\Delta_{n'}) = \emptyset$ *for* $n \neq n'$. *Then* $\mu^{\alpha(H)} = \mu^H$.

Proof. $\alpha$ is strictly linear increasing on each set $\Delta_n$ and it is easy to see that $\alpha^{-1}(x)$ exists on $\alpha\left(\bigcup_{n=1}^{N} \Delta_n\right)$ and has the same form as $\alpha(x)$, i.e. $\alpha^{-1}(x)$ is strictly linear increasing on the Borel sets $\alpha(\Delta_n)$. Obviously, $\alpha$ transforms null sets into null sets and so does $\alpha^{-1} \restriction \alpha\left(\bigcup_{n=1}^{N} \Delta_n\right)$. According to Prop. 3.30 we then get $P^{ac} = P_{\alpha(H)}^{ac}$. Further, since $\alpha^{-1}(\alpha(H)) = H$ (set $\alpha^{-1}(x) = 0$ on $\mathbb{R} \setminus \bigcup_{n=1}^{N} \alpha(\Delta_n)$), we have $\operatorname{com} H = \operatorname{com} \alpha(H)$ and therefore $\operatorname{ima} \mu^H = \operatorname{ima} \mu^{\alpha(H)} = (\operatorname{com} H) \restriction P^{ac}$. For the proof of $\mu^H = \mu^{\alpha(H)}$ it remains to show that $\ker \mu^H = \ker \mu^{\alpha(H)}$.

Set $Q_n = E(\Delta_n) P^{ac}$. Then $Q_n \in \operatorname{com} H$, $Q_n Q_{n'} = Q_{n'} Q_n = 0$ for $n \neq n'$, and $\sum_{n=1}^{N} Q_n = P^{ac}$. Let $A \in \ker \mu^H$. Then $AQ_n \in \ker \mu^H$ for all $n \in \{1, 2, \ldots, N\}$. Now $\|AQ_n \, e^{-it\alpha(H)} P^{ac} u\| = \|A \, e^{-it(a_n H + b)} Q_n u\|$. Since $\mu^H = \mu^{a_n H + b}$, by Prop. 6.40, this equation implies $AQ_n \in \ker \mu^{\alpha(H)}$ for each $n \in \{1, 2, \ldots, N\}$ and thus $A \in \ker \mu^{\alpha(H)}$, according to Prop. 6.44. The converse, namely that $A \in \ker \mu^{\alpha(H)}$ implies $A \in \ker \mu^H$, can be proved in the same manner using the equation $\|AQ_n \, e^{-itH} P^{ac} u\| = \|AQ_n \, e^{-i(t/a_n)(a_n H + b_n)} u\| = \|AQ_n \, e^{-i(t/a_n)\alpha(H)} P^{ac} u\|$. ∎

**Remark 6.** In the course of the proof we have verified a little more than Prop. 5 states, namely that $\ker \mu^H = \ker \mu^{\alpha(H)}$ without the assumption that $\alpha^{-1}$ exists on $\alpha\left(\bigcup_{n=1}^{N} \Delta_n\right)$, i.e. without the assumption $\alpha(\Delta_n) \cap \alpha(\Delta_{n'}) = \emptyset$ for $n \neq n'$. We used only $P^{ac} = P^{ac}_{\alpha(H)}$, but this follows from the special form of $\alpha$ and need not the assumption mentioned above.

From Prop. 5 we easily get that in the class of selfadjoint operators $H'$ such that $\mu^H = \mu^{H'}$ where $H$ is fixed there always exist operators $H'$ which are bounded and operators $H'$ which are unbounded.

**Example 7** (bounded generator). Assume $H$ is unbounded. Define $\alpha$ by $\alpha(x) = \sum \chi_{[n,n+1)}(x) \, [a_n x + b_n]$ for $x \geq 0$ and by $\alpha(x) := -\alpha(-x)$ for $x < 0$, where $a_n = 2^{-n}$ and $b_n = 2 - (n + 2) \, 2^{-n}$. It is not hard to verify that $\alpha$ satisfies the assumptions of Prop. 5 and thus $\mu^H = \mu^{\alpha(H)}$. Furthermore, we have $|\alpha(x)| \leq 2$ as a simple calculation shows. Thus $H' = \alpha(H)$ is a bounded generator of the wave morphism $\mu^H$.

**Example 8** (unbounded generator). Assume $H$ is bounded. Now we can subdivide $\operatorname{spec}^c H$ into infinitely many disjoint Borel sets such that $\bigcup_{n=1}^{\infty} \Delta_n = \operatorname{spec}^c H$ and $E(\Delta_n) > 0$ for each $n \in \mathbb{N}$. Let $c > 2 \, \|H\|$. Then define $\alpha$ by $\alpha(x) = \sum_{n=1}^{\infty} \chi_{\Delta_n}(x) \, [x + cn]$. The function $\alpha$ satisfies the assumptions of Prop. 5 and thus $\mu^H = \mu^{\alpha(H)}$. Furthermore, we have $\|\alpha(H) \, E(\Delta_n)\| = \|E(\Delta_n) \, [H + cn]\| \geq \big| \|HE(\Delta_n)\| - cn \big| \geq c(n - (1/2))$, and this implies that $\alpha(H)$ is an unbounded generator of the wave morphism $\mu^H$.

It is obvious that the class of generators of a given wave ideal $\mathcal{J} = \ker \mu^H$ is larger than the class of generators of the corresponding wave algebra $\mathcal{A} = \operatorname{dom} \mu^H$. This follows already from Remark 6. The following statement on generators of wave ideals does not remain true if we replace wave ideal by wave algebra.

**Proposition 9.** *In the class of generators of a fixed wave ideal $\mathcal{J} = \ker \mu^H$ there is a generator $H'$, i.e. $\mathcal{J} = \ker \mu^{H'}$, which is absolutely continuous and is simple.*

**Proof.** By Corollary 4 we can assume that $P^{ac} = 1$. Further, there is a set of projections $Q_n$, $n \in \{1, 2, \ldots, N \leq \infty\}$, such that $Q_n \in \operatorname{com} H$, $Q_n Q_{n'} = Q_{n'} Q_n = 0$ for $n \neq n'$, $\sum_{n=1}^{N} Q_n = 1$ and $H \restriction Q_n \mathcal{H}$ is simple and bounded with the spectrum $\Delta_n$ (see Chapter 4). Then it is not difficult to construct a set of functions $\alpha_n(x) = a_n x + b_n$, $a_n > 0$ and $b_n \in \mathbb{R}$ such that $\alpha_n(\Delta_n) \cap \alpha_{n'}(\Delta_{n'}) = \emptyset$ if $n \neq n'$. Thus the operator $H' := \sum_{n=1}^{N} \alpha_n(H) \, Q_n = \sum_{n=1}^{N} (a_n H + b_n) \, Q_n$ is clearly absolutely continuous and is simple. It remains to show $\ker \mu^H = \ker \mu^{H'}$. But this follows from the equality

$$\|AQ_n \, e^{-itH} u\| = \|AQ_n \, e^{-i(t/a_n)(a_n H + b_n)} u\| = \|AQ_n \, e^{-i(t/a_n)H'} u\|$$

with the help of Prop. 6.44. ∎

The statements in this section show that the generators of a given wave morphism, wave algebra or wave ideal can considerably differ in their spectral properties. In the next section we shall see that only the multiplicity of the absolutely continuous spectrum is in a certain sense an invariant for the class of generators of a wave algebra.

### 8.1.2. On the class of generators of the wave algebra dom $\mu^H$

In Section 7.3.2 we studied the properties of the wave morphism $\mu^H$ restricted on certain v. Neumann algebras $\mathscr{A}$. Under additional assumptions on $\mathscr{A}$ it could be shown that $\mu^H \mid \mathscr{A}$ is a spatial isomorphism from $\mathscr{A}$ onto $\mu^H(\mathscr{A})$ realized by a partial isometry $W$ which is a limit of $e^{itH}T_t$ where $T_t$ commutes with all elements from $\mathscr{A}$ and is asymptotically isometric.

The following statement presents a necessary condition for the equality of two wave algebras dom $\mu^H$, dom $\mu^{H_0}$ within the framework of Section 7.3.2, where the role of the $W^*$-algebra $\mathscr{A}$ contained in dom $\mu^H$ is played by com $H_0$.

Later, within the framework of the algebraic scattering theory considered in Part III (Chapter 10), the partial isometry $W$ mentioned above is called the *generalized wave* operator of the system $\{H, H_0, \mathscr{A} = \text{com } H_0\}$.

**Theorem 10.** *Let $H$ and $H_0$ be two generators of a wave algebra $\mathscr{A}$, i.e. $\mathscr{A} = \text{dom } \mu^H = \text{dom } \mu^{H_0}$. Then there exist a real-valued, a.e.* (mod $|\cdot|$) *finite Borel function $\alpha$ on $\mathbb{R}$ and an operator-valued function $T_t\colon \mathbb{R} \to \mathscr{C}(\mathscr{H})$ defined on a dense set $\mathscr{D}$ in $H$ such that:*

(i) $(\text{com } \alpha(H_0)) \mid P^{ac}_{H_0} = (\text{com } H_0) \mid P^{ac}_{H_0}$ and $P^{ac}_{H_0} = P^{ac}_{\alpha(H_0)}$,

(ii) *$T_t$ is for each $t \in \mathbb{R}$ a function of $H_0$,*

(iii) *$\mathscr{D} \subseteq \text{dom } T_t$ for all $t \in \mathbb{R}$,*

(iv) $\lim\limits_{t\to\infty} \|T_t u\| = \|u\|$ *for $u \in \mathscr{D}$ in case $\mu^H = \mu_s{}^H$* and $\text{A-}\lim\limits_{t\to\infty} \|T_t u\| = \|u\|$ *for $u \in \mathscr{D}$ in case $\mu^H = \mu_a{}^H$,*

(v) *$T_t = e^{-it\alpha(H_0)}G_t$ with* $\operatorname{Lim}\limits_{t\to\infty} (G_{t+s} - G_t)\, u = 0$ *for $s \in \mathbb{R}$, $u \in \mathscr{D}$,*

(vi) *$W_+ u := \operatorname{Lim}\limits_{t\to\infty} e^{itH}T_t P^{ac}_{H_0} u$ exists for all $u \in \mathscr{D}$, and $W_+{}^*W_+ = P^{ac}_{H_0}$, $W_+W_+{}^* = P^{ac}$ and $W_+ A W_+{}^* = \mu^H(A)$ for all $A \in \text{com } H_0$.*

**Proof.** From dom $\mu^H = \text{dom } \mu^{H_0}$ we get $\ker \mu^H = \ker \mu^{H_0}$ (see Remark 2) and ima $\mu^{H_0} \subset \text{dom } \mu^H$, ima $\mu^H \subset \text{dom } \mu^{H_0}$. This implies, by Prop. 7.20, that $\mu^H$ is an isometric $*$-isomorphism from ima $\mu^{H_0} = \text{com } H_0 \mid P^{ac}_{H_0}$ onto ima $\mu^H = \text{com } H \mid P^{ac}$. Further, ima $\mu^{H_0} \mid P^{ac}_{H_0}\mathscr{H}$ and ima $\mu^H \mid P^{ac}\mathscr{H}$ are v. Neumann algebras and $(\text{com}(\text{ima } \mu^{H_0})) \mid P^{ac}_{H_0} = (\text{com}(\text{com } H_0)) \mid P^{ac}_{H_0}$, $(\text{com}(\text{ima } \mu^H)) \mid P^{ac} = (\text{com}(\text{com } H)) \mid P^{ac}$ are Abelian. Furthermore, $\mu^H(P^{ac}_{H_0}) = \mu^H(1) + \mu^H(P^{ac}_{H_0} - 1) = P^{ac}$, since $P^{ac}_{H_0} - 1 \in \ker \mu^{H_0} = \ker \mu^H$. Thus the assumptions of Theorem 7.22 are fulfilled (with $E = P^{ac}_{H_0}$ and $P = P^{ac}$), and this together with Theorem 7.23 proves all assertions except (i) and (v).

From Prop. 7.24 and its proof we get that $T_t = e^{-itK}G_t$ where $G_t$ satisfies (v), $e^{itK} \in \text{com}(\text{com } H_0)$, $P^{ac}_K = P^{ac}_{H_0}$ and $W_+KW_+{}^* = HP^{ac}$. This implies $W_+ (\text{com } K)\, W_+{}^* = (\text{com } H) \mid P^{ac}$ and, by $(\text{com } H) \mid P^{ac} = W_+(\text{com } H_0)\, W_+{}^*$, that $(\text{com } H_0) \mid P^{ac}_{H_0} = (\text{com } K) \mid P^{ac}_{H_0}$. From this and $K(1 - P^{ac}_{H_0}) = 0$ (or from $e^{itK} \in \text{com}(\text{com } H_0)$) we obtain that $K = \alpha(H_0)$ where $\alpha$ is some real-valued Borel function which is a.e. (mod $|\cdot|$) *finite.* ∎

Evidently, the converse, namely that the existence of $\alpha$, $T_t$ and $W_+$ satisfying (i)—(vi) for $H_0$, $H$ implies dom $\mu^H$ = dom $\mu^{H_0}$ is not true. For instance, take the case $H_0 = -H$. Then all assumptions are satisfied, but we have dom $\mu^H \neq$ dom $\mu^{H_0}$ = dom $\mu^{-H}$ (see the remarks at the beginning of this chapter).

Up to now we have not given any example where dom $\mu^H$ = dom $\mu^{H_0}$ and $G_t$ is not trivial, i.e. $G_t = 1$ or $\operatorname{s-lim}_{t\to\infty} G_t = 1$. The following example provides two generators $H_0$ and $H$ of a wave algebra with a nontrivial $G_t$, that is, where $\operatorname{Lim}_{t\to\infty} e^{itH} e^{-it\alpha(H_0)} P^{ac}_{H_0}$ does not exist for any function $\alpha$.

**Example 11.** Let $H_0 = -\mathrm{i}d/dx$ on $\mathscr{H} = L^2(\mathbb{R}, dx)$ and $H = H_0 + q(\cdot)$ with $q(x) = 1$ for $|x| \leqq 1$ and $q(x) = |x|^{-1}$ for $|x| > 1$. Clearly, $H$ and $H_0$ are selfadjoint operators with the usual definition of the domains, and $e^{-itH} = U e^{-itH_0} U^*$ with $(Uf)(x) = e^{-ip(x)} f(x)$ where $p(x) = x$ for $|x| \leqq 1$ and $p(x) = (\operatorname{sgn} x) \log |x|$ for $|x| \geqq 1$. Now define $T_t := e^{-itH_0} e^{-ip(t)}$. Obviously $T(t)$ is unitary for all $t$ and is a function of $H_0$, for all $t$. Further, for $t > 1$ and $x + t > 1$, we have

$$(e^{itH} T_t u)(x) = (U e^{itH_0} U^* e^{-itH_0} e^{-i\log t} u)(x)$$

$$= e^{-ip(x)} e^{ip(x+t)} e^{-i\log t} u(x) = e^{-ip(x)} e^{i\log(1+x/t)} u(x).$$

Because of $\log(1 + x/t) \to 0$ as $t \to \infty$ for all $x \in \mathbb{R}$, this equality proves that $e^{itH} T_t u \xrightarrow[s]{} v$ with $v(x) = e^{-ip(x)} u(x)$. That is, $U = \operatorname{s-lim}_{t\to\infty} e^{itH} T_t$ exists and is a unitary operator. This implies that also $U^* = \operatorname{s-lim}_{t\to\infty} T_t^* e^{-itH}$ exists (use the equality $\|T_t^* e^{-itH} u - U^* u\|^2 = 2\|u\|^2 - 2 \operatorname{Re}(Uu, e^{itH} T_t u)$ for the proof). From the existence of these strong limits and the equality

$$e^{itH} A e^{-itH} = e^{itH} T_t e^{-itH_0} A e^{-itH_0} T_t^* e^{-itH}$$

we then easily get that $A \in$ dom $\mu^{H_0}$ implies $A \in$ dom $\mu^H$. Similarly, one proves the converse relation and so we get dom $\mu^H$ = dom $\mu^{H_0}$. Here we have proved the existence of the generalized wave operator $W_+ = U$ by construction of $T_t$. It is easy to see that the existence of $\operatorname{s-lim}_{t\to\infty} e^{itH} T_t$, with our special $G_t = e^{-ip(t)}$ implies that $\operatorname{s-lim}_{t\to\infty} e^{itH} e^{-it\alpha(H_0)}$ cannot exist.

From Theorem 10 (vi) we obtain that $W_+(\operatorname{com} H_0) P^{ac}_{H_0} W_+^* = (\operatorname{com} H) \restriction P^{ac}$ follows from dom $\mu^H$ = dom $\mu^{H_0}$. Thus the multiplicities of the absolutely continuous parts of the generators are invariants. For instance, if $H_0 \restriction P^{ac}_{H_0}\mathscr{H}$ has multiplicity $n$, then the same is true for the other generator $H$.

### 8.1.3. The class of generators of the wave ideal $\mathscr{I} = \ker \mu_a^H$

Our aim in this section is to prove that the existence of some limits in terms of integrals over $e^{itH} e^{-isH_0}$ are necessary (and also sufficient under some additional assumptions on $H_0$, $H$) for the equality $\ker \mu_a^H = \ker \mu_a^{H_0}$. One can prove similar statements for the generators of the strong wave ideal $\ker \mu_s^H$, but we omit them because they are more complicated and have not led to useful implications up to this date. That this is no accident show the following paragraphs; the wave morphism $\mu_a^H$ allows more general assertions than the wave morphism $\mu_s^H$, and with respect to this $\mu_a^H$ is a more pleasant mathematical object than $\mu_s^H$.

We begin with a statement true for both kinds of wave ideals.

**Proposition 12.** *Let $H, H_0$ be selfadjoint on $\mathscr{H}$. Assume that* $\ker \mu^{H_0} \subseteq \ker \mu^H$. *Then for all $v \in P^{ac}\mathscr{H}$ and all $u \in \mathscr{M}_\infty(H_0)$*

$$\operatorname*{Lim}_{t\to\infty} \int_{-\infty}^{0} |(e^{-isH_0}u, e^{-itH}v)|^2 \, ds = 0. \tag{1}$$

**Proof.** Because of $\int |(e^{-isH_0}u, f)|^2 \, ds \leq 2\pi \, |||u|||^2_{H_0,\infty} \, \|f\|^2$ (see Prop. 3.22) it is sufficient to prove the assertion for a generating set $\mathscr{D}$ in $\mathscr{M}_\infty(H_0)$ (with respect to $|||\cdot|||_{H_0,\infty}$). As such set $\mathscr{D}$ we chose the generating set given by Lemma 4.27. According to this lemma, it is easy to see that it is no loss of generality to assume that there exist vectors $u \in \mathscr{M}_\infty(H_0)$ such that for each $u$ there is $a > 0$ so that $e^{imaH_0}u$, $m = 0, \pm 1, \pm 2, \ldots$, is an orthonormal system, and that it suffices to consider these vectors $u$. Put $u_m = e^{imaH_0}u$ and $g_t = e^{-itH}v$. Then

$$\int_{-\infty}^{0} |(e^{-isH_0}u, e^{-itH}v)|^2 \, ds = \sum_{n=0}^{\infty} \int_{-a(n+1)}^{-an} |(e^{-isH_0}u, g_t)|^2 \, ds$$

$$= \sum_{n=0}^{\infty} \int_{0}^{a} |(e^{-isH_0}u_{n+1}, g_t)|^2 \, ds = \int_{0}^{a} \left( \sum_{n=0}^{\infty} |(u_{n+1}, e^{isH_0}g_t)|^2 \right) ds.$$

Now we define a projection $Q$ by $Qf = \sum (u_{n+1}, f) \, u_{n+1}$, then $\|Qf\|^2 = \sum_{n\geq 0} |(u_{n+1}, f)|^2$. Thus

$$\int_{-\infty}^{0} |(e^{-isH_0}u, e^{-itH}v)|^2 \, ds = \int_{0}^{a} \|Q \, e^{isH_0}g_t\|^2 \, ds. \tag{2}$$

Further, we have $\|Q \, e^{-imaH_0}f\|^2 = \sum_{n\geq 0} |(u_{n+1+m}, f)|^2 \to 0$, for $f \in \mathscr{H}_0$, as $m \to \infty$. Hence $Q \in \ker \mu_s^{H_0}$ follows by a similar argument as in the proof of Lemma 6.41 ($f$ is substituted by $e^{-i\tau aH_0}f$, $0 \leq \tau \leq 1$, which is a compact set). This implies $Q \, e^{isH_0} \in \ker \mu_s^{H_0} \subseteq \ker \mu^H$. Because of $\|Q \, e^{isH_0}g_t\| \leq \|g_t\| = \|v\|$ and the Lebesgue dominated convergence theorem, (1) follows from (2). ■

**Remark 13.** First, we note that in (1) we can replace $\int_{-\infty}^{0}$ by $\int_{-\infty}^{c}$ where $c$ is an arbitrary real number. Secondly, using the equality

$$(2\pi)^{-1} \int_{-\infty}^{\infty} \|XR_{H_0}(\lambda + i\varepsilon) \, g\|^2 \, d\lambda = \int_{0}^{\infty} e^{-2\varepsilon t} \, \|X \, e^{-itH_0}g\|^2 \, dt$$

(see (11) in Chapter 3), we get from (1) that

$$\lim_{t\to\infty} \lim_{\varepsilon\to+0} \int_{-\infty}^{\infty} |(e^{-itH}v, R_{H_0}(\lambda - i\varepsilon) \, u)|^2 \, d\lambda = 0.$$

Since $\lim_{\varepsilon\to+0} \lim_{t\to\infty} \int \left|(e^{-itH}v, R_{H_0}(\lambda - i\varepsilon) \, u)\right|^2 d\lambda = 0$ is trivially satisfied, by $\lim_{t\to\infty} (f, e^{-itH}v) = 0$ for $v \in P^{ac}_H\mathscr{H}$, this relation says that we can interchange the two limits, $\lim_{t\to\infty}$ and $\lim_{\varepsilon\to+0}$.

**Proposition 14.** *Let $H, H_0$ be selfadjoint on $\mathscr{H}$. Assume* $\ker \mu_a^{H_0} \subseteq \ker \mu_a^H$. *Then for all $v \in P^{ac}\mathscr{H}$ and all $u \in \mathscr{M}_\infty(H_0)$*

$$\lim_{p\to\infty} \sup_{T\geq 1} T^{-1} \int_{0}^{T} dt \int_{pT}^{\infty} |(e^{-isH_0}u, e^{-itH}v)|^2 \, ds = 0. \tag{3}$$

Proof. 1. Reasoning as in the proof of Prop. 12, we get that it is enough to prove (3) for a generating set $\mathscr{D}$ in $\mathscr{M}_\infty(H_0)$ and that we can assume there are vectors $u$ (and it suffices to consider such vectors $u$) so that for each $u$ there is $a > 0$ such that $e^{imaH_0}u$, $m = 0, \pm 1, \ldots$, is an orthonormal basis in clo spa $\{e^{itH_0}u; t \in \mathbb{R}\}$. Now we fix such $u$ (without loss of generality we may assume $a = 1$). In the following it is sufficient to consider the simple (cyclic) part of $H_0$ with the generating vector $u$ and we may assume this part of $H_0$ to be bounded.

2. Now we assume that (3) is not true. Then there exist sequences $\varrho_n \to \infty$ and $T_n \geq 1$ such that $T_n^{-1} \int_0^{T_n} dt \int_{\varrho_n T_n}^{\infty} w(t, s)\, ds \geq \beta > 0$ where $\beta$ is some constant and $w(t, s) = |(e^{-isH_0}u, e^{-itH}v)|^2$. It is easy to see that necessarily $T_n \to \infty$ as $n \to \infty$. Put $R_n = \varrho_n T_n$. Without loss of generality we can assume that $T_n$, $R_n$ are natural numbers.

The next step is the construction of subsequences by the following process which proceeds according to the diagram

$$\begin{array}{ccccccc} T_1 \to T_2' & & T_3' \to T_4' & & \cdots \\ & \downarrow & \uparrow & \downarrow & \\ R_1 & & R_2' \to R_3' & & R_4' \to \end{array}$$

Let $0 < \varepsilon < \beta/2$ be given. Then $T'_{n+1}$ is defined by

$$(T'_{n+1})^{-1} \int_0^{T_n'} dt \int_{R_n'}^{\infty} w(t, s)\, ds < \varepsilon$$

and $R'_{n+2}$ is defined by $(T'_{n+1})^{-1} \int_0^{T'_{n+1}} dt \int_{R'_{n+1}}^{R'_{n+2}} w(t, s)\, ds > (\beta/2 + \varepsilon)$. Then we obtain

$$A_n := (T'_{n+1})^{-1} \int_{T_n'}^{T'_{n+1}} dt \int_{R'_{n+1}}^{R'_{n+2}} w(t, s)\, ds = (T'_{n+1})^{-1} \int_0^{T'_{n+1}} dt \int_{R'_{n+1}}^{R'_{n+2}} w(t, s)\, ds$$
$$- (T'_{n+1})^{-1} \int_0^{T_n'} dt \int_{R'_{n+1}}^{R'_{n+2}} w(t, s)\, ds > (\beta/2 + \varepsilon) - \varepsilon = \beta/2. \tag{4}$$

For convenience we denote the subsequences $T_n'$, $R_n'$ by $T_n$, $R_n$ again.

3. The last step is to calculate $A_n$ and to show that $A_n \to 0$ as $n \to \infty$ which is a contradiction to (4). First we associate to each section $(R_n, R_{n+1}]$ the projection $P_n$ defined by $P_n f = \sum_{k \in \mathscr{I}_n} (u_k, f)\, u_k$ where $\mathscr{I}_n = \{R_n + 1, R_n + 2, \ldots, R_{n+1}\}$ and $u_k = e^{-i(k-1)H_0}u$. Obviously $P_n P_{n'} = \delta_{n,n'} P_n$. Let $Q_n \leq P_n$ be a subprojection and put $Q = \sum Q_n$. We claim $Q \in \ker \mu_a^H$ when $\dim Q_n \mathscr{H} \leq T_n - T_{n-1}$. According to Corollary 6.42 and the fact that $H_0$ is bounded on clo spa $\{e^{itH_0}u; t \in \mathbb{R}\} =$ clo spa $\{e^{inH_0}u; n = 0, \pm 1, \ldots\}$, we have to prove $\lim_{N \to \infty} N^{-1} \sum_{\varrho=1}^{N} \|Qu_\varrho\|^2 = 0$. Now we have

$$N^{-1} \sum_{\varrho=1}^{N} \|Qu_\varrho\|^2 \leq N^{-1} \sum_{n=1}^{R_{n_0}} \sum_{\varrho \in \mathscr{I}_n} \|Q_n u_\varrho\|^2 \leq N^{-1} \sum_{n=1}^{R_{n_0}} \operatorname{tr} Q_n$$
$$\leq N^{-1} \sum_{n=1}^{R_{n_0}} (T_n - T_{n-1}) \leq (R_{n_0})^{-1} T_{n_0}$$

where $R_{n_0}$ is determined by $R_{n_0} < N \leqq R_{n_0+1}$. By definition ($R_n = T_n \varrho_n$), the right side tends to zero as $N \to \infty$ and thus the left side tends to zero as $N \to \infty$.

Further, set $\mathscr{K}_n = \{T_n + 1, T_n + 2, \ldots, T_{n+1}\}$ and $v_m = e^{-i(m-1)H}v$. Then

$$A_n = (T_{n+1})^{-1} \int_0^1 dt \int_0^1 \left( \sum_{m \in \mathscr{K}_n} \sum_{k \in \mathscr{I}_{n+1}} \left|(u_k\, e^{i\sigma H_0}\, e^{-itH} v_m)\right|^2 \right) d\sigma$$

$$= (T_{n+1})^{-1} \int_0^1 dt \int_0^1 \left( \sum_{m \in \mathscr{K}_n} \|P_{n+1}\, e^{i\sigma H_0}\, e^{-itH} v_m\|^2 \right) d\sigma. \tag{5}$$

Now we consider spa $\{P_{n+1}\, e^{i\sigma H_0}\, e^{-itH} v_m;\, m \in \mathscr{K}_n\}$ and denote the projection onto this subspace by $Q^{\sigma,t}_{n+1}$. Then $Q^{\sigma,t}_{n+1} \leqq P_{n+1}$ and $\dim Q^{\sigma,t}_{n+1}\mathscr{H} \leqq T_{n+1} - T_n$. Hence $Q^{\sigma,t} = \sum Q^{\sigma,t}_{n+1}$ belongs to $\ker \mu_a^{H_0} \subseteqq \ker \mu_a^{H}$. Further

$$\|P_n\, e^{i\sigma H_0}\, e^{-itH} v_m\| = \|P_n Q_n^{\sigma,t}\, e^{i\sigma H_0}\, e^{-itH} v_m\| \leqq \|Q^{\sigma,t}\, e^{i\sigma H_0}\, e^{-itH} v_m\|.$$

Clearly, $Q^{\sigma,t}\, e^{i\sigma H_0}\, e^{-itH} \in \ker \mu_a^{H}$ for all $\sigma, t \in [0, 1]$. That is,

$$(T_{n+1})^{-1} \sum_{m \in \mathscr{K}_n} \|P_{n+1}\, e^{i\sigma H_0}\, e^{-itH} v_m\|^2 \leqq (T_{n+1})^{-1} \sum_{m=1}^{T_{n+1}} \|Q^{\sigma,t}\, e^{-i\sigma H_0}\, e^{-itH} v_m\|^2 \to 0,\ n \to \infty,$$

according to Lemma 6.9. This and the estimate $(T_{n+1})^{-1} \sum_{m=1}^{T_{n+1}} \|Q^{\sigma,t}\, e^{i\sigma H_0}\, e^{-itH} v_m\|^2 \leqq \|v\|^2$ imply that $A_n \to 0$ as $n \to \infty$ by the Lebesgue dominated convergence theorem. ∎

With the following statement we finish the discussion on the generators of a given wave ideal $\ker \mu_a^{H}$. At least in case that $H$ and $H_0$ are bounded and they have finite multiplicity the question whether $\ker \mu^{H} = \ker \mu^{H_0}$ is completely solved with

**Theorem 15.** *Let $H_0$ be a bounded selfadjoint operator on $\mathscr{H}$ such that the multiplicites of the homogeneous parts of $H_0 P^{ac}_{H_0}$ are bounded by $m < \infty$ and $P^{ac}_{H_0} \geqq P^{ac}$. Suppose that for all $v \in \mathscr{M}_\infty(H)$ and all $u \in \mathscr{M}_\infty(H_0)$ we have*

(I) $$\lim_{T \to \infty} T^{-1} \int_0^T dt \int_{-\infty}^0 |(e^{-isH_0}u,\, e^{-itH}v)|^2\, ds = 0,$$

(II) $$\lim_{\varrho \to \infty} \sup_{T \geqq 1} T^{-1} \int_0^T dt \int_{\varrho T}^\infty |(e^{-isH_0}u,\, e^{-itH}v)|^2\, ds = 0.$$

*Then* $\ker \mu_a^{H_0} \subseteqq \ker \mu_a^{H}$.

**Proof.** We have $P^{ac}_{H_0}\mathscr{H} = \bigoplus_{s=1}^{m} \operatorname{clo\,spa} \{E_{H_0}(\Delta)\, u_s;\, \Delta \subset \mathbb{R}\}$ where the generating vectors $u_s$, $s = 1, 2, \ldots, m$, can be chosen such that $u_s \in \mathscr{M}_\infty(H_0)$ and $\|g\|^2 = (1/2\pi) \int |(e^{-isH_0}u_\varrho, g)|^2\, ds$ for all $g \in \operatorname{clo\,spa} \{E_{H_0}(\Delta)\, u_\varrho;\, \Delta \subset \mathbb{R}\}$. For $g \in P^{ac}_{H_0}\mathscr{H}$ we get

$$\|g\|^2 = (1/2\pi) \sum_{\varrho=1}^{m} \int_{-\infty}^{\infty} |(e^{-isH_0}u_\varrho, g)|^2\, ds. \tag{6}$$

Now we define $S_\Delta(A, v, T)$ by

$$S_\Delta(A, v, T) = \left( (2\pi T)^{-1} \sum_{\varrho=1}^{m} \int_0^T dt \int_\Delta |(e^{-isH_0}u_\varrho,\, A\, e^{-itH}v)|^2\, ds \right)^{1/2}. \tag{7}$$

$S_\Delta(A, v, T)$ satisfies $S_\Delta(A + B, v, T) \le S_\Delta(A, v, T) + S_\Delta(B, v, T)$. Further

$$S_{\mathbb{R}}(A, v, T) = \left(T^{-1} \int_0^T \|A\, e^{-itH} v\|^2 \, dt\right)^{1/2}. \tag{8}$$

Now we choose $\Delta = [0, pT]$ and fix $\varepsilon > 0$. Then, according to (I), (II), and (6), there is $p(\varepsilon) > 0$ and $T(\varepsilon) > 0$ such that for all $p > p(\varepsilon)$, $T > T(\varepsilon)$

$$\|v\| \ge S_{[0,pT]}(1, v, T) \ge (\|v\|^2 - \varepsilon)^{1/2}. \tag{9}$$

Now let $P \in \ker \mu_a^{H_0}$ be a projection. Then

$$S_{[0,pT]}(1, v, T) \le S_{[0,pT]}(P, v, T) + S_{[0,pT]}(P^\perp, v, T).$$

Using (8), we obtain

$$S_{[0,pT]}(P^\perp, v, T) \le S_{\mathbb{R}}(P^\perp, v, T) = \left(T^{-1} \int_0^T \|P^\perp e^{-itH} v\|^2 \, dt\right)^{1/2}$$

$$= \left(\|v\|^2 - T^{-1} \int_0^T \|P\, e^{-itH} v\|^2 \, dt\right)^{1/2},$$

and on the other hand

$$\sum_{\varrho=1}^m (2\pi)^{-1} \int_0^T |(e^{-isH_0} u_\varrho, P\, e^{-itH} v)|^2 \, dt$$

$$= (2\pi)^{-1} \sum_{\varrho=1}^m \int_0^T |(P\, e^{-isH_0} u_\varrho, e^{-itH} v)|^2 \, dt \le |||v|||_{H,\infty} \sum_{\varrho=1}^m \|P\, e^{-isH_0} u_\varrho\|^2,$$

that is

$$S_{[0,pT]}(P, v, T) \le |||v|||_{H,\infty} \left(\sum_{\varrho=1}^m T^{-1} \int_0^{pT} \|P\, e^{-isH_0} u_\varrho\|^2 \, ds\right)^{1/2}.$$

Thus from (9) and $S_{[0,pT]}(1, v, T) \le S_{[0,pT]}(P, v, T) + S_{[0,pT]}(P^\perp, v, T)$ we get

$$(\|v\|^2 - \varepsilon)^{1/2} \le |||v|||_{H,\infty} \left(\sum_{\varrho=1}^m T^{-1} \int_0^{pT} \|P\, e^{-isH_0} u_\varrho\|^2 \, ds\right)^{1/2}$$

$$+ \left(\|v\|^2 - T^{-1} \int_0^T \|P\, e^{-itH} v\|^2 \, dt\right)^{1/2}.$$

Now we fix $p > p(\varepsilon)$ and take the limit $T \to \infty$. By $P \in \ker \mu_a^{H_0}$, this gives

$$(\|v\|^2 - \varepsilon)^{1/2} \le \liminf_{T\to\infty} \left(\|v\|^2 - T^{-1} \int_0^T \|P\, e^{-itH} v\|^2 \, dt\right)^{1/2}$$

$$= \left(\|v\|^2 - \limsup_{T\to\infty} T^{-1} \int_0^T \|P\, e^{-itH} v\|^2 \, dt\right)^{1/2}.$$

Hence $\limsup_{T\to\infty} T^{-1} \int_0^T \|P\, e^{-itH} v\|^2 \, dt \le \varepsilon$ and this implies $P \in \ker \mu_a^H$ because $\mathcal{M}_\infty(H)$ is dense in $P_H^{ac}\mathcal{H}$. By the continuity in norm of $\mu_a^{H_0}$, we can extend this result to all selfadjoint operators $P \in \ker \mu_a^{H_0}$, and since $\ker \mu_a^{H_0}$ is a left ideal, to all $P \in \ker \mu_a^{H_0}$. Thus $\ker \mu_a^{H_0} \subseteqq \ker \mu_a^{H}$ is proved. ■

## 8.2. The invariance of the wave morphism $\mu_a^H$

### 8.2.1. Admissible functions

First we introduce a class of functions which we can roughly characterize as piecewise strictly increasing functions which are smooth in a certain sense.

**Definition 16.** Let $\alpha$ be a real-valued function on $\mathbb{R}$. Let $\Delta = \bigcup_{n=1}^{N} \mathcal{I}_n \subseteq \mathbb{R}$, $N \leqq \infty$, where the intervals $\mathcal{I}_n = (a_n, b_n)$ are mutually disjoint. Then $\alpha$ is called *admissible* on $\Delta$ if on each $\mathcal{I}_n$:

(I) $\alpha(x)$ is continuously differentiable,
(II) $\alpha'(x) > 0$,
(III) $\alpha'(x)$ is locally of bounded variation.
$\alpha$ is called *admissible* if $\Delta = \mathbb{R}$.

Simple examples of admissible functions are, for all $\Delta \subseteqq \mathbb{R}$, the polynomials $x^{2n+1}$ and, for $\Delta \subseteqq (0, \infty)$ the functions $\sqrt{x}$ and $x^{2n}$.

**Lemma 17.** *Let $\alpha$ be an admissible function on $\Delta$. Then:*

(i) *$\alpha^{-1}$ exists on each set $\alpha(\mathcal{I}_n)$ and is admissible on $\alpha(\mathcal{I}_n)$,*

(ii) *If $\alpha$ is a.e. finite with respect to $E(\cdot)$ and $\Delta$ contains a spectral core of $H \upharpoonright P^c\mathcal{H}$, then* $P^{ac}_{\alpha(H)} = P^{ac}_H$.

**Proof.** (i) The existence of $\alpha^{-1}$ on $\alpha(\mathcal{I}_n)$ follows from the strict monotonicity of $\alpha$ on $\mathcal{I}_n$. Further, $\left(\alpha^{-1}(y)\right)' = \left(\alpha'(x)\right)^{-1}$ with $y = \alpha(x)$ for $x \in \mathcal{I}_n$. This ensures (I) and (II) of Definition 16 for $\alpha^{-1}$ on $\alpha(\mathcal{I}_n)$, and also (III) if we take into account that "$g(x)$ is of bounded variation on $[a, b]$, $g(x) \neq 0$ on $[a, b]$" implies "$1/g(x)$ is of bounded variation on $[a, b]$" (see Natanson [1, p. 243]). Thus, the functions $\beta_n$ defined by $\beta_n(x) = \alpha^{-1}(x)$ on $\alpha(\mathcal{I}_n)$ and $\beta_n(x) = 0$ on $\mathbb{R} \setminus \alpha(\mathcal{I}_n)$ are admissible on $\alpha(\mathcal{I}_n)$.

(ii) First note that $\alpha(H)$ is well-defined and $E(\mathbb{R} \setminus \Delta)\,\mathcal{H}$ belongs to $\mathcal{H}_p(H) \cap \mathcal{H}_p(\alpha(H))$ (see e.g. Weidmann [1, p. 197]). Thus we can restrict ourselves to $E(\Delta)\,\mathcal{H} = \bigotimes_{n=1}^{N} E(\mathcal{I}_n)\,\mathcal{H}$. Clearly, it suffices to show that $\mathcal{H}_{ac}\left(H \upharpoonright E(\mathcal{I}_n)\,\mathcal{H}\right) = \mathcal{H}_{ac}\left(\alpha(H) \upharpoonright E(\mathcal{I}_n)\,\mathcal{H}\right)$ for all $n \in \{1, 2, \ldots, N\}$ in order to prove $P^{ac}_{\alpha(H)} = P^{ac}_H$. But this follows from (I) and (II) of Definition 16, by Remark 3.31. ■

Now we prove some asymptotic estimates of the Fourier transform of $e^{is\alpha(x)}$ for admissible functions $\alpha$.

**Lemma 18.** *Let $\alpha$ be admissible on $\Delta \subseteqq \mathbb{R}$. Then for each finite interval $[c, d] \subset \Delta$ we have*

$$\lim_{p \to \infty} \sup_{T \geq 1} T^{-1} \int_0^T dt \left| \int_{\pm pT}^{\pm\infty} ds \left| \int_c^d e^{isx} e^{-it\alpha(x)}\, dx \right|^2 \right| = 0.$$

**Proof.** Integration by parts provides

$$\left|\int_c^d e^{isx}\,e^{-it\alpha(x)}\,dx\right|^2 = \left|\int_c^d e^{-it\alpha(x)}\,\frac{1}{is}\left(\frac{d}{dx}\,e^{isx}\right)dx\right|^2$$

$$\leq \left|\left[\frac{1}{is}\,e^{isx}\,e^{-it\alpha(x)}\right]_c^d + \frac{t}{s}\int_c^d e^{isx}\,e^{-it\alpha(x)}\alpha'(x)\,dx\right|^2$$

$$\leq 8\,|s|^{-2} + 2\,|s|^{-2}\,t^2\,|f_t(s)|^2$$

where $f_t(s)$ denotes the Fourier transform of $\chi_{[c,d]}(x)\,\alpha'(x)\,e^{-it\alpha(x)}$. Since $\alpha'(x)$ exists and is continuous on $[c, d] \subset \Delta$, $\alpha'(x)\,\chi_{[c,d]}(x)$ belongs to $L^2(\mathbb{R}, dx)$ and therefore $f_t(\cdot) \in L^2(\mathbb{R}, dx)$. Thus

$$\int_{-\infty}^{\infty} |f_t(s)|^2\,ds = \int_c^d |\alpha'(x)|^2\,dx = c_1 < \infty$$

with $c_1$ independent of $t$. Using this and the estimate above, we obtain

$$T^{-1}\int_0^T dt\left|\int_{\pm pT}^{\pm\infty} ds\left|\int_c^d e^{isx}\,e^{-it\alpha(x)}\,dx\right|^2\right|$$

$$\leq T^{-1}\int_0^T dt\left|\int_{\pm pT}^{\pm\infty}(8\,|s|^{-2} + 2\,|f_t(s)|^2\,t^2\,|s|^{-2}\,ds\right|$$

$$\leq T^{-1}\int_0^T\left(8(pT)^{-1} + 2c_1t^2(pT)^{-2}\right)dt \leq p^{-1}(8T^{-1} + c_1p^{-1}).$$

The right side tends to zero as $p \to \infty$ uniformly in $T \geq 1$. This proves the lemma. ∎

**Lemma 19.** *Let $\alpha$ be admissible on $\Delta \subseteq \mathbb{R}$. Then for each finite interval $[c, d] \subset \Delta$ we have*

$$\lim_{t\to\infty}\int_{-\infty}^{0} ds\left|\int_c^d e^{isx}\,e^{-it\alpha(x)}\,dx\right|^2 = 0.$$

**Proof.** On $[c, d]$ the function $\alpha(x)$ is continuously differentiable and $\alpha'(x) > 0$. Thus $c_1 := \min_{x\in[c,d]} \alpha'(x) > 0$. Let $M$ be the total variation of $\alpha'(x)$ on $[c, d]$. Then a simple calculation shows that

$$\int_c^d \left|d\left((s - t\alpha'(x))^{-1}\right)\right| \leq Mt\,|s - tc_1|^{-2} \quad \text{for} \quad t > 0,\ \ s \leq 0. \tag{10}$$

Using this, we get by integration by parts

$$\int_{-\infty}^{0} \mathrm{d}s \left| \int_c^d \mathrm{e}^{\mathrm{i}sx}\, \mathrm{e}^{-\mathrm{i}t\alpha(x)}\, \mathrm{d}x \right|^2 = \int_{-\infty}^{0} \mathrm{d}s \left| \int_c^d \big(s - t\alpha'(x)\big)^{-1} \left( \frac{\mathrm{d}}{\mathrm{d}x}\, \mathrm{e}^{\mathrm{i}(sx - t\alpha(x))} \right) \mathrm{d}x \right|^2$$

$$\leqq \int_{-\infty}^{0} \mathrm{d}s \left| \Big[\big(s - t\alpha'(x)\big)^{-1}\Big]_c^d \right| + \left| \int_c^d \mathrm{e}^{\mathrm{i}(sx - t\alpha(x))}\, \mathrm{d}\Big(\big(s - t\alpha'(x)\big)^{-1}\Big) \right|^2$$

$$\leqq \int_{-\infty}^{0} \mathrm{d}s \left\{ 2\, |s - tc_1|^{-1} + \int_c^d \left| \mathrm{d}\Big(\big(s - t\alpha'(x)\big)^{-1}\Big) \right| \right\}^2$$

$$\leqq \int_{-\infty}^{0} \mathrm{d}s \{8\, |s - tc_1|^{-2} + 2M^2t^2\, |s - tc_1|^{-4}\}.$$

A straightforward calculation shows that the right side tends to zero as $t \to \infty$ which proves the assertion. ∎

**Lemma 20.** *Let $\alpha$ or $-\alpha$ be admissible on $\Delta \subseteq \mathbb{R}$. Let $[c, d]$ be a finite interval from $\Delta$ such that $\alpha'(x)$ is strictly monotone on $[c, d]$. Then*

$$w_t(s) := \int_c^d \mathrm{e}^{\mathrm{i}sx}\, \mathrm{e}^{-\mathrm{i}t\alpha(x)}\, \mathrm{d}x$$

*tends to zero as $t \to \infty$, uniformly in $s \in \mathbb{R}$.*

**Proof.** Since $\alpha'(x)$ is strictly monotone and continuous on $[c, d]$, it follows that for each $\varepsilon > 0$ and $y \in \mathbb{R}$ there is an interval $\Delta_y \subset [c, d]$ with $|\Delta_y| < \varepsilon$ such that

$$a_\varepsilon := \min_{y \in \mathbb{R}} \left( \min_{x \in [c,d] \setminus \Delta_y} |\alpha'(x) + y| \right) > 0.$$

Hence, with $y = -st^{-1}$, we have

$$|w_t(s)| = \left| \int_c^d \mathrm{e}^{\mathrm{i}sx}\, \mathrm{e}^{-\mathrm{i}t\alpha(x)}\, \mathrm{d}x \right| \leqq \varepsilon + \left| \int_{[c,d] \setminus \Delta_y} \mathrm{e}^{-\mathrm{i}t(\alpha(x) + yx)}\, \mathrm{d}x \right|. \tag{11}$$

$[c, d] \setminus \Delta_y$ consists of one or two intervals depending on $\varepsilon$ and $y$. Let $[a, b]$ be such an interval. Then

$$\left| \int_a^b \mathrm{e}^{-\mathrm{i}t(\alpha(x) + xy)}\, \mathrm{d}x \right| = \left| \int_a^b \left( \frac{\mathrm{d}}{\mathrm{d}x}\, \mathrm{e}^{-\mathrm{i}t(\alpha(x) + xy)} \right) \frac{\mathrm{d}x}{\mathrm{i}t\big(\alpha'(x) + y\big)} \right|$$

$$\leqq \left| \Big[\mathrm{e}^{-\mathrm{i}t(\alpha(x) + xy)} \Big(\mathrm{i}t\big(\alpha'(x) + y\big)\Big)^{-1}\Big]_a^b \right| + \left| \int_a^b \mathrm{e}^{-\mathrm{i}t(\alpha(x) + xy)}\, \mathrm{d}\Big(\Big(\mathrm{i}t\big(\alpha'(x) + y\big)\Big)^{-1}\Big) \right|$$

$$\leqq 2(a_\varepsilon t)^{-1} + \int_a^b \left| \mathrm{d}\Big(\Big(t\big(\alpha'(x) + y\big)\Big)^{-1}\Big) \right| \leqq 2|a_\varepsilon t|^{-1} + M(a_\varepsilon^2 t)^{-1}$$

where $M = |\alpha'(b) - \alpha'(a)|$ is the total variation of $\alpha'(x)$ on $[a, b]$ and we have used the estimate (10). (11) and the last estimate prove the lemma since $M$ and $a_s$ are independent of $s$. ∎

### 8.2.2. The invariance of the wave morphism $\mu_a{}^H$

Assume $\mu_a{}^H = \mu_a{}^{H'}$. Then we have $\ker \mu_a{}^H = \ker \mu_a{}^{H'}$ and $(\mathrm{com}\, H) \restriction P^{ac} = (\mathrm{com}\, H') \restriction P^{ac}_{H'}$. The last relation implies $P^{ac}_{H'} = P^{ac}$ and that there exist two functions $\alpha$ and $\beta$ such that $\alpha(H \restriction P^{ac}\mathscr{H}) = H' \restriction P^{ac}\mathscr{H}$ and $\beta(H' \restriction P^{ac}\mathscr{H}) = H \restriction P^{ac}\mathscr{H}$. Clearly, $\alpha$ and $\beta$ are real-valued Borel functions, a.e. finite with respect to $E(\cdot)$ and $E_{H'}(\cdot)$, respectively. On the subspace $(1 - P^{ac})\,\mathscr{H} = (1 - P^{ac}_{H'})\,\mathscr{H}$ the operators $H$ and $H'$ can be arbitrarily different. These considerations show that we can easily find the class of generators of a given wave morphism $\mu_a{}^H$ if the class of functions $\alpha$ is known for which $\ker \mu_a{}^H \subseteq \ker \mu_a{}^{\alpha(H)}$. But this is answered in

**Theorem 21.** *Let $\alpha$ be a real-valued Borel function which is a.e. finite with respect to $E(\cdot)$ (thus $\alpha(H)$ is selfadjoint according to the functional calculus). Assume that $P^{ac}_{\alpha(H)} \leq P^{ac}$. Then* $\ker \mu_a{}^H \subseteq \ker \mu_a{}^{\alpha(H)}$ *iff for all Borel sets $\Delta \subset (a, b) \cap \mathrm{spec}^c(H \restriction P^{ac}\mathscr{H})$, $(a, b)$ an arbitrary finite interval, the following conditions hold:*

$$\lim_{t\to\infty} t^{-1} \int_0^t \mathrm{d}t' \int_{-\infty}^0 \mathrm{d}s \left| \int_\Delta \mathrm{e}^{\mathrm{i}sx}\, \mathrm{e}^{-\mathrm{i}t'\alpha(x)}\, \mathrm{d}x \right|^2 = 0, \tag{12}$$

$$\lim_{\varrho\to\infty} \sup_{t\geq 1} t^{-1} \int_0^t \mathrm{d}t' \int_{\varrho t}^\infty \mathrm{d}s \left| \int_\Delta \mathrm{e}^{\mathrm{i}sx}\, \mathrm{e}^{-\mathrm{i}t'\alpha(x)}\, \mathrm{d}x \right|^2 = 0. \tag{13}$$

**Proof. 1.** Let $\{Q_n\}_1^\infty$ be a sequence of pairwise disjoint projections from com $H$ such that $P^{ac} = \sum Q_n$, and $H \restriction Q_n\mathscr{H}$ is bounded and simple. $Q_n \in \mathrm{com}\, \alpha(H)$ and we define $H_n = H \restriction Q_n\mathscr{H}$, $\tilde{H}_n = \alpha(H) \restriction Q_n\mathscr{H}$. Then obviously, $\ker \mu_a{}^H \subseteq \ker \mu_a{}^{\alpha(H)}$ is equivalent to $\ker \mu_a{}^{H_n} \subseteq \ker \mu_a{}^{\tilde{H}_n}$ for all $n$. Thus it suffices to prove the theorem for the case: $H$ is bounded, absolutely continuous and simple.

2. According to Prop. 12, Prop. 14, and Theorem 15, $\ker \mu_a{}^H \subseteq \ker \mu_a{}^{\alpha(H)}$ is equivalent to (I) and (II) of Theorem 15 with $H$, $\alpha(H)$ in place of $H_0$, $H$. With respect to the spectral representation of $H$, the conditions (I) and (II) of Theorem 15 transform into (12) and (13) without $\Delta$ and with $\mathrm{e}^{\mathrm{i}sx}$ replaced by $\mathrm{e}^{\mathrm{i}sx}\overline{u(x)}\, v(x)$, $u \in \mathscr{M}_\infty(H)$ and $v \in \mathscr{M}_\infty(\alpha(H))$. Because of

$$\int_{-\infty}^{\infty} \mathrm{d}s \left| \int_{-\infty}^{\infty} \mathrm{e}^{\mathrm{i}sx}\, \mathrm{e}^{-\mathrm{i}t'\alpha(x)}\overline{u(x)}\, v(x)\, \mathrm{d}x \right|^2 \leq 2\pi\, |||u|||_{H,\infty}^2\, \|v\|^2$$

these conditions are equivalent to the ones with $u$ and $v$ replaced by generating sets $\mathscr{D}_1$ in $\mathscr{M}_\infty(H)$ (with respect to $|||\cdot|||_{H,\infty}$) and $\mathscr{D}_2$ in $P^{ac}_{\alpha(H)}\mathscr{H}$. Such sets are vectors $u$ and $v$ with $u(x) = \chi_\Delta(x)$ and $v(x) = \chi_{\Delta'}(x)$ where $\Delta, \Delta' \subseteq (a, b) \cap \mathrm{spec}^c(H \restriction P^{ac}\mathscr{H})$. This yields (12) and (13) and shows that (12) and (13) are equivalent to $\ker \mu_a{}^H \subseteq \ker \mu_a{}^{\alpha(H)}$. ∎

**Remark 22.** Under the assumptions of Theorem 21 on $H$ and $\alpha$ we also get that $\ker \mu_a{}^{-H} \subseteq \ker \mu_a{}^{-\alpha(H)}$ is equivalent to (12) and (13). This follows from the fact that the function $\beta(x) = -\alpha(-x)$ satisfies (12) and (13) for $-H$ in place of $H$ when $\alpha$ does this for $H$ and from $\beta(-H) = -\alpha(H)$.

The next statement shows that the assumptions (12) and (13) are fulfilled for a sufficiently wide class of functions.

**Theorem 23.** *Let $\alpha$ be an admissible function on $\Delta \supseteq \operatorname{spec}^c (H \mid P^c_H\mathscr{H})$ which is a.e. finite with respect to $E(\cdot)$. Then* $\ker \mu_a{}^H = \ker \mu_a{}^{\alpha(H)}$

Proof. From Lemma 17 we get $P^{ac} = P^{ac}_{\alpha(H)}$. Further, we have a decomposition $\Delta = \bigcup_{n=1}^{N} \mathscr{I}_n$ for $\alpha$. Set $Q_n = E(\mathscr{I}_n)\, P^{ac}$. Then $P^{ac} = \sum_{n=1}^{N} Q_n$ by the properties of the sets $\mathscr{I}_n$. Clearly, to prove $\ker \mu_a{}^H = \ker \mu_a{}^{\alpha(H)}$, it suffices to prove $\ker \mu_a{}^{H_n} = \ker \mu_a{}^{\alpha(H_n)}$ for all $n$ where $H_n = H \mid Q_n\mathscr{H}$. According to Lemmas 18 and 19, the function $\alpha$ satisfies (12) and (13) for finite intervals $\Delta$. The extension to arbitrary Borel sets $\Delta$ as in Theorem 21 follows from

$$\int_{-\infty}^{\infty} \mathrm{d}s \left| \int_{-\infty}^{\infty} \mathrm{e}^{\mathrm{i}sx}\, \mathrm{e}^{-\mathrm{i}t\alpha(x)} v(x)\, \mathrm{d}x \right|^2 \leq 2\pi \|v\|^2.$$

Thus from Theorem 21 we obtain $\ker \mu_a{}^{H_n} \subseteq \ker \mu_a{}^{\alpha(H_n)}$. On the other hand, the functions $\beta_n$, defined by $\beta_n(x) = \alpha^{-1}(x)$ on $\alpha(\mathscr{I}_n)$ and $\beta_n(x) = 0$ on $\mathbb{R} \setminus \alpha(\mathscr{I}_n)$, are admissible on $\operatorname{spec}^c \alpha(H_n)$, by Lemma 17. Thus we can apply Theorem 21 to $H' = \alpha(H_n)$ and the function $\beta_n$ (replace $H$, $\alpha$ by $H' = \alpha(H_n)$, $\beta_n$ in Theorem 21) and get $\ker \mu_a{}^{\alpha(H_n)} \subseteq \ker \mu_a{}^{\beta_n(\alpha(H_n))} = \ker \mu_a{}^{H_n}$ because of $\beta_n(\alpha_n(H)) = H_n$. This proves $\ker \mu_a{}^{H_n} = \ker \mu_a{}^{\alpha(H_n)}$. ■

**Remark 24.** In general the equality of $\ker \mu_a{}^H$ and $\ker \mu_a{}^{\alpha(H)}$ does not imply the equality of the corresponding wave algebras. However, since $\operatorname{com} H \subseteq \operatorname{com} \alpha(H)$, we have $\operatorname{dom} \mu_a{}^H \subseteq \operatorname{dom} \mu_a{}^{\alpha(H)}$. If additionally $\alpha$ is strictly increasing on $\operatorname{spec}_c H$, then we get the equality $\operatorname{dom} \mu_a{}^H = \operatorname{dom} \mu_a{}^{\alpha(H)}$ and thus $\mu_a{}^H = \mu_a{}^{\alpha(H)}$.

When we have functions $\alpha$ which are not piecewise increasing on the spectrum of $H$, then $\ker \mu_a{}^H \neq \ker \mu_a{}^{\alpha(H)}$ in general. However, we have the following assertion.

**Proposition 25.** *Let $\alpha$ be a real-valued Borel function which is a.e. finite with respect to $E(\cdot)$. Assume there is an open set $\Delta = \Delta_1 \cup \Delta_2 \supseteq \operatorname{spec}^c (H \mid P^c_H\mathscr{H})$ such that $\alpha$ is admissible on $\Delta_1$ and $-\alpha$ is admissible on $\Delta_2$. Then $\operatorname{dom} \mu_a{}^H \cap \operatorname{dom} \mu_a{}^{-H} \subseteq \operatorname{dom} \mu_a{}^{\alpha(H)} \cap \operatorname{dom} \mu_a{}^{-\alpha(H)}$, and for all $A \in \operatorname{dom} \mu_a{}^H \cap \operatorname{dom} \mu_a{}^{-H}$ we have*

$$\mu_a{}^{\alpha(H)}(A) = \mu_a{}^H(A)\, E(\Delta_1) + \mu_a{}^{-H}(A)\, E(\Delta_2), \tag{14}$$

$$\mu_a{}^{-\alpha(H)}(A) = \mu_a{}^{-H}(A)\, E(\Delta_1) + \mu_a{}^H(A)\, E(\Delta_2). \tag{15}$$

Proof. Set $A_+ = \mu_a{}^H(A)$, $A_- = \mu_a{}^{-H}(A)$, $E_+ = E(\Delta_1)\, P^{ac}$ and $E_- = E(\Delta_2)\, P^{ac}$. Then $A = (A - A_+)\, E_+ + (A - A_-)\, E_- + A_+E_+ + A_-E_- + A(1 - P^{ac})$. This gives

$$\mathrm{e}^{\mathrm{i}t\alpha(H)} A\, \mathrm{e}^{-\mathrm{i}t\alpha(H)} P^{ac} = \mathrm{e}^{\mathrm{i}t\alpha(H)} (A - A_+)\, \mathrm{e}^{-\mathrm{i}t\alpha(H)} E_+$$
$$+ \mathrm{e}^{\mathrm{i}t\alpha(H)} (A - A_-)\, \mathrm{e}^{-\mathrm{i}t\alpha(H)} E_- + A_+E_+ - A_-E_-.$$

To prove (14), we therefore have to show $P^{ac} = P^{ac}_{\alpha(H)}$ and

$$|\mathrm{A}|\text{-}\lim_{t\to\pm\infty} \mathrm{e}^{\mathrm{i}t\alpha(H)} (A - A_\pm)\, \mathrm{e}^{-\mathrm{i}t\alpha(H)} E_\pm = 0, \quad \text{for both signs.} \tag{16}$$

$P^{ac} = P^{ac}_{\alpha(H)}$ is an easy consequence of Lemma 17. Clearly, (16) is equivalent to $A - A_+ \in \ker \mu_a^{\alpha(H_+)}$ and $A - A_- \in \ker \mu_a^{\alpha(H_-)}$ with $H_+ = HE_+$ and $H_- = HE_-$. Now $\alpha$ is an admissible function on $\Delta_1 \supseteq \operatorname{spec}^c(H_+ \mid P^c_{H_+}\mathscr{H})$ and $-\alpha$ on $\Delta_2 \supseteq \operatorname{spec}^c(H_- \mid P^c_{H_-}\mathscr{H})$. Thus, by Theorem 23, we have $\ker \mu_a^{H_+} = \ker \mu_a^{\alpha(H_+)}$ and $\ker \mu_a^{H_-} = \ker \mu_a^{-\alpha(H_-)}$. The last equality implies $\ker \mu_a^{-H_-} = \ker \mu_a^{\alpha(H_-)}$ (see Remark 22). Since $A - A_+ \in \ker \mu_a^H \subseteq \ker \mu_a^{H_+}$ and $A - A_- \in \ker \mu_a^{-H} \subseteq \ker \mu_a^{-H_-}$, we get $A - A_+ \in \ker \mu_a^{\alpha(H_+)}$ and $A - A_- \in \ker \mu_a^{\alpha(H_-)}$ and thus (16). This concludes the proof of (14). Similarly we get (15). ■

When $\alpha$ is twice continuously differentiable (piecewise), then there is a decomposition of $\mathbb{R}$, $\mathbb{R} = \bigcup_{n=-N_1}^{N_2} (a_n, a_{n+1}]$ with $a_j < a_{j+1}$ for each $j \in \{-N_1, -N_1 + 1, \ldots, N_2\}$ such that $\alpha$ or $-\alpha$ is admissible on each open interval $(a_j, a_{j+1})$. Thus $\mathbb{R} = \Delta_1 \cup \Delta_2 \cup \Delta_3$ where the sets $\Delta_j$ are mutually disjoint, $\Delta_1$ and $\Delta_2$ are open sets and $\Delta_3$ consists of countably many points. Furthermore, $\alpha$ is admissible on $\Delta_1$ and $-\alpha$ is admissible on $\Delta_2$. Therefore $\alpha$ satisfies the assumptions of Prop. 25 for each operator $H$ provided $\alpha(H)$ is well-defined.

## 8.3. The invariance of the wave morphism $\mu_s^H$

### 8.3.1. A no-go theorem

With respect to the invariance of $\mu_s^H$ under transformations $H \to H'$ the same remarks are true as the ones made at the beginning of Section 2.2. Thus we have only to study the question of the equality $\ker \mu_s^H = \ker \mu_s^{\alpha(H)}$ for a class of functions $\alpha$. In Section 1.1 we saw that this equality is satisfied for piecewise linear increasing functions. Now the question arises whether a stronger result like Theorem 23 is also possible for strong wave ideals. The answer is "no" as the following theorem says.

**Theorem 26.** *Let $\alpha$ or $-\alpha$ be admissible on a finite interval $[a, b]$. Assume $\alpha'(x)$ is strictly monotone on $[a, b]$ and $|[a, b] \cap \operatorname{spec}^c(H \mid P^{ac}\mathscr{H})| > 0$. Then there is a projection $P$ such that $P \in \ker \mu_s^H$ but $P \notin \ker \mu_s^{\alpha(H)}$.*

Proof. 1. By $|[a, b] \cap \operatorname{spec}^c(H \mid P_H^{ac}\mathscr{H})| > 0$, there is a reducing subspace $\mathscr{H}_1$ for $H$ such that $\mathscr{H}_1$, $H \mid \mathscr{H}_1$ can be identified with $L^2(\Delta, \mathrm{d}x)$, $x$ where $\Delta \subseteq [a, b]$. Clearly, it suffices to construct the projection $P$ in the subspace $\mathscr{H}_1$. Thus we restrict ourselves to the special case $\mathscr{H} = L^2(\Delta, \mathrm{d}x)$, $H = x$ with $\Delta \subseteq [a, b]$. For simplicity we assume $[a, b] = [0, 2\pi]$ and $\alpha'(x) > 0$ on $[0, 2\pi]$. The general case can be easily reduced to this special one ($\alpha'(x) < 0$ leads to $\ker \mu_a^H = \ker \mu_a^{-\alpha(H)}$ and thus to $\ker \mu_s^H \neq \ker \mu_s^{\alpha(H)}$) Similarly as in other proofs we first solve the problem in the bigger space $\mathscr{H}' = L^2([0, 2\pi], \mathrm{d}x) \supseteq \mathscr{H}$, where $H'$ denotes the multiplication operator by $x$ in $\mathscr{H}'$.

2. Set $u_t(x) = (2\pi)^{-1/2} \mathrm{e}^{-itx}$ and $v_t(x) = (2\pi)^{-1/2} \mathrm{e}^{-it\alpha(x)}$. The set $\{u_s\}_{-\infty}^{\infty}$ is an orthonormal basis in $\mathscr{H}'$ and we have $\mathrm{e}^{-imH'} u_n = u_{n+m}$. By Lemma 20,

$$a_t(s) := (u_s, v_t) = (2\pi)^{-1} \int_0^{2\pi} \mathrm{e}^{isx}\, \mathrm{e}^{-it\alpha(x)}\, \mathrm{d}x$$

tends to zero as $t \to \infty$ uniformly in $s \in \mathbb{R}$. Thus we can find for each $\varepsilon \in (0, 1/2)$ sequences $t_n \to \infty$ and $p_n \to \infty$ ($p_n \in \mathbb{N}$) such that (use also $\sum_{n=0}^{-\infty} |(u_n, v_t)|^2 \to 0$ as $t \to \infty$, by Lemma 19)

$$\|f_n - v_{t_n}\| \leqq \varepsilon^n \quad \text{with} \quad f_n = \sum_{m=p_n+1}^{p_{n+1}} (u_m, v_{t_n})\, u_m. \tag{17}$$

Define $w_n = f_n/\|f_n\|$. Then $\{w_n\}_1^\infty$ is an orthonormal system and we obtain from (17) that $\|w_n - v_{t_n}\| \leqq 2\varepsilon^n$. Set $P' = \sum w_n(w_n, \cdot)$. $P'$ is a projection and $\|P'u_m\|^2 = |(w_n, u_m)|^2$ where $n$ is given by $m \in \{p_n + 1, \ldots, p_{n+1}\}$ for $m \geqq p_1 + 1$. Since $n \to \infty$ as $m \to \infty$, this implies

$$\|P'u_m\|^2 = |(w_n, u_m)|^2 = \|f_n\|^{-2}\, |(f_n, u_m)|^2 \leqq (1 - \varepsilon^n)^{-2} |\,(u_m, v_{t_n})|^2 \to 0$$

as $m \to \infty$. Because of Corollary 6.42 this proves $P' \in \ker \mu_s^{H'}$. On the other hand, we have

$$\|P'v_{t_n}\| \geqq |(w_n, v_{t_n})|^2 = |(w_n - v_{t_n}, v_{t_n}) + (v_{t_n}, v_{t_n})|^2$$
$$\geqq \|(w_n - v_{t_n}, v_{t_n})\,| - 1|^2 \to 1$$

as $n \to \infty$, by $\|w_n - v_{t_n}\| \leqq 2\varepsilon^n$. Thus $\|P'v_{t_n}\|^2 = \|P'\, \mathrm{e}^{-\mathrm{i}t_n\alpha(H')} u_0\|^2 \to 1$ as $n \to \infty$, and this shows $P' \notin \ker \mu_s^{\alpha(H')}$.

3. The last step is to extend the result in 2. to $H = x$ on $\mathscr{H} = L^2(\Delta, \mathrm{d}x)$. For this aim it suffices to construct a projection $\tilde{P}$ in $\mathscr{H}'$ with $\tilde{P} = \tilde{P}Q = Q\tilde{P}$ and $\tilde{P} \in \ker \mu_s^{H'}$, $\tilde{P} \notin \ker \mu_s^{\alpha(H')}$ where $Q$ denotes the projection onto $\mathscr{H} = L^2(\Delta, \mathrm{d}x)$. Set $F = UP'Q$ where $U$ denotes a partial isometry with $U^*U = 1$ and $UU^* = Q$. Then $F \in \ker \mu_s^{H'}$. Further, we have

$$\|Fv_{t_n}\|^2 = \|P'Qv_{t_n}\|^2 \geqq |(w_n, Qv_{t_n})|^2 \geqq |(w_n - v_{t_n}, Qv_{t_n})| - \|Qv_{t_n}\|^2|^2.$$

Now $\|Qv_{t_n}\|^2 = (2\pi)^{-1} \int_\Delta |\mathrm{e}^{-\mathrm{i}t_n\alpha(x)}|^2 \,\mathrm{d}x = |\Delta|/2\pi > 0$. Using this and $\|w_n - v_{t_n}\| \leqq 2\varepsilon^n$, we get from the estimate above that $\lim_{n\to\infty} \|Fv_{t_n}\|^2 \geqq |\Delta|/2\pi > 0$ which shows that $F \notin \ker \mu_s^{\alpha(H')}$. The same is then true for $|F|$ and also for the projection $\tilde{P} = E_{|F|}((\varepsilon, \infty))$ for a sufficiently small $\varepsilon > 0$. $\tilde{P} \in \ker \mu_s^{H'}$ is also easily verified from $F \in \ker \mu_s^{H'}$. ∎

From this statement we see that if $\alpha$ is a real-valued Borel function which is piecewise twice continuously differentiable, then there are two possibilities. Either $\alpha$ is piecewise linear increasing on $\operatorname{spec}^c(H \mid P^c\mathscr{H})$ and thus $\ker \mu_s^H = \ker \mu_s^{\alpha(H)}$ or the assumptions of Theorem 26 are satisfied and thus $\ker \mu_s^H \neq \ker \mu_s^{\alpha(H)}$. Therefore the essential content of Theorem 26 is that $\ker \mu_s^H$ is not invariant under the transformation $H \to \alpha(H)$ for a sufficiently wide class of functions.

### 8.3.2. Some special classes of operators

First we note an immediate consequence of Theorem 23 and Theorem 6.23.

**Theorem 27.** *Let $\alpha$ be an admissible function on $\Delta \supseteqq \operatorname{spec}^c(H \mid P_H^c\mathscr{H})$ which is a.e. finite with respect to $E(\cdot)$. Let $A \in \operatorname{dom} \mu_s^H \cap \operatorname{dom} \mu_s^{\alpha(H)}$. Then $\mu_s^H(A) = \mu_s^{\alpha(H)}(A)$.*

Proof. It is obvious. ∎

In the preceding section we have seen that $\ker \mu_s^H \neq \ker \mu_s^{\alpha(H)}$ for most of all functions $\alpha$. Thus we are now interested in weaker statements, namely statements providing classes of operators $A$ such that $A \in \bigcap_\alpha \operatorname{dom} \mu_s^{\alpha(H)}$ where $\alpha$ varies in a wide class of functions. A simple class of such operators was already described (see Theorem 6.45 and Remark 6.46), namely the set of all bounded operators $A$ for which there exists a set of pairwise disjoint projections $Q_n$ from com $H$ such that $AQ_n$ is compact and $\sum Q_n \geqq P^{\mathrm{ac}}$. In this case $\alpha$ runs through the set of all real-valued Borel functions which are a.e. finite with respect to $E(\cdot)$ and for which $P^{\mathrm{ac}}_{\alpha(H)} \leqq P^{\mathrm{ac}}$.

The next statement provides another class of such operators.

**Proposition 28.** *Let $\alpha$ be a real-valued Borel function on $R$ which is a.e. finite with respect to $E(\cdot)$. Assume there is a decomposition of $\mathbb{R}$, $\mathbb{R} = \bigcup_{n=1}^{N} \mathscr{I}_n$, where $\mathscr{I}_n = (a_n, a_{n+1}]$ are disjoint intervals, such that $\alpha$ is strictly increasing and twice continuously differentiable on $(a_n, a_{n+1})$. Let $\mathscr{D}$ be generating in $P^{\mathrm{ac}}\mathscr{H}$ such that $\mathscr{D} \subseteqq \mathscr{M}_\infty(H)$ and $u \in \mathscr{D}$ implies $u \in E([c, d])\,\mathscr{H}$ with $[c, d]$ contained in some interval $(a_n, a_{n+1})$, $n \in \{1, 2, \ldots, N\}$. Let $A \in \mathscr{L}(\mathscr{H})$. Suppose $A(t)\,u := \mathrm{e}^{\mathrm{i}tH} A\, \mathrm{e}^{-\mathrm{i}tH} u$, $u \in \mathscr{D}$, is strongly differentiable with*

$$\{\|A'(t)\,u\|\} \in L^2(\mathbb{R}, \mathrm{d}x) \cap L^1(\mathbb{R}, \mathrm{d}x)$$

*and $\{|t|^p\,\|A'(t)\,u\|\} \in L^1(R_+, \mathrm{d}x)$ for some $p > 0$. Then $A \in \operatorname{dom} \mu_s^{\alpha(H)}$.*

For the proof of this statement we need

**Lemma 29.** *Let $f(t)$ be a strongly differentiable function from $\mathbb{R}$ to $\mathscr{H}$ such that:*

(I) $\|f(t)\| \to 0$ *as* $|t| \to \infty$,
(II) $\|f'(\cdot)\| \in L^1(\mathbb{R}, \mathrm{d}t) \cap L^2(\mathbb{R}, \mathrm{d}t)$,
(III) $\{|t|^p\,\|f'(t)\|\} \in L^1(\mathbb{R}, \mathrm{d}t)$ *for some $p > 0$. Then $\hat{f}(\cdot) \in L^1(\mathbb{R}, \mathrm{d}t; \mathscr{H})$.*

**Proof.** We have

$$\int_{-N}^{N} \mathrm{e}^{-\mathrm{i}sx} s f(x)\,\mathrm{d}x = \mathrm{i} \int_{-N}^{N} f(x)\,(\mathrm{d}/\mathrm{d}x\, \mathrm{e}^{-\mathrm{i}sx})\,\mathrm{d}x$$
$$= \mathrm{i}\{f(N)\,\mathrm{e}^{-\mathrm{i}sN} - f(-N)\,\mathrm{e}^{\mathrm{i}sN}\} - \mathrm{i} \int_{-N}^{N} \mathrm{e}^{-\mathrm{i}sx} f'(x)\,\mathrm{d}x.$$

Because of (I) and (II), this implies that the Fourier transform of $f$ exists, for $s \neq 0$, and $\{\|\hat{f}(s)\|\, s\} \in L^2(\mathbb{R}, \mathrm{d}s)$. Further, since $\|f'(\cdot)\| \in L^1(\mathbb{R}, \mathrm{d}x)$, $\hat{f}(s)\, s$ is continuous in $s$. An easy calculation, using (II), (III) and the identity above, shows that the Fourier transform $g(s)$ of $f'(x)$ satisfies the estimate $\|g(s) - g(0)\| \leqq c\,|s|^q$ with $2q \leqq \min\{1, p/2\}$. Further, from (I) and $g(0) = \operatorname*{s-lim}_{N\to\infty} \int_{-N}^{N} f'(x)\,\mathrm{d}x = \operatorname*{s-lim}_{N\to\infty} [f(N) - f(-N)]$, we get $g(0) = 0$. From the assertions on $g(s)$, the identity above and $\{\|\hat{f}(s)\|\, s\} \in L^2(\mathbb{R}, \mathrm{d}s)$ we finally obtain $\hat{f}(\cdot) \in L^1(\mathbb{R}, \mathrm{d}s; \mathscr{H})$. ∎

**Proof of Proposition 28.** 1. First we show $A \in \operatorname{dom} \mu_s^H$. Since $A(t)\,u$ is strongly differentiable and $\|A'(t)\,u\|$ is summable, $A(t)\,u - A(s)\,u = \int_s^t A'(\lambda)\,u\,\mathrm{d}\lambda$ and, by

$\|A'(\cdot)\,u\| \in L^1(\mathbb{R}_+, \mathrm{d}t)$, this implies that the strong limit of $A(t)\,u$ as $t \to \infty$ exists for all $u \in \mathcal{D}$. Thus $A \in \operatorname{dom} \mu_s^H$ and we set $A_+ = \mu_s^H(A)$.

2. By Prop. 3.3, we have

$$\mathrm{e}^{-\mathrm{i}t\alpha(H)} g(H)\, u = (2\pi)^{-1/2} \int_{-\infty}^{\infty} g(t, s)\, \mathrm{e}^{-\mathrm{i}sH} u \,\mathrm{d}s \tag{18}$$

with

$$g(t, s) = (2\pi)^{-1/2} \int_{-\infty}^{\infty} \mathrm{e}^{\mathrm{i}sx}\, \mathrm{e}^{-\mathrm{i}t\alpha(x)} g(x) \,\mathrm{d}x \tag{19}$$

for functions $g(t, \cdot) \in L^1(\mathbb{R}, \mathrm{d}s) \cap L^2(\mathbb{R}, \mathrm{d}s)$. Let $g(x) \geqq 0$ have a support in $[c, d] \subset (a_n, a_{n+1})$ and assume $g(x)$ is twice continuously differentiable. Then, since $\alpha(x)$ is also twice continuously differentiable, we get from (19) that $g(t, \cdot)$ is the Fourier transform of a function with compact support and second derivative in $L^1(\mathbb{R}, \mathrm{d}x)$. Thus a simple calculation shows that $g(t, \cdot) \in L^1(\mathbb{R}, \mathrm{d}s)$ for each $t \in \mathbb{R}$. The fact that $g(t, \cdot) \in L^2(\mathbb{R}, \mathrm{d}s)$ is obvious. Further, we have that

$$g(t, s) \to 0 \quad \text{as} \quad t \to \infty \quad \text{for each} \quad s \in \mathbb{R}. \tag{20}$$

This is an easy consequence of the fact that on the support of $g(x)$, $\alpha$ has an inverse $\alpha^{-1}$ which is also twice continuously differentiable (make the transformation $\alpha(x) = y$ in (19) and use the Riemann-Lebesgue lemma (Prop. 3.23). From Lemma 19 we further get

$$\int_{-\infty}^{\infty} |g(t, s)|^2 \,\mathrm{d}s \to 0 \quad \text{as} \quad t \to \infty. \tag{21}$$

3. Now we consider the expression

$$\begin{aligned} I(t) &:= \left(A\, \mathrm{e}^{-\mathrm{i}t\alpha(H)} - \mathrm{e}^{-\mathrm{i}t\alpha(H)} A_+\right) g(H)\, u \\ &= \int_{-\infty}^{\infty} \left(A g(t, s)\, \mathrm{e}^{-\mathrm{i}sH} - g(t, s)\, \mathrm{e}^{-\mathrm{i}sH} A_+\right) u \,\mathrm{d}s = \int_{-\infty}^{\infty} g(t, s)\, \mathrm{e}^{-\mathrm{i}sH}\left(A(s) - A_+\right) u \,\mathrm{d}s \end{aligned}$$

where we have used (18). Let $k_1$ and $k_2$ be two nonnegative functions such that $k_1(x) + k_2(x) = 1$ for $x \in \mathbb{R}$, $k_1(x) = 1$ on $\mathbb{R}_+$, $k_1(x) = 0$ on $(-\infty, -1)$ and $k_1, k_2$ are infinitely differentiable functions. Then we can write $I(t) = I_1(t) + I_2(t)$ with

$$I_j(t) = \int_{-\infty}^{\infty} g(t, s)\, k_j(s)\, \mathrm{e}^{-\mathrm{i}sH}\left(A(s) - A_+\right) u \,\mathrm{d}s.$$

Now we have, using Prop. 3.22,

$$\begin{aligned} \|I_2(t)\| &= \sup_{\|v\|=1} \left|\left(v, I_2(t)\right)\right| \leqq \sup_{\|v\|=1} \int_{-\infty}^{0} |g(t, s)| \left|\left(v, \mathrm{e}^{-\mathrm{i}sH}\left(A(s) - A_+\right) u\right)\right| \mathrm{d}s \\ &\leqq \sup_{\|v\|=1} \int_{-\infty}^{0} |g(t, s)| \left|\left(v, (A - A_+)\, \mathrm{e}^{-\mathrm{i}sH} u\right)\right| \mathrm{d}s \\ &\leqq \sup_{\|v\|=1} \left\{\int_{-\infty}^{0} |g(t, s)|^2 \,\mathrm{d}s\right\}^{1/2} \left\{\int_{-\infty}^{0} \left|\left((A - A_+)^* v, \mathrm{e}^{-\mathrm{i}sH} u\right)\right|^2 \mathrm{d}s\right\}^{1/2} \\ &\leqq (2\pi)^{1/2} \left\{\int_{-\infty}^{0} |g(t, s)|^2 \,\mathrm{d}s\right\}^{1/2} \|A - A_+\|\; |||u|||_{H,\infty}. \end{aligned}$$

By (21) the first factor on the right side tends to zero as $t \to \infty$, thus $\|I_2(t)\| \to 0$ as $t \to \infty$.

Next, let us consider $I_1(t)$. By our assumptions on $A(t)$ and $k_1(t)$, the vector-valued function $D(t) = k_1(t)\left(A(s) - A_+\right) u$ satisfies the assumptions of Lemma 29. Thus, the Fourier transform $\hat{D}(s)$ of $D(t)$ is from $L^1(\mathbb{R}, ds; \mathscr{H})$. Then

$$I_1(t) = \int_{-\infty}^{\infty} g(t, s)\, e^{-isH} D(s)\, ds$$

exists and we have

$$\left(v, I_1(t)\right) = \int_{-\infty}^{\infty} \left(\overline{g(t, s)}\, e^{isH} v, D(s)\right) ds$$

$$= \int_{-\infty}^{\infty} \left(e^{it\alpha(H-p)} g(H-p)\, v, \hat{D}(p)\right) dp$$

where we have used (18) and the Parseval equation. Thus, by $\|g(H)\| \leqq 1$, we get from this equality $\|I_1(t)\| \leqq \|\hat{D}(\cdot)\|_{L^1(\mathbb{R},dp;\mathscr{H})}$. Now we claim that $\lim_{t\to\infty} \|I_1(t)\| = 0$. By the last estimate it suffices to prove this for special vector-valued functions $\hat{D}_f(\cdot) = f(\cdot)\, u \in L^1(\mathbb{R}, ds; \mathscr{H})$ with $f \in C_0^\infty(\mathbb{R})$ and $u \in \mathscr{H}$ because we can approximate $\hat{D}(s)$ by a linear combination of such elements in $L^1(\mathbb{R}, ds; \mathscr{H})$-norm. Then we get $I_1(t) = w_t(H)\, u$ where

$$w_t(s) = \int_{-\infty}^{\infty} f(p)\, e^{-it\alpha(s-p)} g(s-p)\, dp.$$

Since $f \in L^1(\mathbb{R}, dp)$, we get $|w_t(s)| \leqq \|f\|_{L^1}$ for all $t$ and $s$. By (20) and (19), $w_t(s) \to 0$ as $t \to \infty$ for each $s$. This together with $|w_t(s)| \leqq c$ implies that $w_t(H)\, u$ tends strongly to zero. Thus $I_1(t) \xrightarrow[s]{} 0$ as $t \to \infty$ and so does $I(t)$. Now

$$\|I(t)\| = \|(A\, e^{-it\alpha(H)} - e^{-it\alpha(H)} A_+)\, g(H)\, u\|$$

$$= \|e^{it\alpha(H)}(A - A_+)\, e^{-it\alpha(H)} g(H)\, u\|.$$

Since the set of such vectors $g(H)\, u$ is generating in $P^{ac}\mathscr{H}$, this proves (taking into account $P^{ac}_{\alpha(H)} = P^{ac}$ by Lemma 17) $A - A_+ \in \ker \mu_s^{\alpha(H)}$ and therefore $A \in \operatorname{dom} \mu_s^{\alpha(H)}$. ∎

## Notes and Remarks to Part II

Limits of $e^{itH} X\, e^{-itH}$ formally appeared first in papers on scattering in relativistic and non-relativistic quantum theory, see e.g. Ekstein [1956a], Jauch [1958b], and Kato/Mugibayashi [1963a]. The development of the mathematical scattering theory, in particular the algebraic scattering theory and some central questions of the mathematical scattering theory proposed by Kreĭn [1964a], stimulated the precise study of such limits.

Much of this part is a slight generalization and reformulation of previous results in Baumgärtel [1973a], Obermann [1977a], and Wollenberg [1979a], [1979b]. In general the proofs are simplified and organized as comprehensive as possible.

*Chapter 6.* The basic properties of the wave morphism $\mu_s^H$ contained in Theorems 23, 25, 28 and 35 are partly given by Lavine [1970a] and completely by Baumgärtel [1973a]. Theorem 29 is due to Baumgärtel (it is contained in Baumgärtel "On a recent note of A. W. Saenz and W. W. Zachary" (unpublished)). The corresponding basic properties for $\mu_a^H$ are due to Obermann [1977a]. The notations wave morphism (or wave homomorphism), wave algebra and wave ideal go back to the relation between these notions and the wave operator and were introduced in LAVINE [1970a] and WOLLENBERG [1977b]. The notation asymptotic constant can also be found in THIRRING [1, 2].

The use of the absolute Abelian limit, |A|-lim, goes back to OBERMANN [1977a]. The reason why it is not senseful to take the Abelian limit, A-lim, or the weak limit of the study for expressions like $e^{itH}X\,e^{-itH}$ is that they are not multiplicative. A simple example for this fact is given by $H := x$ on $L^2(\mathbb{R}, dx)$, $X = e^{-a(d/dx)}$, $X^* = e^{a(d/dx)}$ having the properties $XX^* = 1$, $\text{A-}\lim_{t\to\infty} e^{itH}X\,e^{-itH} = 0$, $\text{A-}\lim_{t\to\infty} e^{itH}X^*\,e^{-itH} = 0$.

The equality of the limit processes (Abel mean, Cesaro mean, second Hölder mean) and Corollary 14 are well-known (see e.g. HARDY [1]).

The proofs of Lemma 13 and 17 are essentially taken from v. NEUMANN/KOOPMAN [1932a].

Most of the results of 6.3 go back to WOLLENBERG and are contained in WOLLENBERG [1979a]. Prop. 43 is due to BAUMGÄRTEL. $\bigcap_H \ker \mu_s^H = \mathcal{L}_\infty(\mathcal{H})$ was first proved in DEMUTH/NEIDHARDT/WOLLENBERG [1976a], and dom $\mu_s^H \subset \mathcal{L}(\mathcal{H})$ was shown by HOWLAND [1972a].

The idea of using a topological characterization of dom $\mu_s^H$ as in 6.4 and the results of 6.4.1 and 6.4.2 for $\mu_s^H$ go back to BAUMGÄRTEL [1973a]. The corresponding definition and the results for $\mu_a^H$ were given by OBERMANN [1977a], 6.4.3 is due to WOLLENBERG [1979a].

*6.5.* For selfadjoint operators $H$ (even Schrödinger operators) with $P_H^{sc} \neq 0$ and a vector $u \in P_H^{sc}\mathcal{H}$ such that $\lim_{t\to\infty} (u, e^{-itH}u) = 0$, see SINHA [1977a]. An example for an operator $H$, with $P_H^{sc} \neq 0$ and a vector $u \in P_H^{sc}\mathcal{H}$ such that $\lim_{t\to\infty} (u, e^{-itH}u) \nrightarrow 0$, can be found in v. NEUMANN/KOOPMAN [1932a].

It is possible to generalize the notions of wave algebra, wave ideal etc. in the following sense: Replace $e^{-itH}$ by a unitary representation of $\mathbb{R}^n$ $(n > 1)$. At the place of the half line $\mathbb{R}_+ \ni t$ put an open cone $K \subset \mathbb{R}^n$ and replace the limit $t \to \infty$ by a suitable limit $|x| \to \infty$ with $x \in K$. For instance, if we consider the Poincare group, then the cone $K$ is the forward cone. Many statements similar to the ones in Chapter 6 can then be proved. The notion of absolute continuity requires care in this case. For details we refer to REHBERG [1979a].

*Chapter 7.* The main results of 7.1, Theorems 4 and 8 and Prop. 6, were first proved by WOLLENBERG [1977b]. The idea to use the shift representation for the proof of Theorem 4 goes back to BAUMGÄRTEL. The assertions in 7.2 are slight generalizations of results in BAUMGÄRTEL [1973a], BAUMGÄRTEL "On a recent note of A. W. SAENZ and W. W. ZACHARY" (unpublished) and WOLLENBERG [1979a]. Prop. 18 (ii) goes back to SAENZ/ZACHARY [1976a].

*7.3.* Prop. 21 was first proved in AMREIN/GEORGESCU/MARTIN [1974a]. Theorems 22 and 23 and Prop. 24 are essentially taken from AMREIN/MARTIN/MISRA [1970a].

*7.4.* The definition of the map $\Gamma_H$ and the proof of its essential properties under the restriction on certain integral operators goes back to FRIEDRICHS [1938a], [1948a]. The general case can be found in KATO [1]. Prop. 35 is a reformulation of a theorem of PUTNAM [1] and KATO [1968a] on nonnegative commutators.

Theorem 38 and Prop. 39 and 40 are taken from WOLLENBERG [1979b]. Prop. 43 and 46 are reformulations of statements on gentle perturbations (see FRIEDRICHS [1]).

*Chapter 8.* The results of 8.1.1 and 8.1.2 are taken from WOLLENBERG [1979b]. Prop. 5 is a slight generalization of a theorem in WOLLENBERG [1975a], see also DOLLARD [1976a].

The proof of the invariance of the wave ideal ker $\mu_a^H$ under the transformation $H \to \alpha(H)$, $\alpha$ an admissible function, is due to WOLLENBERG and was first published in OBERMANN/WOLLENBERG [1978b]. For a slightly larger class of admissible functions, $\ker \mu_a^H = \ker \mu_a^{\alpha(H)}$ was proved in WOLLENBERG [1979b].

*8.3* contains results from WOLLENBERG [1975a], [1979b]. Prop. 28 is, essentially, a reformulation of an invariance principle for wave operators given by CHANDLER/GIBSON [1976a], [1979a]. Lemma 29 was also proved in the latter paper.

Concerning Theorem 15 it is not difficult to strengthen the conditions (I) and (II) such that the other condition "$H$ has finite multiplicity" can be dropped. Assume that a system of pairwise orthogonal projections $P_1, P_2, \ldots, \sum P_n = P_{H_0}^{sc}$, reducing $H_0$, is given such that $H_0 \restriction P_n\mathcal{H}$ has finite multiplicity. In conditions (I) and (II) of Theorem 15 replace the expression $|(e^{-isH_0}u, e^{-itH}v)|^2$ by $\sum |(e^{-isH_0}u_n, e^{-itH}v)|^2$ where $u_n \in \mathcal{M}_\infty(H_0 \restriction P_n\mathcal{H}) \subset P_n\mathcal{H}$. Then the assertion $\ker \mu_a^{H_0} \subseteq \ker \mu_a^H$ is valid.

# Part III. Two-Space Wave Operators and Scattering Operators

This part is devoted to the investigation of certain limits of expressions of the form $e^{itH}J\,e^{-itH_0}f$ for $t \to \pm\infty$ where $H$, $H_0$ are selfadjoint operators in Hilbert spaces $\mathscr{H}$, $\mathscr{H}_0$ and $J \in \mathscr{L}(\mathscr{H}_0, \mathscr{H})$. The special case $\mathscr{H} = \mathscr{H}_0$, $H = H_0$ was extensively studied in Part II. The operator $J(t) := e^{itH}J\,e^{-itH_0}$ is called the *time-dependent pre-wave operator*. The limits of this operator are considered on the absolutely continuous subspace of $H_0$, as in Part II. The only exception is Chapter 9 containing the general theory. It will be developed for subspaces defined by universal projections. Further, we restrict ourselves in general to the limit $t \to \infty$ because the limit $t \to -\infty$ for the system $\{H, H_0; J\}$ is equivalent to the limit $t \to \infty$ for the system $\{-H, -H_0; J\}$. Therefore the assertions for $\{H, H_0; J\}$ and $t \to \infty$ can be easily transformed into the corresponding assertions for $\{H, H_0; J\}$ and $t \to -\infty$.

There are several reasons suggesting the study of limits of pre-wave operators, pure mathematical reasons as well as reasons coming from physical considerations within the framework of scattering theory, in particular in quantum mechanics.

1. Recall the physical motivation given in the Introduction. It is based on the interpretation of $\{H, \mathscr{H}\}$ as interacting (perturbed) system and $\{H_0, \mathscr{H}_0\}$ as free (unperturbed) system where $J$ connects the states of the unperturbed and the perturbed system. This interpretation is closely related to the problem to compare asymptotically the evolutions $e^{-itH}$ and $e^{-itH_0}$. Each $u_0 \in \mathscr{H}_0$ defines a time-trajectory $u_0(t) := e^{-itH_0}u_0$. Correspondingly, $Ju_0$ determines the time-trajectory $e^{-itH}Ju_0$. Then $e^{-itH}Ju_0 = J(-t)\,u_0(t)$. If asymptotically $J(-t) \approx J_{-\infty}$, then $e^{-itH}Ju_0 \approx J_{-\infty}\,e^{-itH_0}u_0$ (for $t \approx \infty$). This means that the complicated evolution $e^{-itH}$ can be asymptotically described by the "simple" evolution $e^{-itH_0}$ and the limit $J_{-\infty}$ of the identification operator. The same result appears if one wants to find a vector $u_+$ associated to $u_0$ such that $e^{-itH}u_+ \approx J\,e^{-itH_0}u_0$ asymptotically. The latter relation implies $J(t)\,u_0 \approx u_+$ asymptotically and therefore $J_\infty u_0 = u_+$.

2. A more mathematical motivation results from the fact that limits of pre-wave operators are always intertwining operators for $H, H_0$. The existence of a nontrivial intertwining operator implies the existence of nontrivial parts of $H$ and $H_0$ which are unitarily equivalent. Hence limits of pre-wave operators are sometimes convenient tools in order to decide the question of unitary equivalence of the perturbed and the unperturbed operators (see Part I, Section 4.5.3).

In this part we use the notations $E_0(\cdot) := E_{H_0}(\cdot)$, $E(\cdot) := E_H(\cdot)$, $P_0^{ac} := P_{H_0}^{ac}$, $P^{ac} := P_H^{ac}$, $E_0^{ac}(\cdot) := E_0(\cdot)\,P_0^{ac}$, $E^{ac}(\cdot) := E(\cdot)\,P^{ac}$, $R_0(z) := R_{H_0}(z)$ and $R(z) := R_H(z)$ where $H_0$ and $H$ denote selfadjoint operators on $\mathscr{H}_0$ and $\mathscr{H}$. $J$ denotes always a bounded operator from $\mathscr{H}_0$ into $\mathscr{H}$. We use only direct integrals with separable admissible systems.

# 9. Elementary Theory of Wave and Scattering Operators

## 9.1. Pre-wave operators and their limits

### 9.1.1. Time-dependent and stationary pre-wave operators

In this section we introduce the expression

$$J(t) := e^{itH} J\, e^{-itH_0} \tag{1}$$

which is called the *time-dependent pre-wave operator*. Our main interest in the following sections is the investigation of the existence and general properties of several kinds of limits of $J(t)$ with respect to $t \to \pm\infty$.

Since there exist several topologies in the Hilbert space and several kinds of limiting procedures, there are many possible ways to define the limit of (1). There are, however, only a few that interest us.

For instance, the operator norm limit can be excluded a priori because of

**Proposition 1.** *The operator norm limit* $\text{n-}\lim_{t\to\infty} e^{itH} J e^{itH_0}$ *exists iff $J$ is an intertwining operator with respect to $H$, $H_0$, that is, if*

$$J\, e^{-itH_0} = e^{-itH} J, \qquad t \in \mathbb{R}. \tag{2}$$

**Proof.** We have

$$\|e^{it_1H} J\, e^{-it_1H_0} - e^{it_2H} J\, e^{-it_2H_0}\| = \|J - e^{i(t_2-t_1)H} J\, e^{-i(t_2-t_1)H_0}\|.$$

Thus the norm convergence of (1) implies

$$\|J - e^{itH} J\, e^{-itH_0}\| = 0, \qquad t \in \mathbb{R},$$

that is, $J = e^{itH} J\, e^{-itH_0}$. Hence $J$ is an intertwining operator. The converse is obvious. ■

Similarly as in PartII we are mainly interested in the investigation of

$$\text{s-}\lim_{t\to\infty} e^{itH} J\, e^{itH_0} u =: u_+ \tag{3}$$

and

$$|A|\text{-}\lim_{t\to\infty} e^{itH} J\, e^{-itH_0} u =: u_+ \tag{4}$$

for certain sets of vectors $u \in \mathscr{H}_0$. Recall that the absolute Abelian limit $|A|$-lim in (4) means

$$\lim_{\varepsilon\to+0} \int_0^\infty \omega_\varepsilon(t)\, \|e^{itH} J\, e^{-itH_0} u - u_+\|^2\, dt = 0 \tag{5}$$

where, according to Lemma 6.2, $\omega_\varepsilon(t)$ can be chosen as $\varepsilon\, e^{-\varepsilon t}$, $\varepsilon\chi_{[0,\varepsilon^{-1}]}(t)$ or another equivalent averaging function. In the following we prefer $\varepsilon\, e^{-\varepsilon t}$. Because of the identity

$$\int_0^\infty \varepsilon\, e^{-\varepsilon t} \|e^{itH} J e^{-itH_0} u - u_+\|^2\, dt$$

$$= \|u_+\|^2 + \int_0^\infty \varepsilon\, e^{-\varepsilon t} \|J e^{-itH_0} u\|^2\, dt - 2\operatorname{Re}\left(u_+, \int_0^\infty \varepsilon\, e^{-\varepsilon t} e^{itH} J e^{-itH_0} u\, dt\right),$$

the following limits, whose existence is implied by (4), are of some interest:

$$\Omega_+^w u := \text{w-lim}_{\varepsilon \to +0} \int_0^\infty \varepsilon\, e^{-\varepsilon t} e^{itH} J e^{-itH_0} u\, dt, \tag{7}$$

$$\Omega_+ u := \text{s-lim}_{\varepsilon \to +0} \int_0^\infty \varepsilon\, e^{-\varepsilon t} e^{itH} J e^{-itH_0} u\, dt. \tag{8}$$

The operator

$$\Omega_\varepsilon := \int_0^\infty \varepsilon\, e^{-\varepsilon t} e^{itH} J e^{-itH_0}\, dt \tag{9}$$

is called the *stationary pre-wave operator*. This operator plays a decisive role in the so-called stationary theory of wave operators (see Chap. 13), because it can be easily expressed by means of resolvents and spectral measures only.

### 9.1.2. Limits of pre-wave operators

The first step towards the definition of wave operators is the investigation of the limits of the pre-wave operators, introduced in the preceding section. Considering the set of all vectors $u$ such that $\operatorname{Lim}_{t\to\infty} J(t)\, u$ exists, we obtain in a natural way an operator $J_\infty$, defined by $J_\infty u := \operatorname{Lim}_{t\to\infty} J(t)\, u$ on this set which turns out to be dom $J_\infty$. Sometimes this operator $J_\infty$ is already called wave operator (see Kato [1979a], Weidmann [1]). However, in this book (except Paragraph 10.1) the term wave operator is reserved for a slightly more special concept. In this section we collect simple properties of the operator $J_\infty$.

**Definition 2.** The set of all vectors $u \in \mathscr{H}_0$ such that

$$\operatorname{Lim}_{t\to\infty} e^{itH} J e^{-itH_0} u = \operatorname{Lim}_{t\to\infty} J(t)\, u$$

exists is denoted by $\mathscr{U}^{ex} = \mathscr{U}^{ex}(H, H_0, J)$.

Recall that Lim can be interpreted in two ways: either as |A|-lim or as s-lim. Thus, if it is necessary, we shall write $\mathscr{U}_a^{ex}$ or $\mathscr{U}_s^{ex}$ to indicate the corresponding limit.

**Lemma 3.** *$\mathscr{U}^{ex}$ is a subspace.*

**Proof.** See Prop. 6.19. ∎

The projection onto the subspace $\mathscr{U}^{ex}(H, H_0, J)$ is denoted by $P^{ex} = P^{ex}(H, H_0, J)$. The operator $\operatorname{Lim}_{t\to\infty} J(t)\, P^{ex}$ is denoted by $J_\infty$.

The following relation shows that $|A|\text{-}\lim_{t\to\infty} J(t)\, P^{ex}_a$ is a natural extension of $\text{s-}\lim_{t\to\infty} J(t)\, P^{ex}_s$.

**Proposition 4.** *If* $u_s := \text{s-}\lim_{t\to\infty} J(t)\, u$ *exists, then also* $u_a := |A|\text{-}\lim_{t\to\infty} J(t)\, u$ *exists and* $u_s = u_a$ *is valid.*

**Proof.** It is an immediate consequence of Corollary 6.14. ∎

Prop. 4 says that in any event $P^{ex}_s \leq P^{ex}_a$ is valid and that both operators $J^a_\infty$, $J^s_\infty$ coincide on $P^{ex}_s \mathscr{H}_0$.

**Proposition 5.** *The operator* $J_\infty$ *is bounded,* $J_\infty \in \mathscr{L}(\mathscr{H}_0, \mathscr{H})$, *and* $\|J_\infty\| \leq \|J\|$.

**Proof.** Because of Lemma 6.17 we have in any event $J_\infty = \text{s-}\lim_{n\to\infty} J(t_n)\, P^{ex}$ for a certain sequence $t_n \to \infty$. Since $\|J(t_n)\| \leq \|J\|$ for all $t_n$, we obtain the assertion. ∎

**Proposition 6.** *If* $u \in P^{ex}\mathscr{H}_0$, *then* $\text{s-}\lim_{\varepsilon\to+0} \Omega_\varepsilon u$ *and a fortiori* $\text{w-}\lim_{\varepsilon\to+0} \Omega_\varepsilon u$ *exist and equal* $J_\infty u$.

**Proof.** If $u \in P^{ex}\mathscr{H}_0$, then

$$\left\| \int_0^\infty \varepsilon\, e^{-\varepsilon t} J(t)\, u\, dt - J_\infty u \right\| = \left\| \int_0^\infty \varepsilon\, e^{-\varepsilon t} \big(J(t)\, u - J_\infty u\big)\, dt \right\|$$

$$\leq \left( \int_0^\infty \varepsilon\, e^{-\varepsilon t}\, \|J(t)\, u - J_\infty u\|^2\, dt \right)^{1/2}.$$

Hence $\text{s-}\lim_{\varepsilon\to+0} \Omega_\varepsilon u$ exists and equals $J_\infty u$. The other assertion is obvious. ∎

**Remark 7.** Similarly as $\mathscr{U}^{ex}$ we can introduce the set $\mathscr{U}^{ex}_w$ of all vectors $u \in \mathscr{H}_0$ such that $\text{w-}\lim_{\varepsilon\to+0} \Omega_\varepsilon u$ exists. Then it turns out that also $\mathscr{U}^{ex}_w$ is a subspace. The projection onto $\mathscr{U}^{ex}_w$ is denoted by $P^{ex}_w$. From Prop. 6 we obtain:

(I) $P^{ex} \leq P^{ex}_w$ and

(II) $\text{w-}\lim_{\varepsilon\to+0} \Omega_\varepsilon P^{ex}_w =: \Omega^w_+$ is a natural extension of $J^a_\infty = |A|\text{-}\lim_{t\to\infty} J(t)\, P^{ex}_a$.

In the next example $P^{ex}$ and $J_\infty$ can be calculated explicitly. This example, where $e^{-itH_0}$ is the shift representation of the translation group in $L^2(\mathbb{R}, dx)$, is of principal interest. It will be used often, sometimes with the Hilbert space $L^2(\mathbb{R}, dx; \mathscr{K})$ where $\mathscr{K}$ is a Hilbert space, for the construction of counter-examples.

**Example 8.** Let $\mathscr{H}_0 = L^2(\mathbb{R}, dx)$, $H_0 = -i\, d/dx$, $J = 1$, and $H = H_0 + q(\cdot)$, where $q(x)$ denotes a real-valued function from $L^\infty(\mathbb{R}, dx)$ which is locally $L^1$-integrable. As it is well-known,

$(e^{-itH_0} u)(x) = u(x - t)$ for all $u \in L^2(\mathbb{R}, dx)$. Therefore one easily calculates

$$(e^{itH} e^{-itH_0} u)(x) = e^{i\int_x^{x+t} q(y)dy} u(x),$$

that is, $e^{itH} e^{-itH_0}$ is a multiplication operator. Now we impose on $q$ the condition $q \in L^1(\mathbb{R}, dx)$. Then we obtain

$$(J_\infty u)(x) = \left(\text{s-lim}_{t\to\infty} e^{itH} e^{-itH_0} u\right)(x) = e^{i\int_x^{\infty} q(y)dy} u(x)$$

for all $u \in \mathscr{H}_0$. This means $P_s^{ex} = 1$ in this case.

Finally we present an example illustrating the fact that $P_s^{ex} < P_a^{ex}$ is possible. In other words, there is an example, where $J_\infty u$ exists with respect to $|A|\text{-}\lim_{t\to\infty}$ but not with respect to $\text{s-lim}_{t\to\infty}$ for certain vectors $u \in \mathscr{H}_0$.

**Example 9.** Let $\mathscr{H}_0 = L^2([0, 2\pi], dx)$, $H_0 = x$ (multiplication operator by $x$), $\mathscr{H}_0 = \mathscr{H}$, $J = 1$. We define $H$ by $H = U^*H_0U$ where $U$ is a unitary operator on $\mathscr{H}_0$ of the form $U = 1 - 2Q$ and where $Q$ is a projection. Then

$$e^{itH} e^{itH_0} = U^* e^{itH_0} U e^{-itH_0} = U^*(1 - 2e^{itH_0} Q e^{-itH_0}). \tag{10}$$

Thus the existence of $\operatorname{Lim}_{t\to\infty} e^{itH} e^{-itH_0} u = J_\infty u$ is equivalent to the existence of $\operatorname{Lim}_{t\to\infty} e^{itH_0} Q e^{-itH_0} u$. Because of $P_0^{ac} = 1$ the existence of the latter limit for all $u \in \mathscr{H}_0$ is equivalent to $Q \in \operatorname{dom} \mu^{H_0}$ (see Section 6.2.1). Now in the proof of Theorem 6.50 there was constructed a projection $Q$ (if $H_0$ is given) such that $Q \in \operatorname{dom} \mu_a^{H}$ but $Q \notin \operatorname{dom} \mu_s^{H_0}$. Taking this projection $Q$, we see that the limit of (10) exist in the absolute Abelian sense but not in the strong sense. Thus $|A|\text{-}\lim_{t\to\infty} e^{itH} e^{-itH_0} = J_\infty$ exists (that is $P_a^{ex} = 1$), but $\text{s-lim}_{t\to\infty} e^{itH} e^{-itH_0}$ does not exist and we have $P_s^{ex} < 1$.

### 9.1.3. Simple properties of the limit $J_\infty$ of the pre-wave operator

In this section we collect some properties centered around the intertwining property and related facts. Some of these properties were already presented in Part II (see Section 6.2.1).

**Proposition 10.** *The projection $P^{ex}$ belongs to* com $H_0$, $P^{ex} \in \operatorname{com} H_0$, *and* $J_\infty$ *is an intertwining operator,*

$$J_\infty e^{itH_0} = e^{itH} J_\infty, \qquad t \in \mathbb{R}. \tag{11}$$

Proof. The first assertion is contained in Prop. 6.19. The second assertion is a consequence of

$$J(t) e^{-i\tau H_0} = e^{-i\tau H} J(t + \tau), \qquad t, \tau \in \mathbb{R}, \tag{12}$$

and the translation invariance of the strong limit and the absolute Abelian limit, $\text{s-lim}_{t\to\infty} f(t) = \text{s-lim}_{t\to\infty} f(t + \tau)$, $|A|\text{-}\lim_{t\to\infty} f(t) = |A|\text{-}\lim_{t\to\infty} f(t + \tau)$ (see Lemma 6.8.). ■

A similar proposition can be formulated with respect to $P_w^{ex}$ and its corresponding limit (denoted by $\Omega_+^w$), introduced in Remark 7.

**Proposition 11.** *The projection* $P_{\mathrm{w}}^{\mathrm{ex}}$ *belongs to* com $H_0$, $P_{\mathrm{w}}^{\mathrm{ex}} \in$ com $H_0$, *and* $\Omega_+^{\mathrm{w}}$ *is an intertwining operator,*

$$\Omega_+^{\mathrm{w}} \, \mathrm{e}^{\mathrm{i}tH_0} = \mathrm{e}^{\mathrm{i}tH} \Omega_+^{\mathrm{w}}, \qquad t \in \mathbb{R}. \tag{13}$$

**Proof.** Let $f \in \mathscr{H}_0$, $g \in \mathscr{H}$. From (12) we obtain

$$\int_0^\infty \varepsilon \, \mathrm{e}^{-\varepsilon t} \big(g, J(t) \, \mathrm{e}^{-\mathrm{i}\tau H_0} f\big) \, \mathrm{d}t = \int_0^\infty \varepsilon \, \mathrm{e}^{-\varepsilon t} \big(\mathrm{e}^{\mathrm{i}\tau H} g, J(t+\tau) \, f\big) \, \mathrm{d}t.$$

Hence both assertions follow from the translation invariance of A-lim. ■

In Section 3.2.3 it was pointed out that clo spa $\{E_0(\Delta) \, u; u \in \mathscr{M}, \Delta$ Borel set$\}$ $=$ clo spa $\{\mathrm{e}^{\mathrm{i}tH_0} u; u \in \mathscr{M}, t \in \mathbb{R}\}$, where $\mathscr{M}$ is any subset of $\mathscr{H}_0$. This, together with Prop. 10, leads to the following two lemmas.

**Lemma 12.** *Let* $\mathscr{M}$ *be a subset of* $\mathscr{H}_0$. *If* $\mathscr{M} \subseteq \mathscr{U}^{\mathrm{ex}}$, *then also* clo spa $\{E_0(\Delta) \, u; u \in \mathscr{M}$, $\Delta$ Borel set$\} \subseteq \mathscr{U}^{\mathrm{ex}}$.

**Lemma 13.** *Let* $Q$ *be a projection in* $\mathscr{H}_0$. *Let* $\mathscr{M}$ *be generating with respect to* $Q\mathscr{H}_0$ *and* $H_0$. *If* $\mathscr{M} \subseteq \mathscr{U}^{\mathrm{ex}}$ *then* $Q \leq P^{\mathrm{ex}}$.

These pure technical lemmas are very important for the investigation of the existence of wave operators. They say that in every case it is sufficient to reduce existence considerations to (linear) manifolds which are generating in the considered subspace with respect to $H_0$.

**Example 14.** In general the projection $P^{\mathrm{ex}} \in$ com $H_0$ is not a spectral projection of $H_0$. This can be seen easily by a slight generalization of Example 8. Let $\mathscr{H}_0 = L^2(\mathbb{R}, \mathrm{d}x; \mathscr{K})$ where $\mathscr{K}$ is finite-dimensional, dim $\mathscr{K} < \infty$. Let $H_0 = -\mathrm{i} \, \mathrm{d}/\mathrm{d}x$ as before, hence $(\mathrm{e}^{-\mathrm{i}tH_0} f)(x) = f(x - t)$. Let $\mathscr{H} = \mathscr{H}_0$, $J = 1$ and $H = H_0 + q(\cdot)$, where $q(x)$ denotes a Borel function $q \colon \mathbb{R} \to \mathscr{L}(\mathscr{K})$ with $\|q(x)\| < \infty$ for all $x \in \mathbb{R}$ and $q(x)$ selfadjoint on $\mathscr{K}$ and $q(x)$, $q(x')$ pairwise commuting for all $x, x' \in \mathbb{R}$. Then one calculates as before

$$(\mathrm{e}^{\mathrm{i}tH} \, \mathrm{e}^{-\mathrm{i}tH_0} f)(x) = \mathrm{e}^{\mathrm{i} \int_x^{x+t} q(y) \mathrm{d}y} f(x), \qquad f \in \mathscr{H}_0.$$

Thus $\mathrm{e}^{\mathrm{i}tH} \, \mathrm{e}^{-\mathrm{i}tH_0}$ is a multiplication operator. Hence $P^{\mathrm{ex}}$ is necessarily also a multiplication operator $Q(x) \in \mathscr{L}(\mathscr{K})$. Now $P^{\mathrm{ex}} \in$ com $H_0$, i.e. $Q(x)$ commutes with $\mathrm{e}^{-\mathrm{i}tH_0}$. Thus we have $Q(x) \, f(x - t)$ $= Q(x - t) \, f(x - t)$ for all $f \in L^2(\mathbb{R}, \mathrm{d}x; \mathscr{K})$ and all $t \in \mathbb{R}$, therefore $Q(x) = \Pi$ a.e. (mod $|\cdot|$) follows where $\Pi$ is a constant projection from $\mathscr{L}(\mathscr{K})$. But a projection $P^{\mathrm{ex}}$ of this type is a spectral projection of $H_0$ iff either $\Pi = 0$ or $\Pi = 1_{\mathscr{K}}$. Namely, if $P^{\mathrm{ex}}$ is a spectral projection, then $FP^{\mathrm{ex}}F^* = \chi_\Delta$ for some Borel set $\Delta$ (note that the Fourier transform of the spectral measure of $H_0$ is given by the spectral measure $\chi_\Delta(\cdot)$ on $L^2(\mathbb{R}, \mathrm{d}\lambda; \mathscr{K})$. On the other hand, $FP^{\mathrm{ex}}F^*$ is the constant projection $\Pi$, because $P^{\mathrm{ex}}$ is the constant projection $\Pi$, i.e. we obtain the equation $\chi_\Delta(\lambda) \, g(\lambda)$ $= \Pi g(\lambda)$ for all $g \in L^2(\mathbb{R}, \mathrm{d}\lambda; \mathscr{K})$, but this equation immediately yields either $\Pi = 0$ (and $|\Delta| = 0$) or $\Pi = 1_{\mathscr{K}}$ (and $\Delta = \mathbb{R}$). Now we can choose $q$ such that $0 < \Pi < 1_{\mathscr{K}}$ (a simple example is given by $\mathscr{K} = \mathbb{C}^2$, $q(x) = \{q_1(x), q_2(x)\}$ (diagonal matrix) with $q_1(x) = (1 + |x|)^{-(1+\varepsilon)}$, $\varepsilon > 0$, and $q_2(x) \equiv 1$; in this case an easy calculation shows that $\Pi = \{1, 0\}$.

The intertwining property of $J_\infty$ leads to the following

**Theorem 15.** *Let* $J_\infty = \operatorname{sgn} J_\infty \, |J_\infty|$ *be the polar decomposition of* $J_\infty$ *and let* $E_0 =$ $(\operatorname{sgn} J_\infty)^* \operatorname{sgn} J_\infty$, $E = \operatorname{sgn} J_\infty (\operatorname{sgn} J_\infty)^*$. *Then* $E_0 \leq P^{\mathrm{ex}}$, $E_0 \in$ com $H_0$ and $P^{\mathrm{ex}} - E_0$ *is the projection onto* ker $(J_\infty \restriction P^{\mathrm{ex}} \mathscr{H}_0)$, *while* $E_0$ *is the projection onto* supp $J_\infty$

$= (\ker J_\infty)^\perp$. *Also,* $|J_\infty| \in \operatorname{com} H_0$. *Furthermore,* $E$, $|J_\infty^*| \in \operatorname{com} H$ *and* $E$ *is the projection onto* clo ima $J_\infty$ *which equals* supp $J_\infty^*$. sgn $J_\infty$ *is also intertwining for* $H, H_0$ *and* $H \mid E\mathscr{H}$ *is unitarily equivalent (via* sgn $J_\infty$*) to* $H_0 \mid E_0\mathscr{H}_0$.

**Proof.** The result follows immediately from the intertwining property of $J_\infty$ (see the corresponding assertions in Section 1.3.3). ∎

A further consequence of the intertwining property (and the last theorem) is

**Proposition 16.** *The following relations are valid:*

$$\operatorname*{Lim}_{t\to\infty} \{e^{-itH} J_\infty - J e^{-itH_0} P^{ex}\} = 0, \tag{14}$$

$$\operatorname*{Lim}_{t\to\infty} \{(J_\infty - J) e^{-itH_0} P^{ex}\} = 0, \tag{15}$$

$$\operatorname*{Lim}_{t\to\infty} E^\perp J e^{-itH_0} P^{ex} = 0. \tag{16}$$

**Proof.** (14) follows from the definition of $J_\infty$. (15) follows from (14), $J_\infty = J_\infty P^{ex}$ and the intertwining property. (16) follows from (15) and $E^\perp J_\infty = 0$. ∎

Another important implication of the intertwining property of $J_\infty$ is connected with its representation by a family of operators $\hat{J}_\infty(\lambda)$ with respect to direct integral representations of $\mathscr{H}$ and $\mathscr{H}_0$ (see Section 4.4.2, in particular Corollary 4.19).

**Theorem 17.** *Let* $\Phi_0: \mathscr{H}_0 \to L^2(\operatorname{spec}^c H_0, \mu_0(d\lambda); \mathscr{H}_{0\lambda})$ *and* $\Phi: \mathscr{H} \to L^2(\operatorname{spec}^c H, \mu(d\lambda); \mathscr{H}_\lambda)$ *be isometric maps such that* $E_0(\Delta)$, $E(\Delta)$ *are represented by* $\chi_\Delta(\lambda)$ (direct integral representations of $\mathscr{H}_0$, $\mathscr{H}$ with respect to $E_0(\Delta)$, $E(\Delta)$, where the corresponding admissible systems are separable). *Then there is a function* $\hat{J}_\infty(\lambda)$: $\operatorname{spec}^c H_0 \to \mathscr{L}(\mathscr{H}_{0\lambda}, \mathscr{H}_\lambda)$ *a.e.* (mod $\mu_0$) *defined on* $\operatorname{spec}^c H_0$ *and with essentially bounded norm* $\|\hat{J}_\infty(\lambda)\|$ *such that*

$$(J_\infty f)^\wedge(\lambda) = \hat{J}_\infty(\lambda)\, \hat{f}(\lambda) \tag{17}$$

*where* $\hat{f}$ *denotes the representer of* $f$ *in* $L^2(\operatorname{spec}^c H_0, \mu_0(d\lambda); \mathscr{H}_{0\lambda})$, $\hat{f} = \Phi_0 f$, *and* $(J_\infty f)^\wedge$ *denotes the representer of* $J_\infty f$ *in*

$$L^2(\operatorname{spec}^c H, \mu(d\lambda); \mathscr{H}_\lambda), \quad (J_\infty f)^\wedge = \Phi(J_\infty f).$$

**Proof.** The result follows immediately from Corollary 4.19. ∎

## 9.2. Two-space wave operators

### 9.2.1. Definition of wave operators with respect to universal spectral projections

Recall the concept of a universal spectral projection $P_{H_0}$ (abbreviated: universal projection) with respect to selfadjoint operators $H_0$ in $\mathscr{H}_0$. A universal projection $P_{H_0}$ is a spectral projection $P_{H_0} = E_{H_0}(\Delta_{H_0})$ (where $\Delta_{H_0}$ may vary with $H_0$) which is compatible with unitary transformations of $H_0$ and with the transition to a part $H_0 \mid Q\mathscr{H}_0$ ($Q \in \operatorname{com} H_0$) of $H_0$ (for details see Section 6.5.1). In some sense $P_{H_0}$ represents a universal spectral property of selfadjoint operators such as continuity or absolute continuity or multiplicity not bigger than $m$.

The projection $P^{ex} = P^{ex}(H, H_0, J)$ defined after Lemma 3, in general, depends not only on $H_0$ but also on $H$, $J$. Of course, the advantage in this case is that $J_\infty = J_\infty P^{ex}$ always exist; the difficulties are contained in the complicated structure of $P^{ex}$. In order to establish a useful theory of wave operators it is more convenient to fix a certain $P^{ex} = P^{ex}(H_0)$ (perhaps not the maximal one) and to develop the theory of the limits $\operatorname{Lim}_{t\to\infty} J(t) f$ for the vectors $f$ from the fixed subspace $P^{ex}(H_0)\mathscr{H}_0$. Obviously, there is much freedom for choosing $P^{ex}(H_0)$. Our "ansatz" is to choose a certain universal projection. Thus the wave operator is defined as follows:

**Definition 18.** Let $P_{H_0}$ be a universal projection of $H_0$. Then the *wave operator* $W_+(H, H_0; JP_{H_0})$ is defined by

$$W_+(H, H_0; JP_{H_0}) := \operatorname{Lim}_{t\to\infty} e^{itH} J\, e^{-itH_0} P_{H_0}. \tag{18}$$

If Lim $=$ |A|-lim, we sometimes call $W_+(H, H_0; JP_{H_0})$ *Abel wave operator*; if Lim $=$ s-lim, then we call these operators *strong wave operators.*

We use also the abbreviated notations $W_+(H, H_0; J)$ and $W_+$ for the wave operator. If $\mathscr{H} = \mathscr{H}_0$ and $J = 1$, then we put $W_+(H, H_0) := W_+(H, H_0; 1)$.

The wave operator $W_-(H, H_0; J)$ (for the negative time direction) is defined by

$$W_-(H, H_0; J) := W_+(-H, -H_0; J). \tag{19}$$

Thus it is sufficient for the general investigation of the properties of the wave operators to consider $W_+$ only.

**Remark 19.** Obviously the propositions of Sections 1.2 and 1.3 are valid also for $W_+$ instead of $J_\infty$. The reason is that $W_+(H, H_0; J) = J_\infty P_{H_0}$. For example, $W_+$ is intertwining with respect to $H$, $H_0$, $e^{-itH} W_+ = W_+ e^{-itH_0}$ for $t \in \mathbb{R}$, and also $\Omega_+^w := \operatorname{w-lim}_{\varepsilon\to 0} \Omega_\varepsilon P_{H_0}$ is an intertwining operator; the relations (14)—(16) now are written in the form $\operatorname{Lim}_{t\to\infty} \{e^{-itH} W_+ - J\, e^{-itH_0} P_{H_0}\} = 0$, $\operatorname{Lim}_{t\to\infty} (W_+ - J)\, e^{-itH_0} P_{H_0} = 0$ and $\operatorname{Lim}_{t\to\infty} E^\perp J\, e^{-itH_0} P_{H_0} = 0$, where now $E = \operatorname{sgn} W_+ (\operatorname{sgn} W_+)^*$. Lemma 13 can be formulated as follows: "Let $\mathscr{M} \subseteq P_{H_0}\mathscr{H}_0$ be generating with respect to $P_{H_0}\mathscr{H}_0$ and $H_0$ and assume that $\operatorname{Lim}_{t\to\infty} J(t)\, u$ exists for all $u \in \mathscr{M}$, then $W_+$ exists".

Next we collect two further properties of $W_+$.

**Proposition 20.** *Let* $W_+ = W_+(H, H_0; J)$ *exist and put* $E = \operatorname{sgn} W_+ (\operatorname{sgn} W_+)^*$. *Then* $E \leq P_H$ *is valid.*

**Proof.** According to Theorem 15 we have $H \restriction E\mathscr{H}$ is unitarily equivalent to $H_0 \restriction E_0\mathscr{H}_0$, where $E_0 = (\operatorname{sgn} W_+)^* \operatorname{sgn} W_+$. The operator $H_0 \restriction E_0\mathscr{H}_0$ is a part of $H_0 \restriction P_{H_0}\mathscr{H}_0$. Thus Prop. 6.64 leads to $E \leq P_H$. ■

**Theorem 21** (Chain rule). *Let* $H_0$, $H'$, $H$ *be selfadjoint operators on* $\mathscr{H}_0$, $\mathscr{H}'$, $\mathscr{H}$, *respectively. Let* $J_0 \in \mathscr{L}(\mathscr{H}_0, \mathscr{H}')$, $J' \in \mathscr{L}(\mathscr{H}', \mathscr{H})$ *and put* $J := J'J_0$. *Assume that*

*$W_+(H', H_0; J_0)$ and $W_+(H, H'; J')$ exist. Then also $W_+(H, H_0; J)$ exists and*

$$W_+(H, H_0; J) = W_+(H, H'; J')\, W_+(H', H_0; J_0) \tag{20}$$

*is valid.*

**Proof.** We have

$$e^{itH} J\, e^{-itH_0} P_{H_0} = e^{itH} J'\, e^{-itH'}\, e^{itH'} J_0\, e^{-itH_0} P_{H_0}$$

$$= e^{itH} J'\, e^{-itH'} P_{H'}\, e^{itH'} J_0\, e^{-itH_0} P_{H_0} + e^{itH} J'\, e^{-itH'} P_{H'}^{\perp}\, e^{itH'} J_0\, e^{-itH_0} P_{H_0}.$$

Because of (16) the second term on the right-hand side vanishes for $\operatorname{Lim}_{t\to\infty}$. The first term on the right-hand side tends to $W_+(H, H'; J')\, W_+(H', H_0; J_0)$ since $\operatorname{Lim}_{t\to\infty} A(t)\, B(t) = \operatorname{Lim}_{t\to\infty} A(t) \operatorname{Lim}_{t\to\infty} B(t)$ (see Remark 6.18). Thus (20) is proved. ∎

**Remark 22.** Theorem 17 is also valid for $W_+$ instead of $J_\infty$ (if $W_+$ exists). The operator function $\hat{J}_\infty(\lambda)$ from $\operatorname{spec}^c H_0$ into $\mathscr{L}(\mathscr{H}_{0\lambda}, \mathscr{H}_\lambda)$ in this case is called the *wave matrix* and it is denoted by $\hat{W}_+(\lambda)$. Note that the wave matrix depends on the isometries $\Phi$, $\Phi_0$ (see Theorem 17).

**Remark 23.** Without loss of generality we can assume $P_{H_0} \leq P_{H_0}^c$ in the following. In principle the projection $P_{H_0}^p$ onto the spectrally discrete subspace of $H_0$ can also be used as the universal projection. But this choice leads to trivial results. Let $\operatorname{Lim} e^{itH} J \cdot e^{-itH_0} P_{H_0}^p$ exist. Assume $\lambda_0$ to be an eigenvalue of $H_0$ and let $u_0$ be a corresponding eigenvector. Then $e^{itH} J\, e^{-itH_0} u_0 = e^{-it\lambda_0}\, e^{itH} J u_0$. Further, according to the mean ergodic theorem (Prop. 3.9), Prop. 6 and Lemma 6.2, we get

$$W_+ u_0 = \operatorname*{s-lim}_{\varepsilon\to+0} \int_0^\infty \varepsilon\, e^{-\varepsilon t}\, e^{itH} J\, e^{-itH_0} u_0\, dt = E_H(\{\lambda_0\})\, J u_0.$$

But from (15) we obtain

$$\operatorname*{Lim}_{t\to\infty} (W_+ - J)\, e^{-itH_0} u_0 = \operatorname*{Lim}_{t\to\infty} e^{-it\lambda_0} (E_H(\{\lambda_0\}) - 1_{\mathscr{H}})\, J u_0 = 0.$$

Therefore either $J u_0 = 0$ or $\lambda_0$ is also an eigenvalue of $H$ and $J u_0$ is a corresponding eigenvector.

In particular, if $\mathscr{H} = \mathscr{H}_0$, $\mathscr{J} = 1$ and $\dim \mathscr{H}_0 = n < \infty$ (finite-dimensional case) we obtain: $W_+(H, H_0)$ exists iff $H = H_0$, and in this case obviously we have $W_+ = 1$.

In almost all cases we shall even assume $P_{H_0} \leq P_{H_0}^{ac}$.

### 9.2.2. Completeness and semicompleteness

With respect to a given universal projection $P_{H_0}$ we can define semicompleteness and completeness of wave operators.

**Definition 24.** The wave operator $W_+(H, H_0; J)$ is called *$H_0$-semicomplete* if $E_0 := (\operatorname{sgn} W_+)^* \operatorname{sgn} W_+ = P_{H_0}$. It is called *$H$-semicomplete* if $E := \operatorname{sgn} W_+ (\operatorname{sgn} W_+)^* = P_H$. It is called *complete* if it is $H_0$-semicomplete and H-semicomplete.

The meaning of these definitions results from Theorem 15. If $W_+$ is $H_0$-semicomplete, then the partial isometry sgn $W_+$ has the initial subspace $P_{H_0}\mathscr{H}_0$, that is, the part of $H_0$ with the universal spectral property represented by $P_{H_0}$ is unitarily mapped by means of sgn $W_+$ onto a part of $H \upharpoonright P_H\mathscr{H}$. If $W_+$ is $H$-semicomplete, then the partial isometry sgn $W_+$ has the final subspace $P_H\mathscr{H}$. Completeness of $W_+$ means that sgn $W_+$ yields a unitary equivalence between $H \upharpoonright P_H\mathscr{H}$ and $H_0 \upharpoonright P_{H_0}\mathscr{H}_0$, that is, the parts of $H$ and $H_0$ with the given universal spectral property.

**Example 25** (Completeness). The wave operator of Example 8, considered with respect to the absolutely continuous subspace ($P_{H_0} = P_{H_0}^{ac}$), that is, with respect to the universal spectral property "absolutely continuous", is complete. Namely, in this case $P_{H_0}^{ac} = 1$ and $W_+(H, H_0)$ exists on the whole space $\mathscr{H}_0$; furthermore, $W_+(H, H_0)$ is unitary. Thus $E = W_+W_+^* = 1 = P_H^{ac}$ $= E_0 = W_+^*W_+ = P_{H_0}^{ac}$ and $W_+(H, H_0)$ is complete.

**Example 26** (Non-completeness). Let $\mathscr{H} = \mathscr{H}_0$, $J = 1$ and $H_0$ be selfadjoint with $\dim (P_0^{ac})^\perp = \infty$, $P_0^{ac} > 0$. Further, let $H$ be absolutely continuous, $P_0^{ac} \in \operatorname{com} H$ and $H \upharpoonright P_0^{ac}\mathscr{H}_0 = H_0 \upharpoonright P_0^{ac}\mathscr{H}_0$. In this example we take as universal spectral property "absolutely continuous", i.e. $P_H = P^{ac}$ and $P_{H_0} = P_0^{ac}$. Then we obtain $e^{itH} e^{-itH_0}P_0^{ac} = e^{itH}P_0^{ac} e^{-itH_0}P_0^{ac} = e^{itH_0}P_0^{ac} e^{-itH_0}P_0^{ac} = P_0^{ac}$. Therefore the wave operator $W_+(H, H_0)$ exists and it is $H_0$-semicomplete. But $E = W_+ W_+^*$ $= P_0^{ac} < 1 = P^{ac}$ and this means that $W_+$ is not $H$-semicomplete. Thus $W_+$ is not complete.

Sometimes it is possible to prove non-completeness of a wave operator by the following lemma.

**Lemma 27.** *Let $W_+(H, H_0; J)$ exist and put (as before) $E = \operatorname{sgn} W_+(\operatorname{sgn} W_+)^*$. Assume that a selfadjoint operator $H_0'$ and a bounded operator $J'$ exist with the properties: $W_+(H, H_0; J')$ exists, $E' := \operatorname{sgn} W_+(H, H_0'; J') \left(\operatorname{sgn} W_+(H, H_0'; J')\right)^* \neq 0$ and $EE' = 0$. Then $W_+(H, H_0; J)$ is necessarily non-complete.*

Proof. Obvious. ∎

### 9.2.3. One-dimensional perturbations

This section is intended to give a simple but important example where the existence of the wave operator $W_+(H, H_0)$ with respect to $P_{H_0} = P_0^{ac}$ can be proved rather directly. Furthermore, the wave matrix in this case can be calculated explicitly.

We put $\mathscr{H}_0 = \mathscr{H}$ and $J = 1$. Further let $V$ be a one-dimensional selfadjoint operator. That is, let

$$V = a(h, \cdot)\, h, \quad h \in \mathscr{H} \text{ and } \|h\| = 1, \tag{21}$$

where $a \in \mathbb{R}$, $a \neq 0$. Assume $H_0$ to be an arbitrary selfadjoint operator. If

$$H := H_0 + V \tag{22}$$

then $H$ is also selfadjoint, where dom $H$ = dom $H_0$.

Our aim is to prove the existence of $\operatorname*{s-lim}_{t\to\infty} e^{itH} e^{-itH_0}P_0^{ac}$, that is, the existence of the wave operator $W_+(H, H_0)$ with respect to the universal spectral property "absolutely continuous". First we collect some preliminary lemmas which will be applied in the proof of the main theorem of this section. We introduce the notation

$$V(t) := e^{itH} e^{-itH_0}. \tag{23}$$

Then $V(t)$ is strongly differentiable on dom $H_0$, and with $f \in$ dom $H_0$ we obtain $V'(t) f = \mathrm{i}\, e^{\mathrm{i}tH} V e^{-\mathrm{i}tH_0} f$, hence

$$V(t_2) f - V(t_1) f = \mathrm{i} \int_{t_1}^{t_2} e^{\mathrm{i}\tau H} V e^{-\mathrm{i}\tau H_0} f \, d\tau \tag{24}$$

follows for all $f \in \mathscr{H}_0$.

**Lemma 28.** *Let $Q, Q_0$ be projections in $\mathscr{H}$. Assume $\mathscr{D} \subset Q\mathscr{H}$, $\mathscr{D}_0 \subset Q_0\mathscr{H}$ to be generating in $Q\mathscr{H}$, $Q_0\mathscr{H}$, respectively. If $\lim_{t\to\infty} (g, V(t) f)$ exists for $g \in \mathscr{D}$, $f \in \mathscr{D}_0$, then* $\text{w-lim}_{t\to\infty} QV(t) Q_0$ *exists.*

**Proof.** It is obvious. ■

**Lemma 29.** *Let $H_0$ be selfadjoint in $\mathscr{H}$ and let $V$ and $H$ be defined by* (21) *and* (22). *Then*

$$V_\infty := \text{w-lim}_{t\to\infty} P^{ac} e^{\mathrm{i}tH} e^{-\mathrm{i}tH_0} P_0^{ac} \tag{25}$$

*exists.*

**Proof.** Let $g \in \mathscr{H}$. Then from (24) we obtain

$$(g, V(t_2) f - V(t_1) f) = \mathrm{i}\alpha \int_{t_1}^{t_2} (e^{-\mathrm{i}\tau H} g, h) (h, e^{-\mathrm{i}\tau H_0} f) \, d\tau$$

and

$$|(g, V(t_2) f - V(t_1) f)| \leq |\alpha| \left( \int_{t_1}^{t_2} |(e^{-\mathrm{i}\tau H} g, h)|^2 \, d\tau \right)^{1/2} \left( \int_{t_1}^{t_2} |(h, e^{-\mathrm{i}\tau H_0} f)|^2 \, d\tau \right)^{1/2}. \tag{26}$$

Now we choose $f \in \mathscr{M}_\infty(H_0)$, $g \in \mathscr{M}_\infty(H)$. Note that $\mathscr{M}_\infty(H_0)$, $\mathscr{M}_\infty(H)$ are dense in $P_0^{ac}\mathscr{H}$, $P^{ac}\mathscr{H}$, respectively (see Prop. 3.19). From Prop. 3.22 we immediately obtain

$$\int |(e^{-\mathrm{i}\tau H} g, h)|^2 \, d\tau < \infty, \qquad \int |(h, e^{-\mathrm{i}\tau H_0} f)|^2 \, d\tau < \infty.$$

Hence from (26) the existence of $\lim_{t\to\infty} (g, V(t) f)$ follows. Finally, Lemma 28 (with $\mathscr{D} = \mathscr{M}_\infty(H)$, $\mathscr{D}_0 = \mathscr{M}_\infty(H_0)$, $Q = P^{ac}$, $Q_0 = P^{ac}$) yields the assertion. ■

The next step is to prove that $V_\infty$ is partially isometric with initial subspace $P_0^{ac}\mathscr{H}_0$. If this is true, then it is easy to prove that $V_\infty$ is simultaneously the strong limit of $V(t) P_0^{ac}$.

In this connection it is convenient to assume, without loss of generality, that *$H$ and $H_0$ are simple in $\mathscr{H}$ and that $h$ is a generating vector for both operators.*

**Remark 30.** In the general case, if $V$ and $H$ are defined by (21) and (22), we consider the subspace $\mathscr{K}_0 =$ clo spa $\{E_0(\Delta) h; \Delta$ Borel set$\}$. $\mathscr{K}_0$ is reducing for $H_0$ and contains $h$. It is the smallest subspace with these properties. Correspondingly, $\mathscr{K} =$ clo spa $\{E(\Delta) h; \Delta$ Borel set$\}$ is reducing for $H$, contains $h$ and is the smallest subspace with these properties. Now both subspaces $\mathscr{K}_0$, $\mathscr{K}$ are reducing for $V$, hence they must coincide. We denote it by $\mathscr{K}$. In $\mathscr{K}^\perp$ we have $H = H_0$, in $\mathscr{K}$ both operators $H$, $H_0$ are simple and $h$ is a generating vector.

**Proposition 31.** *The operator $V_\infty$ is partially isometric with initial subspace $P_0^{ac}\mathscr{H}$.*

**Proof.** $V_\infty$ is an intertwining operator between $H$ and $H_0$, as can be easily seen from (25). Further, we have $V_\infty = P^{ac} V_\infty = V_\infty P_0^{ac}$, therefore it is sufficient to verify the isometry property for vectors of the form $f = E_0(\Delta) P_0^{ac} h$ only. Because of

$$V_\infty E_0(\Delta) P_0^{ac} h = E(\Delta) V_\infty P_0^{ac} h, \qquad \Delta \text{ a Borel set}, \tag{27}$$

it is enough to prove the equation $\|E(\Delta) V_\infty P_0^{ac} h\| = \|E_0(\Delta) P_0^{ac} h\|$ for all Borel sets $\Delta$ (the extension to clo spa $\{E_0(\Delta) P_0^{ac} h;\ \Delta$ Borel set$\}$ is obvious). Let $g \in \mathscr{M}_\infty(H)$, $f \in \mathscr{M}_\infty(H_0)$. Then from

$$(g, V(t) f) = (g, f) + \mathrm{i}\alpha \int_0^t (e^{-\mathrm{i}\tau H} g, h) (h, e^{-\mathrm{i}\tau H_0} f) \, d\tau$$

we obtain

$$(g, V_\infty f) = (g, f) + \mathrm{i}a \int_0^\infty (\mathrm{e}^{-\mathrm{i}\tau H} g, h)\,(h, \mathrm{e}^{-\mathrm{i}\tau H_0} f)\,\mathrm{d}\tau. \tag{28}$$

Note that we have very simple spectral representations for $H$, $H_0$ because these operators are simple and $h$ is generating. Let $\Phi$, $\Phi_0$ denote the isometric representation operators (cf. Corollary 4.23)

$$\Phi\colon \mathscr{H} \to L^2(\operatorname{spec}^c H, \mu(\mathrm{d}\lambda))$$

$$\Phi_0\colon \mathscr{H}_0 \to L^2(\operatorname{spec}^c H_0, \mu_0(\mathrm{d}\lambda))$$

where $\mu(\Delta) = (h, E(\Delta)\,h)$, $\mu_0(\Delta) = (h, E_0(\Delta)\,h)$. The representation of $f$ by $(\Phi f)\,(\cdot)$ is given by $f = \int (\Phi f)\,(\lambda)\,E(\mathrm{d}\lambda)\,h$, similarly for $\Phi_0$. Because of $(\mathrm{e}^{-\mathrm{i}\tau H} g, h) = \int \mathrm{e}^{\mathrm{i}\tau\lambda}\overline{(\Phi g)\,(\lambda)}\,\mu(\mathrm{d}\lambda)$ we can write (28) in the form

$$(g, V_\infty f) = (g, f) + \mathrm{i}a \int_{-\infty}^{\infty} \overline{(\Phi g)\,(\lambda)} \left\{\int_0^\infty \mathrm{e}^{\mathrm{i}\tau\lambda}(h, \mathrm{e}^{-\mathrm{i}\tau H_0} f)\,\mathrm{d}\tau\right\} \mu(\mathrm{d}\lambda).$$

From Section 3.5.3 we obtain immediately (in particular, note the formulas between Prop. 24 and Prop. 25)

$$\int_0^\infty \mathrm{e}^{\mathrm{i}\tau\lambda}(h, \mathrm{e}^{-\mathrm{i}\tau H_0} f)\,\mathrm{d}\tau = \mathrm{i}(h, R_0(\lambda + \mathrm{i}0)\,f).$$

Note that $(h, R_0(\lambda + \mathrm{i}0)\,f) := \lim_{\varepsilon \to +0} (h, R_0(\lambda + \mathrm{i}\varepsilon)\,f)$ exists a.e. (mod $|\cdot|$) and the limit is a member of $L^2(\mathbb{R}, \mathrm{d}\lambda)$. Now the formula

$$(g, V_\infty f) = (g, f) - a \int_{-\infty}^{\infty} \overline{(\Phi g)\,(\lambda)}\,(h, R_0(\lambda + \mathrm{i}0)\,f)\,\mu(\mathrm{d}\lambda) \tag{29}$$

easily follows. (29) can be written in the form

$$(g, V_\infty f) = \int_{-\infty}^{\infty} \overline{(\Phi g)\,(\lambda)}\,\{(\Phi f)\,(\lambda) - a(h, R_0(\lambda + \mathrm{i}0)\,f)\}\,\mu(\mathrm{d}\lambda).$$

According to the remarks at the beginning of this proof we need this formula only for $f = h$. By $(\Phi h)\,(\lambda) \equiv 1$ and Prop. 1.2, we obtain

$$(\Phi V_\infty h)\,(\lambda) = 1 - a(h, R_0(\lambda + \mathrm{i}0)\,h), \qquad \text{a.e. mod } |\cdot|. \tag{30}$$

For abbreviation we put

$$\omega(z) := 1 - a(h, R_0(z)\,h) = \big(h, (1 - VR_0(z))\,h\big), \qquad \operatorname{Im} z \neq 0.$$

Later the function $1 - VR_0(z)$ is often denoted by $G_0(z)$. Now by a straightforward calculation (which is left to the reader) we obtain the following identity:

$$(\varepsilon/\pi)\,\|R_0(\lambda + \mathrm{i}\varepsilon)\,h\|^2 = |\omega(\lambda + \mathrm{i}\varepsilon)|^2\,(\varepsilon/\pi)\,\|R(\lambda + \mathrm{i}\varepsilon)\,h\|^2, \qquad \varepsilon > 0. \tag{31}$$

According to Prop. 3.15, from (31) the relation

$$(h, E_0(\mathrm{d}\lambda)\,h)/\mathrm{d}\lambda = |\omega(\lambda + \mathrm{i}0)|^2\,\big((h, E(\mathrm{d}\lambda)\,h)/\mathrm{d}\lambda\big) \tag{32}$$

follows a.e. mod $|\cdot|$, that is $\mu_0'(\lambda) = |\omega(\lambda + \mathrm{i}0)|^2\,\mu'(\lambda)$.

From (30) we obtain

$$\|E(\Delta)\ V_\infty h\|^2 = \|E(\Delta)\ V_\infty P_0^{\mathrm{ac}} h\|^2 = \int_\Delta |\omega(\lambda + \mathrm{i}0)|^2\,\tilde{\mu}(\mathrm{d}\lambda), \quad \tilde{\mu}(\Delta) := (P^{\mathrm{ac}} h, E(\Delta)\, h).$$

Combining this with (32), the equation

$$\|E(\Delta)\ V_\infty P_0^{\mathrm{ac}} h\|^2 = \|E_0(\Delta)\ P_0^{\mathrm{ac}} h\|^2$$

follows. Thus the proof is finished. ■

**Remark 32.** Note that actually we have proved not only the partial isometry of $V_\infty$ with initial subspace $P_0^{\mathrm{ac}}\mathscr{H}$, but we also presented the explicit form of this operator, which is contained in formula (30). For an arbitrary $f \in P_0^{\mathrm{ac}}\mathscr{H}$ we have

$$(\Phi V_\infty f)\ (\lambda) = \omega(\lambda + \mathrm{i}0)\ (\Phi_0 f)\ (\lambda), \qquad \text{a.e. mod } |\cdot|, \tag{33}$$

which can be easily derived from (30).

The last step can now be done, namely to prove the existence of the strong wave operator.

**Theorem 33.** *Let $H_0$ be selfadjoint in $\mathscr{H}$ and let $V$ and $H$ be defined by (21) and (22). Then*

$$W_+(H, H_0) = \operatorname*{s-lim}_{t\to\infty} \mathrm{e}^{\mathrm{i}tH}\, \mathrm{e}^{-\mathrm{i}tH_0} P_0^{\mathrm{ac}} \tag{34}$$

*exists.*

**Proof.** According to Lemma 29 and Prop. 31 we have: $V_\infty = \operatorname*{w-lim}_{t\to\infty} P^{\mathrm{ac}}\, \mathrm{e}^{\mathrm{i}tH}\, \mathrm{e}^{-\mathrm{i}tH_0} P_0^{\mathrm{ac}}$ exists and $V_\infty$ is partially isometric with initial subspace $P_0^{\mathrm{ac}}\mathscr{H}$. Note that $\|V(t)\ P_0^{\mathrm{ac}}\| \leqq 1$. Let $f \in P_0^{\mathrm{ac}}\mathscr{H}$. Then

$$\|V(t)\, f - V_\infty f\|^2 = \|V(t)\, f\|^2 + \|V_\infty f\|^2 - 2\mathrm{Re}\,(V_\infty f, V(t)\, f).$$

Hence,

$$\limsup_{t\to\infty} \|V(t)\, f - V_\infty f\| \leqq 2\, \|f\|^2 - 2\mathrm{Re} \lim_{t\to\infty} (V_\infty f, V(t)\, f)$$

where the latter limit exists because $V_\infty f = P^{\mathrm{ac}} V_\infty f$. Since $\lim_{t\to\infty} (V_\infty f, V(t)\, f) = \|V_\infty f\|^2 = \|f\|^2$, we obtain $\limsup_{t\to\infty} \|V(t)\, f - V_\infty f\|^2 = 0$. ■

**Corollary 34.** *If the assumptions of Theorem 33 are satisfied and if $H, H_0$ are simple, $h$ is generating and $\Phi, \Phi_0$ denote the corresponding isometric operators from $\mathscr{H}$ onto the spectral representation spaces $L^2(\operatorname{spec}^{\mathrm{c}} H, \mu(\mathrm{d}\lambda))$, $L^2(\operatorname{spec}^{\mathrm{c}} H_0, \mu_0(\mathrm{d}\lambda))$, respectively, then the wave matrix of $W_+(H, H_0)$ with respect to $\Phi, \Phi_0$ is given by*

$$\hat{W}_+(\lambda) = \omega(\lambda + \mathrm{i}0) := 1 - a(h, R_0(\lambda + \mathrm{i}0)\, h).$$

**Proof.** It follows immediately from Remark 32. ■

## 9.3. Asymptotically equivalent identification operators

### 9.3.1. Simple properties of asymptotically equivalent identification operators

Let $H_0, \mathscr{H}_0, P_{H_0}$ be given and let $A, B \in \mathscr{L}(\mathscr{H}_0, \mathscr{H})$. First recall that $A$ and $B$ are called asymptotically equivalent with respect to $H_0 P_{H_0}$ if $\operatorname*{Lim}_{t\to\infty} (A - B)\, \mathrm{e}^{-\mathrm{i}tH_0} P_{H_0} = 0$. Note that this equivalence relation is denoted by $A \underset{H_0}{\approx} B$. Another notation (see Section 6.52) is $A - B \in \mathscr{L}_{H_0, P_{H_0}}(\mathscr{H}_0, \mathscr{H})$. In Part II (see especially Chapter 6) we have shown that to a given operator $J$ there exists a lot of operators $J'$ with $J' \underset{H_0}{\approx} J$. The following proposition shows that the wave operators $W_+(H, H_0; J)$ depend only on the equivalence class defined by $J$ via $J' \underset{H_0}{\approx} J$.

**Proposition 35.** *If $J \underset{H_0}{\approx} J'$ and $W_+(H, H_0; J)$ exists, then also $W_+(H, H_0; J')$ exists and $W_+(H, H_0; J) = W_+(H, H_{0+} J')$ is valid. Conversely, if both wave operators exists and are equal, then $J \underset{H_0}{\approx} J'$ follows.*

Proof. From the equation

$$\|e^{itH} J\, e^{-itH_0} P_{H_0} f - e^{itH} J'\, e^{-itH_0} P_{H_0} f\| = \|(J - J')\, e^{-itH_0} P_{H_0} f\|$$

the assertions easily follow. ■

Special equivalent identification operators are the wave operator itself and also $P_H J$ as the following proposition states.

**Proposition 36.** *Let $W_+ = W_+(H, H_0; J)$ exist. Then*

$$W_+ \underset{H_0}{\approx} J, \qquad P_H J \underset{H_0}{\approx} J \tag{35}$$

*are valid.*

Proof. $W_+ \underset{H_0}{\approx} J$ follows from (15). Further we have $E \leq P_H$ according to Prop. 20. Then $P_H J \underset{H_0}{\approx} J$ follows from (16). ■

If $P_{H_0} \leq P^c_{H_0}$ and Lim $=$ |A|-lim, then $\mathscr{L}_\infty(\mathscr{H}_0, \mathscr{H}) \subseteq \mathscr{L}_{H_0 P_{H_0}}\mathscr{H}_0, \mathscr{H})_a$, that is, all compact operators $K$ are asymptotically equivalent to 0. Hence all operators $J + K$ are identification operators with $W_+(H, H_0; J + K) = W_+(H, H_0; J)$ (see Section 6.3.2).

### 9.3.2. Criteria for completeness and semicompleteness

The wave operator $W_+$ maps $P_{H_0}\mathscr{H}_0$ into $P_H\mathscr{H}$. From Theorem 15 it easily follows that $W_+$ is $H_0$-semicomplete iff the map $W_+\colon P_{H_0}\mathscr{H}_0 \to P_H\mathscr{H}$ is injective. $W_+$ is $H$-semicomplete iff the map $W_+{}^*\colon P_H\mathscr{H} \to P_{H_0}\mathscr{H}_0$ is injective or equivalently iff ima $W_+$ is dense in $P_H\mathscr{H}$. Hence $W_+$ is complete iff $W_+\colon P_{H_0}\mathscr{H}_0 \to P_H\mathscr{H}$ is injective and ima $W_+$ is dense in $P_H\mathscr{H}$. These simple assertions lead to the following corollary of the chain rule (Theorem 21).

**Corollary 37.** *If $W_+(H', H_0; J_0)$ is $H_0(H')$-semicomplete and $W_+(H, H'; J')$ is $H'(H)$-semicomplete, then also $W_+ = W_+(H, H_0; J) = W_+(H, H_0; J'J_0) = W_+(H, H'; J')\, W_+ \times (H', H_0; J_0)$ is $H_0(H)$-semicomplete. The completeness of $W_+(H', H_0; J_0)$ and $W_+(H, H'; J')$ implies that of $W_+$.*

With the help of the following definition we are able to formulate some useful criteria for semicompleteness and completeness.

**Definition 38.** Let $J \in \mathscr{L}(\mathscr{H}_0, \mathscr{H})$, $J' \in \mathscr{L}(\mathscr{H}, \mathscr{H}_0)$. $J'$ is called an *asymptotic left inverse* to $J$ if

$$J'J \underset{H_0}{\approx} 1_{\mathscr{H}_0}. \tag{36}$$

For example, if $J$ is bounded invertible, then $J^{-1} \in \mathscr{L}(\mathscr{H}, \mathscr{H}_0)$ is obviously an asymptotic left inverse.

**Proposition 39.** *Let $W_+ = W_+(H, H_0; J)$ exist and let $J'$ be an asymptotic left inverse to $J$. Then $W_+$ is $H_0$-semicomplete and*

$$\operatorname*{Lim}_{t\to\infty} e^{itH_0} J' e^{-itH} W_+ = P_{H_0}, \tag{37}$$

$$\operatorname*{Lim}_{t\to\infty} (JJ' - 1_{\mathscr{H}})\, e^{-itH} W_+ = 0 \tag{38}$$

*are valid.*

**Proof.** According to (14) we have

$$\operatorname*{Lim}_{t\to\infty} (e^{-itH} W_+ - J\, e^{-itH_0} P_{H_0}) = 0. \tag{39}$$

Hence, applying $J'$ and using (36), we obtain

$$\operatorname*{Lim}_{t\to\infty} (J' e^{-itH} W_+ - e^{-itH_0} P_{H_0}) = 0 \tag{40}$$

which implies (37). From (40), applying $J$ and using (39), we obtain

$$\operatorname*{Lim}_{t\to\infty} (JJ' e^{-itH} W_+ - e^{-itH} W_+) = 0$$

which is just (38). From (37) and $W_+ f = 0$ we obtain $P_{H_0} f = 0$; that is, $W_+$ is $H_0$-semicomplete. ■

**Proposition 40.** *Let the assumptions of Prop. 39 be satisfied. Then: $W_+$ is complete iff*
(i) *$J$ is asymptotic left inverse to $J'$ with respect to $HP_H$, that is*

$$JJ' \underset{H}{\approx} 1_{\mathscr{H}}, \tag{41}$$

*and*

(ii) *$W_+' = W_+(H_0, H; J')$ exists.*
*In this case $W_+'$ is also complete and*

$$W_+' W_+ = P_{H_0}, \qquad W_+ W_+' = P_H. \tag{42}$$

**Proof.** Let (i) and (ii) be true. Then from Prop. 39 we obtain that $W_+'$ is $H$-semicomplete. From (37) and $W_+ = P_H W_+$ the relation $W_+' W_+ = P_{H_0}$ immediately follows. Hence clo ima $W_+' = P_{H_0} \mathscr{H}_0$, and $W_+'$ is complete. Now apply Prop. 39 to $W_+'$ and $J$. Then (37) yields $W_+ W_+' = P_H$, and $W_+$ is complete. Conversely, let $W_+$ be complete. Then ima $W_+$ is dense in $P_H \mathscr{H}$ and (37) implies the existence of $W_+'$ with $W_+' W_+ = P_{H_0}$. Similarly (38) implies $JJ' \underset{H}{\approx} 1_{\mathscr{H}}$. ■

## 9.4. Wave operators and limits of pre-wave operators

In this paragraph we discuss the connection between the wave operator (Abel or strong) and the weak limit of the pre-wave operator. Further, conditions for the asymptotic equivalence of $\Omega_+^w$ and $J$ are considered. These considerations are necessary in order to be able to get stationary criteria for the existence of wave operators (see Chap. 13).

### 9.4.1. Connection between $W_+$ and $\Omega_+^{\mathrm{w}}$

Recall the definition of $\Omega_+^{\mathrm{w}}$ given in Remark 19 and in (7). Further recall the identity (6) which already shows the connection between the Abel wave operator $W_+$ and $\Omega_+^{\mathrm{w}}$, stated in the next proposition.

**Proposition 41.** *The Abel wave operator $W_+(H, H_0; J)$ exists iff $\Omega_+^{\mathrm{w}}$ exists and $\Omega_+^{\mathrm{w}}$ is asymptotically equivalent to $J$ with respect to $H_0P_{H_0}$, $\Omega_+^{\mathrm{w}} \underset{H_0}{\approx} J$ (in |A|-lim sense). In this case $W_+ = \Omega_+^{\mathrm{w}}$ is valid.*

**Proof.** Let $\Omega_+^{\mathrm{w}}$ exist and let $\Omega_+^{\mathrm{w}} \underset{H_0}{\approx} J$. Then we obtain, by Prop. 11 (intertwining property of $\Omega_+^{\mathrm{w}}$),

$$0 = \lim_{\varepsilon \to +0} \int_0^\infty \varepsilon\, e^{-\varepsilon t} \|(J - \Omega_+^{\mathrm{w}})\, e^{-itH_0} P_{H_0} f\|^2\, dt$$

$$= \lim_{\varepsilon \to +0} \int_0^\infty \varepsilon\, e^{-\varepsilon t} \|e^{itH} J\, e^{-itH_0} P_{H_0} f - \Omega_+^{\mathrm{w}} P_{H_0} f\|^2\, dt, \tag{43}$$

that is, the existence of $W_+(H, H_0; J)$ and its equality with $\Omega_+^{\mathrm{w}}$ follow. Conversely, if $W_+(H, H_0; J)$ exists, then according to Prop. 36 we have $W_+ \underset{H_0}{\approx} J$. Moreover, from Prop. 6 we obtain the existence of $\operatorname{w-lim}_{\varepsilon \to +0} \int_0^\infty \varepsilon\, e^{-\varepsilon t}\, e^{itH} J\, e^{-itH_0} P_{H_0}\, dt$ and its equality with $W_+$. ∎

It should be noted that in the proof of the "if"-part of Prop. 41 we have only used the intertwining property of $\Omega_+^{\mathrm{w}}$. Thus we can replace the assumption for this part of Prop. 41 by "there exists an intertwining operator $V$ such that $V \underset{H_0}{\approx} J$".
A similar statement holds for the strong wave operator.

**Proposition 42.** *The strong wave operator $W_+(H, H_0; J)$ exists iff $\Omega_+^{\mathrm{w}}$ exists and $\Omega_+^{\mathrm{w}}$ is asymptotically equivalent to $J$ with respect to $H_0P_{H_0}$ in the strong sense, i.e. $\Omega_+^{\mathrm{w}} \underset{H_0}{\approx} J$ (strongly).*

**Proof.** Let $W_+$ exist. Then from Prop. 6, $W_+ = \Omega_+^{\mathrm{w}}$ follows and we have $W_+ \underset{H_0}{\approx} J$ (see Prop. 36). Conversely, let $\Omega_+^{\mathrm{w}}$ exist and assume $\Omega_+^{\mathrm{w}} \underset{H_0}{\approx} J$. Then $\|e^{itH} J\, e^{-itH_0} P_{H_0} f - \Omega_+^{\mathrm{w}} f\| = \|e^{itH} J\, e^{-itH_0} P_{H_0} f - e^{itH} \Omega_+^{\mathrm{w}}\, e^{-itH_0} P_{H_0} f\| = \|(J - \Omega_+^{\mathrm{w}})\, e^{-itH_0} P_{H_0} f\| \to 0$ as $t \to \infty$, hence the assertion follows. ∎

### 9.4.2. Asymptotic equivalence of $\Omega_+^{\mathrm{w}}$ and $J$

For later application we extract a corollary from Prop. 41.

**Corollary 43.** *Let $\Omega_+^{\mathrm{w}}$ exist. Then $\Omega_+^{\mathrm{w}} \underset{H_0}{\approx} J$ (with respect to |A|-lim) iff*

$$\lim_{\varepsilon \to +0} \int_0^\infty \varepsilon\, e^{-\varepsilon t} \|J\, e^{-itH_0} P_{H_0} f\|^2\, dt = \|\Omega_+^{\mathrm{w}} f\|^2, \qquad f \in \mathscr{H}_0, \tag{44}$$

*is valid.*

**Proof.** Let $\Omega_+^w \underset{H_0}{\approx} J$ be satisfied. Then from Prop. 41 we obtain the existence of $W_+(H, H_0; J)$. Thus, by Lemma 6.15, (44) follows. Conversely, let (44) be true. According to (6) we obtain

$$\limsup_{\varepsilon \to +0} \int_0^\infty \varepsilon\, e^{-\varepsilon t} \|e^{itH} J\, e^{-itH_0} P_{H_0} f - \Omega_+^w f\|^2\, dt$$

$$\leqq \limsup_{\varepsilon \to +0} \int_0^\infty \varepsilon\, e^{-\varepsilon t} \|J\, e^{-itH_0} P_{H_0} f\|^2\, dt - \|\Omega_+^w f\|^2. \tag{45}$$

Thus (45) and (44) yield the existence of $W_+$ and its equality with $\Omega_+^w$. Hence $\Omega_+^w \underset{H_0}{\approx} J$ follows from Prop. 36. ∎

Similarly we get a counterpart of Corollary 43 with respect to strong convergence.

**Corollary 44.** *Let* $\operatorname{w-lim}_{t\to\infty} e^{itH} J e^{-itH_0} P_{H_0}$ *exist (then also* $\Omega_+^w$ *exists and equals this limit). Then* $\Omega_+^w \underset{H_0}{\approx} J$ *(with respect to strong convergence) iff*

$$\lim_{t\to\infty} \|J\, e^{-itH_0} P_{H_0} f\| = \|\Omega_+^w f\|, \qquad f \in \mathscr{H}_0, \tag{46}$$

*is valid.*

**Proof.** If (46) is true, then also $\lim_{t\to\infty} \|e^{itH} J\, e^{-itH_0} P_{H_0} f\| = \|\Omega_+^w f\|$, hence we get the existence of $W_+$ and the equality $W_+ = \Omega_+^w$. Thus $\Omega_+^w \underset{H_0}{\approx} J$ follows from Prop. 36. Conversely, if $\Omega_+^w \underset{H_0}{\approx} J$, then because of

$$\begin{aligned} \big|\|\Omega_+^w f\| - \|J\, e^{-itH_0} P_{H_0} f\|\big| &= \big|\|\Omega_+^w\, e^{-itH_0} P_{H_0} f\| - \|J\, e^{-itH_0} P_{H_0} f\|\big| \\ &\leqq \|(\Omega_+^w - J)\, e^{-itH_0} P_{H_0} f\| \end{aligned}$$

the assertion follows. ∎

## 9.5. Partially isometric wave operators

### 9.5.1. Asymptotically partially isometric identification operators

If $J$ is isometric from $\mathscr{H}_0$ onto $\mathscr{H}$, then obviously the wave operator $W_+(H, H_0; J)$, provided it exists, is partially isometric with initial subspace $P_{H_0}\mathscr{H}_0$. This follows from

$$\|W_+ u\| = \operatorname*{Lim}_{t\to\infty} \|e^{itH} J\, e^{-itH_0} P_{H_0} u\| = \operatorname*{Lim}_{t\to\infty} \|J\, e^{-itH_0} P_{H_0} u\| = \|P_{H_0} u\|.$$

But this property of $W_+$ is also satisfied in much more general cases. In this connection recall the notion of asymptotically partially isometric operators with respect to $H_0 P_{H_0}$ (see Chapter 6, Definitions 71 and 72): $J$ is asymptotically partially isometric with respect to $H_0 P_{H_0}$ if $J \underset{H_0}{\approx} V$, where $V$ is partially isometric. With the help of this notion we obtain

**Proposition 45.** *Let $W_+ = W_+(H, H_0; J)$ exist. Then: $W_+$ is partially isometric and $H_0$-semicomplete (that is, the initial subspace is $P_{H_0}\mathscr{H}_0$) iff there is a partial isometry $V \in \mathscr{L}(\mathscr{H}_0, \mathscr{H})$ with initial subspace $P_{H_0}\mathscr{H}_0$* and $J \underset{H_0}{\approx} V$.

Proof. Let $W_+$ be partially isometric and $H_0$-semicomplete. Then, by Prop. 36, $J \underset{H_0}{\approx} W_+$ where $W_+ = V$ is a partial isometry with initial subspace $P_{H_0}\mathscr{H}_0$. Conversely, assume there is a $V$ with $V^*V = P_{H_0}$ and $V \underset{H_0}{\approx} J$. Then, because of $\|V\,e^{-itH_0}P_{H_0}f\| = \|P_{H_0}f\|$, we obtain

$$\|W_+f\| = \operatorname*{Lim}_{t\to\infty} \|J\,e^{-itH_0}P_{H_0}f\| = \operatorname*{Lim}_{t\to\infty} \|V\,e^{-itH_0}P_{H_0}f\| = \|P_{H_0}f\|. \;\blacksquare \tag{47}$$

Formally Prop. 45 may be weakened by means of the following statement.

**Lemma 46.** *Let $W_+ = W_+(H, H_0; J)$ exist. Then $J$ is asymptotically equivalent to $V$ with $V^*V = P_{H_0}$*, $J \underset{H_0}{\approx} V$, iff

$$\operatorname*{Lim}_{t\to\infty} \|J\,e^{-itH_0}P_{H_0}f\| = \|P_{H_0}f\|, \qquad f \in \mathscr{H}_0. \tag{48}$$

Proof. If $J$ is asymptotically partially isometric with "asymptotic initial subspace" $P_{H_0}\mathscr{H}_0$, then (48) immediately follows. If (48) is valid, then we obtain from (47) the partial isometry and the $H_0$-semicompleteness of $W_+$. Hence from Prop. 36 the assertion follows. $\blacksquare$

Note that (48) and also the notion of an asymptotically partial isometry $J$ with "asymptotic initial subspace" $P_{H_0}\mathscr{H}_0$ do not contain any operator $H$. But the equivalence of these two properties is proved under the assumption of the existence of $W_+(H, H_0; J)$ for some $H$.

Further note that all previous considerations are also valid for $t \to -\infty$ (recall (19)). In particular, according to Prop. 45, we obtain: If $W_- = W_-(H, H_0; J)$ exists, then $W_-$ is partially isometric and $H_0$-semicomplete iff there is a partial isometry $V_-$ with initial projection $P_{H_0}$ such that $J \underset{H_0}{\approx} V_-$ with respect to $t \to -\infty$, $J \underset{-H_0}{\approx} V_-$. If both wave operators $W_+$ and $W_-$ exist and are partially isometric and $H_0$-semicomplete, then according to Prop. 36 we obtain $J \underset{H_0}{\approx} W_+$ and $J \underset{-H_0}{\approx} W_-$. Of course, the question arises if it is possible to choose a unique partial isometry $V$ with initial projection $P_{H_0}$ and with $J \underset{H_0}{\approx} V$, $J \underset{-H_0}{\approx} V$ simultaneously in this case. If $P_{H_0} = P^{ac}_{H_0}$, then this question can be answered affirmatively. This will be proved in Chapter 11 (see Theorem 11.12).

**Corollary 47.** *If $W_+(H, H_0; J)$ exists and if it is partially isometric and $H_0$-semicomplete, then necessarily $\|J\| \geqq 1$.*

Proof. Let $\|J\| < 1$. Then $\operatorname*{Lim}_{t\to\infty} \|J\,e^{-itH_0}P_{H_0}f\| \leqq \|J\|\,\|P_{H_0}f\| < \|P_{H_0}f\|$ in contradiction to (48). $\blacksquare$

### 9.5.2. Completeness criteria

**Proposition 48.** *Let $W_+ = W_+(H, H_0; J)$ exist. Further, let*

$$J^*J \underset{H_0}{\approx} 1_{\mathscr{H}_0}, \tag{49}$$

*that is, assume that $J^*$ is asymptotic left inverse to $J$. Then $W_+$ is $H_0$-semicomplete and partially isometric with initial projection $P_{H_0}$. In this case $W_+$ is complete iff*

(i) *$W_+' = W_+(H_0, H; J^*)$ exists and*

(ii) *$J$ is asymptotic left inverse to $J^*$, $JJ^* \underset{H}{\approx} 1_{\mathscr{H}}$.*
*Further $W_+' = W_+^*$ is valid.*

**Proof.** According to Prop. 45 and Lemma 46, from

$$\operatorname*{Lim}_{t\to\infty} \|J\,e^{-itH_0}P_{H_0}f\|^2 = \operatorname*{Lim}_{t\to\infty} (e^{itH_0}J^*J\,e^{-itH_0}P_{H_0}f, P_{H_0}f) = (P_{H_0}f, P_{H_0}f)$$

the $H_0$-semicompleteness and the partial isometry of $W_+$ follow. The second assertion follows from Prop. 40. From this proposition we further obtain $W_+W_+' = P_H$. Hence, $W_+^*W_+W_+' = P_{H_0}W_+' = W_+^*P_H$. Then, since $P_{H_0}W_+' = W_+'$ and $W_+^*P_H = W_+^*$, the last assertion is also true. ∎

The condition (49) is a very strong condition and in many important cases not satisfied (see for example multichannel scattering, Chapter 10). In these cases Prop. 48 is not applicable as a completeness criterion. But sometimes the following variant is useful.

**Proposition 49.** *Let $J$ be asymptotically unitary, that is, let $J \underset{H_0}{\approx} U$ with $U^*U = 1_{\mathscr{H}_0}$, $UU^* = 1_{\mathscr{H}}$ (then for $U$ condition (49) is satisfied). Then, if $W_+(H, H_0; J)$ exists, it is $H_0$-semicomplete and partially isometric. It is complete iff $W_+(H_0, H; U^*)$ exists.*

**Proof.** It follows easily from Prop. 45 and Prop. 48. ∎

Prop. 49 may be useful because it is not necessary that $J$ itself satisfies property (49).

**Proposition 50.** *Let $W_+ = W_+(H, H_0; J)$ be partially isometric and $H_0$-semicomplete. Further let $\|J\| = 1$. Then: $W_+$ is complete iff $W_+(H_0, H; J^*)$ exists and $JJ^* \underset{H}{\approx} 1_{\mathscr{H}}$. In this case $W_+(H_0, H; J^*) = \big(W_+(H, H_0; J)\big)^*$ is valid.*

**Proof.** According to Prop. 48 it is sufficient to prove that in this case (49) is satisfied. First, we note $1_{\mathscr{H}_0} - J^*J \geq 0$. Secondly, from the partial isometry of $W_+$ with initial projection $P_{H_0}$ we obtain $\operatorname*{Lim}_{t\to\infty} \|J\,e^{-itH_0}P_{H_0}f\| = \|P_{H_0}f\|$, $f \in \mathscr{H}_0$, so that

$$\operatorname*{Lim}_{t\to\infty} \big((P_{H_0} - P_{H_0}J^*JP_{H_0})\,e^{-itH_0}f, e^{-itH_0}f\big) = 0.$$

Therefore also

$$\operatorname*{Lim}_{t\to\infty} \|(1_{\mathscr{H}_0} - J^*J)^{1/2}\,e^{-itH_0}P_{H_0}f\| = 0 \quad \text{and} \quad \operatorname*{Lim}_{t\to\infty} \|(1_{\mathscr{H}_0} - J^*J)\,e^{-itH_0}P_{H_0}f\| = 0.$$

That is, we obtain $J^*J \underset{H_0}{\approx} 1_{\mathscr{H}_0}$. ∎

## 9.6. Scattering operator and scattering matrix

### 9.6.1. Scattering operator

Assume that a universal spectral projection $H \to P_H$ is fixed.

**Definition 51.** A triple $\{H, H_0; J\}$ is called a *scattering system* if $W_+(H, H_0; J)$ and $W_-(H, H_0; J)$ exist.

For instance the triple $\{H, H_0; 1\}$ defined in Example 8 is a scattering system, also the triple $\{H_0 + V, H_0; 1\}$ of Section 2.3.

**Definition 52.** Let $\{H, H_0; J\}$ be a scattering system. Then the operator $S := W_+{}^*W_-$ is called the *scattering operator*.

**Proposition 53.** *The scattering operator $S$ of a scattering system $\{H, H_0; J\}$ commutes with $H_0$, $S \in \text{com}\, H_0$.*

Proof. It is obvious because of the intertwining property of the wave operators $W_+$, $W_-$. ■

**Proposition 54.** *Let $\{H, H_0; J\}$ be a scattering system. If the corresponding wave operators $W_+$ and $W_-$ are complete and partially isometric, then*

$$S^*S = SS^* = P_{H_0}, \tag{50}$$

*that is, $S \restriction P_{H_0}\mathscr{H}_0 \in \mathscr{L}(P_{H_0}\mathscr{H}_0)$ is unitary.*

Proof. We have $S^* = W_-{}^*W_+$. Hence

$$SS^* = W_+{}^*W_-W_-{}^*W_+ = W_+{}^*P_HW_+ = W_+{}^*W_+ = P_{H_0},$$

$$S^*S = W_-{}^*W_+W_+{}^*W_- = W_-{}^*P_HW_- = W_-{}^*W_- = P_{H_0}. \ ■$$

Note that the assertion (50) is also true if $W_+$, $W_-$ are partially isometric, $H_0$-semi-complete and if $W_+W_+{}^* = W_-W_-{}^* \leq P_H$.

**Example 55.** We consider once more Example 8 (in the generalized form of Example 14): $\mathscr{H} = \mathscr{H}_0 = L^2(\mathbb{R}, dx; \mathscr{K})$, $J = 1$, $H_0 = -\mathrm{id}/dx$, $H = H_0 + q(\cdot)$ where $q: \mathbb{R} \to \mathscr{L}(\mathscr{K})$ is a Borel function, $\operatorname{ess\,sup}_{x\in\mathbb{R}} \|q(x)\| < \infty$, $\int \|q(x)\|\, dx < \infty$ and the values of $q(x)$ are pairwise commuting selfadjoint operators on $\mathscr{K}$. Then the strong wave operators $W_\pm = W_\pm(H, H_0)$ exist and are given by

$$(W_+f)(x) = \exp\{\mathrm{i} \int_x^\infty q(y)\, dy\}\, f(x) \quad \text{and} \quad (W_-f)(x) = \exp\{-\mathrm{i} \int_{-\infty}^x q(y)\, dy\}\, f(x).$$

$W_+$ and $W_-$ are unitary operators, hence complete. We obtain

$$(Sf)(x) = \exp\{-\mathrm{i} \int_{-\infty}^\infty q(y)\, dy\}\, f(x). \tag{51}$$

**Example 56.** We consider the case of a one-dimensional perturbation $H = H_0 + V$, $V = a(h, \cdot)h$, $\|h\| = 1$, $a \in \mathbb{R}$, of an arbitrary selfadjoint operator $H_0$ in $\mathscr{H}_0$; $\mathscr{H} = \mathscr{H}_0$, $J = 1$, $P_{H_0} = P_0^{ac}$. First, we apply Theorem 33 and obtain: $W_+(H, H_0)$, $W_-(H, H_0)$ exist, thus $\{H, H_0; 1\}$ is a scattering system. Further $W_+{}^*W_+ = W_-{}^*W_- = P_0^{ac}$. Second, we apply Theorem 33 to the system $\{H_0, H, 1\}$ and obtain: $W_+(H_0, H)$, $W_-(H_0, H)$ exist, they are partially isometric and $H$-semi-complete. Third, we apply Prop. 48 and obtain: $W_+(H, H_0)$ and $W_-(H, H_0)$ are complete; furthermore $W_+(H_0, H) = (W_+(H, H_0))^*$, similarly for $W_-$. Thus Prop. 54 yields that $S$ is unitary.

### 9.6.2. Scattering matrix

Let $\{H, H_0; J\}$ be a scattering system. The property $S \in \text{com}\, H_0$ for the scattering operator of this system immediately leads to the existence of a representation of $S$ by an operator-valued function in a direct integral representation of $\mathscr{H}_0$ with respect to $H_0$.

**Proposition 57.** *Let* $\Phi_0\colon \mathscr{H}_0 \to L^2(\mathrm{spec^c}\, H_0, \mu_0(\mathrm{d}\lambda); \mathscr{K}_{0\lambda})$ *be an isometric map such that* $E_0(\Delta)$ *is represented by* $\chi_\Delta(\cdot)$ *(direct integral representation of* $\mathscr{H}_0$ *with respect to* $E_0(\cdot)$*, where the corresponding admissible system* $\mathscr{S}_0$ *is separable). Further, let* $S$ *be the scattering operator of a scattering system* $\{H, H_0; J\}$. *Then there is an operator function* $\hat{S}(\cdot)\colon \mathrm{spec^c}\, H_0 \to \mathscr{L}(\mathscr{K}_{0\lambda})$, *a.e. defined* (mod $\mu_0$), $\|\hat{S}(\lambda)\|_{\mathscr{K}_{0\lambda}}$ *essentially bounded, such that*

$$\big(\Phi_0(Sf)\big)(\lambda) = \hat{S}(\lambda)\,(\Phi_0 f)(\lambda), \qquad f \in \mathscr{H}_0. \tag{52}$$

*The operator function* $\hat{S}(\cdot)$ *is uniquely determined* mod $\mu_0$.

**Proof.** It immediately follows from Prop. 4.18. ■

The operator function $\hat{S}(\lambda)$ is called the *scattering matrix* (S-matrix) of the scattering system $\{H, H_0; J\}$. Note that the scattering matrix depends on $\Phi_0$.

Recall the notion of the wave matrix, introduced in Remark 22 for a wave operator $W_+(H, H_0; J)$. If a scattering system $\{H, H_0; J\}$ is given and if we fix two isometries $\Phi, \Phi_0$ (yielding direct integral representations of $\mathscr{H}, \mathscr{H}_0$ with respect to $E(\cdot), E_0(\cdot)$, respectively), then the two wave matrices $\hat{W}_+(\lambda), \hat{W}_-(\lambda)$ are well-defined. We have

**Proposition 58.** *Let* $\{H, H_0; J\}$ *be a scattering system, assume that two isometries* $\Phi, \Phi_0$ *are fixed, such that the scattering matrix* $\hat{S}(\lambda)$ *and the wave matrices* $\hat{W}_+(\lambda), \hat{W}_-(\lambda)$ *are well-defined. Then*

$$\hat{S}(\lambda) = \hat{W}_+(\lambda)^* \, \hat{W}_-(\lambda) \tag{53}$$

*is valid a.e. on* $\mathrm{spec^c}\, H_0 (\mathrm{mod}\ \mu_0)$.

**Proof.** It immediately follows from Prop. 4.18 and Remark 22. ■

**Proposition 59.** *If* $S \mid P_{H_0}\mathscr{H}_0$ *is unitary, and* $P_{H_0} = E_0(\Delta_0)$ *then* $\hat{S}(\lambda)$ *is unitary in* $\mathscr{L}(\mathscr{K}_{0\lambda})$ *for* $\lambda \in \Delta_0 (\mathrm{mod}\ \mu_0)$, *where* $\Phi_0$ *is the isometric map of* Prop. 57.

**Proof.** It follows from Prop. 4.18. ■

**Example 60.** (I) We continue Example 55. The equation (51) shows that $S$ is represented by a constant operator $Q$ in $\mathscr{L}(\mathscr{K})$, $Q = \int q(y)\,\mathrm{d}y$, such that $(Sf)(x) = \mathrm{e}^{-\mathrm{i}Q} f(x)$. Hence, if we choose $\Phi_0 = F_0$ (Fourier transformation), then $H_0$ is represented in $F_0\mathscr{H}_0 = L^2(\mathbb{R}, \mathrm{d}\lambda; \mathscr{K})$ by the multiplication operator by $\lambda$ and therefore we obtain

$$\hat{S}(\lambda) = \mathrm{e}^{-\mathrm{i}Q}$$

with respect to $F_0$.

(II) We continue Example 56. We choose $\Phi, \Phi_0$ as in Section 2.3 and apply Corollary 34. Then we obtain explicit formulas for the corresponding wave matrices,

$$\hat{W}_+(\lambda) = \omega(\lambda + \mathrm{i}0) := 1 - a\big(h, R_0(\lambda + \mathrm{i}0)\, h\big),$$

$$\hat{W}_-(\lambda) = \omega(\lambda - \mathrm{i}0) := 1 - a\big(h, R_0(\lambda - \mathrm{i}0)\, h\big).$$

Applying Prop. 58, we obtain

$$\hat{S}(\lambda) = \overline{\omega(\lambda + \mathrm{i}0)}\, \omega(\lambda - \mathrm{i}0), \qquad \lambda \in \mathrm{spec}^{\,c}(H_0 \mid P_0^{ac}\mathscr{H}_0)$$

a.e. (mod $|\cdot|$). Note that in this case we have $\overline{\omega(\lambda + \mathrm{i}0)} = \omega(\lambda - \mathrm{i}0)$, hence

$$\hat{S}(\lambda) = \omega(\lambda - \mathrm{i}0)^2 \tag{54}$$

is valid. (54) shows that $\hat{S}(\lambda)$ can be analytically continued into the lower half plane $\operatorname{Im} z < 0$ by

$$\hat{S}(z) = \omega(z)^2, \qquad \operatorname{Im} z < 0.$$

Further note that from $W_+(H, H_0)^* = W_+(H_0, H)$ the equation

$$\overline{\omega(\lambda + i0)} = 1 + a(h, R(\lambda + i0)\, h)$$

follows. But the identity

$$\big(1 - a(h, R_0(z)\, h)\big)\big(1 + a(h, R(z)\, h)\big) = 1$$

yields $\overline{\omega(\lambda + i0)} = (\omega(\lambda + i0))^{-1}$ or $|\omega(\lambda + i0)| = 1$ a.e. (mod $|\cdot|$) on $\operatorname{spec}^c (H_0 \restriction P_0^{ac}\mathscr{H}_0)$, which is obvious because of $\hat{S}(\lambda) = \big(\overline{\omega(\lambda + i0)}\big)^2$ (apply Prop. 59).

## 9.7. Examples

### 9.7.1. Finite-dimensional perturbations

In this section we apply the results of the preceding sections to the case of finite-dimensional perturbations, obtaining generalizations of the results of Section 2.3. We put $\mathscr{H}_0 = \mathscr{H}$, $J = 1$ and $P_{H_0} = P_0^{ac}$ (as in Section 2.3). Let

$$H := H_0 + V \tag{55}$$

be a perturbation of an arbitrary selfadjoint operator $H_0$ by a finite-dimensional selfadjoint operator $V$, dim (ima $V$) $< \infty$. If $e_1, e_2, \ldots, e_n$ denotes a suitable orthonormal basis in ima $V$, then $V$ can be written in the form

$$V = \sum_{\varrho=1}^{n} a_\varrho(e_\varrho, \cdot)\, e_\varrho, \qquad a_1, a_2, \ldots, a_n \in \mathbb{R}. \tag{56}$$

Our aim is to prove the existence of $W_+(H, H_0) = \operatorname{s-lim}_{t\to\infty} e^{itH} e^{-itH_0} P_0^{ac}$.

**Theorem 61.** *Let $V$ and $H$ be defined by* (56) *and* (55). *Then $W_+ = W_+(H, H_0)$ exists, is partially isometric and complete. Further $W_+(H_0, H)$ exists and is equal to $W_+(H, H_0)^*$.*

Proof. We have $V = \sum_{\varrho=1}^{n} V_\varrho$ where $V_\varrho = a_\varrho(e_\varrho, \cdot)\, e_\varrho$. We put $H_1 = H_0 + V_1$, $H_2 = H_1 + V_2, \ldots$, $H_n = H_{n-1} + V_n = H$. Then according to Theorem 33, $W_+(H_\varrho, H_{\varrho-1})$ exists for $\varrho = 1, 2, \ldots, n$. Therefore, applying Theorem 21 (chain rule), we obtain the existence of $W_+(H, H_0)$. Interchanging the role of $H$ and $H_0$ we also obtain the existence of $W_+(H_0, H)$. Hence, according to Prop. 48, the wave operator is complete. The partial isometry of $W_+(H, H_0)$ is obvious. ∎

From Theorem 61 we immediately obtain

**Corollary 62.** *Let the assumptions of Theorem 61 be satisfied. Then the absolutely continuous parts of $H_0$ and $H$ are unitarily equivalent. The unitary equivalence between $H_0 \restriction P_0^{ac}\mathscr{H}_0$ and $H \restriction P^{ac}\mathscr{H}$ may be realized by $W_+$.*

Proof. Apply Theorem 61 and Theorem 15. ∎

Similarly $W_- = W_-(H, H_0)$ exists, is partially isometric and complete. Thus $\{H, H_0; J\}$ is a scattering system; the $S$-operator $S \restriction P_0^{ac}\mathscr{H}_0$ is unitary. Our next aim is the calculation of the wave matrices for $W_\pm(H, H_0)$ and of the scattering matrix. First, we present

**Lemma 63.** *Let the assumptions of Theorem 61 be satisfied. Then*

$$W_+P_0^{ac}f = P_0^{ac}P_0^{ac}f - \int_{-\infty}^{\infty} E(d\lambda)\, P^{ac} V R_0(\lambda + i0)\, P_0^{ac}f, \qquad f \in \mathscr{M}_\infty(H_0), \tag{57}$$

*where the vector spectral integral in this formula is defined by*

$$\int_{-\infty}^{\infty} E(\mathrm{d}\lambda)\, P^{\mathrm{ac}}\{\cdots\} := \operatorname*{s-lim}_{N\to\infty} \int_{\Delta_N} E(\mathrm{d}\lambda)\,\{\cdots\} \tag{58}$$

*and where* $\Delta_1 \subset \Delta_2 \subset \cdots$ *is a special sequence of Borel sets such that* $\bigcup_{N=1}^{\infty} \Delta_N = \operatorname{spec^c}(H \restriction P^{\mathrm{ac}}\mathscr{K})$.

**Proof.** First from

$$e^{itH} e^{-itH_0} P_0^{\mathrm{ac}} f = P_0^{\mathrm{ac}} f + i \int_0^t e^{i\tau H} V e^{-i\tau H_0} P_0^{\mathrm{ac}} f \, d\tau \tag{59}$$

we obtain that $\int_0^{\infty} e^{i\tau H} V\, e^{-i\tau H_0} P_0^{\mathrm{ac}} f \, d\tau = \operatorname*{s-lim}_{t\to\infty} \int_0^t \{\cdots\}\, d\tau$ exists. Then the relation

$$\int_0^{\infty} | e^{itH} V\, e^{-itH_0} P_0^{\mathrm{ac}} f \, dt = \operatorname*{s-lim}_{\varepsilon\to+0} \int_0^{\infty} e^{-\varepsilon t} e^{itH} V\, e^{-itH_0} P_0^{\mathrm{ac}} f \, dt$$

is obvious. Now to $\int_0^{\infty} e^{itH} V e^{-itH_0} e^{-\varepsilon t} P_0^{\mathrm{ac}} f \, dt$ and $e^{i\tau H} = \int_{-\infty}^{\infty} e^{i\tau\lambda} E(\mathrm{d}\lambda)$ we apply Prop. 5.8 (intertwining formula for vector spectral integrals). Then we obtain

$$\int_0^{\infty} e^{-\varepsilon\tau} e^{i\tau H} V\, e^{-i\tau H_0} P_0^{\mathrm{ac}} f \, d\tau = i \int_{-\infty}^{\infty} E(\mathrm{d}\lambda)\, V R_0(\lambda + i\varepsilon)\, P_0^{\mathrm{ac}} f$$

and for all Borel sets $\Delta$

$$E(\Delta) \int_0^{\infty} e^{i\tau H} V e^{-i\tau H_0} P_0^{\mathrm{ac}} f \, d\tau = \operatorname*{s-lim}_{\varepsilon\to+0} i \int_{\Delta} E(\mathrm{d}\lambda)\, V R_0(\lambda + i\varepsilon)\, P_0^{\mathrm{ac}} f$$

is valid. Now for convenience we put $V = BA$ ($A$, $B$ finite-dimensional and selfadjoint) and define $\Delta_N$ by

$$\Delta_N = \{\lambda \in \mathbb{R};\ \|(B^* E(\mathrm{d}\lambda)\, B/\mathrm{d}\lambda)\| \le N,\ \lambda \notin \operatorname{spec^c}(H \restriction (P^{\mathrm{ac}})^{\perp} \mathscr{K}),\ \lambda \in \operatorname{spec^c}(H \restriction P^{\mathrm{ac}}\mathscr{K})\}.$$

Then to $\int_{\Delta_N} E(\mathrm{d}\lambda)\, B\{A R_0(\lambda + i\varepsilon)\, P_0^{\mathrm{ac}} f\}$ we apply Prop. 5.16 where $f \in \mathscr{M}_\infty(H_0)$; note further Remark 3.26. Thus we obtain the following result: $\int_{\Delta_N} E(\mathrm{d}\lambda)\, V R_0(\lambda + i0)\, P_0^{\mathrm{ac}} f$ exists and we have

$$E(\Delta_N) \int_0^{\infty} e^{i\tau H} V\, e^{-i\tau H_0} P_0^{\mathrm{ac}} f \, d\tau = i \int_{\Delta_N} E(\mathrm{d}\lambda)\, V R_0(\lambda + i0)\, P_0^{\mathrm{ac}} f. \tag{60}$$

Since $\operatorname*{s-lim}_{N\to\infty} E(\Delta_N) = P^{\mathrm{ac}}$, from (59) and (60) we obtain (57). ∎

(57) can be written in the form

$$W_+ f = \int_{-\infty}^{\infty} E(\mathrm{d}\lambda)\, P^{\mathrm{ac}} G_0(\lambda + i0)\, f, \qquad f \in \mathscr{M}_\infty(H_0), \tag{61}$$

where $G_0(\lambda + i0) := 1 - V R_0(\lambda + i0)$. Now we choose the isometries $\Phi$, $\Phi_0$ (according to Theorem 17 and Remark 22) such that the wave matrices are well-defined.

We put $\mathscr{K} = \operatorname{ima} V$, then $\dim \mathscr{K} = n < \infty$. We can assume, without loss of generality, that

$$\mathscr{H}_0 = \text{clo spa}\,\{E(\Delta)\, k;\ k \in \mathscr{K}, \Delta \text{ Borel set}\} = \text{clo spa}\,\{E_0(\Delta)\, k;\ k \in \mathscr{K}, \Delta \text{ Borel set}\}. \tag{62}$$

The arguments for (62) are similar to the corresponding arguments used in the case $n = 1$ (see Section 2.3). Because of Corollary 62 we can assume $\Delta_0 := \operatorname{spec^c}(H_0 \restriction P_0^{\mathrm{ac}} \mathscr{H}_0) = \operatorname{spec^c}(H \restriction P^{\mathrm{ac}}\mathscr{K})$.

From (62) we obtain

$$P_0^{ac}\mathscr{H}_0 = \text{clo spa}\,\{E_0(\Delta)\,k;\ k \in P_0^{ac}\mathscr{K},\ \Delta \subset \Delta_0\},$$
$$P^{ac}\mathscr{H}_0 = \text{clo spa}\,\{E(\Delta)\,k;\ k \in P^{ac}\mathscr{K},\ \Delta \subset \Delta_0\}. \tag{63}$$

Without loss of generality $\Delta_0$ can be chosen such that $(k, E_0(\mathrm{d}\lambda)\,k')/\mathrm{d}\lambda$ exists for all $k, k' \in P_0^{ac}\mathscr{K}$, $\lambda \in \Delta_0$ and $(k, E(\mathrm{d}\lambda)\,k')/\mathrm{d}\lambda$ exists for all $k, k' \in P^{ac}\mathscr{K}$, $\lambda \in \Delta_0$ (note that $P_0^{ac}\mathscr{K}$, $P^{ac}\mathscr{K}$ are spectral). Further, (63) says that $P_0^{ac}\mathscr{K}$, $P^{ac}\mathscr{K}$ are generating. Thus we may define $\Phi$, $\Phi_0$, according to Section 4.5.1 and Section 5.2.4 by

$$\Phi_0\colon P_0^{ac}\mathscr{H} \to L^2(\Delta_0, \mathrm{d}\lambda; \mathscr{F}_{0,\lambda}), \tag{64}$$

$$\Phi\colon P^{ac}\mathscr{H}_0 \to L^2(\Delta_0, \mathrm{d}\lambda; \mathscr{F}_\lambda), \tag{65}$$

where we put $\mathscr{F}_{0,\lambda} := (P_0^{ac}\mathscr{K})_\lambda^{\wedge} = \text{clo}_{|\cdot|_\lambda}(D_{E_0,\lambda}P_0^{ac}\mathscr{K})$, $\mathscr{F}_\lambda := (P^{ac}\mathscr{K})_\lambda^{\wedge} = \text{clo}_{|\cdot|_\lambda}(D_{E,\lambda}P^{ac}\mathscr{K})$. Now we are able to calculate the wave matrix. We apply Prop. 5.15 to (61) and obtain

$$(\Phi(W_+f))\,(\lambda) = D_{E,\lambda}\{P^{ac}G_0(\lambda + \mathrm{i}0)\,f\} \in \mathscr{F}_\lambda, \quad f \in \mathscr{M}_\infty(H_0) \tag{66}$$

a.e. (mod $|\cdot|$), in particular for $k \in P_0^{ac}\mathscr{K}$ we have $(\Phi(W_+k))\,(\lambda) = D_{E,\lambda}\{P^{ac}G_0(\lambda + \mathrm{i}0)\,k\}$. We know a priori that $(\Phi(W_+f))\,(\lambda)$ depends only on $(\Phi_0 f)\,(\lambda)$ and that $\hat{W}_+(\lambda)$ is isometric from $\mathscr{F}_{0,\lambda}$ onto $\mathscr{F}_\lambda$. This means the wave matrix $\hat{W}_+(\lambda)$ can be calculated by means of the operator $P^{ac}G_0(\lambda + \mathrm{i}0)\,P_0^{ac}\colon P_0^{ac}\mathscr{K} \to P^{ac}\mathscr{K}$ in the following form.

**Proposition 64.** *Let the assumptions of Theorem* 61 *be satisfied and let* $\Phi$, $\Phi_0$ *defined by* (64), (65). *Then the wave matrix* $\hat{W}_+(\lambda)$ *is given by the map*

$$\mathscr{F}_{0,\lambda} \ni D_{E_0,\lambda}k \to \hat{G}_0(\lambda + \mathrm{i}0)\,D_{E_0,\lambda}k := D_{E,\lambda}\{P^{ac}G_0(\lambda + \mathrm{i}0)\,P_0^{ac}k\} \in \mathscr{F}_\lambda$$

*a.e.* (mod $|\cdot|$), *where* $k \in P_0^{ac}\mathscr{K}$, *thus* $\hat{W}_+(\lambda) = \hat{G}_0(\lambda + \mathrm{i}0)$ *a.e.* (mod $|\cdot|$).

**Proof.** It immediately follows from (66). ■

Correspondingly we obtain $\hat{W}_-(\lambda) = \hat{G}_0(\lambda - \mathrm{i}0)$ where $\hat{G}_0(\lambda - \mathrm{i}0)$ is defined by $P^{ac}G_0(\lambda - \mathrm{i}0)P_0^{ac}$. Finally we have

$$\hat{S}(\lambda) = \hat{G}_0(\lambda + \mathrm{i}0)^*\,\hat{G}_0(\lambda - \mathrm{i}0) \quad \text{a.e. (mod } |\cdot|).$$

## 9.7.2. Special symmetric hyperbolic systems (uniformly propagative systems)

This section is intended to give a further illustration of the framework developed in the preceding sections. Naturally, we cannot present existence proofs for wave operators of the discussed systems; this is possible only in Part IV. Let

$$\mathscr{H}_0 = L^2(\mathbb{R}^n, \mathrm{d}x; \mathbb{C}^p) = L^2(\mathbb{R}^n, \mathrm{d}x) \otimes \mathbb{C}^p, \qquad p \in \mathbb{N},$$

with the norm

$$\|u\|_0^2 = \int_{\mathbb{R}^n} \|u(x)\|_{\mathbb{C}^p}^2\,\mathrm{d}x, \qquad u(x) = \{u_1(x), \ldots, u_p(x)\},$$

$$\|a\|_{\mathbb{C}^p}^2 = \sum_{j=1}^{p} |a_j|^2, \qquad a = \{a_1, \ldots, a_p\} \in \mathbb{C}^p.$$

Further, let $A_1, \ldots, A_n$ be selfadjoint operators in $\mathbb{C}^p$. Then we put

$$H_0 = -\mathrm{i}\sum_{j=1}^{n} A_j\,\mathrm{d}/\mathrm{d}x_j.$$

Obviously, $H_0$ is selfadjoint with dom $H_0 \supset C_0^\infty(\mathbb{R}^n, \mathbb{C}^p)$. Now let $E(x): \mathbb{R}^n \to \mathscr{L}(\mathbb{C}^p)$ be an operator-valued function with selfadjoint values satisfying $0 < E_1 \leq E(x) \leq E_2 < \infty$, i.e. $E(x)$ is strictly positive. Let $\mathscr{H}$ be the Hilbert space whose elements are just the elements of $\mathscr{H}_0$ but whose norm is given by

$$\|u\|^2 = \int_{\mathbb{R}^n} (u(x), E(x)\, u(x))_{\mathbb{C}^p}\, dx, \qquad u \in \mathscr{H}_0.$$

Now we define in $\mathscr{H}$ an operator $H$ by the relation

$$(Hu)(x) = E(x)^{-1} \left(-i \sum_{j=1}^{n} A_j\, d/dx_j\right) u(x). \tag{67}$$

If we introduce the identification $J: \mathscr{H}_0 \to \mathscr{H}$ by $Ju = u$, then it is easy to see that $H = E^{-1} J H_0 J^{-1}$ and dom $H = J$ dom $H_0$ where $E^{-1}$ denotes the bounded multiplication operator by $E(x)^{-1}$ in $\mathscr{H}$. Furthermore, $H$ is clearly selfadjoint. Other possible identification operators from $\mathscr{H}_0$ to $\mathscr{H}$ can be defined by

$$(J_1 u)(x) = E(x)^{-1} u(x), \qquad (J_2 u)(x) = E(x)^{-1/2} u(x).$$

$J_2$ is even isometric from $\mathscr{H}_0$ onto $\mathscr{H}$.

Now we restrict ourselves to operators $H_0$ with matrices $A_j$ satisfying some additional assumptions which can be expressed by the solutions $\lambda(k)$ of the characteristic equation

$$\det \{\lambda 1 - \sum_{j=1}^{n} A_j k_j\} = 0, \qquad k = \{k_1, k_2, \ldots, k_n\} \in \mathbb{R}^n. \tag{68}$$

**Definition 65.** The operator $H_0$ is called *weakly uniformly propagative* if the different characteristics roots $\lambda_q(k)$ of (68), $q = 1, \ldots, l$, satisfy:

(I) If $l = 2m$, $m \in \mathbb{N}$, then there exists an arrangement $\lambda_{\varrho_1}(k) > \cdots > \lambda_{\varrho_m}(k) > 0 > \lambda_{\varrho_{m+1}}(k) > \cdots > \lambda_{\varrho_{2m}}(k)$ of the characteristic roots for all $k \in \mathbb{R}^n \setminus \{0\}$.

(II) If $l = 2m + 1$, $m \in \mathbb{N}$, then there exists an arrangement $\lambda_{\varrho_1}(k) > \cdots > \lambda_{\varrho_m}(k) > \lambda_{\varrho_{m+1}}(k) = 0 > \lambda_{\varrho_{m+2}}(k) > \cdots > \lambda_{\varrho_{2m+1}}(k)$ of the characteristic roots $\lambda_q(k)$ for all $k \in \mathbb{R}^n \setminus \{0\}$.

(III) The roots $\lambda_q(k)$ have constant multiplicity for all $k \in \mathbb{R}^n \setminus \{0\}$.

(IV) $\lambda_q(k) = |k|\, \lambda_q(\hat{k})$ with $\hat{k} = k/|k|$ for all $q$, and $0 < a_1 \leq \lambda_q(\hat{k}) \leq a_2 < \infty$ is valid for all $q$ except the root $\lambda_{\varrho_{m+1}}(k) = 0$ from (II).

First order systems described by $H_0$ with a slightly stronger condition are called *uniformly propagative*. However, the properties (I)–(IV) are the essential ones of uniformly propagative systems. The roots $\lambda_q(k)$ correspond to propagation speeds of the system. This can easily be seen by transition to the Fourier transform $\hat{u}(k)$ of $u(\cdot) \in \mathscr{H}_0$, given by $\hat{u}(k) = (2\pi)^{-n/2} \int_{\mathbb{R}^n} e^{-ikx} u(x)\, dx$. Obviously, $(H_0 u)^\wedge(k) = \sum_{j=1}^{n} A_j k_j \hat{u}(k)$ and from this relation one obtains that projections $P_q$ reducing $H_0$ and satisfying $\sum P_q = 1$ are associated with the roots $\lambda_q(k)$. In the Fourier transformed representation the projections $P_q$ are $\mathscr{L}(\mathbb{C}^p)$-valued functions of $k$, i.e.

$$(P_q u)^\wedge(k) = \hat{P}_q(k)\, \hat{u}(k), \qquad \hat{P}_q(k) = \hat{P}_q(k/|k|).$$

Further we have

$$(H_0 P_q u)^\wedge(k) = \lambda_q(k)\, \hat{P}_q(\hat{k})\, \hat{u}(k) \tag{69}$$

for all $u \in$ dom $H_0$. (References to proofs of these simple relations and to more details are given in the Notes and Remarks.) From (69) we see that the existence of a characteristic root $\lambda_{q_0}(k) = 0$

corresponds to an eigenvalue 0 of $H_0$ called zero energy mode. The other roots correspond to the absolutely continuous spectrum of $H_0$ ($P^{ac}_{H_0} = P^c_{H_0}$ as one easily sees from (69) and the properties of $\lambda_q(k)$).

**Proposition 66.** *If $H_0$ is weakly uniformly propagative and $u \in C_0^\infty(\mathbb{R}^n, \mathbb{C}^p)$, then $u'(t, x) := (-i\, e^{-itH_0} H_0 u)(x)$ tends to 0 as $t \to \infty$ uniformly with respect to $x \in \Omega$, where $\Omega$ denotes an arbitrary compact subset of $\mathbb{R}^n$.*

Proof. The proof is simple and we omit it (see the Notes and Remarks). ∎

**Lemma 67.** *If $H_0$ is weakly uniformly propagative, $0 < E_1 \leq E(x) \leq E_2 < \infty$, and if*

$$\|E(x) - 1\|_{\mathbb{C}^p} \to 0 \quad as \ |x| \to \infty \tag{70}$$

*then $J_2 \underset{H_0}{\approx} J$ (in the strong sense).*

Proof. It is sufficient to prove $\operatorname*{s-lim}_{t\to\infty} (J_2 - J)\, e^{-itH_0} P_0^{ac} u = 0$ for all $u = H_0 w$ with $w \in C_0^\infty(\mathbb{R}^n, \mathbb{C}^p)$
This set of vectors $u$ is dense in $P_0^{ac}\mathscr{H}_0$. Then we have

$$\begin{aligned}
\|(J_2 - J)\, e^{-itH_0} P_0^{ac} u\|^2 &= \int_{\mathbb{R}^n} \Big(E(x)\,(E(x)^{-1/2} - 1)\, w'(t, x), w'(t, x)\Big)_{\mathbb{C}^p}\, dx \\
&= \int_{\mathbb{R}^n} \Big(E(x)^{1/2}\,(1 - E(x)^{1/2})\, w'(t, x), w'(t, x)\Big)_{\mathbb{C}^p}\, dx \\
&\leq \Big(\int_{\Omega} \|w'(t, x)\|^2_{\mathbb{C}^p}\, dx\Big) \sup_x \|E(x)^{1/2}\,(1 - E(x)^{1/2})\|_{\mathbb{C}^p} \\
&\quad + \sup_{x \in \mathbb{R}^n \setminus \Omega} \|E(x)^{1/2}\,(1 - E(x)^{1/2})\|_{\mathbb{C}^p} \int_{\mathbb{R}^n} \|w'(t, x)\|^2_{\mathbb{C}^p}\, dx.
\end{aligned}$$

The first term tends to 0 by Prop. 66 and the second on the right side is smaller than each $\varepsilon > 0$ if we choose $\Omega$ large enough, by (70). This proves $J_2 \underset{H_0}{\approx} J$ (in the strong sense). ∎

Similarly $J \underset{H_0}{\approx} J_1$ can be proved. Concerning existence results for wave operators in this framework for the universal projection $P^{ac}_H$ see Part IV (Section 16.3.3).

# 10. Identifikation Operators

In Chapter 9 the special case $\mathscr{H} = \mathscr{H}_0$, $J = 1$ played a particular role, for example the finite-dimensional perturbations were considered with respect to this case. Recall that the wave operator is simply given by

$$W_+ = \operatorname*{Lim}_{t\to\infty} e^{itH} e^{-itH_0} P_{H_0} \tag{1}$$

in this case. Formula (1) has been the starting point of the so-called one-space scattering theory, which was developed in close connection to the applications of this theory to Schrödinger equations in Quantum Mechanics. The one-space theory is included in the framework of Chapter 9. Despite of the general motivations given in the introduction of Part III, at first sight the two-space theory with identification operator $J$ might appear as a mere generalization of the one-space theory. In order to motivate this generalization as strongly as possible and not just with the help of either very general arguments or special examples (as in Section 9.7.2), we give in this chapter some basic material for this purpose. The motivations we present are from different areas: multichannel (multiparticle) scattering, algebraic scattering theory, second order equations and quantum field theory. The last topic is an example in which unbounded identification operators are of interest. However it should be noted that our later investigations are mainly devoted to the case of bounded identification operators.

## 10.1. Abstract multichannel scattering theory

### 10.1.1. Channel projections and channel Hamiltonians

In the following we adopt the quantum mechanical language. Let the selfadjoint operator $H$ (the Hamiltonian) describe a "real" evolution in the Hilbert space (the state space) $\mathscr{H}$. We assume $H \neq c1$. Further, let there be given a system of commuting selfadjoint operators which describe so-called free (noninteracting) evolutions in $\mathscr{H}$. We first collect a few of the simple properties of such free Hamiltonians.

**Definition 1.** A system $\mathscr{A}$ of selfadjoint operators in $\mathscr{H}$, being essentially selfadjoint on a common dense submanifold $\mathscr{D} \subset \operatorname{dom} A$ for all $A \in \mathscr{A}$, is called a *system of free Hamiltonians* if the following four properties are satisfied:

(I) $\mathscr{A}$ is linear. That is, from $A, B \in \mathscr{A}$; $a, b \in \mathbb{R}$, also $aA + bB \in \mathscr{A}$ follows.

(II) $1 \in \mathscr{A}$.

(III) The operators $A \in \mathscr{A}$ are pairwise commuting.

(IV) If $A \in \mathscr{A}$ and $A \neq c1$, $c$ real, then $A$ is absolutely continuous ($P_A^{ac} = 1$).

**Example 2.** Let $\mathscr{H} = L^2(\mathbb{R}^n, dp)$. Define $\mathscr{A}$ as the system of all multiplication operators given by all polynomials $\pi(p)$ with real coefficients and with a degree not bigger than two. Then the properties (I)–(IV) are obviously satisfied. Thus in this way we obtain a system of free Hamiltonians. Also the set of all polynomials without the condition on the degree is such a system.

The main problem of scattering theory is to compare asymptotically, with respect to $t \to \pm\infty$, the trajectories of the real evolution $e^{-itH}f$ with trajectories $e^{-itA}g$ of free evolutions. In the following we consider only the case $t \to \infty$. The definitions for the case $t \to -\infty$ could be similarly given (see (9.19)).

According to Definition 9.2 and Lemma 9.3, for each $A \in \mathscr{A}$ the projection $P^{ex}(H, A, 1)$ is defined as the projection onto the subspace of all $f \in \mathscr{H}$ such that $\operatorname{Lim}_{t\to\infty} e^{itH} e^{-itA}f$ exists. In this paragraph the corresponding operator $J_\infty$ (see Section 9.1.2) is denoted by

$$W_+ = W_+(H, A) := \operatorname{Lim}_{t\to\infty} e^{itH} e^{-itA}Q_A, \tag{2}$$

where $Q_A := P^{ex}(H, A, 1)$. It is partially isometric and $W_+^*W_+ = Q_A$. The corresponding final projection is denoted by $F_A$. Then

$$F_A := W_+W_+^*. \tag{3}$$

If $A = c1$, then $Q_A = E_H(\{c\})$ (see Remark 9.23). If $A$ is absolutely continuous, then either $Q_A = 0$ or $\dim Q_A = \infty$.

**Definition 3.** The operator $A \in \mathscr{A}$ is called a *channel Hamiltonian* (or an *asymptotic Hamiltonian*) with respect to $H$, and $Q_A$ is called the corresponding *channel projection*, if $Q_A > 0$. In this case $W_+(H, A)$ is called the corresponding *channel wave operator*.

The last part of this definition is just the exception of our general terminology mentioned in Section 9.1.2.

If $A = c1$, then $A$ is a channel Hamiltonian iff $c$ is an eigenvalue of $H$. In this case we call $A = c1$ a *discrete channel Hamiltonian*. In the other case $A$ is called an *absolutely continuous channel Hamiltonian*.

**Proposition 4.** *Let* H *be selfadjoint let* $\mathscr{A}$ *be a system of free Hamiltonians and let* $A, B \in \mathscr{A}$ *be channel Hamiltonians, where* $A \neq B$. *Then*

$$F_AF_B = F_BF_A = 0. \tag{4}$$

Proof. We put

$$f_+ = W_+(H, A)\, f, \qquad g_+ = W_+(H, B)\, g, \qquad f, g \in \mathscr{H}. \tag{5}$$

It is sufficient to show that $(f_+, g_+) = 0$. We have (see (6.2) for the notation $[\cdot]_\varepsilon$ and $[\cdot, \cdot]_\varepsilon$)

$$\lim_{\varepsilon\to+0} [e^{itH} e^{-itA}f - f_+]_\varepsilon = 0, \qquad \lim_{\varepsilon\to+0} [e^{itH} e^{-itB}g - g_+]_\varepsilon = 0.$$

Hence

$$\lim_{\varepsilon\to+0} |[e^{itH} e^{-itA}f, e^{itH} e^{-itB}g]_\varepsilon - (f_+, g_+)| = 0.$$

Further, $[e^{itH} e^{-itA}f, e^{itH} e^{-itB}g]_\varepsilon = [e^{-it(A-B)}f, g]_\varepsilon$. If $A - B$ is absolutely continuous, we obtain, according to *Riemann-Lebesgue's* Lemma (see Prop. 3.23), $\lim_{t\to\infty} (e^{-it(A-B)}f, g) = 0$; therefore also $[e^{-it(A-B)}f, g]_\varepsilon \to 0$ as $\varepsilon \to +0$. Hence we obtain in this case $(f_+, g_+) = 0$.

If $A - B = c1$ (the second case), then $c \neq 0$ and we obtain $(e^{-it(A-B)}f, g) = e^{itc}(f, g)$. Now $A\text{-}\lim_{t\to\infty} e^{itc}(f, g) = 0$, hence also in this case $(f_+, g_+) = 0$. ■

Sometimes property (4) is called the *asymptotic orthogonality of the channels*. The triple $\{A, Q_A, W_A(H, A)\}$ is called a *channel* with respect to the pair $\{H, \mathscr{A}\}$.

**Corollary 5.** *If* H *and* $\mathscr{A}$ *are given as in Prop. 4 and if* $\mathscr{H}$ *is separable., then there is at most a denumerable set of channels corresponding to the pair* $\{H, \mathscr{A}\}$.

Proof. According to Prop. 4 the projections $F_A$, which correspond to channel Hamiltonians $A \in \mathscr{A}$, form a system of pairwise orthogonal projections on $\mathscr{H}$. The cardinality of such a system must be necessarily at most denumerable. ■

### 10.1.2. Reaction channels and asymptotic completeness

Let $A \in \mathscr{A}$ be a channel Hamiltonian. In this section we consider channel Hamiltonians of the form $A + c1$. We have

**Proposition 6.** *Let H and* $\mathscr{A}$ *be given as in Prop. 4. Further, let* $A, B \in \mathscr{A}$ *be channel Hamiltonians with* $A - B = c1$, $c \neq 0$. *Then*

$$Q_A Q_B = Q_B Q_A = 0 \tag{6}$$

*that is, in this case also the channel projections are pairwise orthogonal.*

Proof. We have $A = B + c1$. Now according to (5)

$$[e^{itH} e^{-itB} Q_B g - g_+]_\varepsilon \to 0 \quad \text{as} \quad \varepsilon \to +0$$

and

$$[e^{itH} e^{-itB} e^{-itc} Q_A f - f_+]_\varepsilon \to 0 \quad \text{as} \quad \varepsilon \to +0$$

hold, where $f, g \in \mathscr{H}$. The latter relation can be written in the form $[e^{itH} e^{-itB} Q_A f - e^{itc} f_+]_\varepsilon \to 0$ as $\varepsilon \to +0$. Therefore we obtain

$$|[e^{itH} e^{-itB} Q_B g, e^{itH} e^{-itB} Q_A f]_\varepsilon - [g_+, e^{itc} f_+]_\varepsilon| \to 0 \quad \text{as} \quad \varepsilon \to +0.$$

Now $[g_+, e^{itc} f_+]_\varepsilon = \int_0^\infty \varepsilon\, e^{-\varepsilon t}\, e^{itc}(g_+, f_+)\, dt \to 0$, $\varepsilon \to +0$, hence

$$[e^{itH} e^{-itB} Q_B g, e^{itH} e^{-itB} Q_A f]_\varepsilon \to 0, \qquad \varepsilon \to +0,$$

follows. Because of

$$(e^{itH} e^{-itB} Q_B g, e^{itH} e^{-itB} Q_A f) = (Q_B g, Q_A f)$$

we obtain $(Q_B g, Q_A f) = 0$. ■

The system of all channels corresponding to channel Hamiltonians of the form $A + c1$ ($c$ is varying) is called a *reaction channel* defined by the channel Hamiltonian $A$. It is denoted by $\{A\}$. Because of Prop. 6,

$$Q_{\{A\}} := \sum_c Q_{A+c1} \tag{7}$$

is a well-defined projection belonging to com $A$. It is called the *reaction channel projection* of the reaction channel $\{A\}$.

Note that the system of all discrete channels forms exactly one reaction channel, the *discrete reaction channel*. All other reaction channels consist of absolutely continuous channels. Then from (3), property (IV), and Prop. 4 we immediately obtain

$$\sum_{A \in \mathscr{A} \setminus \{c1\}} F_A \leqq P^{ac}. \tag{8}$$

**Definition 7.** The pair $\{H, \mathscr{A}\}$ is called *asymptotically complete* if

$$\sum_{A \in \mathscr{A} \setminus \{c1\}} F_A = P^{ac}. \tag{9}$$

Property (9) means that the system $\mathscr{A}$ is sufficiently large in order to exhaust the asymptotic possibilities of trajectories $e^{-itH}f$ for $f \in P^{ac}\mathscr{H}$.

### 10.1.3. Example: Multiparticle scattering

As a concrete model for the abstract framework of Sections 1 and 2 we consider the quantum mechanical nonrelativistic $N$-particle problem for distinguishable spinless particles with two-point interactions. The particles are labeled by $1, 2, \ldots, N$. The masses and the Cartesian coordinates of the particles are denoted by $m_j$ and $x_j \in \mathbb{R}^3$.

Let $\Gamma$ be a fragment of the whole system, consisting of $p$ particles, $1 \leqq p \leqq N$. The Hilbert space of the fragment is given by

$$\mathscr{H}_\Gamma = \bigotimes_{j\in\Gamma} L^2(\mathbb{R}^3, dx_j) = L^2(\mathbb{R}^{3p}, \prod_{j\in\Gamma} dx_j),$$

the corresponding Hamiltonian $H_\Gamma$ is formally given by

$$H_\Gamma \sim -\sum_{j\in\Gamma} (2m_j)^{-1} \Delta_{x_j} + \sum_{i<j;\, i,j\in\Gamma} V_{i,j}(x_i - x_j)$$

where $\Delta_{x_j}$ denotes the Laplacian with respect to the coordinate $x_j$ and $V_{i,j}$ denotes the interaction between $i$ and $j$. We put

$$K_\Gamma = -\sum_{j\in\Gamma} (2m_j)^{-1} \Delta_{x_j}, \qquad V_\Gamma = \sum_{i<j;\, i,j\in\Gamma} V_{i,j}(x_i - x_j).$$

Then $H_\Gamma = K_\Gamma + V_\Gamma$. We introduce the center of mass by $X_\Gamma = M_\Gamma^{-1} \sum_{i\in\Gamma} m_i x_i$, $M_\Gamma = \sum_{i\in\Gamma} m_j$. Further there are additionally $3(p-1)$ so-called internal coordinates $X^\Gamma$, such that

$$\mathscr{H}_\Gamma = L^2(\mathbb{R}^3, dX_\Gamma) \otimes L^2(\mathbb{R}^{3p-3}, dX^\Gamma). \tag{10}$$

The Hamiltonian $H_\Gamma$ of the fragment can then be written in the form

$$H_\Gamma = K_{X_\Gamma} \otimes 1_{X^\Gamma} + 1_{X_\Gamma} \otimes \hat{H}_{X^\Gamma}$$

with respect to the resolution (10), where $K_{x_\Gamma} = -(2M_\Gamma)^{-1} \Delta_{x_\Gamma}$ is (up to the factor $-(2M_\Gamma)^{-1}$) the Laplacian with respect to the center of mass. Defining $K_{x^\Gamma}$ by the equation

$$K_\Gamma = K_{X_\Gamma} \otimes 1_{X^\Gamma} + 1_{X_\Gamma} \otimes K_{X^\Gamma} \tag{11}$$

and putting $V_\Gamma = 1_{X_\Gamma} \otimes V_{X^\Gamma}$, we have $\hat{H}_{X^\Gamma} = K_{X^\Gamma} + V_{X^\Gamma}$. Further we have $e^{-itH_\Gamma} = e^{-itK_{X_\Gamma}} \otimes e^{-it\hat{H}_{X^\Gamma}}$ where $e^{-itK_{X_\Gamma}}$ describes the evolution of the center of mass of the fragment and $e^{-it\hat{H}_{X^\Gamma}}$ describes the internal evolution of the fragment.

Now let $\alpha = \{\Gamma_1, \Gamma_2, \ldots, \Gamma_k\}$, $2 \leq k \leq N$, be a clustering of the whole system into a system of $k$ fragments. A clustering $\alpha$ implies a resolution of $\mathscr{H} = \mathscr{H}_{\{1,2,\ldots,N\}}$ into the following tensor product

$$\mathscr{H} = \bigotimes_{\varrho=1}^{k} \{L^2(\mathbb{R}^3, dX_{\Gamma_\varrho}) \otimes L^2(\mathbb{R}^{3p_\varrho-3}, dX^{\Gamma_\varrho})\} \cong L^2(\mathbb{R}^{3k}, dX_{\Gamma_1} \cdots dX_{\Gamma_k}) \otimes \left(\bigotimes_{\varrho=1}^{k} L^2(\mathbb{R}^{3p_\varrho-3}, dX^{\Gamma_\varrho})\right). \tag{12}$$

The special clustering $\{\{1\}, \{2\}, \ldots, \{N\}\}$ is denoted by $\alpha_0$. In this case we have $k = N$ and the internal part has to be omitted. According to the resolution (12) of $\mathscr{H}$ we define

$$H_\alpha = \sum_{\varrho=1}^{k} 1_{(X_{\Gamma_1}, X^{\Gamma_1})} \otimes 1_{(X_{\Gamma_2}, X^{\Gamma_2})} \otimes \cdots \otimes H_{\Gamma_\varrho} \otimes \cdots \otimes 1_{(X_{\Gamma_k}, X^{\Gamma_k})},$$

and

$$K_\alpha = \sum_{\varrho=1}^{k} 1_{(X_{\Gamma_1}, X^{\Gamma_1})} \otimes 1_{(X_{\Gamma_2}, X^{\Gamma_2})} \otimes \cdots \otimes (K_{X_{\Gamma_\varrho}} \otimes 1_{X^{\Gamma_\varrho}}) \otimes \cdots \otimes 1_{(X_{\Gamma_k}, X^{\Gamma_k})}. \tag{13}$$

By $L^2(\mathbb{R}^{3N}, dp_1 \cdots dp_N)$ we denote the Fourier transform of $L^2(\mathbb{R}^{3N}, dx_1 \cdots dx_N)$. Let $\tilde{\mathscr{A}}$ be the system of free Hamiltonians on $L^2(\mathbb{R}^{3N}, dp_1 \cdots dp_N)$ defined in Example 2 (with degree of the polynomial $\pi(p_1, \ldots, p_N)$, $p_j = \{p_j^1, p_j^2, p_j^3\}$, not bigger than 2). Further let $\mathscr{A}$ be the system of free Hamiltonians on $L^2(\mathbb{R}^{3N}, dx_1 \cdots dx_N)$ obtained from $\tilde{\mathscr{A}}$ by inverse Fourier transformation. For example, all operators $K_\alpha$ belong to $\mathscr{A}$, because via Fourier transformation to $K_\alpha$ corresponds the polynomial $\sum_{\varrho=1}^{k} (2M_{\Gamma_\varrho})^{-1} |p_{\Gamma_\varrho}|^2$, where $p_{\Gamma_\varrho} = \sum_{j \in \Gamma_\varrho} p_j$, but this polynomial belongs to $\tilde{\mathscr{A}}$.

We consider the pair $\{H, \mathscr{A}\}$. Our aim is to determine the channels with respect to this pair.

First we introduce the concept of an *asymptotic configuration* of the whole $N$-particle system. Let $\alpha = \{\Gamma_1, \Gamma_2, \ldots, \Gamma_k\}$ be a fixed clustering. An asymptotic configuration is given if each fragment $\Gamma_\varrho$, $\varrho = 1, 2, \ldots, k$, is to be found in a fixed bound state given by an eigenvalue $\lambda_{\Gamma_\varrho}$ of $\hat{H}_{X^{\Gamma_\varrho}}$ and a corresponding state vector from $E_{\hat{H}_{X^{\Gamma_\varrho}}}(\{\lambda_{\Gamma_\varrho}\}) L^2(\mathbb{R}^{3p_\varrho-3}, dX^{\Gamma_\varrho})$. We introduce the projection

$$Q_\alpha(\lambda_{\Gamma_1}, \lambda_{\Gamma_2}, \ldots, \lambda_{\Gamma_k}) := 1_{X_{\Gamma_1} X_{\Gamma_2} \ldots X_{\Gamma_k}} \bigotimes_{\varrho=1}^{k} E_{\hat{H}_{X^{\Gamma_\varrho}}}(\{\lambda_{\Gamma_\varrho}\}). \tag{14}$$

Then we obtain that $Q_\alpha(\lambda_{\Gamma_1}, \ldots, \lambda_{\Gamma_k})$ is reducing with respect to $H_\alpha$ and $K_\alpha$ and that

$$H_\alpha \restriction Q_\alpha(\lambda_{\Gamma_1}, \lambda_{\Gamma_2}, \ldots, \lambda_{\Gamma_k}) \mathscr{H} = \left(K_\alpha + \sum_{\varrho=1}^{k} \lambda_{\Gamma_\varrho}\right) \restriction Q_\alpha(\lambda_{\Gamma_1}, \lambda_{\Gamma_2}, \ldots, \lambda_{\Gamma_k}) \mathscr{H} \tag{15}$$

is valid. In particular we have

$$Q_{\alpha_0} = 1_{\mathscr{H}}, \qquad H_{\alpha_0} = K_{\{1,2,\ldots,N\}} = -\sum_{j=1}^{N} (2m_j)^{-1} \Delta_{x_j}.$$

Note that the operator $A_\alpha(\lambda_{\Gamma_1}, \lambda_{\Gamma_2}, \ldots, \lambda_{\Gamma_k}) := K_\alpha + \sum_{\varrho=1}^{N} \lambda_{\Gamma_\varrho}$ belongs to $\mathscr{A}$. Now we have

**Lemma 8.** *Let $\{H, \mathscr{A}\}$ be defined as above. Assume*

$$Q_\alpha(\lambda_{\Gamma_1}, \lambda_{\Gamma_2}, \ldots, \lambda_{\Gamma_k}) \leq P^{\mathrm{ex}}(H, H_\alpha, 1). \tag{16}$$

*Then*

$$Q_\alpha(\lambda_{\Gamma_1}, \lambda_{\Gamma_2}, \ldots, \lambda_{\Gamma_k}) \leq Q_{A_\alpha(\lambda_{\Gamma_1}, \lambda_{\Gamma_2}, \ldots, \lambda_{\Gamma_k})} := P^{\mathrm{ex}}(H, A_\alpha(\lambda_{\Gamma_1}, \lambda_{\Gamma_2}, \ldots, \lambda_{\Gamma_k}), 1). \tag{17}$$

Proof. It immediately follows from (15) and (16). ∎

If we fix a clustering $\alpha$, then by variation of the eigenvalue system $\{\lambda_{\Gamma_1}, \ldots, \lambda_{\Gamma_k}\}$ we obtain channel Hamiltonians of the form $K_\alpha + \lambda$, where $\lambda = \sum_{\varrho=1}^{k} \lambda_{\Gamma_\varrho}$. This means that the channels obtained in this way belong to the same reaction channel, which is determined by $\alpha$. We put

$$Q_\alpha := \sum_{\{\lambda_{\Gamma_1}, \lambda_{\Gamma_2}, \dots, \lambda_{\Gamma_k}\}} Q_\alpha(\lambda_{\Gamma_1}, \lambda_{\Gamma_2}, \dots, \lambda_{\Gamma_k}).$$

Recall that the reaction channel projection, given by $\alpha$, is denoted by $Q_{\{K_\alpha\}}$. Then we have

**Lemma 9.** *Let $\{H, \mathscr{A}\}$ be as in Lemma 8 and let (16) be satisfied. Then*

$$Q_\alpha \leqq Q_{\{K_\alpha\}}. \tag{18}$$

**Proof.** It immediately follows from (17). ■

**Remark 10.** (I) Note that $Q_\alpha(\lambda_{\Gamma_1}, \lambda_{\Gamma_2}, \dots, \lambda_{\Gamma_k})$ and $Q_\alpha$ are defined in the framework of the spectral theory of $\hat{H}_{X^{\Gamma_\varrho}}$, $\varrho = 1, 2, \dots, k$, without any relation to time asymptotics. On the other hand, $P^{\mathrm{ex}}(H, H_\alpha, 1).Q_{A_\alpha(\lambda_{\Gamma_1}, \lambda_{\Gamma_2}, \dots, \lambda_{\Gamma_k})} = P^{\mathrm{ex}}(H, A_\alpha(\lambda_{\Gamma_1} \lambda_{\Gamma_2}, \dots, \lambda_{\Gamma_k}), 1)$ and $Q_{\{K_\alpha\}}$ are defined with respect to $t \to \infty$. Similarly, we obtain corresponding operators with respect to $t \to -\infty$ (use (9.19)). We use the notations $P^{\mathrm{ex}}_\pm(H, H_\alpha, 1), Q^\pm_{A_\alpha(\lambda_{\Gamma_1}, \dots, \lambda_{\Gamma_k})}, Q^\pm_{\{K_\alpha\}}$. Obviously, $Q_\alpha \leqq Q^\pm_{\{K_\alpha\}}$ is valid if (16) is assumed to be true for $P^{\mathrm{ex}}_\pm(H, H_\alpha, 1)$. (II) Note that the cardinality of the set of all clusterings $\alpha$ is finite.

We put

$$F_\alpha{}^\pm = (W_\pm(H, H_\alpha) Q_\alpha)(W_\pm(H, H_\alpha) Q_\alpha)^*. \tag{19}$$

Then from $F_\alpha{}^\pm \leqq F^\pm_{\{K_\alpha\}}$ and (8) we obtain

$$\sum_\alpha F_\alpha{}^\pm \leqq \sum_\alpha F^\pm_{\{K_\alpha\}} \leqq P^{\mathrm{ac}}.$$

**Lemma 11.** *Let $\{H, \mathscr{A}\}$ be as in Lemma 8 and let (16) be satisfied for $P^{\mathrm{ex}}_\pm(H, H_\alpha, 1)$ (in place of $P^{\mathrm{ex}}(H, H_\alpha, 1)$). Further assume*

$$\sum_\alpha F_\alpha{}^\pm = P^{\mathrm{ac}}. \tag{20}$$

*Then*

$$Q_\alpha = Q^+_{\{K_\alpha\}} = Q^-_{\{K_\alpha\}} \tag{21}$$

*that is, the system of all reaction channels is given by the system of all clusterings $\alpha$, which is a finite set. The reaction channel projections $Q_\alpha$ (and the channel Hamiltonians $K_\alpha + \lambda$) do not depend on the time direction. A channel within a given reaction channel $\alpha$ is specified by fixing of a bound state eigenvalue for each fragment of the clustering. Furthermore, the system $\{H, \mathscr{A}\}$ is asymptotically complete.*

**Proof.** From (17), (20) and Prop. 4 immediately

$$Q_\alpha(\lambda_{\Gamma_1}, \dots, \lambda_{\Gamma_k}) = Q^+_{A_\alpha(\lambda_{\Gamma_1}, \dots, \lambda_{\Gamma_k})} = Q^-_{A_\alpha(\lambda_{\Gamma_1}, \dots, \lambda_{\Gamma_k})}$$

follows. This implies (21). ■

**Remark 12.** (I) In Part IV methods are developed, which allow to verify (16) for concrete systems in many cases. Also (20) can be verified in some cases, for example if $N = 3$ (problem of asymptotic completeness). The results for $N > 3$ are not yet sufficient. For a proof of asymptotic completeness of $N$-body systems see the Notes and Remarks.

(II) By the preceding considerations the following additional assumptions are in some sense justified: (I) $Q_A{}^+ = Q_A{}^-$; (II) There is only a finite set of reaction channels. Sometimes also the existence of a channel $\alpha_0$ with $Q_{\alpha_0} = 1$ is used.

### 10.1.4. Identification operator

Let the pair $\{H, \mathscr{A}\}$ be given in $\mathscr{H}$. We assume:

(I) The number $r$ of reaction channels is finite, $r < \infty$.

(II) The channel projections (and channel Hamiltonians) are independent of the time-direction.

We denote the reaction channel index by $\alpha$, the channel index within a reaction channel by $(\alpha, j)$, $j = 1, 2, \ldots$ The channel Hamiltonians and the corresponding channel projections are denoted by $A_{\alpha,j}$, $Q_{\alpha,j}$. We put

$$A_\alpha = \sum_j A_{\alpha,j} Q_{\alpha,j}. \tag{22}$$

$A_\alpha$ is called the *reaction channel Hamiltonian*. Note that in general $A_\alpha$ does not belong to $\mathscr{A}$ because $A_\alpha \restriction Q_\alpha^\perp \mathscr{H} = 0$, whereas $A_\alpha \restriction Q_\alpha \mathscr{H}$ is absolutely continuous (recall that $Q_\alpha = \sum Q_{\alpha,j}$). Thus, if $Q_\alpha < 1_{\mathscr{H}}$, then $A_\alpha$ is not a member of $\mathscr{A}$. We put

$$\mathscr{H}_{\mathscr{A}} := \mathscr{H} \times \mathscr{H} \times \cdots \times \mathscr{H} \quad (r \text{ times}), \tag{23}$$

$$P_0 := \{Q_1, Q_2, \ldots, Q_r\}, \tag{24}$$

$$H_0 := \{A_1, A_2, \ldots, A_r\}. \tag{25}$$

Then $P_0 = P_{H_0}^{ac}$. We define $J \in \mathscr{L}(\mathscr{H}_{\mathscr{A}}, \mathscr{H})$ by

$$J\hat{f} := \sum_{\varrho=1}^{r} Q_\varrho f_\varrho, \qquad \hat{f} = \{f_1, f_2, \ldots, f_r\} \in \mathscr{H}_{\mathscr{A}}, \tag{26}$$

and obtain the following obvious properties of $J$:

$$\|J\| \leq r^{1/2},$$

$$J^* f = \{Q_1 f, Q_2 f, \ldots, Q_r f\}, \qquad f \in \mathscr{H},$$

$$J^* J = \{Q_\alpha Q_\beta\}_{\alpha,\beta=1}^{r}$$

where the right-hand side means the representation of $J^*J$ as a $r \times r$-matrix of operators $Q_\alpha Q_\beta$ with respect to the resolution (23) of $\mathscr{H}_{\mathscr{A}}$. Further, we have $JJ^* = \sum_{\varrho=1}^{r} Q_\varrho$, hence

$$(JJ^* f, f) = \sum_{\varrho=1}^{r} \|Q_\varrho f\|^2. \tag{27}$$

From (27 we obtain

**Proposition 13.** *If* $Q_{\alpha_0} = 1$ *for some reaction channel* $\alpha_0$, *then* $JJ^*$ *is bounded invertible.*

**Proof.** Obvious. ■

With these notations we can introduce the following two-space wave operator

$$W_+(H, H_0; J) := \operatorname*{Lim}_{t \to \infty} e^{itH} J\, e^{-itH} P_0 \tag{28}$$

where we have chosen the absolutely continuous projection as the universal projection. Then we obtain

**Proposition 14.** *Let* $H$ *be selfadjoint and let* $\mathscr{A}$ *be a system of free Hamiltonians. Assume* (I) *and* (II) *to be satisfied and let* $P_0, H_0, J$ *be defined by* (24), (25), (26). *Then* $W_+(H, H_0; J)$

*exists, it is partially isometric and $H_0$-semicomplete. Similarly $W_-(H, H_0; J)$ exists and the same properties are valid.*

**Proof.** The existence of $W_+(H, H_0; J)$ is obvious because for $f = \{0, 0, \ldots, 0, f_\varrho, \ldots, 0\} \in \mathscr{H}_{\mathscr{A}}$ we obtain

$$e^{itH} J\, e^{-itH_0} P_0 f = e^{itH}\, e^{-itA_\varrho} Q_\varrho f_\varrho .$$

The partial isometry of $W_+(H, H_0; J)$ is a direct consequence of Prop. 4. ■

**Corollary 15.** *Let the assumptions of Prop. 14 be satisfied. Then there are partial isometries $V_+, V_- \in \mathscr{L}(\mathscr{H}_{\mathscr{A}}, \mathscr{H})$ with initial projection $P_0$ such that $J \underset{H_0}{\approx} V_+$, $J \underset{-H_0}{\approx} V_-$. That is, $J$ is asymptotically partially isometric (with asymptotic initial projection $P_0$) for each time direction separately. One can choose $V_+ = W_+$, $V_- = W_-$.*

**Proof.** It immediately follows from Prop. 9.36. ■

**Remark 16.** According to the remarks in Section 9.5.1, under the assumptions of Corollary 15, one can formulate the stronger result that there is *one* partial isometry $V$ with the property $J \underset{\pm H_0}{\approx} V$ for both time directions simultaneously (a proof is given in Chapter 11, see Theorem 11.12).

**Proposition 17.** *Let the assumptions of Prop. 14 be satisfied. Then the pair $\{H, \mathscr{A}\}$ is asymptotically complete (with respect to $t \to \infty$) iff $W_+(H, H_0; J)$ is complete.*

**Proof.** It is obvious because of $W_+ W_+^* = \sum_{\varrho=1}^{r} F_{A_\varrho}^+$. ■

We emphasize that the identification operator $J$ results, by definition (26), from a pair $\{H, \mathscr{A}\}$, thus $J$ depends on $H$. In the next section we discuss some "internal" conditions of the system $\{H_0, P_0\}$ (without using $H$ in the formulation of these conditions) such that the identification operator (26) is asymptotically partially isometric.

### 10.1.5. General channel systems

In this section we consider finite systems of pairwise commuting selfadjoint operators together with corresponding reducing projections which are not a priori given as the system of reaction channel Hamiltonians and projections of a certain system $\{H, \mathscr{A}\}$.

**Definition 18.** A finite system $A_1, A_2, \ldots, A_r$ of pairwise commuting selfadjoint operators in $\mathscr{H}$ together with a corresponding system $Q_1, Q_2, \ldots, Q_r$ of reducing projections, $Q_j \in \operatorname{com} A_j$, where $A_j \restriction Q_j\mathscr{H}$ is absolutely continuous, is called a *general channel system.*

If such a system is given, then the operator $J: \mathscr{H}_0 = \underbrace{\mathscr{H} \times \mathscr{H} \times \cdots \times \mathscr{H}}_{r \text{ times}} \to \mathscr{H}$,

$$J\mathfrak{f} := \sum_{\varrho=1}^{r} Q_\varrho f_\varrho, \qquad \mathfrak{f} = \{f_1, f_2, \ldots, f_r\} \in \mathscr{H}_0, \tag{29}$$

is well-defined. We put $H_0 = \{A_1Q_1, A_2Q_2, \ldots, A_rQ_r\}$ and $P_0 = \{Q_1, Q_2, \ldots, Q_r\}$. Then $P_{H_0}^{ac} = P_0$. In the following we ask for conditions on the general channel system such that the operator $J$, defined by (29), is asymptotically partially isometric with respect to $H_0P_0$ and $-H_0P_0$.

In particular, we formulate a condition such that *one* partial isometry can be chosen for both time directions.

We can assume without loss of generality that $Q_iQ_j \neq 0$ for all $i, j$. Namely, if $Q_iQ_j = 0$, then also $Q_jQ_i = 0$, that is $Q_i + Q_j$ is also an (ortho-) projection and one can consider $A_{i,j} = A_iQ_i + A_jQ_j$ as a single operator on $(Q_i + Q_j)\mathscr{H}$.

For our purpose it is convenient to start with two lemmas. In the following the universal projection is always the projection onto the absolutely continuous subspace.

**Lemma 19.** *Let $J$ be defined by (29). Assume $J \underset{H_0}{\approx} V$, where $V^*V = P_0$. Put $P_\varrho := \{0, \ldots, Q_\varrho, \ldots, 0\}$, $V_\varrho f := VP_\varrho\{0, \ldots 0, \overset{(\varrho)}{f}, 0, \ldots, 0\}$, $f \in \mathscr{H}$ and $F_\varrho := V_\varrho V_\varrho^*$. Then $F_1, F_2, \ldots, F_r$ is a system of pairwise orthogonal projections, $F_\varrho F_\sigma = \delta_{\varrho,\sigma}F_\sigma$, $\varrho, \sigma = 1, 2, \ldots, r$, $F_\varrho Q_\varrho \in \operatorname{dom} \mu^{A_\varrho}$ and $\mu^{A_\varrho}(F_\varrho Q_\varrho) = Q_\varrho$.*

**Proof.** We have $F_\varrho F_\sigma = (VP_\varrho)(VP_\varrho)^*(VP_\sigma)(VP_\sigma)^* = VP_\varrho V^*VP_\sigma V^* = VP_\varrho P_0 P_\sigma V^* = VP_\varrho P_\sigma V^* = \delta_{\varrho,\sigma}VP_\varrho V^* = \delta_{\varrho,\sigma}F_\varrho$. Further, we have by assumption

$$\operatorname*{Lim}_{t\to\infty} (JP_\varrho - V_\varrho)\,\mathrm{e}^{-\mathrm{i}tH_0}P_\varrho = 0, \qquad \varrho = 1, 2, \ldots, r. \tag{30}$$

If $\hat{f} = \{0, \ldots, 0, f_\varrho, 0, \ldots, 0\}$, then $JP_\varrho \hat{f} = Q_\varrho f_\varrho$. Hence from (30) we obtain

$$\operatorname*{Lim}_{t\to\infty} (Q_\varrho - V_\varrho)\,\mathrm{e}^{-\mathrm{i}tA_\varrho}Q_\varrho = 0, \qquad \varrho = 1, 2, \ldots, r.$$

Therefore we have $\mu^{A_\varrho}(Q_\varrho - V_\varrho) = 0$ (note that $V_\varrho = V_\varrho Q_\varrho$). Hence the assertion follows from Lemma 7.16. ■

**Lemma 20.** *Let $F_1, F_2, \ldots, F_r$ be a system of pairwise orthogonal projections with $F_\varrho Q_\varrho \in \operatorname{dom} \mu^{A_\varrho}$ and*

$$\mu^{A_\varrho}(F_\varrho Q_\varrho) = Q_\varrho, \qquad \varrho = 1, 2, \ldots, r. \tag{31}$$

*Then there is a partial isometry $V \in \mathscr{L}(\mathscr{H}_0, \mathscr{H})$ with initial projection $P_0$ such that $J \underset{H_0}{\approx} V$. If (31) is assumed also for $-A_\varrho$ (time direction $t \to -\infty$), then $V$ can be chosen such that $J \underset{H_0}{\approx} V$ and $J \underset{-H_0}{\approx} V$ (the same $V$ for both time directions).*

**Proof.** We apply Lemma 7.17 and put $F = F_\varrho$, $Q = Q_\varrho$, and $H = A_\varrho$ in this lemma. Then we obtain the existence of partial isometries $V_\varrho$ in $\mathscr{H}$ with $V_\varrho^*V_\varrho = Q_\varrho$, $V_\varrho V_\varrho^* = F_\varrho$ and $\mu^{A_\varrho}(V_\varrho) = Q_\varrho$. Now we define $V \in \mathscr{L}(\mathscr{H}_0, \mathscr{H})$ by $V := \sum_{\varrho=1}^{r} V_\varrho P_\varrho$. Then we obtain

$$\operatorname*{Lim}_{t\to\infty} (JP_\varrho - VP_\varrho)\,\mathrm{e}^{-\mathrm{i}tH_0}P_\varrho = \operatorname*{Lim}_{t\to\infty} (Q_\varrho - V_\varrho)\,\mathrm{e}^{-\mathrm{i}tA_\varrho}Q_\varrho = 0,$$

hence also $\operatorname*{Lim}_{t\to\infty} (J - V)\,\mathrm{e}^{-\mathrm{i}tH_0}P_0 = 0$ is valid, that is $J \underset{H_0}{\approx} V$. The second part is also implied by Lemma 7.17. ■

Now we can present the main result of this section.

**Theorem 21.** *Let $A_1, A_2, \ldots, A_r; Q_1, Q_2, \ldots, Q_r$ be a general channel system. Let $Q_i = Q_j$ imply $i = j$. The ordering of the $Q_j$ can be arranged in such a way that from $Q_\alpha < Q_\beta$ also $\alpha < \beta$ follows. Assume*

$$Q_jQ_\varrho \in \operatorname{dom} \mu^{A_\varrho} \quad \text{and} \quad \mu^{A_\varrho}(Q_jQ_\varrho) = 0, \quad 1 \leqq j < \varrho, \quad \varrho = 2, \ldots, r. \tag{32}$$

*Then there is a partial isometry $V_+ \in \mathcal{L}(\mathcal{H}_0, \mathcal{H})$ with initial projection $P_0$ such that $J \underset{H_0}{\approx} V_+$. If (32) is assumed also for $-A_\varrho$ (time direction $t \to -\infty$), then $V_+$ can be chosen such that also $J \underset{-H_0}{\approx} V_+$ holds.*

Proof. According to Lemma 20 it is sufficient to prove the existence of a system $F_1, F_2, \ldots, F_r$ of pairwise orthogonal projections such that $\mu^{A_j}(F_jQ_j) = Q_j$, $j = 1, 2, \ldots, r$. This will be proved by induction. First we put $F_1 = Q_1$, then we have by assumption $\mu^{A_\varrho}(Q_1Q_\varrho) = 0$ for $\varrho = 2, 3, \ldots, r$. Now we assume the existence of a system $F_1, F_2, \ldots, F_n$, $n < r$, of pairwise orthogonal projections such that $\mu^{A_\alpha}(F_\alpha Q_\alpha) = Q_\alpha$, $\alpha = 1, 2, \ldots, n$ and $\mu^{A_\beta}(F_\alpha Q_\beta) = 0$ for $\alpha = 1, 2, \ldots, n$, $\beta = n + 1, \ldots, r$. The problem is to construct a projection $F_{n+1}$ with the properties $F_{n+1}F_\alpha = F_\alpha F_{n+1} = 0$, $\alpha = 1, \ldots, n$; $\mu^{A_{n+1}}(F_{n+1}Q_{n+1}) = Q_{n+1}$ and $\mu^{A_\beta}(F_{n+1}Q_\beta) = 0$, $\beta = n + 2, \ldots, r$. We put $F := \sum_{\alpha=1}^{n} F_\alpha$ and $C := F^\perp Q_{n+1} F^\perp$. Then $\mu^{A_{n+1}}(CQ_{n+1}) = \mu^{A_{n+1}}(F^\perp Q_{n+1} F^\perp Q_{n+1}) = Q_{n+1}$. Therefore, according to Prop. 7.1, there is a nonvanishing spectral projection $F_{n+1} := E_c((\varepsilon, 1])$, $\varepsilon \in (0, 1)$, such that $\mu^{A_{n+1}}(F_{n+1}Q_{n+1}) = Q_{n+1}$. Further, we have $F_{n+1} \perp F_\alpha$, $\alpha = 1, 2, \ldots, n$. Moreover we obtain for $\beta = n + 2, \ldots, r$ the equation $\mu^{A_\beta}(F_{n+1}Q_\beta) = 0$. Namely, from $\mu^{A_\beta}(Q_{n+1}Q_\beta) = 0$ and $\mu^{A_\beta}(F_\alpha Q_\beta) = 0$ for $\alpha = 1, 2, \ldots, n$ we obtain

$$\mu^{A_\beta}(F^\perp Q_{n+1} F^\perp Q_\beta) = \mu^{A_\beta}\left(\left(1 - \sum_{\alpha=1}^{n} F_\alpha\right) Q_{n+1} \left(1 - \sum_{\alpha=1}^{n} F_\alpha\right) Q_\beta\right) = 0,$$

that is $\mu^{A_\beta}(CQ_\beta) = 0$, This implies $\mu^{A_\beta}\big(E_C(\Delta)\, Q_\beta\big) = 0$ for $\Delta = (\varepsilon, 1]$, because of $C \geqq 0$, $E_C(\Delta) \leqq \varepsilon^{-1}C$ and Prop. 6.43. Thus, the "additional condition" is "invariant" with respect to the transition from $Q_{n+1}$ to $F_{n+1}$. Now by induction Theorem 21 is proved. ■

**Remark 22.** Theorem 21 is also valid if $r = \infty$, but in this case $J$ is unbounded and densely defined. In the case of Section 3 the assumptions of Theorem 21 can be verified (see the Notes and Remarks).

## 10.2. Algebraic scattering theory

### 10.2.1. Basic ideas and notions

Let $H$, $H_0$ be selfadjoint operators acting on $\mathcal{H}$. The usual one-space scattering theory is based on the concept of the wave operators $W_\pm(H, H_0)$, $W_\pm(H_0, H)$ with respect to some universal projection $P_{H_0}, P_H$ (see Chapter 9); the subspace $P_H\mathcal{H}$ is sometimes called the set of *scattering states* of $H$. In the following we choose $P_H = P_H^{ac}$ because this is the most interesting case in applications. For some physical interesting cases (e.g. Coulomb scattering) the wave operators $W_\pm(H, H_0)$ do not exist. Therefore one looks for generalizations to manage such situations. In this paragraph we present an approach in order to get generalized wave operators $W_\pm$. The link between generalized wave operators and two-space wave operators (in the sense of Chapter 9) is pointed out.

This approach starts with a $C^*$-subalgebra $\mathscr{A}$ of com $H_0$, in other words one considers special constants of the dynamics described by the group $e^{-itH}$, which represent physically the preparation and measurements of the corresponding scattering experiment. One requires that this algebra $\mathscr{A}$ satisfies some conditions with respect to the group $e^{-itH}$ (note the introduction to Part II in this connection).

First, one requires that the elements of $\mathscr{A}$ are asymptotic constants of the dynamics described by the group $e^{-itH}$ on the set of the scattering states; i.e. $\mathscr{A} \in \text{dom}\,\mu^H \cap \text{dom}\,\mu^{-H}$. The limits of $e^{itH}A\, e^{-itH}P^{ac}$ as $t \to \pm\infty$ are given by $\mu^{\pm H}(A)$, $A \in \mathscr{A}$.

Second, one requires that the algebra $\mathscr{A}$ is large in the sense that $\mu^{\pm H}(\mathscr{A}) \restriction P^{ac}\mathscr{H}$ is cyclic. Physically this means that the algebra $\mathscr{A}$ can be used to characterize completely the scattering states $P^{ac}\mathscr{H}$ (via the time limits $t \to \pm\infty$ for the time evolution of the operators of $\mathscr{A}$).

Third, one requires that the maps $\mu^{\pm H} \restriction \mathscr{A}$ can be extended to spatial isomorphisms from the weak closure of $\mathscr{A}$ onto the weak closures of $\mu^{\pm H}(\mathscr{A})$, i.e. there are two partial isometries $W_+$ and $W_-$ on $\mathscr{H}$ with $W_\pm{}^*W_\pm = P_0^{ac}$, $W_\pm W_\pm{}^* = P^{ac}$ such that $\mu^{\pm H}(A) = W_\pm A W_\pm{}^*$ for all $A \in \mathscr{A}$. These operators are called *generalized wave operators.* The *generalized scattering operator* $S$ for the system $\{H, H_0; \mathscr{A}\}$ is defined by $S := W_+{}^*W_-$.

Clearly, the operators $W_+$, $W_-$, and $S$ are highly nonunique. Even in the case $\mathscr{A} = \text{com}\, H_0 \restriction P_0^{ac}$ we obtain that the operators $\widetilde{W}_\pm = W_\pm U_\pm$, where $U_\pm$ are functions of $H_0$ with $U_\pm{}^*U_\pm = U_\pm U_\pm{}^* = P_0^{ac}$, are also generalized wave operators with the generalized scattering operator $\widetilde{S} = U_+{}^*SU_-$.

### 10.2.2. Algebraic scattering systems

Now we turn to the precise mathematical formulation of the algebraic scattering theory.

**Definition 23.** Let $H, H_0$ be selfadjoint operators acting on $\mathscr{H}$ and let $\mathscr{A}$ be a $C^*$-algebra. The triple $\{H, H_0; \mathscr{A}\}$ is called an *algebraic scattering system* if the following conditions are satisfied:

(I) $\mathscr{A} \subseteqq (\text{com}\, H_0) \restriction P_0^{ac}$, $\mathscr{A} \subset \text{dom}\,\mu^H \cap \text{dom}\,\mu^{-H}$.

(II) $\mathscr{A} \cap \ker \mu^H = \{0\}$, $\mathscr{A} \cap \ker \mu^{-H} = \{0\}$.

(III) The isometric $*$-isomorphisms between the $C^*$-algebras $\mathscr{A} \restriction P_0^{ac}\mathscr{H}$ and $\mu^{\pm H}(\mathscr{A}) \restriction P^{ac}\mathscr{H}$, given by $\mathscr{A} \ni A \to \mu^{\pm H}(A) \in \mu^{\pm H}(\mathscr{A})$ are spatial.

The condition (III) means explicitely the existence of two partial isometries $W_+$ and $W_-$ such that

$$W_+{}^*W_+ = W_-{}^*W_- = P_0^{ac}, \qquad W_+W_+{}^* = W_-W_-{}^* = P^{ac}, \tag{33}$$

$$\mu^H(A) = W_+AW_+{}^*, \qquad \mu^{-H}(A) = W_-AW_-{}^*, \qquad A \in \mathscr{A}. \tag{34}$$

Further note that (I) implies $A = P_0^{ac}AP_0^{ac}$ for all $A \in \mathscr{A}$. The *generalized wave operators* $W_\pm$ are uniquely defined up to multiplication from the right by partial isometries $U_\pm$ from com $\mathscr{A}$ with $U_\pm{}^*U_\pm = U_\pm U_\pm{}^* = P_0^{ac}$. This is the content of

**Proposition 24.** *Let* $W_+$ *satisfy* (33) *and* (34). *Then a second partial isometry* $\widetilde{W}_+$ *satisfies*

(33) *and* (34) *iff* $\tilde{W}_+ = W_+U_+$ *where* $U_+ \in \operatorname{com} \mathscr{A}$ *and* $U_+^*U_+ = U_+U_+^* = P_0^{ac}$. *A similar assertion is true for* $W_-$.

**Proof.** If we define $\tilde{W}_+ = W_+U_+$, where $U_+ \in \operatorname{com} \mathscr{A}$ and $U_+^*U_+ = U_+U_+^* = P_0^{ac}$, then (33) and (34) are obviously satisfied. Conversely, assume $\tilde{W}_+$ satisfies (33) and (34). Then we can write $\tilde{W}_+ = W_+U_+$. From (34) we get $W_+U_+AU_+^*W_+^* = W_+AW_+^*$. This gives $U_+AU_+^* = P_0^{ac}AP_0^{ac} = A$ for all $A \in \mathscr{A}$. Thus $U_+ \in \operatorname{com} \mathscr{A}$. The second condition is obvious. ∎

In principle one could drop the relation of $\mathscr{A}$ to a selfadjoint operator $H_0$ and define algebraic scattering systems without reference to an unperturbed operator $H_0$. But in all applications there is always a natural candidate for $H_0$, and the existence of such an operator $H_0$ facilitates the task of setting up the relations to the usual scattering theory. Now we present some sufficient criteria for the validity of the decisive condition (III).

**Proposition 25.** *Let the C*-algebras* $\mathscr{A} \mid P_0^{ac}\mathscr{H}$ *and* $\mu^{\pm H}(\mathscr{A}) \mid P^{ac}\mathscr{H}$ *be cyclic. Assume the existence of generating vectors* $u_0, u_\pm$ *such that* $(u_0, Au_0) = \left(u_\pm, \mu^{\pm H}(A)\, u_\pm\right)$ *for all* $A \in \mathscr{A}$. *Further let* (I) *of Definition* 23 *be satisfied. Then also* (II) *and* (III) *of Definition* 23 *are valid.*

**Proof.** We have to construct operators $W_\pm$ such that (33) and (34) are true. (33): Set $A_+ := \mu_H(A)$ for $A \in \mathscr{A}$. Define the operator $W_+$ by $W_+Au_0 = A_+u_+$ and $W_+(P_0^{ac})^\perp = 0$. Then $W_+$ is densely defined because $u_0$ is generating in $P_0^{ac}\mathscr{H}$ with respect to $\mathscr{A} \mid P_0^{ac}\mathscr{H}$. Furthermore, since the vectors $A_+u_+$ are dense in $P^{ac}\mathscr{H}$ by assumption, we have ima $W_+ \mid \mathscr{A}u_0$ is dense in $P^{ac}\mathscr{H}$. Since $\mu^H$ restricted to $\mathscr{A}$ is a *-homomorphism, by assumption we obtain

$$(Au_0, Bu_0) = (A_+u_+, B_+u_+), \qquad A, B \in \mathscr{A}.$$

Using the definition of $W_+$ and this relation for $A = B$, we obtain $\|W_+u\|^2 = \|u\|^2$ for $u \in \{Au_0; A \in \mathscr{A}\}$. Hence we may extend $W_+$ by continuity to the whole space $\mathscr{H}$ with the property $W_+^*W_+ = P_0^{ac}$. Because ima $W_+$ is dense in $P^{ac}\mathscr{H}$, also $W_+W_+^* = P^{ac}$ follows. (34): We have $W_+ABu_0 = A_+B_+u_+$, further $W_+Bu_0 = B_+u_+$ and $P_0^{ac}Bu_0 = W_+^*B_+u_+$. But from (I) of Definition 23 the relation $P_0^{ac}B = B$ follows, hence we obtain $Bu_0 = W_+^*B_+u_+$. Inserting this above, we find $W_+AW_+^*B_+u_+ = A_+B_+u_+$. Since the set $\{B_+u_+; B \in \mathscr{A}\}$ is dense in $P^{ac}\mathscr{H}$, we obtain $W_+AW_+^* = A_+$ for all $A \in \mathscr{A}$. This proves (34). Similarly we proceed for $W_-$. ∎

**Proposition 26.** *Assume the triple* $\{H, H_0; \mathscr{A}\}$ *satisfies* (I) *of Definition* 23. *Further let* $\mathscr{A} = C^*(A_1, A_2, \ldots, A_n)$ *and assume* $\mu^{\pm H}(A_j) = U_\pm A_j U_\pm^*$, $j = 1, \ldots, n$, *where* $U_\pm$ *are partially isometric with initial subspace* $P_0^{ac}\mathscr{H}$ *and final subspace* $P^{ac}\mathscr{H}$. *Then* (II) *and* (III) *of Definition* 23 *are satisfied.*

**Proof.** By assumption we immediately obtain $\mu^{\pm H}(A) = U_\pm AU_\pm^*$, $A \in \mathscr{A}$, because of Theorem 6.25. Thus (II) and (III) immediately follow. ∎

**Proposition 27.** *Assume the triple* $\{H, H_0; \mathscr{A}\}$ *satisfies* (I) *and* (II) *of Definition* 23. *Further, assume* $\mathscr{A} \mid P_0^{ac}\mathscr{H}$ *to be a v. Neumann algebra such that* $(\operatorname{com} \mathscr{A}) \mid P_0^{ac}$, $\left(\operatorname{com} \mu_H(\mathscr{A})\right) \mid P^{ac}$ *and* $\left(\operatorname{com} \mu^{-H}(\mathscr{A})\right) \mid P^{ac}$ *are Abelian. Then* (III) *of Definition* 23 *is satisfied.*

Proof. According to Prop. 7.20, $\mu_H(\cdot)$ is, by assumption (II), an isometric *-isomorphism from $\mathscr{A} \restriction P_0^{ac}\mathscr{H}$ onto $\mu_H(\mathscr{A}) \restriction P^{ac}\mathscr{H}$. Further, because of Prop. 7.21, $\mu^H(\mathscr{A})$ is a $W^*$-algebra. Now we apply Theorem 7.22 to obtain the result. ■

A simple example for Prop. 26 is given by

**Example 28.** Let $H$ and $H_0$ be unitarily equivalent operators, $H = UH_0U^*$. Further, assume $H$ to be simple and $H - H_0$ to be compact. Then $R(z) - R_0(z) \in \mathscr{L}_\infty(\mathscr{H}) \subset \ker \mu^H \cap \ker \mu^{-H}$ (see Theorem 6.35). Hence the $C^*$-algebra $\mathscr{A}$ of all continuous bounded functions of $H_0$ belongs to dom $\mu^H \cap$ dom $\mu^{-H}$ and moreover $\mu^{\pm H}(\alpha(H_0)) = \alpha(H) P^{ac}$ for all $\alpha \in C(\mathbb{R})$. Inserting $UH_0U^* = H$ into this relation, we obtain $\mu^{\pm H}(\alpha(H_0)) = U\alpha(H_0)\, U^*P^{ac}$. Thus the assumptions of Prop. 26 are satisfied; therefore $\{H, H_0; \mathscr{A}\}$ with $\mathscr{A} = \{\alpha(H_0)\, P_0^{ac}, \alpha \in C(\mathbb{R})\}$ is an algebraic scattering system.

### 10.2.3. General properties of algebraic scattering systems

Let $\{H, H_0; \mathscr{A}\}$ be an algebraic scattering system with generalized wave operators $W_+$, $W_-$. The generalized scattering operator $S$ is defined by $S := W_+{}^*W_-$. $S$ is not unique because of Prop. 24. As in (7.5) we define two unitary groups $e^{-itK_\pm}$ on $\mathscr{H}$ by

$$e^{-itK_\pm} := W_\pm{}^* e^{-itH} W_\pm + (P_0^{ac})^\perp. \tag{35}$$

By definition, $e^{-itK_\pm}$ are strongly continuous unitary groups with selfadjoint generators $K_\pm$, where $K_\pm(P_0^{ac})^\perp = 0$. Since $H$ commutes with all operators $\mu^{\pm H}(A)$, $A \in \mathscr{A}$, we obtain, by $A = W_\pm{}^*\mu^{\pm H}(A)\, W_\pm$, that $K_\pm$ commutes with all operators $A \in \mathscr{A}$, in particular we have $e^{-itK_\pm} \in$ com $\mathscr{A}$ for all $t \in \mathbb{R}$.

If $\mathscr{A} = (\text{com}\, H_0) \restriction P_0^{ac}$, then $(\text{com}\, \mathscr{A}) \restriction P_0^{ac} = W^*(E_0(\Delta)\, P_0^{ac};\ \Delta \subset \mathscr{B}_{\mathbb{R}})$. In this case $K_+$ and $K_-$ are functions of $H_0$.

In quantum mechanical scattering systems $H_0$ corresponds to the so-called free Hamiltonian and $K_+$, $K_-$ are called *renormalized* free Hamiltonians.

Note that $W_\pm$ are intertwining operators between $K_\pm$ and $H$.

**Proposition 29.** *Let $\{H, H_0; \mathscr{A}\}$ be an algebraic scattering system. Further, let $S$ be a scattering operator for $\{H, H_0; \mathscr{A}\}$. Then*

$$e^{-itK_+}S = S\, e^{-itK_-}, \qquad t \in \mathbb{R}, \tag{36}$$

$$S^*S = SS^* = P_0^{ac}. \tag{37}$$

*Moreover, $S$ is unique up to multiplication from the right or from the left by partial isometries $U \in$ com $\mathscr{A}$ with $U^*U = UU^* = P_0^{ac}$.*

Proof. We have $e^{-itK_+}S = e^{-itK_+}W_+{}^*W_- = W_+{}^*\, e^{-itH}W_- = W_+{}^*W_-\, e^{-itK_-} = S\, e^{-itK_-}$; (37) is also obvious and the last assertion immediately follows from Prop. 24. ■

**Proposition 30.** *Let $\{H, H_0; \mathscr{A}\}$ be an algebraic scattering system. Assume that $\alpha(H_0)\, P_0^{ac} \in \mathscr{A}$ for all $\alpha \in C(\mathbb{R})$ and $\mu^{\pm H}(\alpha(H_0)\, P_0^{ac}) = \alpha(H)\, P^{ac}$. Then $K_+ = K_- = H_0P_0^{ac}$ and $S$ commutes with $H_0$.*

Proof. By assumption, we have $\mu^{\pm H}(e^{-itH_0}P_0^{ac}) = e^{-itH}P^{ac}$. On the other hand, $\mu^{\pm H}(e^{-itH_0}P_0^{ac}) = W_\pm\, e^{-itH_0}P_0^{ac}W_\pm{}^*$, hence we obtain $e^{-itH_0}P_0^{ac} = W_\pm{}^*\, e^{-itH}W_\pm$. Then from (35) the assertion follows. ■

The next assertion is sometimes useful for proving that a system $\{H, H_0; \mathscr{A}\}$ is an algebraic scattering system. Moreover, it shows that each scattering system $\{H, H_0, 1\}$ (see Definition 9.51) is an algebraic scattering system $\{H, H_0; \mathscr{A}\}$, where $\mathscr{A} = (\text{com } H_0) \restriction P_0^{ac}$.

**Proposition 31.** *Let $H'$ be a selfadjoint operator acting on $\mathscr{H}$. Suppose that $\{H', H_0; \mathscr{A}\}$ is an algebraic scattering system. Further, suppose that the wave operators $W_\pm(H, H'; 1)$ exist and are complete. Then also $\{H, H_0; \mathscr{A}\}$ is an algebraic scattering system.*

**Proof.** We consider only the case $t \to \infty$. We have

$$e^{itH} A\, e^{-itH} P^{ac} = e^{itH} e^{-itH'} e^{itH'} A\, e^{-itH'} e^{itH'} e^{-itH} P^{ac}.$$

Hence from the multiplicativity of Lim (recall Lemma 6.13) and from Theorem 6.25 we obtain that $\operatorname{Lim}_{t\to\infty} e^{itH} A\, e^{-itH} P^{ac}$ exists for all $A \in \mathscr{A}$ and

$$\operatorname{Lim}_{t\to\infty} e^{itH} A\, e^{-itH} P^{ac} = W_+(H, H'; 1)\, \mu^{H'}(A)\, W_+(H, H'; 1)^*$$

is valid. By assumption we have $\mu^{H'}(A) = W_+ A W_+^*$ where $W_+$ denotes the generalized wave operator of the system $\{H', H_0; \mathscr{A}\}$. Thus $A \in \text{dom } \mu^H$ and (I) of Definition 23 is satisfied. Furthermore,

$$\mu^H(A) = W_+(H, H'; 1)\, W_+ A W_+^* W_+(H, H'; 1)^*$$

follows. Therefore (II) and (III) of Definition 23 follow and $\{H, H_0; \mathscr{A}\}$ is an algebraic scattering system with the generalized wave operators $W_+(H, H'; 1)\, W_+$ and $W_-(H, H'; 1)\, W_-$. ■

**Remark 32.** (I) Obviously $\{H_0, H_0; (\text{com } H_0) \restriction P_0^{ac}\}$ is an algebraic scattering system with $W_\pm = P_0^{ac}$. Thus, if $\{H, H_0; 1\}$ is a scattering system, that is, if $W_\pm(H, H_0; 1)$ exist and are complete then, according to Prop. 31, $\{H, H_0; (\text{com } H_0) \restriction P_0^{ac}\}$ is an algebraic scattering system with generalized wave operators $W_\pm(H, H_0; 1)$. (II) If the $C^*$-algebra $\mathscr{A} \restriction P_0^{ac} \mathscr{H}_0 \subseteq (\text{com } H_0) \restriction P_0^{ac} \mathscr{H}_0$ is even a v. Neumann algebra, then we obtain stronger results on the system $\{H, H_0; \mathscr{A}\}$, for example with respect to properties and representation of the generalized wave operators. These results can be easily obtained by a simple application of Theorem 7.22, Prop. 7.24 and Theorem 7.23, where it has to be assumed that the assumptions of Prop. 27 are satisfied (with $E$ and P replaced by $P_0^{ac}$ and $P^{ac}$).

On the other hand, Prop. 7.25 yields sufficient criteria for a system $\{H, H_0; \mathscr{A}\}$ to be an algebraic scattering system.

### 10.2.4. Algebraic scattering systems and wave operators

Let $\{H, H_0; \mathscr{A}\}$ be an algebraic scattering system. In some cases the corresponding generalized wave operators can be represented by usual wave operators with a certain identification operator. This possibility is mainly connected with Theorem 7.4.

The crucial point for such a representation is the assumption $K_+ = K_- =: K_0$ because this property is a necessary and sufficient condition for

$$e^{itH} W_\pm = W_\pm e^{itK_0}, \qquad t \in \mathbb{R}, \tag{38}$$

the intertwining property of the generalized wave operators with respect to $H$ and $K_0$ (note Prop. 30 in this connection).

For example, if the assumptions of Prop. 27 are satisfied, then, according to Theorem 7.23, the representation of $W_\pm$ has the form

$$W_\pm u = \operatorname*{Lim}_{t\to\pm\infty} e^{itH} e^{-itK_0} G_\pm(t)\, u$$

where the operator functions $G_\pm(t)$ satisfy the property formulated in Prop. 7.24 and Theorem 7.23.

The link between the generalized wave operators and the usual wave operators (with identification operators) is given by

**Proposition 33.** *Let $\{H, H_0; \mathscr{A}\}$ be an algebraic scattering system and let $K_+ = K_- =: K_0$ Then there is a bounded identification operator $J \in \mathscr{L}(\mathscr{H})$ such that both $W_\pm(H, K_0; J)$ exist and $W_\pm(H, K_0; J) = W_\pm$, where $W_\pm$ are the generalized wave operators of $\{H, H_0; \mathscr{A}\}$.*

Proof. First note that (38) is true. Further, according to Theorem 7.4, there is a bounded operator $C \in \mathscr{L}(\mathscr{H})$ such that $\mu^{K_0}(C) = P^{ac}_{K_0}$, $\mu^{-K_0}(C) = 0$. Now we define $J := W_+C + W_-(1 - C)$. Then we obtain

$$\begin{aligned}\operatorname*{Lim}_{t\to\pm\infty} e^{itH} J\, e^{-itK_0} P^{ac}_{K_0} &= \operatorname*{Lim}_{t\to\pm\infty} e^{itH}\big(W_+C + W_-(1 - C)\big)\, e^{-itK_0} P^{ac}_{K_0}\\ &= \operatorname*{Lim}_{t\to\pm\infty} W_+ e^{itK_0} C\, e^{-itK_0} P^{ac}_{K_0} + \operatorname*{Lim}_{t\to\pm\infty} W_- e^{itK_0}(1 - C)\, e^{-itK_0} P^{ac}_{K_0}\\ &= W_+\mu^{\pm K_0}(C) + W_-\mu^{\pm K_0}(1 - C) = W_\pm. \ \blacksquare\end{aligned}$$

### 10.2.5. Examples

(I) *Schrödinger operator with magnetic potentials.* Let $H_0$ denote the natural selfadjoint extension of the negative Laplacian $-\Delta = \sum_{j=1}^{3} (-i\, \partial/\partial x_j)^2$ on $L^2(\mathbb{R}^3, dx)$ with $x = \{x_1, x_2, x_3\}$. Further, let $H$ be defined by $H = UH_0U^*$ where $U$ denotes the unitary operator given by $(Uv)(x) = e^{ip(x)}v(x)$. We assume that $p(x)$ is a real-valued twice continuously differentiable function in $\mathbb{R}$ which only depends on $r = |x| = (x_1^2 + x_2^2 + x_3^2)^{1/2}$. Set $q_j(x) := (\partial/\partial x_j)\, p(x) = ((d/dr)\, p(r))\, \partial r/\partial x_j = ((d/dr)\, p(r))\, (x_j/r)$. Then an easy calculation shows that

$$(Hu)(x) = \sum_{j=1}^{3} \left(-i\frac{\partial}{\partial x_j} - q_j(x)\right)^2 u(x)$$

and $H$ is selfadjoint on dom $H = U(\text{dom}\, H_0)$. Clearly, $P^{ac}_0 = P^{ac} = 1$. The question is wether generalized wave operators exist for $H$, $H_0$ and some $C^*$-algebra $\mathscr{A} \subseteq$ com $H_0$ under additional restrictions on the function $p(r)$. We assume that the functions $q_j(x)$ are essentially bounded. Further we assume the operators $Q_j$ represented by the functions $q_j(x)$ belong to $\ker \mu_s^{H_0} \cap \ker \mu_s^{-H_0}$. By Theorem 7.38, this means that $\lim_{R\to\infty} R^{-3} \int_{|x|\le R} |q_j(x)|^2\, dx = 0$. To be specific, we can take the function $p(r) = (1 + r)^\alpha$, $\alpha \in (0, 1)$. Then $q_j(x) = \alpha(1 + r)^{\alpha-1}\frac{x_j}{r}$ and it can easily be shown that these functions satisfy the condition mentioned above.

Concerning the $C^*$-algebra $\mathscr{A}$ we first introduce the operators $P_j$ representing the free momenta of the system, given by $H$ and $H_0$, defined by $P_j = -i\frac{\partial}{\partial x_j}$ and the operators $L_j$ representing the angular momenta, defined by $L_1 = -i\left(x_2 \frac{\partial}{\partial x_3} - x_3 \frac{\partial}{\partial x_2}\right)$, $L_2$, $L_3$ (given by cyclic permutation of the indices 1, 2, 3 in the formula for $L_1$). Now let $\mathscr{A}$ be the $C^*$-algebra generated by the continuous bounded functions of $P_j$ and $L_j$, $j = 1, 2, 3$. It is well-known that $P_j$ and $L_j$ commute with $H_0$ and the commutant of this algebra $\mathscr{A}$ is com (com $H_0$) (i.e. the set of all essentially boundeal measurable functions of $H_0$). Thus $\mathscr{A} \subseteq$ com $H_0$ and com $\mathscr{A}$ is Abelian.

We use Prop. 26 to prove that $\{H, H_0; \mathscr{A}\}$ is an algebraic scattering system. Define $\tilde{P}_j := UP_jU^*$. Since $p(x) = p(r)$, $U$ commutes with the angular momenta $L_j$, i.e. $UL_jU^* = L_j$. Let $f \in C(\mathbb{R}^3)$, then $e^{itH}f(L_j)\,e^{-itH} = U\,e^{itH_0}U^*f(L_j)\,U\,e^{-itH_0}U^* = f(L_j)$, $j = 1, 2, 3$, since $L_j$ commutes also with $H_0$. Therefore $f(L_j) \in$ com $H \subset$ dom $\mu^H \cap$ dom $\mu^{-H}$. Thus to show that $\mathscr{A} \subset$ dom $\mu^H \cap$ dom $\mu^{-H}$, it suffices to verify that all bounded continuous functions of the momenta $P_j$ have this property. For this aim we use Prop. 7.27, that is, we prove that $R_{P_j}(z) \in$ dom$^H \cap$ dom $\mu^{-H}$ and $\mu^{\pm H}(R_{P_j}(z)) = R_{\tilde{P}_j}(z)$. We have, using (1.13) and $\tilde{P}_j = P_j - Q_j$,

$$e^{itH}(R_{P_j}(z) - R_{\tilde{P}_j}(z))\,e^{-itH} = e^{itH}R_{P_j}(z)\,(P_j - \tilde{P}_j)\,R_{\tilde{P}_j}(z)\,e^{-itH}$$

$$= e^{itH}R_{P_j}(z)\,Q_jU\,e^{-itH_0}U^*R_{\tilde{P}_j}(z) = e^{itH}R_{P_j}(z)\,UQ_j\,e^{-itH_0}U^*R_{\tilde{P}_j}(z).$$

From this equation we obtain, by the assumption $Q_j \in \ker \mu^{H_0} \cap \ker \mu^{-H_0}$, the relation $R_{P_j}(z) - R_{\tilde{P}_j}(z) \in \ker \mu^H \cap \ker \mu^{-H}$. Hence $R_{P_j}(z) \in$ dom $\mu^H \cap$ dom $\mu^{-H}$ and $\mu^{\pm H}(R_{P_j}(z)) = R_{\tilde{P}_j}(z)$ is proved. This shows that the system $\{H, H_0; \mathscr{A}\}$ satisfies the assumptions of Prop. 26, therefore $\{H, H_0; \mathscr{A}\}$ is a generalized scattering system. The generalized wave operators $W_+$ and $W_-$ are equal to $U$.

(II) *Schrödinger operators with long range potentials.* Let $H_0$ and $\mathscr{H}$ be as in (I). Now assume that $H$ is given by $-\Delta + q(\cdot)$ where $q(x) = c\,|x|^{-\alpha}$, $\alpha \in (0, 1]$. Then $H$ is selfadjoint with dom $H$ = dom $H_0$ because the perturbation $V$, defined by $q(x)$, is $H_0$-bounded with relative bound 0 (see Kato [1, p. 302]). Furthermore, $H_0$ and $H$ have the same absolutely continuous spectrum. More precisely, if one introduces the decomposition $\mathscr{H} = \bigoplus_{l=0}^{\infty} \bigoplus_{|m| \le l} \mathscr{H}_{l,m}$ where $\mathscr{H}_{l,m}$ denotes the set of all functions of the form $r^{-1}u(r)\,Y_{l,m}(\theta, \varphi)$ with $u(\cdot) \in L^2(\mathbb{R}_+, dr)$ and $Y_{l,m}$ denotes the spherical harmonics, then each $\mathscr{H}_{l,m}$ reduces $H$ and $H_0$ since both operators are rotational invariant. Now we use the following result (see Weidmann [1967a]):

Let $H_0 = -\Delta$ and $H = -\Delta + q(x)$ on $L^2(\mathbb{R}^3, dx)$. Assume that $q(|x|)$ satisfies the following assumptions: There is a $c > 0$ and an $r_0 > 0$ such that $|x|^s\,|q(|x|)| \le c$ for $|x| < r_0$ with $s < 3/2$, $|q(|x|)| \le c$ for $|x| \ge r_0$; further, $q(|x|) = q_1(|x|) + q_2(|x|)$ for $|x| > r_0$ with $q_1(r)$ of bounded variation and $q_2(\cdot) \in L^1([r_0, \infty), dr)$. Then, for each $\{l, m\}$, $H \restriction \mathscr{H}_{l,m}$ is simple and has the absolutely continuous spectrum $[0, \infty)$.

The potential $q(x) = c\,|x|^{-\alpha}$, $\alpha > 0$, satisfies the assumptions of this result. Furthermore, it is well-known that $H_0 \restriction \mathscr{H}_{l,m}$ is simple and absolutely continuous on $[0, \infty)$. If we can prove that there is a partial isometry $W$ such that $W^*W = 1$, $WH_0 = HW$ and each $\mathscr{H}_{l,m}$ is a reducing subspace for $W$, then from Lemma 4.28 it easily follows that $WW^* = P^{ac}$.

According to Remark 32 (II), one way to prove that $\{H, H_0;$ com $H_0\}$ is an algebraic scattering system is to apply Prop. 7.25, that is to find unitary families $T_t^{\pm}$ with the properties described in this proposition. The idea of the proof is as follows: First construct unitary families $T_t^{\pm}$ which are functions of $H_0$ for each $t \in \mathbb{R}_{\pm}$ such that

$$W_{\pm}u := \operatorname*{s-lim}_{t \to \pm\infty} e^{itH}T_t^{\pm}u, \qquad u \in \mathscr{H}, \tag{39}$$

exist. Then immediately $W_+{}^*W_+ = W_-{}^*W_- = 1$ follows and, by the properties of $T_t^{\pm}$ and $H$, $W_{\pm}$ are reduced by the subspaces $\mathscr{H}_{l,m}$. Secondly, one shows that $W_{\pm}$ are intertwining operators for $H$, $H_0$, this means $W_{\pm}\,e^{itH_0} = e^{itH}\,W_{\pm}$, $t \in \mathbb{R}$. Then, by the remark given above, $W_{\pm}W_{\pm}^* = P^{ac}$ follows. Then, by Prop. 7.25, $\{H, H_0;$ com $H_0\}$ is an algebraic scattering system and $W_{\pm}$ are generalized wave operators. Therefore, it remains to construct families $T_t^{\pm}$ such that (39) is satisfied and to show that $W_{\pm}$ are intertwining operators for $H$ and $H_0$.

For simplicity we consider only the physical important case that $q(x)$ is the Coulomb potential $c\,|x|^{-1}$, $c \in \mathbb{R}$. In this case we define

$$T_t^{\pm} := e^{-iX_t}; \quad X_t := tH_0 + (1/2) \int_0^t c(H_0^{1/2}|t'|)^{-1} \chi_{[0,\infty)}(4\,|t'|\,H_0 - 1)\,dt', \tag{40}$$

the characteristic function is to avoid singularities of the operator function $X_t$ at $t = 0$. Here $T_t^+ = T_t^- =: T_t$. Let $\mathscr{D}$ be the set of vectors $u(\cdot) \in L^2(\mathbb{R}^3, dx)$ whose Fourier transform $\hat{u}(\cdot) \in C_0^\infty(\mathbb{R}^3 \setminus \{0\})$. Clearly, $\mathscr{D}$ is dense in $\mathscr{H}$. Because the Fourier transform of $H_0$ is multiplication by $|k|^2$, the Fourier transform of $X_t$ is also multiplication by a function $X_t(k)$. For $4t\,|k|^2 > 1$ we obtain

$$X_t(k) = t\,|k|^2 + (1/2)\,c\,|k|^{-1}\,(\log|t| + \log(4\,|k|^2)). \tag{41}$$

A consequence of (41) is that for each vector $u \in \mathscr{D}$ there is a $t_u > 0$ such that for all $t > t_u$

$$X_t u = [tH_0 + (1/2)\,cH_0^{-1/2}(\log|t| + \log(4H_0)]\,u$$

and

$$T_t u = e^{-i(tH_0 + (1/2)cH_0^{-1/2}(\log|t| + \log(4H_0))}\,u. \tag{42}$$

From (41) it easily follows that $u \in \operatorname{dom} X_t'$ for all $t > t_u$ and $T_t u \in \mathscr{D} \subsetneqq \operatorname{dom} H_0 = \operatorname{dom} H$. Thus $T_t u$ is strongly differentiable for all $u \in \mathscr{D}$ and we obtain

$$\frac{d}{dt}\,e^{itH}T_t u = i\,e^{itH}\left(H - \frac{d}{dt}X_t\right)T_t u = i\,e^{itH}\,(H - H_0 - c(2H_0^{1/2}t)^{-1})\,T_t u$$

for all sufficiently large $t$. The existence of (39) will be proved with the help of the estimate

$$\|e^{itH}T_t u - e^{isH}T_s u\| \leqq \int_s^t \|(H - H_0 - c(2H_0^{1/2}t')^{-1})\,T_{t'}u\|\,dt'. \tag{43}$$

In what follows we restrict ourselves to the case of $W_+$. The case $W_-$ is similar. Set $u_1 := H_0^{-1/2}u$, that is $\hat{u}_1(k) = |k|^{-1}\hat{u}(k)$. Then it is a straightforward calculation to show that

$$\left((H - H_0 - c(2H_0^{1/2}t)^{-1})\,T_t u\right)(x) = c\,|x|^{-1}(T_t u)(x) - c(2t)^{-1}\,(T_t u_1)(x)$$

$$= c\,|x|^{-1}\,D_u(x,t) - c(2t)^{-1}\,D_{u_1}(x,t), \tag{44}$$

where $D_u(x, t)$ is defined by

$$D_u(x,t) := (T_t u)(x) - (2it)^{-3/2} \exp\{ix^2/4t - ict\,|x|^{-1}\log(x^2/t)\}\,\hat{u}(x/2t). \tag{45}$$

Now we claim that $D_u(x, t)$ satisfies the estimate

$$|D_u(x,t)| \leqq a(\log|t|)^s\,t^{-5/2}(1 + |x|/t)^2)^{-m} \tag{46}$$

for all $t > t_u$, arbitrary integer $m$ and constants $s \in \mathbb{N}$, $a \in \mathbb{R}$ depending on $m$ and $u$ only. Using this estimate together with (44) and (43), we obtain that (39) exists for $t \to \infty$ and for all $u \in \mathscr{D}$. By the uniform boundedness of $e^{itH}T_t$, (39) exists for all $u \in \mathscr{H}$. To prove (46), we set $u(x, t)$ $:= (e^{-iX_t}\,e^{itH_0}u)(x)$. Because of (42) for $t > t_u$ we obtain

$$\hat{u}(k,t) = \exp\{-(ic/2)\,|k|^{-1}\log(4tk^2)\}\,\hat{u}(k).$$

Using the integral representation of $e^{-itH_0}$ (see Reed/Simon [2, p. 59])

$$(e^{-itH_0}u)(x) = (4i\pi t)^{-3/2} \int_{\mathbb{R}^3} e^{i|x-y|^2/4t}u(y)\,dy,$$

we find

$$(T_t u)(x) = (4i\pi t)^{-3/2} \int_{\mathbb{R}^3} e^{i|x-y|^2/4t}u(y,t)\,dy.$$

This and the elementary formula $e^{i|x-y|^2/4t} = e^{i|x|^2/4t}\, e^{-i(x,y)/2t}\, e^{i|y|^2/4t}$ give

$$D_u(x, t) = (4i\pi t)^{-3/2}\, e^{i|x|^2/4t} \int_{\mathbb{R}^3} e^{-i(x,y)/2t}\, (e^{i|y|^2/4t} - 1)\, u(y, t)\, dy.$$

By assumption $\hat{u}(\cdot, t) \in C_0^\infty(\mathbb{R}^3) \subset \mathscr{S}(\mathbb{R}^3)$ and therefore $u(\cdot, t) \in \mathscr{S}(\mathbb{R}^3)$. Now we use the following result on the Fourier transform on $\mathscr{S}(\mathbb{R}^3)$ (see REED/SIMON [2, p. 2]): The Fourier transform is a continuous linear map from $\mathscr{S}(\mathbb{R}^3)$ onto $\mathscr{S}(\mathbb{R}^3)$ and $\|\hat{u}\|_{\alpha,\beta} \leq \sum_{j=1}^{M} c_j \|u\|_{\alpha_j,\beta_j}$ where $\|\cdot\|_{\alpha,\beta}$ denote the seminorms on $\mathscr{S}$ given by $\sup_{x\in\mathbb{R}^3} |x^\alpha D^\beta u(x)|$, $\alpha$ and $\beta$ are multiindices, that is, $\alpha = \{\alpha_1, \alpha_2, \alpha_3\}$, $\beta = \{\beta_1, \beta_2, \beta_3\}$, $D^\beta = \partial^{\beta_1+\beta_2+\beta_3}/\partial x_1^{\beta_1}\, \partial x_2^{\beta_2}\, \partial x_3^{\beta_3}$ and $x^\alpha = x_1^{\alpha_1} x_2^{\alpha_2} x_3^{\alpha_3}$, $\alpha_j \in \mathbb{N}$ and $\beta_j \in \mathbb{N}$. From the special form of $\hat{u}(k, t)$ we obtain the estimate $\|\hat{u}(\cdot, t)\|_{\alpha,\beta} \leq a_{\alpha,\beta}(\log t)^p$ where $p$ depends on $\alpha$ and $\beta$. By the result just mentioned this implies

$$\|u(\cdot, t)\|_{\alpha,\beta} \leq b_{\alpha,\beta}(\log t)^p \tag{47}$$

with $p$ depending on $\alpha$ and $\beta$. From the formula for $D_u(x, t)$ above and (47) we obtain

$$\left|\left(\frac{x^2}{4t^2}\right)^m D_u(x, t)\right| \leq |4\pi t|^{-3/2} \int_{\mathbb{R}^3} \left|\left(\left(\sum_{j=1}^{3} \frac{\partial^2}{\partial y_j^2}\right)^m\right) e^{-i(x,y)/2t}\, (e^{i|y|^2/4t} - 1)\, u(y, t)\right| dy$$

$$\leq c|t|^{-5/2} \int_{\mathbb{R}^3} dy (1 + |y|)^{-4} \sum_{j=\{0,0,0\}}^{l(m)} \sum_{s=\{0,0,0\}}^{a(m)} \left| y^j \frac{\partial^s}{\partial y^s} u(y, t)\right| \leq d_m\, |t|^{-5/2} (\log |t|)^{q(m)}.$$

Here we have used also the estimate $|(e^{i|y|^2/4t} - 1)| \leq |y|^2/4t$. The estimate for $D_u(x, t)$ leads to (46) proving the existence of (39). It remains to show the intertwining relation $W_+ H_0 = H W_+$. We have

$$e^{isH}\, e^{itH}\, e^{-iX_t}\, e^{-isH_0} = e^{i(t+s)H}\, e^{-iX_{t+s}} (e^{iX_{t+s} - iX_t}\, e^{-isH_0}).$$

The intertwining relation $e^{isH} W_+\, e^{-isH_0} = W_+$ follows from this relation if we can show that $\operatorname{s-lim}_{t\to\infty} (e^{-iX_t}\, e^{-isH_0} - e^{-iX_{t+s}}) = 0$. We prove this for vectors $u$ from the dense set $\mathscr{D}$. By (42), for $t > t_u$ and $t + s > t_u$ we obtain

$$(e^{-iX_{t+s}} - e^{-iX_t}\, e^{-isH_0})\, u$$

$$= e^{-i(t+s)H_0}\, e^{-i(c/2)H_0^{-1/2}\log(4H_0)}\, (e^{-i(c/2)H_0^{-1/2}\log(t+s)} - e^{-i(c/2)H_0^{-1/2}\log})$$

$$= \exp\{-i(t + s)\, H_0 - i(c/2)\, H_0^{-1/2} \log (4H_0)$$
$$- i(c/2)\, H_0^{-1/2} \log t\}\, (\exp\{-i(c/2)\, H_0^{-1/2} \log (1 + s/t)\} - 1)\, u.$$

Since $\log (1 + s/t) \to 0$ as $t \to \infty$ for each $s$, and since $e^{iaH_0^{-1/2}}$ is strongly continuous in $a$, we obtain from the equation above that the right side tends to 0 strongly. This proves the intertwining relation $e^{isH} W_+\, e^{-isH_0} = W_+$, therefore also $W_+ H_0 = H W_+$, and the proof that $W_+$ is a generalized wave operator is finished.

We note that

$$\operatorname{w-lim}_{t\to\infty} e^{itH}\, e^{-itH_0} = \operatorname{w-lim}_{t\to\infty} e^{itH}\, e^{-iX_t}\, e^{iX_t}\, e^{-itH_0} = 0$$

because of (39) and

$$\operatorname{w-lim}_{t\to\infty} e^{iX_t}\, e^{-itH_0} = \operatorname{w-lim}_{t\to\infty} e^{i(c/2)H_0^{-1/2}(\log t + \log(4H_0))} u = 0$$

for $u \in \mathscr{D}$. The latter relation is implied by Prop. 3.23 (note that $H_0^{-1/2}$ is absolutely continuous). Therefore the usual wave operator $\operatorname{Lim}_{t\to\infty} e^{itH}\, e^{-itH_0}$ does not exist in this case.

## 10.3. Equations of second order in $d/dt$

The examples for two-space wave operators with identification operator considered so far were connected with the abstract Schrödinger equations $u'(t) = -iHu(t)$, $v'(t) = -iH_0v(t)$. In other words, the wave operators were connected with the interpretation of $e^{-itH}u$, $e^{-itH_0}v$ as trajectories of the time evolution of given states $u$, $v$ at time $t = 0$ in the state spaces $\mathscr{H}$, $\mathscr{H}_0$. The reason for introducing the identification operator $J$ was the possible existence of many channels (asymptotic configurations) (see Paragraph 1) and — on the other hand — the necessity of modifying $H_0$ (see Paragraph 2).

It is also possible to consider modifications of the underlying equations and to use identification operators $J$ in order to transform the modified equations into a Schrödinger form. As an example we consider second order equation in this paragraph. The universal projection is always $P_H^{ac}$.

### 10.3.1. Reduction to the Schrödinger equation

Let $\mathscr{X}$ be a separable Hilbert space and let $A$ be a selfadjoint nonnegative operator on $\mathscr{X}$, $A \geqq 0$. We consider the initial value problem for the following second order equation:

$$u''(t) + Au(t) = 0, \quad t \in \mathbb{R}, \quad u(0) = u_0 \text{ and } u'(0) = v_0. \tag{48}$$

We put $B = A^{1/2}$, $B \geqq 0$. The operator $B^{-1} \sin(tB) \in \mathscr{L}(\mathscr{X})$ is well-defined, even when $B^{-1}$ does not exist. Then we obtain easily for the solution of (48)

$$\begin{pmatrix} u(t) \\ u'(t) \end{pmatrix} = \begin{pmatrix} \cos(tB) & B^{-1}\sin(tB) \\ -B\sin(tB) & \cos(tB) \end{pmatrix} \begin{pmatrix} u_0 \\ v_0 \end{pmatrix}.$$

We put

$$U(t) = \begin{pmatrix} \cos(tB) & B^{-1}\sin(tB) \\ -B\sin(tB) & \cos(tB) \end{pmatrix}$$

and define a new Hilbert space, consisting of pairs $\{u, v\}$ such that $U(t)$ acts on this space as a unitary group. First we define a scalar product on (dom $B$) mod (ker $B$) by

$$(u, v)_B := (Bu, Bv)_{\mathscr{X}}, \qquad \|u\|_B = \|Bu\|_{\mathscr{X}}.$$

Then $\mathscr{F}_A := \mathrm{clo}_{\|\cdot\|_B}$ (dom $B$) mod (ker $B$) is a well-defined Hilbert space with scalar product $(\cdot, \cdot)_B$. In general this closure contains elements which are not in $\mathscr{X}$. We put

$$\mathscr{H}_A := \mathscr{F}_A \times \mathscr{X}$$

and equip $\mathscr{H}_A$ with the usual scalar product

$$(\hat{u}_1, \hat{u}_2)_A := (u_1, u_2)_B + (k_1, k_2)_{\mathscr{X}}; \qquad \{u_1, k_1\} = \hat{u}_1, \quad \{u_2, k_2\} = \hat{u}_2 \in \mathscr{H}_A.$$

The norm in $\mathscr{H}_A$ is given by

$$\|\hat{u}\|_A^2 = \|u\|_B^2 + \|k\|_{\mathscr{X}}^2, \qquad \{u, k\} = \hat{u} \in \mathscr{H}_A.$$

Now we let $U(t)$ act on the space $\mathscr{H}_A$. We obtain

**Lemma 34.** *$U(t)$ is a unitary group on $\mathscr{H}_A$.*

**Proof.** We have

$$U(t)\begin{pmatrix} u \\ k \end{pmatrix} = \begin{pmatrix} \cos(tB)\,u & +B^{-1}\sin(tB)\,k \\ -B\sin(tB)\,u & +\cos(tB)\,k, \end{pmatrix}, \quad \{u, k\} = \hat{u} \in \mathscr{H}_A.$$

Therefore we obtain

$$\begin{aligned} \|U(t)\,\hat{u}\|_A^2 &= \|B\cos(tB)\,u + \sin(tB)\,k\|_{\mathscr{K}}^2 + \|-B\sin(tB)\,u + \cos(tB)\,k\|_{\mathscr{K}}^2 \\ &= \left(B\cos^2(Bt)\,u, Bu\right) + \left(\sin^2(tB)\,k, k\right) + 2\mathrm{Re}\left(B\cos(tB)\sin(tB)u, k\right) \\ &\quad + \left(B\sin^2(tB)\,u, Bu\right) \\ &\quad + \left(\cos^2(tB)\,k, k\right) - 2\mathrm{Re}\left(B\sin(tB)\cos(tB)\,u, k\right) \\ &= \|Bu\|_{\mathscr{K}}^2 + \|k\|_{\mathscr{K}}^2 = \|u\|_B^2 + \|k\|_{\mathscr{K}}^2 = \|\hat{u}\|_A^2. \quad \blacksquare \end{aligned}$$

### 10.3.2. Connection between $U(t)$ and $e^{-itB}$

First we note the relations

$$\mathscr{K} = \mathrm{clo}\,(\mathrm{ima}\,B) \oplus \ker B$$

and

$$\mathscr{H}_A = \left(\mathscr{F}_A \times \mathrm{clo}\,(\mathrm{ima}\,B)\right) \oplus (\{0\} \times \ker B)$$

where $\oplus$ denotes an orthogonal direct sum (in the latter equation with respect to the scalar product in $\mathscr{H}_A$). Now we have for $k \in \ker B$

$$U(t)\begin{pmatrix} 0 \\ k \end{pmatrix} = \begin{pmatrix} tk \\ k \end{pmatrix} = \begin{pmatrix} 0 \\ k \end{pmatrix},$$

because of $tk = 0 \bmod (\ker B)$, that is $\{0\} \times \ker B$ is an eigenspace for $U(t)$ with eigenvalue 1. Secondly, we define an operator $\tilde{B}$ by

$$\tilde{B}: \mathscr{F}_A \to \mathrm{clo}\,(\mathrm{ima}\,B), \qquad \tilde{B}f = Bf, \qquad f \in \mathrm{dom}\,B.$$

$\tilde{B}$ is isometric from $\mathscr{F}_A$ onto clo (ima $B$). Further we define an operator $T$,

$$T: \mathscr{F}_A \times \mathrm{clo}\,(\mathrm{ima}\,B) \to \mathrm{clo}\,(\mathrm{ima}\,B) \times \mathrm{clo}\,(\mathrm{ima}\,B),$$

by the equation

$$\begin{pmatrix} f \\ g \end{pmatrix} = T\begin{pmatrix} u \\ v \end{pmatrix} = \frac{1}{\sqrt{2}}\begin{pmatrix} \tilde{B} & \mathrm{i} \\ \tilde{B} & -\mathrm{i} \end{pmatrix}\begin{pmatrix} u \\ v \end{pmatrix}. \tag{49}$$

Then $T$ is also isometric from $\mathscr{F}_A \times$ clo (ima $B$) onto clo (ima $B$) $\times$ clo (ima $B$) and

$$T^{-1} = \frac{1}{\sqrt{2}}\begin{pmatrix} \tilde{B}^{-1} & \tilde{B}^{-1} \\ -\mathrm{i} & \mathrm{i} \end{pmatrix} \tag{50}$$

is valid. Now we have

**Lemma 35.** *$U(t)$ is unitarily equivalent to* $e^{-itB} \oplus e^{itB} \oplus 1$.

**Proof.** By a simple calculation we obtain

$$T\big(U(t) \mid \mathcal{F}_A \times \text{clo}\,(\text{ima}\,B)\big)\, T^{-1} = \begin{pmatrix} e^{-itB} & 0 \\ 0 & e^{itB} \end{pmatrix}. \tag{51}$$

This relation together with the fact that $\{0\} \times \ker B$ is an eigenspace for $U(t)$ with eigenvalue 1 proves the assertion. ■

We put $U(t) = e^{-itH}$. The connection between the absolutely continuous subspace of $H$ and the corresponding space of $B$ is given by

**Lemma 36.** *$\{u, v\} \in P_H^{ac}\mathcal{H}_A$ iff $f \in P_B^{ac}\mathcal{X}$ and $g \in P_A^{ac}\mathcal{X}$ where $\{u, v\}$ and $\{f, g\}$ are connected by* (49).

**Proof.** Let $\{u, v\} \in \mathcal{F}_A \times \text{clo}\,(\text{ima}\,B)$, that is, let $v \in \text{clo}\,(\text{ima}\,B)$. Then, according to (49), (50), (51), the equivalence asserted in the lemma is obvious. That is, it is sufficient to show that both conditions imply $v \in \text{clo}\,(\text{ima}\,B)$. First, let $f \in P_B^{ac}\mathcal{X}$, $g \in P_B^{ac}\mathcal{X}$. Because of $P_B^{ac}\mathcal{X} \subseteq (\ker B)^\perp = \text{clo}\,(\text{ima}\,B)$, obviously, $v \in \text{clo}\,(\text{ima}\,B)$ is valid. Second, let $\{u, v\} \in P_H^{ac}\mathcal{H}_A$. Hence $\{u, v\}$ is orthogonal to any eigenspace of $H$, in particular to $\{0\} \times \ker B$. Therefore we obtain $\{u, v\} \perp \{0\} \times \ker B$, that is $v \in (\ker B)^\perp = \text{clo}\,(\text{ima}\,B)$. ■

### 10.3.3. The first identification operator

Suppose we have two second order equations $u'' + A_j u = 0$ in the same Hilbert space $\mathcal{X}$. We put $B_j = A_j^{1/2}$, $U_j(t) = e^{-itH_j}$ in $\mathcal{H}_{A_j} = \mathcal{F}_{A_j} \times \text{clo}\,(\text{ima}\,B_j) \oplus \{0\} \times \ker B_j$. We ask for conditions on $A_j$ such that the wave operators $W_\pm(H_2, H_1; J)$ exist by introducing of an appropriate identification operator $J$. There are different possibilities to choose $J$. First we choose a mathematically convenient one. We denote by $E_j$ the orthoprojection of $\mathcal{X}$ onto clo (ima $B_j$). Then we have $E_j \geq P_{B_j}^{ac}$. Now we define

$$J: \mathcal{H}_{A_1} \to \mathcal{H}_{A_2}, \quad J\{u, v\} = \{\tilde{B}_2^{-1} E_2 \tilde{B}_1 u, v\}, \quad \{u, v\} \in \mathcal{F}_{A_1} \times \mathcal{X} = \mathcal{H}_{A_1}.$$

Then we immediately obtain

$$\|J\| \leq 1, \qquad J^*\{u, v\} = \{\tilde{B}_1^{-1} E_1 \tilde{B}_2 u, v\}.$$

Now we can formulate a criterion for the existence of the wave operators $W_\pm(H_2, H_1; J)$ relating their existence to the existence of the wave operators $W_\pm(B_2, B_1) = \operatorname*{Lim}_{t \to \pm\infty} e^{itB_2} \times e^{-itB_1} P_{B_1}^{ac}$.

**Theorem 37.** *Let both wave operators $W_\pm(B_2, B_1)$ exist (exist and be complete). Then both wave operators $W_\pm(H_2, H_1; J)$ exist and are partially isometric and $H_1$-semicomplete (exist and are partially isometric and complete). In the second case also $W_\pm(H_1, H_2; J^*)$ exist and are complete and*

$$W_\pm(H_1, H_2; J^*) = W_\pm(H_2, H_1; J)^*$$

*is valid.*

**Proof.** Because of Lemma 36 it is sufficient to consider vectors $\{u, v\} \in \mathcal{F}_{A_j} \times \operatorname{clo}(\operatorname{ima} B_j)$. Then we obtain on the set $\operatorname{clo}(\operatorname{ima} B_j) \times \operatorname{clo}(\operatorname{ima} B_j) = T_j(\mathcal{F}_{A_j} \times \operatorname{clo}(\operatorname{ima} B_j))$ the relation

$$T_j\, e^{-itH_j} T_j^{-1} = \begin{pmatrix} e^{-itB_j} & 0 \\ 0 & e^{itB_j} \end{pmatrix} =: e^{-it\hat{B}_j}$$

where $e^{-it\hat{B}_j}$ is acting on $\operatorname{clo}(\operatorname{ima} B_j) \times \operatorname{clo}(\operatorname{ima} B_j)$. According to Lemma 36, $T_j$ maps $P^{ac}_{H_j}\mathcal{H}_{A_j}$ onto $P^{ac}_{\hat{B}_j}(\operatorname{clo}(\operatorname{ima} B_j) \times \operatorname{clo}(\operatorname{ima} B_j))$. We obtain

$$e^{itH_2} J\, e^{-itH_1}\{u, v\} = T_2^{-1} e^{it\hat{B}_2} T_2 J T_1^{-1}\, e^{-it\hat{B}_1} T_1\{u, v\}.$$

Now we have $T_2 J T_1^{-1} = \begin{pmatrix} E_2 & 0 \\ 0 & E_2 \end{pmatrix}$; hence

$$e^{itH_2} J\, e^{-itH_1} \begin{pmatrix} u \\ v \end{pmatrix} = T_2^{-1} e^{it\hat{B}_2} \begin{pmatrix} E_2 & 0 \\ 0 & E_2 \end{pmatrix} e^{-it\hat{B}_1} T_1 \begin{pmatrix} u \\ v \end{pmatrix}$$

$$= T_2^{-1} \begin{pmatrix} e^{itB_2} E_2 e^{-itB_1} & 0 \\ 0 & e^{-itB_2} E_2\, e^{itB_1} \end{pmatrix} T_1 \begin{pmatrix} u \\ v \end{pmatrix}.$$

Using

$$\operatorname*{Lim}_{t\to\pm\infty} e^{itB_2} E_2\, e^{-itB_1} P^{ac}_{B_1} = E_2 \operatorname*{Lim}_{t\to\pm\infty} e^{itB_2}\, e^{itB_1} P^{ac}_{B_1} = E_2 W_\pm(B_2, B_1) = W_\pm(B_2, B_1),$$

we obtain the existence of $W_\pm(H_2, H_1; J)$, its partial isometry and $H_1$-semicompleteness and the equation

$$W_\pm(H_2, H_1; J) = T_2^{-1} \begin{pmatrix} W_\pm(B_2, B_1) & 0 \\ 0 & W_\mp(B_2, B_1) \end{pmatrix} T_1. \tag{52}$$

(52) immediately yields the completeness of $W_\pm(H_2, H_1; J)$ if $W_\pm(B_2, B_1)$ are complete. Similarly we proceed for $W_\pm(H_1, H_2; J^*)$. ∎

**Remark 38.** Recall that the wave operators $W_\pm(H_2, H_1; J)$ are considered with respect to the universal projection $P_0^{ac}$. The partial isometry and $H_1$-semicompleteness of $W(H_2, H_1; J)$, which properties are obvious in view of (52), can be proved also with the help of Prop. 9.48 because of the following lemma.

**Lemma 39.** *Let $W_\pm(B_2, B_1)$ exist. Then $J^*$ is an $H_1$-asymptotic left inverse to $J$.*

**Proof.** Let $\{u, v\} \in \mathcal{F}_{A_1} \times \mathcal{K}$. Then we have

$$\|(1 - J^*J)\{u, v\}\|_{\mathcal{H}_{A_1}} = \|\{(1 - \tilde{B}_1^{-1} E_1 E_2 \tilde{B}_1)\, u, 0\}\|_{\mathcal{H}_{A_1}}$$

$$= \|\tilde{B}_1^{-1}(1 - E_1E_2)\, \tilde{B}_1 u\|_{B_1} = \|(1 - E_1E_2)\, B_1 u\|_{\mathcal{K}}.$$

Therefore it is sufficient to show $\operatorname*{Lim}_{t\to\pm\infty} (1 - E_1E_2)\, B_1 u_1(t) = 0$, where $e^{-itH_1}\{u, v\} = \{u_1(t), \ldots\}$; that is it is sufficient to show $\operatorname*{Lim}_{t\to\pm\infty} \|(1 - E_1E_2)\, e^{-itB_1} f\| = 0$ for $f \in P^{ac}_{B_1}\mathcal{K}$. Now we have $1 - E_1E_2 = (1 - E_1) + E_1(1 - E_2)$, and moreover $(1 - E_1)\, e^{-itB_1} P^{ac}_{B_1} = 0$ because of $P^{ac}_{B_1} \le E_1$. Further, $\operatorname*{Lim}_{t\to\pm\infty} (1 - E_2)\, e^{-itB_1} f = 0$ because of $P^{ac}_{B_2} \le E_2$. This concludes the proof. ∎

**Remark 40.** Since $B_j = A_j^{1/2}$ are defined according to the abstract spectral theory they are not "elementary" operators even when the $A_j$ are. Thus it might appear that the existence of $W_\pm(B_2, B_1)$ is difficult to verify in concrete problems. This is not necessarily so, however. In many cases $W_\pm(B_2, B_1)$ exist and are complete if $W_\pm(A_2, A_1)$ exist and are complete and they coincide (according to the invariance principle of wave operators, see Chapter 11).

### 10.3.4. The second identification operator

The identification operator of Section 3 is mathematically convenient. However it is a different question whether it is a natural choice from the physical point of view. In many cases it would be physically reasonable to identify a pair $\{u, v\} \in \mathscr{H}_{A_1}$ with the same pair $\{u, v\} \in \mathscr{H}_{A_2}$. But this is possible only if $\mathscr{H}_{A_1} = \mathscr{H}_{A_2}$. We consider a more general case. Let

$$\operatorname{dom} B_2 \subseteqq \operatorname{dom} B_1,$$

$$m \, \|B_1 u\| \leqq \|B_2 u\| \leqq M \, \|B_1 u\|, \qquad u \in \operatorname{dom} B_2,$$

where $m, M > 0$. Such a situation may occur for example when $A_2$ is the Friedrichs extension of a symmetric operator $A_0 \geqq 0$ in $\mathscr{K}$ and $A_1$ is another nonnegative extension of $A_0$. Obviously we also obtain $\mathscr{F}_{A_2} \subseteqq \mathscr{F}_{A_1}$ under an obvious identification for ideal elements and

$$m \, \|\tilde{B}_1 u\|_{B_1} \leqq \|\tilde{B}_2 u\|_{B_2} \leqq M \, \|\tilde{B}_1 u\|_{B_1}, \qquad u \in \mathscr{F}_{A_2}.$$

This implies that the two norms $\|\cdot\|_{B_1}$ and $\|\cdot\|_{B_2}$ in $\mathscr{F}_{A_2}$ are equivalent so that $\mathscr{F}_{A_2}$ is a subspace of $\mathscr{F}_{A_1}$. The orthogonal projection of $\mathscr{F}_{A_1}$ onto $\mathscr{F}_{A_2}$ is denoted by $P$. Now we introduce the following identification operator:

$$\hat{J} : \mathscr{H}_{A_1} \to \mathscr{H}_{A_2}, \qquad \hat{J}\{u, v\} = \{Pu, v\}.$$

$\hat{J}$ is nothing else than the orthoprojection of $\mathscr{H}_{A_1}$ onto $\mathscr{H}_{A_2}$, $\mathscr{H}_{A_2}$ regarded as a subspace of $\mathscr{H}_{A_1}$. If $\mathscr{F}_{A_1} = \mathscr{F}_{A_2}$, then $\hat{J}\{u, v\} = \{u, v\}$, that is, in this case we have a simple identification.

In many cases the two identifications $J$ and $\hat{J}$ are asymptotically equivalent with respect to $H_1 P_{H_1}^{ac}$ and $-H_1 P_{H_1}^{ac}$. We consider as an example the following special case.

**Theorem 41.** *Let* $W_\pm(B_2, B_1)$ *exist and further assume* $\operatorname{dom} A_2 \subseteqq \operatorname{dom} A_1$

$$\operatorname*{Lim}_{t \to \pm\infty} (A_1 - A_2) \, e^{-itB_2} v = 0, \qquad v \in \operatorname{dom} A_2 \cap P_{A_2}^{ac} \mathscr{K}. \tag{53}$$

*Then* $J \underset{\pm H_1}{\approx} \hat{J}$, *i.e.* $\operatorname*{Lim}_{t \to \pm\infty} (J - \hat{J}) \, e^{-itH_1} P_{H_1}^{ac} = 0$.

**Proof.** Let $\{f', g'\} \in \mathscr{F}_{A_1} \times \operatorname{clo}(\operatorname{ima} B_1)$ and put $\{u, v\} = T_1\{f', g'\}$ (recall the definition of $T_1$ given by (49)). Then we obtain

$$\|(J - \hat{J}) \, e^{-itH_1}\{f', g'\}\|_{\mathscr{H}_{A_2}} = \|(J - \hat{J}) \, T_1^{-1} \, e^{-it\hat{H}_1} T_1\{f', g'\}\|_{\mathscr{H}_{A_2}}$$

$$= \|(J - \hat{J}) \, T_1^{-1} \, e^{-it\hat{H}_1}\{u, v\}\|_{\mathscr{H}_{A_2}} = \left\|(J - \hat{J}) \left\{e^{-itB_1}(1/\sqrt{2}) \, (\tilde{B}_1^{-1} u + \tilde{B}_1^{-1} v), 0\right\}\right\|_{\mathscr{H}_{A_2}}.$$

We put $u + v = f \in \operatorname{ima} B_1$. Hence $\tilde{B}_1^{-1} f =: g \in \operatorname{dom} B_1$, and

$$\|(J - \hat{J}) e^{-itH_1}\{f', g'\}\|_{\mathscr{H}_{A_2}} = \frac{1}{\sqrt{2}} \|(\tilde{B}_2^{-1} E_2 \tilde{B}_1 - P) e^{-itB_1} g\|_{B_2}$$

$$= \frac{1}{\sqrt{2}} \|\tilde{B}_2^{-1}(E_2 B_1 - \tilde{B}_2 P) e^{-itB_1} g\|_{B_2}$$

follows. Now we have $B_1 e^{-itB_1} g = e^{-itB_1} B_1 g$ and

$$\operatorname*{Lim}_{t \to \pm\infty} (e^{-itB_1} B_1 g - e^{-itB_2} W_\pm B_1 g) = 0, \tag{54}$$

where $e^{-itB_2} W_\pm B_1 g = B_2 e^{-itB_2} W_\pm g$. Further we obtain

$$\operatorname*{Lim}_{t \to \pm\infty} (E_2 B_1 e^{-itB_1} g - B_2 e^{-itB_2} W_\pm g) = 0.$$

Therefore, because of $e^{-itB_2} W_\pm g \in \operatorname{dom} B_2$, we have

$$B_2 e^{-itB_2} W_\pm g - \tilde{B}_2 P e^{-itB_1} g = B_2 P e^{-itB_2} W_\pm g - \tilde{B}_2 P e^{-itB_1} g$$

and it is sufficient to consider $\|P(e^{-itB_2} W_\pm g - e^{-itB_1} g)\|_{B_2}$ for the proof. Since the norms $\|\cdot\|_{B_1}$ and $\|\cdot\|_{B_2}$ are equivalent in $\mathscr{F}_{A_2}$, it is enough to look at

$$\|e^{-itB_2} W_\pm g - e^{-itB_1} g\|_{B_1} = \|B_1 e^{-itB_2} W_\pm g - B_1 e^{-itB_1} g\|_{\mathscr{H}}. \tag{55}$$

Because of (54) the expression to be considered can be slightly transformed into

$$\|(B_1 - B_2) W e^{-itB_2} W_\pm g\|_{\mathscr{H}} = \|e^{itB_1}(B_1 - B_2) e^{-itB_2} W_\pm g\|_{\mathscr{H}}, \quad g \in \operatorname{dom} B_1.$$

Now we have

$$\operatorname*{Lim}_{t \to \pm\infty} e^{itB_1} B_2 e^{-itB_2} W_\pm g = \operatorname*{Lim}_{t \to \pm\infty} e^{itB_1} e^{-itB_2} B_2 W_\pm g = B_1 g$$

so that it is sufficient to consider

$$\|e^{itB_1} B_1 e^{-itB_2} W_\pm g - B_1 g\|_{\mathscr{H}} = \|B_1(e^{itB_1} e^{-itB_2} W_\pm g - g)\|_{\mathscr{H}}.$$

But $\operatorname*{w\text{-}lim}_{t \to \pm\infty} B_1(e^{itB_1} e^{-itB_2} W_\pm g - g) = 0$ because of

$$\operatorname*{Lim}_{t \to \pm\infty} \big(B_1(e^{itB_1} e^{-itB_2} W_\pm g - g), f\big) = \operatorname*{Lim}_{t \to \pm\infty} (e^{itB_1} e^{-itB_2} W_\pm g - g, B_1 f) = 0$$

for $f \in \operatorname{dom} B_1$. Therefore it is enough to prove

$$\operatorname*{Lim}_{t \to \pm\infty} \|B_1 e^{itB_1} e^{-itB_2} W_\pm g\| = \|B_1 g\|, \qquad g \in \operatorname{dom} B_1.$$

It is sufficient to take $g$ from $\operatorname{dom} A_1 \cap P_{A_1}^{ac} \mathscr{H}$ which is dense in $\operatorname{dom} B_1 \cap P_{A_1}^{ac} \mathscr{H}$. Then

$W_\pm g \in \operatorname{dom} A_2 \cap P^{ac}_{A_2}\mathscr{H}$. Now we obtain

$$\begin{aligned}\operatorname*{Lim}_{t\to\pm\infty} \|B_1 e^{itB_1} e^{-itB_2} W_\pm g\|^2 &= \operatorname*{Lim}_{t\to\pm\infty} (B_1 e^{-itB_2} W_\pm g, B_1 e^{-itB_2} W_\pm g)\\ &= \operatorname*{Lim}_{t\to\pm\infty} (e^{itB_2} A_1 e^{-itB_2} W_\pm g, W_\pm g) = \operatorname*{Lim}_{t\to\pm\infty} (e^{itB_2} A_2 e^{-itB_2} W_\pm g, W_\pm g)\\ &= \operatorname*{Lim}_{t\to\pm\infty} \|B_2 e^{-itB_2} W_\pm g\|^2 = \|B_2 W_\pm g\|^2 = \|B_1 g\|^2\end{aligned}$$

where the third step works because of (53). ∎

**Remark 42.** There are other conditions instead of (53) which yield the same result (see the Notes and Remarks).

Since $\mathscr{H}_{A_2} \subseteq \mathscr{H}_{A_1}$, that is $\mathscr{H}_{A_2}$ is a subspace of $\mathscr{H}_{A_1}$, we can define the simple embedding operator $J'$ by

$$J' : \mathscr{H}_{A_2} \to \mathscr{H}_{A_1}; \qquad J'\{u, v\} = \{u, v\}, \qquad \{u, v\} \in \mathscr{H}_{A_2}$$

and we obtain easily that in some cases $J'$ is an asymptotic left inverse for the projection operator $\hat{J}$.

**Lemma 43.** *If $W_\pm(B_2, B_1)$ exist and if (53) is satisfied, then $J'$ is asymptotically left inverse to $J$ (with respect to $H_1 P^{ac}_{H_1}$ and $-H_1 P^{ac}_{H_1}$).*

**Proof.** Similarly as in the proof of Theorem 28 we obtain

$$\|(J'\hat{J} - 1) e^{-itH_1}\{f', g'\}\|_{\mathscr{H}_{A_1}} = \left\|(J'\hat{J} - 1)\left\{(1/\sqrt{2})\, e^{-itB_1} \tilde{B}_1^{-1}(u + v), 0\right\}\right\|_{\mathscr{H}_{A_1}}.$$

Therefore it is sufficient to consider

$$\|(P - 1) e^{-itB_1} g\|_{B_1}, \qquad g \in \operatorname{dom} B_1$$

for $t \to \pm\infty$. Under the assumption of the lemma we have: (55) tends to 0 for $t \to \pm\infty$ with respect to Lim, hence it is sufficient to consider

$$\operatorname*{Lim}_{t\to\pm\infty} \|P e^{-itB_1} g - e^{-itB_2} W_\pm g\|_{B_1}.$$

Now $e^{-itB_2} W_\pm g \in \operatorname{dom} B_2$, hence $P e^{-itB_2} W_\pm g = e^{-itB_2} W_\pm g$ follows. Therefore it is sufficient to consider

$$\operatorname*{Lim}_{t\to\pm\infty} \|B_1(e^{-itB_1} g - e^{-itB_2} W_\pm g)\|_{\mathscr{H}} = \operatorname*{Lim}_{t\to\pm\infty} \|B_1(g - e^{itB_1} e^{-itB_2} W_\pm g)\|_{\mathscr{H}},$$

which equals 0 (cf. proof of Theorem 41). ∎

**Remark 44.** Here we only indicate possible examples, which will be treated later in Part IV (see Chapter 15), because at the present stage of development no strong existence criteria for wave operators are available. The first example refers to the wave equation with a potential. The Hilbert space $\mathscr{H}$ is $L^2(\mathbb{R}^3, dx)$, $A_1 = -\Delta$ (negative Laplacian) and $A_2 = A_1 + q(\cdot)$. If $q(x)$ is the sum of an $L^2$-function and an $L^\infty$-function, then $A_2$ is selfadjoint and $\operatorname{dom} A_2 = \operatorname{dom} A_1$ is true. By strengthening of the assumptions on $q$ all assumptions of the foregoing considerations can be verified (see Section 15.2.4). The second example refers to the wave equation in an exterior domain (scattering by obstacles). The Hilbert space $\mathscr{K}$ is $L^2(\mathbb{R}^n, dx)$, $A_1 = -\Delta$. If $\Omega_0$ is a bounded region in $\mathbb{R}^n$ and $\Omega = \mathbb{R}^n \setminus \Omega_0$, then $L^2(\mathbb{R}^n, dx) \cong L^2(\Omega_0, dx) \times L^2(\Omega, dx)$. The operator $A_2$ is defined by $A_2 = A_2' \times A_2''$, where $A_2'$ and $A_2''$ are the negative Laplacians in $\Omega_0$ and $\Omega$ with Dirichlet boundary conditions. Also in this case several assumptions of the foregoing consideration can be verified (see Section 15.1.3).

## 10.4. Quantum Field Theory

Scattering theory for interacting quantum fields is a remarkable example for the fact that identification operators are, in general, unavoidable for fitting physical scattering theories into the framework of wave operators. Moreover, these identification operators turn out to be unbounded, although densely defined. Thus this example also shows the importance of unbounded identification operators. In this paragraph we present the necessary foundations on the so-called Axiomatic Quantum Field Theory. Further we introduce identification operators between free and interacting fields. The corresponding wave operators are linked with the Haag-Ruelle scattering theory.

### 10.4.1. Abstract quantum fields

In this section we present the necessary abstract material to define Hermitian scalar quantum fields. The following objects $\mathscr{F}$, $\mathfrak{G}$, $V$ are assumed to be given a priori.

Let $\mathscr{F}$ be a locally convex topological $*$-algebra with 1 (identity). Further, let $\mathfrak{G}$ be a locally compact separable group and let $V_g$ be a continuous $*$-representation of $\mathfrak{G}$ in aut $\mathscr{F}$ (group of all continuous automorphisms of $\mathscr{F}$) such that

$$\left.\begin{aligned} &V_g(a_1 f_1 + a_2 f_2) = a_1 V_g f_1 + a_2 V_g f_2,\\ &V_g(f_1 \cdot f_2) = V_g f_1 \cdot V_g f_2,\\ &V_g f^* = (V_g f)^*. \end{aligned}\right\} \tag{56}$$

In the following we denote by $\mathscr{H}$ a separable Hilbert space, by $\mathscr{D}$ a dense (linear) manifold in $\mathscr{H}$ and by $\mathscr{C}^+(\mathscr{D})$ the set of all not necessarily bounded operators $A$ acting on $\mathscr{H}$ with $\mathscr{D} \subseteq \operatorname{dom} A$, $\mathscr{D} \subseteq \operatorname{dom} A^*$, $A\mathscr{D} \subseteq \mathscr{D}$, $A^*\mathscr{D} \subseteq \mathscr{D}$. Further $\omega$ denotes a unit vector from $\mathscr{H}$, $\|\omega\| = 1$.

**Definition 45.** The 5-tuple $\{\mathscr{H}, \mathscr{D}, U_g, \omega, A(f)\}$ is called an *abstract* quantum *field* (with respect to $\mathscr{F}$, $\mathfrak{G}$, $V_g$), if $U_g$ is a unitary strongly continuous representation of $\mathfrak{G}$ in $\mathscr{L}(\mathscr{H})$ such that $U_g\mathscr{D} \subseteq \mathscr{D}$, $\omega \in \mathscr{D}$, $U_g\omega = \omega$ for all $g \in \mathfrak{G}$ and if $A(\cdot)$ is a $*$-representation of $\mathscr{F}$ in $\mathscr{C}^+(\mathscr{D})$, $\mathscr{F} \ni f \to A(f) \in \mathscr{C}^+(\mathscr{D})$ with the following properties (the meaning of the notation $A(f)$ is always $A(f) \restriction \mathscr{D}$):

(I) $A(a_1 f_1 + a_2 f_2) = a_1 A(f_1) + a_2 A(f_2)$ (linearity),

(II) $A(f_1 \cdot f_2) = A(f_1) \cdot A(f_2)$ (multiplicativity),

(III) $A(f)^* = A(f^*)$,

(IV) $A(1) = 1$,

(V) $\omega$ is generating with respect to $A(\cdot)$, $\text{clo}\,\{A(f)\,\omega; f \in \mathscr{F}\} = \mathscr{H}$,
(VI) $(u, A(f)\,v)$, $u, v \in \mathscr{D}$, is a continuous functional on $\mathscr{F}$,
(VII) $A(V_g f) = U_g A(f)\, U_g^{-1}$ (covariance).

If an abstract quantum field is given, then the corresponding ***Wightman functional*** $W(f)$ is defined by

$$W(f) := (\omega, A(f)\,\omega), \qquad f \in \mathscr{F}. \tag{57}$$

It satisfies the following properties, collected in

**Proposition 46.** *Let $W(f)$ be the Wightman functional of an abstract quantum field. Then*

(i) $W(a_1 f_1 + a_2 f_2) = a_1 W(f_1) + a_2 W(f_2)$ *(linearity)*,
(ii) $W(f)$ *is continuous*,
(iii) $W(1) = 1$,
(iv) $W(f^* f) \geqq 0$ *(positivity)*,
(v) $W(V_g f) = W(f)$ *($\mathfrak{G}$-invariance)*.

**Proof.** It is obvious. ∎

The first important property of the Wightman functional is connected with the fact that it is the only unitary invariant of an abstract quantum field. More precisely we have

**Proposition 47.** *By the Wightman functional $W(f)$ the abstract quantum field $\{\mathscr{H}, \mathscr{D}, U_g, \omega, A(f)\}$ is uniquely defined up to unitary equivalence.*

**Proof.** Let two fields $\{\mathscr{H}_j, \mathscr{D}_j, U_{j,g}, \omega_j, A_j(f)\}$, $j = 1, 2$, be given and assume $(\omega_1, A_1(f)\,\omega_1) = (\omega_2, A_2(f)\,\omega_2) =: W(f)$ for all $f \in \mathscr{F}$. We define an operator $U$ from $\mathscr{H}_1$ to $\mathscr{H}_2$ by $U(A_1(f)\,\omega_1) = A_2(f)\,\omega_2$. Then we obtain $\|U(A_1(f)\,\omega_1)\|^2 = (A_2(f)\,\omega_2, A_2(f)\,\omega_2) = (\omega_2, A_2(f^* f)\,\omega_2) = W(f^* f) = \|A_1(f)\,\omega_1\|^2$. Hence $U(A_1(f)\,\omega_1)$ is independent of the special representation of the vector $A_1(f)\,\omega_1$, it is densely defined with dense image and isometric. By continuous extension we obtain an isometric operator $U$ from $\mathscr{H}_1$ onto $\mathscr{H}_2$. Now it is easy to verify $U_{2g} = U U_{1g} U^*$, $A_2(f) = U A_1(f)\, U^*$ on $\{A_2(f)\,\omega_2; f \in \mathscr{F}\}$, hence on $\mathscr{D}_2$. ∎

The next theorem (the so-called ***reconstruction theorem***) shows that if a functional $W(f)$ on $\mathscr{F}$ has properties (i)—(v) of Prop. 46, then it is the Wightman functional of a certain abstract quantum field.

**Theorem 48.** *Let $\mathscr{F}$, $\mathfrak{G}$, $V$ satisfy* (56). *Assume $W(f)$ to be a functional on $\mathscr{F}$ with the properties* (i)—(v) *of Prop. 46. Then there is an abstract quantum field $\{\mathscr{H}, \mathscr{D}, U_g, \omega, A(f)\}$ such that $W(f)$ is the Wightman functional of this abstract quantum field.*

**Proof.** First we define a semiscalar product on $\mathscr{F}$ by the equation

$$(f_1, f_2) := W(f_1^* f_2), \qquad \|f\|^2 = W(f^* f). \tag{58}$$

From Schwarz' inequality

$$|W(f_1^* f_2)|^2 \leqq W(f_1^* f_1)\, W(f_2^* f_2)$$

it easily follows that the left kernel lker $W$, defined by

$$\text{lker } W := \{f;\ W(f^*f) = 0\}$$

(which is equivalently the set of all elements $f$ from $\mathscr{F}$ which satisfy $\|f\| = 0$), is a (closed) subspace, is a left ideal and is $\mathfrak{G}$-invariant. We put

$$\tilde{\mathscr{F}} := \mathscr{F} \bmod (\text{ lker } W).$$

We define a scalar product on $\tilde{\mathscr{F}}$ by the equation

$$(\tilde{f}_1, \tilde{f}_2)_{\tilde{\mathscr{F}}} := (f_1, f_2)$$

where $f_1, f_2$ are representers (elements) of the equivalence classes $\tilde{f}_1, \tilde{f}_2$. Then $\tilde{\mathscr{F}}$ becomes a pre-Hilbert space. The corresponding Hilbert space $\text{clo}_{\|\cdot\|_{\tilde{\mathscr{F}}}} \tilde{\mathscr{F}}$ is denoted by $\mathscr{H}_W$. The operator of the canonical embedding of $\mathscr{F}$ into $\mathscr{H}_w$ is denoted by $J$,

$$J\colon \mathscr{F} \to \mathscr{H}_w, \qquad Jf = \tilde{f} \in \tilde{\mathscr{F}}.$$

$J$ is linear. Further from

$$\|Jf\|^2_{H_w} = W(f^*f), \tag{59}$$

(ii) and the continuity of $f \cdot g$ and $f^*$ in $\mathscr{F}$ the continuity of $J$ follows. Also from (59) it can be seen that ker $J =$ lker $W$. Now we put $\mathscr{D} := \tilde{\mathscr{F}} =$ ima $J$ and we define

$$\omega := J(1), \tag{60}$$

$$U_g(Jf) := J(V_gf), \qquad f \in \mathscr{F}, \tag{61}$$

$$A(f)\,(Ju) := J(f \cdot u), \qquad f, u \in \mathscr{F}. \tag{62}$$

Now we have to check the properties of $U_g$ and $A(\cdot)$. First we begin with $U_g$. We have $\|U_gJu\|^2 = \big(J(V_gu), J(V_gu)\big) = W\big((V_gu)^*\,(V_gu)\big) = W\big(V_g(u^*u)\big) = W(u^*u) = \|Ju\|^2$,

so that $U_g$ is isometric and densely defined. Because $V_g \in \text{aut}\,\mathscr{F}$ we obtain ima $(U_g \restriction \text{ima } J)$ $=$ ima $J$. Therefore $U_g$ can be extended by continuity on $\mathscr{H}_W$ as a unitary operator. Obviously, $U_g$ is a strongly continuous representation of $\mathfrak{G}$ and $U_g\mathscr{D} \subseteq \mathscr{D}$ is valid. Further we have

$$U_g\omega = U_gJ(1) = J(V_g1) = J(1) = \omega.$$

Second, we check the properties (I)—(VII) of $A(\cdot)$. From (62) $A(f)\,\mathscr{D} \subseteq \mathscr{D}$ immediately follows. (I) follows from the linearity of $J$. (II) follows from

$$A(f_1f_2)\,Ju = J\big((f_1 \cdot f_2)\,u\big) = A(f_1)\,J(f_2u) = A(f_1)\,A(f_2)\,Ju.$$

(III) follows from

$$\begin{aligned}\big(Ju_1, A(f)\,Ju_2\big) &= \big(Ju_1, J(fu_2)\big) = W(u_1{}^*fu_2) = W\big((f^*u_1)^*\,u_2\big)\\ &= \big(J(f^*u_1), Ju_2\big) = \big(A(f^*)\,Ju_1, Ju_2\big),\end{aligned}$$

hence also $A(f)^* \mathscr{D} \subseteq \mathscr{D}$ is valid. (IV) is obvious. (V) follows from $A(f)\,\omega = J(f\cdot 1) = Jf$ which yields the dense set ima $J$ if $f$ runs through $\mathscr{F}$. (VI): The continuity of

$$\left(Ju_1, A(f)\,Ju_2\right) = \left(Ju_1, J(fu_2)\right) = W(u_1{}^*fu_2)$$

follows from the continuity of $W$. (VII) follows from

$$A(V_gf)\,Ju = J(V_gf\cdot u) = J(V_gf\cdot V_gV_g{}^{-1}u) = J\left(V_g(f\cdot V_g{}^{-1}u)\right) = U_gJ(f\cdot V_g{}^{-1}u)$$
$$= U_gA(f)\,J(V_g{}^{-1}u) = U_gA(f)\,U_g{}^{-1}Ju.$$

Finally from (58), (60) and (62), $\left(\omega, A(f)\ \omega\right) = (J1, Jf)_{\mathscr{F}} = (1, f) = W(f)$ follows. ■

The importance of the reconstruction theorem lies in the fact that one can work with the objects $\mathscr{F}$, $\mathfrak{G}$, $V_g$, $W(\cdot)$ only, in principle. The operator $A(f)$ on $\mathscr{H}_W$ is called the (abstract) *field operator*. For our purposes the identification operator $J$ is very important. It is defined if $W$ is given. By $J$ the embedding of $\mathscr{F}$ into the Hilbert space is defined and simultaneously the field operator is defined by means of the multiplication in $\mathscr{F}$.

### 10.4.2. Tensor algebra of test functions, Poincaré group, and regular representation

In this section we consider special cases of the objects $\mathscr{F}$, $\mathfrak{G}$, $V_g$ which are of fundamental interest for the definition of quantum fields.

(I) Let $\mathscr{F}$ be the tensor algebra on $\mathscr{S}(\mathbb{R}^4)$, the Schwartz space on $\mathbb{R}^4$. Sometimes the functions from $\mathscr{S}(\mathbb{R}^4)$ are called test functions. Thus

$$\mathscr{F} \ni f = \{f_0, f_1(x_1), f_2(x_1, x_2), \ldots, f_n(x_1, x_2, \ldots, x_n), 0, 0, \ldots\}$$

where $f_0 \in \mathbb{C}$, $f_1 \in \mathscr{S}(\mathbb{R}^4)$, $f_2 \in \mathscr{S}(\mathbb{R}^8)$, ... and where $n$ may depend on $f$. The $k$-th coordinate of $f$ is denoted by $f_k$. The operations $f^*$ and $f\cdot g$ are defined as follows:

$$(f^*)_n\,(x_1, x_2, \ldots, x_n) := \overline{f_n(x_n, x_{n-1}, \ldots, x_1)},$$

$$(f\cdot g)_n\,(x_1, \ldots, x_n) := \sum_{\alpha+\beta=n} f_\alpha(x_1, \ldots, x_\alpha)\,g_\beta(x_{\alpha+1}, \ldots, x_{\alpha+\beta}).$$

The topology in $\mathscr{F}$ is defined by a system of seminorms given by $\|f\|_{\gamma_n,\nu_n} := \sum\limits_{n=0}^{\infty} \gamma_n \|f_n\|_{\nu_n}$, $\gamma_n > 0$, $\nu_n$ nonnegative integers, where $\|f_n\|_\nu$, $\nu = 0, 1, 2, \ldots$ denotes a denumerable system of seminorms in $\mathscr{S}_n = \mathscr{S} \otimes \mathscr{S} \otimes \cdots \otimes \mathscr{S}$ ($n$ times) defining the topology of the tensor product of the Schwarz spaces $\mathscr{S}$. Then $\mathscr{F}$ becomes a locally convex topological $*$-algebra. Note that $1 = \{1, 0, 0, \ldots\}$.

(II) Let $\mathfrak{G} = \mathfrak{P}_+^\uparrow = \mathscr{L}_+^\uparrow \overset{s}{\otimes} \mathbb{R}^4$ be the Poincaré group. $\mathscr{L}_+^\uparrow$ denotes the proper Lorentz group, $\mathbb{R}^4$ is considered as its translation group, s means semi-direct product. Thus the elements $g \in \mathfrak{G}$ are given by pairs $g = \{\Lambda, a\}$ where $a \in \mathbb{R}^4$ and $\Lambda \in \mathscr{L}_+^\uparrow$.

(III) There is a special representation of $\mathfrak{P}_+^\uparrow$ as a group of automorphisms on $\mathscr{F}$, the so-called regular representation, defined by

$$(V_g f)(x) = f(g^{-1}x),\ g^{-1}\{x_1, x_2, \ldots, x_n\} = \{g^{-1}x_1, g^{-1}x_2, \ldots, g^{-1}x_n\}$$

If $g = \{1, a\}$, then

$$(V_{\{1,a\}}f)_n(x_1, \ldots, x_n) = f_n(x_1 - a, \ldots, x_n - a).$$

For convenience we write $V_a$ instead of $V_{\{1,a\}}$. Now we note some properties of the special settings just introduced.

In $\mathscr{F}$ the Fourier transformation is defined by

$$(Ff)_n(p_1, p_2, \ldots, p_n) = (2\pi)^{-2n} \int\limits_{\mathbb{R}^{4n}} e^{-i\{(p_1,x_1)+(p_2,x_2)+\cdots+(p_n,x_n)\}} f_n(x_1, x_2, \ldots, x_n)\, dx_1 \ldots dx_n.$$

As is well-known, $F$ is bijective and bicontinuous on $\mathscr{F}$. We put $Ff = \hat{f}$ and we will distinguish between the space $\mathscr{F}$ and its Fourier transform $\hat{\mathscr{F}}$. Obviously, the Fourier transform $\hat{V}_a = FV_aF^*$ of $V_a$ is given by

$$(\hat{V}_a\hat{f})_n(p) = e^{-i\left(a\sum\limits_{j=1}^{n} p_j\right)} \hat{f}_n(p). \tag{63}$$

Let $\mathfrak{G} = \mathfrak{P}_+{}^\uparrow$ and assume $U_g$ to be a unitary strongly continuous representation of $\mathfrak{G}$ in a Hilbert space $\mathscr{H}$. Then by restriction of this representation to $\mathbb{R}^4$ (translation group) we obtain a unitary strongly continuous representation of the $\mathbb{R}^4$. It is denoted by $U_a$, $a \in \mathbb{R}^4$, for convenience. Then according to the SNAG-Theorem (see for example Neĭmark [1, p. 487]) there is a spectral measure $E(\cdot)$, defined on the Borel sets of $\mathbb{R}^4$ (where $\mathbb{R}^4$ is now considered as the dual space of the translation group of the $\mathbb{R}^4$; in this connection we denote the points of $\mathbb{R}^4$ by $p \in \mathbb{R}^4$, sometimes we write $p = \{p_0, \mathfrak{p}\}$, $\mathfrak{p} = \{p^1, p^2, p^3\}$) such that

$$U_a = \int\limits_{\mathbb{R}^4} e^{-i(a,p)} E(dp). \tag{64}$$

Note that this formula is a generalization of Stone's theorem (cf. Section 3.2.1). The *spectrum* of the representation $U_g$ is defined by supp $E(\cdot)$, we write spec $U :=$ supp $E(\cdot)$. Note that $0 \in$ supp $E$ because of $U_a\omega = \omega$.

According to (59), ker $J$ is given by the equation $W(f^*f) = 0$. The functional $W(f^*g)$ is constitutive for the Hilbert space $\mathscr{H}_W$. The representation (64), hence also the spectral measure $E(\cdot)$, is defined on $\mathscr{H}_W$ via (61) and (63). Therefore it is obvious that there are intimate connections between ker $J$ and supp $E$. For our purposes we need only partial insights which depend essentially on the following lemma.

Note that together with $J$, defined on $\mathscr{F}$, the operator $\hat{J}$, defined by $\hat{J}\hat{f} := Jf$ is given on $\hat{\mathscr{F}}$. Let $k \in \mathscr{S}(\mathbb{R}^4)$. We define the following operator in $\hat{\mathscr{F}}$:

$$\hat{\mathscr{F}} \ni \hat{f} \to k\hat{f} := \{k(0)\,\hat{f}_0,\ k(p_1)\,\hat{f}_1(p_1),\ k(p_1 + p_2)\,\hat{f}_2(p_1, p_2), \ldots\} \tag{65}$$

where $\hat{f} = \{\hat{f}_0, \hat{f}_1(p_1), \hat{f}_2(p_1, p_2), \ldots\}$. Now we have

**Lemma 49.** *Let* $k \in \mathscr{S}(\mathbb{R}^4)$ *and let* $J$ *be the identification operator of an abstract quantum field, where* $\mathscr{F}$, $\mathfrak{G}$, $V_g$ *are defined by the conditions* (I)—(III) *of this section. Then*

$$\hat{J}(k\hat{f}) = \int\limits_{\mathbb{R}^4} k(\lambda)\, E(d\lambda)\, \hat{J}\hat{f}, \qquad \hat{f} \in \hat{\mathscr{F}}. \tag{66}$$

**Proof.** We have

$$\int_{\mathbb{R}^4} \tilde{k}(\lambda)\, E(\mathrm{d}\lambda)\, \hat{J}\hat{f} = \int_{\mathbb{R}^4} (2\pi)^{-2} \left( \int_{\mathbb{R}^4} k(a)\, \mathrm{e}^{-\mathrm{i}(a,\lambda)}\, \mathrm{d}a \right) E(\mathrm{d}\lambda)\, \hat{J}\hat{f}$$

$$= (2\pi)^{-2} \int_{\mathbb{R}^4} \mathrm{d}a k(a) \int_{\mathbb{R}^4} \mathrm{e}^{-\mathrm{i}(a,\lambda)} E(\mathrm{d}\lambda)\, \hat{J}\hat{f} = (2\pi)^{-2} \int_{\mathbb{R}^4} \mathrm{d}a k(a)\, U_a \hat{J}\hat{f}$$

$$= (2\pi)^{-2} \int_{\mathbb{R}^4} k(a)\, \hat{J}(\hat{V}_a \hat{f})\, \mathrm{d}a = \hat{J}\left( (2\pi)^{-2} \int_{\mathbb{R}^4} k(a)\, \hat{V}_a \hat{f}\, \mathrm{d}a \right).$$

Further we obtain

$$\left( (2\pi)^{-2} \int_{\mathbb{R}^4} k(a)\, \hat{V}_a \hat{f}\, \mathrm{d}a \right)_n = (2\pi)^{-2} \int_{\mathbb{R}^4} k(a)\, \mathrm{e}^{-\mathrm{i}\left(a, \sum_{j=1}^{n} p_j\right)} \hat{f}_n(p)\, \mathrm{d}a$$

$$= \tilde{k}(p_1 + p_2 + \cdots + p_n)\, \hat{f}_n(p) = (\tilde{k}\hat{f})_n.$$

The equality of $\hat{J}(\int_{\mathbb{R}^4} k(a)\, \hat{V}_a \hat{f}\, \mathrm{d}a)$ and $\int_{\mathbb{R}^4} k(a)\, \hat{J}(\hat{V}_a \hat{f})\, \mathrm{d}a$ used above, is valid because the Riemannian sums of the integral $\int_{|a| \leq R} k(a)\, \hat{V}_a \hat{f}\, \mathrm{d}a$ converge to this integral with respect to the topology of $\mathscr{F}$ and because $J$ is continuous from $\mathscr{F}$ into $\mathscr{H}_W$. ■

Now from this lemma we derive a first connection between supp $E$ und ker $J$.

**Corollary 50.** *If* $\mathscr{F} \ni \hat{f} = \{0, \hat{f}_1(p_1), 0, 0, \ldots\}$, supp $\hat{f}_1$ *is compact and*

$$\operatorname{supp} \hat{f}_1 \cap \operatorname{supp} E = \emptyset, \tag{67}$$

*then* $\hat{J}\hat{f} = 0$ *follows.*

**Proof.** We choose a function $\tilde{k} \in \mathscr{S}(\mathbb{R}^4)$ such that $\tilde{k}(\lambda) = 1$ on supp $\hat{f}_1$ and $\tilde{k}(\lambda) = 0$ on supp $E$. Then we obtain $\tilde{k}\hat{f} = \hat{f}$ and from (66) immediately $\hat{J}\hat{f} = 0$ follows. ■

### 10.4.3. Quantum fields

In this section we augment the framework of Section 1 with additional assumptions which refer to the specifications of Section 2, that is, in the following $\mathscr{F}$ is the tensor algebra of test functions, $\mathfrak{G}$ is the Poincaré group and $V_g$ is the regular representation of $\mathfrak{G}$ in $\mathscr{F}$. These additional assumptions have physical motivation, but this will be not discussed here.

(I) *Uniqueness of the vacuum.* There exists exactly one linear independent vector $u \in \mathscr{H}$ with the property $U_g u = u$ for all $g \in \mathfrak{G}$.

We use the notation $\langle p, q \rangle := p_0 q_0 - \sum_{j=1}^{3} p_j q_j$ where $p, q \in \mathbb{R}^4$. Then the so-called *closed forward cone* in $\mathbb{R}^4$, clo $V_+$, is defined by

$$\operatorname{clo} V_+ := \{p \in \mathbb{R}^4;\ \langle p, p \rangle = p_0^2 - |\mathfrak{p}|^2 \geq 0,\ p_0 \geq 0\}.$$

The second condition is formulated with the help of this notion.

(II) *Spectrality.* spec $U = \operatorname{supp} E(\cdot) \subseteqq \operatorname{clo} V_+$.

For the formulation of the third condition we need the notion of space-like disjointness of elements of $\mathcal{F}$. Let $f, g \in \mathcal{F}$. First, we call supp $f_n$ and supp $g_n$ *space-like disjoint* if

$$\langle x_l - y_l, x_j - y_j \rangle < 0, \qquad l, j = 1, 2, \ldots, n,$$

for all $\{x_1, x_2, \ldots, x_n\} \in \operatorname{supp} f_n$, $\{y_1, y_2, \ldots, y_n\} \in \operatorname{supp} g_n$. Second, we call $f$ and $g$ *space-like disjoint* if supp $f_n$ and supp $g_n$ are space-like disjoint for all $n = 1, 2, 3, \ldots$ Now we can formulate the concept of locality.

(III) *Locality.* If $f \in \mathcal{F}$ and $g \in \mathcal{F}$ are space-like disjoint, then $A(f)\, A(g) = A(g)\, A(f)$.

Note that it is sufficient to require this condition for the special elements $f = \{0, f_1, 0, 0, \ldots\}$ and $g = \{0, g_1, 0, 0, \ldots\}$ only. The set $\mathcal{F}_0 := \operatorname{spa} \{f_0, f_1(x_1), f_2^{(1)}(x_1) f_2^{(2)}(x_2), \ldots\}$ is dense in $\mathcal{F}$ with respect to its topology, further $J\big((f \cdot g - g \cdot f)\, u\big) = \big(A(f)\, A(g) - A(g)\, A(f)\big)\, Ju$ is continuous for $\{f, g\} \in \mathcal{F} \times \mathcal{F}$. If the locality is true for the special elements mentioned above, then it is true for elements of $\mathcal{F}_0$. Thus it is true for arbitrary elements of $\mathcal{F}$.

For simplicity in the following an abstract quantum field (with $\mathcal{F}$, $\mathfrak{G}$, $V_g$ as in Section 2) is called a *quantum field* if it satisfies the conditions (I)—(III) of this section (uniqueness of the vacuum, spectrality; locality).

The condition of spectrality may be strengthened as follows (for the theory of unitary strongly continuous representations of the Poincaré group; in particular its irreducible representations, we refer to MACKEY [1]). First, the set of all $p = \{p_0, \mathfrak{p}\} \in \mathbb{R}^4$ with $p_0^2 - |\mathfrak{p}|^2 = m_0^2$, $m_0 \geqq 0$ a constant, is called a *mass hyperboloid* with mass $m_0$, it is denoted by $H_{m_0}$. Second, let us introduce the concept of the mass spectrum of a quantum field, more precisely the mass spectrum of the unitary representation $U$ connected with a quantum field. It is defined by

$$\operatorname{spec^m} U := \{m;\, m \geqq 0,\, m^2 = p_0^2 - |\mathfrak{p}|^2, \{p_0, \mathfrak{p}\} \in \operatorname{supp} E(\cdot)\}. \tag{68}$$

Because $\{0, 0\} \in \operatorname{supp} E(\cdot)$ we have always $0 \in \operatorname{spec^m} U$. Further note that if $\{p_0, \mathfrak{p}\} \in \operatorname{spec} U$, then also $H_m \subseteqq \operatorname{spec} U$, where $m^2 = p_0^2 - |\mathfrak{p}|^2$. Now the strengthening of the spectrality mentioned above can be formulated.

(II') *Upper and lower mass gap.* spec $U$ contains a topological discrete mass hyperboloid $H_{m_0}$, $H_{m_0} \subset \operatorname{spec} U$, $m_0 > 0$.

According to the definition of the mass spectrum, $\operatorname{spec^m} U$, condition (II') means that $\operatorname{spec^m} U$ has an upper and a lower gap, that is, $m_0$ is a topological discrete point of $\operatorname{spec^m} U$, in other words

$$\operatorname{spec^m} U = \{0\} \cup \{m_0\} \cup \Lambda_0, \qquad \Lambda_0 \subseteqq [m_0 + \varepsilon, \infty), \qquad \varepsilon > 0. \tag{69}$$

Now we state some implications of (II') and we formulate a further additional condition of a quantum field (the so-called coupling of the vacuum to the one-particle states).

**Proposition 51.** *Let $J$ be the identification operator of a quantum field and assume (II') to be satisfied. Let $\mathcal{F} \ni \mathring{f} = \{0, \mathring{f}_1(p_1), 0, \ldots\}$, supp $\mathring{f}_1$ compact, and supp $\mathring{f}_1 \cap$ supp $E \subseteqq H_{m_0}$. Then*

$$E(\{H_{m_0}\})\, J\mathring{f} = J\mathring{f}, \tag{70}$$

*that is, $\hat{J}\hat{f} \in E(\{H_{m_0}\})\mathscr{H}$, in other words, $\hat{J}\hat{f}$ belongs to that invariant subspace of the representation $U_g$ which is characterized by the mass $m_0$.*

**Proof.** We pick up a function $\hat{k} \in \mathscr{S}(\mathbb{R}^4)$ with the properties $\hat{k}(\lambda) = 1$ on supp $\hat{f}_1$ and $\hat{k}(\lambda) = 0$ on (supp $E$) $\setminus \{H_{m_0}\}$. Then, obviously $\hat{k}\hat{f} = \hat{f}$ and we obtain

$$\hat{J}\hat{f} = \int_{\mathbb{R}^4} \hat{k}(\lambda)\, E(\mathrm{d}\lambda)\, \hat{J}\hat{f} = \int_{\operatorname{supp}\hat{k} \cap \operatorname{supp} E} \hat{k}(\lambda)\, E(\mathrm{d}\lambda)\, \hat{J}\hat{f} = \int_{H_{m_0}} \hat{k}(\lambda)\, E(\mathrm{d}\lambda)\, \hat{J}\hat{f} = E(\{H_{m_0}\})\, \hat{J}\hat{f}. \blacksquare$$

Note that the conditions (I) and (III) are not necessary for the validity of Prop. 51. If $\hat{J}\hat{f}$ would be always zero for $\hat{f} = \{0, \hat{f}_1(p_1), 0, \ldots\}$ with supp $\hat{f}_1$ compact, supp $\hat{f}_1 \cap$ supp $E \subseteq H_{m_0}$, then (70) would be trivial. In this connection we have

**Lemma 52.** *Let the assumptions of Prop. 51 be satisfied. Assume $\hat{J}\hat{f} = 0$ for all $\{0, \hat{f}_1(p_1), 0, \ldots\} = \hat{f} \in \mathscr{F}$, supp $\hat{f}_1$ compact and supp $\hat{f}_1 \cap$ supp $E \subseteq H_{m_0}$. Then $E(\{H_{m_0}\})\, \hat{J}\hat{g} = 0$ for all $\hat{g} = \{0, \hat{g}_1(p_1), 0, \ldots\} \in \mathscr{F}$ without any assumption on supp $\hat{g}_1$.*

**Proof.** First let supp $\hat{g}_1$ be compact and supp $\hat{g}_1 \cap H_{m_0} = \emptyset$. Then we pick up a function $\hat{k} \in \mathscr{S}(\mathbb{R}^4)$ with $\hat{k}(\lambda) = 0$ on $H_{m_0}$ and $\hat{k}(\lambda) = 1$ on supp $\hat{g}_1$. Hence $\hat{g} = \hat{k}\hat{g}$ and $E(\{H_{m_0}\})\, \hat{J}\hat{g} = \int_{H_{m_0}} \hat{k}(\lambda)\, E(\mathrm{d}\lambda)\, \hat{J}\hat{g} = 0$. This means that such vectors $\hat{J}\hat{g}$ cannot contribute to the subspace $E(\{H_{m_0}\})\mathscr{H}$. Furthermore, if $\hat{g} = \hat{f}$ with supp $\hat{f}_1$ compact and supp $\hat{f}_1 \cap$ supp $E \subseteq H_{m_0}$, then, by assumption, $\hat{J}\hat{f} = 0$. Thus we obtain: $E(\{H_{m_0}\})\, \hat{J}\hat{g} = 0$ in both cases. But the span of the elements $\hat{g}$ with one of these two properties is dense in $\mathscr{S}(\mathbb{R}^4)$ with respect to its topology. Therefore the assertion follows. $\blacksquare$

(II') implies $E(\{H_{m_0}\})\mathscr{H} \supset \{0\}$. Because $\{Jf; f \in \mathscr{F}\}$ is dense in $\mathscr{H}$, in any event there is an $f \in \mathscr{F}$ with $E(\{H_{m_0}\})\, Jf \neq 0$. $f$ can be chosen in the form $f = \{0, 0, \ldots, 0, f_k, 0, \ldots\}$, where $k$ is minimal. The subspace $E(\{H_{m_0}\})\mathscr{H}$ is called the subspace of *one-particle states*. Then, because of $Jf = A(f)\,\omega$ and $E(\{H_{m_0}\})\, A(f)\,\omega \neq 0$, by $f$ the vacuum $\omega$ is "coupled" to the one-particle states. Now we can formulate the additional condition mentioned above.

(IV) *Coupling of the vacuum to the one-particle states.* There is an $f = \{0, f_1, 0, \ldots\}$ from $\mathscr{F}$ with $E(\{H_{m_0}\})\, Jf \neq 0$ (in other words, $k = 1$ is assumed).

An equivalent formulation of (IV) can be obtained as follows. First note that

$$W(f) = \sum_{\varrho=0}^{\infty} W_\varrho(f_\varrho),\ f = \{f_0, f_1, \ldots\} \in \mathscr{F},\ W_\varrho(f_\varrho) := W(\{0, 0, \ldots, f_\varrho, 0, \ldots\}). \tag{71}$$

For $W_2(\cdot)$ there is a representation, the so-called Källen-Lehmann representation (see Reed/Simon [2, p. 70]), which yields for $f = \{0, f_1, 0, \ldots\}$

$$W(f^* f) = \|\hat{J}\hat{f}\|^2 = c\, |\hat{f}_1(0)|^2 + \int_0^\infty \int_{H_m} |\hat{f}_1(p)|^2\, \mu_m(\mathrm{d}p)\, \varrho(\mathrm{d}m), \tag{72}$$

where $\mu_m(\cdot)$ denotes the Borel measure on $H_m$,

$$\mu_m(\Delta) := \int_{j(\Delta)} (m^2 + |\mathfrak{p}|^2)^{-1/2}\, \mathrm{d}\mathfrak{p}, \tag{73}$$

and $j(\cdot)$ denotes the homeomorphism of $H_m$ onto $\mathbb{R}^3$ given by

$$j(p) := j\big(((m^2 + |\mathfrak{p}|^2)^{1/2}, \mathfrak{p})\big) = \mathfrak{p} \tag{74}$$

Further $\varrho(\cdot)$ denotes a nonnegative Borel measure on $\mathbb{R}_+$ which is uniquely determined by $W_2$. Now we have

**Proposition 53.** *Let* (II') *be satisfied. Then* (IV) *is valid iff*

$$\{m_0\} \subseteqq \operatorname{supp} \varrho(\cdot) \subseteqq \operatorname{spec}^{\mathrm{m}} U. \tag{75}$$

Proof. The "if"-part is obvious (note Prop. 51). Thus we assume (IV) to be satisfied. We consider $f = \{0, f_1, 0, \ldots\}$ where $\operatorname{supp} f_1$ is compact and $\operatorname{supp} f_1 \cap \operatorname{supp} E = \emptyset$. Then from Corollary 50 and (72) we obtain

$$0 = \int_0^\infty \int_{H_m} |f_1(p)|^2 \mu_m(dp)\, \varrho(dm) = \int_{\operatorname{supp}\varrho} \int_{H_m} |f_1(p)|^2 \mu_m(dp)\, \varrho(dm) = \int_{\bigcup_{m \in \operatorname{supp}\varrho} H_m} |f_1(p)|^2 \mu_m(dp)\, \varrho(dm),$$

that is $f_1 \equiv 0$ on $\bigcup_{m \in \operatorname{supp}\varrho} H_m$ or, equivalently, $\bigcup_{m \in \operatorname{supp}\varrho} H_m \cap \operatorname{supp} f_1 = \emptyset$. In other words, we obtain $\bigcup_{m \in \operatorname{supp}\varrho} H_m \subseteqq \operatorname{supp} E$ or, equivalently, $\operatorname{supp} \varrho \subseteqq \operatorname{spec}^{\mathrm{m}} U$. Furthermore, if $\mathrm{m}_0 \notin \operatorname{supp} \varrho$, then $\mathfrak{J}f = 0$ for all $f$ such that $\operatorname{supp} f_1 \cap \operatorname{supp} E \subseteqq H_{m_0}$ (this follows from (72)). Applying now Lemma 52, we obtain a contradiction to (IV). ∎

### 10.4.4. Example: The free scalar field of mass $m_0 > 0$

In this section we present one of the simplest examples satisfying all assumptions of the previous sections. This example refers to a so-called free field. It is of principal interest for the following considerations. Let $\mathscr{F}$, $\mathfrak{G}$, $V_g$ as in Section 2. According to Theorem 48, in order to define a quantum field it is sufficient to define a functional $W(\cdot)$ on $\mathscr{F}$ with the properties (i)—(v) of Prop. 46. Then one obtains an abstract quantum field and the other conditions (spectrality, locality etc.) are to verify. Since (71) is valid and $\hat{W}_0(f_0) = f_0$ it is sufficient to define the functionals $\hat{W}_\varrho(\cdot)$, $\varrho = 1, 2, \ldots$ Let $m_0 > 0$. We put

$$\hat{W}_2(f_2(\cdot,\cdot)) := \int_{H_{m_0}} f_2(-q, q)\, \mu_{m_0}(dq), \tag{76}$$

$$\hat{W}_{2n}(f_1 f_2 \cdots f_{2n}) := \sum \hat{W}_2(f_{i_1} f_{j_1}) \cdots \hat{W}_2(f_{i_n} f_{j_n}) \tag{77}$$

where the sum has to be extended over all $(2n)!/2^n n!$ permutations of $\{i_1, \ldots, i_n, j_1, \ldots, j_n\}$ such that $i_1 < i_2 < \cdots < i_n$ and $i_1 < j_1, i_2 < j_2, \ldots, i_n < j_n$,

$$\hat{W}_\varrho(f_\varrho) := 0, \qquad \varrho \text{ odd}. \tag{78}$$

The proof that these definitions (76)—(78) define a quantum field is omitted (see Simon [2, § II.5] and further references there). Also (II') is true (with the topological discrete mass hyperboloid $H_{m_0}$). This quantum field is called the *free* (*scalar*) *field of mass* $m_0$.

For our purpose it is very useful to have a certain explicit formula for lker $W = \ker J_0$ and the identification operator $J_0$ itself in this case. Although the calculation is straightforward, it is a little complicated and is therefore omitted here. We restrict ourselves to describe the result (we refer the reader to BAUMGÄRTEL/NEIDHARDT/REHBERG [1983a] for details).

First, the symmetrization operator in $\mathscr{F}$ is denoted by $S$,

$$(S\hat{f})_n\,(p_1, p_2, \ldots, p_n) = (1/n!) \sum_{\pi} \hat{f}_n(p_{\pi(1)}, p_{\pi(2)}, \ldots, p_{\pi(n)})$$

where $\pi$ runs through all permutations of $\{1, 2, 3, \ldots, n\}$. By $\mathscr{F}_s$ we denote the submanifold of $\mathscr{F}$ consisting of all symmetric elements $\hat{f} = S\hat{f}$. Second, by $\hat{f} \mid H$ we denote the element $\hat{f}$; where all variables $p_j$ are restricted to $H_{m_0}$,

$$\hat{f} \mid H := \{\hat{f}_0, \hat{f}_1 \mid H_{m_0}, \hat{f}_2 \mid H_{m_0} \times H_{m_0}, \ldots\}.$$

Then the identification operator $J_0$ can be described by

$$J_0 f = S(T\hat{f}) \mid H \tag{79}$$

where $T$ is a continuous linear operator in $\mathscr{F}$ of the structure

$$T\hat{f} := \{\hat{W}(\hat{f}), \hat{W}^{(1)}(\hat{f}), \hat{W}^{(2)}(\hat{f}), \ldots\}.$$

In this formula $\hat{W}$ is the Wightman functional, whereas $\hat{W}^{(n)}: \mathscr{F} \mapsto \mathscr{S}(\mathbb{R}^{4n})$, $n = 1, 2, \ldots$, denote continuous linear maps, closely connected with $\hat{W}$.

An important fact is the simple behavior of $T$ on the symmetric submanifold $\mathscr{F}_s$. In this case it turns out that

$$\hat{W}^{(n)} g = \sum_{\varrho=0}^{\infty} \binom{n+\varrho}{\varrho} \hat{W}_{\varrho} g_{n+\varrho} = g_n + \cdots, \qquad n = 0, 1, 2, \ldots, g \in \mathscr{F}_s$$

where $(\hat{W}_{\varrho} g_{n+\varrho})\,(p_1; \ldots, p_n) = \hat{W}_{\varrho} g_{n+\varrho}(p_1, \ldots, p_n, \cdot, \ldots, \cdot)$ and $\hat{W}_{\varrho}$ operates on the last $\varrho$ variables.

Note that $T$ restricted to $\mathscr{F}_s$ is one-to-one and onto. Namely let $\hat{f} = \{\hat{f}_0, \hat{f}_1, \ldots, \hat{f}_N, 0, 0, \ldots\} \in \mathscr{F}_s$ and $T\hat{f} = \hat{g}$. The latter equation means explicitely

$$\begin{aligned}
\hat{W}_0(\hat{f}_0) + \hat{W}_1(\hat{f}_1) + \hat{W}_2(\hat{f}_2) + \cdots + \hat{W}_N(\hat{f}_N) &= \hat{g}_0, \\
\hat{W}_0(\hat{f}_1) + \binom{2}{1} \hat{W}_1(\hat{f}_2) + \cdots + \binom{N}{N-1} \hat{W}_{N-1}(\hat{f}_N) &= \hat{g}_1, \\
&\cdots \\
\hat{W}_0(\hat{f}_{N-1}) + \binom{N}{1} \hat{W}_1(\hat{f}_N) &= \hat{g}_{N-1}, \\
\hat{W}_0(\hat{f}_N) &= \hat{g}_N.
\end{aligned} \tag{80}$$

Hence $\hat{f} \in \mathscr{F}_s$ implies $\hat{g} \in \mathscr{F}_s$; further, if $\hat{g} \in \mathscr{F}_s$ is given, then we obtain $\hat{f}_N = \hat{g}_N$, $\hat{f}_{N-1} = \hat{g}_{N-1} - \binom{N}{1} \hat{W}_1(\hat{g}_N), \ldots$ where $\hat{f} \in \mathscr{F}_s$ is uniquely determined, in particular $T\hat{f} = 0$ for $\hat{f} \in \mathscr{F}_s$ implies $\hat{f} = 0$.

Note that in this case the Hilbert space $\mathscr{H}_W$ can be identified with the so-called Fock space

$$\mathbb{C} \times \mathscr{H}_0 \times S_2(\mathscr{H}_0 \otimes \mathscr{H}_0) \times S_3(\mathscr{H}_0 \otimes \mathscr{H}_0 \otimes \mathscr{H}_0) \times \cdots$$

where $S_n$ denotes the projection of the tensor product $\otimes \mathscr{H}_a$ ($n$ times) onto the symmetric tensor product. Further, $\mathscr{H}_0$ denotes the Hilbert space $L^2(H_{m_0}, \mu_{m_0})$ which is a representation space of the irreducible representation of the Poincaré group belonging to the mass $m_0 > 0$ and spin $s = 0$. $U_g$ is acting on this space by $(U_g f)(p) := e^{-i(a,p)} f(\Lambda^{-1}p)$, $g = \{\Lambda, a\}$. Note also

$$\operatorname{spec} U = \{0\} \cup H_{m_0} \cup \{p;\, p_0^2 - |\mathfrak{p}|^2 \geqq (2m_0)^2\}.$$

### 10.4.5. Identification operators between free and interacting fields

Let $J_0$ be the identification operator of the free scalar field of mass $m_0 > 0$ given by $W$ with the properties (76)—(78). Our next aim is to define a certain submanifold $\hat{\mathscr{F}}_{H_{m_0}}$ of $\hat{\mathscr{F}}$ such that this submanifold contains exactly one member of each equivalence class mod (ker $\hat{J}_0$) and such that ima $\hat{J}_0$ = ima $\hat{J}_0 \restriction \hat{\mathscr{F}}_{H_{m_0}}$.

For this purpose we pick up a fixed function $h \in C^\infty(\mathbb{R}^4)$ with supp $h \subseteqq \bigcup\limits_{m_0 - \varepsilon \leqq m \leqq m_0 + \varepsilon} H_m$, $\varepsilon$ fixed and sufficiently small, and assume $h$ to be constant on $H_m$, that is, $h(p) = h_0(m)$, $h_0 \in \mathscr{S}(\mathbb{R}^1)$, supp $h_0 \subseteqq [m_0 - \varepsilon, m_0 + \varepsilon]$. Further let $h_0(m_0) = 1$.

Now we define $\hat{\mathscr{F}}_{H_{m_0}}$ to be the set of all symmetric elements $g \in \hat{\mathscr{F}}_s$ such that

$$g_n(p_1, p_2, \ldots; p_n) = h(p_1)\, h(p_2) \cdots h(p_n)\, \varphi_n(\mathfrak{p}_1, \mathfrak{p}_2, \ldots, \mathfrak{p}_n) \tag{81}$$

where $\varphi_n \in \mathscr{S}(\mathbb{R}^{3n})$ and $\varphi_n$ is symmetric in $\mathfrak{p}_1, \mathfrak{p}_2, \ldots, \mathfrak{p}_n$. Obviously, $\hat{\mathscr{F}}_{H_{m_0}}$ is a (linear) submanifold of $\hat{\mathscr{F}}$. If $\hat{f} \in \hat{\mathscr{F}}_{H_{m_0}}$ then supp $(\hat{f})_n \subseteqq H_\varepsilon \times H_\varepsilon \times \cdots \times H_\varepsilon$ ($n$ times), where $H_\varepsilon$ denotes the union of all mass hyperboloids with $m \in [m_0 - \varepsilon, m_0 + \varepsilon]$. Further, $T$ is one-to-one on $\hat{\mathscr{F}}_{H_{m_0}}$ and maps $\hat{\mathscr{F}}_{H_{m_0}}$ onto $\hat{\mathscr{F}}_{H_{m_0}}$, as it can be easily seen from (80). Now we prove

**Proposition 54.** *The manifold $\hat{\mathscr{F}}_{H_{m_0}}$ contains precisely one member of each equivalence class* mod (ker $J_0$). *That is,*

$$\text{ima } J_0 \restriction \hat{\mathscr{F}}_{H_{m_0}} = \text{ima } J_0, \tag{82}$$

*and $\hat{f} \in \hat{\mathscr{F}}_{H_{m_0}}$ and $\hat{J}_0\hat{f} = 0$ imply $\hat{f} = 0$.*

Proof. Let $\hat{f} \in \hat{\mathscr{F}}_{H_{m_0}}$, $\hat{f}_n(p_1, p_2, \ldots; p_n) = \prod\limits_{j=1}^{n} h(p_j)\, \gamma_n(\mathfrak{p}_1; \mathfrak{p}_2, \ldots, \mathfrak{p}_n)$. We put

$$(T\hat{f})_n\,(p_1, p_2, \ldots, p_n) = \prod_{j=1}^{n} h(p_j)\, \gamma_n'(\mathfrak{p}_1, \mathfrak{p}_2, \ldots, \mathfrak{p}_n)$$

and obtain $\gamma_n'(\mathfrak{p}_1, \mathfrak{p}_2, \ldots, \mathfrak{p}_n) = (T\hat{f})_n \restriction H_{m_0} \times \cdots \times H_{m_0}$. Now let $\hat{J}\hat{f} = 0$, then we get from (79) that $S(T\hat{f}) \restriction H = T\hat{f} \restriction H = 0$, hence $\gamma_n'(\mathfrak{p}_1, \mathfrak{p}_2, \ldots, \mathfrak{p}_n) \equiv 0$ which implies $T\hat{f} = 0$ and $\hat{f} = 0$. That is, the manifold $\hat{\mathscr{F}}_{H_{m_0}}$ contains at most one member from each equivalence classe mod (ker $J_0$). Now we prove (82). Let $\hat{f} \in \hat{\mathscr{F}}$. Then $\hat{J}_0\hat{f} = S(T\hat{f}) \restriction H$. Put $T\hat{f} = \hat{g}$. $S\hat{g}$ is symmetric, $S\hat{g} =: \hat{g}_s$, and we have $\hat{J}_0\hat{f} = \hat{g}_s \restriction H$. Now $(\hat{g}_s)_n\,(p_1, p_2; \ldots, p_n)$

defines after eliminating of all variables $p_j^{(0)}$ via $p_j^{(0)} = (m_0^2 + |\mathfrak{p}_j|^2)^{1/2}$ a certain symmetric function $(\varphi_s)_n(\mathfrak{p}_1; \mathfrak{p}_2, \ldots, \mathfrak{p}_n)$ from $\mathscr{S}(\mathbb{R}^{3n})$. Then the element $\mathfrak{k}_s$ with the coordinates

$$\prod_{j=1}^{n} h(p_j)\,(\varphi_s)_n(\mathfrak{p}_1, \mathfrak{p}_2, \ldots, \mathfrak{p}_n)$$

belongs to $\mathscr{F}_{Hm_0}$ and we have $\mathfrak{k}_s \restriction H = \mathfrak{g}_s \restriction H$. Let $\mathfrak{k}_s = T\hat{\mathfrak{l}}_s$ with $\hat{\mathfrak{l}}_s \in \mathscr{F}_{Hm_0}$. Then $\mathfrak{J}_0\mathfrak{f} = \mathfrak{k}_s \restriction H = T\hat{\mathfrak{l}}_s \restriction H$, that is $\mathfrak{J}_0\mathfrak{f} \in \operatorname{ima} \mathfrak{J}_0 \restriction \mathscr{F}_{Hm_0}$. ■

**Remark 55.** Note that $\mathscr{F}_{Hm_0}$ is not invariant under $\mathfrak{V}_a$. Let $\mathfrak{f} \in \mathscr{F}_{Hm_0}$, $\mathfrak{f} = \{f_0, h(p_1) \times \varphi_1(\mathfrak{p}_1), h(p_1)\,h(p_2)\,\varphi_2(\mathfrak{p}_1, \mathfrak{p}_2), \ldots\} = \{f_0, f_1, f_2, \ldots\}$, then

$$\mathfrak{V}_a\mathfrak{f} = \{f_0, e^{-i(a,(p_1^{(0)},\mathfrak{p}_1))}f_1(p_1), e^{-i(a,(p_1^{(0)}+p_2^{(0)},\mathfrak{p}_1+\mathfrak{p}_2))}f_2(p_1, p_2), \ldots\}. \tag{83}$$

This element does not belong to $\mathscr{F}_{Hm_0}$, in general, because, for example, $e^{-ia_0p_1^{(0)}}$ cannot be expressed as a function of $\mathfrak{p}_1$ only. But, on the other hand, because of Prop. 54, there is precisely one element in $\mathscr{F}_{Hm_0}$, which is a member of the equivalence class mod ($\ker J_0$), defined by the element $\mathfrak{V}_a\mathfrak{f}$; $\mathfrak{f} \in \mathscr{F}_{Hm_0}$. This element is simply

$$\mathfrak{V}_a^{(m_0)}\mathfrak{f} = \{f_0, e^{-i(a,(\mu(\mathfrak{p}_1),\mathfrak{p}_1))}f_1(p_1), e^{-i(a,(\mu(\mathfrak{p}_1)+\mu(\mathfrak{p}_2),\mathfrak{p}_1+\mathfrak{p}_2))}f_2(p_1, p_2), \ldots\} \tag{84}$$

where $\mu(\mathfrak{p}) = (m_0^2 + |\mathfrak{p}|^2)^{1/2}$.

Obviously, $\operatorname{ima} \mathfrak{J}_0 \restriction \mathscr{F}_{Hm_0}$ is invariant under $U_a$ because it coincides with $\operatorname{ima} \mathfrak{J}_0$ which is invariant.

Now we proceed to one simple but important implication of Prop. 54. It states that an identification operator $J$ of an arbitrary quantum field; restricted to $\mathscr{F}_{Hm_0}$ (which is defined with respect to a free field with mass $m_0 > 0$), can be factored in the form $J = KJ_0$, where $K$ is an identification operator between the Hilbert spaces of the two given quantum fields.

**Proposition 56.** *Let $J_0$ be the identification operator of a free quantum field of mass $m_0 > 0$. Further, let $J$ be the identification operator of an arbitrary quantum field. Then there is an operator $K$ from $\mathscr{H}_0$ into $\mathscr{H}$ (where $\mathscr{H}_0$, $\mathscr{H}$ are the image spaces of $J_0$, $J$, respectively) with dom $K$ = ima $J_0$, i.e. $K$ is densely defined, such that*

$$\mathfrak{J}\mathfrak{f} = K\mathfrak{J}_0\mathfrak{f}\,, \qquad \mathfrak{f} \in \mathscr{F}_{Hm_0}. \tag{85}$$

*$K$ is uniquely determined by* (85).

**Proof.** We define $K$ by $K(\mathfrak{J}_0\mathfrak{f}) := \mathfrak{J}\mathfrak{f}$, $\mathfrak{f} \in \mathscr{F}_{Hm_0}$, but we have to show that this is a unique definition. For this reason let $\mathfrak{f}' \in \mathscr{F}_{Hm_0}$, $\mathfrak{J}_0\mathfrak{f} = \mathfrak{J}_0\mathfrak{f}'$ that is, $\mathfrak{J}_0(\mathfrak{f} - \mathfrak{f}') = 0$. Now $\mathfrak{f} - \mathfrak{f}' \in \mathscr{F}_{Hm_0}$ because $\mathscr{F}_{Hm_0}$ is linear. Therefore, according to Prop. 54, $\mathfrak{J}_0(\mathfrak{f} - \mathfrak{f}') = 0$ implies $\mathfrak{f} - \mathfrak{f}' = 0$, hence $\mathfrak{J}\mathfrak{f} = \mathfrak{J}\mathfrak{f}'$. ■

**Remark 57.** Despite of the kind of the definition of $K$ this operator is by no means an intertwining operator of $U_a$, $U_a^{(0)}$ where $U_a$, $U_a^{(0)}$ denote the unitary representations of $\mathbb{R}^4$ belonging to the fields $J$, $J_0$, respectively. This results simply from Remark 55. Namely, on the one hand we have

$$KU_a^{(0)}u = KU_a^{(0)}\mathfrak{J}_0\mathfrak{f} = K(\mathfrak{J}_0(\mathfrak{V}_a\mathfrak{f}) = \mathfrak{J}(\mathfrak{V}_a^{(m_0)}\mathfrak{f})$$

where $u = \hat{J}_0 \hat{f}$ and $\hat{f} \in \mathscr{F}_{H_{m_0}}$. On the other hand we have

$$U_a K u = U_a K \hat{J}_0 \hat{f} = U_a \hat{J} \hat{f} = \hat{J}(\hat{V}_a \hat{f}).$$

But in general $\hat{J}(\hat{V}_a \hat{f}) \neq \hat{J}(\hat{V}_a^{(m_0)} \hat{f})$.

### 10.4.6. Wave operators for quantum fields

By means of the identification operator $K: \mathscr{H}_0 \to \mathscr{H}$ of Prop. 56 we are able to define wave operators with respect to two given fields defined by $J$, $J_0$, where $J_0$ is the identification operator of the free field of mass $m_0 > 0$. Let $\mathscr{D}_0 = \operatorname{ima} J_0$. We put

$$W_+ f_0 := \operatorname*{Lim}_{t \to \infty} e^{itH} K \, e^{-itH_0} f_0, \qquad f_0 \in \mathscr{D}_0, \tag{86}$$

where $e^{-itH} = U_t$ and $e^{-itH_0} = U_t^{(0)}$ denote the time evolutions of the representations $U_a$, $U_a^{(0)}$ belonging to the fields defined by $J$, $J_0$, respectively. Note that $\mathscr{D}_0$ is invariant under $U_a^{(0)}$ and that $\mathscr{D}_0 = \operatorname{dom} K$ according to Prop. 56. Similarly $W_-$ can be defined.

The wave operators (86) can be written in another form which is convenient for some purposes.

**Proposition 58.** *Let $\hat{f} \in \mathscr{F}_{H_{m_0}}$. Then*

$$e^{itH} K \, e^{-itH_0} \hat{J}_0 \hat{f} = \hat{J}(\hat{V}_{-t} \hat{V}_t^{(m_0)} \hat{f}). \tag{87}$$

Proof. We have

$$\hat{J}(\hat{V}_{-t} \hat{V}_t^{(m_0)} \hat{f}) = e^{itH} \hat{J}(\hat{V}_t^{(m_0)} \hat{f}).$$

Because $\hat{f} \in \mathscr{F}_{H_{m_0}}$, $\hat{V}_t^{(m_0)} \hat{f} \in \mathscr{F}_{H_{m_0}}$ is valid. Thus, we obtain by Prop. 56

$$\hat{J}(\hat{V}_t^{(m_0)} \hat{f}) = K\big(\hat{J}_0(\hat{V}_t^{(m_0)} \hat{f})\big).$$

Since $\hat{J}_0(\hat{V}_t \hat{f}) = \hat{J}_0(\hat{V}_t^{(m_0)} \hat{f})$, we obtain

$$\hat{J}(\hat{V}_{-t} \hat{V}_t^{(m_0)} \hat{f}) = e^{itH} K \hat{J}_0(\hat{V}_t \hat{f}) = e^{itH} K \, e^{-itH_0} \hat{J}_0 \hat{f}. \ \blacksquare$$

If a quantum field defined by $J$ is given, then the question arises under what circumstances there is a corresponding free field of a certain mass, defined by $J_0$, such that the wave operators exist. There is a famous result which states the existence and isometry of these wave operators under very general circumstances. Now we formulate this theorem.

**Theorem 59.** *Let $J$ be the identification operator of a quantum field with properties* (II′) *and* (IV). *The special mass of condition* (II′) *is denoted by $m_0$. Further assume the restriction of the representation $U_g$ to $E(\{H_{m_0}\}) \mathscr{H}$ to be irreducible and of spin $s = 0$ (thus this restriction is the irreducible representation characterized by $m_0$, $s = 0$). Further, let $J_0$ be the identification operator of the free scalar field of mass $m_0$. Finally, let the operator $K$ be defined by* (85). *Then the wave operators $W_\pm$ defined by* (86), *exist on $\mathscr{D}_0$ and they are isometric.*

The proof of this deep theorem goes far beyond our framework and cannot be given here. It can be found in many books (see for example Jost [1], Bogoljubov/Logunov/Todorov [1] and Reed/Simon [3]).

# 11. Structural Properties of Wave and Scattering Operators

## 11.1. Structure of wave operators

According to Prop. 9.10 (see also Remark 9.19) a wave operator is necessarily also an intertwining operator. But the converse is not true in general if $J$ is a fixed identification operator. In this paragraph, which can be considered as a continuation of Paragraphs 9.3 and 9.5, we study some structural properties of wave operators. As in Paragraph 9.2 and the following paragraphs we choose a fixed universal projection $P_H$. Moreover, the identification operator $J$ is fixed in this paragraph.

### 11.1.1. Structure of wave operators for arbitrary universal projections

Let $H, H_0$ be selfadjoint operators in $\mathscr{H}$, $\mathscr{H}_0$, respectively, with spectral measures $E(\cdot)$, $E_0(\cdot)$. We consider the class of all intertwining operators $U$ between $H \restriction P_H\mathscr{H}$ and $H_0 \restriction P_{H_0}\mathscr{H}_0$, i.e. the class of all $U \in \mathscr{L}(\mathscr{H}_0, \mathscr{H})$ with the properties

$$U = UP_{H_0}, \qquad P_H U = U, \tag{1}$$

$$UE_0(\Delta) = E(\Delta)\, U \qquad \text{for each Borel set } \Delta \subseteq \mathbb{R}. \tag{2}$$

(1) may be expressed equivalently by $\operatorname{supp} U \subseteq P_{H_0}\mathscr{H}_0$, $\operatorname{ima} U \subseteq P_H\mathscr{H}$. Then it is easy to characterize the wave operator $U = W_+(H, H_0; J)$, $J \in \mathscr{L}(\mathscr{H}_0, \mathscr{H})$ as a special member of this class, if it exists.

**Proposition 1.** *Assume $U$ to be an operator with the properties* (1), (2). *Then the wave operator $W_+(H, H_0; J)$ exists and equals $U$ iff*

$$U = J + C, \qquad \operatorname*{Lim}_{t\to\infty} C\, e^{-itH_0}P_{H_0} = 0 \tag{3}$$

*is valid.*

**Proof.** Let $W_+$ exist. Then by Prop. 9.10 and Theorem 9.15 we obtain (1) and (2). Furthermore, by (9.15), relation (3) follows. Conversely, assume $U$ to be an operator with properties (1), (2) and let (3) be satisfied. Then we obtain

$$\begin{aligned} e^{itH}J\, e^{-itH_0}P_{H_0} &= e^{itH}U\, e^{-itH_0}P_{H_0} - e^{itH}C\, e^{-itH_0}P_{H_0} \\ &= UP_{H_0} - e^{itH}C\, e^{-itH_0}P_{H_0} = U - e^{itH}C\, e^{-itH_0}P_{H_0}. \end{aligned}$$

Hence $\operatorname*{Lim}_{t\to\infty} e^{itH}J\, e^{-itH_0}P_{H_0}$ exists and equals $U$. ∎

A further property of wave operators distinguishing them from mere intertwining operators is the following.

**Proposition 2.** *Let $P$, $P_0$ be projections with $P \in \operatorname{com} H$ and $P_0 \in \operatorname{com} H_0$. Assume $J$ to be intertwining with respect to $P$; $P_0$; $PJ = JP_0$. Then $W_+(H, H_0; J)$ is also intertwining with respect to $P$; $P_0$; $PW_+ = W_+P_0$ (if it exists).*

Proof. It immediately follows from the definition of $W_+$. ■

In the case $\mathscr{H} = \mathscr{H}_0$, $J = 1$ from Prop. 2 we obtain that $P \in \operatorname{com} H \cap \operatorname{com} H_0$ always implies $P \in \operatorname{com} W_+$. The next example shows that this is not true for arbitrary intertwining operators in general.

**Example 3.** Let $H$ be selfadjoint acting on $\mathscr{H}$ and let $P \in \operatorname{com} H$, where $\dim P\mathscr{H} = \dim P^\perp\mathscr{H} = \infty$. Pick up a partial isometry $V$ with $V^*V = P$, $VV^* = P^\perp$ and define $U = V + V^*$. $U$ is unitary and

$$UP = P^\perp U, \qquad P^\perp U^* = U^*P \tag{4}$$

are valid. Define $H'$ by $H' := U^*HU$. Then $UE_{H'}(\Delta) = E_H(\Delta)\, U$, i.e. $U$ is intertwining and $P$ is reducing for $H'$ because of

$$PE_{H'}(\Delta) = PU^*E_H(\Delta)U = U^*P^\perp E_H(\Delta)\, U = U^*E_H(\Delta)\, P^\perp U = U^*E_H(\Delta)\, UP = E_{H'}(\Delta)\, P,$$

but $P \notin \operatorname{com} U$ (cf. (4)).

### 11.1.2. The special cases $P_H = P_H^c$ and $P_H = P_H^{ac}$

Prop. 1. essentially reduces the study of structural properties of wave operators to the investigation of the structure of the corresponding generalized wave ideal $\mathscr{L}_{H_0,P_{H_0}}(\mathscr{H}_0, \mathscr{H})_a$ (or $\mathscr{L}_{H_0,P_{H_0}}(\mathscr{H}_0, \mathscr{H})_s$); note $\mathscr{L}_{H_0,P_{H_0}^{ac}}(\mathscr{H}_0, \mathscr{H}_0) = \ker \mu^{H_0}$ (cf. Definition 6.69 and Prop. 6.70). Concerning structural properties of $\mathscr{L}_{H_0,P_{H_0}}(\mathscr{H}_0; \mathscr{H})$ for special universal projections, see Sections 6.4.2 and 6.5.2. For example, recall that if $P_H = P_H^c$, then $\mathscr{L}_{H_0,P_{H_0}}(\mathscr{H}_0, \mathscr{H})_a$ is the closure of the set of all compact (even finite-dimensional) operators with respect to the (correspondingly defined) $\mu_a^{H_0}$-topology (cf. Definition 6.55). If $P_H = P_H^{ac}$, then the result just cited remains true, moreover in this case $\mathscr{L}_{H_0,P_{H_0}}(\mathscr{H}_0, \mathscr{H})_s$ is the closure of the set of all compact operators with respect to the (correspondingly defined) $\mu_s^{H_0}$-topology (cf. Definition 6.55). Therefore we obtain

**Theorem 4.** *Let $P_H = P_H^c$. Assume $U$ be an operator with properties* (1), (2). *Then the wave operator $W_+ = |\mathrm{A}|\text{-}\lim_{t\to\infty} e^{itH} J\, e^{-itH_0} P_{H_0}^c$ exists and equals $U$ iff*

$$U = J + C$$

*where $C$ is the limit of a sequence $K_n$ of compact operators with respect to the (locally convex) topology defined by the class*

$$p_a(X; H_0, f) := \sup_{\varepsilon>0} \varepsilon \int_0^\infty e^{-\varepsilon t} \|X\, e^{-itH_0} P_{H_0}^c f\|\, dt + \|X(P_{H_0}^c)^\perp f\| \tag{5}$$

*of seminorms on $\mathscr{L}(\mathscr{H}_0, \mathscr{H}) \ni X$.*

Proof. It is obvious because of Prop. 1 and Section 6.5.2. ■

Note that if $U = J + C$, where $C$ is compact, is intertwining with respect to $H, H_0$ and if (1) is satisfied, then $J + C$ is the (existing) wave operator $W_+(H, H_0; J)$. Correspondingly, we obtain

**Theorem 5.** *Let $P_H = P_H^{ac}$. Assume $U$ to be an operator with properties* (1), (2). *Then the wave operator $W_+ = \text{s-}\lim_{t\to\infty} e^{itH} J\, e^{-itH_0} P_{H_0}^{ac}$ exists and equals $U$ iff*

$$U = J + C$$

*where $C$ is the limit of a sequence of compact operators with respect to the (locally convex) topology defined by the class*

$$p_s(X; H_0, f) := \sup_{t\geq 0} \|X\, e^{-itH_0} P_{H_0}^{ac} f\| + \|X(P_{H_0}^{ac})^{\perp} f\| \tag{6}$$

*of seminorms on $\mathscr{L}(\mathscr{H}_0, \mathscr{H}) \ni X$.*

Proof. It is obvious because of Prop. 1, Theorem 6.59 and Section 6.5.2. ■

### 11.1.3. Structure of $H_0$-semicomplete partially isometric wave operators

In this section we restrict the class of intertwining operators between $H \restriction P_H\mathscr{H}$ and $H_0 \restriction P_{H_0}\mathscr{H}_0$ to the special class of intertwining operators with the additional property

$$U^*U = P_{H_0}, \tag{7}$$

that is, the intertwining operators are assumed to be partially isometric with initial projection $P_{H_0}$.

**Proposition 6.** *Let $U \in \mathscr{L}(\mathscr{H}_0, \mathscr{H})$ and assume that conditions* (2) *and* (7) *are satisfied. Then the wave operator $W_+(H, H_0; J)$ exists and equals $U$ iff*

$$U^*J = P_{H_0} + C, \qquad C \in \mathscr{L}_{H_0,P_{H_0}}(\mathscr{H}_0, \mathscr{H}_0) \tag{8}$$

*and*

$$EJ - J \in \mathscr{L}_{H_0,P_{H_0}}(\mathscr{H}_0, \mathscr{H}) \tag{9}$$

*are valid, where $E = UU^*$. Obviously, in this case the wave operator is partially isometric and $H_0$-semicomplete.*

Proof. First, let the wave operator $W_+$ exist and let $W_+ = U$. Then, according to Prop. 1, we have $U = J + C'$, where $\operatorname{Lim}_{t\to\infty} C'\, e^{-itH_0} P_{H_0} = 0$. Therefore we obtain

$$U^*U = P_{H_0} = U^*J + U^*C', \tag{10}$$

that is $U^*J = P_{H_0} - U^*C'$ and $U^*C' \in \mathscr{L}_{H_0,P_{H_0}}(\mathscr{H}_0, \mathscr{H}_0)$. Thus (8) is proved. Moreover, from (10) we obtain $U = EJ + EC'$. By $U = J + C'$ it follows $J \underset{H_0}{\approx} EJ$, that is (9) is true. Conversely, let (8) and (9) be satisfied. From (8) we obtain $U = EJ - UC$. Hence, because of (9),

$$U = J + C', \qquad C' \in \mathscr{L}_{H_0,P_{H_0}}(\mathscr{H}_0, \mathscr{H})$$

follows. Therefore, by Prop. 1, the wave operator exists and equals $U$. ■

Note that condition (9) of Prop. 6 can be replaced by a property of $J$ only, namely its partial isometry with an initial projection larger than $P_{H_0}$. More precisely, we have

**Lemma 7.** *Let $J$ be partially isometric with $J^*J \geqq P_{H_0}$ and assume that* (8) *is satisfied. Then* (9) *is valid.*

Proof. We have

$$\|P_{H_0}u\|^2 = \|J\, e^{-itH_0}P_{H_0}u\|^2$$

$$= \|EJ\, e^{-itH_0}P_{H_0}u\|^2 + \|E^{\perp}J\, e^{-itH_0}P_{H_0}u\|^2$$

$$= \|U^*J\, e^{-itH_0}P_{H_0}\|^2 + \|E^{\perp}J\, e^{-itH_0}P_{H_0}u\|^2.$$

According to (8) the first term on the right-hand side tends to $\|P_{H_0}u\|^2$ with respect to $\operatorname*{Lim}_{t\to\infty}$. Hence the second term tends to 0, that is the assertion is proved. ∎

### 11.1.4. Structure of complete partially isometric wave operators and wave operator equivalence

Now we restrict the class of intertwining operators between $H \mid P_H\mathscr{H}$ and $H_0 \mid P_{H_0}\mathscr{H}_0$ to the special class with the additional property (7) and

$$UU^* = P_H. \tag{11}$$

That is, $H \mid P_H\mathscr{H}$ and $H_0 \mid P_{H_0}\mathscr{H}_0$ are unitarily equivalent iff this class is non-empty.

**Proposition 8.** *Let $U \in \mathscr{L}(\mathscr{H}_0, \mathscr{H})$ and assume that conditions* (2) (7) *and* (11) *are satisfied. Then the wave operator $W_+(H, H_0; J)$ exists and equals $U$ iff* (8) *and $P_HJ \underset{H_0}{\approx} J$ are valid. In this case the wave operator is partially isometric and complete.*

Proof. It immediately follows from Prop. 6. ∎

Now we turn to the definition of an equivalence relation between selfadjoint operators using the concept of a complete wave operator. According to the preceding results, it turns out that this equivalence relation is a strengthening of the unitary equivalence between the parts $H \mid P_H\mathscr{H}$ and $H_0 \mid P_{H_0}\mathscr{H}_0$. As already stated above the latter equivalence means the existence of an operator $U \in \mathscr{L}(\mathscr{H}_0, \mathscr{H})$ with the properties (2), (7) (11). We denote this equivalence by $H_0 \sim H$.

Recall that the identification operator $J$ is fixed in this paragraph. It depends on $\mathscr{H}_0$ and $\mathscr{H}$, that is, $J = J_{\mathscr{H}.\mathscr{H}_0}$. We assume the following properties to be satisfied:

$$J^*_{\mathscr{H}.\mathscr{H}_0}J_{\mathscr{H}.\mathscr{H}_0} = 1_{\mathscr{H}_0}, \quad J_{\mathscr{H}.\mathscr{H}_0}J^*_{\mathscr{H}.\mathscr{H}_0} = 1_{\mathscr{H}},$$

$$J^*_{\mathscr{H}.\mathscr{H}_0} = J_{\mathscr{H}_0.\mathscr{H}}, \qquad J_{\mathscr{H}_1.\mathscr{H}_2}J_{\mathscr{H}_2.\mathscr{H}_1} = J_{\mathscr{H}_2.\mathscr{H}_1}.$$

For example fix a Hilbert space $\mathscr{K}$ and put $J_{\mathscr{H}.\mathscr{K}} = V_{\mathscr{H}}$, $J_{\mathscr{K}.\mathscr{H}} = V_{\mathscr{H}}{}^*$ with $V_{\mathscr{H}}{}^*V_{\mathscr{H}} = 1_{\mathscr{K}}$, $V_{\mathscr{H}}V^*_{\mathscr{H}} = 1_{\mathscr{H}}$ and $V_{\mathscr{K}} = 1_{\mathscr{K}}$. Then the operator $J_{\mathscr{H}_2.\mathscr{H}_1} = V_{\mathscr{H}_2}V^*_{\mathscr{H}_1}$ $\in \mathscr{L}(\mathscr{H}_1, \mathscr{H}_2)$ satisfies these conditions. Note that the following definition depends on the fact that $J$ is fixed.

**Definition 9.** The selfadjoint operators $H_0$, $H$ are called *wave operator equivalent*, if $W_+(H, H_0; J_{\mathscr{H},\mathscr{H}_0})$ exists and is complete. This relation is denoted by $H_0 \approx H$.

The wave operator equivalence $H_0 \approx H$ is an equivalence relation which is stronger than the equivalence $H_0 \sim H$. This is the content of

**Proposition 10.** *The relation $H_0 \approx H$ is an equivalence relation, i.e. $H_0 \approx H_0$; if $H_0 \approx H$, then $H \approx H_0$; if $H_1 \approx H_2$ and $H_2 \approx H_3$, then $H_1 \approx H_3$. If $H_0 \approx H$, then $H_0 \sim H$.*

**Proof.** The first part is obvious because of Prop. 9.48, Theorem 9.21 (chain rule) and the properties of $J_{\mathscr{H},\mathscr{H}_0}$. The second part is obvious because $W_+(H, H_0; J_{\mathscr{H},\mathscr{H}_0})$ satisfies conditions (2),(7) and (11). ■

The connection between $H_0 \sim H$ and $H_0 \approx H$ is given by Prop. 8. Note that according to Lemma 7 and Definition 9 in this case the condition (9) can be omitted. Therefore we have

**Proposition 11.** *Let $H_0 \sim H$. Then $H_0 \approx H$ iff there is an operator $U \in \mathscr{L}(\mathscr{H}_0, \mathscr{H})$ with the properties* (2), (7), (11) *such that*

$$U^*J = P_{H_0} + C, \qquad C \in \mathscr{L}_{H,P_{H_0}}(\mathscr{H}_0, \mathscr{H}_0), \tag{12}$$

*is valid.*

**Proof.** Obvious because of Prop. 8. ■

Note that the operator $U$ in Prop. 11 is nothing else than the wave operator $W_+(H; H_0; J)$. Note further that the condition (12) of Prop. 11, which is formally not symmetric between $H_0$ and $H$, actually is symmetric. Namely from (12) we obtain by left multiplication with $U$

$$P_HJ = U + UC. \tag{13}$$

Hence $P_HJ \underset{H_0}{\approx} U$ and also $J \underset{H_0}{\approx} U$. By left multiplication with $J^*$ we obtain $1_{\mathscr{H}_0} \underset{H_0}{\approx} J^*U$ or $J^*U - P_{H_0} \in \mathscr{L}_{H_0,P_{H_0}}(\mathscr{H}_0, \mathscr{H}_0)$; that is

$$C^* \in \mathscr{L}_{H_0P_{H_0}}(\mathscr{H}_0, \mathscr{H}_0). \tag{14}$$

Further from (13) we obtain $J = U + UC + P_H{}^{\perp}J$ or $J^* = U^* + C^*U^* + J^*P_H{}^{\perp}$. Hence $UJ^* = P_H + UC^*U^* + UJ^*P_H{}^{\perp}$.

But from (14) we obtain $C^*U^* \underset{H}{\approx} 0$, hence with $D = UC^*U^* + UJ^*P_H{}^{\perp}$,

$$(U^*)^* J^* = P_H + D; \qquad D \in \mathscr{L}_{H,P_H}(\mathscr{H}, \mathscr{H}) \tag{15}$$

follows. (15) is of type (12), where the roles of $H$ and $H_0$ are interchanged.

Finally, we mention briefly the case $\mathscr{H} = \mathscr{H}_0$, $J = 1$. In this case we obtain the following result from the preceeding considerations: Let $H_0 \sim H$; that is let the class of operators with the properties

$$UE_0(\Delta) = E(\Delta)\, U, \qquad U^*U = P_{H_0}, \qquad UU^* = P_H$$

be non-empty. Then $H_0 \approx H$ iff there is an operator $U$ from this class such that

$$U^* - P_{H_0} \in \mathscr{L}_{H_0, P_{H_0}}(\mathscr{H}_0, \mathscr{H}_0) \tag{16}$$

is valid. In this case from (16) immediately (note (14))

$$U - P_{H_0} \in \mathscr{L}_{H_0, P_{H_0}}(\mathscr{H}_0, \mathscr{H}_0) \tag{17}$$

and

$$U^* - P_H \in \mathscr{L}_{H, P_H}(\mathscr{H}_0, \mathscr{H}_0), \qquad U - P_H \in \mathscr{L}_{H, P_H}(\mathscr{H}_0, \mathscr{H}_0) \tag{18}$$

follow. Therefore also

$$P_H - P_{H_0} \in \mathscr{L}_{H, P_H}(\mathscr{H}_0, \mathscr{H}_0), \qquad P_H - P_{H_0} \in \mathscr{L}_{H_0, P_{H_0}}(\mathscr{H}_0, \mathscr{H}_0) \tag{19}$$

are valid.

## 11.2. Structure of scattering operators (the inverse problem of scattering theory)

In this paragraph we restrict ourselves by the following assumptions. The universal projection $P_H$ is assumed to be $P_H^{ac}$ and the wave operators to be considered are assumed to be partially isometric and $H_0$-semicomplete. As we shall see, these assumptions imply that $J$ is asymptotically equivalent to *one* fixed partial isometry (with initial projection $P_0^{ac}$) with respect to $H_0P_0^{ac}$ and $-H_0P_0^{ac}$. Surprisingly it turns out that the structural properties of a possible scattering operator are exhausted by

$$S \in \operatorname{com} H_0, \qquad S^*S = SS^* = P_{H_0}^{ac}.$$

Moreover, the condition on $J$ mentioned above is sufficient for the construction of an operator $H$, for fixed $\{J, H_0, S\}$ such that the corresponding wave operators exist and lead to a scattering operator $S(H, H_0, J)$ which coincides with $S$.

### 11.2.1. Asymptotic partial isometry of identification operators

The aim of this section is to prove an important implication of the assumption that wave operators are partially isometric and $H_0$-semicomplete. This implication refers to the identification operator and it was already announced in Remark 9.46 (II).

**Theorem 12.** *Assume that $W_+ = W_+(H, H_0; J)$ and $W_- = W_-(H, H_0; J)$ exist. Further let $W_+$ and $W_-$ be partially isometric with initial projection $P_0^{ac}$. Then there is a partial isometry $V \in \mathscr{L}(\mathscr{H}_0, \mathscr{H})$ with initial projection $P_0^{ac}$ such that $J \underset{H_0}{\approx} V$ and $J \underset{-H_0}{\approx} V$ simultaneously.*

**Proof.** According to Theorem 7.4, there is a projection $P \leqq P_0^{ac}$ in $\mathscr{H}_0$ with the properties $P - P_0^{ac} \in \ker \mu^{H_0}$, $P \in \ker \mu^{-H_0}$. Then we have $P^\perp \in \ker \mu^{H_0}$, $P^\perp - P^{ac} \in \ker \mu^{-H_0}$. We put

$$\tilde{J} := (W_+P + W_-P^\perp)\, P_0^{ac},$$

and obtain immediately $\hat{J} \underset{H_0}{\approx} W_+$ and $\hat{J} \underset{-H_0}{\approx} W_-$. Therefore, because of $J \underset{H_0}{\approx} W_+$ and $J \underset{-H_0}{\approx} W_-$, we obtain $\hat{J} \underset{\pm H_0}{\approx} J$. From the definition of $\hat{J}$, $\hat{J}^* = P_0^{ac}(PW_+^* + P^\perp W_-^*)$ follows. Hence we obtain

$$\hat{J}^*\hat{J} = P_0^{ac}(1 + P^\perp S^* P + PSP^\perp)\, P_0^{ac},$$

where, as usual, $S = W_+^* W_-$. We have $S \in \operatorname{com} H_0$. Therefore the relations

$$\hat{J}^*\hat{J} \in \operatorname{dom} \mu^{H_0} \cap \operatorname{dom} \mu^{-H_0}, \qquad \mu^{H_0}(\hat{J}^*\hat{J}) = \mu^{-H_0}(\hat{J}^*\hat{J}) = P_0^{ac}$$

follow. According to Theorem 6.28 from these relations we also obtain

$$(\hat{J}^*\hat{J})^{1/2} = |\hat{J}| \in \operatorname{dom} \mu^{H_0} \cap \operatorname{dom} \mu^{-H_0}, \qquad \mu^{H_0}(|\hat{J}|) = \mu^{-H_0}(|\hat{J}|) = P_0^{ac}.$$

Now the operator $\hat{J}$ posesses a polar decomposition $\hat{J} = \operatorname{sgn} \hat{J} \cdot |\hat{J}|$. We show that the partial isometry $\operatorname{sgn} \hat{J}$ has the property $\operatorname{sgn} \hat{J} \underset{\pm H_0}{\approx} \hat{J}$. Namely, from $|\hat{J}| \underset{\pm H_0}{\approx} P_0^{ac}$ we obtain $\hat{J} = (\operatorname{sgn} \hat{J})\, |\hat{J}| \underset{\pm H_0}{\approx} (\operatorname{sgn} \hat{J})\, P_0^{ac}$, but $(\operatorname{sgn} \hat{J})\, P_0^{ac} = \operatorname{sgn} \hat{J}$.

We define $Q$ by $Q := (\operatorname{sgn} \hat{J})^* (\operatorname{sgn} \hat{J})$ and get $Q \leqq P_0^{ac}$. From $|\hat{J}| \underset{\pm H_0}{\approx} P_0^{ac}$ immediately $Q\, |\hat{J}| \underset{\pm H_0}{\approx} QP_0^{ac} = Q$ follows. Because of $|\hat{J}| = Q\, |\hat{J}|$ also $Q \underset{\pm H_0}{\approx} P_0^{ac}$ follows. In other words we have $Q - P_0^{ac} \in \ker \mu^{H_0} \cap \ker \mu^{-H_0}$. We put $F := (\operatorname{sgn} \hat{J})\, (\operatorname{sgn} \hat{J})^*$. Obviously $F \leqq 1_{\mathscr{H}}$. Now we consider two cases. First we assume

$$\dim (P_0^{ac} - Q)\, \mathscr{H}_0 \leqq \dim F^\perp \mathscr{H}.$$

Then we choose a partial isometry $V_1$ with initial projection $P_0^{ac} - Q$ and a certain final projection $F_1 \leqq F^\perp$. Because of $V_1 = V_1(P_0^{ac} - Q)$ we have $V_1 \underset{\pm H_0}{\approx} 0$. Therefore

$$V := \operatorname{sgn} \hat{J} + V_1$$

is a partial isometry with initial projection $P_0^{ac}$ such that the property $V \underset{\pm H_0}{\approx} J$ is satisfied.

Now we consider the second case and assume

$$\dim (P_0^{ac} - Q)\, \mathscr{H}_0 > \dim F^\perp \mathscr{H}.$$

In this case we choose a projection $E < Q$ such that $\dim (Q - E)\, \mathscr{H}_0 = \infty$ and $Q - E \in \ker \mu^{H_0} \cap \ker \mu^{-H_0}$ (the existence of such projections is proved in Prop. 7.6, cf. also Remark 7.7.). We put $U := (\operatorname{sgn} \hat{J})\, E$. Then $U$ is partially isometric and $U \underset{\pm H_0}{\approx} \operatorname{sgn} \hat{J}$, hence $U \underset{\pm H_0}{\approx} \hat{J}$. The initial projection of $U$ is $E$. Because of $\dim (Q - E)\, \mathscr{H}_0 \leqq \dim (P_0^{ac} - E)\, \mathscr{H}_0$ we obtain $\dim (P_0^{ac} - E)\, \mathscr{H}_0 = \infty$. On the other hand, $\dim (Q - E)\, \mathscr{H}_0 = \infty$ implies $\dim (1_{\mathscr{H}} - UU^*)\, \mathscr{H} = \infty$. Therefore we obtain

$$\dim (P_0^{ac} - U^*U)\, \mathscr{H}_0 = \dim (1_{\mathscr{H}} - UU^*)\, \mathscr{H},$$

and we can perform once more the construction of $V_1$ described just before (in the first case), but now with respect to the projections $P_0^{ac} - U^*U$ and $1_{\mathscr{H}} - UU^*$. Then

$V := U + V_1$ is partially isometric with initial projection $P_0^{ac}$ and with the property $V \underset{\pm H_0}{\approx} U$. Therefore we obtain $V \underset{\pm H_0}{\approx} \hat{J}$. ∎

### 11.2.2. Formulation of the inverse problem

As was already indicated in the introduction to this paragraph, the question of structural properties of scattering operators can be answered by solving the inverse problem of scattering theory. In this section the corresponding definition is presented. Let $\mathscr{H}$, $\mathscr{H}_0$ $J$, $H_0$ denote the objects which are given a priori. Further, the operator $S \in \mathscr{L}(\mathscr{H}_0)$ is assumed to satisfy the condition

$$S \in \operatorname{com} H_0, \qquad S^*S = SS^* = P_0^{ac}. \tag{20}$$

**Definition 13.** A selfadjoint operator $H$ in $\mathscr{H}$ is called a *solution of the inverse problem* with respect to the data $\{J, H_0, S\}$ if $H$ satisfies the following conditions:

(I) $W_+ = W_+(H, H_0; J)$ and $W_- = W_-(H, H_0; J)$ exist (*existence of the wave operators*),

(II) $W_+^*W_+ = W_-^*W_- = P_0^{ac}$ (*partial isometry and $H_0$-semicompleteness*),

(III) $W_+W_+^* = W_-W_-^*$ (*final projection independent of* $\pm$),

(IV) $W_+^*W_- = S$ (*the scattering operator coincides with* $S$).

Recall that the projection $E = W_+W_+^* = W_-W_-^*$ is a reducing projection of $H$. It is not necessary to require $E = P^{ac}$, but the solution of the inverse problem can be given also under this stronger condition. For this reason we may modify (III) to

(III′) $W_+W_+^* = W_-W_-^* = P^{ac}$ (*completeness*).

That is, the solution of the inverse problem means the construction of a selfadjoint operator $H$ such that both wave operators exist, are partially isometric and complete, and realize the given operator $S$ as scattering operator.

### 11.2.3. The key problem

In order to prepare the solution of the inverse problem, in this section a problem is investigated which turns out to be the key problem. The framework of this problem is the space $\mathscr{H}_0$ and the operators $H_0$ and $S$ only. The direct connection between this problem and the original inverse problem is discussed in the next section. Assume that $H_0$ and $S$ satisfy (20). Then we pose the problem to construct an operator $V \in \mathscr{L}(\mathscr{H}_0)$ with the following properties:

$$V^*V = VV^* = P_0^{ac}, \tag{21}$$

$$V, V^* \in \operatorname{dom} \mu^{H_0} \cap \operatorname{dom} \mu^{-H_0}, \tag{22}$$

$$\mu^{H_0}(V) = P_0^{ac}, \qquad \mu^{-H_0}(V) = SP_0^{ac}. \tag{23}$$

(23) immediately implies

$$\mu^{H_0}(V^*) = P_0^{ac}, \qquad \mu^{-H_0}(V^*) = S. \tag{24}$$

A crucial step for solving this problem is the following lemma, which will be proved at first.

**Lemma 14.** *Let* $A$, $U_+$, $U_- \in \mathscr{L}(\mathscr{H}_0)$. *Assume that:*

(I) $U_+^*U_+ = U_+U_+^* = U_-^*U_- = U_-U_-^* = P_0^{ac}$,

(II) $A, A^* \in \operatorname{dom} \mu^{H_0} \cap \operatorname{dom} \mu^{-H_0}$ *and* $\mu^{H_0}(A) = U_+$, $\mu^{-H_0}(A) = U_-$.

*Then there exists an operator* $V \in \mathscr{L}(\mathscr{H}_0)$ *satisfying:*

(i) $V^*V = VV^* = P_0^{ac}$,

(ii) $V, V^* \in \operatorname{dom} \mu^{H_0} \cap \operatorname{dom} \mu^{-H_0}$,

(iii) $\mu^{H_0}(V) = U_+$, $\mu^{-H_0}(V) = U_-$ *(then also* $\mu^{H_0}(V^*) = U_+^*$, $\mu^{-H_0}(V^*) = U_-^*$*).*

**Proof.** For simplicity let $P_0^{ac} = 1$. Then we have $\mu^{\pm H_0}(A^*A) = 1$, that is $\mu^{\pm H_0}(|A|^2) = 1$. Hence also $\mu^{\pm H_0}(|A|) = 1$. Let $A = \operatorname{sgn} A\, |A|$ be the polar decomposition of $A$. Recall that $\operatorname{sgn} A$ is partially isometric. Further we have $(\operatorname{sgn} A)^* A = |A|$. Because of

$$\operatorname{sgn} A = \operatorname{sgn} A\, |A| + \operatorname{sgn} A(1 - |A|),$$

$$(\operatorname{sgn} A)^* = (\operatorname{sgn} A)^* AU_\pm^* + (\operatorname{sgn} A)^* (1 - AU_\pm^*),$$

and $\mu^{\pm H_0}(AU_\pm^*) = 1$, we obtain $\mu^{\pm H_0}(\operatorname{sgn} A) = \mu^{\pm H_0}(A) = U_\pm$ and $\mu^{\pm H_0}((\operatorname{sgn} A)^*) = \mu^{\pm H_0}(|A|\, U_\pm^*) = U_\pm^*$. We put $(\operatorname{sgn} A)^* (\operatorname{sgn} A) = P$; $(\operatorname{sgn} A)(\operatorname{sgn} A)^* = Q$. Then we have $\mu^{\pm H_0}(P) = \mu^{\pm H_0}(Q) = 1$.

By Theorem 7.8 and Remark 7.9, there are partial isometries $V_P$, $V_Q$ such that $V_P, V_Q, V_P^*, V_Q^* \in \operatorname{dom} \mu^{H_0} \cap \operatorname{dom} \mu^{-H_0}$, $V_P^*V_P = V_QV_Q^* = 1$, $V_PV_P^* = P$, $V_Q^*V_Q = Q$ and $\mu^{\pm H_0}(V_P) = \mu^{\pm H_0}(V_Q) = 1$. Therefore $V = V_Q(\operatorname{sgn} A)\, V_P$ has the desired properties. ∎

Using Lemma 14 it is easy to solve the key problem.

**Theorem 15.** *Assume that* $H_0$ *and* $S$ *satisfy* (20). *Then there is an operator* $V \in \mathscr{L}(\mathscr{H}_0)$ *such that properties* (21)–(23) *are valid.*

**Proof.** According to Theorem 7.4, there is a projection $P$ with the properties $P \in \operatorname{dom} \mu^{H_0} \cap \operatorname{dom} \mu^{-H_0}$, $P \leq P_0^{ac}$ and $\mu^{H_0}(P) = 0$; $\mu^{-H_0}(P) = P_0^{ac}$. With the help of $P$ we form the following operator

$$X = P_0^{ac} - P + PS.$$

Then we obtain immediately $X, X^* \in \operatorname{dom} \mu^{H_0} \cap \operatorname{dom} \mu^{-H_0}$ and

$$\mu^{H_0}(X) = P_0^{ac} - \mu^{H_0}(P) + \mu^{H_0}(PS) = P_0^{ac} + \mu^{H_0}(P)\, S = P_0^{ac}$$

$$\mu^{-H_0}(X) = P_0^{ac} - \mu^{-H_0}(P) + \mu^{-H_0}(PS) = \mu^{-H_0}(P)\, S = P_0^{ac}S = S.$$

That is, $X$ is an operator satisfying the assumptions for the operator $A$ in Lemma 14 with $U_+ = P_0^{ac}$ and $U_- = S$. Hence, according to Lemma 14, we obtain an operator $V$ with the desired properties, (21)–(23). ∎

### 11.2.4. Solution of the inverse problem

Now we apply the results of the preceding section to solve the inverse problem. According to the aim in this section the operator $S$ is assumed to satisfy the condition (20). First, we collect the necessary conditions for the solvability of the inverse problem.

**Theorem 16.** *Let $H$ be a solution of the inverse problem. Then there is a partial isometry $V \in \mathscr{L}(\mathscr{H}_0; \mathscr{H})$ with the properties $V^*V = P_0^{ac}$, $VV^* = E$, where $E = W_+W_+^* = W_-W_-^*$, such that $J \underset{H_0}{\approx} V$ and $J \underset{-H_0}{\approx} V$.*

Proof. According to Theorem 12 and the conditions (I), (II) for a solution $H$ of the inverse problem, we have only to prove that the partial isometry $V$ can be chosen such that $V^*V = E$ is valid. The existence of a partial isometry $V$ with initial projection $P_0^{ac}$ and with the property $J \underset{\pm H_0}{\approx} V$ is already assured by Theorem 12. For this purpose we choose a partial isometry $V_0 \in \mathscr{L}(\mathscr{H}_0)$, according to Theorem 15, with the properties $V_0 \in \operatorname{dom} \mu^{H_0} \cap \operatorname{dom} \mu^{-H_0}$, $\mu^{H_0}(V_0) = P_0^{ac}$, $\mu^{-H_0}(V_0) = S$ and $V_0^*V_0 = V_0V_0^* = P_0^{ac}$. We put $V := W_+V_0 \in \mathscr{L}(\mathscr{H}_0, \mathscr{H})$. Clearly, $V^*V = P_0^{ac}$ and $VV^* = W_+V_0V_0^*W_+^* = W_+W_+^* = E = W_-W_-^*$. It remains to prove $J \underset{H_0}{\approx} V$ and $J \underset{-H_0}{\approx} V$. Because of the existence of the wave operators $W_\pm$ and the properties of $V_0$ we obtain that

$$\operatorname*{Lim}_{t \to \pm\infty} e^{itH}(J - W_+V_0)\, e^{-itH_0}P_0^{ac} = W_\pm - W_+\mu^{\pm H_0}(V_0)$$

exists (for both signs). Moreover, $\mu^{H_0}(V_0) = P_0^{ac}$ implies $W_+ - W_+\mu^{H_0}(V_0) = 0$, and $\mu^{-H_0}(V_0) = S$ implies $W_- - W_+\mu^{-H_0}(V_0) = 0$ (recall that $W_+S = W_-$). Hence we obtain $J \underset{\pm H_0}{\approx} V$. ■

Now we proceed to the converse of Theorem 16. This converse asserts the solvability of the inverse problem provided $J$ is asymptotically equivalent to $V$ with $V^*V = P_0^{ac}$. The solution to be constructed satisfies the completeness condition (III').

**Theorem 17.** *Let $\{J, H_0, S\}$ be given where $S$ satisfies condition* (20). *Assume $J$ to be asymptotically equivalent to $V_0 \in \mathscr{L}(\mathscr{H}_0, \mathscr{H})$, that is*

$$\operatorname*{Lim}_{t \to \pm\infty} (J - V_0)\, e^{-itH_0}P_0^{ac} = 0 \tag{25}$$

*where $V_0^*V_0 = P_0^{ac}$. We put $V_0V_0^* = E$. Then there is a solution $H$ of the inverse problem with $P^{ac} = E$ and with complete wave operators $W_\pm(H, H_0; J)$. The absolutely continuous part $H \upharpoonright P^{ac}\mathscr{H}$ of a special solution can be defined by*

$$H \upharpoonright P^{ac}\mathscr{H} := WH_0W^* \upharpoonright P^{ac}\mathscr{H},$$

*where $W = V_0V^*$ and where $V$ denotes an arbitrary but fixed operator given by Theorem 15.*

Proof. We have $V_0^*(EJ - V_0) = V_0^*J - P_0^{ac}$. From $J \underset{\pm H_0}{\approx} V_0$, that is from (25) immediately $EJ \underset{\pm H_0}{\approx} EV_0 = V_0$ follows. Hence we obtain $V_0^*J - P_0^{ac} \in \ker \mu^{\pm H_0}$.

Because of $W^*J = VV_0^*J$ we get $W^*J \in \operatorname{dom} \mu^{\pm H_0}$ and

$$\mu^{H_0}(W^*J) = \mu^{H_0}(VV_0^*J) = \mu^{H_0}(V)\,\mu^{H_0}(V_0^*J) = P_0^{ac},$$

$$\mu^{-H_0}(W^*J) = \mu^{-H_0}(VV_0^*J) = \mu^{-H_0}(V)\,\mu^{-H_0}(V_0^*J) = S.$$

Therefore we obtain the existence of the wave operators $W_\pm$ by means of the relation

$$W_\pm = \operatorname*{Lim}_{t\to\pm\infty} e^{itH} J\, e^{-itH_0} P_0^{ac} = \operatorname*{Lim}_{t\to\pm\infty} e^{itH} EJ\, e^{-itH_0} P_0^{ac}$$

$$= \operatorname*{Lim}_{t\to\pm\infty} W e^{itH_0} W^*J\, e^{-itH_0} P_0^{ac} = W\mu^{\pm H_0}(W^*J).$$

Moreover, we have $W_+ = W$ and $W_- = WS$, hence we get $W_+^*W_- = S$. The part $H \restriction E^\perp\mathscr{H}$ can be chosen arbitrarily. ■

**Remark 18.** Let $\mathscr{H} = \mathscr{H}_0$ and $J = 1$, that is, assume that the one-space situation is given. In this case the inverse problem is always solvable if $S$ satisfies condition (20). Because of the possibility to choose $V_0 = 1$ in this case, the operator $W$ to be constructed in Theorem 17 is simply $W = V^*$. Further, the wave operators $W_+$ and $W_-$ are given by $W_+ = V^*$ and $W_- = V^*S$. In other words, the defining properties of the operator $W$ in the one-space theory are just the conditions of the key problem in Section 3, namely $\mu^{H_0}(W) = P_0^{ac}$, $\mu^{-H_0}(W) = S^*$ where $W$ is assumed to be unitary on $P_0^{ac}\mathscr{H}_0$. Hence the operator $H$ to be constructed in this case satisfies $P^{ac} = P_0^{ac}$. Let for simplicity $P_0^{ac} = 1$. Then the wave operators obviously have the form

$$W_+ = G_-S^* + G_+, \qquad W_- = G_- + G_+S,$$

where $G_+, G_- \in \operatorname{dom} \mu^{H_0} \cap \operatorname{dom} \mu^{-H_0}$ and where $\mu^{\pm H_0}(G_\pm) = 1$, $\mu^{\pm H_0}(G_\mp) = 0$ is valid. The difficulty, which is expressed by the key problem, is to choose $G_\pm$ such that $W$ becomes unitary. That is, the content of Section 3, where the key problem is solved, can be considered essentially as the solution of the inverse problem for the one-space theory. Thus the crucial difficulty already lies in the one-space problem. The difficulty is not enlarged by the transition to the two-space framework.

### 11.2.5. Description of all solutions

In this section we give an abstract characterization of all solutions of the inverse problem if one special solution is given. We restrict ourselves to solutions of the inverse problem with complete wave operators and for simplicity we assume $P_0^{ac} = 1$, $P^{ac} = 1$. The inverse problem is assumed to be solvable, that is, $J$ is assumed to be asymptotically equivalent to $V$ with $V^*V = 1_{\mathscr{H}_0}$, $VV^* = 1_{\mathscr{H}}$. According to this assumption we choose a fixed isometric operator $V \in \mathscr{L}(\mathscr{H}_0, \mathscr{H})$ with $V \underset{\pm H_0}{\approx} J$ and with $V^*V = 1_{\mathscr{H}_0}$, $VV^* = 1_{\mathscr{H}}$. Further, let $H_1$ be a special solution of the inverse problem. Then

$$H_1 = VU_1^*H_0U_1V^* \tag{26}$$

where $U_1$ is a special unitary operator in $\mathscr{H}_0$, whose existence is ensured by Theorem 17. Now we can formulate

**Theorem 19.** *Let $H_1$ be a special solution of the inverse problem given by (26). Then the set of all solutions $H$ of the inverse problem with $P_H^{ac} = 1$ and with complete wave operators can be characterized by*

$$H = XH_1X^*$$

*where $X$ is unitary on $\mathscr{H}$, $X, X^* \in \operatorname{dom} \mu^{-H_1} \cap \operatorname{dom} \mu^{H_1}$ and*

$$\mu^{\pm H_1}(X) = \mu^{\pm H_1}(X^*) = 1 \quad (\textit{for both signs}) \tag{27}$$

*is valid.*

**Proof.** First, let $H$ be an arbitrary solution. Then $H = WH_0W^*$ with $W = W_+$ follows. Defining $U$ by $U := W^*V$, we easily obtain that $U$ satisfies conditions (21)—(23). Then we have $H = VU^*H_0UV^* = XH_1X^*$ where $X = VU^*U_1V^*$. A simple calculation yields $\mu^{\pm H_1}(X) = \mu^{\pm H_1}(X^*) = 1$. Conversely, let $H_1$ and $X$ satisfy the assumptions of the theorem, in particular let (27) be satisfied. Then we have to verify that $W_\pm(H, H_0; J)$ exist, are unitary and $W_+(H, H_0; J)^* W_-(H, H_0; J) = S$, where $H = XH_1X^*$. Using the existence of the wave operators $W_\pm(H_1, H_0; J)$ and (27) we obtain that

$$\begin{aligned}\operatorname*{Lim}_{t\to\pm\infty} e^{itH} J\, e^{-itH_0} &= \operatorname*{Lim}_{t\to\pm\infty} X\, e^{itH_1} X^*\, e^{-itH_1}\, e^{itH_1} J\, e^{-itH_0} \\ &= X\mu^{\pm H_1}(X^*)\, W_\pm(H_1, H_0; J) = XW_\pm(H_1, H_0; J)\end{aligned} \tag{28}$$

exist. Further, this gives

$$\begin{aligned}W_+(H, H_0; J)^*\, W_-(H, H_0; J) &= W_+(H_1, H_0; J)^*\, X^*XW_-(H_1, H_0; J) \\ &= W_+(H_1, H_0; J)^*\, W_-(H_1, H_0; J) = S. \ \blacksquare\end{aligned}$$

The condition (27) can be also formulated with the help of $H_0$. For this purpose let $H$ be a special solution of the inverse problem with complete wave operators $W_\pm$. Then $W_\pm = W_\pm(H, H_0; V)$ and $W_\pm{}^* = W_\pm(H_0, H; V^*)$ follow. We use

**Lemma 20.** *Let $X \in \mathscr{L}(\mathscr{H}_0)$. Then $X \in \operatorname{dom} \mu^{H_0}$ iff $VXV^* \in \operatorname{dom} \mu^{H}$ and in this case we have*

$$\mu^{H}(VXV^*) = W_+\mu^{H_0}(X)\, W_+{}^*.$$

*Similarly for $\mu^{-H_0}$, $\mu^{-H}$ and $W_-$.*

**Proof.** It is a straightforward calculation. ■

In particular we obtain: $\mu^{\pm H}(Y) = \mu^{\pm H}(Y^*) = 1$ is valid iff $\mu^{\pm H_0}(V^*YV) = \mu^{\pm H_0}(V^*Y^*V) = 1$ is valid. Therefore the following corollary of Theorem 19 is true.

**Corollary 21.** *Let $H_1$ be a special solution of the inverse problem with the data $\{J \approx V, H_0, S\}$ and with complete wave operators. Then the set of all solutions $H$ with complete wave operators and with $P_H^{ac} = 1$ is given by*

$$H = XH_1X^*$$

*where $X$ is unitary on $\mathscr{H}$ and where*

$$\mu^{\pm H_0}(V^*XV) = \mu^{\pm H_0}(V^*X^*V) = 1$$

*is valid.*

**Proof.** It immediately follows from Theorem 19 and Lemma 20. ∎

**Remark 22.** We consider a special case from the one-space theory. Let $\mathscr{H} = \mathscr{H}_0, J = 1$. Further, let $H_1$, $H_0$ be absolutely continuous. The wave operators are assumed to exist and to be complete, that is $W_\pm(H_1, H_0)$ exist and are unitary. Put $W_+^* W_- = S$. In this case we have $V = 1$. Thus Corollary 21 says: If $\mu^{\pm H_0}(X) = \mu^{\pm H_0}(X^*) = 1$ where $X$ is unitary, and if we put $H = XH_1X^*$, then the pair $\{H, H_0\}$ yields the same $S$-operator $S$ as the pair $\{H_1, H_0\}$. The following example shows that the condition $\mu^{\pm H_0}(X) = \mu^{\pm H_0}(X^*) = 1$ is not implied by the property "$\{XH_1X^*, H_0\}$ and $\{H_1, H_0\}$ yield the same S-operator". Let $X \in \text{com com } H_0$, $X$ unitary and let $H_1 = H_0 + V$. Then $H = XH_1X^* = H_0 + XVX^*$ and $W_\pm(H; H_0) = XW_\pm(H_1, H_0)\, X^*$. Hence $S(H, H_0) = XS(H_1, H_0)\, X^* = S(H_1, H_0)$, but $\mu^{\pm H_0}(X) = X$, $\mu^{\pm H_0}(X^*) = X^*$.

### 11.2.6. The inverse problem for multichannel identification operators

Recall the definition of a general channel system in a Hilbert space $\mathscr{H}$ (Definition 10.18). Let the system $\{A_1, A_2, \ldots, A_r; Q_1, Q_2, \ldots, Q_r\}$ be a general channel system. Then the corresponding multichannel identification operator $J$ is well-defined (see (10.29)). In Theorem 10.21 conditions are formulated which ensure that the multichannel identification operator is asymptotically partially isometric. Combining this result with the results of Section 4, we obtain the following result concerning the solution of the inverse problem in this case.

**Theorem 23.** *Let $\{A_1, A_2, \ldots, A_r; Q_1, Q_2, \ldots, Q_r\}$ be a general channel system. Let $Q_i = Q_j$ imply $i = j$. Without loss of generality $Q_iQ_j \neq 0$ can be assumed. The ordering of the $Q_j$ can be arranged in such a way that from $Q_i < Q_j$ also $i < j$ follows. Assume*

$$\operatorname*{Lim}_{t\to\pm\infty} e^{itA_\alpha} Q_j Q_\alpha e^{-itA_\alpha} = 0, \qquad 1 \leqq j < \alpha, \qquad \alpha = 2, 3, \ldots, r.$$

*Further put $\mathscr{H}_0 := \mathscr{H} \times \mathscr{H} \times \cdots \times \mathscr{H}$ ($r$ times) and $H_0 := \{A_1Q_1, A_2Q_2, \ldots, A_rQ_r\}$; $J\hat{f} := \sum_{\varrho=1}^{r} Q_\varrho f_\varrho$ where $\hat{f} = \{f_1, f_2, \ldots, f_r\}$. Assume $S \in \text{com } H_0$ and $S^*S = SS^* = P_0^{ac}$. Then there is a solution $H$ in $\mathscr{H}$ of the inverse problem with data $\{J, H_0, S\}$ and with complete wave operators.*

**Proof.** It is a direct implication of Theorem 10.21 and Theorem 17. ∎

**Remark 24.** In some special cases the solution of the inverse problem can be chosen uniquely from a certain restricted class of operators such that additional a priori assumptions must be satisfied. A well-known example in this direction is the case: $\mathscr{H} = \mathscr{H}_0 = L^2([0, \infty), dx)$; $H_0$ corresponds to $-d^2/dx^2$ and where the following a priori condition for $H$ is used: $V := H - H_0$ is a multiplication operator (potential). For more details see the references cited in the Notes and Remarks.

## 11.3. The invariance principle for wave operators

In this paragraph we investigate the behavior of wave operators $W_+(H; H_0; J)$ if $H$ and $H_0$ are replaced by functions $\alpha(H)$ and $\alpha(H_0)$ of $H$ and $H_0$. For example, we pose the problem of the existence of $W_+(\alpha(H), \alpha(H_0); J)$ and its properties. Since for a large class of functions $\alpha$ the wave operators do not change in the case of such a substitution, one usually speaks of the invariance principle of wave operators in this connection. In the following always $P_H = P_H^{ac}$.

### 11.3.1. Formulation of the invariance principle

First we fix the class of functions $\alpha$ for which the invariance principle will be formulated and proved. The functions $\alpha$ are assumed to be admissible according to Definition 8.16, i.e. an admissible function $\alpha$ is piecewise continuously differentiable, where $\alpha'(x)$ is positive and locally of bounded variation. Sometimes this class of functions is called the *Kato class.*

As already noted in the introduction to this paragraph, in the following the universal projection $P_H$ is always assumed to be $P_H^{ac}$. Only for this universal projection the invariance principle will be proved. Note that for admissible functions $\alpha$ the subspace of absolute continuity is invariant, i.e. $P_{\alpha(H)}^{ac} = P_H^{ac}$ if $\alpha$ is a.e. finite with respect to $E(\cdot)$. Note that for admissible functions $\alpha$ also $\alpha^{-1}$ exists piecewise and is admissible (cf. Lemma 8.17).

Two forms of the invariance principle may be distinguished. The weak form states that if both wave operators $W_+(H, H_0; J)$ and $W_+(\alpha(H), \alpha(H_0); J)$ exist and if $\alpha$ is an admissible function, then

$$W_+(\alpha(H), \alpha(H_0); J) = W_+(H, H_0; J) \tag{29}$$

is valid. The strong form of the invariance principle says that if $W_+(H, H_0; J)$ exists and if $\alpha$ is an admissible function, then also $W_+(\alpha(H), \alpha(H_0); J)$ exists and (29) is valid.

In the following it will be shown that in a scattering theory, where the wave operators are assumed to exist in the strong sense (s-lim), only the weak form of the invariance principle is true for admissible functions. However, in a scattering theory, where the wave operators are assumed to exist only in the absolute Abelian sense (|A|-lim), the strong form of the invariance principle is true for admissible functions.

Clearly, in this connection the results of Section 8.2.2 on the invariance of $\ker \mu_s{}^H$ are crucial, in particular Theorem 8.23 which says that $\ker \mu_s{}^H = \ker \mu^{\alpha(H)}$ if $\alpha$ is admissible and a.e. finite with respect to $E(\cdot)$.

### 11.3.2. Invariance principle, strong form

In this section the limit $\underset{t\to\infty}{\text{Lim}}$ is interpreted in the sense of $\underset{t\to\infty}{|A|\text{-lim}}$, the absolute Abelian limit. The strong form of the invariance principle is an implication from Theorem 8.23. We state it as follows.

**Theorem 25.** *Let the wave operator* $W_+(H, H_0; J) = \underset{t\to\infty}{|A|\text{-lim}}\, e^{itH} J\, e^{-itH_0} P_0^{ac}$ *exist. Further, assume $\alpha$ to be admissible and a.e. finite with respect to $E(\cdot)$ and $E_0(\cdot)$. Then also* $W_+(\alpha(H), \alpha(H_0); J) = \underset{t\to\infty}{|A|\text{-lim}}\, e^{it\alpha(H)} J\, e^{-it\alpha(H_0)} P_{\alpha(H_0)}^{ac}$ *exists and the equality*

$$W_+(H, H_0; J) = W_+(\alpha(H), \alpha(H_0); J) \tag{30}$$

*is valid (similary for $W_-$).*

**Proof.** For brevity we put $W_+ := W_+(H, H_0; J)$. Because of $(\text{ima } W_+^*)^\perp = \ker W_+$ we have

$$\mathscr{H}_0 = \text{clo}(\text{ima } W_+^*) \oplus \ker W_+. \tag{31}$$

These two subspaces are treated separately. First let $u \in \text{ima } W_+^*$, that is $u = W_+^* f$. Thus we have to consider

$$e^{itH} J\, e^{-itH_0} W_+^* = e^{itH} J W_+^*\, e^{-itH}$$

and

$$e^{it\alpha(H)} J\, e^{-it\alpha(H_0)} W_+^* = e^{it\alpha(H)} J W_+^*\, e^{-it\alpha(H)}.$$

Because of $W_+ P_0^{ac} = W_+$ and $P^{ac} W_+ = W_+$, hence $P_0^{ac} W_+^* = W_+^*$ and $W_+^* = W_+^* P^{ac}$, we get

$$e^{itH} J\, e^{-itH_0} P_0^{ac} W_+^* = e^{itH} J W_+^*\, e^{-itH} P^{ac}$$

and

$$e^{it\alpha(H)} J\, e^{-it\alpha(H_0)} P_0^{ac} W_+^* = e^{it\alpha(H_0)} J W_+^*\, e^{-it\alpha(H)} P^{ac}.$$

Now according to the assumption

$$B := \mu_a{}^H(JW_+^*) = |A|\text{-}\lim_{t\to\infty} e^{itH} J W_+^*\, e^{-itH} P^{ac}$$

exists. Then $\mu_a{}^H(JW_+^* - B) = 0$. Note that $B \in \text{com } H$. From Theorem 8.23 and $JW_+^* - B \in \ker \mu_a{}^H$ we obtain $JW_+^* - B \in \ker \mu_a^{\alpha(H)}$, i.e.

$$|A|\text{-}\lim_{t\to\infty} e^{it\alpha(H)} J\, e^{-it\alpha(H_0)} P_0^{ac} W_+^*$$

exists and this limit equals $|A|\text{-}\lim_{t\to\infty} e^{itH} J\, e^{-itH_0} P_0^{ac} W_+^*$. In other words, for $u \in \text{ima } W_+^*$, hence also for $u \in \text{clo}(\text{ima } W_+^*)$, the theorem is already proved. In the second case we proceed as follows: Let $K_+$ denote the projection onto $\ker W_+$. Then we have $|A|\text{-}\lim_{t\to\infty} e^{itH} J\, e^{-itH_0} P_0^{ac} K_+ = 0$. $K_+$ reduces $H_0$. Thus we obtain $|A|\text{-}\lim_{t\to\infty} e^{itH} J K_+\, e^{-itH_0} P_0^{ac} = 0$ and also $|A|\text{-}\lim_{t\to\infty} J K_+\, e^{-itH_0} P_0^{ac} = 0$. Therefore, from Theorem 8.23 it follows that

$$|A|\text{-}\lim_{t\to\infty} J K_+\, e^{-it\alpha(H_0)} P_0^{ac} = 0,$$

hence also

$$|A|\text{-}\lim_{t\to\infty} e^{it\alpha(H)} J\, e^{-it\alpha(H_0)} P_0^{ac} K_+ = 0.$$

In other words, also in the case $u \in \ker W_+$ the existence of $|A|\text{-}\lim_{t\to\infty} e^{it\alpha(H)} J\, e^{-it\alpha(H_0)} P_0^{ac} u$ and the equality with $W_+(H, H_0; J)\, u$ follow. Hence (31) yields the full assertion. ∎

### 11.3.3. Invariance principle, weak form

The weak form of the invariance principle for strong wave operators immediately follows from the result of Section 11.3.2.

**Corollary 26.** *Let the triple $\{H, H_0; J\}$ be given. Assume $\alpha$ to be admissible and a.e. finite with respect to $E(\cdot)$, $E_0(\cdot)$. Further, let both wave operators $W_+(H, H_0; J) = \text{s-lim}_{t\to\infty} e^{itH}J \times e^{-itH_0}P_0^{ac}$ and $W_+(\alpha(H), \alpha(H_0; J) = \text{s-lim}_{t\to\infty} e^{it\alpha(H)}J\, e^{-it\alpha(H_0)}P^{ac}_{\alpha(H_0)}$ exist. Then the equality*

$$W_+(\alpha(H), \alpha(H_0); J) = W_+(H, H_0; J) \tag{32}$$

*is valid.*

**Proof.** The existence of the strong wave operator $W_+(H, H_0; J)$ implies the existence of $W_+(H, H_0; J)$ in the sense of the absolute Abelian limit and its equality with the strong wave operator. Similarly for $W_+(\alpha(H), \alpha(H_0); J)$. Hence (32) immediately follows from Theorem 25. ■

### 11.3.4. A counterexample and piecewise linear functions

In general, if one works with the strong wave operator, it is not true that the existence of $W_+(H, H_0; J)$ implies the existence of $W_+(\alpha(H), \alpha(H_0); J)$. This can be seen by the following

**Proposition 27.** *Let the triple $\{H, H_0, J\}$ be given and let $J \underset{\alpha(H_0)}{\approx} V$, $J \underset{H_0}{\approx} V$ with $V^*V = 1_{\mathscr{H}_0}$, $VV^* = 1_{\mathscr{H}}$. Further, let $\alpha$ be admissible on a finite interval $[a, b]$. Assume $\alpha'(x)$ is strictly increasing on $[a, b]$ and $|[a, b] \cap \operatorname{spec}^c (H_0 \mid P_0^{ac}\mathscr{H}_0)| > 0$. Then there exists always an operator $H$ such that $W_+(H, H_0; J)$ exists but $W_+(\alpha(H), \alpha(H_0); J)$ does not exist (in the strong sense).*

**Proof.** First note that under the assumptions of Prop. 27 there is a projection $P$ such that $P \in \ker \mu_s^{H_0}$ but $P \notin \operatorname{dom} \mu_s^{\alpha(H_0)}$ (see Theorem 8.26 for the construction of such a projection). Now we consider $U = 1 - 2P$, which is unitary in $\mathscr{H}_0$ because of $U^*U = UU^* = (1 - 2P)(1 - 2P) = 1 - 4P + 4P^2 = 1$. We put $W = VU^*$. Then $W \in \mathscr{L}(\mathscr{H}_0, \mathscr{H})$ and it is isometric from $\mathscr{H}_0$ onto $\mathscr{H}$. Now we define $H$ by $H := WH_0W^*$. Then we obtain

$$\text{s-lim}_{t\to\infty} e^{itH}J\, e^{-itH_0}P_0^{ac} = \text{s-lim}_{t\to\infty} W\, e^{itH_0}W^*V\, e^{-itH_0}P_0^{ac} = W\mu_s^{H_0}(U) = W$$

exists, but

$$\text{s-lim}_{t\to\infty} e^{it\alpha(H)}J\, e^{-it\alpha(H_0)}P_0^{ac} = \text{s-lim}_{t\to\infty} W\, e^{it\alpha(H_0)}W^*V\, e^{-it\alpha(H_0)}P_0^{ac} = W\mu_s^{\alpha(H_0)}(U)$$

does not exist. ■

In Prop. 27 functions $\alpha$, which are piecewise linear, are excluded. Actually, one can show that for such functions the invariance principle for strong wave operators is also true in the strong form.

**Theorem 28.** *Let the triple $\{H, H_0; J\}$ be given. Assume the function $\alpha$ is strictly linear increasing on each Borel set $\Delta_n$, $n = 1, 2, \ldots$, where the Borel sets $\Delta_n$ are mutually disjoint and where $\bigcup_n \Delta_n \supset \operatorname{spec}^c H_0$ is valid. Further assume that $W_0(H, H_0; J)$ exists in the strong sense. Then also $W_+(\alpha(H), \alpha(H_0); J)$ exists in the strong sense and it is equal to $W_+(H, H_0; J)$.*

**Proof.** It proceeds similarly to the proof of Theorem 25 where Theorem 8.23 was used. In the present case the relation $\ker \mu_s^{H_0} = \ker \mu_s^{\alpha(H_0)}$ is used. This relation immediately follows from Prop. 8.5. ∎

**Remark 29.** A simple consequence of Theorem 25 and Theorem 28 is that it is sufficient to consider bounded operators $H$, $H_0$ in scattering theories with strong or Abel wave operators. This follows from the fact that for unbounded selfadjoint operators $H$, $H_0$ we can find admissible functions $\alpha$ such that $\alpha(H)$ and $\alpha(H_0)$ are bounded selfadjoint operators and such that the existence of $W_+(H, H_0; J)$ is equivalent (in the corresponding limiting sense) to the existence of $W_+(\alpha(H), \alpha(H_0); J)$. Furthermore, the corresponding wave operators are equal. It is simple to construct such functions $\alpha$, one example is $\alpha(\lambda) = \arctan \lambda$, for Abel wave operators, and $\alpha(x) = \sum_{n=-\infty}^{\infty} \chi_{[n,n+1)}(x)\,(a_n x + b_n)$ (with certain constants $a_n > 0$ and $b_n \in \mathbb{R}$, see Example 8.7) for strong wave operators.

For further results concerning the invariance principle, see the Notes and Remarks to this part and Part IV.

# 12. Lax-Phillips Evolutions and Two-Space Wave Operators

In this chapter the so-called Lax-Phillips scattering theory is linked with the two-space scattering theory developed in Chapter 9. It turns out that the special properties of the evolution of the Lax-Phillips-theory can be used for the construction of the second (unperturbed) evolution in the second space. The S-operator within this framework has very nice properties.

## 12.1. Lax-Phillips evolutions

### 12.1.1. Incoming and outgoing subspaces

Let $\mathscr{H}$ be a separable Hilbert space, let $U(t)$, $t \in \mathbb{R}$, be a unitary strongly continuous group in $\mathscr{H}$ and let $\mathscr{D}_+$, $\mathscr{D}_-$ be subspaces in $\mathscr{H}$.

**Definition 1.** A subspace $\mathscr{D}_+$ is called an *outgoing subspace* with respect to $U(t)$ if the following conditions are satisfied:

(I) $U(t)\,\mathscr{D}_+ \subseteq \mathscr{D}_+, \quad t \geq 0,$

(II) $\bigcap_{t\in\mathbb{R}} U(t)\,\mathscr{D}_+ = \{0\},$

(III) $\operatorname{clo}\left\{\bigcup_{t\in\mathbb{R}} U(t)\,\mathscr{D}_+\right\} = \mathscr{H}.$

A subspace $\mathscr{D}_-$ is called an *incoming subspace*, if (II) and (III) are satisfied and if (I) is replaced by

(I') $U(t)\,\mathscr{D}_- \subseteq \mathscr{D}_-, \quad t \leq 0.$

### 12.1.2. Examples

(a) Let $\mathscr{H} = L^2(\mathbb{R}, dx; \mathscr{K})$ where $\mathscr{K}$ denotes a separable Hilbert space, and let $(U(t)f)(x) = f(x-t)$, $f \in L^2(\mathbb{R}, dx; \mathscr{K})$. Then, as it can be easily verified, the subspace $\mathscr{D}_+ = L^2([0, \infty), dx; \mathscr{K})$ is outgoing and the subspace $\mathscr{D}_- = L^2((-\infty, 0), dx; \mathscr{K})$ is incoming.

(b) Let $f_1, f_2 \in C_0^\infty(\mathbb{R}^3)$. We consider pairs $f = \{f_1, f_2\}$ and define a norm (which comes from a scalar-product) by

$$\|f\|^2 := (1/2) \int_{\mathbb{R}^3} \{|\operatorname{grad} f_1|^2 + |f_2|^2\}\, dx.$$

We define a Hilbert space $\mathscr{H}$ by $\mathscr{H} := \operatorname{clo}_{\|\cdot\|} \{C_0^\infty(\mathbb{R}^3) \times C_0^\infty(\mathbb{R}^3)\}$. We consider the wave equation $\partial^2 u/\partial t^2 = \Delta u$ in $\mathbb{R}^3$ and in particular the initial value problem

$$u(x, 0) = f_1(x), \qquad (\partial u/\partial t)\,(x, 0) = f_2(x).$$

Then the map

$$U(t)\colon \mathscr{H} \to \mathscr{H}, \qquad U(t)\,\{f_1, f_2\} = \{u, \partial u/\partial t\}$$

defines a unitary group on $\mathscr{H}$. Recall that the solution $u(x, t)$ is uniquely determined by its initial data $f = \{f_1, f_2\}$, that is, we have an operator $W$, $u = Wf$. Now we consider the set $\mathscr{D}_+$ of all initial data $f$, such that

$$(Wf)\,(x, t) \equiv 0, \qquad |x| \leq t,$$

that is, we assume that $u(x, t) \equiv 0$ in the so-called forward cone. Then $\mathscr{D}_+$ is outgoing with respect to $U(t)$, as it can easily be shown. Note that in this case the properties (I), (II), (III) correspond to simple geometrical properties of the cone and its time evolution.

### 12.1.3. Structure theorem for Lax-Phillips evolutions

Let $U(t)$ be a strongly continuous unitary group in $\mathscr{H}$, defined on the real line (unitary evolution group in $\mathscr{H}$).

**Definition 2.** A unitary evolution group $U(t)$ in $\mathscr{H}$ is called a *Lax-Phillips evolution* if there are incoming and outgoing subspaces in $\mathscr{H}$ with respect to $U(t)$.

According to this definition the evolutions of Examples (a) and (b) are Lax-Phillips evolutions. Example (a) refers to the translation group in the Hilbert space $L^2(\mathbb{R}, \mathrm{d}x; \mathscr{K})$. That is, this translation group is a Lax-Phillips evolution. This example is highly interesting because essentially all Lax-Phillips evolutions are of this form. In particular, essentially all outgoing and all incoming subspaces are of the form in this example. This is the content of the following

**Theorem 3.** *Let $U(t)$ be a Lax-Phillips evolution in $\mathscr{H}$ with outgoing subspace $\mathscr{D}_+$ and incoming subspace $\mathscr{D}_-$. Then there is an auxiliary Hilbert space $\mathscr{K}$ and an isometric operator $V_+$ from $\mathscr{H}$ onto $L^2(\mathbb{R}, \mathrm{d}x; \mathscr{K})$ such that $V_+\mathscr{D}_+ = L^2\big([0, \infty), \mathrm{d}x; \mathscr{K}\big)$ and*

$$\big(U_+(t)\, f\big)\,(x) := \big(V_+U(t)\, V_+{}^*f\big)\,(x) = f(x - t), \qquad f \in L^2(\mathbb{R}, \mathrm{d}x; \mathscr{K}).$$

*Similarly, there is an isometric operator $V_-$ from $\mathscr{H}$ onto $L^2(\mathbb{R}, \mathrm{d}x; \mathscr{K}')$, where $\mathscr{K}'$ is an auxiliary Hilbert space, such that $V_-\mathscr{D}_- = L^2\big((-\infty, 0], \mathrm{d}x; \mathscr{K}'\big)$ and*

$$\big(U_-(t)\, f\big)\,(x) := \big(V_-U(t)\, V_-{}^*f\big)\,(x) = f(x - t), \qquad f \in L^2(\mathbb{R}, \mathrm{d}x; \mathscr{K}').$$

*One can always choose $\mathscr{K} = \mathscr{K}'$, but in general $V_+ \neq V_-$ is valid. These representations of $U(t)$, given by $V_+$ and $V_-$, are unique up to isomorphisms of $\mathscr{K}$.*

Proof. There are many proofs of this theorem, we give a proof which depends on the spectral multiplicity theory. First we state a lemma, which is used in the course of the proof and which is proved for example in Reed/Simon [3, p. 215].

**Lemma 4.** *Let $\mu$ be a nontrivial Borel measure on $\mathbb{R}$ which is quasitranslation invariant, that is $\mu$ is equipped with the property that $\mu(\cdot + a)$ is equivalent to $\mu(\cdot)$ for all $a \in \mathbb{R}$. Then $\mu$ is equivalent to the Lebesgue measure $|\cdot|$.*

Now we define

$$\mathscr{D}_+(t) := U(t)\,\mathscr{D}_+, \qquad t \in \mathbb{R},$$

$$\mathscr{D}_+(\infty) := \{0\}, \qquad \mathscr{D}_+(-\infty) := \mathscr{H},$$

$$\mathscr{N}_{a,b} := \mathscr{D}_+(a) \cap \mathscr{D}_+(b)^\perp, \qquad a < b.$$

Let $E\big((a, b]\big)$ be the projection onto $\mathscr{N}_{a,b}$. According to the properties of $\mathscr{D}_+$ and the (strong) continuity of $U(t)$ it can easily be seen that $E\big((a, b]\big)$ generates a spectral measure $E(\Delta)$ on the real line with the property

$$U(t)\,E(\Delta)\,U(-t) = E(\Delta + t), \qquad t \in \mathbb{R}, \qquad \Delta \text{ Borel set.} \tag{1}$$

Sometimes (1) is called an imprimitivity relation. We define the selfadjoint operator $X = \int \lambda E(\mathrm{d}\lambda)$. Then from (1) we obtain

$$U(t)\,XU(-t) = X - t1_{\mathscr{H}}.$$

According to (1) the spectral measure classes of $X$ are invariant under translations (see Lemma 4.20 and Prop. 4.24 for spectral measure classes). Applying Lemma 4, we obtain: There is only one spectral measure class in this case which contains the Lebesgue measure. Therefore we obtain an isometric operator $\Omega_+$ from $\mathscr{H}$ onto $L^2(\mathbb{R}, \mathrm{d}x; \mathscr{K})$ such that

$$\Omega_+ E(\Delta)\,\Omega_+^* = \chi_\Delta \tag{2}$$

is valid, where $\chi_\Delta$ denotes the multiplication operator by $\chi_\Delta(\cdot)$. $\Omega_+$ already satisfies the first property of the operator $V_+$ of the theorem because $E\big((0, \infty)\big)$ is the projection onto $\mathscr{D}_+$. But $\Omega_+$ has to be modified in order to satisfy the second property, too. We define

$$W(t) := \Omega_+ U(t)\,\Omega_+^*; \tag{3}$$

$$\big(T(t)\,f\big)\,(x) := f(x - t), \tag{4}$$

that is, $W(t)$ and $T(t)$ are both unitary groups in $L^2(\mathbb{R}, \mathrm{d}x; \mathscr{K})$. We obtain

$$T(t)\,\chi_\Delta(\cdot)\,T(-t) = \chi_{\Delta+t}(\cdot), \qquad W(t)\,\chi_\Delta(\cdot)\,W(-t) = \chi_{\Delta+t}(\cdot),$$

hence

$$W(t)\,T(-t)\,\chi_\Delta = \chi_\Delta W(t)\,T(-t)$$

for all Borel sets $\Delta$. Therefore $V(t) := W(t)\,T(-t)$ is a unitary operator in $L^2(\mathbb{R}, \mathrm{d}x; \mathscr{K})$, which is a multiplication operator. Thus the operator $V(t)$ corresponds to an operator function $V(t; x)$ where $V(t; x)$ is a unitary operator in $\mathscr{K}$ defined a.e. (mod $|\cdot|$) in $x \in \mathbb{R}$. We have $W(t) = V(t)\,T(t)$. Now let $B$ be an arbitrary unitary multiplication operator in $L^2(\mathbb{R}, \mathrm{d}x; \mathscr{K})$, realized by a unitary operator function $B(x)$ in $\mathscr{K}$, a.e. (mod $|\cdot|$) defined on $\mathbb{R}$. Then

$$BW(t)\,B^* = B(\cdot)\,V(t; \cdot)\,T(t)\,B(\cdot)^* = B(\cdot)\,V(t; \cdot)\,B(\cdot - t)^*\,T(t)$$

because of $T(t)\,B(x)^* = B(x - t)^*\,T(t)$. Now the aim is to choose $B(x)$ such that

$$B(x)\,V(t; x)\,B(x - t)^* = 1_{\mathscr{K}}$$

is valid. Therefore it is sufficient to choose $B(x)$ such that

$$B(x-t) = B(x)\, V(t; x) \tag{5}$$

is valid for almost all $x$ and all $t$. If we fix $x = x_0$ and if we put $B(x_0) = 1_{\mathscr{K}}$, then we get $B(x_0 - t) = V(t; x_0)$. Therefore, the question is if there is $x_0$ such that $V(t; x_0)$ is defined for almost all $t$ and $V(t; x_0)$ is admissible in the sense of Section 4.4.2. Note that if $t$ is arbitrary but fixed, the representer $V(t; x)$ of $V(t)$ may be changed on a null set without changing the operator $V(t)$. Obviously it is sufficient to show that one can choose a representer $V(t; x)$ such that $V(t; x)$ is weakly measurable with respect to $|\cdot|$ and $(t, x) \in \mathbb{R}^2$ and it is defined a.e. mod $|\cdot|$ in $\mathbb{R}^2$. To this aim we consider the operator $C_h(t, x)$ defined by $C_h(t, x)\, k := h^{-1} \int_x^{x+h} V(t; y)\, k \,\mathrm{d}y$, $k \in \mathscr{K}$, which is a member of $\mathscr{L}(\mathscr{K})$ and which is strongly continuous with respect to $(t, x) \in \mathbb{R}^2$. Hence $C_+(t, x) := \operatorname{s-lim}_{h\to 0} C_h(t, x)$ is weakly measurable. Moreover, $C_+(t, x)$ is defined a.e. mod $|\cdot|$ in $\mathbb{R}^2$ and, if $t$ is fixed, $C_+(t, x)$ is a representer of $V(t)$ because of $C_+(t, x)\, k = V(t; x)\, k$ a.e. mod $|\cdot|$ (with respect to $x$) for all $k \in \mathscr{K}$. Now we define $B(\cdot)$ by

$$B(\xi) := C_+(x_0 - \xi, x_0)$$

where $x_0$ is suitable chosen. Then (5) is satisfied and we obtain $BW(t)\, B^* = T(t)$. Defining $V_+$ by $V_+ = B\Omega_+$ according to (3), we obtain the assertion. ■

**Remark 5.** The representation $U_+(t)$, $L^2([0, \infty), \mathrm{d}x; \mathscr{K})$, $L^2(\mathbb{R}, \mathrm{d}x; \mathscr{K})$ is called an *outgoing translation representation* of $U(t)$, defined by $V_+$; the representation $U_-(t)$, $L^2((-\infty, 0), \mathrm{d}x; \mathscr{K})$, $L^2(\mathbb{R}, \mathrm{d}x; \mathscr{K})$ is called an *incoming translation representation* of $U(t)$, defined by $V_-$.

**Remark 6.** If a unitary strongly continuous group $U(t)$ in $\mathscr{H}$ is given, then, by Stone's theorem, one has always a representation of $\mathscr{H}$ by a direct integral

$$\mathscr{H} \ni f \to \hat{f}(\lambda) \in L^2(\mathbb{R}, \mu(\mathrm{d}\lambda); \mathscr{K}_\lambda; \mathscr{S})$$

such that $U(t)$ is represented in $L^2(\mathbb{R}, \mu(\mathrm{d}\lambda); \mathscr{K}_\lambda; \mathscr{S})$ by the multiplication operator $\mathrm{e}^{-it\lambda}$. Recalling the connection of the translation group $f(x) \to f(x-t)$ in $L^2(\mathbb{R}, \mathrm{d}x; \mathscr{K})$ and the multiplication group $\hat{f}(\lambda) \to \mathrm{e}^{-it\lambda}\hat{f}(\lambda)$ in $L^2(\mathbb{R}, \mathrm{d}\lambda; \mathscr{K})$, given by Fourier transformation, we can state the following, in some sense a surprising fact: If there is an outgoing subspace $\mathscr{D}_+$, corresponding to $U(t)$, then one can choose $\mathscr{K}_\lambda \equiv \mathscr{K}$ (that is the Hilbert space $\mathscr{K}_\lambda$ in the direct integral can be fixed) and $\mu(\cdot) = |\cdot|$ (Lebesgue measure); in particular we have $U(t) = \mathrm{e}^{-itH}$ where $H$ is absolutely continuous, is of uniform multiplicity and has the property $\operatorname{spec}^c H = \mathbb{R}$.

In the following Lax-Phillips evolutions are considered which have the following additional property

$$\mathscr{D}_+ \perp \mathscr{D}_-. \tag{7}$$

The link between Lax-Phillips scattering and two-space scattering will be developed for Lax-Phillips evolutions with property (7) (see 12.2). Moreover, property (7) leads to nice properties of the $S$-matrix (see 12.3).

### 12.1.4. Spectral representation of Lax-Phillips evolutions

Let $U(t)$ be a Lax-Phillips evolution in $\mathscr{H}$ and let $V_+$ be chosen according to Theorem 3. By $F$ we denote the Fourier transformation from $L^2(\mathbb{R}, dx; \mathscr{K})$ onto $L^2(\mathbb{R}, d\lambda; \mathscr{K})$. We consider the isometric operator $FV_+$. Then

$$(FV_+)\, \mathscr{D}_+ = F\big(L_2(\mathbb{R}_+, dx; \mathscr{K})\big) = \mathscr{H}_-^2(\mathbb{R}, d\lambda; \mathscr{K})$$

where $\mathscr{H}_-^2(\mathbb{R}, d\lambda; \mathscr{K})$ denotes the Hardy class with respect to the lower half plane, and we get

$$FV_+U(t)\,(FV_+)^* = e^{-itH_0}$$

where $e^{-itH_0}$ means the multiplication operator by $e^{-it\lambda}$ in $L^2(\mathbb{R}, d\lambda; \mathscr{K})$. This representation of $U(t)$ in $L^2(\mathbb{R}, d\lambda; \mathscr{K})$, given by $FV_+$, is called an *outgoing spectral representation* of $U(t)$.

Correspondingly by $FV_-$ we obtain an *incoming spectral representation* with respect to $\mathscr{D}_-$ of $U(t)$ in $L^2(\mathbb{R}, d\lambda; \mathscr{K})$.

### 12.1.5. The Lax-Phillips scattering operator

Let $V_+$ and $V_-$ be given according to Theorem 3. We call $f_+ = V_+f$ the representer of $f$ in the outgoing translation representation of $U(t)$ and $f_- = V_-f$ the representer of $f$ in the incoming translation representation of $f$. Then the so-called *Lax-Phillips scattering operator* $S_{LP}$ is defined by $S_{LP}f_- := f_+$ as an operator from $L^2(\mathbb{R}, dx; \mathscr{K})$ into $L^2(\mathbb{R}; dx; \mathscr{K})$). Hence $S_{LP}V_-f = V_+f$ and we obtain

$$S_{LP} = V_+V_-^*. \tag{8}$$

$S_{LP}$ is unitary in $L^2(\mathbb{R}, dx; \mathscr{K})$ and commutes with the translation group $T(t)$ in this space. By Fourier transformation $F$ we define the operator $\hat{S}_{LP}$ by

$$\hat{S}_{LP} := FS_{LP}F^* \tag{9}$$

and obtain: $\hat{S}_{LP}$ commutes with the multiplication group $e^{-it\lambda}$ in $L^2(\mathbb{R}, d\lambda; \mathscr{K})$, hence $\hat{S}_{LP}$ is a multiplication operator, that is

$$(\hat{S}_{LP}f)\,(\lambda) = \hat{S}_{LP}(\lambda)\, f(\lambda), \qquad f \in L^2(\mathbb{R}, d\lambda; \mathscr{K}),$$

where $\hat{S}_{LP}(\lambda)$ is unitary in $\mathscr{K}$ and a.e. (mod $|\cdot|$) defined on $\mathbb{R}$. The operator function $\hat{S}_{LP}(\lambda)$ is sometimes called the *Lax-Phillips scattering matrix.*

## 12.2. Two-space formulation of the Lax-Phillips scattering theory

In this paragraph the link between the scattering theory of Lax-Phillips evolutions and the general two-space scattering theory is developed. It turns out that if a Lax-Phillips evolution $U(t)$ with property (7) is given, then it is always possible to construct a second evolution $U_0(t)$ in a second space $\mathscr{H}_0$ and an identification operator $J \in \mathscr{L}(\mathscr{H}_0, \mathscr{H})$ such that the two-space wave operators exist and up to an unessential factor the two-space scattering operator and the Lax-Phillips scattering operator coincide.

### 12.2.1. Construction of special incoming and outgoing translation representations

Let $U(t)$ in $\mathscr{H}$ and $U_0(t)$ in $\mathscr{H}_0$ be unitary strongly continuous evolutions. Further, let $J \in \mathscr{L}(\mathscr{H}_0; \mathscr{H})$. We assume the following conditions:

(I) There are subspaces $\mathscr{E}_+$, $\mathscr{E}_-$ of $\mathscr{H}_0$ which are orthogonal, $\mathscr{E}_+ \perp \mathscr{E}_-$, and such that $J$ is isometric on $\mathscr{E}_+ \oplus \mathscr{E}_-$ and $J$ maps $(\mathscr{E}_+ \oplus \mathscr{E}_-)^\perp$ into $\big(J(\mathscr{E}_+ \oplus \mathscr{E}_-)\big)^\perp$. We put $\mathscr{D}_+ := J\mathscr{E}_+$ and $\mathscr{D}_- := J\mathscr{E}_-$.

(II) $\mathscr{E}_+$ and $\mathscr{E}_-$ are outgoing and incoming subspaces for $U_0(t)$. $\mathscr{D}_+$ and $\mathscr{D}_-$ are outgoing and incoming subspaces for $U(t)$.

(III) The intertwining relations $U(t)\, Jf = JU_0(t)\, f$, $f \in \mathscr{E}_+$, $t \geqq 0$, and $U(t)\, Jf = JU_0(t)\, f$, $f \in \mathscr{E}_-$, $t \leqq 0$, are valid.

(IV) There is a representation of $\mathscr{H}_0$ by $L^2(\mathbb{R}, dx; \mathscr{K})$ such that this representation is (essentially) both incoming and outgoing for $U_0(t)$, that is, there exists an isometric operator $V_0$ from $\mathscr{H}_0$ onto $L^2(\mathbb{R}, dx; \mathscr{K})$ such that for some $a \geqq 0$

$$V_0\mathscr{E}_+ = L^2\big([a, \infty), dx; \mathscr{K}\big), \tag{10}$$

$$V_0\mathscr{E}_- = L^2\big((-\infty, -a], dx; \mathscr{K}\big), \tag{11}$$

$$V_0U_0(t)\, V_0^* = T(t), \qquad \big(T(t)\, f\big)(x) = f(x - t), \tag{12}$$

is valid.

**Remark 7.** It is always possible to fit a given Lax-Phillips evolution with property (7) into this framework. Let $U(t)$ be such an evolution with outgoing subspace $\mathscr{D}_+$ and incoming subspace $\mathscr{D}_-$. Further let $V_+$, $V_-$ be chosen according to Theorem 3. We define

$$\mathscr{E}_+ := L^2\big([a, \infty), dx; \mathscr{K}\big), \qquad \mathscr{E}_- := L^2\big((-\infty, -a], dx; \mathscr{K}\big),$$

$$Jf := \begin{cases} V_+^*T(-a)\, f, & f \in \mathscr{E}_+, \\ V_-^*T(a)\, f, & f \in \mathscr{E}_-, \\ Jf \in \big(J(\mathscr{E}_+ \oplus \mathscr{E}_-)\big)^\perp, & f \in L^2([-a, a], dx; \mathscr{K}). \end{cases}$$

Further, put $U_0(t) = T(t)$, $V_0 = 1$. Then obviously (I), (II), (IV) are satisfied. We have to check (III) only. But for $f \in \mathscr{E}_+$ we obtain $U(t)\, Jf = U(t)\, V_+^*T(-a)\, f = V_+^*T(t) \times T(-a)\, f = JU_0(t)\, f$ for $t \geqq 0$. Similarly, we can verify the second relation.

In general it is not possible to define the free dynamics $U_0(t)$ on the same Hilbert space as $U(t)$ without modifying the subspaces $\mathscr{D}_+$, $\mathscr{D}_-$. The reason is that the subspace $(\mathscr{D}_+ \oplus \mathscr{D}_-)^\perp$ can have arbitrary dimension.

Now under the conditions (I)—(IV) we can construct a special outgoing (incoming) translation representation. The crucial point is the construction of the representation; the general existence of outgoing (incoming) translation representations is already proved in Theorem 3.

Let $f \in \mathscr{D}_+$. Then there is $f_0 \in \mathscr{E}_+$, which is uniquely determined, such that $Jf_0 = f$. Now we define an operator $V_+$ by

$$V_+U(t)\, f := T(t)\, V_0 f_0.$$

The operator $V_+$ is isometric because of

$$\|U(t)\, f\| = \|f\| = \|f_0\| = \|T(t)\, V_0 f_0\|$$

and it is densely defined because $\mathscr{D}_+$ is an outgoing subspace. Further, ima $V_+$ is dense in $L^2(\mathbb{R}, dx; \mathscr{K})$ because of (10) and the fact that if translations are applied to $L^2([a, \infty), dx; \mathscr{K})$, then the whole space can be densely exhausted. Therefore, by continuous extension we finally get an isometric operator $V_+$ from $\mathscr{H}$ onto $L^2(\mathbb{R}, dx; \mathscr{K})$ with the property

$$V_+U(t)\, f = T(t)\, V_0 f_0, \qquad f \in \mathscr{D}_+,$$

and

$$V_+U(t)\, f = T(t)\, V_+ f, \qquad f \in \mathscr{H}. \tag{13}$$

Further we have $V_+\mathscr{D}_+ = L^2([a, \infty), dx; \mathscr{K})$. Therefore $\tilde{V}_+ := T(-a)\, V_+$ has the desired property

$$\tilde{V}_+\mathscr{D}_+ = L^2(\mathbb{R}_+, dx; \mathscr{K}).$$

Moreover, because of (13) $U(t)$ is represented via $\tilde{V}_+$ by $T(t)$, that is, $\tilde{V}_+$ yields an outgoing translation representation for $U(t)$. Similarly, we proceed for the construction of an incoming translation representation. Note that property (I) automatically implies $\mathscr{D}_+ \perp \mathscr{D}_-$.

### 12.2.2. Wave operators and scattering operator

In this section the existence of complete wave operators within the framework of Section 1 is proved.

**Proposition 8.** *Under the conditions* (I)—(IV) *the two-space strong wave operators*

$$W_\pm f = \operatorname*{s-lim}_{t \to \pm\infty} U(-t)\, JU_0(t)\, f, \qquad f \in \mathscr{H}_0$$

*exist and are isometric on* $\mathscr{H}_0$. *Moreover*

$$W_+ f = Jf \quad \text{for} \quad f \in \mathscr{E}_+, \qquad W_- f = Jf \quad \text{for} \quad f \in \mathscr{E}_-. \tag{14}$$

**Proof.** We define a dense set $\mathscr{D}$ of $\mathscr{H}_0$ by the property that $f \in \mathscr{D}$ iff the function $\hat{f}(x) = (V_0 f)\,(x)$ has compact support on $\mathbb{R}$. Because of $\big(T(s)\, \hat{f}\big)\,(x) = \hat{f}(x - s)$ we obtain supp $T(s)\, \hat{f} = s + \operatorname{supp} \hat{f}$, that is, if $s$ is sufficiently large, then in every case we have supp $T(s)\, \hat{f} \in [a, \infty)$ or

$$T(s)\, V_0 f \in V_0\mathscr{E}_+, \qquad f \in \mathscr{D}.$$

In other words, we obtain $V_0{}^*T(s)\,V_0 f \in \mathscr{E}_{+}$, that is $U_0(s)\,f \in \mathscr{E}_+$ for all $f \in \mathscr{D}$. Therefore, because of (III), we obtain for $t \geqq 0$

$$U(-(t+s))\,JU_0(t+s)\,f = U(-s)\,U(-t)\,JU_0(t)\,U_0(s)\,f$$
$$= U(-s)\,U(-t)\,U(t)\,JU_0(s)\,f = U(-s)\,JU_0(s)\,f.$$

Thus the wave operator $W_+$ exists obviously and it is isometric. Relation (14) is immediately implied by (III). Similarly we proceed in the case $t \to -\infty$. ∎

Now we prove the completeness of the wave operators $W_\pm$ and then we clarify the relation between the scattering operator defined by $W_+$ and $W_-$ and the Lax-Phillips scattering operator $S_{\text{LP}}$.

**Proposition 9.** *The wave operators $W_+$ and $W_-$ are complete.*

Proof. Let $\mathscr{D}$ be defined as in the foregoing proof. Assume $f \in \mathscr{D}$. Then we get $U_0(s)\,f \in \mathscr{E}_+$ for some $s$. Hence, because of (14), we obtain

$$W_+U_0(s)\,f = JU_0(s)\,f \in \mathscr{D}_+. \tag{15}$$

Now recall the operator $\tilde{V}_+ = T(-a)\,V_+$ which realizes an outgoing translation representation for $U(t)$. Therefore the operator $\tilde{V}_+W_+$ is intertwining with respect to $U_0(t)$ and $T(t)$ and we obtain

$$\tilde{V}_+W_+f = T(-s)\,\tilde{V}_+W_+U_0(s)\,f = T(-s)\,\tilde{V}_+JU_0(s)\,f = T(-s)\,T(-a)\,V_+JU_0(s)\,f.$$

Because of (15) and $V_+Jf_0 = V_0f_0$ for all $f_0 \in \mathscr{E}_+$, we have $V_+(JU_0(s)\,f) = V_0U_0(s)\,f$. Hence we further obtain $\tilde{V}_+W_+f = T(-s)\,T(-a)\,V_0U_0(s)\,f = T(-a)\,V_0f$ since $V_0$ is intertwining between $U_0(t)$ and $T(t)$. But $\mathscr{D}$ is dense, hence $\tilde{V}_+W_+ = T(-a)\,V_0$ or

$$V_+W_+ = V_0 \tag{16}$$

follows. From (16) we obtain $W_+ = V_+{}^*V_0$ and $W_+{}^* = V_0{}^*V_+$, and finally $W_+W_+{}^* = 1_{\mathscr{H}} = P_H^{\text{ac}}$ because the generator $H$ of $U(t)$ is absolutely continuous. Similarly, we proceed for $W_-$. Hence $W_-W_-{}^* = 1_{\mathscr{H}} = P_H^{\text{ac}}$ is also valid, that is, $W_+$ and $W_-$ are complete wave operators. ∎

**Theorem 10.** *The S-operator corresponding to the wave operators of Prop.* 8 *is given by*

$$S = W_+{}^*W_- = V_0{}^*T(2a)\,S_{\text{LP}}V_0, \tag{17}$$

*that is, except for the unessential factor $T(2a)$ which commutes with $S_{\text{LP}}$, $S$ is just the Lax-Phillips scattering operator $S_{\text{LP}}$ pulled back to $\mathscr{H}_0$.*

Proof. We have $W_+ = \tilde{V}_+{}^*T(-a)\,V_0$ and $W_- = \tilde{V}_-{}^*T(a)\,V_0$, hence we obtain

$$W_+{}^*W_- = V_0{}^*T(a)\,\tilde{V}_+\tilde{V}_-{}^*T(a)\,V_0 = V_0{}^*T(2a)\,S_{\text{LP}}V_0. \;∎$$

Finally we note the connection between the corresponding $S$-matrices, for which the connection is even more intimate.

**Proposition 11.** *The S-matrices $\hat{S}(\lambda)$, $\hat{S}_{LP}(\lambda)$ of $S$, $S_{LP}$ respectively, are unitary operator functions from $\mathbb{R}$ into $\mathscr{L}(\mathscr{K})$ defined a.e.* (mod $|\cdot|$) *and satisfy the relation*

$$\hat{S}(\lambda) = e^{-2ia\lambda}\hat{S}_{LP}(\lambda). \tag{18}$$

Proof. $S$ commutes with $U_0(t)$, hence $V_0SV_0^*$ commutes with $T(t)$, and also $T(2a)\, S_{LP}$ commutes with $T(t)$. By Fourier transformation we obtain: The space $L^2(\mathbb{R}, d\lambda; \mathscr{K})$ may be used as a spectral representation space for $U_0(t)$, such that $S$ is represented by $\hat{S}(\lambda)$, $U_0(t)$ by $e^{-it\lambda}$. Then the assertion follows from (17). ∎

## 12.3. Analytic properties of the Lax-Phillips scattering matrix

The special properties of Lax-Phillips evolutions with condition (7) lead to nice analytic properties of $\hat{S}_{LP}(\lambda)$ with respect to the upper half plane, which will be derived in the following.

### 12.3.1. Analyticity in the upper half plane

In order to show the analytic continuability of $\hat{S}_{LP}(\lambda)$ into the upper half plane, we begin with two propositions.

**Proposition 12.** *The operator $S_{LP}$ is unitary in $L^2(\mathbb{R}, dx; \mathscr{K})$, it commutes with $T(t)$ and*

$$S_{LP}\big(L^2(\mathbb{R}_-, dx; \mathscr{K})\big) \subseteq L^2(\mathbb{R}_-, dx; \mathscr{K}) \tag{19}$$

*is valid.*

Proof. Only (19) is to be proved. For this reason let $f \in L^2(\mathbb{R}_-, dx; \mathscr{K})$. Then $f = V_-d_-$ where $d_- \in \mathscr{D}_-$. Hence we get

$$S_{LP}f = V_+V_-^*V_-d_- = V_+d_-.$$

Now we have $d_- \perp \mathscr{D}_+$, hence we obtain $(V_+d_-) \perp V_+\mathscr{D}_+ = L^2(\mathbb{R}_+, dx; \mathscr{K})$. Therefore $V_+d_- \in L^2(\mathbb{R}_-, dx; \mathscr{K})$ results, that is $S_{LP}f \in L^2(\mathbb{R}_-, dx; \mathscr{K})$. ∎

From Prop. 12 we easily obtain by Fourier transformation

**Proposition 13.** *The operator $\hat{S}_{LP} = FS_{LP}F^*$ is unitary in $L^2(\mathbb{R}, d\lambda; \mathscr{K})$, it commutes with the multiplication group $e^{-it\lambda}$ and*

$$\hat{S}_{LP}\mathscr{H}_+^2(\mathbb{R}, d\lambda; \mathscr{K}) \subseteq \mathscr{H}_+^2(\mathbb{R}, d\lambda; \mathscr{K}) \tag{20}$$

*is valid.*

Proof. The assertion is the Fourier transformed version of Prop. 12. ∎

Property (20) has the following interesting consequence:

**Theorem 14.** *Let $U(t)$ be a Lax-Phillips evolution in $\mathscr{H}$ with property (7) and with the Lax-Phillips scattering matrix $\hat{S}_{LP}(\lambda)$. Then there is a holomorphic operator function $\hat{S}(z)$, defined on the upper half plane, $\operatorname{Im} z > 0$, with values in $\mathscr{L}(\mathscr{K})$ and with the following properties:*

$$\|\hat{S}(z)\|_{\mathscr{L}(\mathscr{K})} \le 1, \tag{21}$$

*the strong limit* s-$\lim_{\varepsilon\to+0} \hat{S}(\lambda + i\varepsilon)$ *exists a.e.* (mod $|\cdot|$) *on* $\mathbb{R}$ *and*

$$\text{s-}\lim_{\varepsilon\to+0} \hat{S}(\lambda + i\varepsilon) = \hat{S}_{LP}(\lambda) \tag{22}$$

*is valid a.e.* (mod $|\cdot|$) *on* $\mathbb{R}$; *in other words,* $\hat{S}_{LP}(\lambda)$ *can be analytically continued into the whole upper half plane.*

**Proof.** Let $k \in \mathscr{K}$ and $\alpha(\lambda)$ be a fixed scalar function from $\mathscr{H}_+{}^2(\mathbb{R}, d\lambda)$, for example $\alpha(\lambda) = (\lambda + i)^{-1}$. Then, because of (20), we have

$$\hat{S}_{LP}(\lambda) \left(\alpha(\lambda)\, k\right) = \alpha(\lambda)\, \hat{S}_{LP}(\lambda)\, k =: g(\lambda; k) \in \mathscr{H}_+{}^2(\mathbb{R}, d\lambda; \mathscr{K}),$$

hence $\hat{S}_{LP}(\lambda)\, k = (\lambda + i)\, g(\lambda; k)$. We define $\hat{S}(z)\, k := (z + i)\, g(z; k)$, $\operatorname{Im} z > 0$. $\hat{S}(z)$ is a linear operator. Because of the Hardy class properties (see Section 1.2.3) we immediately get s-$\lim_{\varepsilon\to+0} \hat{S}(\lambda + i\varepsilon)\, k = \hat{S}_{LP}(\lambda)\, k$ a.e. (mod $|\cdot|$) on $\mathbb{R}$ follows. The exceptional set may depend on $k$. Further $\|\hat{S}(z)\, k\|_{\mathscr{K}} \leq \|\hat{S}_{LP}(\lambda)\, k\| = \|k\|$ follows from the maximum principle, hence we obtain (21). Finally, with the help of (21), the dependence of the exceptional set from $k$ may be removed and (22) follows. ∎

According to (20) we can form the subspace $\mathscr{H}_+{}^2(\mathbb{R}, d\lambda; \mathscr{K}) \ominus \hat{S}_{LP}\mathscr{H}_+{}^2(\mathbb{R}, d\lambda; \mathscr{K})$. In the following we need a characterization of this subspace. First we consider the (inverse) Fourier transformed space $L^2(\mathbb{R}_-, dx; \mathscr{K}) \ominus S_{LP}L^2(\mathbb{R}_-, dx; \mathscr{K})$.

**Lemma 15.** *The equation*

$$L^2(\mathbb{R}_-, dx; \mathscr{K}) \ominus S_{LP}L^2(\mathbb{R}_-, dx; \mathscr{K}) = V_+(\mathscr{D}_+ \oplus \mathscr{D}_-)^\perp \tag{23}$$

*is valid.*

**Proof.** We have $V_+\mathscr{D}_+{}^\perp = L^2(\mathbb{R}_-, dx; \mathscr{K})$ and $(\mathscr{D}_+ \oplus \mathscr{D}_-)^\perp = \mathscr{D}_+{}^\perp \ominus \mathscr{D}_-$. Further, $V_+\mathscr{D}_- = S_{LP}V_-\mathscr{D}_- = S_{LP}L^2(\mathbb{R}_-, dx; \mathscr{K})$ is valid. Hence (23) is proved. ∎

**Lemma 16.** *The equation*

$$\mathscr{H}_+{}^2(\mathbb{R}, d\lambda; \mathscr{K}) \ominus \hat{S}_{LP}\mathscr{H}_+{}^2(\mathbb{R}, d\lambda; \mathscr{K}) = (FV_+)\, (\mathscr{D}_+ \oplus \mathscr{D}_-)^\perp \tag{24}$$

*is valid.*

**Proof.** Fourier transformation of (23). ∎

Another form of (24) is expressed by

**Proposition 17.** *The subspace* $(FV_+)\, (\mathscr{D}_+ \oplus \mathscr{D}_-)^\perp$ *is given by*

$$(FV_+)\, (\mathscr{D}_+ \oplus \mathscr{D}_-)^\perp = \mathscr{H}_+{}^2(\mathbb{R}, d\lambda; \mathscr{K}) \cap \hat{S}_{LP}\mathscr{H}_-{}^2(\mathbb{R}, d\lambda; \mathscr{K}) \tag{25}$$

*or, in other words,* $f \in (\mathscr{D}_+ \oplus \mathscr{D}_-)^\perp$ *iff the outgoing spectral representer* $\hat{f}(\lambda)$ *of* $f$ *has the properties*

$$\hat{f} \in \mathscr{H}_+{}^2(\mathbb{R}, d\lambda; \mathscr{K}), \qquad \hat{S}^*_{LP}\hat{f} \in \mathscr{H}_-{}^2(\mathbb{R}, d\lambda; \mathscr{K}). \tag{26}$$

**Proof.** (25) is obvious since $\mathscr{H}_+{}^2 = (\mathscr{H}_-{}^2)^\perp$ and $\hat{S}_{LP}$ is unitary. But $\hat{f} = \hat{S}_{LP}\hat{g}$ with a $\hat{g} \in \mathscr{H}_-{}^2$ is equivalent with $\hat{S}^*_{LP}\hat{f} = \hat{g}$, hence (26) is proved. ∎

## 12.3.2. Analytic continuitation of $\hat{S}(z)$ into the lower half plane and the Lax-Phillips semigroup

According to Section 3.1, it is natural to ask for conditions of analytic continuation of $\hat{S}(z)$ into the lower half plane. Since $\hat{S}(z)$ is unitary on the real axis, analytic continuation is simply given by the formula

$$\hat{S}(z) = \left(\hat{S}(\bar{z})^*\right)^{-1}, \qquad \operatorname{Im} z < 0. \tag{27}$$

But in order to apply (27), we have to get information on the existence of $\hat{S}(\bar{z})^{-1}$ as a bounded operator, in other words, we need to know $0 \in \operatorname{res} \hat{S}(\bar{z})$. The investigation of this property can be connected with the study of the following semigroup $Z(t)$, which corresponds to a given Lax-Phillips evolution $U(t)$.
Let $P_+$ denote the projection onto $\mathcal{D}_+^\perp$, $P_-$ the projection onto $\mathcal{D}_-^\perp$ and $P_{\mathcal{N}}$ the projection onto $\mathcal{N} := (\mathcal{D}_+ \oplus \mathcal{D}_-)^\perp$. Then we define

$$Z(t) := P_{\mathcal{N}} U(t) P_{\mathcal{N}}, \qquad t \geqq 0. \tag{28}$$

We note the following properties of $Z(t)$:

**Proposition 18.** *The family $Z(t)$, $t \geqq 0$, is a strongly continuous semigroup of contractions. The relations*

$$Z(t) = P_+ U(t) P_- \tag{29}$$

*and*

$$\operatorname*{s-lim}_{t\to\infty} Z(t) = 0 \tag{30}$$

*are valid.*

**Proof.** Note that $P_+^\perp$ is the projection onto $\mathcal{D}_+$. Note further that $U(t)$ maps $\mathcal{D}_+$ into $\mathcal{D}_+$ for $t \geqq 0$. Therefore $P_+^\perp U(t) P_+^\perp = U(t) P_+^\perp$ or

$$P_+ U(t) = P_+ U(t) P_+, \qquad t \geqq 0.$$

Further, $U(-t)$ maps $\mathcal{D}_-$ into $\mathcal{D}_-$ for $t \geqq 0$. Therefore $P_-^\perp U(-t) P_-^\perp = U(-t) P_-^\perp$ or

$$P_- U(t) P_- = U(t) P_-, \qquad t \geqq 0.$$

Moreover, we have $P_{\mathcal{N}} = P_- P_+ = P_+ P_-$. Hence

$$P_{\mathcal{N}} U(t) P_{\mathcal{N}} = P_+ P_- U(t) P_- P_+ = P_+ U(t) P_- P_+ = P_+ U(t) P_-$$

follows, that is, (29) is proved. From (29) and $P_+ P_- = P_- P_+$ we get

$$Z(t)\, Z(s) = P_+ U(t) P_- P_+ U(s) P_- = P_+ U(t)\, U(s) P_- = P_+ U(t + s) P_- = Z(t + s),$$

that is, $Z(t)$ is a semigroup. Obviously, this semigroup is strongly continuous and $Z(t)$ is a contraction for all $t \geqq 0$. Now we prove (30). Let $u \in \mathcal{N}$ and $\varepsilon > 0$ be given. According to property (III) of Definition 1, there are a $T = T(\varepsilon)$ and a vector $f \in \mathcal{D}_+$ such that

$$\|u - U(-T) f\| < \varepsilon.$$

Further, $P_+U(t)\,U(-T)\,f = P_+U(t-T)\,f = 0$ for $t > T$ because of $U(t-T)\,f \in \mathscr{D}_+$. Therefore

$$\|Z(t)\,u\| = \|P_+U(t)\,u\| = \|P_+U(t)\,u - P_+U(t)\,U(-T)\,f\| \leqq \|u - U(-T)\,f\| < \varepsilon$$

for $t > T$, that is, (30) is proved. ■

Prop. 18 implies the existence of an accretive operator $B \in \mathscr{L}(\mathscr{N})$ with spec $B \subseteqq \{z; \operatorname{Re} z \geqq 0\}$ such that $Z(t) \restriction \mathscr{N} = e^{-tB}$. $B$ is the generator of the semigroup $Z(t) \restriction \mathscr{N}$ (see Yosida [1, p. 250], where the theorem is formulated for the dissipative operator $-B$). Now we calculate $Z(t)$ in the outgoing translation representation. We put $Z_+(t) := V_+Z(t)\,V_+^*$. Let $f_+ = V_+f$, $f \in \mathscr{N}$, that is, according to Lemma 15, $f_+ \in L^2(\mathbb{R}_-, ds; \mathscr{K}) \ominus S_{LP}L^2(\mathbb{R}_-, ds; \mathscr{K})$. Then we obtain

**Lemma 19.** *$Z_+(\cdot)$ is given by*

$$\big(Z_+(t)\,f_+\big)(s) = \chi_{\mathbb{R}_-}(s)\,f_+(s-t), \qquad t \geqq 0. \tag{31}$$

Proof. We have $Z_+(t) = V_+P_+U(t)\,P_-V_+^* = V_+P_+V_+^*T(t)\,V_+P_-V_+^*$. $V_+P_+V_+^*$ is the projection onto $L^2(\mathbb{R}_-, ds; \mathscr{K})$, that is, multiplication by $\chi_{\mathbb{R}_-}(s)$. Further we have $(V_+P_-V_+^*)\,V_+f = V_+P_-f = V_+f$ for $f \in \mathscr{N}$. Hence (31) is proved. ■

Now we can prove the following interesting result which states a first link between $Z(t)$ and $\hat{S}_{LP}$.

**Proposition 20.** *Let* $\operatorname{Re}\xi > 0$. *Then*

$$\dim\big(\ker(\xi 1 - B)\big) = \dim\big(\ker \hat{S}(i\bar{\xi})^*\big). \tag{32}$$

Proof. Let $f \in \ker(\xi 1 - B)$, that is, let $f \in \mathscr{N}$ and $Bf = \xi f$ or

$$Z(t)\,f = e^{-t\xi} f, \qquad t \geqq 0. \tag{33}$$

The representer of $f$ in the outgoing translation representation is denoted by $f_+$. Then, by Lemma 19, (33) is equivalent to

$$\chi_{\mathbb{R}_-}(s)\,f_+(s-t) = e^{-t\xi} f_+(s), \qquad t \geqq 0. \tag{34}$$

Since $f \in \operatorname{dom} B$, the function $f_+(s)$ is continuous for $s \to -0$. Hence (34) is valid iff there is a vector $k \in \mathscr{K}$ such that

$$f_+(s) = e^{s\xi}\chi_{\mathbb{R}_-}(s)\,k. \tag{35}$$

The outgoing spectral representer of $f$ is denoted by $\hat{f}_+$, $\hat{f}_+ \in L^2(\mathbb{R}, d\lambda; \mathscr{K})$. From (35) we obtain

$$\hat{f}_+(\lambda) = (2\pi)^{-1/2} \int_{-\infty}^{\infty} e^{-is\lambda} f_+(s)\,ds = -(2\pi)^{-1/2}\,(i\lambda - \xi)^{-1}\,k.$$

According to Prop. 17, $f \in \mathscr{N}$ is equivalent to $\hat{f}_+ \in \mathscr{H}_+{}^2(\mathbb{R}, d\lambda; \mathscr{K})$ and $\hat{S}_{LP}\hat{f}_+ \in \mathscr{H}_-{}^2(\mathbb{R}, d\lambda; \mathscr{K})$. In our case $\hat{f}_+ \in \mathscr{H}_+{}^2(\mathbb{R}, d\lambda; \mathscr{K})$ is obvious. Therefore in this case the condition $f \in \mathscr{N}$ is equivalent to

$$\hat{S}_{LP}(\lambda)^*\big((i\lambda - \xi)^{-1}\,k\big) = (i\lambda - \xi)^{-1}\,\hat{S}_{LP}(\lambda)^*\,k \in \mathscr{H}_-{}^2(\mathbb{R}, d\lambda; \mathscr{K}). \tag{36}$$

Since $\hat{S}(z)$ is the analytic continuation of $\hat{S}_{LP}(\lambda)$ into the upper half plane, $\hat{S}(\bar{z})^*$ is the analytic continuation of $\hat{S}_{LP}(\lambda)^*$ into the lower half plane. Hence the continuation of (36) is given by

$$(iz - \xi)^{-1} \hat{S}(\bar{z})^* k.$$

Because of the condition $\hat{S}^*_{LP}\hat{f}_+ \in \mathscr{H}_-^2(\mathbb{R}, d\lambda; \mathscr{K})$ this function has to be holomorphic in the lower half plane. But this is the case if and only if

$$\hat{S}(-i\bar{\xi})^* k = \hat{S}(i\bar{\xi})^* k = 0,$$

that is, if $k \in \ker \hat{S}(i\bar{\xi})^*$. If two vectors $f_1$, $f_2$ from $\ker(\xi 1 - B)$ are given, where $f_1 \perp f_2$, then we immediately obtain for the corresponding vectors $k_1$, $k_2$ also $k_1 \perp k_2$. Hence (32) is proved. ■

Similarly

$$\dim\left(\ker\left(\bar{\xi} - B^*\right)\right) = \dim\left(\ker \hat{S}(i\bar{\xi})\right) \tag{37}$$

can be proved. Finally, we study the connection, mentioned above, between $Z(t)$ and the analytic continuation of $\hat{S}(z)$ into the lower half plane. First we state two lemmas.

By Lemma 19, the generator $B_+$ of $Z_+(t)$ is characterized by

$$(B_+f_+)(s) = \chi_{\mathbb{R}_-}(s) f_+'(s)$$

where dom $B_+$ consists of all $f_+ \in \mathscr{N}_+ = L^2(\mathbb{R}_-, ds; \mathscr{K}) \ominus S_{LP}L^2(\mathbb{R}_-, ds; \mathscr{K})$ such that $f_+' \in \mathscr{N}$. For the Fourier transformed generator $\hat{B}_+ = FB_+F^*$ we obtain the following characterization.

**Lemma 21.** *Let* $\hat{f}_+ \in \hat{\mathscr{N}}_+$. *Then* $\hat{f}_+ \in$ dom $\hat{B}_+$ *iff there is a vector* $k \in \mathscr{K}$ *such that*

$$i\lambda\hat{f}_+(\lambda) - k \in \hat{\mathscr{N}}_+ = \mathscr{H}_+^2(\mathbb{R}, d\lambda; \mathscr{K}) \ominus \hat{S}_{LP}\mathscr{H}_+^2(\mathbb{R}, d\lambda; \mathscr{K}).$$

*In this case*

$$(\hat{B}_+\hat{f}_+)(\lambda) = i\lambda\hat{f}_+(\lambda) - k$$

*is valid.*

Proof. We obtain for all Schwartz space functions $g \in L^2(\mathbb{R}, ds; \mathscr{K})$ the relation

$$\int_{-\infty}^{\infty} \left(g(s), (B_+f_+)(s)\right)_{\mathscr{K}} ds = \int_{-\infty}^{\infty} \left(\hat{g}(\lambda), i\lambda\hat{f}_+(\lambda) - k\right)_{\mathscr{K}} d\lambda, \tag{38}$$

where $k = -(2\pi)^{-1/2} f_+(0)$. From (38) the assertion follows. ■

**Lemma 22.** *Let* $\operatorname{Re}\xi > 0$ *and* $k \in \mathscr{K}$, $\|k\|_{\mathscr{K}} = 1$. *Then there is a vector* $g \in \mathscr{N}$, $g \neq 0$, *such that*

$$\|Z(t) g - e^{-\xi t} g\| \leq 4 \|\hat{S}(i\bar{\xi})^* k\|_{\mathscr{K}} \|g\| \tag{39}$$

Proof. See Lax/Phillips [1, p. 78]. ■

Now we can formulate the theorem. It states essentially that the unitary operator function $\hat{S}_{LP}(\lambda)$, defined a.e. (mod $|\cdot|$) on $\mathbb{R}$, can be holomorphically continued into the region $\{z; z = -i\xi, \xi \in \operatorname{res} B\}$ of the lower half plane. Moreover, this region is the maximal region of holomorphic continuation for $\hat{S}_{LP}(\lambda)$. Note that the holomorphic continuations of $\hat{S}_{LP}(\lambda)$ into the lower and the upper half plane are not necessarily branches of a unique analytic function. In order to decide this question one has to investigate $\hat{S}_{LP}(\lambda)$ on the real axis. According to (27) we can transform the question of holomorphic continuability of $\hat{S}_{LP}(\lambda)$ into the lower half plane into the question of bounded invertibility of $\hat{S}(\bar{z})$ or $\hat{S}(\bar{z})^*$. Therefore we can state the theorem as follows.

**Theorem 23.** *Let* $\operatorname{Re} \xi > 0$. *Then* $\xi \in \operatorname{res} B$ *iff* $\hat{S}(i\bar{\xi})$ *is bounded invertible.*

Proof. First, we assume $\xi \in \operatorname{res} B$. Then, by Prop. 20, we obtain $\ker \hat{S}(i\xi)^* = \{0\}$. Further, we have $\bar{\xi} \in \operatorname{res} B^*$, and applying (37); we obtain $\ker \hat{S}(i\bar{\xi}) = \{0\}$. Hence $\hat{S}(i\bar{\xi})^*$ is one-to-one and $\operatorname{ima} \hat{S}(i\bar{\xi})^*$ is dense in $\mathscr{K}$. Now we consider the operator

$$(1 + B)^{-1} - (1 + \xi)^{-1} = \int_0^\infty e^{-t}\left(Z(t) - e^{-\xi t}\right) dt.$$

Since $(1 + B)^{-1} - (1 + \xi)^{-1} = (\xi - B)(1 + B)^{-1}(1 + \xi)^{-1}$ this operator is bounded invertible. Applying Lemma 22 we obtain

$$\left\|\left((1 + B)^{-1} - (1 + \xi)^{-1}\right) g\right\| \leq 4 \,\|\hat{S}(i\bar{\xi})^* k\|\, \|g\|,$$

hence $\|\hat{S}(i\bar{\xi})^* k\| \geq (1/4) \|g\|^{-1} \left\|\left((1 + B)^{-1} - (1 + \xi)^{-1}\right) g\right\|$. Therefore $\left(\hat{S}(i\bar{\xi})^*\right)^{-1}$ is bounded, hence $\hat{S}(i\bar{\xi})^{-1}$ is also bounded.

Second, we assume $\hat{S}(i\bar{\xi})^*$ is bounded invertible. Then, by Prop. 20, we have $\ker (\xi - B) = \{0\}$ and we assert that $(\xi - B)^{-1}$ is bounded. According to the closed graph theorem it is sufficient to prove $\operatorname{ima} (\xi - B) = \mathscr{N}$, or equivalently $\operatorname{ima} (\xi - \hat{B}_+) = \hat{\mathscr{N}}_+$. For this purpose let $\hat{f}$ be arbitrary from $\hat{\mathscr{N}}_+$. Then, according to Lemma 21, we have to prove the existence of a function $\hat{h} \in \hat{\mathscr{N}}_+$ and a vector $k \in \mathscr{K}$ such that

$$\xi \hat{h}(\lambda) - \left(i\lambda \hat{h}(\lambda) - k\right) = \hat{f}(\lambda),$$

or

$$\hat{h}(\lambda) = (\xi - i\lambda)^{-1} \left(\hat{f}(\lambda) - k\right). \tag{40}$$

In other words, we assert that one can choose $k \in \mathscr{K}$ such that $\hat{h} \in \hat{\mathscr{N}}_+$. By Prop. 17, we have to check the properties: $\hat{h} \in \mathscr{H}_+^2(\mathbb{R}, d\lambda; \mathscr{K})$ and $\hat{S}_{LP}(\lambda)^* \hat{h} \in \mathscr{H}_-^2(\mathbb{R}, d\lambda; \mathscr{K})$. The first property is obvious: First, we have $\hat{f} \in \mathscr{H}_+^2(\mathbb{R}, d\lambda; \mathscr{K})$, second, $(\xi - iz)^{-1}$ is holomorphic in the upper half plane and from $\mathscr{H}_+^2(\mathbb{R}, d\lambda; \mathscr{K})$. Next we consider

$$\hat{S}_{LP}(\lambda)^* \hat{h}(\lambda) = (\xi - i\lambda)^{-1} \left(\hat{S}_{LP}(\lambda)^* \hat{f}(\lambda) - \hat{S}_{LP}(\lambda)^* k\right).$$

We put $\hat{S}_{LP}(\lambda)^* \hat{f}(\lambda) =: \hat{g}(\lambda)$. Since $\hat{f} \in \hat{\mathscr{N}}_+$, $\hat{g}(\lambda)$ is from $\mathscr{H}_-^2(\mathbb{R}, d\lambda; \mathscr{K})$. Therefore $\hat{S}_{LP}(\lambda)^* \hat{h}(\lambda)$ is analytic continuable into the lower half plane by $(\xi - iz)^{-1} \left(\hat{g}(z) - \hat{S}(\bar{z})^* k\right)$. But this function has to be holomorphic in the lower half plane, hence necessarily $\hat{g}(-i\xi) = \hat{S}(i\bar{\xi})^* k$ and we obtain a unique $k$,

$$k = \left(\hat{S}(\bar{\xi} i)^*\right)^{-1} \hat{g}(-i\xi).$$

For this $k$ the function $\hat{S}_{LP}(\lambda)^* \hat{h}(\lambda)$ is from $\mathscr{H}_-^2$. This proves the assertion. ■

# 13. Stationary Theory

The revolvent $R(z) = (z - H)^{-1}$ and the spectral measure $E(\cdot)$ of a selfadjoint operator $H$, together with the spectral parameter $\lambda$ are called *stationary* terms with respect to $H$. On the other hand, the unitary evolution group $e^{-itH}$, where $t$ is interpreted as time, is called a *time-dependent* term with respect to $H$.

The principal aim of a stationary scattering theory is to express the objects of scattering theory, introduced in Chap. 9, for example the wave operators, and their properties in stationary terms. Thus the first problem of a stationary scattering theory is to transform the stationary pre-wave operator (see Section 9.1.1) into a pure stationary form. Further, the limit of the stationary pre-wave operator has to be interpreted in a stationary language. The second problem is to continue the study of the connections between the wave operator and the limit of the stationary pre-wave operator (see Paragraph 9.4).

According to the two essentially different kinds of stationary transformations of $\Omega_\varepsilon$ to be developed in 13.1, we obtain on the one hand a stationary theory within the framework of revolvents only (Paragraph 3) and on the other hand a stationary theory where the theory of operator spectral integrals (developed in Chapter 5) plays the decisive role as the main instrument of investigation (Paragraph 4). It turns out that the absolute Abelian limit is most appropriate for the definition of wave operators in order to develop a well-rounded stationary theory.

The universal projection is $P_A^{ac}$ in this chapter.

## 13.1. The stationary pre-wave operator

### 13.1.1. Integral representations

The stationary pre-wave operator $\Omega_\varepsilon$, defined by (9.9), can be expressed in weak sense as an integral with respect to the spectral parameter $\lambda$ and also as an operator spectral integral with respect to the spectral measures $E(\cdot)$ and $E_0(\cdot)$ of $H$ and $H_0$. Recall $R(z) = (z - H)^{-1}$ and $R_0(z) = (z - H_0)^{-1}$.

**Proposition 1.** *Let $\Omega_\varepsilon$ be the pre-wave operator given by* (9.9). *Then the representations*

$$(g, \Omega_\varepsilon f) = (\varepsilon/2\pi) \int_{-\infty}^{\infty} \big(g, R(\lambda - i\varepsilon/2)\, JR_0(\lambda + i\varepsilon/2)\, f\big)\, d\lambda, \quad f \in \mathscr{H}_0, \quad g \in \mathscr{H}, \tag{1}$$

$$\Omega_\varepsilon = i \int_{-\infty}^{\infty} E(d\lambda)\, \varepsilon J R_0(\lambda + i\varepsilon), \tag{2}$$

$$\Omega_\varepsilon = -i \int_{-\infty}^{\infty} \varepsilon R(\lambda - i\varepsilon)\, J E_0(d\lambda) \tag{3}$$

*are valid for all* $\varepsilon > 0$.

**Proof.** First we prove (1). We put

$$f_{H_0}(t) = \chi_{[0,\infty)}(t)\, e^{-\varepsilon t} J\, e^{-itH_0} f, \qquad f \in \mathscr{H}_0,$$

$$g_H(t) = \chi_{[0,\infty)}(t)\, e^{-\varepsilon t}\, e^{-itH} g, \qquad g \in \mathscr{H}.$$

Then we get

$$\hat{f}_{H_0}(\lambda) := (2\pi)^{-1/2} \int_{-\infty}^{\infty} e^{it\lambda} f_{H_0}(t)\, dt = (2\pi)^{-1/2}\, iJR_0(\lambda + i\varepsilon)\, f,$$

$$\hat{g}_H(\lambda) := (2\pi)^{-1/2} \int_{-\infty}^{\infty} e^{it\lambda} g_H(t)\, dt = (2\pi)^{-1/2}\, iR(\lambda + i\varepsilon)\, g.$$

Because of Parseval's equation $\int \big(g_H(t), f_{H_0}(t)\big)_{\mathscr{H}}\, dt = \int \big(\hat{g}_H(\lambda), \hat{f}_{H_0}(\lambda)\big)_{\mathscr{H}}\, d\lambda$ we obtain

$$(2\pi)^{-1} \int_{-\infty}^{\infty} \big(R(\lambda + i\varepsilon)\, g, JR_0(\lambda + i\varepsilon)\, f\big)\, d\lambda = \int_0^{\infty} e^{-2\varepsilon t} (e^{-itH} g, J\, e^{-itH_0} f)\, dt$$

and

$$\varepsilon\pi^{-1} \int_{-\infty}^{\infty} \big(g, R(\lambda - i\varepsilon)\, JR_0(\lambda + i\varepsilon)\, f\big)\, d\lambda = (g, \Omega_{2\varepsilon} f).$$

Now we prove (2). First we consider $E\big([a, b)\big)\, \Omega_\varepsilon$, where $[a, b)$ denotes a finite interval. Since $E\big([a, b)\big)\, e^{itH} = \int_a^b e^{it\lambda} E(d\lambda)$, we obtain

$$E\big([a, b)\big)\, \Omega_\varepsilon = \varepsilon \int_0^{\infty} e^{-\varepsilon t} \left\{ \int_a^b e^{it\lambda} E(d\lambda) \right\} J\, e^{-itH_0}\, dt.$$

Now we apply Prop. 5.8. In the present case we have $[c, d) = [0, \infty)$, $A(t) = \varepsilon J\, e^{-\varepsilon t}\, e^{-itH_0}$ and $\alpha(\lambda, t) = e^{it\lambda}$. The operator function $A(t)$ has the properties $\int_0^{\infty} \|A(t)\|\, dt < \infty$ and $\int_0^{\infty} t\, \|A(t)\|\, dt < \infty$. Therefore the operator function $F(\lambda)$, defined by the Bochner integral

$$F(\lambda) = \int_0^{\infty} e^{i\lambda t} A(t)\, dt,$$

is continuously differentiable (with respect to the operator norm) and we get

$$F'(\lambda) = i \int_0^{\infty} e^{i\lambda t}\, t A(t)\, dt.$$

Further, we have $F(\lambda) = i\varepsilon J(\lambda + i\varepsilon - H_0)^{-1}$. Now, by Prop. 5.4, $\int_a^b E(d\lambda)\, F(\lambda)$ exists. From Prop. 5.8 we obtain

$$\int_a^b E(d\lambda)\, F(\lambda) = \int_0^{\infty} \left\{ \int_a^b e^{i\lambda t} E(d\lambda) \right\} A(t)\, dt = E\big([a, b)\big)\, \Omega_\varepsilon,$$

that is $E([a, b))\, \Omega_\varepsilon = \mathrm{i} \int_a^b E(\mathrm{d}\lambda)\, \varepsilon J R_0(\lambda + \mathrm{i}\varepsilon)$. Because of $\Omega_\varepsilon = \text{s-}\lim_{a\to-\infty, b\to\infty} E([a, b))\, \Omega_\varepsilon$ we immediately obtain (2) (recall the definition of $\int E(\mathrm{d}\lambda)\, F(\lambda)$ as a strong limit of integrals over finite intervals). (3) can be proved similarly. ■

If the operator $V := HJ - JH_0$ can be defined, then we can change the representations (3), (4) slightly such that $V$ appears explicitly. For example we have

**Proposition 2.** *Let $J \operatorname{dom} H_0 \subseteq \operatorname{dom} H$. Then $V$ is defined on $\operatorname{dom} H_0$. In this case we obtain*

$$\Omega_\varepsilon f = Jf - \int_{-\infty}^{\infty} E(\mathrm{d}\lambda)\, V R_0(\lambda + \mathrm{i}\varepsilon)\, f, \qquad f \in \mathscr{H}_0, \tag{4}$$

$$\Omega_\varepsilon f = Jf + \int_{-\infty}^{\infty} R(\lambda - \mathrm{i}\varepsilon)\, V E_0(\mathrm{d}\lambda)\, f, \qquad f \in \operatorname{dom} H_0, \tag{5}$$

*for all $\varepsilon > 0$, where the operator spectral integrals on the right-hand side exist.*

Proof. According to (2) we have

$$\begin{aligned}
(\Omega_\varepsilon - J)\, f &= \mathrm{i} \int_{-\infty}^{\infty} E(\mathrm{d}\lambda)\, \varepsilon J R_0(\lambda + \mathrm{i}\varepsilon)\, f - \int_{-\infty}^{\infty} E(\mathrm{d}\lambda)\, Jf \\
&= \int_{-\infty}^{\infty} E(\mathrm{d}\lambda)\, J\big(\mathrm{i}\varepsilon R_0(\lambda + \mathrm{i}\varepsilon) - 1\big)\, f = \int_{-\infty}^{\infty} E(\mathrm{d}\lambda)\, J(-\lambda + H_0)\, R_0(\lambda + \mathrm{i}\varepsilon)\, f \\
&= -\int_{-\infty}^{\infty} E(\mathrm{d}\lambda)\, V R_0(\lambda + \mathrm{i}\varepsilon)\, f + \int_{-\infty}^{\infty} E(\mathrm{d}\lambda)\, (-\lambda + H)\, J R_0(\lambda + \mathrm{i}\varepsilon)\, f,
\end{aligned}$$

but the last integral vanishes according to Prop. 5.11. Thus (4) is proved. (5) can be proved similarly. ■

Formula (1) will become the central formula for the stationary theory to be developed in Paragraph 3. Formula (2) and also (4) turn out be decisive for the other stationary theory to be developed in Paragraph 4. The main question in these approaches will be in which sense one can interchange the limit $\varepsilon \to +0$ with the corresponding integrals, for example with the integral $\int E(\mathrm{d}\lambda)\, \{\cdot\}$ in the latter approach. As a preliminary to Paragraph 4 we note some further properties in connection with the representations (2), (4) of $\Omega_\varepsilon$ as an operator spectral integral, which are based essentially on the study of such integrals in Chapter 5. We put

$$G_\varepsilon(\lambda) := \mathrm{i}\varepsilon J R_0(\lambda + \mathrm{i}\varepsilon). \tag{6}$$

Then from Prop. 1 it follows that

$$\Omega_\varepsilon = \int_{-\infty}^{\infty} E(\mathrm{d}\lambda)\, G_\varepsilon(\lambda). \tag{7}$$

It is also possible to define $G_\varepsilon(\lambda)$ by

$$G_\varepsilon(\lambda) := J - V R_0(\lambda + \mathrm{i}\varepsilon). \tag{8}$$

Then according to Prop. 2 formula (7) is also valid. In the following let $G_\varepsilon(\lambda)$ be defined by (6) or by (8). From Chapter 5 we obtain

**Proposition 3.** *Let $f \in \mathscr{H}_0$, $g \in \mathscr{H}$. Then the expression*

$$\big(E(\mathrm{d}\lambda)\, G_\varepsilon(x)\, f, G_\varepsilon(x)\, f\big)/\mathrm{d}\lambda|_{x=\lambda}, \qquad \varepsilon > 0, \tag{9}$$

*exists a.e.* (mod $|\cdot|$) *on* $\mathbb{R}$, *and the relations*

$$\big(E(\mathrm{d}\lambda)\, G_\varepsilon(x)\, f, G_\varepsilon(x)\, f\big)/\mathrm{d}\lambda|_{x=\lambda} = \big(E(\mathrm{d}\lambda)\, \Omega_\varepsilon f, \Omega_\varepsilon f\big)/\mathrm{d}\lambda \tag{10}$$

*a.e.* (mod $|\cdot|$) *on* $\mathbb{R}$ *and*

$$\left\| P^{\mathrm{ac}} \int_{-\infty}^{\infty} E(\mathrm{d}\lambda)\, G_\varepsilon(\lambda)\, f \right\|^2 = \int_{-\infty}^{\infty} \big(E(\mathrm{d}\lambda)\, G_\varepsilon(x)\, f, G_\varepsilon(x)\, f\big)/\mathrm{d}\lambda|_{x=\lambda}\, \mathrm{d}\lambda \tag{11}$$

*are valid. Further,*

$$\big(g, E(\mathrm{d}\lambda)\, G_\varepsilon(x)\, f\big)/\mathrm{d}\lambda|_{x=\lambda}, \qquad \varepsilon > 0,$$

*exists a.e.* (mod $|\cdot|$) *on* $\mathbb{R}$, *and the relations*

$$\big(g, E(\mathrm{d}\lambda)\, G_\varepsilon(x)\, f\big)/\mathrm{d}\lambda|_{x=\lambda} = \big(g, E(\mathrm{d}\lambda)\, \Omega_\varepsilon f\big)/\mathrm{d}\lambda \tag{12}$$

*a.e.* (mod $|\cdot|$) *on* $\mathbb{R}$ *and*

$$\left(g, P^{\mathrm{ac}} \int_{-\infty}^{\infty} E(\mathrm{d}\lambda)\, G_\varepsilon(\lambda)\, f \right) = \int_{-\infty}^{\infty} \big(g, E(\mathrm{d}\lambda)\, G_\varepsilon(x)\, f\big)/\mathrm{d}\lambda|_{x=\lambda}\, \mathrm{d}\lambda$$

*are valid.*

**Proof.** The existence of $\big(E(\mathrm{d}\lambda)\, G_\varepsilon(x)\, f, G_\varepsilon(x)\, f\big)/\mathrm{d}\lambda|_{x=\lambda}$ and $\big(g, E(\mathrm{d}\lambda)\, G_\varepsilon(x)\, f\big)/\mathrm{d}\lambda|_{x=\lambda}$ follows by extending Prop. 5.9 to arbitrary spectral measures $E(\cdot)$ (the proof of Prop. 5.9 works also in this case). Further (11) follows from Prop. 5.10. By Corollary 5.12, this implies (10) a.e. on $\Delta^{\mathrm{ac}}$, where $P^{\mathrm{ac}} = E(\Delta^{\mathrm{ac}})$, hence a.e. (mod $|\cdot|$) on $\mathbb{R}$. By a corresponding modification of Prop. 5.10 one obtains the last relation and finally (12) follows. ■

### 13.1.2. Some properties of the stationary pre-wave operators

In this section we collect some properties of $\Omega_\varepsilon$ which are of technical nature but useful in the following. First we introduce

$$\Omega_\varepsilon(\Delta) := (\varepsilon/2\pi) \int_{\Delta} R(\lambda - \mathrm{i}\varepsilon/2)\, J R_0(\lambda + \mathrm{i}\varepsilon/2)\, \mathrm{d}\lambda, \qquad \Delta \text{ a Borel set.} \tag{13}$$

If $\Delta$ is compact, then this integral is a usual Bochner integral. If $\Delta$ is non-compact, then the integral is defined in weak sense as in (1). We have $\Omega_\varepsilon(\mathbb{R}) = \Omega_\varepsilon$.

**Proposition 4.** *If $\Delta$, $\Delta'$ are Borel sets with $\Delta \cap \Delta' = \emptyset$, then*

$$\operatorname*{s-lim}_{\varepsilon \to +0} \Omega_\varepsilon(\Delta)\, E_{H_0}(\Delta')\, P_0^{\mathrm{ac}} = 0.$$

**Proof.** Let $u \in E_0(\Delta') P_0^{ac}\mathscr{H}_0$ and $v \in \mathscr{H}$. Then we have

$$\big(v, \Omega_{2\varepsilon}(\Delta)\, u\big) = (\varepsilon/\pi) \int_\Delta \big(R(\lambda + i\varepsilon)\, v, JR_0(\lambda + i\varepsilon)\, u\big)\, d\lambda.$$

Hence

$$|(v, \Omega_{2\varepsilon}(\Delta)\, u)| \leqq \|J\| \int_\Delta (\varepsilon/\pi)^{1/2}\, \|R_0(\lambda + i\varepsilon)\, u\|\, (\varepsilon/\pi)^{1/2}\, \|R(\lambda + i\varepsilon)\, v\|\, d\lambda$$

and

$$|(v, \Omega_{2\varepsilon}(\Delta)\, u)| \leqq \|J\| \Big(\int_\Delta (\varepsilon/\pi)\, \|R_0(\lambda + i\varepsilon)\, u\|^2\, d\lambda\Big)^{1/2} \Big(\int_\Delta (\varepsilon/\pi)\, \|R(\lambda + i\varepsilon)\, v\|^2\, d\lambda\Big)^{1/2}$$

follow. Since

$$(\varepsilon/\pi)\, \|R_0(\lambda + i\varepsilon)\, u\|^2 = (\varepsilon/\pi) \int_{-\infty}^{\infty} \big((\lambda - \mu)^2 + \varepsilon^2\big)^{-1} \big(u, E_0(d\mu)\, u\big)$$

(and also the corresponding formula with respect to $R(\lambda + i\varepsilon)$ and $E(d\mu)$ is valid, see (3.17)), we obtain

$$(v, \Omega_{2\varepsilon}(\Delta)\, u)| \leqq$$

$$\leqq \|J\| \Big(\int_{-\infty}^{\infty} \big(u, E_0(d\mu)\, u\big) \int_\Delta \frac{\varepsilon/\pi}{(\lambda - \mu)^2 + \varepsilon^2}\, d\lambda\Big)^{1/2} \Big(\int_{-\infty}^{\infty} \big(v, E(d\mu)\, v\big) \int_\Delta \frac{\varepsilon/\pi}{(\lambda - \mu)^2 + \varepsilon^2}\, d\lambda\Big)^{1/2}. \tag{14}$$

Now

$$\int_\Delta \frac{\varepsilon/\pi}{(\lambda - \mu)^2 + \varepsilon^2}\, d\lambda = \int_{-\infty}^{\infty} \chi_\Delta(\lambda)\, \frac{\varepsilon/\pi}{(\lambda - \mu)^2 + \varepsilon^2}\, d\lambda = \int_{-\infty}^{\infty} \chi_\Delta(\lambda)\, \delta_\varepsilon(\mu - \lambda)\, d\lambda =: (\delta_\varepsilon * \chi_\Delta)\, (\mu)$$

is valid (recall Definition (1.6) of $\delta_\varepsilon(\cdot)$). We have $(\delta_\varepsilon * \chi_\Delta)\, (\mu) \leqq 1$, therefore we get

$$\big|\big(v, \Omega_{2\varepsilon}(\Delta)\, u\big)\big| \leqq \|J\| \Big(\int_{-\infty}^{\infty} (\delta_\varepsilon * \chi_\Delta)\, (\mu)\, \big(u, E_0(d\mu)\, u\big)\Big)^{1/2} \Big(\int_{-\infty}^{\infty} (\delta_\varepsilon * \chi_\Delta)\, (\mu)\, \big(v, E(d\mu)\, v\big)\Big)^{1/2}$$

$$\leqq \|J\| \Big(\int_{-\infty}^{\infty} (\delta_\varepsilon * \chi_\Delta)\, (\mu)\, \big(u, E_0(d\mu)\, u\big)\Big)^{1/2} \|v\|,$$

hence

$$\|\Omega_{2\varepsilon}(\Delta)\, u\| \leqq \|J\| \Big(\int_{-\infty}^{\infty} (\delta_\varepsilon * \chi_\Delta)\, (\mu)\, \big(u, E_0(d\mu)\, u\big)\Big)^{1/2}.$$

Now $\lim_{\varepsilon \to +0} (\delta_\varepsilon * \chi_\Delta)\, (\mu) = \chi_\Delta(\mu)$ a.e. on $\mathbb{R}$ and $\chi_\Delta(\mu) = 0$ a.e. on $\mathbb{R}$ with respect to $\big(u, E_0(\cdot) u\big)$. Hence the assertion follows from the Lebesgue dominated convergence theorem. ∎

The next lemma is connected with the intertwining property of the limit of the stationary pre-wave operator (cf. Prop. 9.11 and Remark 9.19).

**Lemma 5.** *Let $\Delta$ be a compact Borel set of $\mathbb{R}$. Then*

$$\lim_{\varepsilon \to +0} \sup_{t \in \Delta} \|e^{itH} \Omega_\varepsilon - \Omega_\varepsilon\, e^{itH_0}\| = 0. \tag{15}$$

**Proof.** From (9.9) we obtain

$$e^{-isH}\Omega_{2\varepsilon} - \Omega_{2\varepsilon}\, e^{-isH_0}$$

$$= 2\varepsilon\, e^{-isH}\left(\int_0^s e^{-2\varepsilon t}\, e^{itH} J\, e^{-itH_0}\, dt + (1 - e^{2\varepsilon s}) \int_s^\infty e^{-2\varepsilon t}\, e^{itH} J\, e^{-itH_0}\, dt\right).$$

Hence

$$\|e^{-isH}\Omega_{2\varepsilon} - \Omega_{2\varepsilon}\, e^{-isH_0}\| \leqq 2\varepsilon(|s| \cdot \|J\| + |1 - e^{2\varepsilon s}|\, e^{-2\varepsilon s}\, \|J\|)$$
$$= 2\varepsilon\, \|J\|\, (|s| + |e^{-2\varepsilon s} - 1|),$$

so that the left-hand side tends to 0 as $\varepsilon \to +0$ uniformly with respect to $s$, where $s$ is from a fixed compact set. ∎

The next lemma is also related to the intertwining property of the limit of the stationary pre-wave operator, but may also be considered as a counterpart to Prop. 4. Pay attention to the fact that this lemma is formulated for closed sets only, whereas Prop. 4 is valid for arbitrary Borel sets. It seems that Lemma 6 is not true in general for arbitrary Borel sets, but we are not able to present a counterexample.

**Lemma 6.** *Let $\Lambda$ be a closed set of $\mathbb{R}$. Then*

$$\operatorname*{s\text{-}lim}_{\varepsilon\to+0} E(\Lambda)\, \Omega_\varepsilon P_0^{ac} E_0(\mathbb{R} \setminus \Lambda)\, f = 0, \qquad f \in \mathscr{H}_0. \tag{16}$$

**Proof.** Note that $\mathbb{R} \setminus \Lambda$ is open. Let $[a, b]$ be a finite closed interval from $\mathbb{R} \setminus \Lambda$. First we prove

$$\operatorname*{s\text{-}lim}_{\varepsilon\to+0} E(\Lambda)\, \Omega_\varepsilon P_0^{ac} E_0([a, b])\, f = 0. \tag{17}$$

Since $\|\Omega_\varepsilon\| \leqq \|J\|$ and clo spa $\{P_0^{ac} E_0([a, b])\, f;\ [a, b] \subset \mathbb{R} \setminus \Lambda, f \in \mathscr{H}_0\} = P_0^{ac} \times E_0(\mathbb{R} \setminus \Lambda)\mathscr{H}_0$, (16) follows immediately from (17). For the proof of (17) we use the integral representation (3). Then we obtain

$$E(\Lambda)\, \Omega_\varepsilon P_0^{ac} E_0([a, b])\, f = -i\varepsilon \int_a^b E(\Lambda)\, R(\lambda - i\varepsilon)\, J E_0(d\lambda)\, P_0^{ac} f. \tag{18}$$

Now we integrate by parts (see Section 5.1.3.) and get

$$\int_a^b E(\Lambda)\ R(\lambda - i\varepsilon)\, J E_0(d\lambda)\, P_0^{ac} f = E(\Lambda)\, R(\lambda - i\varepsilon)\, J E_0(\lambda)\, P_0^{ac} f\big|_a^b$$

$$- \int_a^b E(\Lambda)\, R'(\lambda - i\varepsilon)\, J E_0(\lambda)\, P_0^{ac} f\, d\lambda. \tag{19}$$

The functions $E(\Lambda)\, R(\lambda - i\varepsilon)$ and $E(\Lambda)\, R'(\lambda - i\varepsilon)$ are holomorphic for $\lambda \in [a, b]$, $\varepsilon \geqq 0$, because spec $\{H \restriction E(\Lambda)\, \mathscr{H}\} \subseteq \Lambda$. Hence from (18) and (19) the relation (17) follows. ∎

**Remark 7.** In particular (16) is valid if $\Lambda = \{a\}$ is a one-point set. Since $P_0^{ac} E_0(\{a\})\, f = 0$, we obtain in this case from (16) that

$$\operatorname*{s\text{-}lim}_{\varepsilon\to+0} E(\{a\})\, \Omega_\varepsilon P_0^{ac} f = 0, \qquad a \in \mathbb{R}, \qquad f \in \mathscr{H}_0. \tag{20}$$

From (20) and (16) the validity of

$$\operatorname*{s-lim}_{\varepsilon\to+0} E(\mathbb{R} \setminus \Delta)\, \Omega_\varepsilon P_0^{\mathrm{ac}} E_0(\Delta)\, f = 0, \qquad f \in \mathscr{H}_0, \tag{21}$$

for arbitrary intervals $\Delta$ (bounded, non-bounded, open, closed, semiclosed) follows.

In the following paragraphs we need some properties on equiabsolute continuity of $E(\Delta)\, P^{\mathrm{ac}}\Omega_\varepsilon f$. We have to distinguish between weak and strong equiabsolute continuity.

**Lemma 8.** *For any $f \in \mathscr{H}_0$ the vector function $E(\Delta)\, P^{\mathrm{ac}}\Omega_\varepsilon f$ is weakly equiabsolutely continuous with respect to $\varepsilon > 0$, that is, for all $g \in \mathscr{H}$ the expression*

$$\sup_{\varepsilon>0} |(g, E(\Delta)\, P^{\mathrm{ac}}\Omega_\varepsilon f)|$$

*is absolutely continuous.*

**Proof.** We have $\big(g, E(\Delta)\, P^{\mathrm{ac}}\Omega_\varepsilon f\big) = \big(E(\Delta)\, P^{\mathrm{ac}}g, \Omega_\varepsilon f\big)$. Hence

$$\sup_{\varepsilon>0} |(g, E(\Delta)\, P^{\mathrm{ac}}\Omega_\varepsilon f)| \leq \sup_{\varepsilon>0} \|\Omega_\varepsilon f\| \cdot \|E(\Delta)\, P^{\mathrm{ac}}g\| \leq \|J\|\, \|f\| \cdot \|E(\Delta)\, P^{\mathrm{ac}}g\|.$$

Since $\|E(\Delta)\, P^{\mathrm{ac}}g\|$ is absolutely continuous, the assertion follows. ∎

**Lemma 9.** *Let $0 < a < b$, $f \in \mathscr{H}_0$. Then $E(\Delta)\, P^{\mathrm{ac}}\Omega_\varepsilon f$ is strongly equiabsolutely continuous with respect to $\varepsilon \in [a, b]$, that is*

$$\sup_{\varepsilon\in[a,b]} \|E(\Delta)\, P^{\mathrm{ac}}\Omega_\varepsilon f\|$$

*is absolutely continuous.*

**Proof.** This is an easy consequence of the facts that $f_\varepsilon := P^{\mathrm{ac}}\Omega_\varepsilon f$ is holomorphic in $\varepsilon$ for $\varepsilon \in [a, b]$ and $\|E(\Delta)\, f_\varepsilon\|$ is absolutely continuous for each fixed $\varepsilon > 0$. ∎

Finally we present a uniform estimate with respect to $\varepsilon$ of $E\big(\mathbb{R} \setminus (-N, N)\big)\, \Omega_\varepsilon P_0^{\mathrm{ac}} f$. This estimate will complete the equiabsolute continuity of $E(\Delta)\, P^{\mathrm{ac}}\Omega_\varepsilon P_0^{\mathrm{ac}} f$. Both properties together correspond to the so-called uniform integrability of the function $f_\varepsilon(\lambda) = G_\varepsilon(\lambda)\, P_0^{\mathrm{ac}} f$ in (8) (see Dieudonné [1, 13.12. Example 4]).

**Lemma 10.** *The relation*

$$\lim_{N\to\infty} \sup_{0<\varepsilon\leq 1} \big\|E\big(\mathbb{R} \setminus (-N, N)\big)\, \Omega_\varepsilon P_0^{\mathrm{ac}} f\big\| = 0, \qquad f \in \mathscr{H}_0, \tag{22}$$

*is valid.*

**Proof.** The relation

$$\lim_{N\to\infty} \sup_{\delta\leq\varepsilon<1} \big\|E\big(\mathbb{R} \setminus (-N, N)\big)\, \Omega_\varepsilon P_0^{\mathrm{ac}} f\big\| = 0 \tag{23}$$

for each $\delta \in (0, 1)$ follows from the facts that $\Omega_\varepsilon P_0^{\mathrm{ac}} f$ is holomorphic in $\varepsilon \in [\delta, 1]$ and that for each fixed $\varepsilon > 0$ one has $\lim\limits_{N\to\infty} \big\|E\big(\mathbb{R} \setminus (-N, N)\big)\, \Omega_\varepsilon P_0^{\mathrm{ac}} f\big\| = 0$. Now let $\Delta$ be an arbitrary bounded interval. We choose $N_0$ such that $\operatorname{clo} \Delta \subset (-N_0, N_0)$. Then, applying (21), we obtain

$$\operatorname*{s-lim}_{\varepsilon\to+0} E\big(\mathbb{R} \setminus (-N, N)\big)\, \Omega_\varepsilon P_0^{\mathrm{ac}} E_0(\Delta)\, f = 0, \tag{24}$$

uniformly with respect to $N \geqq N_0$. From (23) and (24) we obtain

$$\lim_{N\to\infty} \sup_{0<\varepsilon\leq 1} \|E(\mathbb{R} \setminus (-N, N))\, \Omega_\varepsilon P_0^{ac} E_0(\Delta) f\| = 0.$$

But $\|\Omega_\varepsilon\| \leqq \|J\|$ and clo spa $\{P_0^{ac}E_0(\Delta) f;\ \Delta$ bounded interval, $f \in \mathscr{H}_0\} = P_0^{ac}\mathscr{H}_0$. Therefore (22) follows. ■

**Remark 11.** Lemma 10 says essentially that it is sufficient to prove the convergence of $E(\Delta)\, \Omega_\varepsilon P_0^{ac} f$ as $\varepsilon \to +0$ for each bounded interval $\Delta$ to get the convergence of $\Omega_\varepsilon P_0^{ac} f$.

## 13.2. Limits of the stationary pre-wave operator

### 13.2.1. Weak limits

In this section properties of the weak limit (cf. Remark 9.19)

$$\Omega_+^w := \operatorname*{w-lim}_{\varepsilon\to+0} \Omega_\varepsilon P_0^{ac} \tag{25}$$

and simple conditions for its existence are presented. $\Omega_+^w$ is bounded and we have $\|\Omega_+^w\| \leqq \|J\|$. We begin with a simple consequence of Lemma 5.

**Lemma 12.** *The relation*

$$\operatorname*{w-lim}_{\varepsilon\to+0} \{\Omega_\varepsilon E_0(\Delta) - E(\Delta)\, \Omega_\varepsilon\} = 0 \tag{26}$$

*is valid for all Borel sets $\Delta$.*

**Proof.** Let $f \in \mathscr{H}_0$, $g \in \mathscr{H}$. Then we obtain immediately from Lemma 5

$$\lim_{\varepsilon\to+0} \{(g, e^{-isH}\Omega_\varepsilon f) - (g, \Omega_\varepsilon e^{-isH_0} f)\} = 0, \qquad s \in \mathbb{R};$$

or

$$\lim_{\varepsilon\to+0} \{(e^{isH} g, \Omega_\varepsilon f) - (g, \Omega_\varepsilon e^{-isH_0} f)\} = 0.$$

Thus, if $t(\lambda) = \sum_{\varrho=1}^{m} \alpha_\varrho e^{is_\varrho\lambda}$ is an arbitrary real trigonometrical polynomial, then

$$\lim_{\varepsilon\to+0} \{(t(H)\, g, \Omega_\varepsilon f) - (g, \Omega_\varepsilon\, t(H_0)\, f)\} = 0.$$

Now let $\Delta$ be a compact interval. Then, according to Section 3.2.3, we can choose a sequence $t_n(\lambda)$ of real trigonometrical polynomials such that $\operatorname*{s-lim}_{n\to\infty} t_n(H_0)\, f = E_0(\Delta)\, f$ and $\operatorname*{s-lim}_{n\to\infty} t_n(H)\, g = E(\Delta)\, g$ simultaneously. Hence the assertion follows for compact intervals and finally for all Borel sets. ■

From Lemma 12 we immediately obtain the intertwining property of $\Omega_+^w$, which was already proved in Prop. 9.11.

**Proposition 13.** *Let the weak limit* $\Omega_+^{\mathrm{w}} = \operatorname*{w-lim}_{\varepsilon\to+0} \Omega_\varepsilon P_0^{\mathrm{ac}}$ *exist. Then it is an intertwining operator with respect to* $H$, $H_0$, *that*

$$E(\Delta)\,\Omega_+^{\mathrm{w}} = \Omega_+^{\mathrm{w}} E_0(\Delta) \tag{27}$$

*is valid for all Borel sets* $\Delta$.

Proof. Let $f \in P_0^{\mathrm{ac}}\mathscr{H}_0$. Then also $E_0(\Delta) f \in P_0^{\mathrm{ac}}\mathscr{H}_0$. Hence the assertion follows immediately from (26). ∎

Another consequence of Lemma 12, directly related to the intertwining property, is

**Proposition 14.** *The relations*

$$\operatorname*{w-lim}_{\varepsilon\to+0} (P^{\mathrm{ac}})^\perp\, \Omega_\varepsilon P_0^{\mathrm{ac}} = 0, \qquad \operatorname*{w-lim}_{\varepsilon\to+0} P^{\mathrm{ac}} \Omega_\varepsilon (P_0^{\mathrm{ac}})^\perp = 0 \tag{28}$$

*are valid.*

Proof. We have $(P^{\mathrm{ac}})^\perp = E(\Delta_0)$, where $|\Delta_0| = 0$. Namely, $(P^{\mathrm{ac}})^\perp = P^{\mathrm{sc}} + P^{\mathrm{p}}$. Further, $P^{\mathrm{sc}} = E(\Delta^{\mathrm{sc}})$, where $|\Delta^{\mathrm{sc}}| = 0$, and $P^{\mathrm{p}} = E(\Delta^{\mathrm{d}})$, where $\Delta^{\mathrm{d}} = \bigcup_{\varrho=1}^{\infty} \lambda_\varrho$ and $\lambda_1, \lambda_2, \ldots$ denotes the sequence of all eigenvalues of $H$. Obviously $|\Delta^{\mathrm{d}}| = 0$. Then $\Delta_0$ can be chosen as $\Delta^{\mathrm{sc}} \cup \Delta^{\mathrm{d}}$. Now $E_0(\Delta_0)\, P_0^{\mathrm{ac}} = 0$, hence the first assertion follows from (26). The second assertion follows from the first by replacing $\Omega_\varepsilon$ by $\Omega_\varepsilon^*$ and taking the adjoint. ∎

The next result follows directly from Prop. 14.

**Lemma 15.** *The weak limit* $\operatorname*{w-lim}_{\varepsilon\to+0} \Omega_\varepsilon P_0^{\mathrm{ac}} = \Omega_+^{\mathrm{w}}$ *exists iff* $\operatorname*{w-lim}_{\varepsilon\to+0} P^{\mathrm{ac}}\Omega_\varepsilon P_0^{\mathrm{ac}}$ *exists and in this case we have* $\Omega_+^{\mathrm{w}} = \operatorname*{w-lim}_{\varepsilon\to+0} P^{\mathrm{ac}}\Omega_\varepsilon P_0^{\mathrm{ac}}$.

Proof. Obvious. ∎

A further consequence of Lemma 12 is

**Lemma 16.** *Let* $\mathscr{K}_0 \subset \mathscr{H}_0$, $\mathscr{K} \subset \mathscr{H}$ *be linear manifolds so that* $P_0^{\mathrm{ac}}\mathscr{K}_0$, $P^{\mathrm{ac}}\mathscr{K}$ *are generating with respect to* $P_0^{\mathrm{ac}}\mathscr{H}_0$, $P^{\mathrm{ac}}\mathscr{H}$ *and* $H_0$, $H$, *respectively. Further assume the existence of* $\lim_{\varepsilon\to+0} \big(P^{\mathrm{ac}}k, E(\Lambda)\, \Omega_\varepsilon P_0^{\mathrm{ac}} k_0\big)$ *for all* $k_0 \in \mathscr{K}_0$, $k \in \mathscr{K}$ *and all bounded intervals* $\Lambda$. *Then* $\Omega_+^{\mathrm{w}}$ *exists.*

Proof. From the assumptions and $\|\Omega_\varepsilon\| \leqq \|J\|$ the weak convergence of $E(\Lambda)\, P^{\mathrm{ac}}\Omega_\varepsilon \times P_0^{\mathrm{ac}} k_0$ follows for all $k_0 \in \mathscr{K}_0$. Applying Lemma 12, we obtain the weak convergence of $P^{\mathrm{ac}}E(\Lambda)\, \Omega_\varepsilon E_0(\Delta)\, P_0^{\mathrm{ac}} k_0$ for all Borel sets $\Delta$. Then from $\|\Omega_\varepsilon\| \leqq \|J\|$ and from the fact that $P_0^{\mathrm{ac}}\mathscr{K}_0$ is generating with respect to $P_0^{\mathrm{ac}}\mathscr{H}_0$ and $H_0$, we get the weak convergence of $E(\Lambda)\, P^{\mathrm{ac}}\Omega_\varepsilon P_0^{\mathrm{ac}}$. Applying Lemma 10, we obtain the weak convergence of $P^{\mathrm{ac}}\Omega_\varepsilon P_0^{\mathrm{ac}}$. Hence, because of Lemma 15, the assertion follows. ∎

Now we can prove a first sufficient condition for the existence of $\Omega_+^{\mathrm{w}}$.

**Proposition 17.** *Let* $\mathscr{K}_0 \subset \mathscr{H}_0$, $\mathscr{K} \subset \mathscr{H}$ *be linear manifolds so that* $P_0^{\mathrm{ac}}\mathscr{K}_0$, $P^{\mathrm{ac}}\mathscr{K}$ *are generating with respect to* $P_0^{\mathrm{ac}}\mathscr{H}_0$, $P^{\mathrm{ac}}\mathscr{H}$ *and* $H_0$, $H$ *respectively. Further, assume that for all* $k_0 \in \mathscr{K}_0$, $k \in \mathscr{K}$

$$\big(k, E(\mathrm{d}\lambda)\, G_\varepsilon(\mu)\, k_0\big)/\mathrm{d}\lambda\big|_{\mu=\lambda} \tag{29}$$

*is a Cauchy net as $\varepsilon \to +0$ with respect to the convergence in measure on $\mathbb{R}$. Then $\Omega_+^w$ exists.*

**Proof.** Because of Prop. 3 we have

$$(k, E(\Delta)\, P^{ac}\Omega_\varepsilon k_0) = \int_\Delta (k, E(d\lambda)\, G_\varepsilon(\mu)\, k_0)/d\lambda|_{\mu=\lambda}\, d\lambda \tag{30}$$

for all bounded intervals $\Delta$. Further, because of Lemma 8, the integral on the right-hand side is equiabsolutely continuous with respect to $\varepsilon$. Therefore the assumption on the convergence in measure of (29) together with the Vitali convergence theorem (see NATANSON [1, p. 169]) imply the existence of $\lim_{\varepsilon\to+0} (k, E(\Delta)\, P^{ac}\Omega_\varepsilon k_0)$. Because of (28) we have $\lim_{\varepsilon\to+0} (k, E(\Delta)\, P^{ac}\Omega_\varepsilon (P_0^{ac})^\perp\, k_0) = 0$. Hence we obtain the existence of $\lim_{\varepsilon\to+0} (k, E(\Delta)\, P^{ac}\Omega_\varepsilon P_0^{ac} k_0)$. Applying now Lemma 16, the assertion follows. ■

Moreover, we obtain the equation

$$\lim_{\varepsilon\to+0} (k, E(\Delta)\, P^{ac}\Omega_\varepsilon k_0) = \int_\Delta \alpha_+(\lambda; k, k_0)\, d\lambda, \tag{31}$$

where

$$\alpha_+(\lambda; k, k_0) := \operatorname*{m-lim}_{\varepsilon\to+0} (k, E(d\lambda)\, G_\varepsilon(\mu)\, k_0)/d\lambda|_{\mu=\lambda} \tag{32}$$

denotes the limit in measure of the Cauchy net (29).

According to Prop. 13 the existence of the weak limit $\Omega_+^w$ implies the intertwining property of $\Omega_+^w$. Therefore there is a corresponding weak wave matrix $\hat{\Omega}_+^w(\lambda)$ if direct integral representations of $P_0^{ac}\mathscr{H}_0$, $P^{ac}\mathscr{H}$ are given with respect to $H_0$, $H$, respectively. Let

$$\Phi_0: P_0^{ac}\mathscr{H}_0 \mapsto L^2(\Lambda_0, d\lambda; \mathscr{N}_{0\lambda}), \qquad \Phi: P^{ac}\mathscr{H} \to L^2(\Lambda, d\lambda; \mathscr{N}_\lambda)$$

be isometric operators realizing two direct integral representations ($\Lambda_0$, $\Lambda$ are spectral cores of $\operatorname{spec}_{ac} H_0$, $\operatorname{spec}_{ac} H$ respectively). Then it is easy to link the weak wave matrix $\hat{\Omega}_+^w(\lambda)$ with the function $\alpha_+(\lambda; k, k_0)$. This is the content of

**Proposition 18.** *Assume that the assumptions of Prop. 17 are satisfied. Then*

$$\alpha_+(\lambda; k, k_0) = (\hat{k}(\lambda), \hat{\Omega}_+^w(\lambda)\, \hat{k}_0(\lambda)) \tag{33}$$

*is valid for all $k \in \mathscr{K}$, $k_0 \in \mathscr{K}_0$, a.e. on $\mathbb{R}$, where $\hat{k}(\cdot)$, $\hat{k}_0(\cdot)$ denote the representers of $P^{ac}k$, $P_0^{ac}k_0$ with respect to the direct integral representations $L^2(\Lambda, d\lambda; \mathscr{N}_\lambda)$, $L^2(\Lambda_0, d\lambda; \mathscr{N}_{0\lambda})$ of $P^{ac}\mathscr{H}$, $P_0^{ac}\mathscr{H}_0$.*

**Proof.** Obviously, we have

$$(k, E(\Delta)\, \Omega_+^w k_0) = \int_\Delta (\hat{k}(\lambda), \hat{\Omega}_+^w(\lambda)\, \hat{k}_0(\lambda))_{\mathscr{N}_\lambda}\, d\lambda. \tag{34}$$

Therefore from (31) we obtain

$$\int_\Delta (\hat{k}(\lambda), \hat{\Omega}_+^w(\lambda)\, \hat{k}_0(\lambda))_{\mathscr{N}_\lambda}\, d\lambda = \int_\Delta \alpha_+(\lambda; k; k_0)\, d\lambda$$

for all Borel sets $\Delta$. Hence (33) follows a.e. (mod $|\cdot|$) on $\mathbb{R}$. ■

**Remark 19.** It seems that the conditions of Prop. 17, in general, are not necessary conditions, at least in the case $P_0^{ac}\mathscr{K}_0 = P_0^{ac}\mathscr{H}_0$, $P^{ac}\mathscr{K} = P^{ac}\mathscr{H}$. Namely, if $\Omega_+^w$ exists, then according to (30) we obtain the existence of $\lim_{\varepsilon\to+0} \int_\Delta (g, E(d\lambda)\, G_\varepsilon(\mu) f)/d\lambda|_{\mu=\lambda}\, d\lambda$ for all $g \in P^{ac}\mathscr{H}$, $f \in P_0^{ac}\mathscr{H}_0$ and for all Borel sets $\Delta$ and, according to (34), the equation

$$\lim_{\varepsilon\to+0} \int_\Delta (g, E(d\lambda)\, G_\varepsilon(\mu) f)/d\lambda|_{\mu=\lambda}\, d\lambda = \int_\Delta (\hat{g}(\lambda), \hat{\Omega}_+^w(\lambda)\, \hat{f}(\lambda))_{\mathscr{N}_\lambda}\, d\lambda.$$

But from this equation, in general, it is not possible to conclude that the integrand in the left-hand integral is a Cauchy net with respect to measure convergence.

### 13.2.2. Strong limits

The investigation of $\Omega_+^s := \text{s-lim}_{\varepsilon\to+0} \Omega_\varepsilon P_0^{ac}$ is in some sense more complicated than that of $\Omega_+^w$. By Lemma 15, the investigation of $\Omega_+^w$ is reduced to the study of $\text{w-lim}_{\varepsilon\to+0} P^{ac}\Omega_\varepsilon P_0^{ac}$. If $\text{s-lim}_{\varepsilon\to+0} (P^{ac})^\perp\, \Omega_\varepsilon P_0^{ac}$ exists, then, according to Prop. 14, it is necessarily zero, because the strong limit equals the weak limit. Thus the crucial question for a similar reduction in the strong limit case is the question of the existence of $\text{s-lim}_{\varepsilon\to+0} (P^{ac})^\perp\, \Omega_\varepsilon P_0^{ac}$. It seems that this strong limit in general does not exist (we were not able to decide this question), but the existence can be shown if additional a priori information about the singular spectrum of $H$ is available (see Prop. 20). Despite of this difficulty we consider in this section

$$\text{s-lim}_{\varepsilon\to+0} P^{ac}\Omega_\varepsilon P_0^{ac} \tag{35}$$

and we present conditions for its existence. In some contrast to the situation in the weak limit case in the present (strong limit) case we are able to prove a simple necessary and sufficient criterion.

**Proposition 20.** *If* $|\text{spec}_{sc} H \cup \text{spec}_p H| = 0$, *then*

$$\text{s-lim}_{\varepsilon\to+0} (P^{ac})^\perp\, \Omega_\varepsilon P_0^{ac} = 0. \tag{36}$$

**Proof.** We apply Lemma 6 and put $\Lambda = \text{spec}_{sc} H \cup \text{spec}_p H$ in this lemma. $\Lambda$ is closed and $|\Lambda| = 0$. Therefore $P_0^{ac}E_0(\Lambda) = 0$. Hence (36) follows from (16). ∎

**Lemma 21.** *Let $\mathscr{K}_0 \subset \mathscr{H}_0$ be a linear manifold such that $P_0^{ac}\mathscr{K}_0$ is generating with respect to $P_0^{ac}\mathscr{H}_0$ and $H_0$. Suppose that* $\text{s-lim}_{\varepsilon\to+0} E(\Delta)\, P^{ac}\Omega_\varepsilon P_0^{ac} k_0$ *exists for all bounded intervals $\Delta$ and all $k_0 \in \mathscr{K}_0$. Then* $\text{s-lim}_{\varepsilon\to+0} P^{ac}\Omega_\varepsilon P_0^{ac}$ *exists.*

**Proof.** Obviously from the assumptions of Lemma 21 the existence of $\text{s-lim}_{\varepsilon\to+0} E(\Delta) \times P^{ac}\Omega_\varepsilon P_0^{ac}$ follows. Applying Lemma 10 the assertion follows. ∎

Now we prepare the mentioned criterion. First recall Section 3.5.5 and the notations

$$(g, f)_{E,\lambda} := (g, E(d\lambda)\, f)/d\lambda,$$

$$|f|_{E,\lambda} := (f, f)_{E,\lambda}^{1/2}$$

for vectors $f$, $g$ from a spectral manifold $\mathscr{K} \subset P^{ac}\mathscr{H}$ and for $\lambda$ from a corresponding spectral core $\Gamma$. In particular recall that if $\lambda = \lambda_0 \in \Gamma$ is fixed, then $(g, f)_{E,\lambda_0}$ is a semi-scalar product on $\mathscr{K}$; $|f|_{E,\lambda_0}$ is the corresponding seminorm.

Recall further that according to Prop. 3 the expression

$$|G_{\varepsilon_1}(\lambda) f - G_{\varepsilon_2}(\lambda) f|_{E,\lambda}$$

exists for all $f \in \mathscr{H}_0$, $\varepsilon_1, \varepsilon_2 > 0$, a.e. (mod $|\cdot|$) on $\mathbb{R}$. Now we have

**Theorem 22.** (I) *Let $\mathscr{K}_0$ be a linear manifold in $\mathscr{H}_0$ and let $P_0^{ac}\mathscr{K}_0$ be generating with respect to $P_0^{ac}\mathscr{H}_0$ and $H_0$.*

(II) *Further let $E(\Delta)\, P^{ac}\Omega_\varepsilon P_0^{ac} k_0$ be strongly equiabsolutely continuous with respect to $\varepsilon \in (0, 1]$ for all $k_0 \in \mathscr{K}_0$.*

(III) *Assume that*

$$|G_{\varepsilon_1}(\lambda)\, P_0^{ac} k_0 - G_{\varepsilon_2}(\lambda)\, P_0^{ac} k_0|_{E,\lambda}, \qquad \varepsilon_1, \varepsilon_2 > 0, \tag{37}$$

*is a Cauchy net as $\varepsilon_1, \varepsilon_2 \to +0$ with respect to convergence in measure on $\mathbb{R}$ for all $k_0 \in \mathscr{K}_0$. Then $\operatorname{s-lim}_{\varepsilon\to+0} P^{ac}\Omega_\varepsilon P_0^{ac}$ exists. Conversely if the latter strong limit exists, then* (II) *and* (III) *are valid even for all $f \in P_0^{ac}\mathscr{H}_0$.*

**Proof.** (A) Let (I), (II), (III) be satisfied and let $\Delta$ be a bounded interval. Then according to Prop. 3 we have

$$\|E(\Delta)\, P^{ac}\Omega_{\varepsilon_1} P_0^{ac} k_0 - E(\Delta)\, P^{ac}\Omega_{\varepsilon_2} P_0^{ac} k_0\|^2 = \int_\Delta |G_{\varepsilon_1}(\lambda)\, P_0^{ac} k_0 - G_{\varepsilon_2}(\lambda)\, P_0^{ac} k_0|_{E,\lambda}^2 \, d\lambda. \tag{38}$$

Now applying the Vitali convergence theorem (see Natanson [1, p. 169]) to this integral the existence of $\operatorname{s-lim}_{\varepsilon\to+0} E(\Delta)\, P^{ac}\Omega_\varepsilon P_0^{ac} k_0$ follows. Then by Lemma 21, the assertion is proved. (B) Assume the existence of $\operatorname{s-lim}_{\varepsilon\to+0} P^{ac}\Omega_\varepsilon P_0^{ac} =: \Omega_+^{ac}$. Then $P^{ac}\Omega_\varepsilon P_0^{ac} f$ is strongly continuous with respect to $\varepsilon$ on $[0, 1]$, where the value of the function at $\varepsilon = 0$ is defined by $\Omega_+^{ac} f$. Then (II) follows from this fact and Lemma 9. (III) easily follows from (38), where $k_0$ has to be replaced by $f$. ∎

We recall that convergence in measure as $\varepsilon \to +0$ implies the existence of a sequence $\varepsilon_n \to +0$ such that a.e.-convergence is valid. Therefore we obtain from Theorem 22

**Corollary 23.** *If $\operatorname{s-lim}_{\varepsilon\to+0} P^{ac}\Omega_\varepsilon P_0^{ac}$ exists, then for every $f \in P_0^{ac}\mathscr{H}_0$ there is a sequence $\varepsilon_n \to +0$ (which depends on $f$) such that*

$$\lim_{n,m\to\infty} |G_{\varepsilon_n}(\lambda) f - G_{\varepsilon_m}(\lambda) f|_{E,\lambda} = 0 \tag{39}$$

*is valid a.e.* (mod $|\cdot|$) *on $\mathbb{R}$; that is if $f$ is fixed, then $G_{\varepsilon_n}(\lambda) f$ is strongly convergent with respect to the seminorm $|\cdot|_{E,\lambda}$ for almost all $\lambda \in \mathbb{R}$*(mod $|\cdot|$).

Further we recall that a.e.-convergence as $\varepsilon \to +0$ implies measure convergence as $\varepsilon \to +0$. Therefore from Theorem 22 we obtain

**Corollary 24.** *Let the conditions* (I) *and* (II) *of Theorem 22 be satisfied. If $G_\varepsilon(\lambda)\, P_0^{ac} k_0$ is a Cauchy net as $\varepsilon \to +0$ with respect to the seminorm $|\cdot|_{E,\lambda}$ for all $k_0 \in \mathscr{K}_0$ and almost all $\lambda \in \mathbb{R}$*(mod $|\cdot|$) *then $\operatorname{s-lim}_{\varepsilon\to+0} P^{ac}\Omega_\varepsilon P_0^{ac}$ exists.*

There is a condition on $H$, which can be formulated in terms of the $\mu_a^{H_0}$-topology in $\mathscr{L}(\mathscr{H}_0, \mathscr{H})$ which is sufficient for the validity of condition (II) in Theorem 22 (for the $\mu_a^{H_0}$-topology see Definition 6.55).

**Proposition 25.** *If $P^{ac}E(\Delta) J$ is absolutely continuous with respect to the $\mu_a^{H_0}$-topology, then condition* (II) *of Theorem 22 is satisfied.*

Proof. We have $P^{ac}E(\Delta)\, \Omega_\varepsilon P_0^{ac} f = \varepsilon \int_0^\infty e^{-\varepsilon t} P^{ac}E(\Delta)\, e^{itH} J\, e^{-itH_0} P_0^{ac} f\, dt$. Hence

$$\|P^{ac}E(\Delta)\, \Omega_\varepsilon P_0^{ac} f\| \leqq \varepsilon \int_0^\infty e^{-\varepsilon t} \|P^{ac}E(\Delta)\, J\, e^{-itH_0} P_0^{ac} f\|\, dt$$

$$\leqq p_a\big(P^{ac}E(\Delta)\, J; H_0, f\big),$$

that is we obtain

$$\sup_{\varepsilon > 0} \|E(\Delta)\, P^{ac}\Omega_\varepsilon P_0^{ac} f\| \leqq p_a\big(P^{ac}E(\Delta)\, J; H_0, f\big)$$

and the assertion follows. ∎

## 13.3. Stationary theory of the Abel wave operator in terms of resolvents only

In this paragraph we develop the stationary theory of the Abel wave operator in connection with the representation (1) of the stationary pre-wave operator $\dot{\Omega}_\varepsilon$.

### 13.3.1. Stationary characterization of the Abel wave operator

First we state two propositions on the stationary characterization of operators being asymptotically equivalent to zero, which are slight generalizations of Prop. 6.32 and Prop. 6.33.

**Proposition 26.** *An operator $X \in \mathscr{L}(\mathscr{H}_0, \mathscr{H})$ is asymptotically equivalent to 0 with respect to $H_0 P_0^{ac}$, $X \underset{H_0}{\approx} 0$ (|A|-lim) iff*

$$\lim_{\varepsilon \to +0} \varepsilon \int_{-\infty}^{\infty} \|X R_0(\lambda + i\varepsilon)\, P_0^{ac} f\|^2\, d\lambda = 0, \qquad f \in \mathscr{H}_0. \tag{40}$$

Proof. It follows from Prop. 6.32. ∎

**Proposition 27.** *An operator $X \in \mathscr{L}(\mathscr{H}_0, \mathscr{H})$ is asymptotically equivalent to 0 with respect to $H_0 P_0^{ac}$, $X \underset{H_0}{\approx} 0$ (|A|-lim) iff for each $f \in \mathscr{H}_0$ the function $\varepsilon^{1/2} X R_0(\lambda + i\varepsilon)\, P_0^{ac} f$ is strongly convergent in measure to 0 as $\varepsilon \to +0$.*

Proof. It follows from Prop. 6.33. ∎

If $\mathscr{H}_0 = \mathscr{H}$, then the characterizations of $\ker \mu_s^{H_0}$ (Prop. 6.32 and Prop. 6.33) are obtained. A corresponding simple characterization of $\ker \mu_s^{H_0}$ does not seem to exists. If the universal projection $P_0^{ac}$ is replaced by $P_0^c$, then, remarkably enough, Prop. 26 and Prop. 27 remain true (for the absolute Abelian limit).

**Theorem 28.** *The Abel wave operator $W_+(H, H_0; J)$ exists iff there is an operator $\Omega \in \mathscr{L}(\mathscr{H}_0, \mathscr{H})$ such that*

$$\lim_{\varepsilon \to +0} \varepsilon \int_{-\infty}^{\infty} \|JR_0(\lambda + i\varepsilon) P_0^{ac} f - R(\lambda + i\varepsilon) \Omega f\|^2 \, d\lambda = 0, \quad f \in \mathscr{H}_0, \tag{41}$$

*is valid. In this case $W_+(H, H_0; J) = \Omega$.*

**Proof.** We have

$$\int_0^{\infty} e^{-2\varepsilon t} \|e^{itH} J e^{-itH_0} P_0^{ac} f - W_+ f\|^2 \, dt$$

$$= \int_{-\infty}^{\infty} \|\chi_{[0,\infty)}(t) e^{-\varepsilon t} (J e^{-itH_0} P_0^{ac} f - e^{-itH} W_+ f)\|^2 \, dt.$$

Now we use the fact that $i(2\pi)^{-1/2} JR_0(\lambda + i\varepsilon) u$ is the inverse Fourier transform of $\chi_{[0,\infty)}(t) e^{-\varepsilon t} J e^{-itH_0} u$ and $i(2\pi)^{-1/2} R(\lambda + i\varepsilon) v$ is the inverse Fourier transform of $\chi_{[0,\infty)}(t) e^{-\varepsilon t} e^{-itH} v$. Further, we apply Parseval's equation and obtain

$$\int_0^{\infty} e^{-2\varepsilon t} \|e^{itH} J e^{-itH_0} P_0^{ac} f - W_+ f\|^2 \, dt$$

$$= (2\pi)^{-1} \int_{-\infty}^{\infty} \|JR_0(\lambda + i\varepsilon) P_0^{ac} f - R(\lambda + i\varepsilon) W_+ f\|^2 \, d\lambda.$$

Hence (41) is equivalent to the existence of $W_+$ and the equality $W_+ = \Omega$. ∎

A consequence of Theorem 28 is

**Corollary 29.** *The Abel wave operator $W_+(H, H_0; J)$ exists iff there is an operator $\Omega \in \mathscr{L}(\mathscr{H}_0, \mathscr{H})$ satisfying the following conditions:* (I) $\Omega = P^{ac} \Omega P_0^{ac}$, (II) *for all $u \in \mathscr{H}_0$ the function*

$$\varepsilon^{1/2} \big(JR_0(\lambda + i\varepsilon) P_0^{ac} u - R(\lambda + i\varepsilon) \Omega u\big) \tag{42}$$

*is (strongly) measure convergent to 0 as $\varepsilon \to +0$.*

**Proof.** The "only if"-part is obvious (see NATANSON [1, p. 185]). Now assume (I) and (II) to be satisfied and let $u \in \mathscr{H}_0$ be fixed. Recall that convergence in measure of (42) as $\varepsilon \to +0$ implies the existence of a sequence $\varepsilon_n \to +0$ (which depends on $u$) such that

$$\operatorname*{s-lim}_{n \to \infty} \varepsilon_n^{1/2} \big(JR_0(\lambda + i\varepsilon_n) P_0^{ac} u - R(\lambda + i\varepsilon_n) \Omega u\big) = 0$$

a.e. (mod $|\cdot|$) on $\mathbb{R}$. First we prove that $\Omega$ is an intertwining operator with respect to $H$, $H_0$, that is we have to prove $E(\Delta) \Omega = \Omega E_0(\Delta)$ for arbitrary Borel sets. We put $u = E_0(\Delta) f$. Then $\operatorname*{s-lim}_{n \to \infty} \varepsilon_n^{1/2} R_0(\lambda + i\varepsilon_n) P_0^{ac} u = 0$ for all $\lambda \notin \operatorname{clo} \Delta$ and therefore also

$\operatorname{s-lim}_{n\to\infty} \varepsilon_n^{1/2} R(\lambda + i\varepsilon) \Omega u = 0$ for almost all $\lambda \notin \operatorname{clo} \Delta$; that is $\lim_{n\to\infty} \varepsilon_n \|R(\lambda + i\varepsilon_n) \Omega u\|^2 = 0$ for almost all $\lambda \notin \operatorname{clo} \Delta$. Hence $(E(d\lambda) \Omega u, \Omega u)/d\lambda = 0$ follows for almost all $\lambda \notin \operatorname{clo} \Delta$, that is we get $\Omega u = E(\operatorname{clo} \Delta) \Omega u$. Therefore we obtain for closed sets $\Lambda$: $\Omega E_0(\Lambda) = E(\Lambda) \Omega E_0(\Lambda)$. Recall that $E_0(\{a\}) g = 0$ if $g \in P_0^{ac}\mathscr{H}_0$. Now let $\Lambda = [a, b]$. Then $E(\Lambda) \Omega g = E(\Lambda) \Omega E_0(\Lambda) g + E(\Lambda) \Omega E_0(\mathbb{R} \setminus \Lambda) g$. For the second term on the right-hand side we obtain

$$E(\Lambda) \Omega E_0(\mathbb{R} \setminus \Lambda) g = E(\Lambda) \Omega E_0(\mathbb{R} \setminus (a, b)) g$$
$$= E(\Lambda) E(\mathbb{R} \setminus (a, b)) \Omega E_0(\mathbb{R} \setminus (a, b)) g = E(\{a\} \cup \{b\}) \Omega E_0(\mathbb{R} \setminus (a, b)) g = 0,$$

because $\Omega E_0(\mathbb{R} \setminus (a, b)) g \in P^{ac}\mathscr{H}$. Therefore we obtain $\Omega E_0(\Lambda) = E(\Lambda) \Omega$ for arbitrary closed intervals $\Lambda$, hence this relation is valid also for open intervals and finally also for arbitrary Borel sets. Now (42) can be written in the form

$$\varepsilon^{1/2}(J - \Omega) R_0(\lambda + i\varepsilon) P_0^{ac}u. \tag{43}$$

To this vector function we can apply Prop. 27 and obtain $J \underset{H_0}{\approx} \Omega$. This implies the assertion and the equality $\Omega = W_+$. ∎

### 13.3.2. The wave matrix

In this section we connect the integrand of the representation (1) with the wave matrix of the corresponding Abel wave operator. To prepare this connection, we introduce direct integral representations $\Phi: P^{ac}\mathscr{H} \mapsto L^2(\Lambda; d\lambda; \mathscr{K}_\lambda)$ and $\Phi_0: P_0^{ac}\mathscr{H}_0 \mapsto L^2(\Lambda_0, d\lambda; \hat{\mathscr{K}}_{0\lambda})$ with respect to $H$, $H_0$; the representers of $g \in P^{ac}\mathscr{H}$; $f \in P_0^{ac}\mathscr{H}_0$ are denoted by $\hat{g}(\cdot)$, $\hat{f}(\cdot)$ respectively.

**Theorem 30.** *Suppose that the Abel wave operator $W_+ = W_+(H, H_0; J)$ exists. Then there is a dense set $\mathscr{D} \subset \mathscr{H}$ such that for all $f \in \mathscr{H}_0$, $g \in \mathscr{D}$ the function*

$$\alpha_\varepsilon(\lambda; g, f) := \varepsilon(g, R(\lambda - i\varepsilon) JR_0(\lambda + i\varepsilon) P_0^{ac}f)$$

*is $L^1$-convergent with respect to $L^1(\mathbb{R}, d\lambda)$ as $\varepsilon \to +0$ with the limit*

$$\alpha_+(\lambda; g, f) := (\hat{g}(\lambda), \hat{W}_+(\lambda) \hat{f}(\lambda))_{\mathscr{K}_\lambda},$$

*where $\hat{W}_+(\lambda)$ denotes the wave matrix corresponding to $\Phi$, $\Phi_0$, and where $\hat{f}(\cdot)$, $\hat{g}(\cdot)$ denote the representers of $P_0^{ac}f$, $P^{ac}g$, respectively.*

Proof. First recall $|A|\text{-}\lim_{t\to\infty} E^\perp J e^{-itH_0} P_0^{ac} = 0$ (see (9.16)), where $E$ denotes the projection onto $\operatorname{clo} \operatorname{ima} W_+$. Hence we have

$$\lim_{\varepsilon\to+0} \varepsilon \int_{-\infty}^{\infty} |(E^\perp g, R(\lambda - i\varepsilon) JR_0(\lambda + i\varepsilon) P_0^{ac}f)| \, d\lambda$$

$$= \lim_{\varepsilon\to+0} \varepsilon \int_{-\infty}^{\infty} |(R(\lambda + i\varepsilon) g, E^\perp JR_0(\lambda + i\varepsilon) P_0^{ac}f)| \, d\lambda$$

$$\le \lim_{\varepsilon\to+0} \left\{\varepsilon \int_{-\infty}^{\infty} \|E^\perp JR_0(\lambda + i\varepsilon) P_0^{ac}f\|^2 \, d\lambda\right\}^{1/2} \sup_{\varepsilon>0} \left\{\varepsilon \int_{-\infty}^{\infty} \|R(\lambda + i\varepsilon) g\|^2 \, d\lambda\right\}^{1/2} = 0.$$

Furthermore,

$$\int_{-\infty}^{\infty} |\alpha_\varepsilon(\lambda; g, f) - \alpha_+(\lambda; g, f)| \, d\lambda \leqq \int_{-\infty}^{\infty} \Big|\big(Eg, \varepsilon R(\lambda - i\varepsilon) \, JR_0(\lambda + i\varepsilon) \, P_0^{ac} f\big)$$

$$- \alpha_+(\lambda; g, f)\Big| \, d\lambda + \int_{-\infty}^{\infty} \Big|\big(E^\perp g, \varepsilon R(\lambda - i\varepsilon) \, JR_0(\lambda + i\varepsilon) \, P_0^{ac} f\big)\Big| \, d\lambda$$

is valid. We put $\mathscr{D} := \operatorname{ima} W_+ \oplus E^\perp \mathscr{H}$ and $g \in \mathscr{D}$, $Eg = W_+ f_0$, $f_0 \in P_0^{ac} \mathscr{H}_0$. Then we obtain

$$\int_{-\infty}^{\infty} \Big|\big(W_+ f_0, \varepsilon R(\lambda - i\varepsilon) \, JR_0(\lambda + i\varepsilon) \, P_0^{ac} f\big) - \alpha_+(\lambda; g, f)\Big| \, d\lambda$$

$$= \int_{-\infty}^{\infty} \Big|\varepsilon\big(R(\lambda + i\varepsilon) \, W_+ f_0, JR_0(\lambda + i\varepsilon) \, P_0^{ac} f\big) - \alpha_+(\lambda; g, f)\Big| \, d\lambda. \tag{44}$$

According to (9.15) we have $W_+ \underset{H_0}{\approx} J$. Therefore by Prop. 26 $\lim\limits_{\varepsilon \to +0} \varepsilon \int_{-\infty}^{\infty} \|(W_+ - J) R_0(\lambda + i\varepsilon) \, P_0^{ac} u\|^2 \, d\lambda = 0$ is implied for all $u \in \mathscr{H}_0$. Thus on the right-hand side of (44) we can replace $JR_0(\lambda + i\varepsilon) \, P_0^{ac} f$ by $W_+ R_0(\lambda + i\varepsilon) \, P_0^{ac} f$ as $\varepsilon \to +0$ without changing the limit properties of that expression. That is, we have to consider

$$\int_{-\infty}^{\infty} \Big|\varepsilon\big(R(\lambda + i\varepsilon) \, W_+ f_0, W_+ R_0(\lambda + i\varepsilon) \, P_0^{ac} f\big) - \alpha_+(\lambda; g, f)\Big| \, d\lambda$$

$$= \int_{-\infty}^{\infty} \Big|\varepsilon\big(R(\lambda + i\varepsilon) \, W_+ f_0, R(\lambda + i\varepsilon) \, W_+ f\big) - \alpha_+(\lambda; g, f)\Big| \, d\lambda.$$

Recall that

$$\begin{aligned} \alpha_+(\lambda; g, f) &= (g, W_+ f)_{E,\lambda} = \big(g, E(d\lambda) \, W_+ f\big)/d\lambda = \big(Eg, E(d\lambda) \, W_+ f\big)/d\lambda \\ &= \big(W_+ f_0, E(d\lambda) \, W_+ f\big)/d\lambda = (W_+ f_0, W_+ f)_{E,\lambda}, \end{aligned}$$

that is we have to prove

$$\lim_{\varepsilon \to +0} \int_{-\infty}^{\infty} \Big|\varepsilon\big(R(\lambda + i\varepsilon) \, W_+ f_0, R(\lambda + i\varepsilon) \, W_+ f\big) - \big(W_+ f_0, E(d\lambda) \, W_+ f\big)/d\lambda\Big| \, d\lambda = 0.$$

But this is true by Prop. 3.18. ∎

**Corollary 31.** *Suppose that the Abel wave operator $W_+ = W_+(H, H_0; J)$ exists. Then for all $f \in \mathscr{H}_0$, $g \in \mathscr{D}$ the expression*

$$\varepsilon\big(g, R(\lambda - i\varepsilon) \, JR_0(\lambda + i\varepsilon) \, P_0^{ac} f\big) - \big(\hat{g}(\lambda), \hat{W}_+(\lambda) \hat{f}(\lambda)\big)_{\mathscr{K}_\lambda} \tag{45}$$

*is measure-convergent to 0 as $\varepsilon \to +0$. Moreover there is a sequence $\varepsilon_n \to +0$ (which depends on $f, g$) such that*

$$\lim_{n \to \infty} \varepsilon_n\big(g, R(\lambda - i\varepsilon_n) \, JR_0(\lambda + i\varepsilon_n) \, P_0^{ac} f\big) = \big(\hat{g}(\lambda), \hat{W}_+(\lambda) \, \hat{f}(\lambda)\big)_{\mathscr{K}_\lambda} \tag{46}$$

*is valid a.e.* (mod $|\cdot|$) *on* $\mathbb{R}$.

**Proof.** It is obvious, because $L^1$-convergence implies convergence in measure and this implies the existence of a sequence $\varepsilon_n \to +0$ such that convergence a.e. is true. ■

**Remark 32.** It may be useful to compare Corollary 31 and Prop. 18. In the latter case

$$\underset{\varepsilon\to+0}{\text{m-lim}} \left(k, E(\mathrm{d}\lambda)\, G_\varepsilon(\mu)\, k_0\right)/\mathrm{d}\lambda|_{\mu=\lambda} = \left(\hat{k}(\lambda), \hat{\Omega}_+^{\mathrm{w}}(\lambda)\, \hat{k}_0(\lambda)\right)_{\mathscr{K}_\lambda} \tag{47}$$

was proved, in the preceding case

$$\underset{\varepsilon\to+0}{\text{m-lim}} \left(g, \varepsilon R(\lambda - \mathrm{i}\varepsilon)\, JR_0(\lambda + \mathrm{i}\varepsilon)\, P_0^{\mathrm{ac}} f\right) = \left(\hat{g}(\lambda), \hat{W}_+(\lambda)\, \hat{f}(\lambda)\right)_{\mathscr{K}_\lambda}. \tag{48}$$

But in (47) the existence of the measure limit on the left-hand side has to be assumed (and the existence of $\Omega_+^{\mathrm{w}}$ appears as a conclusion), whereas in (48) the existence of $W_+$ is assumed and the existence of the measure limit on the left-hand side can be proved. It should be noted that a weaker proposition on the relation between $\left(g, \varepsilon R(\lambda - \mathrm{i}\varepsilon)\, J \times R_0(\lambda + \mathrm{i}\varepsilon)\, P_0^{\mathrm{ac}} f\right)$ and $\left(\hat{g}(\lambda), \hat{\Omega}_+^{\mathrm{w}}(\lambda)\, \hat{f}(\lambda)\right)_{\mathscr{K}_\lambda}$ than that given in Corollary 31 can be proved if only the existence of $\Omega_+^{\mathrm{w}}$ is assumed.

**Proposition 32.** *Suppose that $\Omega_+^{\mathrm{w}}$ exists. Then*

$$\lim_{\varepsilon\to+0} \int_\Delta \left(g, \varepsilon R(\lambda - \mathrm{i}\varepsilon)\, JR_0(\lambda + \mathrm{i}\varepsilon)\, P_0^{\mathrm{ac}} f\right) \mathrm{d}\lambda = \int_\Delta \left(\hat{g}(\lambda), \hat{\Omega}_+^{\mathrm{w}}(\lambda)\, \hat{f}(\lambda)\right)_{\mathscr{K}_\lambda} \mathrm{d}\lambda \tag{49}$$

*is valid for all $f \in \mathscr{H}_0$, $g \in \mathscr{H}$ and for all Borel sets $\Delta$, where $\hat{f}(\cdot)$, $\hat{g}(\cdot)$ denote the representers of $P_0^{\mathrm{ac}} f$, $P^{\mathrm{ac}} g$, respectively.*

**Proof.** We have $\Omega_\varepsilon = \Omega_\varepsilon(\Delta) + \Omega_\varepsilon(\mathbb{R} \setminus \Delta)$, hence $\Omega_\varepsilon P_0^{\mathrm{ac}} E_0(\Delta) = \Omega_\varepsilon(\Delta)\, P_0^{\mathrm{ac}} E_0(\Delta) + \Omega_\varepsilon(\mathbb{R} \setminus \Delta)\, P_0^{\mathrm{ac}} E_0(\Delta)$. According to Prop. 4, we have $\underset{\varepsilon\to+0}{\text{s-lim}}\, \Omega_\varepsilon(\mathbb{R} \setminus \Delta)\, P_0^{\mathrm{ac}} E_0(\Delta) = 0$. Therefore $\underset{\varepsilon\to+0}{\text{w-lim}}\, \Omega_\varepsilon(\Delta)\, P_0^{\mathrm{ac}} E_0(\Delta)$ exists and equals $\Omega_+^{\mathrm{w}} E_0(\Delta)$. Now we have $\Omega_\varepsilon(\Delta)\, P_0^{\mathrm{ac}} f = \Omega_\varepsilon(\Delta)\, P_0^{\mathrm{ac}} E_0(\Delta)\, f + \Omega_\varepsilon(\Delta)\, P_0^{\mathrm{ac}} E_0(\mathbb{R} \setminus \Delta)\, f$. Hence $\underset{\varepsilon\to+0}{\text{w-lim}}\, \Omega_\varepsilon(\Delta)\, P_0^{\mathrm{ac}} f$ exists for all $f \in \mathscr{H}_0$ and we have

$$\underset{\varepsilon\to+0}{\text{w-lim}}\, \Omega_\varepsilon(\Delta)\, P_0^{\mathrm{ac}} f = \Omega_+^{\mathrm{w}} E_0(\Delta)\, f.$$

Because of $\left(g, \Omega_+^{\mathrm{w}} E_0(\Delta)\, f\right) = \int_\Delta \left(\hat{g}(\lambda), \hat{\Omega}_+^{\mathrm{w}}(\lambda)\, \hat{f}(\lambda)\right)_{\mathscr{K}_\lambda} \mathrm{d}\lambda$ the relation (49) follows. ■

**Remark 33.** Compare Prop. 32 with Remark 19 where the corresponding relation with respect to representation (2) is presented.

## 13.4. Stationary theory of the Abel wave operator using operator spectral integrals

In this paragraph we develop the stationary theory of the Abel wave operator in connection with representation (2) of the stationary pre-wave operator.

### 13.4.1. The general case

In this section we collect some results from Paragraph 2 and from Paragraph 9.4 to obtain a necessary and sufficient stationary criterion for the existence of the Abel wave operator. The weak point in this criterion is that an assumption like (9.44) linking $J$ and $\Omega_+^w$ seems to be unavoidable.

**Theorem 34.** (I) *Let $\mathscr{K}_0 \subset \mathscr{H}_0$, $\mathscr{K} \subset \mathscr{H}$ be linear manifolds and assume that $P_0^{ac}\mathscr{K}_0$, $P^{ac}\mathscr{K}$ are generating with respect to $P_0^{ac}\mathscr{H}_0$, $P^{ac}\mathscr{H}$ and $H_0$, $H$, respectively.*

(II) *Assume that $\left(k, E(\mathrm{d}\lambda)\, G_\varepsilon(\mu)\, k_0\right)/\mathrm{d}\lambda|_{\mu=\lambda}$ is a Cauchy net as $\varepsilon \to +0$ with respect to convergence in measure on $\mathbb{R}$ for all $k_0 \in \mathscr{K}_0$, $k \in \mathscr{K}$* (then $\Omega_+^w$ exists).

(III) *Further assume*

$$\lim_{\varepsilon\to+0} \int_0^\infty \varepsilon\, \mathrm{e}^{-\varepsilon t}\, \|J\, \mathrm{e}^{-\mathrm{i}tH_0} P_0^{ac} f\|^2\, \mathrm{d}t = \|\Omega_+^w f\|^2, \qquad f \in \mathscr{H}_0. \tag{50}$$

*Then the Abel wave operator $W_+ = W_+(H, H_0; J)$ exists. Conversely, if the Abel wave operator exists, then* (II) *is true for $\mathscr{K}_0 = P_0^{ac}\mathscr{H}_0$, $\mathscr{K} = P^{ac}\mathscr{H}$ and also* (III) *is satisfied.*

Proof. (A) Let (I)—(III) be satisfied. Then the existence of $\Omega_+^w$ follows from Prop. 17. Further, from Corollary 9.43 we obtain $\Omega_+^w \underset{H_0}{\approx} J$. Finally, Prop. 9.41 yields the existence of $W_+(H, H_0; J)$. (B) Conversely, from the existence of $W_+$ we obtain (III) via Prop. 9.41 and Corollary 9.43. Further, from Prop. 9.6. the existence of $\operatorname{s-lim}_{\varepsilon\to+0} \Omega_\varepsilon P_0^{ac}$ follows. Now we apply Theorem 22. Therefore $|G_{\varepsilon_1}(\lambda)\, P_0^{ac} f - G_{\varepsilon_2}(\lambda)\, P_0^{ac} f|_{E,\lambda}$ is a Cauchy net as $\varepsilon_1, \varepsilon_2 \to +0$ with respect to convergence in measure for all $f \in \mathscr{H}_0$. Now we have

$$\begin{aligned} &\left|\left(g, E(\mathrm{d}\lambda)\, G_{\varepsilon_1}(\mu)\, P_0^{ac} f\right)/\mathrm{d}\lambda|_{\mu=\lambda} - \left(g, E(\mathrm{d}\lambda)\, G_{\varepsilon_2}(\mu)\, P_0^{ac} f\right)/\mathrm{d}\lambda|_{\mu=\lambda}\right| \\ &\le |g|_{E,\lambda}\, |G_{\varepsilon_1}(\lambda)\, P_0^{ac} f - G_{\varepsilon_2}(\lambda)\, P_0^{ac} f|_{E,\lambda} \end{aligned} \tag{51}$$

a.e. (mod $|\cdot|$) on $\mathbb{R}$. Thus (II) is satisfied for $\mathscr{K}_0 = P_0^{ac}\mathscr{H}_0$, $\mathscr{K} = P^{ac}\mathscr{H}$. ■

### 13.4.2. The case of partially isometric Abel wave operators

In this section we present another variant of a sufficient and necessary criterion. Namely, we split up the condition (50) into two assumptions on $J$ and $\Omega_+^w$ separately, which together yield (50). The condition on $J$ is expressed by

$$\lim_{\varepsilon\to+0} \int_0^\infty \varepsilon\, \mathrm{e}^{-\varepsilon t}\, \|J\, \mathrm{e}^{-\mathrm{i}tH_0} P_0^{ac} f\|^2\, \mathrm{d}t = \|P_0^{ac} f\|^2, \qquad f \in \mathscr{H}_0. \tag{52}$$

This condition was already considered in Chapter 9.5. A slightly stronger condition is $J \underset{H_0}{\approx} V$, where $V$ is partially isometric with initial subspace $P_0^{ac}\mathscr{H}_0$ (cf. Lemma 9.46). According to Lemma 9.46 and Prop. 9.45 the condition (52) yields necessarily partially isometric and $H_0$-semicomplete wave operators.

**Theorem 35.** *Let condition* (52) *be satisfied.*

(I) *Let $\mathscr{K}_0 \subset \mathscr{H}_0$ be a linear manifold such that $P_0^{ac}\mathscr{K}_0$ is generating with respect to $P_0^{ac}\mathscr{H}_0$ and $H_0$.*

(II) *Assume that $G_\varepsilon(\lambda)\, k_0$ is a Cauchy net as $\varepsilon \to +0$ with respect to convergence in measure on $\mathbb{R}$ and $|\cdot|_{E,\lambda}$ for all $k_0 \in \mathscr{K}_0$, that is, let*

$$\operatorname*{m-lim}_{\varepsilon,\varepsilon' \to +0} |G_\varepsilon(\lambda)\, k_0 - G_{\varepsilon'}(\lambda)\, k_0|_{E,\lambda} = 0, \qquad k_0 \in \mathscr{K}_0, \tag{53}$$

(note that if (53) is valid, then also $\operatorname*{m-lim}_{\varepsilon \to +0} |G_\varepsilon(\lambda)\, k_0|_{E,\lambda}$ exists).

(III) *Further assume*

$$\operatorname*{m-lim}_{\varepsilon \to +0} |G_\varepsilon(\lambda)\, k_0|_{E,\lambda} = |k_0|_{E_0,\lambda}, \qquad k_0 \in \mathscr{K}_0. \tag{54}$$

*Then the Abel wave operator $W_+ = W_+(H, H_0; J)$ exists and is partially isometric and $H_0$-semicomplete. Conversely, if $W_+$ exists, then* (II) *and* (III) *are satisfied for $\mathscr{K}_0 = P_0^{ac}\mathscr{H}_0$.*

Proof. According to (51) (where $P_0^{ac}f$ is replaced by $k_0$ and where $g \in P^{ac}\mathscr{H}$) and Prop. 17 we obtain the existence of $\Omega_+^w$. Now we prove that $\Omega_+^w$ is partially isometric with initial subspace $P_0^{ac}\mathscr{H}_0$. Since $\Omega_+^w f = 0$ for $f \in (P_0^{ac})^\perp \mathscr{H}_0$, we must prove $\|\Omega_+^w f\| = \|f\|$ for all $f \in P_0^{ac}\mathscr{H}_0$. Because of $\|\Omega_+^w\| \leqq \|J\|$ it is sufficient to prove this equation for a dense set in $P_0^{ac}\mathscr{H}_0$. We choose as a dense set spa $\{E_0(\Delta)\, P_0^{ac}k_0;\ k_0 \in \mathscr{K}_0,$ $\Delta$ bounded Borel set$\}$. First let $k_0 \in \mathscr{K}_0$ be fixed. Then from (III) we obtain the existence of a sequence $\varepsilon_n \to +0$ and the existence of a Borel set $\Lambda_0$ with $|\Lambda_0| = 0$ such that

$$\lim_{n\to\infty} |G_{\varepsilon_n}(\lambda)\, k_0|_{E,\lambda} = |k_0|_{E_0,\lambda}, \qquad \lambda \in \mathbb{R} \setminus \Lambda_0 \tag{55}$$

is valid. Now we put $\alpha(\lambda) := \sup_{n,m} |G_{\varepsilon_n}(\lambda)\, k_0 - G_{\varepsilon_m}(\lambda)\, k_0|_{E,\lambda}$. Then $\alpha(\lambda)$ is a Borel function and $0 \leqq \alpha(\lambda) < \infty$ for $\lambda \in \mathbb{R} \setminus \Lambda_0$. Hence $\Lambda_N := \{\lambda;\ \alpha(\lambda) < N\}$ is a Borel set and we have $\Lambda_1 \subset \Lambda_2 \subset \cdots \subset \mathbb{R} \setminus \Lambda_0$, $\bigcup_{N=1}^{\infty} \Lambda_N = \mathbb{R} \setminus \Lambda_0$. Further, let $\Delta$ be an arbitrary bounded Borel set. Then we obtain

$$\|P^{ac}E(\Lambda_N)\, E(\Delta)\, (\Omega_{\varepsilon_n}k_0 - \Omega_{\varepsilon_m}k_0)\|^2 = \int_{\Lambda_N \cap \Delta} |G_{\varepsilon_n}(\lambda)\, k_0 - G_{\varepsilon_m}(\lambda)\, k_0|_{E,\lambda}^2\, d\lambda.$$

Hence, according to the Vitali convergence theorem, $\|P^{ac}E(\Lambda_N)\, E(\Delta)\, (\Omega_{\varepsilon_n}k_0 - \Omega_{\varepsilon_m}k_0)\|$ is a Cauchy sequence, that is, $\operatorname*{s-lim}_{n\to\infty} P^{ac}E(\Lambda_N)\, E(\Delta)\, \Omega_{\varepsilon_n}k_0$ exists and moreover we have

$$\begin{aligned}\operatorname*{s-lim}_{n\to\infty} E(\Lambda_N)\, E(\Delta)\, P^{ac}\Omega_{\varepsilon_n}k_0 &= E(\Lambda_N)\, E(\Delta)\, P^{ac}\Omega_+^w k_0 = E(\Lambda_N)\, E(\Delta)\, P^{ac}\Omega_+^w P_0^{ac}k_0\\ &= E(\Lambda_N)\, E(\Delta)\, \Omega_+^w P_0^{ac}k_0.\end{aligned}$$

Thus also

$$\lim_{n\to\infty} \|E(\Lambda_N)\, E(\Delta)\, P^{ac}\Omega_{\varepsilon_n}k_0\|^2 = \|E(\Lambda_N)\, E(\Delta)\, \Omega_+^w P_0^{ac}k_0\|^2$$

follows. On the other hand, we obtain, by Vitali's convergence theorem, that

$$\begin{aligned}\lim_{n\to\infty} \|E(\Lambda_N \cap \Delta)\, P^{ac}\Omega_{\varepsilon_n}k_0\|^2 &= \lim_{n\to\infty} \int_{\Lambda_N\cap\Delta} |G_{\varepsilon_n}(\lambda)\, k_0|_{E,\lambda}^2\, d\lambda = \int_{\Lambda_N\cap\Delta} \lim_{n\to\infty} |G_{\varepsilon_n}(\lambda)\, k_0|_{E,\lambda}^2\, d\lambda\\ &= \int_{\Lambda_N\cap\Delta} |k_0|_{E_0,\lambda}^2\, d\lambda = \|E_0(\Lambda_N \cap \Delta)\, P_0^{ac}k_0\|^2 = \|E_0(\Lambda_n)\, E_0(\Delta)\, P_0^{ac}k_0\|^2.\end{aligned}$$

Therefore we have the equation

$$\|E(\Lambda_N)\, E(\Delta)\, \Omega_+^{\mathrm{w}} P_0^{\mathrm{ac}} k_0\| = \|E_0(\Lambda_N)\, E_0(\Delta)\, P_0^{\mathrm{ac}} k_0\|,$$

or, because of the intertwining property of $\Omega_+^{\mathrm{w}}$,

$$\|\Omega_+^{\mathrm{w}} E_0(\Lambda_N)\, E_0(\Delta)\, P_0^{\mathrm{ac}} k_0\| = \|E_0(\Lambda_N)\, E_0(\Delta)\, P_0^{\mathrm{ac}} k_0\|.$$

Now $\operatorname*{s-lim}_{N\to\infty} E_0(\Lambda_N)\, E_0(\Delta)\, P_0^{\mathrm{ac}} k_0 = E_0(\Delta)\, P_0^{\mathrm{ac}} k_0$ is valid, hence

$$\|\Omega_+^{\mathrm{w}} E_0(\Delta)\, P_0^{\mathrm{ac}} k_0\| = \|E_0(\Delta)\, P_0^{\mathrm{ac}} k_0\| \tag{56}$$

follows for all $k_0 \in \mathscr{K}_0$ and for all bounded Borel sets $\Delta$. From (56) and from the intertwining property of $\Omega_+^{\mathrm{w}}$ we easily obtain

$$\left\|\Omega_+^{\mathrm{w}} \sum_{\varrho=1}^{m} E_0(\Delta_\varrho)\, P_0^{\mathrm{ac}} k_\varrho\right\| = \left\|\sum_{\varrho=1}^{m} E_0(\Delta_\varrho)\, P_0^{\mathrm{ac}} k_\varrho\right\|,$$

where the bounded Borel sets $\Delta_\varrho$ are pairwise disjoint. Now we can conclude: $\Omega_+^{\mathrm{w}}$ is isometric on $P_0^{\mathrm{ac}}\mathscr{H}_0$. Therefore because of (52) the relation (50) is satisfied. Thus according to Theorem 34 the Abel wave operator $W_+$ exists and equals $\Omega_+^{\mathrm{w}}$. Conversely, let $W_+$ exist. Then $\operatorname*{s-lim}_{\varepsilon\to+0} \Omega_\varepsilon P_0^{\mathrm{ac}} f$ exists for all $f \in \mathscr{H}_0$. By Theorem 22, we obtain (II) for $\mathscr{K}_0 = P_0^{\mathrm{ac}}\mathscr{H}_0$. Further we obtain for all Borel sets $\Delta$ (use Vitali's convergence theorem)

$$\lim_{\varepsilon\to+0} \|E(\Delta)\, P^{\mathrm{ac}} \Omega_\varepsilon P_0^{\mathrm{ac}} f\|^2 = \lim_{\varepsilon\to+0} \int_\Delta |G_\varepsilon(\lambda)\, P_0^{\mathrm{ac}} f|_{E,\lambda}^2 \, \mathrm{d}\lambda = \int_\Delta \operatorname*{m-lim}_{\varepsilon\to+0} |G_\varepsilon(\lambda)\, P_0^{\mathrm{ac}} f|_{E,\lambda}^2 \, \mathrm{d}\lambda$$

$$= \|E(\Delta)\, W_+ P_0^{\mathrm{ac}} f\|^2 = \|W_+ E_0(\Delta)\, P_0^{\mathrm{ac}} f\|^2 = \|E_0(\Delta)\, P_0^{\mathrm{ac}} f\|^2 = \int_\Delta |f|_{E_0,\lambda}^2 \, \mathrm{d}\lambda.$$

Hence

$$\operatorname*{m-lim}_{\varepsilon\to+0} |G_\varepsilon(\lambda)\, P_0^{\mathrm{ac}} f|_{E,\lambda} = |f|_{E_0,\lambda} = |P_0^{\mathrm{ac}} f|_{E_0,\lambda}, \qquad f \in \mathscr{K}_0,$$

follows a.e. (mod $|\cdot|$) on $\mathbb{R}$. Thus (III) is satisfied for $\mathscr{K}_0 = P_0^{\mathrm{ac}}\mathscr{H}_0$. ∎

**Remark 36.** If we replace the limits in measure in conditions (II) and (III) of Theorem 35 by limits a.e., that is, if $G_\varepsilon(\lambda)\, k_0$ is a Cauchy net as $\varepsilon \to +0$ with respect to the seminorm $|\cdot|_{E,\lambda}$ for almost all $\lambda \in \mathbb{R}$ (mod $|\cdot|$) and if $\lim_{\varepsilon\to+0} |G_\varepsilon(\lambda)\, k_0|_{E,\lambda} = |k_0|_{E_0,\lambda}$ a.e. (mod $|\cdot|$) on $\mathbb{R}$; then conditions (II) and (III) of Theorem 35 are satisfied. That is, in this way we obtain sufficient conditions for the existence of the Abel wave operator $W_+$. Applications of Theorem 35 will be given in Chapter 14.

### 13.4.3. Convergence to the wave matrix

In this section we refer to Theorem 22 and in particular to the expression (37). The problem to be considered is to determine the limit function of the Cauchy net $|G_\varepsilon(\lambda)\, f - G_{\varepsilon'}(\lambda)\, f|_{E,\lambda}$; $f \in P_0^{\mathrm{ac}}\mathscr{H}_0$, which is a Cauchy net with respect to the convergence in measure, explicitly and to link this limit with the wave matrix. First, from

$$P^{ac}\Omega_\varepsilon P_0^{ac}f - \Omega_+^w f = \int_{-\infty}^{\infty} E(d\lambda)\, P^{ac}\big(G_\varepsilon(\lambda)\, P_0^{ac}f - \Omega_+^w f\big), \qquad f \in P_0^{ac}\mathscr{H}_0, \tag{57}$$

we immediately obtain

$$\|P^{ac}\Omega_\varepsilon P_0^{ac}f - \Omega_+^w f\|^2 = \int_{-\infty}^{\infty} |G_\varepsilon(\lambda)\, P_0^{ac}f - \Omega_+^w f|_{E,\lambda}^2\, d\lambda.$$

Hence

$$\operatorname*{m-lim}_{\varepsilon\to+0} |G_\varepsilon(\lambda)\, P_0^{ac} - \Omega_+^w f|_{E,\lambda} = 0, \qquad f \in P_0^{ac}\mathscr{H}_0, \tag{58}$$

follows. Thus the crucial point is to link this relation with the wave matrix $\hat{\Omega}_+^w(\lambda)$, if direct integral representations of $P_0^{ac}\mathscr{H}_0$ and $P_0^{ac}\mathscr{H}$ are given.

In order to establish this link we recall the definitions and facts from Section 3.5.5 and Section 5.2.4.

First, we recall the concept of a spectral and generating manifold $\mathscr{X} \subset P^{ac}\mathscr{H}$ with respect to $H$ and the corresponding direct integral representation of $P^{ac}\mathscr{H}$, given by $\Phi_{\mathscr{X}}: P^{ac}\mathscr{H} \mapsto L^2(\Lambda, d\lambda; \hat{\mathscr{X}}_\lambda)$ ($\Lambda$ spectral core of $H \mid P^{ac}\mathscr{H}$). Second, we recall the manifold $\mathscr{D}_\lambda$ and the Banach space $\hat{\mathscr{D}}_\lambda$, further the representation of $f = \int E(d\lambda)\, u(\lambda) \in P^{ac}\mathscr{H}$ by $f \leftrightarrow D_\lambda u(\lambda) \in \hat{\mathscr{D}}_\lambda$ ("almost" a spectral representation) and the fact that $(\Phi_{\mathscr{X}}f)(\lambda) = D_\lambda u(\lambda)$ a.e. (mod $|\cdot|$) on $\mathbb{R}$. With the help of these concepts we are able to express the desired connection by

**Proposition 37.** *Let* $\operatorname*{s-lim}_{\varepsilon\to+0} \Omega_\varepsilon P_0^{ac} = \Omega_+^w$ *exist. Then*

$$\operatorname*{m-lim}_{\varepsilon\to+0} \big[D_\lambda\big(G_\varepsilon(\lambda)\, f\big) - \hat{\Omega}_+^w(\lambda)\, (\Phi_0 f)\, (\lambda)\big]_{E,\lambda} = 0, \qquad f \in P_0^{ac}\mathscr{H}_0, \tag{59}$$

*where the wave matrix* $\hat{\Omega}_+^w(\lambda)$ *of* $\Omega_+^w$ *is defined with respect to an arbitrary direct integral representation of* $P_0^{ac}\mathscr{H}_0$ *with respect to* $H_0$*, given by the map* $\Phi_0$ *and a direct integral representation of* $P^{ac}\mathscr{H}$ *with respect to* $H$ *given by* $\Phi_{\mathscr{X}}$*, where* $\mathscr{X} \subset P^{ac}\mathscr{H}$ *is spectral and generating.*

**Proof.** According to (58) we have

$$\big[D_\lambda\big(G_\varepsilon(\lambda)\, f\big) - D_\lambda(\Omega_+^w f)\big]_{E,\lambda} \to 0, \qquad \varepsilon \to +0, \quad f \in P_0^{ac}\mathscr{H}_0,$$

with respect to convergence in measure. Thus (59) follows, because

$$D_\lambda(\Omega_+^w f) = (\Phi_{\mathscr{X}}\Omega_+^w f)\, (\lambda) = \hat{\Omega}_+^w(\lambda)\, (\Phi_0 f)\, (\lambda) \text{ a.e. (mod } |\cdot|) \text{ on } \mathbb{R}. \blacksquare$$

## 13.5. A completeness criterion

In this paragraph we present a stationary criterion for the completeness of partially isometric and $H_0$-complete wave operators. It is related to Paragraph 3 on the one hand and uses so-called approximate operator spectral integrals (see Paragraph 5.3) on the other hand. The formula

$$R(z) = JR_0(z)\, F(z), \qquad z = \lambda + i\varepsilon, \qquad \varepsilon > 0, \tag{60}$$

can be considered as a starting point in this framework where the function $F(z) \in \mathscr{L}(\mathscr{H}, \mathscr{H}_0)$ is assumed to satisfy smoothness properties as $\varepsilon \to +0$. This "ansatz" is often used in connection with completeness considerations (see the Notes and Remarks). The crucial points in *our* approach are the following: (I) the condition (60) is weakened to an asymptotic relation

$$\lim_{\varepsilon \to +0} \varepsilon \int_{-\infty}^{\infty} \|R(\lambda + i\varepsilon) f - JR_0(\lambda + i\varepsilon) F_\varepsilon(\lambda) f\|^2 \,d\lambda = 0, \qquad f \in \mathscr{H}, \tag{61}$$

similar to the asymptotic relations considered in Paragraph 3. The relation (61) is a counterpart to formula (41) in Theorem 28. (II) The smoothness property of $F_\varepsilon(\lambda)$ as $\varepsilon \to +0$ is not assumed in a pointwise sense but in an averaged sense with respect to $\lambda$. It can be best expressed by an asymptotic property of an approximate operator spectral integral of the type

$$\int_{-\infty}^{\infty} E_{0,\varepsilon}(d\lambda)\, F_\varepsilon(\lambda), \qquad E_{0,\varepsilon}(\Delta) = (\varepsilon/\pi) \int_{\Delta} R_0(\lambda - i\varepsilon)\, R_0(\lambda + i\varepsilon)\, d\lambda.$$

### 13.5.1. A property in connection with the adjoint of the wave operator

In this section the decisive preparation of the criterion is presented. In the following the condition (52) is assumed which ensures partially isometric and $H_0$-semicomplete wave operators. Further we assume the existence of the Abel wave operator $W_+ = W_+(H, H_0; J)$. First, we prove the following lemma.

**Lemma 38.** *Let $\mathscr{D}$ be a linear manifold from $\mathscr{H}$ and let $F_\varepsilon(\lambda) \in \mathscr{L}(\mathscr{H}, \mathscr{H}_0)$, $\varepsilon > 0$, $\lambda \in \mathbb{R}$. Assume $F_\varepsilon(\lambda) f$ to be strongly continuous with respect to $\lambda$ for all $\varepsilon > 0$ and all $f \in \mathscr{D}$. Suppose the existence of $\int_{-\infty}^{\infty} E_{0,\varepsilon}(d\lambda)\, F_\varepsilon(\lambda) f$ for all $\varepsilon > 0$ and that for all $f \in \mathscr{D}$ the following conditions are satisfied:*

$$Z_+ f := \operatorname*{w-lim}_{\varepsilon \to +0} P_0^{ac} \int_{-\infty}^{\infty} E_{0,\varepsilon}(d\lambda)\, F_\varepsilon(\lambda) f \tag{62}$$

*exists.*

$$\lim_{\varepsilon \to +0} \int_{-\infty}^{\infty} \big(F_\varepsilon(\lambda) f, E_{0,\varepsilon}(d\lambda)\, F_\varepsilon(\lambda) f\big) = \|Z_+ f\|^2. \tag{63}$$

$$\lim_{\varepsilon \to +0} \varepsilon \int_{-\infty}^{\infty} \|R(\lambda + i\varepsilon) f - JR_0(\lambda + i\varepsilon) F_\varepsilon(\lambda) f\|^2 \,d\lambda = 0. \tag{64}$$

*Then $Z_+ f = W_+^* f$ for all $f \in \mathscr{D}$.*

Proof. First we prove

$$\lim_{\varepsilon \to +0} (\varepsilon/\pi) \int_{-\infty}^{\infty} \big\|R_0(\lambda + i\varepsilon) \big(F_\varepsilon(\lambda) f - Z_+ f\big)\big\|^2 \,d\lambda = 0. \tag{65}$$

In terms of approximate spectral integrals (65) means

$$\lim_{\varepsilon\to+0} \int_{-\infty}^{\infty} \left(F_\varepsilon(\lambda) f - Z_+f, E_{0,\varepsilon}(\mathrm{d}\lambda) \left(F_\varepsilon(\lambda) f - Z_+f\right)\right) = 0. \tag{66}$$

We have

$$\int_{-\infty}^{\infty} \left(F_\varepsilon(\lambda) f - Z_+f, E_{0,\varepsilon}(\mathrm{d}\lambda) \left(F_\varepsilon(\lambda) f - Z_+f\right)\right) = \int_{-\infty}^{\infty} \left(F_\varepsilon(\lambda) f, E_{0,\varepsilon}(\mathrm{d}\lambda) F_\varepsilon(\lambda) f\right)$$

$$- \left(Z_+f, \int_{-\infty}^{\infty} E_{0,\varepsilon}(\mathrm{d}\lambda) F_\varepsilon(\lambda) f\right) - \left(\int_{-\infty}^{\infty} E_{0,\varepsilon}(\mathrm{d}\lambda) F_\varepsilon(\lambda) f, Z_+f\right)$$

$$+ \left(Z_+f, \int_{-\infty}^{\infty} E_{0,\varepsilon}(\mathrm{d}\lambda) Z_+f\right).$$

Since $\left(Z_+f, \int_{-\infty}^{\infty} E_{0,\varepsilon}(\mathrm{d}\lambda) Z_+f\right) = (\varepsilon/\pi) \int_{-\infty}^{\infty} \|R_0(\lambda + \mathrm{i}\varepsilon) Z_+f\|^2 \,\mathrm{d}\lambda = \|Z_+f\|^2$, because of (62) and (63) we obtain relation (66), hence also (65). If $A \in \mathscr{L}(\mathscr{H}_0, \mathscr{H})$, from (65) we immediately obtain

$$\lim_{\varepsilon\to+0} (\varepsilon/\pi) \int_{-\infty}^{\infty} \left\|AR_0(\lambda + \mathrm{i}\varepsilon) \left(F_\varepsilon(\lambda) f - Z_+f\right)\right\|^2 \mathrm{d}\lambda = 0. \tag{67}$$

Now we assume even $|A|$-$\lim_{t\to\infty} A\, \mathrm{e}^{-\mathrm{i}tH_0} P_0^{\mathrm{ac}} = 0$. Then we are able to prove

$$\lim_{\varepsilon\to+0} (\varepsilon/\pi) \int_{-\infty}^{\infty} \|AR_0(\lambda + \mathrm{i}\varepsilon) F_\varepsilon(\lambda) f\|^2 \,\mathrm{d}\lambda = 0. \tag{68}$$

Namely, according to Prop. 26 the condition $A \underset{H_0}{\approx} 0$ is equivalent to

$$\lim_{\varepsilon\to+0} (\varepsilon/\pi) \int_{-\infty}^{\infty} \|AR_0(\lambda + \mathrm{i}\varepsilon) P_0^{\mathrm{ac}} g\|^2 \,\mathrm{d}\lambda = 0, \qquad g \in \mathscr{H}_0. \tag{69}$$

Now we have

$$\left((\varepsilon/\pi) \int_{-\infty}^{\infty} AR_0(\lambda + \mathrm{i}\varepsilon) F_\varepsilon(\lambda) f\|^2 \,\mathrm{d}\lambda\right)^{1/2}$$

$$\leq \left((\varepsilon/\pi) \int_{-\infty}^{\infty} \|AR_0(\lambda + \mathrm{i}\varepsilon) Z_+f\|^2 \,\mathrm{d}\lambda\right)^{1/2} + \left((\varepsilon/\pi) \int_{-\infty}^{\infty} \left\|AR_0(\lambda + \mathrm{i}\varepsilon) \left(F_\varepsilon(\lambda) f - Z_+f\right)\right\|^2 \mathrm{d}\lambda\right)^{1/2}.$$

The first term on the right-hand side tends to 0 as $\varepsilon \to +0$ because of (69), the second term also tends to 0 as $\varepsilon \to +0$ since (67) is valid. Thus (68) is proved.

Now we choose a special operator $A = J - W_+$. By (9.15) we have $J \underset{H_0}{\approx} W_+$, hence we obtain

$$\lim_{\varepsilon\to+0} (\varepsilon/\pi) \int_{-\infty}^{\infty} \|(J - W_+) R_0(\lambda + \mathrm{i}\varepsilon) F_\varepsilon(\lambda) f\|^2 \,\mathrm{d}\lambda = 0. \tag{70}$$

From (70) and (64) we obtain

$$\lim_{\varepsilon\to+0} (\varepsilon/\pi) \int_{-\infty}^{\infty} \|R(\lambda + i\varepsilon) f - W_+R_0(\lambda + i\varepsilon) F_\varepsilon(\lambda) f\|^2 \, d\lambda = 0, \tag{71}$$

using the estimate

$$\left((\varepsilon/\pi) \int_{-\infty}^{\infty} \|R(\lambda + i\varepsilon) f - W_+R_0(\lambda + i\varepsilon) F_\varepsilon(\lambda) f\|^2 \, d\lambda\right)^{1/2}$$

$$\leqq \left((\varepsilon/\pi) \int_{-\infty}^{\infty} \|R(\lambda + i\varepsilon) f - JR_0(\lambda + i\varepsilon) F_\varepsilon(\lambda) f\|^2 \, d\lambda\right)^{1/2}$$

$$+ \left((\varepsilon/\pi) \int_{-\infty}^{\infty} \|(J - W_+) R_0(\lambda + i\varepsilon) F_\varepsilon(\lambda) f\|^2 \, d\lambda\right)^{1/2}.$$

Now we are able to prove

$$\lim_{\varepsilon\to+0} (\varepsilon/\pi) \int_{-\infty}^{\infty} \big(JR_0(\lambda + i\varepsilon) g, R(\lambda + i\varepsilon) f\big)_{\mathscr{H}} \, d\lambda = (g, Z_+f) \tag{72}$$

for all $g \in P_0^{ac}\mathscr{H}_0$, $f \in \mathscr{D}$. Since

$$(g, Z_+f) = \lim_{\varepsilon\to+0} \left(g, \int_{-\infty}^{\infty} E_{0,\varepsilon}(d\lambda) \; F_\varepsilon(\lambda) f\right) = \lim_{\varepsilon\to+0} \int_{-\infty}^{\infty} \big(g, E_{0,\varepsilon}(d\lambda) \; F_\varepsilon(\lambda) f\big)$$

$$= \lim_{\varepsilon\to+0} (\varepsilon/\pi) \int_{-\infty}^{\infty} \big(R_0(\lambda + i\varepsilon) g, R_0(\lambda + i\varepsilon) F_\varepsilon(\lambda) f\big)_{\mathscr{H}_0} \, d\lambda,$$

it is sufficient to prove

$$\lim_{\varepsilon\to+0} (\varepsilon/\pi) \int_{-\infty}^{\infty} \Big|\big(JR_0(\lambda + i\varepsilon) g, R(\lambda + i\varepsilon) f\big)_{\mathscr{H}}$$

$$- \big(R_0(\lambda + i\varepsilon) g, R_0(\lambda + i\varepsilon) F_\varepsilon(\lambda) f\big)_{\mathscr{H}_0}\Big| \, d\lambda = 0. \tag{73}$$

Note that $\big(R_0(\lambda + i\varepsilon) g, R_0(\lambda + i\varepsilon) F_\varepsilon(\lambda) f\big)_{\mathscr{H}_0} = \big(W_+R_0(\lambda + i\varepsilon) g, W_+R_0(\lambda + i\varepsilon) F_\varepsilon(\lambda) f\big)_{\mathscr{H}}$ is valid because of $g \in P_0^{ac}\mathscr{H}_0$. Now we have

$$\big(JR_0(\lambda + i\varepsilon) g, R(\lambda + i\varepsilon) f\big)_{\mathscr{H}} - \big(W_+R_0(\lambda + i\varepsilon) g, W_+R_0(\lambda + i\varepsilon) F_\varepsilon(\lambda) f\big)_{\mathscr{H}}$$

$$= \big((J - W_+) R_0(\lambda + i\varepsilon) g, R(\lambda + i\varepsilon) f\big)_{\mathscr{H}}$$

$$+ \big(W_+R_0(\lambda + i\varepsilon) g, R(\lambda + i\varepsilon) f - W_+R_0(\lambda + i\varepsilon) F_\varepsilon(\lambda) f\big)_{\mathscr{H}}.$$

Hence we obtain

$$\Big|\big(JR_0(\lambda + i\varepsilon) g, R(\lambda + i\varepsilon) f\big)_{\mathscr{H}} - \big(W_+R_0(\lambda + i\varepsilon) g, W_+R_0(\lambda + i\varepsilon) F_\varepsilon(\lambda) f\big)_{\mathscr{H}}\Big|$$

$$\leqq \|(J - W_+) R_0(\lambda + i\varepsilon) g\| \cdot \|R(\lambda + i\varepsilon) f\|$$

$$+ \|W_+R_0(\lambda + i\varepsilon) g\| \cdot \|R(\lambda + i\varepsilon) f - W_+R_0 (\lambda + i\varepsilon) F_\varepsilon(\lambda) f\|$$

and

$$(\varepsilon/\pi) \int_{-\infty}^{\infty} \left|\left(JR_0(\lambda + i\varepsilon)\, g, R(\lambda + i\varepsilon)\, f\right)_{\mathscr{H}} - \left(R_0(\lambda + i\varepsilon)\, g, R_0(\lambda + i\varepsilon)\, F_\varepsilon(\lambda)\, f\right)_{\mathscr{H}_0}\right| d\lambda$$

$$\leq \left((\varepsilon/\pi) \int_{-\infty}^{\infty} \|(J - W_+)\, R_0(\lambda + i\varepsilon)\, g\|^2\, d\lambda\right)^{1/2} \left((\varepsilon/\pi) \int_{-\infty}^{\infty} \|R(\lambda + i\varepsilon)\, f\|^2\, d\lambda\right)^{1/2}$$

$$+ \left((\varepsilon/\pi) \int_{-\infty}^{\infty} \|W_+R_0(\lambda + i\varepsilon)\, g\|^2\, d\lambda\right)^{1/2}$$

$$\times \left((\varepsilon/\pi) \int_{-\infty}^{\infty} \|R(\lambda + i\varepsilon)\, f - W_+R_0(\lambda + i\varepsilon)\, F_\varepsilon(\lambda)\, f\|^2\, d\lambda\right)^{1/2}.$$

Thus the assertion (73) follows from (71), (69) (where $A$ is replaced by $J - W_+$) and from the identities

$$(\varepsilon/\pi) \int_{-\infty}^{\infty} \|R(\lambda + i\varepsilon)\, f\|^2\, d\lambda = \|f\|^2, \quad (\varepsilon/\pi) \int_{-\infty}^{\infty} \|R_0(\lambda + i\varepsilon)\, g\|^2\, d\lambda = \|g\|^2.$$

(72) can also be written in the form

$$\lim_{\varepsilon\to+0} (\varepsilon/\pi) \int_{-\infty}^{\infty} \left(R(\lambda - i\varepsilon)\, JR_0(\lambda + i\varepsilon)\, g, f\right)_{\mathscr{H}} d\lambda = (g, Z_+f). \tag{74}$$

Hence, according to representation (1), formula (74) means that

$$(W_+g, f) = (g, Z_+f), \qquad g \in P_0^{\mathrm{ac}}\mathscr{H}_0, \qquad f \in \mathscr{D}.$$

Since $Z_+f \in P_0^{\mathrm{ac}}\mathscr{H}_0$, we obtain

$$(g, Z_+f) = (g, P_0^{\mathrm{ac}}Z_+f) = (P_0^{\mathrm{ac}}g, Z_+f) = (W_+P_0^{\mathrm{ac}}g, f) = (g, W_+^*f)$$

for all $g \in \mathscr{H}$. Hence $Z_+f = W_+^*f$ is valid for all $f \in \mathscr{D}$, that is, the assertion is proved. ■

Note that it is also possible to formulate Lemma 38 in a "localized" form, that is, with the help of a fixed (compact) Borel set $\Lambda_0$:

**Lemma 39.** *Let $\mathscr{D} \subset \mathscr{H}$ be a linear manifold and let $F_\varepsilon(\lambda) \in \mathscr{L}(\mathscr{H}, \mathscr{H}_0)$, $\varepsilon > 0$, $\lambda \in \Lambda_0$, where $\Lambda_0 \subset \mathbb{R}$ is a fixed compact set. Assume $F_\varepsilon(\lambda)\, f$ to be strongly continuous with respect to $\lambda \in \Lambda_0$ for all $\varepsilon > 0$ and all $f \in \mathscr{D}$. Suppose that for all $f \in \mathscr{D}$ the following conditions are satisfied:*

$$Z_+f := \operatorname*{w-lim}_{\varepsilon\to+0} P_0^{\mathrm{ac}} \int_{\Lambda_0} E_{0,\varepsilon}(d\lambda)\, F_\varepsilon(\lambda)\, f \tag{75}$$

*exists.*

$$\lim_{\varepsilon\to+0} \int_{\Lambda_0} \left(F_\varepsilon(\lambda)\, f, E_{0,\varepsilon}(d\lambda)\, F_\varepsilon(\lambda)\, f\right) = \|Z_+f\|^2. \tag{76}$$

$$\lim_{\varepsilon\to+0} \int_{\Lambda_0} \|R(\lambda + i\varepsilon)\, f - JR_0(\lambda + i\varepsilon)\, F_\varepsilon(\lambda)\, f\|^2\, d\lambda = 0. \tag{77}$$

*Then* $Z_+f = E_0(\Lambda_0) W_+^*f$ *for all* $f \in \mathcal{D}$.

**Proof.** We note only the points which have to be changed. Instead of (72) one proves

$$\lim_{\varepsilon\to+0} (\varepsilon/\pi) \int_{\Lambda_0} \left(JR_0(\lambda + i\varepsilon)\, g, R(\lambda + i\varepsilon)\, f\right)_{\mathcal{H}} d\lambda = (g, Z_+f) \tag{78}$$

for all $g \in E_0(\Lambda_0)\, P_0^{ac}\mathcal{H}_0$, $f \in \mathcal{D}$, or

$$\lim_{\varepsilon\to+0} (\varepsilon/\pi) \int_{\Lambda_0} \left(R(\lambda - i\varepsilon)\, JR_0(\lambda + i\varepsilon)\, g, f\right)_{\mathcal{H}} d\lambda = (g, Z_+f).$$

Then, because of Prop. 4, also

$$\lim_{\varepsilon\to+0} (\varepsilon/\pi) \int_{-\infty}^{\infty} \left(R(\lambda - i\varepsilon)\, JR_0(\lambda + i\varepsilon)\, g, f\right)_{\mathcal{H}} d\lambda = (g, Z_+f)$$

is valid. Therefore we obtain

$$\left(W_+E_0(\Lambda_0)\, P_0^{ac}g, f\right) = \left(E_0(\Lambda_0)\, P_0^{ac}g, Z_+f\right)$$

for all $g \in \mathcal{H}_0$ $f \in \mathcal{D}$, hence

$$Z_+f = E_0(\Lambda_0)\, W_+^*f$$

follows for all $f \in \mathcal{D}$. ∎

## 13.5.2. Extension of the operator $Z_+$ and formulation of the criterion

According to the lemma proved in the preceding section the operator $Z_+$, defined on $\mathcal{D}$, is an intertwining operator with respect to $H$, $H_0$ in the following sense: If $f \in \mathcal{D}$ and also $E(\Delta)\, f \in \mathcal{D}$, then $E_0(\Delta)\, Z_+f = Z_+E(\Delta)\, f$. Namely, we have $E_0(\Delta)\, Z_+f = E_0(\Delta)\, W_+^*f = W_+^*E(\Delta)\, f = Z_+E(\Delta)\, f$. For this reason it is possible to extend the operator $Z_+$ to spa $\{E(\Delta)\, f; f \in \mathcal{D}\}$ by *requiring* the intertwining property, that is, by the following definition:

$$\hat{Z}_+E(\Delta)\, f := E_0(\Delta)\, Z_+f, \qquad f \in \mathcal{D}. \tag{79}$$

If $E(\Delta)\, f \in \mathcal{D}$, then we obtain $\hat{Z}_+E(\Delta)\, f = Z_+E(\Delta)\, f$, thus $\hat{Z}_+$ is an extension of $Z_+$. Now we prove

**Lemma 40.** *The operator* $\hat{Z}_+$ *is isometric on* $P^{ac}\left(\text{spa}\, \{E(\Delta)\, f; f \in \mathcal{D}\}\right)$ *and the intertwining property*

$$\hat{Z}_+E(\Delta)\, g = E_0(\Delta)\, \hat{Z}_+g; \qquad g \in \text{spa}\, \{E(\Delta)\, f; f \in \mathcal{D}\}$$

*is satisfied.*

**Proof.** The intertwining property is obvious. Therefore it suffices to prove that

$$\|\hat{Z}_+E(\Delta)\, f\| = \|P^{ac}E(\Delta)\, f\|; \qquad f \in \mathcal{D},\ \Delta \text{ Borel set},$$

or, equivalently, that

$$\|E_0(\Delta)\, Z_+f\| = \|E(\Delta)\, P^{ac}f\|. \tag{80}$$

From (71) we obtain $\lim_{\varepsilon\to+0} (\varepsilon/\pi) \int_\Delta \|P^{ac}(R(\lambda + i\varepsilon) f - W_+R_0(\lambda + i\varepsilon) F_\varepsilon(\lambda) f)\|^2 \, d\lambda = 0$ hence also

$$\lim_{\varepsilon\to+0} (\varepsilon/\pi) \int_\Delta \|R(\lambda + i\varepsilon) P^{ac}f\|^2 \, d\lambda = \lim_{\varepsilon\to+0} (\varepsilon/\pi) \int_\Delta \|W_+R_0(\lambda + i\varepsilon) F_\varepsilon(\lambda) f\|^2 \, d\lambda. \quad (81)$$

Because of $\|E(\Delta) g\|^2 = \lim_{\varepsilon\to+0} (\varepsilon/\pi) \int_\Delta \|R(\lambda + i\varepsilon) g\|^2 \, d\lambda$ for all $g \in P^{ac}\mathscr{H}$ and the isometry of $W_+$ on $P_0^{ac}\mathscr{H}_0$, we get from (81) that

$$\|E(\Delta) P^{ac}f\|^2 = \lim_{\varepsilon\to+0} (\varepsilon/\pi) \int_\Delta \|P_0^{ac}R_0(\lambda + i\varepsilon) F_\varepsilon(\lambda) f\|^2 \, d\lambda. \quad (82)$$

By (65) the right-hand side of (82) is nothing else than

$$\lim_{\varepsilon\to+0} (\varepsilon/\pi) \int_\Delta \|P_0^{ac}R_0(\lambda + i\varepsilon) Z_+f\|^2 \, d\lambda = \|E_0(\Delta) Z_+f\|^2.$$

Thus (80) is proved. ∎

The "localized" form of Lemma 40 is given by

**Corollary 41.** *Let $\hat{Z}_+$ be the extension* (79) *of the localized operator $Z_+$ of Lemma* 39. *Then: $\hat{Z}_+$ is isometric on $E(\Lambda_0) P^{ac}$* spa $\{E(\Delta) f; f \in \mathscr{D}\}$ *and the intertwining property*

$$\hat{Z}_+E(\Delta) g = E_0(\Delta) \hat{Z}_+g, \qquad g \in \text{spa}\,\{E(\Delta) f; f \in \mathscr{D}\}$$

*is satisfied.*

**Proof.** It proceeds as the proof of Lemma 40. ∎

According to Lemma 40 we can extend $\hat{Z}_+$ to the subspace $P^{ac}$ clo spa $\{E(\Delta) f; f \in \mathscr{D}\}$ as an isometric operator by continuous extension. Now we can formulate the completeness criterion.

**Theorem 42.** *Let $\mathscr{D} \subset \mathscr{H}$ and $F_\varepsilon(\lambda)$ be given as in Lemma* 38, *in particular let* (62), (63), (64) *be satisfied. Assume further that $P^{ac}\mathscr{D}$ is generating with respect to $P^{ac}\mathscr{H}$ and $H$. Then $W_+$ is complete.*

**Proof.** Let $Z_+$ be the operator from Lemma 38 and let $\hat{Z}_+$ be the extension of $Z_+$ to $P^{ac}$ clo spa $\{E(\Delta) f; f \in \mathscr{D}\}$. Since $P^{ac}\mathscr{D}$ is generating in $P^{ac}\mathscr{H}$, this subspace coincides with $P^{ac}\mathscr{H}$. According to Lemma 40, we have $\hat{Z}_+^*\hat{Z}_+ = P^{ac}$. On the other hand by Lemma 38 we have $Z_+ = W_+^*$. Thus $W_+W_+^* = P^{ac}$ is obtained, which proves the assertion. ∎

Correspondingly we obtain a "localized" criterion:

**Corollary 43.** *Let $\mathscr{D} \subset \mathscr{H}$ and $F_\varepsilon(\lambda)$ be given as in Lemma* 39, *in particular let* (75), (76), (77) *be satisfied. Assume further that $P^{ac}E(\Lambda_0) \mathscr{D}$ is generating with respect to $P^{ac}E(\Lambda_0) \mathscr{H}$ and $H$. Then $E(\Lambda_0) W_+W_+^* = E(\Lambda_0) P^{ac}$ or, in other words, $W_+$ is complete with respect to $\Lambda_0$.*

**Proof.** It proceeds as the proof of Theorem 42. ∎

Interestingly enough; the sufficient conditions of Theorem 42 for the completeness of $W_+$ are also necessary, even in the stronger form that one may take $\mathscr{D} = P^{ac}\mathscr{H}$.

**Theorem 44.** *Let the Abel wave operator $W_+$ exist and suppose $W_+$ to be complete. Then: If we put $\mathcal{D} = P^{ac}\mathcal{H}$ and $F_\varepsilon(\lambda) := W_+^*$ (a constant operator function), then the conditions (62), (63), (64) are satisfied.*

**Proof.** First we have

$$\lim_{\varepsilon\to+0} (\varepsilon/\pi) \int_{-\infty}^{\infty} (R_0(\lambda + i\varepsilon)\, g, R_0(\lambda + i\varepsilon)\, W_+^* f)\, d\lambda = (g, W_+^* f), \qquad g \in \mathcal{H}_0,\quad f \in P^{ac}\mathcal{H},$$

or

$$\text{w-lim}_{\varepsilon\to+0} \int_{-\infty}^{\infty} E_{0,\varepsilon}(d\lambda)\, W_+^* f = W_+^* f,$$

which proves (62). This also implies that

$$\lim_{\varepsilon\to+0} \int_{-\infty}^{\infty} \left(W_+^* f, E_{0,\varepsilon}(d\lambda)\, W_+^* f\right) = \lim_{\varepsilon\to+0} (\varepsilon/\pi) \int_{-\infty}^{\infty} \|R_0(\lambda + i\varepsilon)\, W_+^* f\|^2\, d\lambda = \|W_+^* f\|^2,$$

which proves (63). Finally we have $|\mathrm{A}|\text{-}\lim_{t\to\infty} (W_+ - J)\, e^{-itH_0} P_0^{ac} = 0$. Now $P_0^{ac} = W_+^* W_+$ is valid, that is, we get

$$|\mathrm{A}|\text{-}\lim_{t\to\infty} \{W_+ e^{-itH_0} W_+^* W_+ - J e^{-itH_0} W_+^* W_+\} = 0.$$

Since $W_+ W_+^* = P^{ac}$, we obtain

$$|\mathrm{A}|\text{-}\lim_{t\to\infty} \{e^{-itH} W_+ - J e^{-itH_0} W_+^* W_+\} = 0.$$

We put $f = W_+ g$, where $f$ is an arbitrary vector from $P^{ac}\mathcal{H}$ and obtain

$$\lim_{\varepsilon\to+0} 2\varepsilon \int_{-\infty}^{\infty} e^{-2\varepsilon t}\, \|e^{-itH} f - J e^{-itH_0} W_+^* f\|^2\, dt = 0. \tag{83}$$

From (83) we obtain by Parseval's equation

$$\lim_{\varepsilon\to+0} (\varepsilon/\pi) \int_{-\infty}^{\infty} \|R(\lambda + i\varepsilon)\, f - J R_0(\lambda + i\varepsilon)\, W_+^* f\|^2\, d\lambda = 0,$$

which proves (64). ■

## Notes and Remarks to Part III

For the physical background of scattering theory we refer the reader to Amrein/Jauch/Sinha [1] and Thirring [2] whose books are written in a precise and modern mathematical language. See also Newton [1] and Goldberger/Watson [1].

*Chapter 9.* The elementary properties of wave operators (in case $J = 1$ and $P_H = P_H^{ac}$) described in Section 2.1 can already be found in the first papers on mathematical scattering theory (see Cook [1957a], Kato [1957a] (chain rule), Jauch [1958a]). The extension to two-space wave operators was given by Kato [1967a]. In this paper he also introduced the notions of semi-completeness, completeness, asymptotically equivalent identification operators, and asymptotic left inverse. The criteria for semicompleteness, completeness, and partial isometry of the two-space wave operators given in Paragraphs 3 and 5 are essentially taken from KATO [1967a].

That the wave operators $W_\pm$ $(H, H_0)$ are not always complete was first proved by KATO/KURODA [1959a] in the case: $H_0 = -\Delta$, $H = -\Delta + q(x)$ where $q(x)$ behaves bad at infinity. PEARSON [1975a] gave an example in the same situation where $q(x)$ has compact support and $q(x)$ oscillates very strongly at $x = 0$.

The first example showing that the Abel wave operator may exist for a certain pair $H, H_0$ but not the strong one goes back to M. ROSENBLUM (see HOWLAND [1967a]). For arbitrary operators $H_0$, an example was given by OBERMANN/WOLLENBERG [1978a]. The motivation for study of the Abel wave operator came from the physical literature in which the "stationary" wave operators

$$\Omega_\pm = \operatorname*{s-lim}_{\varepsilon \to \pm 0} \pm \varepsilon \int_0^{\pm\infty} e^{-\varepsilon t} e^{itH} e^{-itH_0} P_0^{ac} \, dt$$

were considered. In general, these operators are not isometric on $P_0^{ac}\mathscr{H}_0$ (take as a simple example $H = H_0 + c$, $c > 0$, then $\Omega_\pm = 0$). Since the isometry of $\Omega_\pm$ on $P_0^{ac}\mathscr{H}_0$ is necessary for the unitarity of the scattering operator $S$ on $P_0^{ac}\mathscr{H}_0$, JORDAN [1962a] required that $\Omega_\pm$ exist *and* are isometric on $P_0^{ac}\mathscr{H}_0$. OBERMANN [1977a] first showed that these two requirements are equivalent to the existence of the limits $W_\pm = |A|\text{-}\lim_{t\to\infty} e^{\pm itH} e^{-itH_0} P_0^{ac}$.

***Finite-dimensional perturbations.*** The results on one- and finite-dimensional perturbations presented in this chapter can be found (partly in a slightly different form) in KATO [1957a] and KURODA [1963a]. Our treatment of this subject uses partly the theory of vector spectral integrals.

***Uniformly propagative systems.*** Scattering problems for general first order systems of the form (67) were initiated by WILCOX [1966a]. He also introduced the name uniformly propagative systems and proved Prop. 66 and the existence of wave operators for such systems. This paper of WILCOX contains also the other details on weakly uniformly propagative systems described in Section 7.2. The asymptotic equivalence of different identification operators (Lemma 67) is due to KATO [1967a]. Existence and completeness of the strong wave operators for such systems were proved in MOCHIZUKI [1969a], BIRMAN [1969b], DEIČ [1971a], SCHULENBERGER/WILCOX [1971a], [1972a], SCHULENBERGER [1972a], DEIFT [1976a], and AVILA/COSTA [1980a]. For further literature see especially DEIFT [1976a] and also the Notes and Remarks to Chapter 12.

***Scattering states.*** The choice of the universal projection (or possibly other projections not having the properties of universal projections) in the definition of the wave operators is not unique and depends on the viewpoint. KATO [1957a], [1957b] chose $P_H^{ac}$ in the definition of the wave operators. JAUCH [1958a] chose $P_H^c$, that is, he identified $P_H^c\mathscr{H}$ with the set of scattering states. In most physical applications $P_H^{ac} = P_H^c$ such that there is no difference. This and the mathematical convenience of the choice of $P_H^{ac}\mathscr{H}$ is the reason that in the following years most papers on scattering theory have been treated the case $P_H = P_H^{ac}$. In potential scattering ($H_0 = -\Delta$, $H = -\Delta + q(x)$ and $\mathscr{H} = L^2(\mathbb{R}^n, dx)$) there is a natural way to define scattering states and bound states via their asymptotically geometric behavior. More precisely, let $F_r$ denote the multiplication operators by $\chi_{|x|\le r}(x)$ on $\mathscr{H} = L^2(\mathbb{R}^n, dx)$ and set

$$\mathfrak{M}_0^\pm := \{u \in \mathscr{H};\ \operatorname*{Lim}_{r\to\infty} \sup_{t\in[0,\pm\infty]} \|(1 - F_r)\, e^{-itH}u\| = 0\},$$

$$\mathfrak{M}_\infty^\pm := \{u \in \mathscr{H};\ \operatorname*{Lim}_{t\to\pm\infty} F_r\, e^{-itH}u = 0,\ r \in (0, \infty)\},$$

where both limiting methods are used.

The vectors $u \in \mathfrak{M}_\infty^\pm$ are called $\pm$ *scattering states* and vectors $u \in \mathfrak{M}_0^\pm$ are called $\pm$ *bound states.* One has always the relations $P_H^p\mathscr{H} \subseteq \mathfrak{M}_0^\pm$, $P_H^c\mathscr{H} \supseteq \mathfrak{M}_\infty^\pm$. For a large class of potentials RUELLE [1969a], AMREIN/GEORGESCU [1973a], [1974a], PEARSON [1974a], [1975b], and COMBESCURE/GINIBRE [1978a] proved that $\mathfrak{M}_0^+ = \mathfrak{M}_0^- = P_H^p\mathscr{H}$ and $P_H^c\mathscr{H} = P_H^{ac}\mathscr{H} = \mathfrak{M}_\infty^+ = \mathfrak{M}_\infty^-$. However, PEARSON [1975a], [1975b] showed that there exist singular potentials with subspaces $\mathfrak{M}_\Omega^\pm$ of $P_H^{ac}\mathscr{H}$ such that the vectors $e^{-itH}u$, $u \in \mathfrak{M}_\Omega^\pm$ approach the singularities of the potential $q(x)$ ($q$ is sufficiently regular outside a compact region $\Omega \subset \mathbb{R}^n$ of measure zero) as $t \to \pm\infty$ and

$P_H^{ac}\mathscr{H} = \mathfrak{M}_\infty^\pm \oplus \mathfrak{M}_Q\pm$, That is, the set of scattering states is property contained in the set of absolutely continuous vectors of $H$. His results were extended to larger classes of potentials by COMBESCURE/GINIBRE [1978a]. See also AMREIN/PEARSON/WOLLENBERG [1980a].

WILCOX [1973a] and AMREIN/GEORGESCU [1974a] gave an abstraction of the above definition of scattering states for arbitrary selfadjoint operators $H$: The operators $F_r$ above are replaced by a fixed set of projections $Q_r$ with the property $\text{s-lim}_{r\to\infty} Q_r = 1$. This method can be considered as a special case of defining universal projections by systems $\mathscr{A}_H$ of bounded operators with certain properties (cf. Section 6.5.1, in particular Definition 6.65 and Prop. 6.66). In the present case the a priori given set of projections $Q_r$ of $\mathscr{H}$ has to be associated to a fixed selfadjoint operator $H$. Then the system $\{Q_r\}$ generates a system $\mathscr{A}_H$ (as the minimal system with the defining properties of $\mathscr{A}_H$ containing all $Q_r$). $\mathscr{A}_H$ defines a universal projection via Prop. 6.66. Some asymptotic properties (with respect to the evolution of absolutely and singularly continuous vectors can be found in GUSTAFSON [1974a], SINHA [1977a] and AMREIN/PEARSON/WOLLENBERG [1980a]. See also the Notes and Remarks to Chapter 6.

*Paragraph 10.1.* The definition of an abstract multichannel system with the help of a system of free Hamiltonians and the results in Sections 1 and 2 is due to BAUMGÄRTEL [1977a]. The idea of introducing a new Hilbert space to formulate the multiparticle scattering in a more transparent and economic way is more or less explicitly contained in the papers of EKSTEIN [1956b], FADDEEV [1963a], and others. A first precise formulation was given by CHANDLER/GIBSON [1973a]. See also PRUGOVECKI [1973a]. This idea was further exploited in GIBSON/CHANDLER [1974a], CHANDLER/GIBSON [1977a], [1978a], [1980a], HOWLAND [1976a], JAFAEV [1978a] (in this paper he used the asymptotic equivalence of the multichannel identification operator with another identification operator), and KATO [1977a]. The new Hilbert space $\breve{\mathscr{H}}$ is sometimes defined in a slightly other manner, namely $\breve{\mathscr{H}} = \mathop{\times}\limits_{j=1}^{r} Q_j\mathscr{H}$ instead of $\breve{\mathscr{H}} = \mathop{\times}\limits_{j=1}^{r} \mathscr{H}$. Theorem 21 goes back to WOLLENBERG [1977b].

*Multiparticle scattering.* Unfortunately, because proofs for existence and completeness of wave operators for $N$-body systems are space-consuming, we are not able to present any details in this book. Besides the papers mentioned above and in the Introduction we cite some further relevant works: v. WINTER [1965a], HUNZIKER [1967a], JAKUBOVSKIĬ [1967a], COMBES [1969b], HEPP [1969a], GINIBRE/MOULIN [1974a], THOMAS [1975a], MOURRE [1977a], GINIBRE [1977a], DEIFT/SIMON [1977a], SIMON [1977b], SIGAL [1978a], [1978b], and HAGEDORN [1978a], [1979a], [1980a].

*Paragraph 10.2.* The algebraic approaches to scattering problems were developed at the end of the sixties and at the beginning of the seventies. One approach working with $C^*$-subalgebras of the commutant of the unperturbed operator $H_0$ was originated by LAVINE [1970a], COMBES [1969a] and CORBETT [1970a], and was further developed by LAVINE [1971a], [1972a], [1973b], THOMAS [1974a], and CORBETT [1975a].

The other approach working with $W^*$-subalgebras of the commutant of $H_0$ is due to AMREIN/MARTIN/MISRA [1970a], and was further developed by MOURRE [1973a] and AMREIN/GEORGESCU/MARTIN [1974a]; see also ZACHARY [1973a].

Since the only quantities of physical significance are the expection values of observables, one would like to replace the strong limit of $e^{itH} X e^{-itH}$ by the weak one in the corresponding expressions. A discussion of this standpoint and some results can be found in PRUGOVECKI [1971a] and GEORGESCU [1977a]; see also BAUMGÄRTEL/DEMUTH/WOLLENBERG [1978b].

To avoid the nonuniqueness of the generalized wave operators $W_\pm$ (and that of the scattering operator), one can try to define them by

$$(e^{itH}X\, e^{-itH}P^{ac} - W_\pm\, e^{itH_0}X\, e^{-itH_0}W_\pm) \to 0 \quad \text{as} \quad t \to \pm\infty, \tag{1}$$

where $X$ runs through a sufficiently large set $\mathscr{A}$ of operators not all contained in the commutant of $H_0$. This point of view was advocated by PRUGOVECKI [1971a]. However, the details of such a theory have not been well worked out up to this date, and, furthermore, one has to remark that in physically interesting cases (potential scattering) the indeterminess of the S-operator has no effect on many scattering quantities (see DOLLARD [1969a], AMREIN/MARTIN/MISRA [1970a], P. MARTIN [1972a]).

For another approach to scattering problems using (1) with $X \in \mathscr{L}_1(\mathscr{H})$ and where the limit is to be taken in trace norm see JAUCH/MISRA/GIBSON [1968a].

The results of Sections 3 and 4 are essentially based on the papers of AMREIN/MARTIN/MISRA [1970a] and THOMAS [1974a].

Usually one is only able to show that a certain $C^*$-subalgebra $\mathscr{A}$ of com $H_0$ belongs to dom $\mu^{H} \cap$ dom $\mu^{-H}$ and not a $W^*$-algebra. Then, the proof of the existence of generalized wave operators can use additional knowledge on the unitary equivalence of $\mathscr{A}_0$ and $\mu^{\pm H}(\mathscr{A}_0)$ where $\mathscr{A}_0$ is some dense subset of $\mathscr{A}$. This is the method used by THOMAS [1974a] who proved the existence and completeness of generalized wave operators for $H_0 = -\Delta$, $H = -\Delta + V(x)$ and $\mathscr{A} = \{$set of all continuous bounded functions of the momenta $P_j$ and angular momenta $L_j$ on $L^2(\mathbb{R}^3, dx)\}$ with $V(x) = V_1(x) + V_2(x)$ real-valued functions satisfying:

(A) $V_1(x) = V_1(r)$, $r = |x|$, is continuously differentiable with $V_1(r) \to 0$ as $r \to \infty$ and $|dV(r)/dr| \leqq c(1 + r)^{-\gamma}$, $\gamma > 1$,

(B) $V_2(x) = (1 + r)^{-\gamma} (f_p(x) + f_\infty(x))$, $\gamma > 1$, $f_\infty \in L^\infty(\mathbb{R}^3, dx)$ and $f_p \in L^p(\mathbb{R}^3, dx)$ for some $p > 3/2$.

The proof uses results of LAVINE [1973b].

In the $W^*$-approach one usually uses the fact that it is sufficient to construct modified free evolutions $T_t^\pm$ (see Section 4) for the proof of the existence of generalized wave operators for $\{H, H_0,$ com $H_0\}$. For Schrödinger operators $H = -\Delta + V(x)$ with long-range potentials $V(x)$ there is a procedure for constructing $T_t^\pm$ firstly developed by DOLLARD [1964a] who proved the assertions in Section 2.5 (II) concerning $H = -\Delta + c/|x|$. The proof we used is essentially taken from ALSHOLM/KATO [1973a]. Potentials $V(x) = c\,|x|^{-\gamma}$, $\gamma > 3/4$ were studied by AMREIN/MARTIN/MISRA [1970a], and potentials $V(x) = c\,|x|^{-\gamma}$, $\gamma > 0$ were considered in BUSLAEV/MATVEEV [1970a]. SAHNOVIČ [1970a] also constructed the modified free evolutions $T_t^\pm$ in the case that $H_0$ is multiplication by $x$ on $L^2(\Delta, dx)$ and $H - H_0$ is an integral operator.

For Schrödinger operators and more general differential operators with long-range potentials there is an extended literature on the existence and completeness of generalized wave operators using the approach with the modified free evolutions $T_t^\pm$. See in particular, ALSHOLM/KATO [1973a], HÖRMANDER [1976a], ALSHOLM [1976a], BERTHIER/COLLET [1977b], and ENSS [1979a]. There are also other approaches to the same problems for these classes of operators $H$ and $H_0$. See e.g. IKEBE [1975a], KITADA [1976a], [1977a], [1978a], SAITO [1], and IKEBE/ISOZAKI [1979a].

The idea to use an identification operator $J$ to treat scattering problems with long-range potentials was introduced by MULHERIN/ZINNES [1970a] in case $V(x) = c/|x|$ (Coulomb scattering). For other long-range potentials this approach was used by MATVEEV/SKRIGANOV [1972a], GEORGESCU [1974a], CHANDLER/GIBSON [1974a] (multichannel Coulomb scattering), KITADA [1977a] [1978a], and KAKO [1978a].

The Coulomb scattering was also treated by other methods in the literature:

1. Another algebraic approach to Coulomb scattering was given by GROSSE/GRÜMM/NARNHOFER/THIRRING [1974a] using results of LAVINE [1970a].

2. An interesting method to solve the Coulomb scattering problem is to consider the Coulomb potential as a limit of screened potentials. For details see ZORBAS [1974a], SEMON/TAYLOR [1975a], [1976a], GOODMANSON/TAYLOR [1980a] and literature cited there.

One can also ask if the existence of generalized wave operators $W_\pm$ for $\{H, H_0, \mathscr{A}\}$ or $\{H, H_0, T_t^\pm\}$ implies the existence and equality with $W_\pm$ of the corresponding generalized wave operators for $\{\alpha(H), \alpha(H_0), \mathscr{A}\}$ or $\{\alpha(H), \alpha(H_0), T_{t,\alpha}^\pm\}$ for certain functions $\alpha$. From the statements of Section 8.2.2 this follows immediately if we consider the $|A|$-lim and the class of admissible functions (see Definition 8.16). For the strong limit there exist some sufficient criteria. The strongest results are contained in CHANDLER/GIBSON [1976a], [1979a]. See also the earlier papers of MATVEEV [1971a] and SAHNOVIČ [1971a].

*Paragraph 10.3.* Scattering theory for the perturbed wave equation was studied in the framework of Hilbert spaces by LAX/PHILLIPS [1962a], [1964a], [1], BIRMAN [1963a], WILCOX [1966a], and KATO [1967a]. KATO [1967a] introduced the approach to second order equations described in Paragraph 3. In the text we follow partly KATO [1967a] and partly REED/SIMON [1977a] who simplified and improved this approach. Further literature to scattering of classical waves the reader can find in the Notes and Remarks to Chapter 12 and to the Klein-Gordon equation (Part IV), and in DEIFT [1976a].

*Paragraph 10.4.* For general references on quantum field theory, see: SEGAL [1], JOST [1], BOGOLJUBOV/LOGUNOV/TODOROV [1] and REED/SIMON [2, 3]. There one can also find historical remarks and further references. The Haag-Ruelle scattering theory was developed by HAAG [1958a] and RUELLE [1962a]. A presentation can also be found in the books cited above (except SEGAL [1]). An algebraic formulation of the Haag-Ruelle scattering theory appeared in ARAKI/HAAG [1967a].

A central problem of the quantum field theory is to find a non-trivial quantum field, that is a quantum field with $S \neq 1$. This was succesfully solved for some models with two space-time dimensions; for details see SIMON [2]. Another approach uses the description of the quantum fields in terms of positive functionals, going back to BORCHERS [1962a] and UHLMANN [1962a], [1968a], to solve this problem. Details can be found in HOFMAN/LASSNER [1975a].

The idea to formulate the Haag-Ruelle scattering theory in the framework of a kind of a two-space scattering theory already appeared in KOŠMANENKO [1974a]. See also KOŠMANENKO [1975a], [1979a]. REED/SIMON [3, p. 321] gave a description closer to the usual two-space scattering theory. The two-space formulation of the Haag-Ruelle scattering theory given in the text is due to BAUMGÄRTEL. The details can be found in BAUMGÄRTEL/NEIDHARDT/REHBERG [1983a]. The difference to the other two-space formulations is that the formulation in the text provides a natural factorization $J = KJ_0$ where $J$ and $J_0$ correspond to the interacting and free quantum fields.

It is also possible to rewrite the $N$-body nonrelativistic quantum theory as a field theory and to rewrite its corresponding scattering theory in a form analogous to the Haag-Ruelle scattering theory. This was shown by HEPP [1965a] and SANDHAS [1966a].

Note that there are approaches to relativistic quantum scattering problems without quantum fields. Some references are: JORDAN [1964a], COESTER [1965a] and RUIJSENAARS [1980a] (and references cited therein).

*Chapter 11.* The problem to characterize the class of strong wave operators for a pair $H_0$, $H$ within the more general class of intertwining operators for $H_0$, $H$ was proposed by KREĬN [1964a] and solved by BAUMGÄRTEL [1973a]. The latter paper was also the main starting point for the development of the theory of asymptotic constants. The obvious generalization of the characterization of strong wave operators to Abel ones was given by OBERMANN [1977a].

*Inverse problem.* The essential results of Paragraph 2 go back to WOLLENBERG [1977b], [1978a]. There are extensions to a more general scattering theory for unitary representations of $\mathbb{R}^n$ $(n > 1)$: see the Notes and Remarks to Chapter 7; for details we refer to REHBERG [1979a] and BAUMGÄRTEL/REHBERG/WOLLENBERG [1982a]. A characterization of the wave operators in terms of $S$ and elements of the wave algebra using the solution of the inverse problem can be found in WOLLENBERG/KOŠMANENKO [1979a]. A description of the class of selfadjoint operators $H$ leading to the same scattering operator $S$ for fixed $H_0$ was already given by EKSTEIN [1960a], COESTER [1965a], COESTER/COHEN/DAY/VINCENT [1970a], and SAENZ/ZACHARY [1976a]. Earlier KOŠMANENKO [1976a] was able to show that the inverse problem has always a solution $H$ for given $H_0$, $S$ (one-space theory), but in his scattering theory the solution $H$ is in general a bilinear functional and not an operator.

That the assumptions of Theorem 23 are satisfied for many-body systems with finitely many scattering channels was proved in WOLLENBERG [1977b]. For two-body channels $\alpha$, $\beta$, COMBES [1969b] already proved that $Q_\alpha Q_\beta$ are compact operators and thus $\mu^{\pm H_\alpha}(Q_\beta Q_\alpha) = 0$ (under slight additional assumptions on the Hamiltonians).

Inverse scattering problems under the additional assumption that $H_0$ is a differential operator and $H - H_0$ is a multiplication operator has been extensively studied in the literature. For a survey see AGRANOVIČ/MARČENKO [1], CHADAN/SABATIER [1], FADDEJEV [1974a] and DEIFT/TRUBOWITZ [1979a].

*Invariance principle.* The main points of the history of the invariance principle for wave operators were already noted in the Introduction to this book. The proof of the strong invariance principle (Theorem 25) goes back to WOLLENBERG and was first published, in an improved form, in OBERMANN/WOLLENBERG [1978b]. An extension of Theorem 25 to a slightly more general class of admissible functions can be found in WOLLENBERG [1979b]. The weak form of this principle for two-space wave operators (Corollary 26) was first proved, under slight additional assumptions on $J$, in WOLLENBERG [1977d]. There are many papers which try to extend the strong invariance principle to strong wave operators under additional assumptions on $\{H, H_0, J\}$. The following list includes papers of this nature: SHENK [1968a], BIRMAN [1968a], KATO/

Kuroda [1971a], Donaldson/Gibson/Hersh [1973a], Veselic/Weidmann [1974a], Chandler/Gibson [1976a], [1979a], Mullin [1976a], Dollard [1976a], Dollard/Friedman [1979a], and Schechter [1979b]. For nonselfadjoint perturbations, see Goldstein [1971a], Kako/Yajima [1976a], and for generalized wave operators see the Notes and Remarks to Chapter 10. See also the Notes and Remarks to Part IV.

*Chapter 12.* All essential definitions and results go back to Lax and Phillips. In particular, see their book Lax/Phillipps [1] which contains further details and some applications to scattering problems of waves and to quantum scattering.

Theorem 3 was first proved by Sinaĭ [1961a].

In the literature scattering of waves by obstacles has been extensively studied with the help of the Lax-Phillips approach by Lax/Phillips [1], [1977a], [1978a], Majda [1976a], [1977a], and others.

Adamjan/Arov [1965a], [1965b], Lax/Phillips [1973a], Arov [1974a], and Foias [1975a] extended the Lax-Phillips theory to contractive semigroups. These works have relations to the scattering theory for nonselfadjoint operators. Special problems and applications of the Lax-Phillips theory were considered in Lax/Phillips [1978a], [1978b], Pavlov [1979a] and Amrein/Wollenberg [1981a].

There are also scattering problems connected with automorphic functions where the Lax-Phillips approach can be used. For details see Pavlov/Faddeev [1972a] and Lax/Phillips [2].

For scattering problems with acoustical and electromagnetic waves the reader may consult the following books and papers: Lax/Phillips [1], Wilcox [1], Reed/Simon [3], Deift [1976a], Helton/Ralston [1976a], Picard [1977a], Deift [1978a], Chen/Yang [1978a], Lyford [1978a], Strauss [1979a].

*Chapter 13.* The development of the stationary theory of the wave operator is tightly connected with the development of special stationary methods for proving the existence of and constructing the wave operators (for literature on stationary methods, see the Notes and Remarks to Chapter 14). Some elements of the stationary theory are already contained in the first papers on scattering problems: Kato [1957a], de Branges [1962a], Birman [1963a] and Kuroda [1963a]. Later Baumgärtel [1967a], [1969a], [1974b], Amrein/Georgescu/Jauch [1971a], and Pearson [1971a] derived stationary expressions of wave operators in terms of spectral integrals. This were the first steps towards a pure stationary description of the wave operators. The solution of this problem goes back to Baumgärtel and it was first published in Baumgärtel/Obermann/Wollenberg [1979a]. The presentation of the stationary theory in the text follows essentially the last mentioned paper. The stationary description of the completeness property (Paragraph 5) is also due to Baumgärtel [1982a]. We remark that the presented theory uses some ideas from Kato/Kuroda [1970a], [1971a], Howland [1967a], [1976a] and Kato [1977a], [1977b]. In particular, Prop. 4, Lemma 5, Lemma 6 (with w-lim) and Lemma 12 are due to Howland [1976a].

We close the Notes and Remarks to this part with a brief description of some of the scattering theories which we have not treated in the text:

(1) *Time-dependent perturbations.* If the selfadjoint operators $H_0$ and $H$, acting on $\mathscr{H}_0$ and $\mathscr{H}$, depend on the parameter $t$ (time), then it is also possible to define wave and scattering operators. We briefly describe the essential ideas. Let $u_0(t)$ and $u(t)$ be the solutions of the equations

$$u_0'(t) = \mathrm{i}H_0(t)\,u_0(t), \qquad u'(t) = \mathrm{i}H(t)\,u(t)$$

with the initial values $u_0(s)$ and $u(s)$ at $t = s$. Associated with these equations are 2-parameter families $U_0(s, t)$ and $U(s, t)$ of evolution operators, sometimes called propagators, defined by $U_0(s, t)\,u_0(s) = u_0(t)$ and $U(s, t)\,u(s) = u(t)$ for all solutions $u_0(t)$, $u(t)$. The propagators obey the identities $U_0(s, s) = U(s, s) = 1$, $U_0(r, t)\,U_0(s, r) = U_0(s, t)$ and $U(r, t)\,U(s, r) = U(s, t)$. The wave operators are defined by

$$W_{\pm,s}(H, H_0; J) := |\mathrm{A}|\text{-}\lim_{t\to\pm\infty} U(t, s)\,J U_0(s, t)\,P,$$

when these limits exist. They depend, in general, on the initial time $s$. Further, in general, it is not clear which projections $P$ is to be taken because the absolutely continuous part of $H_0(t)$ may depend on $t$. If $H_0(t)$ is independent of $t$, which is usually satisfied, then one takes $P = P_0^{\mathrm{ac}}$. In particular, see Schmidt [1975a] for more details.

Usually one studies two kinds of time-dependent perturbations. The first considers the case that $H_0(t)$ is independent of $t$ and $V(t) = H(t)\,J - JH_0$ goes to zero as $|t| \to \infty$ in any sense. The second investigates the case that $V(t)$ is periodic in time. For general properties of wave and scattering operators in these cases and existence criteria, in particular see the papers of HOWLAND [1974a], [1979a] and SCHMIDT [1975a].

HOWLAND [1974a] introduced the method of using the Hilbert spaces $\mathscr{K}_0 := L^2(\mathbb{R}, dt; \mathscr{H}_0)$ and $\mathscr{K} := L^2(\mathbb{R}, dt; \mathscr{H})$ to change time-dependent operators $H_0(t)$ and $H(t)$ into time-independent ones. Define the selfadjoint operators $K_0 := -i(d/dt) + H_0(t)$ and $K := -i(d/dt) + H(t)$ on $\mathscr{K}_0$ and $\mathscr{K}$, respectively. Then the unitary groups $V_0(t) = e^{-itK_0}$ and $V(t) = e^{-itK}$ are connected with the propagators of $H_0(t)$ and $H(t)$ by the relations $(V_0(s)\,u_0)\,(t) = U_0(t, t-s)\,u_0(t-s$ and $(V(s)\,u)\,(t) = U(t, t-s)\,u(t-s)$. HOWLAND proved (in case $J = 1$) that the wave operators $W_{\pm,s}(H, H_0)$ exist if the wave operators $W_\pm(K, K_0) := \operatorname*{s-lim}_{t\to\pm\infty} e^{itK}\,e^{-itK_0}P_0^{ac}$ exist. For more details about the connection between propagators and unitary groups we refer to NEIDHARDT [1981b], [1981c]. For additional work on scattering theory for time-dependent operators, we cite the following: INOUE [1972a], [1973a], [1974a], MONLEZUN [1974a], DAVIES [1974a], J. GOLDSTEIN [1975a], J. GOLDSTEIN/MONLEZUN [1976a], EVANS [1976b], YAJIMA [1977a], KURODA/MORITA [1977a], SOHR [1978a] and KOROTJAEV [1979a]. See also PRUGOVECKI/TIP [1974a], [1974b], and HENDRICKSON [1975a], [1976a]. Note also the work on the adiabatic wave operator: DOLLARD [1966a] and OBERMANN [1982a].

(2) *Scattering for particles in classical mechanics.* We shortly describe the scattering theory for a classical particle moving in an external force field $F(x)$. The states of such a particle are points in the phase space, that means pairs $\{x, v\} \in \mathbb{R}^6$ where $x$ represents the position and $v$ the velocity of the particle. The particle obeys the Newtonian equation $\ddot{x}(t) = F(x(t))$. If $\{x_0, v_0\} = \{x(0), v(0)\}$, then the free dynamics $T_t^0$ is defined by $T_t^0\{x_0, v_0\} = \{x_0 + v_0 t, v_0\}$ and the interacting dynamics $T_t$ by $T_t\{x_0, v_0\} = \{x(t), v(t)\}$ where $x(t), v(t)$ are solutions of the Newtonian equation with the initial values $x_0, v_0$. Now let $\Omega = \mathbb{R}^6$ and $\Omega_0 = \mathbb{R}^6 \setminus \{\{a, b\} \in \mathbb{R}^6; b = 0\}$. Then the classical wave operators $W_\pm$ are defined by

$$W_\pm u := \lim_{t\to\pm\infty} T_{-t}T_t^0 u, \qquad u \in \Omega_0.$$

This approach defining the wave operators as maps on the phase space was used by SIMON [1971a]. See also REED/SIMON [3] and THIRRING [1]. Under continuity and falloff conditions on $F(x)$ one can show that $W_\pm$ exist and have the usual properties as the wave operators in Hilbert spaces. COOK [1967a], HUNZIKER [1968a], and PROSSER [1972a] introduced wave operators in a way similar to the one described in this book. They work in the Hilbert space $\mathscr{H}_0 = L^2(\Omega, dx\,dv)$ and define unitary groups $U_0(t)$, $U(t)$ by $(U_0(t)\,f)\,(x, v) := f(T_t^0\{x, v\})$, $(U(t)\,f)\,(x, v) := f(T_t\{x, v\})$. Then the wave operators are defined as usual in the Hilbert space framework.

For a third type of defining wave operators in classical mechanics, slightly different from the latter, we refer to SOKOLOV [1979a]. See also SHIROKOV [1979a] and ANTONEČ [1978a].

(3) *Nonlinear scattering.* There exists also a scattering theory for pairs $H_0$, $H$ where $H - H_0$ is now a nonlinear operator. This scattering theory is usually connected with concrete equations as $u_{tt} - \Delta u + m^2 u = gu^n$, $u \in L^2(\mathbb{R}^n, dx)$ (nonlinear wave equation,) or $iu_t = (-\Delta + m)\,u + f(u)$ (nonlinear Schrödinger equation). The following list includes papers on this subject: SEGAL [1965a], CHADAM [1972a], [1972b], MORAWETZ/STRAUSS [1972a], [1973a], REED [1976a], WICHNOWSKI [1976a], LIN/STRAUSS [1978a], GINIBRE/VELO [1978a], [1978b], [1979a], [1980a], and GROSSE [1979a]. In particular, see STRAUSS [1974a] and REED [1].

(4) *Nonselfadjoint operators.* In principle the two-space scattering theory is also a theory for certain nonselfadjoint perturbations of the selfadjoint operator $H_0$. Namely assume that $H$ is similar to a selfadjoint operator $\tilde{H}$. Then there is a bounded invertible operator $J_1$ such that $J_1^{-1}\tilde{H}J_1 = H$ and this gives

$$e^{itH}J\,e^{-itH_0} = J_1^{-1}\,e^{-it\tilde{H}}(J_1J)\,e^{-itH_0},$$

that is the scattering theory for $\{H, H_0, J\}$ with $H$ nonselfadjoint is equivalent to the one for $\{\tilde{H}, H_0, J_1J\}$ with $\tilde{H}$ selfadjoint. The literature on scattering theory for nonselfadjoint perturbations can be found in the Notes and Remarks to Part IV.

**(5)** ***Scattering theory for groups on a Hilbert space and for automorphisms on $C^*$-algebras.*** **The first scattering theory compares two unitary group representations of $\mathbb{R}^n$ ($n > 1$) asymptotically, and the second studies limits of $\alpha_{-t} \circ \alpha_t^0(A)$ on $C^*$-algebras $\mathscr{A} \ni A$ where $\alpha_t, \alpha_t^0$ are two automorphism groups on $\mathscr{A}$. See Rehberg [1979a], Voigt [1977a], Narnhofer [1979a], and Notes and Remarks to Paragraph 10.2.**

# Part IV. Existence and Completeness of Wave Operators

**Whereas in the preceding part the general properties of wave operators and scattering operators were derived, in general only under the assumption that these objects exist (except a part of Chapter 13), the aim of this part is to develop several methods leading to simple and powerful criteria for the existence and/or completeness of wave operators. The two basic methods are the stationary theory, working with the resolvents of $H_0$ and $H$, and the time-falloff method, using the integrability of $|(u, e^{itH} V e^{-itH_0} v)|$, where $V := HJ - JH_0$ is the perturbation of $H_0$. Naturally there are lots of systems $\{H, H_0, J\}$ where both methods work.**

**Furthermore, trace class perturbations and smooth perturbations are treated. In particular, the trace class method, using that some function of $H$, $H_0$ and $J$ is a trace operator, is very effective and wide ranging. It has the great advantage that it leads to the existence of wave operators without detailed knowledge of the operators $H$, $H_0$ and $J$ (clearly, the verification of the trace class property of some function of $H$, $H_0$ and $J$ requires some information about $H_0$ and $V$, but this is often not very hard). The smooth perturbations can also be handled with time-falloff methods, in principle. But the smoothness method gives additional information about the spectrum of $H$ that can not be extracted from the time-falloff one in this case.**

**If one restricts oneself to the special case in which $J$ is a unitary operator and the absolutely continuous spectrum of $H_0$ has not infinite multiplicity, then one obtains the following picture concerning the existence and completeness of wave operators $W_\pm(H, H_0; J)$:**

**Stationary Methods** ↓ **Trace Class Methods** ↓ **Smoothness Methods**

**Time-Falloff-Methods** ↓ **Smoothness Methods**

**The arrow points from the more general to the more special method.**

**Except for the stationary ones, these methods always provide criteria for the wave operators to exist even in the strong sense. This is not very surprising because we have already seen in Part II (see Lemma 6.17) that the difference between the strong limit and the absolute Abelian one (time-average limit) is not very "large".**

**If we look at the formula (see e.g. (9.23))**

$$(e^{itH} J e^{-itH_0} - J)\, u = i \int_0^t e^{isH} V e^{-isH_0} u \, ds,$$

then we see that the difference between the strong limit and the time-average limit (Cesaro-limit) of the left side arises only from "oscillations in $s$" of the integrand which cancel each other.

In this part we mostly restrict ourselves to the case in which the wave operators are defined with the projection $P_A^{ac}$ onto the absolutely continuous subspace, that is, the universal projection is always $P_A^{ac}$ unless otherwise stated. It is easy to change the statements into statements with other universal projections, in particular with $P_A^{ac}$ replaced by $E_A(\Delta)\,P_A^{ac}$, $\Delta$ some subset of $\mathbb{R}$, or by $P_A^c$ (in time-falloff methods).

Applications of the methods developed in this part are given to perturbations of the multiplication operator by integral operators, to potential scattering and to some other scattering problems. As was already remarked in the Preface we prefer simplicity to generality in the applications.

In the present part $H_0$ and $H$ always denote selfadjoint operators acting on separable Hilbert spaces $\mathscr{H}_0$ and $\mathscr{H}$, and $J$ denotes a bounded operator from $\mathscr{H}_0$ to $\mathscr{H}$. Further, for convenience, we use the notations $E_0(\cdot) := E_{H_0}(\cdot)$, $E(\cdot) := E_H(\cdot)$, $P_0^{ac} := P_{H_0}^{ac}$, $P^{ac} := P_H^{ac}$, $R_0(z) := R_{H_0}(z)$ and $R(z) := R_H(z)$.

# 14. Stationary methods

## 14.1. Stationary methods with auxiliary manifolds

**In this chapter our starting point are the stationary necessary and sufficient criteria for the existence and for the completeness of the (partially isometric and $H_0$-semicomplete) wave operator $W_\pm(H; H_0; J)$. These criteria (see Theorem 13.35 and Theorem 13.42) refer to the seminorm $|\cdot|_{E,\lambda} = (\cdot, E(d\lambda)\cdot)/d\lambda$ and to approximate spectral integrals. The universal projection is $P_\Lambda^{ac}$. For example, the essential assumption of Theorem 13.35 is that $(J - (HJ - JH_0) R_0(z))\, u$ tends to a limit as $z$ goes to the real axis in the sense of convergence in measure and with respect to $|\cdot|_{E,\lambda}$ for a certain set of vectors $u \in \mathscr{H}_0$. Sometimes the verification of this condition is possible by the following method: First, one introduces an auxiliary manifold $\mathfrak{X}$ with its own topology and shows the existence of the limit of the expression mentioned above with respect to the topology of $\mathfrak{X}$. Second, one shows that the seminorm $|\cdot|_{E,\lambda}$ is continuous with respect to this topology.**

**The properties centered around the concept of such an auxiliary manifold are closely connected with the concepts of spectral forms and spectral manifolds.**

### 14.1.1. Spectral forms and spectral manifolds

**Recall the concept of a spectral form (Section 4.5.1) and of a spectral manifold (Sections 3.5.5 and 5.2.4). In the present case we are interested only in $P_\Lambda^{ac}\mathscr{H}$ and $\mu = |\cdot|$. But for applications it is convenient to start with an arbitrary manifold $\mathfrak{X} \subset \mathscr{H}$. For this reason we give a slightly changed definition of a spectral form for $\Gamma = \mathbb{R}$ and for a given spectral measure $E(\cdot)$ on $\mathscr{H}$ $\left(H = \int \lambda E(d\lambda)\right)$.**

**Definition 1.** **Let $\mathfrak{X} \subset \mathscr{H}$ be a linear manifold. A family of semiscalar products $s(\lambda; f, g)$, defined on $\mathbb{R} \times \mathfrak{X} \times \mathfrak{X}$, is called a *spectral form* with respect to $E(\cdot)$ if $s(\lambda; f, g)$ is integrable on $\mathbb{R}$ with respect to $|\cdot|$ for all $f, g \in \mathfrak{X}$ and if**

$$\int_\Delta s(\lambda; f, g)\, d\lambda = (f, E(\Delta)\, P^{ac} g), \qquad \Delta \subset \mathbb{R} \text{ Borel set},\ f, g \in \mathfrak{X}, \tag{1}$$

**is valid.**

**Note that an equivalent description of $s$ is given by**

$$s(\lambda; f, g) = (f, E(d\lambda)\, g)/d\lambda \qquad \text{a.e. mod } |\cdot|, \quad f, g \in \mathfrak{X}, \tag{2}$$

**where the exceptional set for the derivative on the right-hand side may vary with $f, g$. Then Prop. 4.21 has to be changed into the following form:**

***If $P^{ac}\mathfrak{X} \subset P^{ac}\mathscr{H}$ is generating with respect to $P^{ac}\mathscr{H}$ and $H$ and if $s(\lambda; \cdot, \cdot)$, $\lambda \in \Delta$ ($\Delta \subset \mathbb{R}$ Borel set) is a spectral form on $\mathfrak{X}$ with respect to $E(\cdot)$ in the sense of Definition 1, then $E(\Delta)\, P^{ac}\mathscr{H} \cong L^2(\Delta, d\lambda; \hat{\mathfrak{X}}_\lambda)$, where $\hat{\mathfrak{X}}_\lambda$ is defined as in Section 3.5.5.***

If $s(\lambda; f, f) = 0$ for $f \in \mathscr{X} \cap (P^{ac})^{\perp} \mathscr{H}$, then one can define on $P^{ac}\mathscr{X}$ the expression $s_0(\lambda; P^{ac}f, P^{ac}g) := s(\lambda; f, g)$, $f, g \in \mathscr{X}$, which turns out to be a spectral form in the sense of Section 4.5.1 on $\Gamma = \operatorname{spec}^c(H \restriction P^{ac}\mathscr{H})$, if $P^{ac}\mathscr{X}$ is generating with respect to $P^{ac}\mathscr{H}$ and $H$.

Further recall the concept of a spectral manifold $\mathscr{X}$ (note that in Section 3.5.5 the definition is restricted to the case $\mathscr{X} \subset P^{ac}\mathscr{H}$):

**Definition 2.** Let $\mathscr{X} \subset \mathscr{H}$ be a linear manifold. $\mathscr{X}$ is called *spectral* with respect to $E(\cdot)$ if there is a Borel set $\Lambda_0$, $|\Lambda_0| = 0$, such that $(f, E(\mathrm{d}\lambda)\, g)/\mathrm{d}\lambda$ exists for all $f, g \in \mathscr{X}$ if $\lambda \in \mathbb{R} \setminus \Lambda_0$.

If $\mathscr{X}$ is spectral, then one can define on $\mathscr{X}$ a spectral form by

$$s(\lambda; f, g) := \big(f, E(\mathrm{d}\lambda)\, g\big)/\mathrm{d}\lambda, \qquad \lambda \in \mathbb{R} \setminus \Lambda_0, \qquad f, g \in \mathscr{X}. \tag{3}$$

Sometimes in this case we refer to $s$ as the canonical spectral form on a spectral manifold. Note that if $f \in \mathscr{X} \cap (P^{ac})^{\perp} \mathscr{H}$, then $\big(f, E(\mathrm{d}\lambda)\, f\big)/\mathrm{d}\lambda$ exists on $\mathbb{R} \setminus \Lambda_0$ and equals 0 a.e. (mod $|\cdot|$).

Note further that not necessarily $P^{ac}\mathscr{X}$ is a spectral manifold if $\mathscr{X}$ is a spectral manifold.

In the following the linear manifold $\mathscr{X} \subset \mathscr{H}$ is assumed to be spectral, where $s$ is the canonical spectral form.

### 14.1.2. Approximate spectral forms and auxiliary manifolds

Let $\mathscr{X} \subset \mathscr{H}$ be a spectral manifold with exceptional set $\Lambda_0 \subset \mathbb{R}$, $|\Lambda_0| = 0$.

**Definition 3.** A function $f_\varepsilon(\lambda; u, v)\colon (0, \varepsilon_0) \times \mathbb{R} \times \mathscr{X} \times \mathscr{X} \to \mathbb{C}$, $\varepsilon_0 > 0$, is called an *approximate spectral form* for $\{H, \mathscr{X}, \Lambda_0\}$ if $\lim\limits_{\varepsilon \to +0} f_\varepsilon(\lambda; u, v) =: f_+(\lambda; u, v)$ exists and equals $(u, v)_{E,\lambda} := \big(u, E(\mathrm{d}\lambda)\, v\big)/\mathrm{d}\lambda$ for all $u, v \in \mathscr{X}$ and $\lambda \in \mathbb{R} \setminus \Lambda_0$.

An example for an approximate spectral form for $\{H, \mathscr{X}, \emptyset\}$ is given by: $\mathscr{H} = L^2(\mathbb{R}, \mathrm{d}x)$, $H$ is the multiplication operator by $x$,

$$\mathscr{X} = L^2(\mathbb{R}, \mathrm{d}x) \cap C(\mathbb{R}), \qquad f_\varepsilon(\lambda; u, v) := (\varepsilon/\pi)\, \big(R(\lambda + \mathrm{i}\varepsilon)\, u; R(\lambda + \mathrm{i}\varepsilon)\, v\big).$$

First, $\mathscr{X}$ is spectral and $\big(f, E(\mathrm{d}\lambda)\, g\big)/\mathrm{d}\lambda = (f, g)_{E,\lambda} = \overline{f(\lambda)}\, g(\lambda)$. That is, the canonical spectral form is simply given by $\overline{f(\lambda)}\, g(\lambda)$. Second,

$$\lim_{\varepsilon \to +0} (\varepsilon/\pi)\, \big(R(\lambda + \mathrm{i}\varepsilon)\, u, R(\lambda + \mathrm{i}\varepsilon)\, v\big) = \lim_{\varepsilon \to +0} \int_{-\infty}^{\infty} \delta_\varepsilon(\lambda - x)\, \overline{u(x)}\, v(x)\, \mathrm{d}x$$
$$= \overline{u(\lambda)}\, v(\lambda) \quad \text{for all} \quad \lambda \in \mathbb{R},$$

according to Prop. 3.16.

Note that the function

$$f_\varepsilon^H(\lambda; u, v) := (\varepsilon/\pi)\, \big(R(\lambda + \mathrm{i}\varepsilon)\, u, R(\lambda + \mathrm{i}\varepsilon)\, v\big) \tag{4}$$

is always an approximate spectral form for any spectral manifold $\mathscr{X} \subset \mathscr{H}$. This follows from Prop. 3.16. Sometimes this function is called the *canonical approximate spectral form*. Note that $f_\varepsilon^H(\lambda; \cdot, \cdot)$ itself is a scalar product for all $\varepsilon > 0$, $\lambda \in \mathbb{R}$.

Finally, we assume the spectral manifold $\mathscr{X} \subset \mathscr{H}$ to be a topological space with its own topology. If there is an approximate spectral form $f_\varepsilon(\lambda; \cdot, \cdot)$ on $\mathscr{X}$ which is equicontinuous with respect to $\varepsilon > 0$ on $\mathscr{X} \times \mathscr{X}$ for each $\lambda \in \mathbb{R} \setminus \Lambda_0$, then the canonical spectral form is continuous on $\mathscr{X} \times \mathscr{X}$ with respect to the topology of $\mathscr{X}$.

**Proposition 4.** *Let $\mathscr{X}$ be spectral and a topological space. Further, let $f_\varepsilon(\lambda; \cdot, \cdot)$ be an equicontinuous approximate spectral form for $\{H, \mathscr{X}; \Lambda_0\}$ with respect to $\varepsilon > 0$ for each $\lambda \in \mathbb{R} \setminus \Lambda_0$. Then the canonical spectral form $(\cdot; \cdot)_{E,\lambda}$ is continuous on $\mathscr{X} \times \mathscr{X}$ for each $\lambda \in \mathbb{R} \setminus \Lambda_0$.*

**Proof.** Obvious. ∎

In the example considered above the canonical approximate spectral form is equicontinuous with respect to $\varepsilon > 0$ if the topology in $\mathscr{X}$ is defined by $\|u\|_{\mathscr{X}} := \sup_{x \in \mathbb{R}} |u(x)|$. In this case we obtain

$$(\varepsilon/\pi) \, \|R(\lambda + i\varepsilon) \, u\|^2 = (\varepsilon/\pi) \int_{-\infty}^{\infty} \left((\lambda - x)^2 + \varepsilon^2\right)^{-1} |u(x)|^2 \, dx \leq \|u\|_{\mathscr{X}}^2, \qquad \varepsilon > 0,$$

hence the equicontinuity in $\varepsilon > 0$ follows.

In the following we use spectral manifolds which are topological spaces such that an equicontinuous approximate spectral form can be introduced. We call them *auxiliary manifolds*.

Note that not only the canonical approximate spectral form is of interest. In this connection we present an implication of one of our main assumptions

$$J \underset{H_0}{\approx} W; \qquad W^*W = P_0^{ac}$$

(which implies partial isometry and $H_0$-semicompleteness of the wave operator) with respect to $g_\varepsilon^0(\lambda; u; v) := (\varepsilon/\pi) \left(JR_0(\lambda + i\varepsilon) \, u, JR_0(\lambda + i\varepsilon) \, v\right)$.

**Lemma 5.** *Let $J \in \mathscr{L}(\mathscr{H}_0, \mathscr{H})$ be asymptotically equivalent (with respect to $H_0P_0^{ac}$) to a partial isometry* W *with initial subspace $P_0^{ac}\mathscr{H}_0$;* i.e. $J \underset{H_0}{\approx} W$, $W^*W = P_0^{ac}$. *Then*

$$\underset{\varepsilon \to +0}{\text{m-lim}} \, \{f_\varepsilon^{H_0}(\lambda; u, u) - g_\varepsilon^0(\lambda; u, u)\} = 0 \tag{5}$$

*for all $u \in \mathscr{H}_0$* (m-lim with respect to $|\cdot|$).

**Proof.** First we note that

$$(\varepsilon/\pi) \, \|JR_0(\lambda + i\varepsilon) \, (P_0^{ac})^\perp \, u\|^2 \leq \|J\|^2 \, (\varepsilon/\pi) \, \|R_0(\lambda + i\varepsilon) \, (P_0^{ac})^\perp \, u\|^2 \to 0 \tag{6}$$

as $\varepsilon \to +0$ a.e. (mod $|\cdot|$) because of Prop. 3.16 (note that $\left(f, E_0(d\lambda) \, f\right)/d\lambda = 0$ a.e. (mod $|\cdot|$) if $f \in (P_0^{ac})^\perp \mathscr{H}_0$). Thus it suffices to consider $(\varepsilon/\pi) \, \|JR_0(\lambda + i\varepsilon) \, P_0^{ac}u\|^2$

instead of $g_\varepsilon^0(\lambda; u, u)$. According to NATANSON [1, p. 177] now (5) follows from

$$\lim_{\varepsilon\to+0} \int_{-\infty}^{\infty} |f_\varepsilon^{H_0}(\lambda; P_0^{ac}u; P_0^{ac}u) - g_\varepsilon^0(\lambda; P_0^{ac}u, P_0^{ac}u)| \, d\lambda = 0, \qquad u \in \mathscr{H}_0.$$

But this follows from $\left(f_\varepsilon^{H_0}(\lambda; v) := f_\varepsilon^{H_0}(\lambda; v, v),\ g_\varepsilon^0(\lambda; v) := g_\varepsilon^0(\lambda; v, v)\right)$

$$\int_{-\infty}^{\infty} |f_\varepsilon^{H_0}(\lambda; P_0^{ac}u) - g_\varepsilon^0(\lambda; P_0^{ac}u)| \, d\lambda \leqq \int_{-\infty}^{\infty} |f_\varepsilon^{H_0}(\lambda; P_0^{ac}u)^{1/2} - g_\varepsilon^0(\lambda; P_0^{ac}u)^{1/2}|$$

$$\times \left(f_\varepsilon^{H_0}(\lambda; P_0^{ac}u)^{1/2} + g_\varepsilon^0(\lambda; P_0^{ac}u)^{1/2}\right) d\lambda \leqq 2^{1/2} \left( \int_{-\infty}^{\infty} |f_\varepsilon^{H_0}(\lambda; P_0^{ac}u)^{1/2} - g_\varepsilon^0(\lambda; P_0^{ac}u)^{1/2}|^2 \, d\lambda \right)^{1/2}$$

$$\times \left( \int_{-\infty}^{\infty} \left(f_\varepsilon^{H_0}(\lambda; P_0^{ac}u) + g_\varepsilon^0(\lambda; P_0^{ac}u)\right) d\lambda \right)^{1/2} \leqq 2^{1/2} \|u\| \, (1 + \|J\|^2)^{1/2}$$

$$\times \left( \int_{-\infty}^{\infty} |(\varepsilon/\pi)^{1/2} \|R_0(\lambda + i\varepsilon) P_0^{ac}u\| - (\varepsilon/\pi)^{1/2} \|JR_0(\lambda + i\varepsilon) P_0^{ac}u\||^2 \, d\lambda \right)^{1/2}$$

$$\leqq 2^{1/2} \|u\| \, (1 + \|J\|^2)^{1/2} \left( (\varepsilon/\pi) \int_{-\infty}^{\infty} \|(W - J) \, R_0(\lambda + i\varepsilon) \, P_0^{ac}u\|^2 \, d\lambda \right)^{1/2}$$

and Prop. 13.26. ∎

### 14.1.3. Existence and completeness of wave operators

In this section we first prove the existence of $\mathrm{W}_+(H; H_0; J)$ by means of auxiliary manifolds under continuity assumptions on $G_\varepsilon(\lambda) := G(\lambda + i\varepsilon) := J - VR_0(\lambda + i\varepsilon)$, $V = HJ - JH_0$; $J \operatorname{dom} H_0 \subseteqq \operatorname{dom} H$, and under the assumption that the canonical approximate spectral form $f_\varepsilon^H(\lambda; \cdot, \cdot)$ is equicontinuous. Note that

$$G(z) = J + JH_0R_0(z) - HJR_0(z) = J(z - H_0) \, R_0(z)$$
$$+ JH_0R_0(z) - HJR_0(z) = (z - H) \, JR_0(z),$$

hence we obtain

$$JR_0(\lambda + i\varepsilon) = R(\lambda + i\varepsilon) \, G(\lambda + i\varepsilon) \tag{7}$$

and therefore

$$g_\varepsilon^0(\lambda; u, u) = f_\varepsilon^H\left(\lambda; G_\varepsilon(\lambda) \, u, G_\varepsilon(\lambda) \, u\right). \tag{8}$$

**Theorem 6.** *Let $\mathscr{X}_0 \subset \mathscr{H}_0$, $\mathscr{X} \subset \mathscr{H}$ be linear manifolds equipped with their own topologies. Suppose that:*

(I) *$J \underset{H_0}{\approx} W$ where $W^*W = P_0^{ac}$.*

(II) *$P_0^{ac}\mathscr{X}_0$ is generating with respect to $P_0^{ac}\mathscr{H}_0$ and $H_0$, $\mathscr{X}$ is spectral with respect to $E(\cdot)$, and $\mathscr{X}_0$ is spectral with respect to $E_0(\cdot)$.*

(III) $f_\varepsilon^H(\lambda; \cdot, \cdot)$ *is equicontinuous with respect to* $\varepsilon > 0$ *for* $\{H, \mathscr{X}, \Lambda\}$ *(where* $\Lambda$ *denotes the exceptional set* $\Lambda \subset \mathbb{R}$ *of the spectral manifold* $\mathscr{X}$*).*

(IV) $G_\varepsilon(\lambda)$ *maps* $\mathscr{X}_0$ *into* $\mathscr{X}$ *continuously for all* $\varepsilon > 0$, $\lambda \in \mathbb{R} \setminus \Lambda$, *and* $G_\varepsilon(\lambda)$ *is equicontinuous with respect to* $\varepsilon > 0$ *for all* $\lambda \in \mathbb{R} \setminus \Lambda$.

(V) *For every* $\lambda \in \mathbb{R} \setminus \Lambda$ *and* $u \in \mathscr{X}_0$, $G_\varepsilon(\lambda)\, u$ *is a Cauchy net in* $\mathscr{X}$ *as* $\varepsilon \to +0$.

*Then the Abel wave operator* $W_+(H, H_0; J)$ *exists, is partially isometric and* $H_0$*-semicomplete.*

**Proof.** The assertion is proved if we are able to verify the conditions of Theorem 13.35. First we note that

$$\lim_{\varepsilon \to +0} f_\varepsilon^H(\lambda; u, v) = (u, v)_{E,\lambda}, \qquad u, v \in \mathscr{X}, \qquad \lambda \in \mathbb{R} \setminus \Lambda. \tag{9}$$

Then (III) ensures the continuity of $(u, v)_{E,\lambda}$ for all $\lambda \in \mathbb{R} \setminus \Lambda$ with respect to the $\mathscr{X}$-topology. Therefore $(u, v)_{E,\lambda}$ may be extended by continuity to clo $\mathscr{X}$, i.e. it is well-defined also for the ideal elements (limits of Cauchy nets) of clo $\mathscr{X}$, which do not belong to $\mathscr{X}$ (and $\mathscr{H}$). Furthermore, the relation (9) is true also for the ideal elements of clo $\mathscr{X}$. Now let $u \in \mathscr{X}_0$, $\lambda \in \mathbb{R} \setminus \Lambda$. Then, according to (V), $G_\varepsilon(\lambda)\, u$ is a Cauchy net with respect to the $\mathscr{X}$-topology, hence it is also a Cauchy net with respect to $|\cdot|_{E,\lambda}$. Therefore $\lim_{\varepsilon, \varepsilon' \to +0} |G_\varepsilon(\lambda)\, u - G_{\varepsilon'}(\lambda)\, u|_{E,\lambda} = 0$ for all $\lambda \in \mathbb{R} \setminus \Lambda$ and this implies $\operatorname{m-lim}_{\varepsilon, \varepsilon' \to +0} |G_\varepsilon(\lambda)\, u - G_{\varepsilon'}(\lambda)\, u|_{E,\lambda} = 0$. Furthermore, (IV) and (V) imply the existence of $G_+(\lambda)\, u := \lim_{\varepsilon \to +0} G_\varepsilon(\lambda)\, u$, $u \in \mathscr{X}_0$, $\lambda \in \mathbb{R} \setminus \Gamma$ with respect to the $\mathscr{X}$-topology as an element of clo $\mathscr{X}$, and $G_+(\lambda)$ is a continuous operator from $\mathscr{X}_0$ into clo $\mathscr{X}$. Moreover, $\lim_{\varepsilon \to +0} |G_\varepsilon(\lambda)\, u|_{E,\lambda}$ exists and we have $\lim_{\varepsilon \to +0} |G_\varepsilon(\lambda)\, u|_{E,\lambda} = |G_+(\lambda)\, u|_{E,\lambda}$. Now we obtain the existence of $\lim_{\to +0} f_\varepsilon^H(\lambda; G_\varepsilon(\lambda)\, u, G_\varepsilon(\lambda)\, u)$ for all $\lambda \in \mathbb{R} \setminus \Lambda$, $u \in \mathscr{X}_0$ and its equality with $|G_+(\lambda)\, u|^2_{E,\lambda}$ from the estimate

$$\begin{aligned} |f_\varepsilon^H(\lambda; G_\varepsilon(\lambda)\, u, G_\varepsilon(\lambda)\, u)^{1/2} - |G_+(\lambda)\, u|_{E,\lambda}| &\le f_\varepsilon^H(\lambda; G_\varepsilon(\lambda)\, u - G_+(\lambda)\, u, G_\varepsilon(\lambda)\, u - G_+(\lambda)\, u)^{1/2} \\ &\quad + |f_\varepsilon^H(\lambda; G_+(\lambda)\, u, G_+(\lambda)\, u)^{1/2} - |G_+(\lambda)\, u|_{E,\lambda}|. \end{aligned}$$

The first term on the right-hand side tends to 0 as $\varepsilon \to +0$ for $\lambda \in \mathbb{R} \setminus \Lambda$ because of (III) and (V), the second term tends to 0 as $\varepsilon \to +0$ for $\lambda \in \mathbb{R} \setminus \Lambda$ because of (9). Now (8) implies $\lim_{\varepsilon \to +0} g_\varepsilon^0(\lambda; u, u) = |G_+(\lambda)\, u|^2_{E,\lambda}$, and the spectrality of $\mathscr{X}_0$ implies $\lim_{\varepsilon \to +0} f_\varepsilon^{H_0}(\lambda; u, u) = |u|^2_{E_0,\lambda}$ for all $\lambda \in \mathbb{R} \setminus \Lambda_0$ ($\Lambda_0$ the exceptional set for $\mathscr{X}_0$). Applying Lemma 5, we obtain $\operatorname{m-lim}_{\varepsilon \to +0} |G_\varepsilon(\lambda)\, u|_{E,\lambda} = |u|_{E_0,\lambda}$, i.e. all conditions of Theorem 13.35 are verified and this concludes the proof. ■

**Remark 7.** It can be easily seen that if we replace $f_\varepsilon^H(\lambda; \cdot, \cdot)$ and $G_\varepsilon(\lambda)$ by $-f_{-\varepsilon}^H(\lambda; \cdot, \cdot)$ and $G_{-\varepsilon}(\lambda)$ respectively, and (I) by $J \underset{-H_0}{\approx} W$, $W^*W = P_0^{ac}$, then the modified assumptions of Theorem 6 lead to the corresponding assertions for $W_-(H, H_0; J)$.

Preparing the next theorem on the completeness of wave operators, we state the following formula: Let $u, v \in \mathscr{H}_0$, $\alpha \in L^\infty(\mathbb{R}, E_0)$. Then

$$\lim_{\varepsilon \to +0} (\varepsilon/\pi) \int_{-\infty}^{\infty} \alpha(\lambda) \big(R_0(\lambda + i\varepsilon) u, R_0(\lambda + i\varepsilon) v\big) \, d\lambda = \big(u, \alpha(H_0) v\big). \tag{10}$$

The proof uses similar estimates as the proof of Prop. 13.4. Note that (10) implies

$$\lim_{\varepsilon \to +0} \int_{-\infty}^{\infty} f_\varepsilon^{H_0}\big(\lambda; \alpha(H_0) v - \alpha(\lambda) v\big) \, d\lambda = 0 \tag{11}$$

where we use the notation $f_\varepsilon^{H_0}(\lambda; u) := f_\varepsilon^{H_0}(\lambda; u, u)$. This follows from

$$f_\varepsilon^{H_0}\big(\lambda; \alpha(H_0) v - \alpha(\lambda) v\big) = f_\varepsilon^{H_0}\big(\lambda; \alpha(H_0) v\big) + |\alpha(\lambda)|^2 f_\varepsilon^{H_0}(\lambda; v)$$
$$- \alpha(\lambda) f_\varepsilon^{H_0}\big(\lambda; \alpha(H_0) v, v\big) - \overline{\alpha(\lambda)} f_\varepsilon^{H_0}\big(\lambda; v, \alpha(H_0) v\big)$$

and $\int f_\varepsilon^{H_0}\big(\lambda; \alpha(H_0) v\big) \, d\lambda = \|\alpha(H_0) v\|^2$.

If one wants to prove the completeness of the wave operator, one way to do this is to add to the assumptions of Theorem 6 the corresponding ones for $\{H_0, H; J^*\}$. However, in applications the situation is often so that either it is difficult to verify these additional assumptions or even they are not true. For this reason the following completeness theorem may be useful, which can be proved by application of Theorem 13.42 (or Corollary 13.43).

**Theorem 8.** *Assume that the Abel wave operator $W_+(H, H_0; J)$ exists, is partially isometric and $H_0$-semicomplete. Let $\mathscr{X}_0 \subset \mathscr{H}_0$ and $\mathscr{X} \subset \mathscr{H}$ be linear manifolds equipped with their own norms. Suppose:*

(I) *$P^{ac}\mathscr{X}$ is generating with respect to $P^{ac}\mathscr{H}$ and $H$, $\mathscr{X}_0$ is spectral with respect to $E_0(\cdot)$ (with exceptional set $\Lambda_0 \subset \mathbb{R}$, $|\Lambda_0| = 0$) and $P_0^{ac}\mathscr{X}_0$ is generating with respect to $P_0^{ac}\mathscr{H}_0$ and $H_0$.*

(II) *$f_\varepsilon^{H_0}(\lambda; \cdot, \cdot)$ is equicontinuous with respect to $\varepsilon > 0$ for $\{H_0, \mathscr{X}_0, \Lambda_0\}$, moreover it satisfies the uniform estimate*

$$|f_\varepsilon^{H_0}(\lambda; u, v)| \le c_\Delta \|u\|_{\mathscr{X}_0} \|v\|_{\mathscr{X}_0}, \qquad \lambda \in \Delta \subset \mathbb{R} \setminus \Lambda_0$$

*where $\Delta$ is an arbitrary compact set.*

(III) *There is a family $F_\varepsilon(\lambda): \mathscr{X} \to \mathscr{X}_0$ of continuous operators for $\lambda \in \mathbb{R}$, $\varepsilon \in (0, \varepsilon_0)$, $\varepsilon_0 > 0$, such that $F_\varepsilon(\lambda) \in \mathscr{L}(\mathscr{H}, \mathscr{H}_0)$ and $F_\varepsilon(\lambda) u$ is strongly continuous in $\lambda$ for all $\varepsilon > 0$ with respect to $\mathscr{H}_0$. Further $F_\varepsilon(\lambda) u$ is a Cauchy net with respect to $\|\cdot\|_{\mathscr{X}_0}$ for all $\lambda \in \mathbb{R} \setminus \Lambda_0$ and all $u \in \mathscr{X}$, it is continuous in $\lambda$ with respect to $\|\cdot\|_{\mathscr{X}_0}$ for all $\varepsilon \in (0, \varepsilon_0)$ and satisfies the uniform estimate*

$$\|F_\varepsilon(\lambda) u\|_{\mathscr{X}_0} \le d_\Delta \|u\|_{\mathscr{X}}, \qquad \lambda \in \Delta \subset \mathbb{R} \setminus \Lambda_0,$$

*where $\Delta$ is an arbitrary compact set. Finally*

$$R(\lambda + i\varepsilon) u = J R_0(\lambda + i\varepsilon) F_\varepsilon(\lambda) u, \qquad u \in \mathscr{X}, \quad \lambda \in \mathbb{R}, \quad \varepsilon > 0.$$

*Then the wave operator $W_+(H, H_0; J)$ is complete.*

**Proof.** The assertion is proved if we are able to verify the conditions of Corollary 13.43 for each compact set $\Delta \subset \mathbb{R} \setminus \Delta_0$. First we show that $\left\| \int_\Delta E_{0,\varepsilon}(\mathrm{d}\lambda) F_\varepsilon(\lambda) u \right\|$ is uniformly bounded with respect to $\varepsilon > 0$ for all $u \in \mathscr{X}$. Let $g \in \mathscr{H}_0$. Then

$$\left| \left( g, \int_\Delta E_{0,\varepsilon}(\mathrm{d}\lambda) F_\varepsilon(\lambda) u \right) \right| = \left| \int_\Delta (\varepsilon/\pi) \left( R_0(\lambda + i\varepsilon) g, R_0(\lambda + i\varepsilon) F_\varepsilon(\lambda) u \right) \mathrm{d}\lambda \right|$$
$$= \left| \int_\Delta f_\varepsilon^{H_0}(\lambda; g, F_\varepsilon(\lambda) u) \,\mathrm{d}\lambda \right| \leqq \int_\Delta |f_\varepsilon^{H_0}(\lambda; g, F_\varepsilon(\lambda) u)| \,\mathrm{d}\lambda$$
$$\leqq \int_\Delta f_\varepsilon^{H_0}(\lambda; g, g)^{1/2} f_\varepsilon^{H_0}(\lambda; F_\varepsilon(\lambda) u, F_\varepsilon(\lambda) u)^{1/2} \,\mathrm{d}\lambda$$
$$\leqq \left( \int_\Delta f_\varepsilon^{H_0}(\lambda; g, g) \,\mathrm{d}\lambda \right)^{1/2} \left( \int_\Delta f_\varepsilon^{H_0}(\lambda; F_\varepsilon(\lambda) u, F_\varepsilon(\lambda) u) \,\mathrm{d}\lambda \right)^{1/2}.$$

But $\int_\Delta f_\varepsilon^{H_0}(\lambda; g, g) \,\mathrm{d}\lambda \leqq \|g\|^2$ and $\int_\Delta f_\varepsilon^{H_0}(\lambda; F_\varepsilon(\lambda) u, F_\varepsilon(\lambda) u) \,\mathrm{d}\lambda \leqq c_\Delta d_\Delta^2 |\Delta|, \|u\|_{\mathscr{X}}^2$ hence $\left\| \int_\Delta E_{0,\varepsilon}(\mathrm{d}\lambda) F_\varepsilon(\lambda) u \right\|^2 \leqq c_\Delta d_\Delta^2 |\Delta| \, \|u\|_{\mathscr{X}}^2$. That is, in order to prove the weak convergence of $P_0^{\mathrm{ac}} \int_\Delta E_{0,\varepsilon}(\mathrm{d}\lambda) F_\varepsilon(\lambda) u$ it is sufficient to prove the convergence of $\left( g, \int_\Delta E_{0,\varepsilon}(\mathrm{d}\lambda) F_\varepsilon(\lambda) u \right)$ for a generating set of vectors $g$ of $P_0^{\mathrm{ac}} \mathscr{H}_0$. We take the set $\{E_0(B) P_0^{\mathrm{ac}} v; v \in \mathscr{X}_0, B$ Borel set$\}$ as such a generating set. That is, we have to consider $(P_0^{\mathrm{ac}} = E_0(\Delta_0^{\mathrm{ac}}))$

$$\left( E_0(B) E_0(\Delta_0^{\mathrm{ac}}) v, \int_\Delta E_{0,\varepsilon}(\mathrm{d}\lambda) F_\varepsilon(\lambda) u \right) = \int_\Delta f_\varepsilon^{H_0}\left(\lambda; E_0(B \cap \Delta_0^{\mathrm{ac}}) v, F_\varepsilon(\lambda) u\right) \mathrm{d}\lambda \tag{12}$$

and we have to show the convergence of (12) as $\varepsilon \to +0$. We compare (12) with the expression

$$\int_\Delta \chi_{B \cap \Delta_0^{\mathrm{ac}}}(\lambda) f_\varepsilon^{H_0}\left(\lambda; v, F_\varepsilon(\lambda) u\right) \mathrm{d}\lambda = \int_\Delta f_\varepsilon^{H_0}\left(\lambda; \chi_{B \cap \Delta_0^{\mathrm{ac}}}(\lambda) v, F_\varepsilon(\lambda) u\right) \mathrm{d}\lambda \tag{13}$$

and show that the difference tends to 0 as $\varepsilon \to +0$ (this means that it is sufficient to prove the convergence of (13)). We obtain by an easy calculation, using Schwarz' inequality

$$\left| \int_\Delta f_\varepsilon^{H_0}\left(\lambda; E_0(B \cap \Delta_0^{\mathrm{ac}}) v, F_\varepsilon(\lambda) u\right) \mathrm{d}\lambda - \int_\Delta f_\varepsilon^{H_0}\left(\lambda; \chi_{B \cap \Delta_0^{\mathrm{ac}}}(\lambda) v, F_\varepsilon(\lambda) u\right) \mathrm{d}\lambda \right|$$
$$\leqq \int_\Delta \left| f_\varepsilon^{H_0}\left(\lambda; E_0 (B \cap \Delta_0^{\mathrm{ac}}) v - \chi_{B \cap \Delta_0^{\mathrm{ac}}}(\lambda) v, F_\varepsilon(\lambda) u\right) \right| \mathrm{d}\lambda$$
$$\leqq \left( \int_\Delta f_\varepsilon^{H_0}\left(\lambda; E_0(B \cap \Delta_0^{\mathrm{ac}}) v - \chi_{B \cap \Delta_0^{\mathrm{ac}}}(\lambda) v\right) \mathrm{d}\lambda \right)^{1/2} \left( \int_\Delta f_\varepsilon^{H_0}\left(\lambda; F_\varepsilon(\lambda) u\right) \mathrm{d}\lambda \right)^{1/2}.$$

Since the second factor is uniformly bounded with respect to $\varepsilon > 0$, it is enough to show that the first factor tends to 0 as $\varepsilon \to +0$. But this is true because of (11).

Now we prove the convergence of (13) as $\varepsilon \to +0$. First, (III) implies the existence of $\operatorname{s-lim}_{\|\cdot\|_{\mathscr{X}_0}, \varepsilon \to +0} F_\varepsilon(\lambda) u =: F_+(\lambda) u \in \mathrm{clo}_{\|\cdot\|_{\mathscr{X}_0}} \mathscr{X}_0$. By a similar extension argument as in the proof of Theorem 6 the semiscalar product $(u, v)_{E_0,\lambda}$ can be extended to $\mathrm{clo}_{\|\cdot\|_{\mathscr{X}_0}} \mathscr{X}_0$ and

$(u, v)_{E_0,\lambda} = \lim_{\varepsilon\to+0} f_\varepsilon^{H_0}(\lambda; u, v)$ turns out to be true for all elements of $\mathrm{clo}_{\|\cdot\|_{\mathscr{X}_0}} \mathscr{X}_0$. Therefore the existence of $\lim_{\varepsilon\to+0} f_\varepsilon^{H_0}(\lambda; v, F_\varepsilon(\lambda) u)$ and its equality with $(v, F + (\lambda) u)_{E_0,\lambda}$ follows from the estimate

$$\left|f_\varepsilon^{H_0}(\lambda; v, F_\varepsilon(\lambda) u) - (v, F_+(\lambda) u)_{E_0,\lambda}\right| \leqq \left|f_\varepsilon^{H_0}(\lambda; v, F_\varepsilon(\lambda) u) - f_\varepsilon^{H_0}(\lambda; v, F_+(\lambda) u)\right| + \left|f_\varepsilon^{H_0}(\lambda; v, F_+(\lambda) u) - (v, F + (\lambda) u)_{E_0,\lambda}\right|$$

and from (II). Second, $|f_\varepsilon^{H_0}(\lambda; v, F_\varepsilon(\lambda) u)| \leqq c_\Delta d_\Delta \|v\|_{\mathscr{X}_0} \|u\|_{\mathscr{X}_0}$, hence the existence of the limit of (13) follows from Lebesgue's dominated convergence theorem and we obtain

$$\lim_{\varepsilon\to+0}\left(E_0(B \cap \Delta_0^{\mathrm{ac}}) v, \int_\Delta E_{0,\varepsilon}(\mathrm{d}\lambda) F_\varepsilon(\lambda) u\right) = \int_\Delta \chi_{B\cap\Delta_0^{\mathrm{ac}}}(\lambda) (v, F_+(\lambda) u)_{E_0,\lambda} \,\mathrm{d}\lambda.$$

Now the last step of the proof proceeds as follows: $(u, v)_{E_0,\lambda}$ is the canonical spectral form on $\mathscr{X}_0$ (in the sense of Section 1). Therefore, according to Section 1 (cf. Sections 4.5.1 and 4.1.2), we have the isomorphy $E_0(\Delta) P_0^{\mathrm{ac}} \mathscr{H}_0 \cong L^2(\Delta, \mathrm{d}\lambda; \hat{\mathscr{X}}_{0,\lambda})$.

Note that the ideal element $\text{s-lim}_{\|\cdot\|_{\mathscr{X}_0}, n\to\infty} v_n \in \mathrm{clo}_{\|\cdot\|_{\mathscr{X}_0}} \mathscr{X}_0$ ($v_n \in \mathscr{X}_0$ a Cauchy sequence) can be identified with $\text{s-lim}_{|\cdot|_{E_0,\lambda}, n\to\infty} v_n \pmod{\ker |\cdot|_{E_0,\lambda}}$ from $\hat{\mathscr{X}}_{0,\lambda}$ because of $|v|_{E_0,\lambda} \leqq C_\Delta \|v\|_{\mathscr{X}_0}$, $v \in \mathscr{X}_0$. Hence $F_+(\lambda) u$ can be identified with the limit of the Cauchy net $F_\varepsilon(\lambda) u$ with respect to $|\cdot|_{E_0,\lambda} \pmod{\ker |\cdot|_{E_0,\lambda}}$, i.e. $F_+(\lambda) u \in \hat{\mathscr{X}}_{0,\lambda}$ for all $u \in \mathscr{X}_0$ and all $\lambda \in \mathbb{R} \setminus \Lambda_0$. Further we obtain

$$\lim_{\varepsilon\to+0} \int_\Delta \left(F_\varepsilon(\lambda) u, E_{0,\varepsilon}(\mathrm{d}\lambda) F_\varepsilon(\lambda) u\right) = \lim_{\varepsilon\to+0} \int_\Delta f_\varepsilon^{H_0}\left(\lambda; F_\varepsilon(\lambda) u, F_\varepsilon(\lambda) u\right) \mathrm{d}\lambda = \int_\Delta \left(F_+(\lambda) u, F_+(\lambda) u\right)_{E_0,\lambda} \mathrm{d}\lambda.$$

This follows from $\lim_{n\to\infty} f_\varepsilon^{H_0}(\lambda; F_\varepsilon(\lambda) u) = (F_+(\lambda) u, F_+(\lambda) u)_{E_0,\lambda}$ for all $\lambda \in \mathbb{R} \setminus \Lambda_0$, which is implied by the estimate

$$\begin{aligned}\left|f_\varepsilon^{H_0}(\lambda; F_\varepsilon(\lambda) u)^{1/2} - |F_+(\lambda) u|_{E_0,\lambda}\right| &\leqq \left|f_\varepsilon^{H_0}(\lambda; F_\varepsilon(\lambda) u)^{1/2} - f_\varepsilon^{H_0}(\lambda; F_+(\lambda) u)\right|^{1/2} \\ &\quad + \left|f_\varepsilon^{H_0}(\lambda; F_+(\lambda) u)^{1/2} - |F_+(\lambda) u|_{E_0,\lambda}\right| \\ &\leqq f_\varepsilon^{H_0}(\lambda; F_\varepsilon(\lambda) u - F_+(\lambda) u)^{1/2} + \left|f_\varepsilon^{H_0}(\lambda; F_+(\lambda) u)^{1/2} - |F_+(\lambda) u|_{E_0,\lambda}\right|,\end{aligned}$$

and from the estimate $f_\varepsilon^{H_0}(\lambda; F_\varepsilon(\lambda) u) \leqq c_\Delta d_\Delta^2 \|u\|_{\mathscr{X}}^2$ (use Lebesgue's dominated convergence theorem).

Simultaneoulsy we have shown $\int_\Delta (F_+(\lambda) u, F_+(\lambda) u)_{E_0,\lambda} \,\mathrm{d}\lambda < \infty$, that is $F_+(\cdot)$ belongs to $L^2(\Delta, \mathrm{d}\lambda; \hat{\mathscr{X}}_{0,\lambda})$ so that this function represents a vector from $P_0^{\mathrm{ac}} E_0(\Delta) \mathscr{H}_0$, which we denote by $Z_+ u$. Thus we have

$$\underset{\varepsilon\to+0}{\text{w-lim}}\, P_0^{\mathrm{ac}} \int_\Delta E_{0,\varepsilon}(\mathrm{d}\lambda) F_\varepsilon(\lambda) = Z_+$$

and

$$\lim_{\varepsilon\to+0} \int_\Delta \left(F_\varepsilon(\lambda) u, E_{0,\varepsilon}(\mathrm{d}\lambda) F_\varepsilon(\lambda) u\right) = \|Z_+ u\|^2.$$

The third condition of Corollary 13.43 (Formula (13.77)) is obvious. This concludes the proof. ■

**Remark 9.** (I) The assumption that $P_0^{ac}\mathscr{X}_0$ is generating with respect to $P_0^{ac}\mathscr{H}_0$ and $H_0$ is not essential. If $P_0^{ac}\mathscr{X}_0$ has not this property, then we can always extend $\mathscr{X}_0$ to a normed linear manifold $\tilde{\mathscr{X}}_0 \subset \mathscr{H}_0$ which is generating, without changing the other assumptions of this theorem. However, the assumption that $P^{ac}\mathscr{X}$ is generating with respect to $P^{ac}\mathscr{H}$ and $H$ is essential as can be seen from Corollary 13.43 (or Theorem 13.42).

(II) The identity $R(z) = JR_0(z)\, F(z)$ is the key assumption in this theorem. Unfortunately, it is rather implicit, in fact, $F(z)$ is not uniquely determined because $J$ need not to be injective. Only the relation $G(z)\, F(z) = 1$ is true but not the converse, in general. When $J$ is bijective, then $F(z)$ is unique and given by

$$F(z) = J^{-1}\big(1 + (H - JH_0J^{-1})\big)\, R(z) = (z - H_0)\, J^{-1}R(z).$$

The next statement is a combination of Theorem 6 and Theorem 8 giving the existence and completeness of the wave operator. The reason for this combined statement is to eliminate assumption

(III) in Theorem 6 on the equicontinuity of $f_\varepsilon^H(\lambda; \cdot, \cdot)$ which is sometimes not so convenient in applications.

**Proposition 10.** *Let $\mathscr{X}_0 \subset \mathscr{H}_0$, $\mathscr{X} \subset \mathscr{H}$ be linear manifolds equipped with their own norms $\|\cdot\|_{\mathscr{X}_0}$, $\|\cdot\|_{\mathscr{X}}$. Let $\mathscr{X}_0$, $\mathscr{X}$ be spectral with respect to $E_0(\cdot)$, $E(\cdot)$ and let $P_0^{ac}\mathscr{X}_0$, $P^{ac}\mathscr{X}$ be generating with respect to $P_0^{ac}\mathscr{H}_0$, $P^{ac}\mathscr{H}$ and $H_0$, $H$ respectively. ($\Lambda = \Lambda_0$ is assumed without loss of generality.) Suppose that assumptions* (I), (IV), *and* (V) *of Theorem 6 are satisfied. Further, suppose that assumptions* (II) *and* (III) *of Theorem 8 are satisfied. Then the wave operator $W_+(H, H_0; J)$ exists, is partially isometric and complete.*

**Proof.** The assertion is proved if we are able to verify the condition (III) of Theorem 6. Because of $R(\lambda + i\varepsilon)\, u = JR_0(\lambda + i\varepsilon)\, F_\varepsilon(\lambda)\, u$, $u \in \mathscr{X}$, we obtain

$$g_\varepsilon^0\big(\lambda; F_\varepsilon(\lambda)\, u, F_\varepsilon(\lambda)\, v\big) = f_\varepsilon^H(\lambda; u, v), \qquad u, v \in \mathscr{X}.$$

It is sufficient to show the equicontinuity for $f_\varepsilon^H(\lambda; u, u)$. We have

$$\begin{aligned} f_\varepsilon^H(\lambda; u, u) &= g_\varepsilon^0\big(\lambda; F_\varepsilon(\lambda)\, u, F_\varepsilon(\lambda)\, u\big) = (\varepsilon/\pi)\, \|JR_0(\lambda + i\varepsilon)\, F_\varepsilon(\lambda)\, u\|^2 \\ &\leqq \|J\|^2\, (\varepsilon/\pi)\, \|R_0(\lambda + i\varepsilon)\, F_\varepsilon(\lambda)\, u\|^2 = \|J\|^2\, f_\varepsilon^{H_0}\big(\lambda; F_\varepsilon(\lambda)\, u, F_\varepsilon(\lambda)\, u\big) \\ &\leqq \|J\|^2\, c_\Delta\, \|F_\varepsilon(\lambda)\, u\|_{\mathscr{X}_0}^2 \leqq \|J\|^2\, c_\Delta\, d_\Delta^2\, \|u\|_{\mathscr{X}}^2 \end{aligned}$$

if $\lambda \in \Delta$ ($\Delta$ compact), $\Delta \subset \mathbb{R} \setminus \Lambda_0$; that is for each $\lambda \in \mathbb{R} \setminus \Lambda_0$, and thus condition (III) of Theorem 6 is verified. This concludes the proof. ■

**Remark 11.** Note that it is also possible to omit the assumption "$\mathscr{X}$ is spectral with respect to $E(\cdot)$" and to work with $\lim_{\varepsilon \to +0} f_\varepsilon^H(\lambda; u, v)$, $u, v \in \mathscr{X}$, which exists for all $\lambda \in \mathbb{R} \setminus \Lambda_0$ and equals $(u, v)_{E,\lambda}$ a.e. (mod $|\cdot|$) (where the exceptional set may vary with $u$ and $v$). In this case the application of Theorem 13.35 is also possible.

### 14.1.4. Existence of strong wave operators

In this section we complete the assumptions made in Section 1.3 in such a way that the wave operators exist even in the strong sense. The essential assumptions are: (1) The auxiliary space $\mathscr{X}$ is even a separable Hilbert space, (2) $G_\varepsilon(\lambda)$ has limits with respect to the $\mathscr{X}_0$, $\mathscr{X}$-topologies as $\varepsilon \to +0$ *and* as $\varepsilon \to -0$. Before coming to the theorem, we need some formulas. Let $u, v \in \mathscr{H}_0$, $\alpha \in L^\infty(\mathbb{R}, \mathrm{d}\lambda)$, $\alpha(H_0) v \in P_0^{\mathrm{ac}}\mathscr{H}_0$. Replacing $\alpha(\lambda)$ by $\chi_\Delta(\lambda)\, \alpha(\lambda)$ in formula (10), where $\Delta$ is a Borel set, we obtain from (10)

$$\lim_{\varepsilon\to+0} (\varepsilon/\pi) \int_\Delta \alpha(\lambda) \big(R_0(\lambda + \mathrm{i}\varepsilon)\, u, R_0(\lambda + \mathrm{i}\varepsilon)\, v\big)\, \mathrm{d}\lambda = \int_\Delta \alpha(\lambda) \frac{\big(u, E_0(\mathrm{d}\lambda)\, v\big)}{\mathrm{d}\lambda}\, \mathrm{d}\lambda. \tag{14}$$

Note that $\alpha(\lambda) \big[\big(u, E_0(\mathrm{d}\lambda)\, v\big)/\mathrm{d}\lambda\big]$ is absolutely integrable. From this fact and (14) we obtain, using a famous statement of Lebesgue (see NATANSON [1, p. 174]), that the functions $\alpha(\lambda)\, (\varepsilon/\pi)\, (R_0(\lambda + \mathrm{i}\varepsilon)\, u, R_0(\lambda + \mathrm{i}\varepsilon)\, v)$ have equiabsolutely continuous integrals, that is, for each $\delta > 0$ there is $\eta = \eta(\delta) > 0$ such that

$$\sup_{\varepsilon>0} \left| \int_\Delta (\varepsilon/\pi)\, \alpha(\lambda) \big(R_0(\lambda + \mathrm{i}\varepsilon)\, u, R_0(\lambda + \mathrm{i}\varepsilon)\, v\big)\, \mathrm{d}\lambda \right| < \delta; \quad \text{if } |\Delta| < \eta. \tag{15}$$

**Theorem 12.** *Let $\mathscr{X}_0 \subset \mathscr{H}_0$, $\mathscr{X} \subset \mathscr{H}$ be linear manifolds equipped with their own topologies. Suppose that:*

(I) $J \operatorname{dom} H_0 \subseteq \operatorname{dom} H$.

(II) $J \underset{\pm H_0}{\approx} W$ *(for both signs), where* $W^*W = P_0^{\mathrm{ac}}$.

(III) *Let $\Phi$ be an admissible function* (see Section 8.2.1) *with the property:* $\lim_{t\to\pm\infty} \|J\, \mathrm{e}^{-\mathrm{i}t\Phi(H_0)} P_0^{\mathrm{ac}} u\|$ *exists and equals* $\|P_0^{\mathrm{ac}} u\|$ *for all* $u \in \mathscr{H}_0$ *(and both signs).*

(IV) $P_0^{\mathrm{ac}}\mathscr{X}_0$, $P^{\mathrm{ac}}\mathscr{X}$ *are generating with respect to* $P_0^{\mathrm{ac}}\mathscr{H}_0$, $P^{\mathrm{ac}}\mathscr{H}$ *and* $H_0$, $H$ *respectively.*

(V) $\mathscr{X}$ *is spectral with respect to* $E(\cdot)$ *(exceptional set $\Lambda$) and $\mathscr{X}$ is a separable Hilbert space.*

(VI) $f_\varepsilon^H(\lambda; \cdot, \cdot)$ *and* $\tilde{f}_\varepsilon^H(\lambda; \cdot, \cdot) = -f_{-\varepsilon}^H(\lambda; \cdot; \cdot)$ *are equicontinuous with respect to* $\varepsilon > 0$ *for* $\{H; \mathscr{X}, \Lambda\}$.

(VII) $G_\varepsilon(\lambda)$ *and* $G_{-\varepsilon}(\lambda)$ *map $\mathscr{X}_0$ into $\mathscr{X}$ continuously for all* $\varepsilon > 0$, $\lambda \in \mathbb{R} \setminus \Lambda$ *and are equicontinuous with respect to* $\varepsilon > 0$ *for all* $\lambda \in \mathbb{R} \setminus \Lambda$.

(VIII) *For every* $\lambda \in \mathbb{R} \setminus \Lambda$ *and* $u \in \mathscr{X}_0$, $G_\varepsilon(\lambda)\, u$ *and* $G_{-\varepsilon}(\lambda)\, u$ *are Cauchy nets in $\mathscr{X}$ as* $\varepsilon \to +0$.

*Then the strong wave operators* $W_+\big(\Phi(H), \Phi(H_0); J\big)$ *and* $W_-\big(\Phi(H), \Phi(H_0); J\big)$ *exist, are partially isometric and $H_0$-semicomplete.*

**Proof.** 1. By Theorem 6 and Remark 7, the assumptions imply that the wave operators $W_+ = W_+(H, H_0; J)$, $W_- = W_-(H, H_0; J)$ exist, are partial isometries and $H_0$-semicomplete. Then the invariance principle (see Section 11.3.2) says that also $W_\pm\big(\Phi(H), \Phi(H_0); J\big)$ exist and coincide with $W_\pm$. Now we consider the relation

$$\begin{aligned}\|\mathrm{e}^{\mathrm{i}t\Phi(H)} J\, \mathrm{e}^{-\mathrm{i}t\Phi(H)} P_0^{\mathrm{ac}} f - W_+ f\|^2 &= \|\mathrm{e}^{\mathrm{i}t\Phi(H)} J\, \mathrm{e}^{-\mathrm{i}t\Phi(H_0)} P_0^{\mathrm{ac}} f\|^2 + \|W_+ f\|^2 \\ &\quad - (\mathrm{e}^{\mathrm{i}t\Phi(H)} J\, \mathrm{e}^{-\mathrm{i}t\Phi(H_0)} P_0^{\mathrm{ac}} f, W_+ f) - (W_+ f, \mathrm{e}^{\mathrm{i}t\Phi(H)} J\, \mathrm{e}^{-\mathrm{i}t\Phi(H_0)} P_0^{\mathrm{ac}} f) \\ &= \|J\, \mathrm{e}^{-\mathrm{i}t\Phi(H_0)} P_0^{\mathrm{ac}} f\|^2 + \|P_0^{\mathrm{ac}} f\|^2 - 2 \operatorname{Re} (\mathrm{e}^{\mathrm{i}t\Phi(H)} J\, \mathrm{e}^{-\mathrm{i}t\Phi(H_0)} P_0^{\mathrm{ac}} f, W_+ f).\end{aligned}$$

Using (III), it follows that it suffices to prove that the limits

$$\lim_{t\to\pm\infty} (e^{it\phi(H)}J\, e^{-it\phi(H_0)}f, g) = (W_+f, g)$$

exist for a set of vectors $f$, $g$ being generating in $P_0^{ac}\mathscr{H}_0$, $P^{ac}\mathscr{H}$ respectively. In the following we prove this only for $t\to\infty$. As generating sets we take the sets $f = \alpha(H_0)\, P_0^{ac}u$, $g = \beta(H)\, P^{ac}v$, $u \in \mathscr{X}_0$, $v \in \mathscr{X}$, $\alpha, \beta \in L^\infty(\mathbb{R}, dx)$ with compact supports (note that $P_0^{ac} = E_0(\Delta_0^{ac})$, $P^{ac} = E(\Delta^{ac})$, hence $f = (\alpha\chi_{\Delta_0^{ac}})(H_0)\, u$, $g = (\beta\chi_{\Delta^{ac}})(H)\, v$).

2. Next we derive a formula for $\big(g, (J - W_+)\, \alpha(H_0)\, u\big)$; $g \in \mathscr{H}$, $u \in \mathscr{X}_0$; $\alpha \in L^\infty(\mathbb{R}, d\lambda)$ with compact support. Using the relation $JR_0(z) = R(z)\, G(z)$, we get

$$J\big(R_0(z) - R_0(\bar z)\big) - \big(R(z) - R(\bar z)\big)\, G(z) = R(\bar z)\, \big(G(z) - G(\bar z)\big).$$

This implies

$$\begin{aligned}&\big(g, (\varepsilon/\pi)\, JR_0(\lambda - i\varepsilon)\, R_0(\lambda + i\varepsilon)\, u\big) - \big(g, (\varepsilon/\pi)\, R(\lambda - i\varepsilon)\, JR_0(\lambda + i\varepsilon)\, u\big)\\ &= -(1/2i\pi)\, \big(g, R(\lambda - i\varepsilon)\, D(\lambda + i\varepsilon)\, u\big)\end{aligned}$$

where $D(\lambda + i\varepsilon) := G(\lambda + i\varepsilon) - G(\lambda - i\varepsilon)$. Multiplying this equation by $\alpha(\lambda)$ and integrating in $\lambda$ over $\mathbb{R}$ we obtain

$$\begin{aligned}&(\varepsilon/\pi) \int_{-\infty}^{\infty} \alpha(\lambda)\, \big(R_0(\lambda + i\varepsilon)\, J^*g, R_0(\lambda + i\varepsilon)\, u\big)\, d\lambda\\ &- (\varepsilon/\pi) \int_{-\infty}^{\infty} \alpha(\lambda)\, \big(g, R(\lambda - i\varepsilon)\, JR_0(\lambda + i\varepsilon)\, u\big)\, d\lambda\\ &= -(1/2i\pi) \int_{-\infty}^{\infty} \alpha(\lambda)\, \big(R(\lambda + i\varepsilon)\, g, D(\lambda + i\varepsilon)\, u\big)\, d\lambda. \qquad (16)\end{aligned}$$

From (10) we get that the first term on the left side tends to $\big(J^*g, \alpha(H_0)\, u\big)$ as $\varepsilon \to +0$. Concerning the second term on the left side we assert; for $\alpha(\lambda)$ with $\alpha(\lambda) = \alpha(\lambda)\, \chi_{\Delta_0^{ac}}(\lambda)$.

$$\lim_{\varepsilon\to+0} (\varepsilon/\pi) \int_{-\infty}^{\infty} \alpha(\lambda)\, \big(g, R(\lambda - i\varepsilon)\, JR_0(\lambda + i\varepsilon)\, u\big)\, d\lambda = \big(g, W_+\alpha(H_0)\, u\big).$$

Namely, we have $\big(g, W_+\alpha(H_0)\, u\big) = \lim_{\varepsilon\to+0} (\varepsilon/\pi) \int_{-\infty}^{\infty} \big(g, R(\lambda - i\varepsilon)\, JR_0(\lambda + i\varepsilon)\, \alpha(H_0)\, u\big)\, d\lambda$ and

$$\begin{aligned}&(\varepsilon/\pi) \int_{-\infty}^{\infty} \big|\big(R(\lambda + i\varepsilon)\, g, JR_0(\lambda + i\varepsilon)\, (\alpha(\lambda) - \alpha(H_0))\, u\big)\big|\, d\lambda\\ &\le (\varepsilon/\pi) \int_{-\infty}^{\infty} \|R(\lambda + i\varepsilon)\, g\|\, \|J\|\, \big\|R_0(\lambda + i\varepsilon)\, \big(\alpha(\lambda) - \alpha(H_0)\big)\, u\big\|\, d\lambda\\ &\le \|J\| \left((\varepsilon/\pi) \int_{-\infty}^{\infty} \|R(\lambda + i\varepsilon)\, g\|^2\, d\lambda\right)^{1/2} \left((\varepsilon/\pi) \int_{-\infty}^{\infty} \big\|R_0(\lambda + i\varepsilon)\, \big(\alpha(\lambda) - \alpha(H_0)\big)\, u\big\|^2\, d\lambda\right)^{1/2}\\ &= \|J\|\, \|g\| \left((\varepsilon/\pi) \int_{-\infty}^{\infty} \big\|R_0(\lambda + i\varepsilon)\, \big(\alpha(\lambda) - \alpha(H_0)\big)\, u\big\|^2\, d\lambda\right)^{1/2},\end{aligned}$$

but this tends to 0 as $\varepsilon \to +0$ according to (11). Therefore we obtain the existence of the limit on the right side of (16), if $\alpha(\lambda) = \alpha(\lambda)\,\chi_{\Lambda_0 \infty}(\lambda)$, and the equation

$$\lim_{\varepsilon\to+0} -(1/2i\pi)\int_{-\infty}^{\infty} \alpha(\lambda)\left(R(\lambda+i\varepsilon)\,g,\, D(\lambda+i\varepsilon)\,u\right) d\lambda = \left(g, (J - W_+)\,\alpha(H_0)\,u\right). \quad (17)$$

Moreover, from $\left(J^*g, \alpha(H_0)\,u\right) = \int_{-\infty}^{\infty} \alpha(\lambda)\left[\left(J^*g, E_0(d\lambda)\,u\right)/d\lambda\right] d\lambda$, $\left(W_+{}^*g, \alpha(H_0)\,u\right) = \int_{-\infty}^{\infty} \alpha(\lambda)\left[\left(W_+{}^*g, E_0(d\lambda)\,u\right)/d\lambda\right] d\lambda$, we obtain that the integrands on the left side of (16) have equiabsolutely continuous integrals with respect to $\varepsilon$. Thus also the integrand on the right side of (16) obeys for each $\delta > 0$

$$\left|(1/2i\pi)\int_{\Delta} \alpha(\lambda)\left(R(\lambda+i\varepsilon)\,g,\, D(\lambda+i\varepsilon)\,u\right) d\lambda\right| < \delta \quad \text{if} \quad |\Delta| < \eta(\delta). \quad (18)$$

Now we consider the integrand of the left side of (17) for $g = \beta(H)\,v$, $v \in \mathscr{X}$, $\beta \in L^\infty(\mathbb{R}, d\lambda)$ with compact support and with the property $\beta(\lambda) = \beta(\lambda)\,\chi_{\Lambda \infty}(\lambda)$. First we have

$$\left(R(\lambda+i\varepsilon)\,\beta(H)\,v,\, D(\lambda+i\varepsilon)\,u\right) = \int_{-\infty}^{\infty} \frac{\overline{\beta(x)}}{\lambda - i\varepsilon - x}\left(E(dx)\,v,\, D(\lambda+i\varepsilon)\,u\right). \quad (19)$$

According to the assumptions (V) and (VI), $\left(w, E(d\lambda)\,w'\right)/d\lambda$, $w, w' \in \mathscr{X}$, $\lambda \in \mathbb{R}\setminus\Lambda$, is a continuous sesquilinear form on $\mathscr{X}\times\mathscr{X}$; hence we have $\left(w, E(d\lambda)\,w'\right)/d\lambda = \left(w, H(\lambda)\,w'\right)_{\mathscr{X}}$ where $H(x) \in \mathscr{L}(\mathscr{X})$ and $H(x) \geqq 0$ for all $x \in \mathbb{R}\setminus\Lambda$. Inserting this into (19), we obtain

$$\left(R(\lambda+i\varepsilon)\,\beta(H)\,v,\, D(\lambda+i\varepsilon)\,u\right) = \left(\int_{-\infty}^{\infty} \frac{\beta(x)\,H(x)\,v}{\lambda + i\varepsilon - x}\,dx,\, D(\lambda+i\varepsilon)\,u\right)_{\mathscr{X}}. \quad (20)$$

We put $f_\varepsilon(\lambda) := f(\lambda+i\varepsilon) := \int_{-\infty}^{\infty} (\lambda + i\varepsilon - x)^{-1}\,\beta(x)\,H(x)\,v\,dx$. Now without loss of the assumption for the vectors $\beta(H)\,v$ to be generating in $P^{ac}\mathscr{H}$, we can restrict ourselves to functions $\beta$ such that $\beta(x)\,\|H(x)\|_{\mathscr{L}(\mathscr{X})}$ is uniformly bounded in $x$. Then $\varphi(x) := \beta(x)\,H(x)\,v$ is a member of $L^2(\mathbb{R}, dx; \mathscr{X})$. From Prop. 1.2 we obtain that $f_+(\lambda) := \operatorname{s-lim}_{\varepsilon\to-0} \int_{-\infty}^{\infty} (\lambda + i\varepsilon - x)^{-1}\,\varphi(x)\,dx$ exists a.e. (mod $|\cdot|$). Then, using assumption (VIII), it follows that $\lim_{\varepsilon\to+0}\left(R(\lambda+i\varepsilon)\,\beta(H)\,v,\, D(\lambda+i\varepsilon)\,u\right)$ exists a.e. (mod $|\cdot|$) and equals $\left(f_+(\lambda); D(\lambda+i0)\,u\right)_{\mathscr{X}}$. Therefore, applying (18) and the Vitali convergence theorem, from (17) we obtain

$$\left(\beta(H)\,v, (J - W_+)\,\alpha(H_0)\,u\right) = -(1/2i\pi)\int_{-\infty}^{\infty} \alpha(\lambda)\left(f_+(\lambda), D(\lambda+i0)\,u\right)_{\mathscr{X}} d\lambda \quad (21)$$

3. Next we estimate the right side of (21). First note that according to Section 1.2.5 (in particular Prop. 1.6), $f(\lambda + i\varepsilon)$ is from the Hardy class $\mathscr{H}_+{}^2(\mathbb{R}, d\lambda; \mathscr{X})$ and the $L^2$-norm of $f_+(\cdot)$ is given by

$$\|f_+(\cdot)\|_{L^2(\mathbb{R}, d\lambda; \mathscr{X})} = 2\pi\left(\int_0^{\infty} \|\phi(s)\|_{\mathscr{X}}^2\,ds\right)^{1/2}.$$

Without loss of the assumption for the vectors $\alpha(H_0)\,u$ to be generating in $P_0^{ac}\mathscr{H}_0$, we can restrict ourselves to functions $\alpha$ with the property $|\alpha(\lambda)|\,\|D(\lambda+i0)\,u\|_{\mathfrak{X}}\leq c_0$, uniformly in $\lambda\in\mathbb{R}$. Putting $c_0\,|\mathrm{supp}\,\alpha=c|^{1/2}$ we obtain

$$|(\beta(H)\,v,(J-W_+)\,\alpha(H_0)\,u)|^2\leq c^2\int_{-\infty}^{\infty}\|f_+(\lambda)\|_{\mathfrak{X}}^2\,d\lambda=2\pi c^2\int_0^{\infty}ds\left\|\int_{-\infty}^{\infty}e^{-isx}\varphi(x)\,dx\right\|_{\mathfrak{X}}^2,$$

or, replacing $v$ by $e^{-it\phi(H)}v$, $u$ by $e^{-it\phi(H_0)}u$, $s$ by $-s$,

$$|(\beta(H)\,v,e^{it\phi(H)}(J-W_+)\,e^{-it\phi(H_0)}\alpha(H_0)\,u)|^2$$
$$\leq 2\pi c^2\int_{-\infty}^{0}ds\left\|\int_{-\infty}^{\infty}e^{isx}\,e^{-it\phi(x)}\varphi(x)\,dx\right\|_{\mathfrak{X}}^2. \tag{22}$$

Because of Lemma 8.19 and assumption (III) the limit of the right side of (22) tends to 0 as $t\to\infty$ if $\varphi(x)=\chi_{[c,d]}(x)$, $\Delta$ is a finite interval. Since

$$\int_{-\infty}^{0}ds\left\|\int_{-\infty}^{\infty}e^{isx}\,e^{-it\phi(x)}\varphi(x)\,dx\right\|_{\mathfrak{X}}^2=2\pi\,\|EF^*\,e^{-it\phi(A)}\varphi\|^2_{L^2(\mathbb{R},ds;\mathfrak{X})}$$
$$\leq 2\pi\,\|\varphi\|^2_{L^2(\mathbb{R},dx;\mathfrak{X})},$$

where $A$ is the multiplication operator with $x$, $F$ is the Fourier transformation and $E$ the projection of multiplication by $\chi_{(-\infty,0]}(s)$, the result extends to all $\varphi\in L^2(\mathbb{R},dx;\mathfrak{X})$. This concludes the proof. ∎

**Remark 13.** (I) The assumption that $P^{ac}\mathfrak{X}$ is generating with respect to $P^{ac}\mathfrak{X}$ and $H$ is not essential. If this property is not satisfied, then one can find an extension $\tilde{\mathfrak{X}}$ such that $\tilde{\mathfrak{X}}$ has this property and $f_\varepsilon^H(\lambda;\cdot,\cdot)$, $-f_{-\varepsilon}^H(\lambda;\cdot,\cdot)$ are also equicontinuous approximate spectral forms on $\tilde{\mathfrak{X}}\times\tilde{\mathfrak{X}}$. The other assumptions do not depend on this extension.

(II) Also the assumption "$\mathfrak{X}$ is spectral with respect to $E(\cdot)$" is not essential (see Remark 11). Moreover it is not essential that we start with the canonical approximate spectral forms $\pm f_{\pm\varepsilon}^H(\lambda;\cdot,\cdot)$ to be equicontinuous for $\{H,\mathfrak{X},\Lambda\}$. The heart of the proof is the fact that there is a family $g(\lambda;\cdot,\cdot)$ of continuous sequilinear forms on $\mathfrak{X}\times\mathfrak{X}$ ($\lambda\in\mathbb{R}\setminus\Lambda$, $|\Lambda|=0$) such that $g(\lambda;u,v)=(u,v)_{E,\lambda}$ a.e. (mod $|\cdot|$) (where the exceptional set may vary with $u$ and $v$). For example, it would be also possible to start with the approximate spectral forms $\pm f_{\pm\varepsilon}^{H_0}(\lambda;F_\varepsilon(\lambda)\cdot,F_\varepsilon(\lambda)\cdot)$ where $F_\varepsilon(\lambda)$ is the family of operators from Theorem 8.

Thus, the following proposition is true, its proof is left to the reader.

**Proposition 14.** *Let the assumptions of Theorem* 12 *except* (VI) *be satisfied. Instead of* (VI) *suppose that:*

(VI′) *$\mathfrak{X}_0$ is a linear normed space and it is spectral with respect to $E_0(\cdot)$ (exceptional set $\Lambda_0$); $f_\varepsilon^{H_0}(\lambda;\cdot,\cdot)$ and $-f_{-\varepsilon}^{H_0}(\lambda;\cdot,\cdot)$ are equicontinuous approximate spectral forms for $\{H_0,\mathfrak{X}_0,\Lambda_0\}$ satisfying the uniform estimate $|f_{\pm\varepsilon}^{H_0}(\lambda;u,v)|\leq c_\Delta\,\|u\|_{\mathfrak{X}_0}\,\|v\|_{\mathfrak{X}_0}$ for all $\lambda\in\Delta$ where $\Delta$ is an arbitrary compact subset of $\mathbb{R}\setminus\Lambda_0$. There are two families $F_{\pm\varepsilon}(\lambda)\colon\mathfrak{X}\to\mathfrak{X}_0$ of continuous operators for $\lambda\in\mathbb{R}$, $\varepsilon\in(0,\varepsilon_0)$, $\varepsilon_0>0$, $F_{\pm\varepsilon}\in\mathscr{L}(\mathscr{H},\mathscr{H}_0)$ such that $F_{\pm\varepsilon}(\lambda)$*

*are strongly continuous for all* $\varepsilon \in (0, \varepsilon_0)$ *with respect to* $\mathscr{H}_0$, $F_{\pm\varepsilon}(\lambda)$ *are Cauchy nets with respect to* $\|\cdot\|_{\mathfrak{X}_0}$ *as* $\varepsilon \to +0$ *for all* $\lambda \in \mathbb{R} \setminus \Lambda_0$ *and all* $u \in \mathfrak{X}$ *and satisfy the uniform estimate* $\|F_{\pm\varepsilon}(\lambda)\, u\|_{\mathfrak{X}_0} \leqq d_\Delta \|u\|_{\mathfrak{X}}$, $\lambda \in \Delta \subset \mathbb{R} \setminus \Lambda_0$ *where* $\Delta$ *is an arbitrary compact set. Furthermore,* $R(\lambda \pm i\varepsilon) = JR_0(\lambda \pm i\varepsilon)\, F_{\pm\varepsilon}(\lambda)\, u$, $u \in \mathfrak{X}$.

*Then the strong wave operators* $W_\pm\big(\phi(H), \phi(H_0); J\big)$ *exist, are partial isometries and complete.*

In the next paragraph we present examples where $\mathfrak{X}$ is a separable Hilbert space.

## 14.2. Factorization method

### 14.2.1. The perturbation $V$ and auxiliary manifolds

In Paragraph 1 we did not specify the linear manifolds $\mathfrak{X}_0$, $\mathfrak{X}$ in dependence on the operators $H$, $H_0$ and $J$. In this paragraph we deal with special constructions of the auxiliary manifolds $\mathfrak{X}_0$, $\mathfrak{X}$ with the help of the perturbation $V := HJ - JH_0$ under the assumption $J \operatorname{dom} H_0 \subseteqq \operatorname{dom} H$ such that $V$ is defined on $\operatorname{dom} H_0$.

If one wants to define $\mathfrak{X}_0$ and $\mathfrak{X}$ by

$$\mathfrak{X} := \operatorname{ima} V, \tag{23}$$

$$\mathfrak{X}_0 \subseteqq \operatorname{supp} V, \qquad \operatorname{clo}_{\|\cdot\|_{\mathscr{H}_0}} \mathfrak{X}_0 = \operatorname{supp} V, \tag{24}$$

then it is immediately obvious that the requirements "$P_0^{ac}\mathfrak{X}_0$ is generating with respect to $P_0^{ac}\mathscr{H}_0$ and $H_0$" and/or "$P^{ac}\mathfrak{X}$ is generating with respect to $P^{ac}\mathscr{H}$ and $H$" are, in general, not satisfied.

In the following it is pointed out that, without loss of generality, we can restrict $H$ and $H_0$ to suitable subspaces of $\mathscr{H}$ and $\mathscr{H}_0$ respectively, such that these requirements are satisfied.

We denote by $P_0$ the projection onto clo spa $\{E_0(\Delta)\, v;\ v \in \operatorname{supp} V, \Delta \subset \mathbb{R}\}$ and by $P$ the projection onto clo spa $\{E(\Delta)\, v;\ v \in \operatorname{ima} V, \Delta \subset \mathbb{R}\}$. Then obviously $P_0 \in \operatorname{com} H_0$, $P \in \operatorname{com} H$ and

$$PVP_0u = Vu, \qquad u \in \operatorname{dom} H_0. \tag{25}$$

**Lemma 15.** $W_+(H, H_0; J)$ *exists iff* $W_+(H, H_0; PJP_0)$ *exists, and in this case the equation*

$$W_+(H, H_0; J) = W_+(H, H_0; PJP_0) + (J - PJP_0)\, P_0^{ac} \tag{26}$$

*is valid; in particular* $JP_0{}^\perp P_0^{ac}$ *is intertwining for* $H, H_0$.

Proof. For $f \in \operatorname{dom} H_0$ we have

$$\frac{d}{dt}\left(e^{itH} J\, e^{-itH_0} f\right) = i\, e^{itH} V\, e^{-itH_0} f = i\, e^{itH} PVP_0\, e^{-itH_0} f,$$

hence

$$e^{it_2H} J\, e^{-it_2H_0} f - e^{it_1H} J\, e^{-it_1H_0} f = e^{it_2H} PJP_0\, e^{-it_2H_0} f - e^{it_1H} PJP_0\, e^{-it_1H_0} f.$$

Thus the first part of Lemma 15 is proved. (26) is obtained by setting $t_1 = 0$, $t_2 \to +\infty$, $f \in P_0^{ac}\mathscr{H}_0$. $JP_0{}^{\perp}P_0^{ac}$ is intertwining because of $W_+(H, H_0; J)\, P_0{}^{\perp} = JP_0{}^{\perp}P_0^{ac}$. ∎

Concerning semicompleteness and completeness we obtain the following statement.

**Lemma 16.** *Let* $J \underset{H_0}{\approx} W$ *where* $W^*W = P_0^{ac}$ *and let* $W_+(HP, HP_0; PJP_0)$ *exist. Then:* $JP_0{}^{\perp}P_0{}^{ac}$ *is partially isometric with initial projection* $P_0{}^{\perp}P_0^{ac}$ *and the* $H_0P_0$*-semicompleteness of* $W_+(HP, H_0P_0;\, PJP_0)$ *implies the* $H_0$*-semicompleteness of* $W_+(H, H_0; J)$. *Furthermore, if* $P^{ac}P^{\perp}JP_0P_0^{ac} = 0$ *and the final projection of* $JP_0{}^{\perp}P_0^{ac}$ *is* $P^{\perp}P^{ac}$, *then the completeness of* $W_+(HP, H_0P_0; PJP_0)$ *implies the completeness of* $W_+(H, H_0; J)$.

**Proof.** The first part follows from $PJP_0 \underset{H_0}{\approx} PWP_0$ and (26). If $W_+(HP, H_0P_0; PJP_0)$ is complete, then its final projection is $PP^{ac}$. Further, the final projection of $JP_0{}^{\perp}P_0^{ac}$ is $P^{\perp}P^{ac}$ and $P^{\perp}P^{ac}JP_0P_0^{ac} = 0$. Hence $W_+(H, H_0; J)$ is complete. ∎

Now, if we restrict ourselves to the consideration of $H_0$ in $P_0\mathscr{H}_0$ and $H$ in $P\mathscr{H}$, then, by construction, $P^{ac}\mathscr{X}$ is generating with respect to $P^{ac}P\mathscr{H}$ and $H$ and $P_0^{ac}\mathscr{X}_0$ is generating with respect to $P_0{}^{ac}P_0\mathscr{H}_0$ and $H_0$. In other words, without loss of generality, we may assume $P = 1_{\mathscr{H}}$, $P_0 = 1_{\mathscr{H}_0}$ if $\mathscr{X}_0$ and $\mathscr{X}$ are defined by (23) and (24).

Of course, there are many other possibilities for the definition of $\mathscr{X}_0$ and $\mathscr{X}$.

For example, if $\mathscr{X}$ is defined by (23), then another possibility to define $\mathscr{X}_0$ is given by

$$\mathscr{X}_0 := J^{-1}\mathscr{X}, \tag{27}$$

(where $J^{-1}\mathscr{X}$ denotes the inverse image of $\mathscr{X}$ with respect to $J$) provided ima $J \cap$ ima $V$ is large enough. In particular, this is the case if the inverse operator $J^{-1}$ exists and is bounded (then ima $J = \mathscr{H}$).

If $V$ is bounded, then the norm in $\mathscr{X}$ can be defined by

$$\|u\|_{\mathscr{X}} := \|f\|_{\mathscr{H}}, \qquad u \in \mathscr{X}, \tag{28}$$

where $u = Vf$ and $f \in$ supp $V$. In this case $\mathscr{X}$ is a separable Hilbert space. If $\mathscr{X}_0$ is defined by (27), then the norm on $\mathscr{X}_0$ may be defined by

$$\|u_0\|_{\mathscr{X}_0} := \|Ju_0\|_{\mathscr{X}}, \qquad u_0 \in \mathscr{X}_0. \tag{29}$$

Thus $\mathscr{X}_0$ is also a separable Hilbert space.

If $V$ is not bounded, one may assume that $V$ is factored in the form

$$V = B^*A \tag{30}$$

with $B^* \in \mathscr{L}(\mathscr{K}, \mathscr{H})$, $\mathscr{K}$ a third separable Hilbert space, where $A: \mathscr{H}_0 \to \mathscr{K}$ has the property dom $H_0 \subseteq$ dom $A$. In such a case one can put

$$\mathscr{X} := \text{ima } B^*, \tag{31}$$

and it is possible to define the norm in $\mathscr{X}$ by

$$\|u\|_{\mathscr{X}} := \|f\|_{\mathscr{K}}, \qquad u \in \mathscr{X}, \tag{32}$$

where $u = B^*f$, $f \in$ supp $B^*$. Also in this case $\mathscr{X}$ is a separable Hilbert space.

Of course, in all such cases a similar discussion like that at the beginning of this section on suitable restrictions of $H$ and $H_0$ such that the assumptions regarding the generating properties of $\mathfrak{X}_0$, $\mathfrak{X}$ are satisfied is required (in the case (23), (24) this discussion leads to Lemma 15 and Lemma 16).

### 14.2.2. Existence of wave operators under factorization assumptions

In this section we use Theorem 12 and the concepts of Section 1 to prove an existence result for strong wave operators. For simplicity we assume $\mathscr{H} = \mathscr{H}_0$, $J = 1$.

**Theorem 17.** *Let $A$, $B$ be defined by* (30). *Suppose that:*
(I) $\operatorname{w-lim}_{\varepsilon\to\pm 0} (\varepsilon/\pi)\, BR(\lambda - i\varepsilon)\, R(\lambda + i\varepsilon)\, B^*$ *exist for all* $\lambda \in \mathbb{R} \setminus \Lambda$ *and both signs,* $+$ *and* $-$, *where $\Lambda$ is a null set,* $|\Lambda| = 0$.
(II) $\operatorname{s-lim}_{\varepsilon\to\pm 0} AR_0(\lambda + i\varepsilon)\, B^* =: Q_\pm(\lambda)$ *exist for all* $\lambda \in \mathbb{R} \setminus \Lambda$ *and both signs* $+$ *and* $-$.
*Then the strong wave operators* $W_+(H, H_0)$, $W_-(H, H_0)$ *exist.*

Proof. 1. We apply Theorem 12. In the present case we have

$$H = H_0 + V \tag{33}$$

and we define $\mathfrak{X}$ by (31), $\mathfrak{X}_0$ by (27), that is we put $\mathfrak{X}_0 = \mathfrak{X}$, $P_0$, $P$ are defined as in Section 1. In this case it turns out that $P_0\mathscr{H}_0$ is the smallest subspace reducing $H_0$ and containing ima $V$ and that $P\mathscr{H}_0$ is the smallest subspace reducing $H$ and containing ima $V$. Hence, because of (33), it follows $P = P_0$ and we may restrict $H$ and $H_0$ to $P_0\mathscr{H}_0$. That means (IV) is satisfied. $\mathfrak{X}$ is a separable Hilbert space, that is (V) is satisfied (note that the assumption $\mathfrak{X}$ is spectral can be omitted according to Remark 13). (I) is satisfied because of $V = B^*A$ and $\operatorname{dom} H_0 \subseteqq \operatorname{dom} A$. (II) and (III) are obvious because of $J = 1$. Thus is remains to prove (VI), (VII) and (VIII) of Theorem 12.

2. Let $u, v \in \mathfrak{X}$, that is $u = B^*f$, $v = B^*g$, $f, g \in \mathscr{H}_0$. Then

$$\begin{aligned} f_\varepsilon^H(\lambda; u, v) &= (\varepsilon/\pi)\, \big(R(\lambda - i\varepsilon)\, R(\lambda + i\varepsilon)\, u, v\big) \\ &= (\varepsilon/\pi)\, \big(BR(\lambda - i\varepsilon)\, R(\lambda + i\varepsilon)\, B^*f, g\big). \end{aligned}$$

From (I) we obtain: the family of operators $(\varepsilon/\pi)\, BR(\lambda - i\varepsilon)\, R(\lambda + i\varepsilon)\, B^*$ is uniformly bounded in $\varepsilon > 0$ for all $\lambda \in \mathbb{R} \setminus \Lambda$ by the principle of uniform boundedness. This implies the equicontinuity of $f_\varepsilon^H(\lambda; \cdot, \cdot)$ on $\mathfrak{X} \times \mathfrak{X}$ in $\varepsilon > 0$. Further, by (I), we obtain the existence of the limits of $f_\varepsilon(\lambda; u, v)$ as $\varepsilon \to +0$ for all $\lambda \in \mathbb{R} \setminus \Lambda$. Similarly we get the same result for $-f_{-\varepsilon}^H(\lambda; u, v)$. The equality of these limits with $(u, v)_{E,\lambda}$ a.e. (mod $|\cdot|$) is obvious. This ensures that $f_\varepsilon^H(\lambda; \cdot, \cdot)$ and $-f_{-\varepsilon}^H(\lambda; \cdot, \cdot)$ are approximate spectral forms for $\{H, \mathfrak{X}, \Lambda\}$.

3. By definition, $G_\varepsilon(\lambda) = 1 - VR_0(\lambda + i\varepsilon) = 1 - B^*AR_0(\lambda + i\varepsilon)$. Then we have, with $\mathfrak{X} \ni u = B^*f$,

$$\begin{aligned} \|G_\varepsilon(\lambda)\, u - G_{\varepsilon'}(\lambda)\, u\|_{\mathfrak{X}} &= \|AR_0(\lambda + i\varepsilon)\, u - AR_0(\lambda + i\varepsilon')\, u\| \\ &= \|AR_0(\lambda + i\varepsilon)\, B^*f - AR_0(\lambda + i\varepsilon')\, B^*f\|. \end{aligned}$$

This implies, by (II), that the right side tends to 0 as $\varepsilon, \varepsilon' \to \pm 0$ for all $\lambda \in \mathbb{R} \setminus \Lambda$. Thus (VIII) of Theorem 12 is satisfied. The equicontinuity (VII) follows from the estimate

$$\|G_\varepsilon(\lambda)\, u\|_{\mathfrak{X}} \leq \|u\|_{\mathfrak{X}_0} + \|AR_0(\lambda + i\varepsilon)\, B^*\|\, \|u\|_{\mathfrak{X}_0}$$

where we have used $\|f\| = \|u\|_{\mathfrak{X}} = \|u\|_{\mathfrak{X}_0}$ and the fact that $\|AR_0(\lambda + i\varepsilon)\, B^*\|$ is uniformly bounded in $\varepsilon \gtrless 0$ because of the existence of the strong limits and the principle of uniform boundedness. This concludes the proof that the assumptions of Theorem 12 are satisfied. ∎

If the assumptions (I) and (II) of Theorem 17 are assumed only for $\varepsilon \to +0$, then we obtain, according to Theorem 6, the existence of the Abel wave operator $W_+(H, H_0)$ (correspondingly $W_-(H, H_0)$ for $\varepsilon \to -0$).

### 14.2.3. Completeness of wave operators under smallness and compactness assumptions

Often it is necessary to have criteria which make no explicit use of the perturbed operator $H$. Naturally, this requires stronger hypotheses on the perturbation V. First we consider a smallness condition and assume that the limits of $AR_0(\lambda + i\varepsilon)\, B^*$ exist and are sufficiently small. Again we put $\mathscr{H} = \mathscr{H}_0$, $J = 1$.

**Theorem 18.** *Let $A$, $B$ be defined by* (30). *Suppose that:*

(I) $\operatorname*{w-lim}_{\varepsilon \to \pm 0} (\varepsilon/\pi)\, BR_0(\lambda - i\varepsilon)\, R_0(\lambda + i\varepsilon)\, B^*$ *exist for all $\lambda \in \mathbb{R} \setminus \Lambda$, uniformly on compact subsets of $\mathbb{R} \setminus \Lambda$, where $\Lambda$ is a null set, and both signs, $+$ and $-$.*

(II) *There is $1 > \eta > 0$ and $\varepsilon_0 > 0$ such that $\|AR_0(\lambda \pm i\varepsilon)\, B^*\| \leq \eta$ for all $\lambda \in \mathbb{R} \setminus \Lambda$, $\varepsilon \in (0, \varepsilon_0)$ and both signs.*

(III) $Q_\pm(\lambda) := \operatorname*{s-lim}_{\varepsilon \to \pm 0} AR_0(\lambda + i\varepsilon)\, B^*$ *exists for all $\lambda \in \mathbb{R} \setminus \Lambda$ and both signs.*

*Then the strong wave operators $W_+(H, H_0)$ and $W_-(H, H_0)$ exist and are complete.*

**Proof.** 1. For the proof we use Prop. 14. $\mathfrak{X}$ is defined by (31), $\mathfrak{X}_0$ by (27), $V = B^*A$ according to (30). Thus assumption (I) of Prop. 14 is satisfied. (II) and (III) are obvious. (IV) and (V) are satisfied (cf. the corresponding arguments in the proof of Theorem 17). Therefore it remains to prove the assumptions (VI′), (VII), and (VIII).

The proof that $G_\varepsilon(\lambda)$ map $\mathfrak{X}_0$ into $\mathfrak{X}$ continuously for $\varepsilon > 0$ and $G_\varepsilon(\lambda)\, u$, $G_{-\varepsilon}(\lambda)\, u$ are equicontinuous and Cauchy nets in $\mathfrak{X}$ as $\varepsilon \to +0$ for all $\lambda \in \mathbb{R} \setminus \Lambda$ is the same as in Theorem 17 (use (III)). Hence (VII) and (VIII) are satisfied.

2. Now we prove (VI′). Let $u, v \in \mathfrak{X}_0$ and let $u = B^*f$, $v = B^*g$ with $f, g \in \mathscr{H}_0$. Then

$$\begin{aligned} f_\varepsilon^{H_0}(\lambda; u, v) &= (\varepsilon/\pi)\, \big(R_0(\lambda - i\varepsilon)\, R_0(\lambda + i\varepsilon)\, u, v\big) \\ &= (\varepsilon/\pi)\, \big(BR_0(\lambda - i\varepsilon)\, R_0(\lambda + i\varepsilon)\, B^*f, g\big). \end{aligned}$$

This relation and assumption (I) imply that the limit of $f_\varepsilon^{H_0}(\lambda; u, v)$ exists as $\varepsilon \to +0$ for all $\lambda \in \mathbb{R} \setminus \Lambda$. $f_\varepsilon^{H_0}(\lambda; \cdot, \cdot)$ is equicontinuous on $\mathfrak{X}_0 \times \mathfrak{X}_0$ for $\lambda \in \mathbb{R} \setminus \Lambda$, and $|f_\varepsilon^{H_0}(\lambda; u, v)| \leq c_\Delta \|u\|_{\mathfrak{X}_0} \|b\|_{\mathfrak{X}_0}$ on compact subsets $\Delta$ of $\mathbb{R} \setminus \Lambda$ (use similar argument as in the proof of Theorem 17). Similarly we obtain the corresponding assertion for $-f^{H_0}_{-\varepsilon}(\lambda; \cdot, \cdot)$.

It remains to show the existence of the operator functions $F_\varepsilon(\lambda)$, $F_{-\varepsilon}(\lambda)$ with the properties of (VI′). We have $G_{\pm\varepsilon}(\lambda) = 1 - B^*AR_0(\lambda \pm i\varepsilon)$. By

$$\|B^*AR_0(\lambda \pm i\varepsilon)\, u\|_{\mathscr{X}} = \|AR_0(\lambda \pm i\varepsilon)\, B^*f\| \leqq \|AR_0(\lambda \pm i\varepsilon)\, B^*\|\, \|f\|$$
$$= \|AR_0(\lambda \pm i\varepsilon)\, B^*\|\, \|u\|_{\mathscr{X}},$$

and (II) we get $\|B^*AR_0(\lambda \pm i\varepsilon)\|_{\mathscr{X}} \leqq \eta$ for all $\lambda \in \mathbb{R} \setminus \Lambda$. That is, the operators $1 - B^*AR_0(\lambda \pm i\varepsilon)$ are bounded invertible for all $\varepsilon \in (-\varepsilon_0, 0) \cup (0, \varepsilon_0)$ and $\lambda \in \mathbb{R} \setminus \Lambda$, and the limits $1 - Q_\pm(\lambda)$ also define bounded invertible operators from $\mathscr{X}_0$ into $\mathscr{X}$ for all $\lambda \in \mathbb{R} \setminus \Lambda$. From $R(\lambda + i\varepsilon)\, G_\varepsilon(\lambda) = R_0(\lambda + i\varepsilon)$ we find

$$R(\lambda \pm i\varepsilon) = R_0(\lambda \pm i\varepsilon)\, G_{\pm\varepsilon}(\lambda)^{-1}$$

where the operators $G_{\pm\varepsilon}(\lambda)^{-1}$ are bounded from $\mathscr{X}$ into $\mathscr{X}_0$ (note that in our case $\mathscr{X} = \mathscr{X}_0$). The properties of $G_{\pm\varepsilon}(\lambda)$ imply that the operators $G_{\pm\varepsilon}(\lambda)^{-1}$ are uniformly bounded in $\varepsilon \in (-\varepsilon_0, 0) \cup (0, \varepsilon_0)$ with respect to the $\mathscr{X}(\mathscr{X}_0)$-norms and that they are Cauchy nets as $\varepsilon \to +0$ for all $\lambda \in \mathbb{R} \setminus \Lambda$. This proves that the operators $F_{\pm\varepsilon}(\lambda) := G_{\pm\varepsilon}(\lambda)^{-1}$ have the desired properties, i.e. (VI′) is satisfied. This concludes the proof. ∎

Second we consider the case that the limits of $AR_0(\lambda \pm i\varepsilon)\, B^*$ exist even in norm sense and that these operators are compact. This allows to avoid the smallness assumption in Theorem 18. For preparation of the corresponding theorem we need a technical lemma.

**Lemma 19.** *Let $\mathscr{Y}$ be a Banach space and let $A(z) \in \mathscr{L}_\infty(\mathscr{Y})$, $\operatorname{Im} z \geqq 0$, satisfy the following conditions:*

(I) *$A(z)$ is holomorphic in $z$, $\operatorname{Im} z > 0$, and norm continuous in $z$, $\operatorname{Im} z \geqq 0$.*

(II) *$1 + A(z)$ is bounded invertible for every $z$, $\operatorname{Im} z > 0$. Then the set $\Lambda = \{\lambda \in \mathbb{R};\ 1 + A(\lambda)$ is not invertible$\}$ is a closed null set.*

**Proof.** If $1 + A(\lambda)$ is invertible; $(1 + A(\lambda))^{-1}$ is bounded; since the inversion is continuous, $\Lambda$ is closed. Thus it remains to prove that $|\Lambda| = 0$. Now let $\lambda \in \mathbb{R}$ and $\varrho_\lambda > 0$ such that $(\operatorname{spec} A(\lambda)) \cap \{w; |w + 1| < \varrho_\lambda\}$ contains at most the point $-1$ itself. Then the Riesz' projection

$$P(z) = (1/2i\pi) \int\limits_{|w+1| = \varrho_\lambda} (w - A(z))^{-1}\, dw$$

is well-defined for some neighborhood $\mathscr{U}_\lambda = \{z; |z - \lambda| < \eta_\lambda, \operatorname{Im} z \geqq 0\}$ of $\lambda$. Then for $z \in \mathscr{U}_\lambda$ the operator $1 + A(z)$ is invertible iff $(1 + A(z)) \restriction P(z)\, \mathscr{Y}$ is invertible. But this is the case iff $\det (1 + A(z)) \restriction P(z)\, \mathscr{Y} \neq 0$. Now we have $\det (1 + A(z)) \restriction P(z)\, \mathscr{Y} \neq 0$ for $\operatorname{Im} z > 0$ and it has continuous limits on the border. Therefore (see Muschelischwili [1, I.A. § 10]) this function can vanish only on a closed null set on the border, that means $\det (1 + A(\lambda)) \restriction P(\lambda)\, \mathscr{Y}$ can only vanish on a null set $\Delta_\lambda$. Now let $\lambda \in [a, b]$ from a compact interval, then there is a finite set $\mathscr{U}_{\lambda_1}, \mathscr{U}_{\lambda_2}, \ldots, \mathscr{U}_{\lambda_n}$ which covers $[a, b]$, hence $\Lambda \cap [a, b] = \bigcup_{\varrho=1}^{n} \Delta_{\lambda_\varrho}$ and a fortiori $\Lambda$ itself are null sets. ∎

**Theorem 20.** *Suppose the assumptions of Theorem 18 except (II) and (III) are satisfied with $\Lambda = \emptyset$. Further suppose:*

(II') $AR_0(\lambda + i\varepsilon)\, B^*$ *is compact for* $\varepsilon \neq 0$ *and all* $\lambda \in \mathbb{R}$,

(III') *the limits* $\underset{\varepsilon\to\pm0}{\text{n-lim}}\, AR_0(\lambda + i\varepsilon)\, B^*$ *exist uniformly on compact subsets of* $\mathbb{R}$ *for both signs,* $+$ *and* $-$.

*Then the strong wave operators* $W_+(H, H_0)$ *and* $W_-(H, H_0)$ *exist and are complete.*

**Proof.** The only point which is different from the proof of Theorem 18 is the existence of the limits of $F_{\pm\varepsilon}(\lambda) := G_{\pm\varepsilon}(\lambda)^{-1}$. From (II') we obtain that $B^*AR_0(z)$ is compact on $\mathscr{X}_0$ with the same properties as $AR_0(z)\, B$. This follows from the fact that the operator $B^*$ acts as an isometric map from supp $B^*$ onto $\mathscr{X}_0$. Now we can apply Lemma 19 to $G_\varepsilon(\lambda) = 1 - B^*AR_0(\lambda + i\varepsilon) = 1 - VR_0(\lambda + i\varepsilon)$. It is easy to see that $G_\varepsilon(\lambda)$ is bounded invertible for all $\varepsilon \neq 0$ from the relation $R(\lambda + i\varepsilon)\, G_\varepsilon(\lambda) = R_0(\lambda + i\varepsilon)$. Thus from Lemma 19 we get that $G_\pm(\lambda)^{-1}$ exist for all $\lambda \in \mathbb{R} \setminus \Lambda$ where $\Lambda$ is a closed null set and, by the continuity of the inverse operation this proves that $G_\varepsilon(\lambda)^{-1}$ tends continuously with respect to the $\mathscr{X}_0$-norm to $G_\pm(\lambda)^{-1}$ for all $\lambda \in \mathbb{R} \setminus \Lambda$. Because of $R(\lambda + i\varepsilon) = R_0(\lambda + i\varepsilon)\, G_\varepsilon(\lambda)^{-1}$ we therefore get that $F_\varepsilon(\lambda) := G_\varepsilon(\lambda)^{-1}$ has the desired properties (see assumption (VI') of Prop. 14). ■

## 14.3. Applications

### 14.3.1. Multiplication operator perturbed by integral operators (small gentle perturbations)

Let $H_0$ be the multiplication by $x$ on $\mathscr{H}_0 = L^2(\Delta, dx; \mathscr{K}_0)$ where $\Delta$ is an interval. Let $J$ be bijective and assume that $J - W \in \mathscr{L}_{H_0, P_0^{ac}}(\mathscr{H}_0, \mathscr{H})_a \cap \mathscr{L}_{-H_0, P_0^{ac}}(\mathscr{H}_0, \mathscr{H})_a$ (cf. Section 6.5.2) with $W^*W = P_0^{ac}$. Further, suppose that $\tilde{V} := J^{-1}V = J^{-1}HJ - H_0$ is an integral operator in $\mathscr{H}_0$ with the kernel $v(x, x')$ satisfying a certain smoothness and smallness condition. Then we want to show that the wave operators $W_\pm(H, H_0: J)$ exist, are partially isometric and complete. For the proof we shall use Prop. 10.

1. First we say precisely what we mean with smoothness and smallness of $V$. Recall the definition of the Hölder continuous kernels with the so-called gentleness norm $\|\cdot\|_{\alpha,\beta}$ (see 7.4.4). Here we only consider the special case that $\alpha = \beta$. Then the $\|\cdot\|_\gamma$-norm of a function $v(\cdot, \cdot)\colon \mathbb{R} \times \mathbb{R} \to \mathscr{L}(\mathscr{K}_0)$ is defined dy

$$\|v\|_\gamma := \sup_{\infty \geq h_1, h_2 > 0}\ \sup_{\infty > s_1, s_2 > -\infty} \|(A^1_{h_1} A^2_{h_2} v)\,(s_1, s_2)\| \tag{34}$$

with

$$(A^1_{h_1} v)\,(s_1, s_2) = h_1^{-\gamma}\big(v(s_1 + h_1, s_2) - v(s_1, s_2)\big),$$
$$(A^2_{h_2} v)\,(s_1, s_2) = h_2^{-\gamma}\big(v(s_1, s_2 + h_2) - v(s_1, s_2)\big)$$

for $h_1, h_2 \neq \infty$ and

$$(A^1_\infty v)\,(s_1, s_2) = (1 + |s_1|^\gamma)\, v(s_1, s_2),$$
$$(A^2_\infty v)\,(s_1, s_2) = (1 + |s_2|^\gamma)\, v(s_1, s_2).$$

Now the smoothness of $v(x, x')$ means that $\|v\|_\gamma < \infty$ (if $\Delta \subset \mathbb{R}$, then we set $v(x, x') = 0$ on $\mathbb{R} \times \mathbb{R} \setminus \Delta \times \Delta$), and the smallness means that $\|v\|_\gamma$ is sufficiently small.

According to this norm of $\mathscr{L}(\mathscr{K}_0)$-valued functions we introduce the norm $|\cdot|_\gamma$ for functions $u(\cdot)\colon \mathbb{R} \to \mathscr{K}_0$ by

$$|u|_\gamma := \sup_{x\in\mathbb{R}} (1 + |x|^\gamma)\, \|u(x)\| + \sup_{x\in\mathbb{R}, h>0} h^{-\gamma}\, \|u(x + h) - u(x)\|. \tag{35}$$

It is easy to see that $|\cdot|_\gamma$ is a norm. If $\gamma > 1/2$, then $|u|_\gamma < \infty$ implies $u \in L^2(\mathbb{R}, dx; \mathscr{K}_0)$. In the following we assume that $\gamma > 1/2$ and $|||v(\cdot,\cdot)|||_\gamma < \infty$, and prove only the assertions for $W_+(H, H_+; J)$. The proof for $W_-(H, H_0; J)$ is the same.

2. Next we introduce the auxiliary spaces $\mathscr{X}_0$, $\mathscr{X}$ and consider the expression $f_\varepsilon^{H_0}(\lambda; \cdot, \cdot)$. Let $\mathscr{X}_0$ denote the set of functions $u\colon \mathbb{R} \to \mathscr{K}_0$ which are continuously differentiable and have compact support in $\Delta$. Then $\mathscr{X}_0 \subset L^2(\Delta, dx; \mathscr{K}_0)$ and we get that $u \in \mathscr{X}_0$ implies $|u|_\gamma < \infty$. The completion clo $\mathscr{X}_0$ of $\mathscr{X}_0$ with respect to the $|\cdot|_\gamma$-norm belongs also to $L^2(\Delta, dx; \mathscr{K}_0)$ (see the remark after (35)). $\mathscr{X}$ is defined by $\mathscr{X} := J\mathscr{X}_0$ and the $\mathscr{X}$-norm is given by $\|u\|_{\mathscr{X}} := \|J^{-1}u\|_{\mathscr{X}_0} = |J^{-1}u|_\gamma$. Since $\mathscr{X}_0$ is dense in $\mathscr{H}_0$ and $\mathscr{X}$ is dense in $\mathscr{H}$ ($J$ is bijective), the assumptions about the generating property of these sets in Prop. 10 are satisfied. Now we show that $f_\varepsilon^{H_0}(\lambda; \cdot, \cdot)$ is an equicontinuous approximate spectral form for $\{H_0, \mathscr{X}_0, \mathscr{B}\}$ and satisfies the estimate $|f_\varepsilon^{H_0}(\lambda; u, v)| \leq c_{\tilde{\Delta}}\, |u|_\gamma\, |v|_\gamma$ on compact subsets $\tilde{\Delta}$ of $\Delta$ for $u, v \in \mathscr{X}_0$. We have

$$
\begin{aligned}
f_\varepsilon^{H_0}(\lambda; u, v) &= (1/2i\pi)\left([R_0(\lambda + i\varepsilon) - R_0(\lambda - i\varepsilon)]\, u, v\right) \\
&= (1/2i\pi) \int_{-\infty}^{\infty} [(\lambda - i\varepsilon - x)^{-1} - (\lambda + i\varepsilon - x)^{-1}]\, (u(x), v(x))\, dx \\
&= \int_{-\infty}^{\infty} \delta_\varepsilon(\lambda - x)\, (u(x), v(x))\, dx.
\end{aligned} \tag{36}
$$

Hence

$$
|f_\varepsilon^{H_0}(\lambda; u, v)| \leq |(u(\cdot), v(\cdot))|_{L^\infty} \leq |(u(\cdot), v(\cdot))|_\gamma \leq |u(\cdot)|_\gamma\, |v(\cdot)|_\gamma
$$

for all $\lambda \in \mathbb{R}$, $\varepsilon > 0$, where the last inequality is easily obtained from (35) for $|u|_\gamma < \infty$, $|v|_\gamma < \infty$.

Thus $f_\varepsilon^{H_0}(\lambda; \cdot, \cdot)$ is an equicontinuous approximate spectral form with respect to $\varepsilon > 0$ for $\{H_0, \mathscr{X}_0, \mathscr{B}\}$.

Moreover, $f_+^{H_0}(\lambda; u, v)$ is Hölder continuous in $\lambda$.

3. Now we prove that the operator function $G_\varepsilon(\lambda)$ is equicontinuous on $\mathscr{X}_0$ in $\varepsilon$ and has strong limits. By $G_\varepsilon(\lambda) = J - J\tilde{V}R_0(\lambda + i\varepsilon)$, we have to study the properties of the clo $\mathscr{X}_0$-valued function

$$
v_{\lambda,\varepsilon}(x) := \int_{-\infty}^{\infty} (\lambda - x' + i\varepsilon)^{-1}\, v(x, x')\, u(x')\, dx', \qquad u \in \mathscr{X}_0,
$$

that is, we must show that $|v_{\lambda,\varepsilon}(\cdot)|_\gamma < \infty$ and $v_{\lambda,\varepsilon}$ tends to limits in clo $\mathscr{X}_0$. Again we split up the integral in two terms, one over the region $|\lambda - x'| > 1$ and the other over the region $|\lambda - x'| \leq 1$. Then

$$
\begin{aligned}
&\left\| \int_{|\lambda - x'| > 1} (\lambda - x' + i\varepsilon)^{-1}\, v(x, x')\, u(x')\, dx' \right\|_{\mathscr{K}_0} \\
&\leq |||v|||_\gamma\, (1 + |x|^\gamma)^{-1}\, |u|_\gamma \int_{|\lambda - x'| > 1} |\lambda - x'|^{-1}\, (1 + |x'|^{2\gamma})^{-1}\, dx' \leq c\, |||v|||_\gamma\, |u|_\gamma\, (1 + |x|^\gamma)^{-1}
\end{aligned}
$$

with $c$ independent of $\lambda$, $\varepsilon$. The other term can be estimated by

$$
\begin{aligned}
&\left\| \int_{|\lambda - x'| \leq 1} \frac{v(x, x')\, u(x') - v(x, \lambda)\, u(\lambda)}{\lambda - x' + i\varepsilon}\, dx' + \int_{|\lambda - x'| \leq 1} \frac{v(x, \lambda)\, u(\lambda)}{\lambda - x' + i\varepsilon}\, dx' \right\|_{\mathscr{K}_0} \\
&\leq |u|_\gamma\, |||v|||_\gamma\, (1 + |x|^\gamma)^{-1} \left\{ \int_{|\lambda - x'| \leq 1} \frac{2\, |x' - \lambda|^\gamma}{|\lambda - x'|}\, dx' + \int_{|\lambda - x'| \leq 1} \frac{1}{\lambda - x' + i\varepsilon}\, dx' \right\} \\
&\leq c'\, |u|_\gamma\, |||v|||_\gamma\, (1 + |x|^\gamma)^{-1}.
\end{aligned}
$$

Thus the vector $v_{\lambda,\varepsilon}(x)$ satisfies the uniform estimate in $\lambda$ and $\varepsilon$

$$
\sup_{x \in \mathbb{R}} (1 + |x|^\gamma)\, \|v_{\lambda,\varepsilon}(x)\|_{\mathscr{K}_0} \leq c_1 < \infty. \tag{37}
$$

In a similar way we can estimate

$$\|v_{\lambda,\varepsilon}(x) - v_{\lambda,\varepsilon}(x+h)\|_{\mathscr{K}_0} \leqq c_2 |h|^\gamma \|v\|_\gamma |u|_\gamma \tag{38}$$

with $c_2$ independent of $\lambda \in \mathbb{R}$ and $\varepsilon \geqq 0$. (37) and (38) show that $\|G_\varepsilon(\lambda) u\|_{\mathscr{X}} \leqq a |u|_\gamma$, that is, $G_\varepsilon(\lambda)$ is equicontinuous in $\varepsilon$ and moreover, $\|(G_\varepsilon(\lambda) - J) u\|_{\mathscr{X}} \leqq a_1 \|v\|_\gamma |u|_\gamma$ where the constants $a_1$ and $a$ are independent of $\lambda$ and $u$. By using similar estimates one can verify that $|v_{\lambda,\varepsilon}(\cdot) - v_{\lambda,\varepsilon'}(\cdot)|_\gamma \to 0$ as $\varepsilon, \varepsilon' \to +0$ proving that $G_\varepsilon(\lambda) u$ has a strong limit as $\varepsilon \to +0$.

4. It remains to show the existence of an operator-valued function $F_\varepsilon(\lambda)$ with the properties described in Theorem 8. That is, we must verify that $R(\lambda + i\varepsilon) = JR_0(\lambda + i\varepsilon) F_\varepsilon(\lambda)$ for all $\varepsilon \in (0, \varepsilon_0)$ with some $\varepsilon_0 > 0$, that $F_\varepsilon(\lambda)$ maps clo $\mathscr{X}$ into clo $\mathscr{X}_0$ and is strongly continuous for all $\varepsilon \in (0, \varepsilon_0)$ and $\lambda \in \mathbb{R}$, and that $F_\varepsilon(\lambda) u$ has limits in $\mathscr{X}_0$-norm as $\varepsilon \to +0$ for all $u \in$ clo $\mathscr{X}$.

From $\|G_\varepsilon(\lambda) - Ju\|_{\mathscr{X}} \leqq a_1 \|v\|_\gamma |u|_\gamma$ and the fact that $J$ is bijective we obtain that for all sufficiently small $\|v\|_\gamma$, $G_\varepsilon(\lambda)$ is bounded invertible for all $\lambda \in \mathbb{R}$ and $\varepsilon \in (0, \infty)$. Because of $R(\lambda + i\varepsilon) G_\varepsilon(\lambda) = JR_0(\lambda + i\varepsilon)$ we find that $F_\varepsilon(\lambda) := G_\varepsilon(\lambda)^{-1}$ satisfies the relation $R(\lambda + i\varepsilon) = JR_0(\lambda + i\varepsilon) F_\varepsilon(\lambda)$. Since $G_\varepsilon(\lambda)$ is equicontinuous in $\lambda \in \mathbb{R}$ and $\varepsilon > 0$ on clo $\mathscr{X}_0$, we see that $F_\varepsilon(\lambda)$ maps clo $\mathscr{X}$ into clo $\mathscr{X}_0$. The existence of the limits of $G_\varepsilon(\lambda) u$ in $\mathscr{X}$-norm and the uniform boundedness of $G_\varepsilon(\lambda)^{-1}$ imply that also the limits of $F_\varepsilon(\lambda) u = G_\varepsilon(\lambda)^{-1} u$ exist in $\mathscr{X}_0$-norm for all $\lambda \in \mathbb{R}$ and $u \in$ clo $\mathscr{X}$ (for the proof of all these properties of $G_\varepsilon(\lambda)^{-1}$ one can use that $G_\varepsilon(\lambda)^{-1}$ can be developed in a $\mathscr{L}(\mathscr{X}_0)$-norm convergent power series of $J^{-1}G_\varepsilon(\lambda) - 1$.

### 14.3.2. One-dimensional perturbation

Let $\mathscr{H}_0 = \mathscr{H}$ and assume that $V = H - H_0$ is a one-dimensional selfadjoint perturbation, i.e. $V$ has the form $V = c\varphi(\varphi, \cdot)$ with $c \in \mathbb{R}$, $\varphi \in \mathscr{H}_0$ and $\|\varphi\| = 1$. We shall use Theorem 17 to prove that the strong wave operators exist and are complete. Set $A = \varphi(\varphi, \cdot)$ and $B = B^* = c\varphi(\varphi, \cdot)$. We must show that

$$\operatorname*{w-lim}_{\varepsilon \to \pm 0} (\varepsilon/\pi) BR(\lambda - i\varepsilon) R(\lambda + i\varepsilon) B \tag{39}$$

exist for all $\lambda \in \mathbb{R} \setminus \Lambda$ with $|\Lambda| = 0$ and that

$$\operatorname*{s-lim}_{\varepsilon \to \pm 0} AR_0(\lambda + i\varepsilon) B \tag{40}$$

exist for all $\lambda \in \mathbb{R} \setminus \Lambda$. In this case (39) is a consequence of

$$\operatorname*{s-lim}_{\varepsilon \to \pm 0} AR(\lambda + i\varepsilon) B \quad \text{exist for all } \lambda \in \mathbb{R} \setminus \Lambda \tag{41}$$

because of $B = cA$. Furthermore, (40) and (41) follow from Prop. 3.15. This concludes the proof of the assumptions of Theorem 17. Therefore the strong wave operators $W_\pm(H, H_0)$ exist. Since the assumptions are symmetric with respect to $H$ and $H_0$, we get also the existence of $W_\pm(H_0, H)$. By Prop. 9.40 this implies the completeness of the wave operators.

# 15. Time-Falloff Methods

## 15.1. Existence of wave operators

### 15.1.1. Some abstract criteria

Our goal is to prove some statements about the existence of wave operators $W_+(H, H_0; J)$ under falloff conditions on $(v, e^{itH}(HJ - JH_0) e^{-itH_0}u)$ in $t$. First we derive a technical lemma.

**Lemma 1.** *Let $f(t)$ be a nonnegative function on $[a, \infty)$, $a \in \mathbb{R}$. Suppose $f(t)$ is uniformly continuous in $t \in [a, \infty)$ and $\int_a^\infty f(t)\, dt < \infty$. Then $f(t)$ goes to zero as $t \to \infty$.*

Proof. Assume $f(t) \nrightarrow 0$ as $t \to \infty$. Then there exist a number $c > 0$ and a sequence $t_n$ such that $t_n \to \infty$ as $n \to \infty$, $t_n \in (a, \infty)$, $t_n > t_{n-1} + 1$ and $f(t_n) > c > 0$ for all $n \in \mathbb{N}$. $f(t)$ is uniformly continuous. Thus $|f(t) - f(t')| < \varepsilon$, $\varepsilon > 0$, for all $|t - t'| < \eta$ with $t, t' \in (a, \infty)$. This implies that $f(t) \geqq c/2$ for all $t \in \Delta := \bigcup_{n=1}^{\infty} [t_n - \eta, t_n + \eta]$ with a suitable positive number $\eta$. Further, we may choose $\eta$ small enough so that $t_1 - \eta > a$ and $t_n + \eta < t_{n+1} - \eta$. Then

$$\int_a^\infty f(t)\, dt \geqq \int_\Delta f(t)\, dt \geqq (c/2) \int_\Delta dt = (c/2)\, |\Delta| = \infty$$

shows that the assumption $f(t) \nrightarrow 0$ as $t \to \infty$ leads to a contradiction. Thus the lemma is proved. ■

**Theorem 2.** *Suppose there exist two sets $\mathscr{D}_0 \subseteqq \mathscr{H}_0$ and $\mathscr{D} \subseteqq \mathscr{H}$ such that:*
*(I) $\mathscr{D}_0 \subseteqq \operatorname{dom} H_0$ and $\mathscr{D}_0$ is generating with respect to $P_0^{ac}\mathscr{H}_0$ and $H_0$,*
*(II) $\mathscr{D} \subseteqq \operatorname{dom} H$ and $\mathscr{D}$ is dense in $\mathscr{H}$,*
*(III) Defining $\left(v, e^{isH}(HJ - JH_0) e^{-isH_0}u\right) := (H e^{-isH}v, J e^{-isH_0}u) - (e^{-isH}v, JH_0 \times e^{-isH_0}u)$ the integrals $\int_0^\infty \left|\left(v, e^{isH}(HJ - JH_0) e^{-isH_0}u\right)\right| ds$, $u \in \mathscr{D}_0$, $v \in \mathscr{D}$, exist and are uniformly convergent with respect to $v \in \mathscr{D}$, $\|v\| = 1$, that is*

$$\sup_{v \in \mathscr{D}, \|v\|=1} \int_s^t \left|\left(v^{i\tau H}, e^{i\tau H}(HJ - JH_0) e^{-i\tau H_0}u\right)\right| d\tau \to 0$$

*for $t > s$, $s \to \infty$.*
*Then the strong wave operator $W_+(H, H_0; J)$ exists.*

Proof. For $u \in \mathscr{D}_0 \subseteqq \operatorname{dom} H_0$ and $v \in \mathscr{D} \subseteqq \operatorname{dom} H$, $e^{itH}v$ and $e^{itH_0}u$ are strongly differentiable in $t$ and the derivatives are uniformly bounded. This gives the existence of

$$\frac{d}{dt}(v, e^{itH}J\,e^{-itH_0}u) = \frac{d}{dt}(e^{-itH}v, J\,e^{-itH_0}u) = (v, e^{itH}\,i(HJ - JH_0)\,e^{-itH_0}u)$$

and its uniform boundedness. Thus

$$(v, e^{itH}J\,e^{-itH_0}u - e^{isH}J\,e^{-isH_0}u) = i\int_s^t (v, e^{it'H}(HJ - JH_0)\,e^{-it'H_0}u)\,dt'.$$

Then

$$\|e^{itH}J\,e^{-itH_0}u - e^{isH}J\,e^{-isH_0}u\| = \sup_{v\in\mathscr{D},\|v\|=1} |(v, e^{itH}J\,e^{-itH_0}u - e^{isH}J\,e^{-isH_0}u)|$$

$$\leqq \sup_{v\in\mathscr{D},\|v\|=1} \int_s^t |(v, e^{it'H}(HJ - JH_0)\,e^{-it'H_0}u)|\,dt'$$

because $\mathscr{D}$ is dense in $\mathscr{H}$. This inequality and (III) imply that $\operatorname{s-lim}_{t\to\infty} e^{itH}J\,e^{-itH_0}u$ exists for all $u \in \mathscr{D}_0$. By Lemma 9.13 this leads to the existence of $W_+(H, H_0; J)$ in the strong sense. ∎

This assertion has the disadvantage that it includes the perturbed unitary group $e^{itH}$ in its assumptions. To avoid this one can take the norm of the vector function $e^{itH}(HJ - JH_0)\,e^{-itH_0}u$ under the integral in (III) instead of the scalar product. This leads immediately to

**Corollary 3.** *Suppose that $HJ - JH_0$ is well-defined on a set $\mathscr{D}_0 \subseteqq \operatorname{dom} H_0$ such that*
(I) $e^{-itH_0}\mathscr{D}_0 \subseteqq \mathscr{D}_0$ *for all $t \in \mathbb{R}_+$ and $\mathscr{D}_0$ is generating with respect to $P_0^{ac}\mathscr{H}_0$ and $H_0$,*
(II) $\int_0^\infty \|(HJ - JH_0)\,e^{-itH_0}u\|\,dt < \infty$, $u \in \mathscr{D}_0$.
*Then the strong wave operator $W_+(H, H_0; J)$ exists.*

The following theorem is in particular useful if the perturbation $HJ - JH_0$ is unbounded or not well-defined.

**Theorem 4.** *Suppose there exist a set $\mathscr{D} \subseteqq \mathscr{H}_0$ and $z_0 \in \operatorname{res} H_0 \cap \operatorname{res} H$ such that*
(I) *$\mathscr{D}$ is generating with respect to $P_0^{ac}\mathscr{H}_0$ and $H_0$,*
(II) $\int_0^\infty \|(R(z_0)\,J - JR_0(z_0))\,e^{-itH_0}u\|\,dt < \infty$, $u \in \mathscr{D}$.
*Then the strong wave operator $W_+(H, H_0; J)$ exists.*

**Proof.** Set $J' := R(z_0)\,J$ and $\mathscr{D}_0 := \operatorname{spa}\{R_0(z_0)\,e^{-itH_0}u, u \in \mathscr{D}, t \in \mathbb{R}_+\}$. Further we have $R(z_0)\,J - JR_0(z_0) = R(z_0)\,(HJ - JH_0)\,R_0(z_0) = (HJ' - J'H_0)\,R_0(z_0)$. From Corollary 3 it then follows that $\operatorname{s-lim}_{t\to\infty} e^{itH}J'\,e^{-itH_0}P_0^{ac} = \operatorname{s-lim}_{t\to\infty} e^{itH}R(z_0)\,J\,e^{-itH_0}P_0^{ac}$ exists. Now we have the identity

$$R(z_0)\,e^{itH}J\,e^{-itH_0} - e^{itH}J\,e^{-itH_0}R_0(z_0) = e^{itH}(R(z_0)\,J - JR_0(z_0))\,e^{-itH_0}.$$

From (II) and Lemma 1 it follows that the right side of this identity tends strongly to zero as $t \to \infty$ for all $u \in \mathscr{D}$. This and the existence of $\operatorname{s-lim}_{t\to\infty} R(z_0)\,e^{itH}J\,e^{-itH_0}P_0^{ac}$

prove that

$$\text{s-}\lim_{t\to\infty} e^{itH} J e^{-itH_0} R_0(z_0) u$$

exists for all $u \in \mathcal{D}$, thus the strong wave operator $W_+(H, H_0; J)$ exists since $\tilde{\mathcal{D}}_0 = R_0(z_0) \times \mathcal{D}$ is generating with respect to $P_0^{ac}\mathcal{H}_0$ and $H_0$, by (I). ∎

Sometimes it is useful to write this theorem in a form in which the perturbation $HJ - JH_0$ is factored. For example, set $A := \big(R(z_0)\, J - JR_0(z_0)\big)\,(z_0 - H_0)$ and $B := (\bar{z}_0 - H)$. Then we have $B^*A = HJ - JH_0$ *in the form sense* (this means that there exist two sets $\mathcal{D}_0 \subseteq \operatorname{dom} H_0$, $\mathcal{D} \subseteq \operatorname{dom} H$ dense in $\mathcal{H}_0$, $\mathcal{H}$, respectively, such that

$$(Au, Bv) = (Ju, Hv) - (JH_0u, v)$$

for all $u \in \mathcal{D}_0$, $v \in \mathcal{D}$) and $R(z_0)\, B^*A = A$. Thus, for $u \in \operatorname{dom} H_0 \cap P_0^{ac}\mathcal{H}_0$,

$$\begin{aligned}\|A\, e^{-itH_0}u\| &= \big\|\big(R(z_0)\, J - JR_0(z_0)\big)\, e^{-itH_0}(z_0 - H_0)\, u\big\| \\ &= \big\|\big(R(z_0)\, J - JR_0(z_0)\big)\, e^{-itH_0}u_0\big\|\end{aligned}$$

with $u_0 = (z_0 - H_0)\, u$. This shows that (II) of Theorem 4 implies the integrability of $\|A\, e^{-itH_0}u\|$ for a set $\mathcal{D}_0 \ni u$ which is generating with respect to $P_0^{ac}\mathcal{H}_0$ and $H_0$. Conversely, we obtain that the integrability of $\|A\, e^{-itH_0}u\|$ implies (II) of Theorem 4. We extend this to the following statement.

**Proposition 5.** *Let* $(Ju, Hv) - (JH_0u, v) = (Au, Bv)$ *for every* $u \in \operatorname{dom} H_0$ *and* $v \in \operatorname{dom} H$ *where* $A$ *denotes an operator from* $\mathcal{H}_0$ *to a Banach space* $\mathcal{X}$ *with* $\operatorname{dom} A \supseteq \operatorname{dom} H_0$ *and* $B$ *denotes an operator from* $\mathcal{H}$ *to* $\mathcal{X}^*$ *(the adjoint space of* $\mathcal{X}$*) with* $\operatorname{dom} B \supseteq \operatorname{dom} H$. *Further suppose that*

*(I)* $B$ *is* $H$*-bounded,*

*(II) there is a set* $\mathcal{D}_0 \subseteq \operatorname{dom} H_0$*, which is generating with respect to* $P_0^{ac}\mathcal{H}_0$ *and* $H_0$*, such that for each* $u \in \mathcal{D}_0$ *there is* $t_u > 0$ *so that* $\int_{t_u}^{\infty} \|A\, e^{-itH}u\|\, dt < \infty$.

*Then the strong wave operator* $W_+(H, H_0; J)$ *exists.*

**Proof.** From the definition of $A$ and $B$ we get

$$R(z_0)\, J - JR_0(z_0) = \big(BR(z_0)^*\big)^*\, AR_0(z_0). \tag{1}$$

Since $B$ is $H$-bounded, $\big(BR(z_0)^*\big)^*$ is a bounded operator from $\mathcal{X}^{**}$ to $\mathcal{H}$. $A$ has range in $\mathcal{X}$ and we have the canonical embedding of $\mathcal{X}$ in $\mathcal{X}^{**}$. Thus the expression (1) makes sense. Let $u \in \mathcal{D}_0$. Then we obtain, by the boundedness of $\big(BR(z_0)^*\big)^*$,

$$\int_{t_u}^{\infty} \big\|\big(R(z_0)\, J - JR_0(z_0)\big)\, e^{-itH_0}(z_0 - H_0)\, u\big\|\, dt$$

$$= \int_{t_u}^{\infty} \big\|\big(BR(z_0)^*\big)^*\, A\, e^{-itH_0}u\big\|\, dt \le \big\|\big(BR(z_0)^*\big)^*\big\| \int_{t_u}^{\infty} \|A\, e^{-itH_0}\, u\| dt.$$

This shows that for $u' = (z_0 - H_0)\, u$

$$\int_0^\infty \|(R(z_0)\, J - JR_0(z_0))\, e^{-itH_0}u'\|\, dt \leq t_u\, \|R(z_0)\, J - JR_0(z_0)\|\, \|u'\|$$

$$+ \|(BR(z_0)^*)^*\| \int_{t_u}^\infty \|A\, e^{-itH_0}u\|\, dt < \infty$$

because of (II) and the boundedness of $R(z_0)\, J - JR_0(z_0)$. Thus (II) of Theorem 4 is fulfilled for all vectors $u' \in \mathscr{D} := (z_0 - H_0)\, \mathscr{D}_0$. Since $\mathscr{D}$ satisfies (I) of Theorem 4, we get that the wave operator $W_+(H, H_0; J)$ exists in the strong sense. ■

An especially simple factorization is given by $B = 1$ and $A = HJ - JH_0$. dom $A \supseteq$ dom $H_0$ is then satisfied if $J(\text{dom}\, H_0) \subseteq \text{dom}\, H$ or if we know that $A$ is $H_0$-bounded. This provides Corollary 3 again.

**Remark 6.** Because $A$ is not assumed to be $H_0$-bounded or closable, $\|A\, e^{-itH_0}\, u\|$ may not be $|\cdot|$-measurable in $t$. Therefore, in general, the integral in (II) should be taken in the sense of an upper integral.

**Remark 7.** It is easily to extend Prop. 5 to the case where $(Ju, Hv) - (JH_0u, v) = \sum_{i=1}^n (A_iu, B_iv)$, $A_i$ are operators from $\mathscr{H}_0$ to Banach spaces $\mathscr{K}_i$ with dom $A_i \supseteq$ dom $H_0$ and $B_i$ are operators from $\mathscr{H}$ to $\mathscr{K}_i$ with dom $B_i \supseteq$ dom $H$. Assumptions (I) and (II) are to replace by the corresponding ones for the operators $A_i$ and $B_i$.

It is not hard to show that all statements of this section remain true if we take another universal projection $P_{H_0} \leq P^c_{H_0}$ in place of $P^{ac}_0$ because we did not need the absolute continuity of the vectors $u$ in the proofs. Furthermore, we see that the statements give the existence of the strong wave operators $W_-(H, H_0; J)$ if we only replace the integrability of the corresponding expressions over $(a, \infty)$ by the one over $(-\infty, -a)$.

### 15.1.2. Application: Short range perturbations of the Laplacian

Let $H_0$ denote the natural selfadjoint extension of $-\Delta = \sum_{j=1}^n \left(-i \frac{\partial}{\partial x_j}\right)^2$ on $\mathscr{H}_0 = L^2(\mathbb{R}^n, dx)$, $n \geq 3$. Let $H$ denote a selfadjoint operator on the same space given by

$$(Hu)\,(x) = \left\{\sum_{j,k=1}^n \left(-i \frac{\partial}{\partial x_j} + b_j(x)\right) a_{j,k}(x) \left(-i \frac{\partial}{\partial x_k} + b_k(x)\right) + q(x)\right\} u(x).$$

Thus $H = H_0 + V$ with $V$ defined by

$$(Vu)\,(x) = \sum_{j,k=1}^n (\delta_{j,k} - a_{j,k}(x)) \frac{\partial^2}{\partial x_j\, \partial x_k}\, u(x) + \sum_{j=1}^n d_j(x) \left(-i \frac{\partial}{\partial x_j}\, u(x)\right) + c(x)\, u(x)$$

with

$$d_j(x) = \sum_{k=1}^n \left\{-i \frac{\partial}{\partial x_k}\, a_{j,k}(x) + (a_{j,k}(x) + a_{k,j}(x))\, b_k(x)\right\}$$

and

$$c(x) = \sum_{j,k=1}^n \left\{-i \frac{\partial}{\partial x_j}\, a_{j,k}(x)\, b_k(x) + a_{j,k}(x)\, b_j(x)\, b_k(x)\right\} + q(x).$$

We assume that the functions $q$, $b_k$, and $(a_{j,k} - \delta_{jk})$ are vanishing sufficiently rapidly at infinity. More precisely, we require that the functions $d_j(x)$, $(a_{j,k}(x) - \delta_{jk})$ and $c(x)$ are uniformly bounded and square integrable outside a finite ball $\mathscr{K}_R := \{x \in \mathbb{R}^n; |x| \leqq R\}$ with some $R > 0$.

Our goal is to prove the existence of the strong wave operators $W_\pm(H, H_0)$. For this aim we choose a suitable identification operator $J$ which is asymptotically equivalent to 1 in the strong sense, and for which we can prove the existence of the strong wave operators $W_\pm(H, H_0; J)$ with the help of Prop. 5. Then, by Prop. 9.45 we get the desired result for $W_\pm(H, H_0)$.

Let $J$ be a multiplication operator by $j(x)$ on $L^2(\mathbb{R}^n\, dx)$ with the following properties: 1. $j(x)$ is infinitely differentiable, 2. $j(x) = 0$ on $\mathscr{K}_R$, 3. $j(x) = 1$ for all $x$ with $|x| > R' > R$. From Theorem 7.38 we get that $J - 1 \in \ker \mu_s^{H_0} \cap \ker \mu_s^{-H_0}$. Thus it remains to prove the existence of $W_\pm(H, H_0; J)$ in the strong sense.

Set $B = 1$ and $A = HJ - JH_0$. From the form of $V$ and the properties of $J$ it follows that $A$ has the form

$$(Au)(x) = \left\{\sum_{j,k=1}^{n} e_{j,k}(x) \frac{\partial^2}{\partial x_j\, \partial x_k} + \sum_{j=1}^{n} f_j(x) \frac{\partial}{\partial x_j} + g(x)\right\} u(x)$$

where all functions $e_{j,k}$, $f_j$ and $g$ are uniformly bounded and square integrable. From this it is not hard to show that $A$ is $H_0$-bounded. Thus dom $A \supseteqq$ dom $H_0$. Next we prove the integrability of $\|A\, \mathrm{e}^{-\mathrm{i}tH_0}u\|$ for all $u \in \mathscr{D}_0 = C_0^\infty(\mathbb{R}^n)$. Clearly, $\mathscr{D}_0$ is dense in $L^2(\mathbb{R}^n, dx)$ and furthermore $\left(\mathrm{i}\frac{\partial}{\partial x_j}\right)^m \mathscr{D}_0 \subseteqq \mathscr{D}_0$ for all $m \in \mathbb{N}$ and $j = 1, 2, \ldots, n$. Using this and that $\left(\mathrm{i}\frac{\partial}{\partial x_j}\right)$ commute with $\mathrm{e}^{-\mathrm{i}tH_0}$, we see that in order to prove the integrability of $\|A\, \mathrm{e}^{-\mathrm{i}tH_0}u\|$, $u \in \mathscr{D}_0$, it is sufficient to verify that $\|G\, \mathrm{e}^{-\mathrm{i}tH_0}u\|$ is integrable in $t$ for all $u \in \mathscr{D}_0$, where $G$ denotes a multiplication operator by a function $g(x)$ from $L^2(\mathbb{R}^n, dx)$.

For the unitary group $\mathrm{e}^{-\mathrm{i}tH_0}$ we have the integral representation (see Reed/Simon [2, p. 59])

$$(\mathrm{e}^{-\mathrm{i}tH_0}u)(x) = (4\mathrm{i}\pi t)^{-n/2} \int_{\mathbb{R}^n} \mathrm{e}^{\mathrm{i}|x-y|^2/4t} u(y)\, dy. \tag{2}$$

This gives, for $u \in \mathscr{D}_0$,

$$\begin{aligned}\|G\, \mathrm{e}^{-\mathrm{i}tH_0}u\|^2 &= \int_{\mathbb{R}^n} dx\, |g(x)|^2\, |4\pi t|^{-n} \left|\int_{\mathbb{R}^n} dy\, \mathrm{e}^{\mathrm{i}|x-y|^2/4t} u(y)\right|^2 \\ &\leqq |4\pi t|^{-n} \int_{\mathbb{R}^n} dx\, |g(x)|^2 \left(\int_{\mathbb{R}^n} |u(y)|\, dy\right)^2 = |4\pi t|^{-n}\, \|g\|^2\, \|u\|^2_{L^1(\mathbb{R}^n, dx)}.\end{aligned} \tag{3}$$

Since $u$ belongs to $\mathscr{D}_0 \subset L^1(\mathbb{R}^n, dx)$, this proves that $\|G\, \mathrm{e}^{-\mathrm{i}tH_0}u\|$ is integrable in $t$ for $n \geqq 3$. This concludes the proof that $A$ satisfies the assumptions of Prop. 5 and therefore that the strong wave operator $W_+(H, H_0; J)$ exists. Since $\|A\, \mathrm{e}^{-\mathrm{i}tH_0}u\|$ is also integrable over $(-\infty, 0)$, we get also the existence of the strong wave operator $W_-(H, H_0; J)$.

In particular, this result shows that for the scattering of a quantum mechanical particle by a short range potential with local singularities the strong wave operators exist (under some additional technical assumptions). The assumption that the functions are bounded outside a finite ball is not essential. We have used it only for the proof that dom $A \supseteqq$ dom $H_0$. But this can be shown under weaker assumptions.

### 15.1.3. Application: Schrödinger operators with hard core potential

Let $\mathfrak{E} \subset \mathbb{R}^3$ be an open set such that $\mathbb{R}^3 \setminus \mathfrak{E}$ is compact and connected. Further assume that the boundary $\partial\mathfrak{E}$ is smooth. Let $-H_0$ on $L^2(\mathbb{R}^3, dx)$ and $-H$ on $L^2(\mathfrak{E}, dx)$ denote selfadjoint extensions of the Laplacian $\Delta$ defined on $\mathscr{S}(\mathbb{R}^3)$ and $\mathscr{S}(\mathfrak{E})$ respectively (for example, assume the boundary condition $u(x) = 0$ on $\partial\mathfrak{E}$ for $-\Delta$ on $L^2(\mathfrak{E}, dx)$) (see e.g. Kato [1, p. 299]). Further, let $J$ be a bounded operator from $\mathscr{H}_0 = L^2(\mathbb{R}^3, dx)$ to $\mathscr{H} = L^2(\mathfrak{E}, dx)$ defined by

$$(Ju)(x) = \chi_{\mathfrak{E}}(x)\, u(x).$$

We want to prove that the strong wave operators $W_\pm(H, H_0; J)$ exist, are partially isometric and $H_0$-semicomplete.

Let $\mathfrak{E}' \subset \mathfrak{E}$ where dis $(\partial\mathfrak{E}, \mathfrak{E}') > 0$ and $\mathbb{R}^3 \setminus \mathfrak{E}'$ is compact. Define an operator $\mathcal{J}$ from $\mathscr{H}_0$ to $\mathscr{H}$ by

$$(\mathcal{J}u)(x) = j(x)\, u(x)$$

where $j(x)$ is infinitely differentiable, $j(x) = 0$ on $\mathbb{R}^3 \setminus \mathfrak{E}$ and $j(x) = 1$ on $\mathfrak{E}'$. From Theorem 7.38 we get that

$$\operatorname*{s-lim}_{t\to\pm\infty} (J - \mathcal{J})\, e^{-itH_0} = 0.$$

Therefore, by Prop. 9.35 and Prop. 9.45, it is sufficient to prove the existence of $W_+(H, H_0; \mathcal{J})$ in the strong sense to show that $\mathcal{J}$ is asymptotically equivalent to a unitary operator $U$ (in the strong sense).

For vectors $u \in \mathscr{H}_0$ such that $u(x)$ has support in $\mathfrak{E}'$ one has $Hu = H_0 u$. Using this we obtain that $V := H\mathcal{J} - \mathcal{J}H_0$ has the form

$$v_0(x) + \sum_{j=1}^{3} v_j(x) \frac{\partial}{\partial x_j}$$

where the functions $v_j$ are bounded and vanish outside the bounded region $\mathbb{R}^3 \setminus \mathfrak{E}'$. It is easy to show that $V$ is $H_0$-bounded ($\frac{\partial}{\partial x_j}$ is $H_0$-bounded) and thus dom $V \supseteq$ dom $H_0$. For an operator of the form of $V$ and vectors $u \in C_0^\infty(\mathbb{R}^3)$ we have already proved in Section 2 that $\|V e^{-itH_0}u\|$ is integrable in $t$. Then Prop. 5 with $B = 1$ and $A = V$ shows that the strong wave operators $W_\pm(H, H_0; \mathcal{J})$ exist. It remains to verify that $\mathcal{J}$ is asymptotically isometric.

Let $\mathfrak{E}_1 \subset \mathfrak{E}'$ with dis $(\partial\mathfrak{E}', \mathfrak{E}_1) > 0$ and $\mathbb{R}^3 \setminus \mathfrak{E}_1$ compact. Define a unitary operator $U$ from $\mathscr{H}_0$ onto $\mathscr{H}$ by $Uu = \mathcal{J}u$ for all $u \in L^2(\mathfrak{E}_1, dx) \subset L^2(\mathbb{R}^3, dx)$ and let $U \restriction L^2(\mathbb{R}^3 \setminus \mathfrak{E}_1, dx)$ be an isometry with the final subspace $L^2(\mathfrak{E} \setminus \mathfrak{E}_1, dx) \subset L^2(\mathfrak{E}, dx)$. By definition, $U$ is unitary and

$$U - \mathcal{J} = (U - \mathcal{J})\, \chi_\Delta \tag{4}$$

where $\chi_\Delta$ is the multiplication operator by the function $\chi_\Delta(x)$ on $\mathscr{H}_0 = L^2(\mathbb{R}^3, dx)$ with $\Delta = \mathbb{R}^3 \setminus \mathfrak{E}_1$. $\Delta$ is compact and thus, by Theorem 7.38, $\chi_\Delta \in \ker \mu_s^{H_0} \cap \ker \mu_s^{-H_0}$. Then, as a consequence of (4), we find that $\mathcal{J}$ is asymptotically equivalent to $U$, i.e. $\mathcal{J} \underset{\pm H_0}{\approx} U$. This concludes the proof of the $H_0$-semicompleteness and partial isometry of $W_\pm(H, H_0; J)$.

Physically, this example is used to describe the scattering of a quantum mechanical particle by a potential that is very strong (in the limit infinitely large) in a small space region and approximately zero outside it.

### 15.1.4. Application: Schrödinger operators with rapidly oscillating potentials

This example is to illustrate again the use of asymptotically equivalent identification operators in proving the existence of wave operators with the help of Prop. 5. Furthermore, it shows that for rapidly oscillating potentials the falloff in $|x|$ may be arbitrarily slow.

Let $H_0$ and $H$ be selfadjoint operators on $\mathscr{H}_0 = L^2(\mathbb{R}^3, dx)$ defined by

$$(H_0 u)(x) = -(\Delta u)(x), \qquad (Hu)(x) = -(\Delta u)(x) + q(x)\, u(x)$$

where $q(x)$ is a real-valued function given by

$$q(x) = c|x|^{-\alpha} \sin(a|x|^\gamma), \qquad c \in \mathbb{R}, \qquad \alpha \in [0, 1] \text{ and } \alpha + \gamma > 2, \qquad a > 0.$$

Clearly, then $\gamma > 1$. We have dom $H$ = dom $H_0 \supseteq \mathscr{S}(\mathbb{R}^3)$ because $q(x)$ is the sum of a function from $L^2(\mathbb{R}^3, dx)$ and a function from $L^\infty(\mathbb{R}^3, dx)$ (see Kato [1, p. 303]). Our goal is to show that the strong wave operators $W_\pm(H, H_0)$ exist.

For this we prove the existence of $W_\pm(H, H_0; J)$ in the strong sense with the help of Prop. 5 where $J$ has the form

$$(Ju)(x) = e^{g(x)} j(x)\, u(x).$$

$j$ and $g$ are functions from $C^\infty(\mathbb{R}^3)$ such that $j(x) = 0$ in a neighborhood of $x = 0$, $j(x) = 1$ for $|x| \geqq 1$, and $g(x) = b\,|x|^{-\beta} \sin(a\,|x|^\gamma)$ for $|x| \geqq 1$ with $b = -c(a\gamma)^{-2}$ and $\beta = \alpha + 2\gamma - 2 > \gamma > 1$. Since $|g(x)| \to 0$ as $|x| \to \infty$, $e^{g(x)} j(x) \to 1$ as $|x| \to \infty$. Thus, by Theorem 7.38,

$$\operatorname*{s-lim}_{t\to\pm\infty} (J - 1)\, e^{-itH_0} = 0.$$

This proves that $W_\pm(H, H_0; J) = W_\pm(H, H_0)$ and that the existence of the left side in the strong sense implies the same for the right side.

Next we calculate $V := HJ - JH_0$. First we note that $V$ is an operator of the form

$$\sum_{j=1}^{3} v_j(x) \frac{\partial}{\partial x_j} + v_0(x)$$

where the functions $v_s(x)$ are continuous and bounded. For $|x| \geqq 1$ these functions can be explicitly calculated. We have

$$((HJ - JH_0)\, u)(x)$$
$$= e^{g(x)} \left\{ q(x) - \sum_{j=1}^{3} \frac{\partial^2}{\partial x_j^2} g(x) - 2 \sum_{j=1}^{3} \left(\frac{\partial}{\partial x_j} g(x)\right) \frac{\partial}{\partial x_j} - \left(\frac{\partial}{\partial x_j} g(x)\right)^2 u(x) \right\} \tag{5}$$

for $|x| \geqq 1$. Since $g(x) = g(|x|)$ for $|x| \geqq 1$, we get, with $|x| = r$,

$$\frac{\partial}{\partial x_j} g(r) = g'(r)\, x_j r^{-1} \quad \text{and} \quad (\Delta g)(x) = g''(r) + 2r^{-1} g'(r).$$

Further we have

$$g'(r) = ba\gamma r^{\gamma-\beta-1} \cos(ar^\gamma) - b\beta r^{-\beta-1} \sin(ar^\gamma), \tag{6}$$

$$g''(r) = -b(a\gamma)^2 r^{2\gamma-2-\beta} \sin(ar^\gamma) - ab\gamma(\beta + 1 - \gamma)\, r^{\gamma-2-\beta} \cos(ar^\gamma)$$
$$-ab\beta\gamma r^{\gamma-2-\beta} \cos(ar^\gamma) + b\beta(\beta + 1)\, r^{-2-\beta} \sin(ar^\gamma). \tag{7}$$

By $\beta = \alpha + 2\gamma - 2$, we have $\beta + 1 - \gamma = \alpha + \gamma - 1 > 1$ ($\alpha + \gamma > 2$ by assumption), $\beta + 2 - \gamma > 2$ and $\beta + 2 - 2\gamma = \alpha$. Using this, (6), and (7), we obtain from (5) that the term

$$-b(a\gamma)^2 r^{2\gamma-2-\beta} \sin(ar^\gamma) = cr^{-\alpha} \sin(ar^\gamma)$$

from $(\Delta g)(x)$ cancels exactly the term $q(x)$. All other terms give contributions to $v_s(x)$ which vanish like $|x^{-(1+\varepsilon)}|$, $\varepsilon > 0$. Thus $v_s(x)$, $s = 0, 1, 2, 3$, are continuous functions of $x$ with the asymptotic behavior $|v_s(x)| \leqq c_s\, |x|^{-(1+\varepsilon)}$, $\varepsilon > 0$. From these properties of the functions $v_s(x)$ it is easy to see that $V$ is $H_0$-bounded and $\operatorname{dom} V \supseteqq \operatorname{dom} H_0$. Therefore Prop. 5 is applicable to the system $\{H, H_0, J\}$ with $B = 1$, $A = V$ and $\mathcal{D}_0 = C_0^\infty(\mathbb{R}^3)$ if we can show that

$$\int_{-\infty}^{\infty} \|V\, e^{-itH_0} u\|\, dt < \infty, \qquad u \in C_0^\infty(\mathbb{R}^3).$$

From the concrete form of $V$, the commutativity of $\frac{\partial}{\partial x_j}$ with $H_0$ and $\frac{\partial}{\partial x_j}\, C_0^\infty(\mathbb{R}^3) \subseteqq C_0^\infty(\mathbb{R}^3)$, we see that it suffices to prove

$$\int_{-\infty}^{\infty} \|W\, e^{-itH_0} u\|\, dt < \infty$$

for a multiplication operator $W$ by $w(x)$ where $w(x)$ is bounded and vanishes at infinity as $|x|^{-(1+\varepsilon)}$ at infinity.

For this aim we first prove an estimate for the unitary group $e^{-itH_0}$. Set

$$(V_t u)(x) := (2it)^{-3/2} e^{ix^2/4t} \hat{u}(x/2t).$$

It is an easy calculation to show that $V_t$ is unitary for each $t \in \mathbb{R}$. By (2), we have, for $u \in C_0^\infty(\mathbb{R}^3)$,

$$(e^{-itH_0}u - V_t u)(x) = (4i\pi t)^{-3/2} e^{ix^2/4t} \int_{\mathbb{R}^3} (e^{iy^2/4t} - 1) e^{-ixy/2t} u(y) \, dy.$$

Set $v_t(y) = (e^{iy^2/4t} - 1) u(y)$. Then

$$\begin{aligned} \|W(e^{-itH_0}u - V_t u)\|^2 &= \int_{\mathbb{R}^3} dx \, |w(x)|^2 (2t)^{-3} |\hat{v}_t(x/2t)|^2 \\ &= \int_{\mathbb{R}^3} dx \, |w(2xt)|^2 |\hat{v}_t(x)|^2 \\ &= \int_{|x|>1} |w(2xt)|^2 |\hat{v}_t(x)|^2 \, dx + \int_{|x|<1} dx \, |w(2xt)|^2 |\hat{v}_t(x)|^2. \end{aligned} \tag{8}$$

Now we use that $|w(x)| \leq c_1 |x|^{-(1+\varepsilon)}$, $\varepsilon > 0$, $|\hat{v}_t(y)|$ is uniformly bounded in $t$, and $\int_{\mathbb{R}^3} dy \, |\hat{v}_t(y)|^2 \leq c_2 < \infty$ with $c_2$ independent of $t$. These properties of $v_t$ follow from its special form and from the assumption that $u(x)$ is a $C_0^\infty$-function. This gives

$$\begin{aligned} \|W(e^{-itH_0}u - V_t u)\|^2 &\leq \int_{|x|\geq 1} c_1^2 |2t|^{-2(1+\varepsilon)} |\hat{v}_t(x)|^2 \, dx \\ &\quad + \int_{|x|<1} c_3 c_1^2 |2x|^{-2(1+\varepsilon)} |t|^{-2(1+\varepsilon)} \, dx \\ &\leq c_2 c_1^2 |2t|^{-2(1+\varepsilon)} + \int_{r<1} r^{-2\varepsilon} 4\pi c_3 c_1^2 |2t|^{-2(1+\varepsilon)} = K |t|^{-2(1+\varepsilon)} \end{aligned} \tag{9}$$

where $r = |x|$ and $|\hat{v}_t(x)| \leq c_3$. Using this estimate and the boundedness of $W$, we get

$$\begin{aligned} \|W e^{-itH_0}u\| &\leq \|W(e^{-itH_0}u - V_t u)\| + \|W V_t u\| \\ &\leq \sqrt{K} |t|^{-(1+\varepsilon)} + \|W V_t u\|. \end{aligned} \tag{10}$$

Further, we have

$$\begin{aligned} \|W V_t u\|^2 &= \int_{\mathbb{R}^3} dx \, |w(x)|^2 (2t)^{-3} |\hat{u}(x/2t)|^2 \\ &\leq \int_{\mathbb{R}^3} dx \, |w(x)|^2 (2t)^{-(2+\varepsilon)} (b_{1-\varepsilon})^2 |x|^{-(1-\varepsilon)} \leq b' |t|^{-(2+\varepsilon)} \end{aligned}$$

where we have used that $u \in \mathcal{S}(\mathbb{R}^3)$ and therefore $|\hat{u}(x/2t)| \leq b_\alpha (|2t/x|)^\alpha$ for arbitrary $\alpha > 0$ (here $2\alpha = 1 - \varepsilon$) and that $w(x)$ is bounded and $|w(x)|^2 |x|^{-(1-\varepsilon)} \leq c' |x|^{-(3+\varepsilon')}$, $\varepsilon' > 0$. This together with (10) proves that $\|W e^{-itH_0}u\|$ is integrable in $t$, and thus concludes the proof of the existence of the strong wave operators.

We note that the last step in the proof, namely that $\int \|W e^{-itH_0}u\| \, dt < \infty$, also shows that the strong wave operators $W_\pm(H, H_0)$ exist for $H = H_0 + W$. This extends the result of Section 2.

## 15.2. Completeness of wave operators

### 15.2.1. Completeness from spectral assumptions on $H$

In Section 1.1 we proved the existence of strong wave operators $W_\pm(H, H_0; J)$ under the assumption that $\|(HJ - JH_0)\, e^{-itH_0}u\|$ or similar expressions are integrable in $t$. Now we are interested in the completeness problem. That is, we want to get information about the image (range) and the support of the wave operators. In principle one could solve it by proving that the wave operators $W_\pm(H_0, H; J^*)$ exist, provided that $J^*J \underset{\pm H_0}{\approx} 1_{\mathscr{H}_0}$ and $JJ^* \underset{H}{\approx} 1_{\mathscr{H}}$ (see Prop. 9.48). In applications the asymptotic properties of $J^*J$ and $JJ^*$ are often easy to verify. But the proof of the existence of $W_\pm(H_0, H; J^*)$ with the help of the method of Section 1.1 requires knowledge about the unitary group $e^{-itH}$, and clearly this is in general not available. Sometimes one has additional information on the spectrum of $H$. Then it is possible to get completeness of wave operators by means of

**Proposition 8.** *Assume the wave operator $W_+(H, H_0; J)$ exists, is partially isometric and $H_0$-semicomplete. Further, suppose there exist decompositions $\mathscr{H}_0 = \bigoplus_{l=1}^{N} \mathscr{H}_{0l}$ and $\mathscr{H} = \bigoplus_{l=1}^{N} \mathscr{H}_l$, $N \leqq \infty$, such that for all $l \in \{1, 2, \ldots, N\}$*

(I) *$\mathscr{H}_{0l}$ reduces $H_0$ and $\mathscr{H}_l$ reduces $H$,*

(II) *$J\mathscr{H}_{0l} \subseteqq \mathscr{H}_l$,*

(III) *$\operatorname{spec}^c (H_0 \mid P_0^{ac}\mathscr{H}_{0l}) \supseteqq \operatorname{spec}^c (H \mid P^{ac}\mathscr{H}_l)$*

(IV) *The multiplicity of $H_0 \mid \mathscr{H}_{0l}P_0^{ac}$ is finite and it is not smaller than the multiplicity of $H \mid P^{ac}\mathscr{H}_l$.*

*Then the wave operator $W_+(H; H_0; J)$ is complete.*

**Proof.** From (I) and (II) it follows that $W_+\mathscr{H}_{0l} \subseteqq \mathscr{H}_l$. Thus $W_l = W_+Q_{l0}$ ($Q_{l0}$ denotes the projection from $\mathscr{H}_0$ onto $\mathscr{H}_{0l}$) is an intertwining operator for $H_0 \mid \mathscr{H}_{0l}$ and $H \mid \mathscr{H}_l$ with $W_l{}^*W_l = P_0^{ac}Q_{l0}$ and $P_l := W_lW_l{}^* \leqq P^{ac}Q_l$ ($Q_l$ denotes the projection from $\mathscr{H}$ onto $\mathscr{H}_l$). Hence

$$\operatorname{spec}^c (H_0 \mid P_0^{ac}\mathscr{H}_{0l}) \subseteqq \operatorname{spec}^c (H \mid P^{ac}\mathscr{H}_l)$$

and because of (III) we obtain

$$\operatorname{spec}^c (H_0 \mid P_0^{ac}\mathscr{H}_{0l}) = \operatorname{spec}^c (H_l \mid P^{ac}\mathscr{H}_l) =: \Lambda_0.$$

Further: $H_0 \mid P_0^{ac}\mathscr{H}_{0l}$ is isomorphic to the part $H \mid P_l\mathscr{H}$ of $H \mid P^{ac}\mathscr{H}_l$. Therefore from (IV) we obtain

$$\operatorname{mul} (H_0 \mid P_0^{ac}\mathscr{H}_{0l}) = \operatorname{mul} (H \mid P^{ac}\mathscr{H}_l) =: n < \infty$$

and $H \mid P_l\mathscr{H}_l \cong H \mid P^{ac}\mathscr{H}_l$ because both operators are isomorphic to the multiplication by $\lambda$ in $L^2(\Lambda_0, d\lambda; \mathscr{K})$, where $\dim \mathscr{K} = n$. From $n < \infty$ now we obtain $P_l\mathscr{H}_l = P^{ac}\mathscr{H}_l$, i.e. $P_l = P^{ac}Q_l$ and $W_lW_l{}^* \mid \mathscr{H}_l = P^{ac}\mathscr{H}_l$. This proves $W_+W_+{}^* = P^{ac}$. ■

In a similar way as we have proved the completeness of the generalized wave operators for $H_0 = -\Delta$, $H = -\Delta + q(x)$ where $q(x)$ is the Coulomb potential we can verify the completeness of wave operators for scattering with short range potentials.

**Example 9.** Let $H_0$ and $H$ denote the natural selfadjoint extensions of $-\Delta$ and $-\Delta + q(x)$ on $L^2(\mathbb{R}^3, dx)$ where $q(x) = q(|x|)$ is bounded outside a finite ball $\mathscr{K}_R = \{x \in \mathbb{R}^3; |x| \leqq R\}$ and $q \in L^2(\mathbb{R}^3, dx)$. Since $q(x)$ is spherically symmetric, we can use the decomposition $\mathscr{H} = \bigoplus_{l=0}^{\infty} \bigoplus_{|m| \leqq l} \mathscr{H}_{l,m}$ (see Section 10.2.5) where the subspaces $\mathscr{H}_{l,m}$ belong to the spherical harmonics $Y_{l,m}(\theta, \varphi)$. Each $\mathscr{H}_{l,m}$ reduces $H_0$ and $H$. Further, we know that $H_0 \restriction \mathscr{H}_{l,m}$ is absolutely continuous on $(0, \infty)$ and is simple. By the result of WEIDMANN cited in Section 10.2.5, $H \restriction \mathscr{H}_{l,m}$ has also simple spectrum in $(0, \infty)$ and is absolutely continuous in $(0, \infty)$. Thus we can apply Prop. 8 and obtain that the wave operators $W_\pm(H, H_0)$ (which exist and are partially isometric, see Section 1.2) are complete.

### 15.2.2. Completeness by time-falloff conditions

In this section we strenghten the assumptions on expressions like $\|(HJ - JH_0)\, e^{-itH_0}u\|$ to get additional information about the image of the wave operators and therefore also about the spectrum of the perturbed operator $H$. First we introduce special classes of functions. By $C_\infty(\mathbb{R} \setminus \Gamma)$ we denote the set of all functions $f \in C_\infty(\mathbb{R})$ with $f(x) = 0$ for all $x \in \Gamma$.

**Definition 10.** Let $\Gamma$ be a closed countable set. Then a set $\mathscr{A}_\Gamma \subseteq C(\mathbb{R})$ of functions is called *multiplicative-generating* for $C_\infty(\mathbb{R} \setminus \Gamma)$ if the set spa $\{f; f(x) = a(x)\, g(x),\ a \in \mathscr{A}_\Gamma,\ g \in C_0^\infty(\mathbb{R} \setminus \Gamma)\}$ is dense in $C_\infty(\mathbb{R} \setminus \Gamma)$ (with respect to $|\cdot|$).

The following lemma provides examples for such multiplicative-generating sets.

**Lemma 11.** (i) *If $\varphi \in C(\mathbb{R})$ and $\varphi(x) \neq 0$ for all $x \in \mathbb{R} \setminus \Gamma$, then $\mathscr{A}_\Gamma := \{\varphi\}$ is multiplicative generating.*

(ii) *If $\varphi \in C$ ($\mathbb{R}$ and $\varphi(x) \neq 0$ for all $x \in \mathbb{R} \setminus \Gamma$, then the set $\{f; f(x) = \varphi(x)\, a(x),\ a \in \mathscr{A}_\Gamma\}$ is multiplicative generating, where $\mathscr{A}_\Gamma$ denotes a multiplicative-generating set of functions.*

**Proof.** (i) follows from (ii) if we can show that the function $a(x) \equiv 1$ generates a set $\mathscr{A}_\Gamma$ for arbitrary $\Gamma$. By Definition 10, this means to show that $C_0^\infty(\mathbb{R} \setminus \Gamma)$ is dense in $C_\infty(\mathbb{R} \setminus \Gamma)$ with respect to $|\cdot|$, but this is a well-known fact. The proof of (ii) is obvious. ∎

An important consequence of Definition 10 is

**Lemma 12.** *Let $\mathscr{A}_\Gamma$ be multiplicative generating. Then*

$$P_H^c \mathscr{H} = \text{clo spa}\, \{\varphi(H)\, u;\ u \in P_H^c \mathscr{H},\ \varphi \in \mathscr{A}_\Gamma\}.$$

**Proof.** We have to show that spa $\{\varphi(H)\, u; u \in P_H^c \mathscr{H}, \varphi \in \mathscr{A}_\Gamma\}$ is dense in $P_H^c \mathscr{H}$. Now we can take vectors $u$ such that $u = \psi(H)\, u'$ with $\psi \in C_0^\infty(\mathbb{R} \setminus \Gamma)$ and $u' \in P_H^c \mathscr{H}$. This and Definition 10 show that it suffices to verify that $\{\alpha(H)\, u; u \in P_H^c \mathscr{H}, \alpha \in \Psi\}$ is dense in $P_H^c \mathscr{H}$ where $\Psi$ is some set of functions dense in $C_\infty(\mathbb{R} \setminus \Gamma)$ (with respect to $|\cdot|$).

Now there is a sequence $\varphi_n \in \Psi$ such that $\lim_{n\to\infty} \varphi_n(x) = 1$ for all $x \in \mathbb{R} \setminus \Gamma$ and $|\varphi_n(x)| \leqq 2$ for all $x \in \mathbb{R}$ and $n \in \mathbb{N}$. Thus, it is enough to show that for each $u \in P_H^c \mathscr{H}$ the relation

$\lim_{n\to\infty} \|u - \varphi_n(H)\, u\| = 0$ holds. We have

$$\|\varphi_n(H)\, u - u\|^2 = \int_{-\infty}^{\infty} |1 - \varphi_n(x)|^2 \left(u, E(\mathrm{d}x)\, u\right).$$

Since $|\varphi_n(x)| \leqq 2$, the integrand is uniformly bounded in $x$ with respect to the measure $\left(u, E(\Delta)\, u\right)$. Since $\Gamma$ is countable, $|1 - \varphi_n(x)|^2$ tends to zero a.e. with respect to the continuous measure $\left(u, E(\Delta)\, u\right)$. Then the dominated convergence theorem proves that $\|u - \varphi_n(H)\, u\| \to 0$ as $n \to \infty$. ∎

As a preliminary to the completeness statements we need .

**Lemma 13.** (i) *Let $K \in \mathscr{L}(\mathscr{H}, \mathscr{H}_0)$ and $R_0(\mathrm{i})\, K - KR(\mathrm{i}) \in \mathscr{L}_\infty(\mathscr{H}, \mathscr{H}_0)$. Then for each $\alpha \in C_\infty(\mathbb{R})$, we have*

$$\alpha(H_0)\, K - K\alpha(H) \in \mathscr{L}_\infty(\mathscr{H}, \mathscr{H}_0).$$

(ii) *Let $K \in \mathscr{L}(\mathscr{H}, \mathscr{H}_0)$ and $R_0(\mathrm{i})\, K - KR(\mathrm{i}) \in \mathscr{L}_{H,P_H^c}(\mathscr{H}, \mathscr{H}_0)_\mathrm{a}$. Then for each $\alpha \in C(\mathbb{R})$ we have*

$$\alpha(H_0)\, K - K\alpha(H) \in \mathscr{L}_{H,P_H^c}(\mathscr{H}, \mathscr{H}_0)_\mathrm{a}.$$

**Proof.** The proof is the same in both casesfor $C_\infty$-functions. One uses that $\mathscr{L}_\infty(\mathscr{H}, \mathscr{H}_0)$ and $\mathscr{L}_{H,P_H^c}(\mathscr{H}, \mathscr{H}_0)$ are left invariant under $\mathscr{L}(\mathscr{H}_0)$, right invariant under com $H$, and norm closed (see Section 1.3.4 and Prop. 6.70). The extension to all continuous functions in case (ii) is possible by a density argument. We prove only (ii).

Let $\operatorname{Im} z \neq 0$. Then $R_0(z)\, (H_0K - KH)$ is well-defined on dom $H$ if it is interpreted as $H_0R_0(z)\, K - R_0(z)\, KH$. From the above mentioned properties of $\mathscr{L}_{H,P_H^c}(\mathscr{H}, \mathscr{H}_0)_\mathrm{a}$ and the identity

$$R_0(z)\, (H_0K - KH)\, R(z) = [1 + (\mathrm{i} - z)\, R_0(z)]\, R_0(\mathrm{i})\, (H_0K - KH)\, R(\mathrm{i})\, [1 + (\mathrm{i} - z)\, R(z)] \tag{11}$$

we obtain that $R_0(z)\, (H_0K - KH)\, R(z) \in \mathscr{L}_{H,P_H^c}(\mathscr{H}, \mathscr{H}_0)_\mathrm{a}$ for all $z$ with $\operatorname{Im} z \neq 0$. Hence

$$R_0(z)\, K - KR(z) = R_0(z)\, (H_0K - KH)\, R(z) \in \mathscr{L}_{H,P_H^c}(\mathscr{H}, \mathscr{H}_0)_\mathrm{a}. \tag{12}$$

Furthermore, one has the identity

$$\begin{aligned} &KR(z)^m\, R(\xi)^n - R_0(z)^m\, R_0(\xi)^n\, K \\ &= \left(KR(z)^m - R_0(z)^m\, K\right) R(\xi)^n + R_0(z)^m \left(KR(\xi)^n - R_0(\xi)^n\, K\right). \end{aligned} \tag{13}$$

Setting in (13) $z = \xi$ and $n = N - 1$, we obtain by induction that $KR(z)^N - R_0(z)^N\, K \in \mathscr{L}_{H,P_H^c}(\mathscr{H}, \mathscr{H}_0)_\mathrm{a}$ for each $N = 0, 1, 2, \ldots$ Therefore from (13) one concludes that

$$KR(\mathrm{i})^m\, R(-\mathrm{i})^n - R_0(\mathrm{i})^m\, R_0(-\mathrm{i})^n\, K \in \mathscr{L}_{H,P_H^c}(\mathscr{H}, \mathscr{H}_0)_\mathrm{a}$$

for all $m, n = 0, 1, 2, \ldots$ By the complex Stone-Weierstrass theorem (see the formulation in the proof of Prop. 7.27), we get that each $\alpha \in C_\infty(\mathbb{R})$ is the uniform limit of a sequence of polynomials in $(\mathrm{i} - x)^{-1}$ and $(-\mathrm{i} - x)^{-1}$. Hence $K\alpha(H) - \alpha(H_0)\, K$ is the norm limit of a sequence of operators from $\mathscr{L}_{H,P_H^c}(\mathscr{H},\mathscr{H}_0)_a$. This proves (ii) for all $\alpha \in C_\infty(\mathbb{R})$.

Now let $\alpha \in \dot{C}(\mathbb{R})$. Then $\alpha(x)\,(\mathrm{i}-x)^{-1}$ is a function from $C_\infty(\mathbb{R})$. We have

$$\begin{aligned}\big(\alpha(H_0)\,K - K\alpha(H)\big)\,R(\mathrm{i}) &= \big(\alpha(H_0)\,KR(\mathrm{i}) - \alpha(H_0)\,R_0(\mathrm{i})\,K\big)\\ &\quad + \big(\alpha(H_0)\,R_0(\mathrm{i})\,K - K\alpha(H)\,R(\mathrm{i})\big)\\ &= \alpha(H_0)\,\big(KR(\mathrm{i}) - R_0(\mathrm{i})\,K\big) + \big(\alpha(H_0)\,R_0(\mathrm{i})\,K - K\alpha(H)\,R(\mathrm{i})\big).\end{aligned}$$

The second term on the right side belongs to $\mathscr{L}_{H,P_H^c}(\mathscr{H},\mathscr{H}_0)_a$ because $\alpha(x)\,(\mathrm{i}-x)^{-1}$ belongs to $C_\infty(\mathbb{R})$. The first term on the right side is an element of $\mathscr{L}_{H,P_H^c}(\mathscr{H},\mathscr{H}_0)_a$ since $KR(\mathrm{i}) - R_0(\mathrm{i})\,K$ has this property. This proves that the left side belongs to $\mathscr{L}_{H,P_H^c}(\mathscr{H},\mathscr{H}_0)_a$ and, since $R(\mathrm{i})\,\mathscr{H}$ is dense in $\mathscr{H}$, that $\alpha(H_0)\,K - K\alpha(H) \in \mathscr{L}_{H,P_H^c}(\mathscr{H},\mathscr{H}_0)_a$. This concludes the proof. ■

**Remark 14.** With a similar argumentation as in the last step of the proof of Lemma 13 one can show: If $\alpha(H_0)\,K - K\alpha(H) \in \mathscr{L}_{H,P_H^c}(\mathscr{H},\mathscr{H}_0)_a$ for all functions $\alpha \in C_0^\infty(\mathbb{R}\setminus\Gamma)$ where $\Gamma$ is a countable closed set, then $\alpha(H_0)\,K - K\alpha(H) \in \mathscr{L}_{H,P_H^c}(\mathscr{H},\mathscr{H}_0)_a$ is true for all functions $\alpha \in C(\mathbb{R})$.

Now we come to the main result of this section.

**Theorem 15.** *Suppose there exist two selfadjoint operators $F_+, F_- \in \mathscr{L}(\mathscr{H}_0)$, an operator $Z \in \mathscr{L}(\mathscr{H},\mathscr{H}_0)$ with a bounded inverse and a set $\mathscr{A}_\Gamma$ of multiplicative-generating functions such that the following conditions are satisfied:*

(I) $F_+ + F_- = P_0^{ac}$, $F_+ \in \ker \mu_a^{-H_0}$ *and* $F_- \in \ker \mu_a^{H_0}$, $F_+ = F_+P_0^{ac} = P_0^{ac}F_+$,

(II) $(1 - P_0^{ac})\,\alpha(H_0) \in \mathscr{L}_\infty(\mathscr{H}_0)$ *for all* $\alpha \in \mathscr{A}_\Gamma$,

(III) $\left|\int_0^{\pm\infty} \big\|\big(R(\mathrm{i})\,J - JR_0(\mathrm{i})\big)\,\alpha(H_0)\,\mathrm{e}^{-\mathrm{i}tH_0}F_\pm\big\|\,\mathrm{d}t\right| < \infty$ *for all* $\alpha \in \mathscr{A}_\Gamma$ *and both signs,*

(IV) $R_0(\mathrm{i})\,J^* - J^*R(\mathrm{i}) \in \mathscr{L}_{H,P_H^c}(\mathscr{H},\mathscr{H}_0)_a \cap \mathscr{L}_{-H,P_H^c}(\mathscr{H},\mathscr{H}_0)_a$,

(V) $(J^* - Z) \in \mathscr{L}_{H,P_H^c}(\mathscr{H},\mathscr{H}_0)_a \cap \mathscr{L}_{-H,P_H^c}(\mathscr{H},\mathscr{H}_0)_a$.

*Then the strong wave operators $W_+(H, H_0; J)$, $W_-(H, H_0; J)$ exist, $H$ and $H_0$ have no singularly continuous spectrum,* and $\mathrm{clo}\,(W_+\mathscr{H}_0) = \mathrm{clo}\,(W_-\mathscr{H}_0) = P^{ac}\mathscr{H}$. *If $J$ is asymptotically equivalent to a partial isometry $V$ with $V^*V = P_0^{ac}$ (with respect to $H_0P_0^{ac}$ and $-H_0P_0^{ac}$) i.e. $J - V \in \mathscr{L}_{H,P_H^c}(\mathscr{H}_0,\mathscr{H})_a \cap \mathscr{L}_{-H,P_H^c}(\mathscr{H}_0,\mathscr{H})_a$, then the wave operators are partially isometric and complete.*

**Proof. 1.** First we prove the existence of the wave operators in the strong sense. This is already a consequence of (I) and (II). We begin by showing that the set

$$\mathscr{D}_+ := \{\alpha(H_0)\,\mathrm{e}^{\mathrm{i}sH_0}F_+\,\mathrm{e}^{-\mathrm{i}sH_0}u;\ u \in P_0^{ac}\mathscr{H}_0,\ s \geqq 0,\ \alpha \in \mathscr{A}_\Gamma\}$$

is generating in $P_0^{ac}\mathscr{H}_0$. Let $u \in P_0^{ac}\mathscr{H}_0$ and set

$$f_{+s} := \mathrm{e}^{\mathrm{i}sH_0}F_+\,\mathrm{e}^{-\mathrm{i}sH_0}u.$$

By (I), $F_+ - P_0^{ac} \in \ker \mu_a^{H_0}$ and this implies that, by Lemma 6.17, there is a sequence $s_n \to \infty$ such that $f_{+s_n} \to u$ strongly. Thus, the set of vectors $\{\mathrm{e}^{\mathrm{i}sH_0}F_+\,\mathrm{e}^{-\mathrm{i}sH_0}u;\ u \in P_0^{ac}\mathscr{H}_0,\ s \geqq 0\}$ is dense in $P_0^{ac}\mathscr{H}_0$. From Lemma 12 we immediately obtain that the set $\{\alpha(H_0)\,u;\ u \in P_0^{ac}\mathscr{H}_0,\ \alpha \in \mathscr{A}_\Gamma\}$ is generating in $P_0^{ac}\mathscr{H}_0$. Both together imply that $\mathscr{D}_+$ is also generating in $P_0^{ac}\mathscr{H}_0$.

Now let $v \in \mathscr{D}_+$. Then $v = \alpha(H_0)\, e^{isH_0} F_+ e^{-isH_0} u$ with $u \in P_0^{ac}\mathscr{H}_0$. Thus

$$\int_0^\infty dt\, \|(R(i)\, J - JR_0(i))\, e^{-itH_0} v\| = \int_{-s}^\infty d\tau\, \|(R(i)\, J - JR_0(i))\, e^{-i\tau H_0} \alpha(H_0)\, F_+ e^{-isH_0} u\|$$

$$\leq \|u\| \left\{ \int_0^\infty d\tau\, \|(R(i)\, J - JR_0(i))\, e^{-i\tau H_0} \alpha(H_0)\, F_+\| + s\, \|R(i)\, J - JR_0(i)\|\, \|\alpha(H_0)\, F_+\| \right\}.$$

By (III) and the boundedness of $F_+$ and $R(i)\, J - JR_0(i)$, the left side is finite. Therefore, by Theorem 4, the strong wave operator $W_+(H, H_0; J)$ exists. Similarly, we can prove the existence of the strong wave operator $W_-(H, H_0; J)$.

2. The next point is to show that

$$F\alpha_\pm(H_0)\, (J^* - W_\pm^*) \in \mathscr{L}_{\pm H, P_H^c}(\mathscr{H}, \mathscr{H}_0)_a \tag{14}$$

for all $\alpha \in \tilde{\mathscr{A}}_\Gamma$ where $\tilde{\mathscr{A}}_\Gamma := \{\alpha;\, \alpha(x) = (i + x)^{-1}\, \tilde{\alpha}(x),\, \overline{\tilde{\alpha}(\cdot)} \in \mathscr{A}_\Gamma\}$. By Lemma 12, we have (use that $\tilde{\mathscr{A}}_\Gamma$ is multiplicative generating as a simple calculation proves)

$$P_{H'}^c \mathscr{H}' = \text{clo spa}\, \{\alpha(H')\, u;\, u \in P_{H'}^c \mathscr{H}',\, \alpha \in \tilde{\mathscr{A}}_\Gamma\}$$

for each selfadjoint operator $H'$ on $\mathscr{H}'$, in particular for $H' = H$ and $H' = H_0$. Because of (IV) and Lemma 13,

$$\begin{aligned} A_\pm(s) &:= F_\pm \alpha(H_0)\, e^{isH_0} (e^{-isH_0} J^* - J^* e^{-isH}) \\ &= F_\pm \alpha(H_0)\, (J^* - e^{isH_0} J^* e^{-isH}) \in \mathscr{L}_{H, P_H^c}(\mathscr{H}, \mathscr{H}_0)_a \cap \mathscr{L}_{-H, P_H^c}(\mathscr{H}, \mathscr{H}_0)_a \end{aligned}$$

for each function $\alpha \in C(\mathbb{R})$. Clearly, the weak limits of $A_\pm(t)$ exist as $t \to \pm\infty$ and coincide with the operators in (14) since the strong wave operators $W_\pm$ exist. Thus it remains to show that the limits of $A_\pm(t)$ as $t \to \pm\infty$ also belong to $\mathscr{L}_{\pm H, P_H^c}(\mathscr{H}, \mathscr{H}_0)_a$ respectively. For this it is sufficient to verify that $A_\pm(t)\, R(-i)$ tend in norm to limits as $t \to \pm\infty$.

Assume $\alpha(x) = (i + x)^{-1}\, \alpha_1(x)$ with $\bar{\alpha}_1 \in \mathscr{A}_\Gamma$ and thus $\alpha \in \tilde{\mathscr{A}}_\Gamma$ and $\alpha \in C(\mathbb{R})$. Then from (III) we get the estimate

$$\begin{aligned} \|(A_\pm(t) - A_\pm(s))\, R(-i)\, u\| &= \|F_\pm \alpha_1(H_0)\, R_0(-i)\, (e^{itH_0} J^* e^{-itH} \\ &\qquad - e^{isH_0} J^* e^{-isH})\, R(-i)\, u\| \\ &\leq \left\| \int_s^t d\tau \{F_\pm \alpha_1(H_0)\, e^{i\tau H_0} [R_0(-i)\, J^* - J^* R(-i)]\, e^{-i\tau H}\, u\} \right\| \\ &\leq \int_s^t d\tau\, \|[R(i)\, J - JR_0(i)]\, \bar{\alpha}_1(H_0)\, e^{-i\tau H_0} F_\pm\|\, \|u\| < c < \infty \end{aligned}$$

for all $s, t > 0$, or $s, t < 0$, respectively. This proves that $A_\pm(t)\, R(-i)$ tend in norm to limits as $t \to \pm\infty$ for all functions $\alpha \in \tilde{\mathscr{A}}_\Gamma$.

3. The last point is to show that $\text{clo}\, (W_\pm \mathscr{H}_0) = P_H^c \mathscr{H}$. Because of $\text{clo}\, (W_\pm \mathscr{H}_0) \subseteq P^{ac}\mathscr{H}$ (see Prop. 9.20) this implies that the singular continuous spectrum of $H$ is empty; $P_{H_0}^{sc} = 0$ follows from (II), because according to (II) we have $(P_0^{ac})^\perp \alpha(H_0) \in \mathscr{L}_\infty(\mathscr{H}_0)$ for all $\alpha \in \mathscr{A}_\Gamma$, hence $(P_0^{ac})^\perp \alpha(H_0)\, g(H_0) \in \mathscr{L}_\infty(\mathscr{H}_0)$ for all $g \in C_0^\infty(\mathbb{R} \setminus \Gamma)$. This implies

$(P_0^{ac})^\perp E_{H_0}(\Delta \setminus \Gamma) \in \mathscr{L}_\infty(\mathscr{H}_0)$ for each compact set $\Delta$, hence $(P_0^{ac})^\perp E_{H_0}(\Delta \setminus \Gamma)$ and a fortiori $(P_0^{ac})^\perp E_{H_0}(\Delta \setminus \Gamma)$ is finite-dimensional for each compact set $\Delta$, which implies $P_{H_0}^{sc} = 0$.

Assume $u \in P_H^c\mathscr{H}$ and $u \perp$ ima $W_+$. Then we must verify that $u = 0$. For this we use

$$\|u\| = \|e^{-itH}u\| = \lim_{T\to\infty} T^{-1} \int_0^T dt \, \|Z^{-1}Z \, e^{-itH}u\|$$

$$\leq \|Z^{-1}\| \left\{ \lim_{T\to\infty} T^{-1} \left[ \int_0^T dt \, \|(Z - J^*) \, e^{-itH}u\| + \int_0^T dt \, \|J^* \, e^{-itH}u\| \right] \right\}.$$

By (V), $u = 0$ if $\lim_{T\to\infty} T^{-1} \int_0^T dt \, \|J^* \, e^{-itH}u\| = 0$. From the fact that the set $\{\alpha(H)\, u;$ $u \in P_H^c\mathscr{H}, \alpha \in \tilde{\mathscr{A}}_\Gamma\}$ is generating in $P_H^c\mathscr{H}$ (see the remark at the beginning of 2.) it follows that it is enough to prove

$$\lim_{T\to\infty} T^{-1} \int_0^T dt \, \|J^* \, e^{-itH}\alpha(H)\, u\|^2 = 0, \qquad \alpha \in \tilde{\mathscr{A}}_\Gamma. \tag{15}$$

We have the identity

$$\begin{aligned} J^* \, e^{-itH}\alpha(H)\, u &= \{J^*\alpha(H) - \alpha(H_0)\, J^*\} \, e^{-itH}u \\ &+ (1 - F_+ - F_-)\, \alpha(H_0)\, J^* \, e^{-itH}u + F_+\alpha(H_0)\, (J^* - W_+^*) \, e^{-itH}u \\ &+ F_-\alpha(H_0)\, (J^* - W_-^*) \, e^{-itH}u + F_+\alpha(H_0) \, e^{-itH_0}W_+^*u \\ &+ F_-\alpha(H_0) \, e^{-itH_0}W_-^*u. \end{aligned}$$

The first four terms on the right side vanish in the $|A|\text{-}\lim_{t\to\infty}$ because they contain operators belonging to $\mathscr{L}_{H,P_H^c}(\mathscr{H}, \mathscr{H}_0)_a$ by (IV) and Lemma 13, by (I) and (II), and by (14). The fifth term is identically zero since $u \perp W_+\mathscr{H}$, by assumption. The sixth term vanishes in $|A|\text{-}\lim_{t\to\infty}$ since $F_- \in \ker \mu_a^{H_0}$ and $W_-^*u \in P_0^{ac}\mathscr{H}_0$. Thus the right side goes to zero in the absolute Abelian limit and this proves (15). This concludes the proof of the theorem. ∎

**Remark 16.** The assumption that the operators $F_\pm$ are selfadjoint is not necessary. One sees from the proof that it suffices to assume $(F_+ + F_- - P_0^{ac})\, \alpha(H_0) \in \mathscr{L}_\infty(\mathscr{H}_0)$, $F_+$ and $F_+^*$ are elements of $\ker \mu_a^{H_0}$, and $F_-$ and $F_-^*$ are elements of $\ker \mu_a^{-H_0}$.

**Remark 17.** If we can factor the perturbation $V = HJ - JH_0 = B^*A$ such that $B$ is $H$-bounded and $A$ is $H_0$-bounded, then

$$R(i)\, J - JR_0(i) = R(i)\, B^*AR_0(i) = \big(BR(-i)\big)^* \, AR_0(i)$$

and $\big(BR(-i)\big)^*$ is bounded. Thus assumption (III) can be replaced by the other

$$\left| \int_0^{\pm\infty} dt \, \|A\alpha(H_0)\, R_0(i) \, e^{-itH_0}F_\pm\| \right| < \infty$$

where $\alpha \in \mathscr{A}_\Gamma$. (In general the integrals here and in (III) are upper integrals.)

**Remark 18.** In practice, (IV) and (V) are often verified via compactness arguments. For example, (IV) is a consequence of the assumption: $\alpha(H_0) J^* - J^*\alpha(H) \in \mathscr{L}_\infty(\mathscr{H}, \mathscr{H}_0)$ for all $\alpha \in C_0^\infty(\mathbb{R} \setminus \Gamma)$ where $\Gamma$ is any countable closed set (see Remark 14).

At first sight this theorem looks a little complicated since it includes some auxiliary objects such as the operators $F_\pm$ and $Z$ and the functions $\alpha \in \mathscr{A}_\Gamma$. Obviously, the use of the freedom in the choice of the set $\mathscr{A}_\Gamma$ and the operators $F_\pm$, $Z$ is to make this statement flexible. That assumption (III) must contain some operators $F_\pm$ is clear because without these operators (III) is only true in the trivial case that $J$ is an intertwining operator for $H$ and $H_0$. Such operators $F_\pm$ were already introduced in Section 7.1.2. They are asymptotically equivalent to $P_0^{ac}$ with respect to $\pm H_0 P_0^{ac}$. Without (V) we would get only the existence of the strong wave operators $W_\pm(H, H_0; J)$ and that $|A|\text{-}\lim_{t\to\pm\infty} J^* e^{-itH}u = 0$ for all $u \in P_H^c\mathscr{H} \cap (W_\pm\mathscr{H}_0)^\perp$. In applications (V) is mostly harmless. Assumption (IV) is not very restrictive. For instance, consider the case $\mathscr{H} = \mathscr{H}_0$ and $J = 1$. Then (IV) is a necessary condition for the existence of the complete wave operators $W_\pm(H, H_0)$ (see the remark at the end of Section 7.3.3).

Under compactness conditions and under the restriction to the special case $\mathscr{H} = \mathscr{H}_0$ and $J = 1$ we obtain from Theorem 15 and its proof a somewhat stronger result.

**Corollary 19.** *Assume $\mathscr{H} = \mathscr{H}_0$ and $J = 1$. Further suppose that there exist two self-adjoint operators $F_+$, $F_-$ and a set $\mathscr{A}_\Gamma$ of multiplicative-generating functions with the following properties:*

(I) $F_+ + F_- = P_0^{ac}$, $F_+ \in \ker \mu_a^{-H_0}$ *and* $F_- \in \ker \mu_a^{H_0}$,

(II) $(1 - P_0^{ac})\, \alpha(H_0) \in \mathscr{L}_\infty(\mathscr{H}_0)$ *for all* $\alpha \in \mathscr{A}_\Gamma$,

(III) $R_0(\mathrm{i}) - R(\mathrm{i}) \in \mathscr{L}_\infty(\mathscr{H}_0)$

(IV) $\left| \int_0^{\pm\infty} \|(R_0(\mathrm{i}) - R(\mathrm{i}))\, e^{-\mathrm{i}tH_0}\alpha(H_0)\, F_\pm\|\, \mathrm{d}t \right| < \infty$ *for all* $\alpha \in \mathscr{A}_\Gamma$ *and both signs.*

*Then the strong wave operators $W_+(H, H_0)$, $W_-(H, H_0)$ exist and are complete. Further, $H$ and $H_0$ have no singularly continuous spectrum and each eigenvalue of $H$ and $H_0$ in $\mathbb{R} \setminus \Gamma$ is of finite multiplicity and these eigenvalues accumulate at most at points of $\Gamma$ or at $\pm\infty$.*

**Proof.** From Theorem 13 we get all results except the assertion on the eigenvalues of $H$. In the course of the proof of Theorem 15 we have verified that

$$F_\pm\alpha(H_0)\,(1 - W_\pm^*) \in \mathscr{L}_{\pm H, P_H^c}(\mathscr{H}_0, \mathscr{H}_0)_a$$

for $\alpha \in \tilde{\mathscr{A}}_\Gamma := \{\alpha;\, \alpha(x) = (\mathrm{i} + x)^{-1}\, \alpha_1(x),\, \overline{\alpha_1(\cdot)} \in \mathscr{A}_\Gamma\}$. In the same way we can show under the compactness assumption (III) that

$$F_\pm\alpha(H_0)\,(1 - W_\pm^*)\, R(-\mathrm{i}) \in \mathscr{L}_\infty(\mathscr{H}_0).$$

Now let $\lambda \in \mathbb{R} \setminus \Gamma$ and assume there is an infinite orthonormal sequence of eigenvectors of $H$, $\{f_n\}_1^\infty$, with $Hf_n = \lambda_n f_n$ $\lambda_n \in \mathbb{R} \setminus \Gamma$, where $\lambda_n \to \lambda$. Choose $\alpha \in \tilde{\mathscr{A}}_\Gamma$ so that $|\alpha(\lambda)| > 0$ (such function exists because of Definition 10). One has $\operatorname{w-lim}_{n\to\infty} f_n = 0$ and $\lim_{n\to\infty} \|\alpha(H)\, R(-\mathrm{i})\, f_n\| = |\alpha(\lambda)\, (\mathrm{i} + \lambda)^{-1}| > 0$. Since $W_\pm^* f_n = 0$, we obtain

$$\alpha(H)\, R(-\mathrm{i})\, f_n = \big(\alpha(H) - \alpha(H_0)\big)\, R(-\mathrm{i})\, f_n + (1 - F_+ - F_-)\, \alpha(H_0)\, R(-\mathrm{i})\, f_n$$
$$+ F_+\alpha(H_0)\,(1 - W_+^*)\, R(-\mathrm{i})\, f_n + F_-\alpha(H_0)\,(1 - W_-^*)\, R(-\mathrm{i})\, f_n.$$

Since all operators appearing on the right side are compact (we have

$$(1 - F_+ - F_-)\,\alpha(H_0) = (1 - P_0^{ac})\,R_0(\mathrm{i})\,\alpha_1(H_0) = \left(\overline{\alpha}_1(H_0)\,(1 - P_0^{ac})\right)^* R_0(\mathrm{i})$$

with $\overline{\alpha}_1 \in \mathscr{A}_\Gamma$) and $\text{w-lim}_{n\to\infty} f_n = 0$, we get $\lim_{n\to\infty} \|\alpha(H)\,R(-\mathrm{i})\,f_n\| = 0$. But this contradicts (see above)

$$\lim_{n\to\infty} \|\alpha(H)\,R(-\mathrm{i})\,f_n\| = |\alpha(\lambda)\,(\mathrm{i} + \lambda)^{-1}| > 0.$$

Thus there does not exist such a sequence of eigenvectors. This proves the assertion on the eigenvectors of $H$. ∎

### 15.2.3. Application: Differential operators of first order

Let $H_0$ and $H$ denote the natural selfadjoint extensions of $-\mathrm{i}(\mathrm{d}/\mathrm{d}x)$ and of $\overline{a(x)}\,(-\mathrm{i}\mathrm{d}/\mathrm{d}x)\,a(x) + b(x)$ respectively, where $a(x)$ is continuously differentiable and $b(x)$ is continuous and real-valued. Further, assume that $a(x) = 1$ for $|x| > R > 0$ and $b(x) = 0$ for $|x| > R$, and assume that $\operatorname{dom} H = \operatorname{dom} H_0$ (which holds for instance if $\sup_{x\in\mathbb{R}} (|a(x)|^2 - 1) < 1$).

Now we study the question if complete strong wave operators exist for these simple operators $H$ and $H_0$. We shall do this with the help of Theorem 15. In our case $J = 1$ and $\mathscr{H} = \mathscr{H}_0 = L^2(\mathbb{R}, \mathrm{d}x)$ and $P_0^{ac} = 1$. We choose $\mathscr{A}_\Gamma = \{\alpha; \alpha(x) \equiv 1\}$. $F_+$ and $F_-$ are defined to be the multiplication operators by $\chi_{[0,\infty)}(x)$ and $\chi_{(-\infty,0]}(x)$ respectively. From the proof of Theorem 7.4 we get that $F_+$ satisfies (I) of Theorem 15. The perturbation $V = H - H_0$ is given by

$$\mathrm{i}a_1(x)\,\frac{\mathrm{d}}{\mathrm{d}x} + b_1(x) = a_1(x)\left(\mathrm{i} + \mathrm{i}\,\frac{\mathrm{d}}{\mathrm{d}x}\right) + b_2(x)$$

where the functions $a_1(x)$, $b_1(x)$ and $b_2(x)$ are continuous and vanish for $|x| > R$. From this form of $V$ it is easy to see that $\operatorname{dom} V \supseteq \operatorname{dom} H_0$ and $V$ is $H_0$-bounded (the assumption $\operatorname{dom} H = \operatorname{dom} H_0$ was not used for the proof of this assertion).

First we show that $V$ is $H_0{}^2$-compact or what is the same that $VR_0(\mathrm{i})^2$ is compact. This implies also that $VR(\mathrm{i})^2$ is compact because of $\operatorname{dom} H = \operatorname{dom} H_0$ (see WEIDMANN [1, p. 261]). Clearly, $VR_0(\mathrm{i})^2$ is compact if we can prove that $AR_0(\mathrm{i}) \in \mathscr{L}_2(\mathscr{H}_0)$ where $A$ is multiplication by a continuous function $g(x)$ with compact support in $(-R, R)$.

Now the resolvent $R_0(\mathrm{i}) = (\mathrm{i} + \mathrm{i}(\mathrm{d}/\mathrm{d}x))^{-1}$ has the form

$$(R_0(\mathrm{i})\,u)\,(x) = -\mathrm{i} \int_{-\infty}^{x} \mathrm{e}^{y-x} u(y)\,\mathrm{d}y \tag{16}$$

as a straightforward calculation proves. Thus

$$\|AR_0(\mathrm{i})\|_2{}^2 = \int_{\mathbb{R}} \mathrm{d}x\,|a(x)|^2 \left| \int_{-\infty}^{x} \mathrm{d}y\,\mathrm{e}^{y-x} \right|^2$$

$$\leq \int_{-R}^{R} \mathrm{d}x\,\mathrm{e}^{-2x}\,|a(x)|^2 \left| \int_{-\infty}^{R} \mathrm{d}y\,\mathrm{e}^{y} \right|^2 < \infty$$

because of the support property of the function $a(x)$. Hence $VR(\mathrm{i})^2$ is compact and therefore

$$(R_0(\mathrm{i}) - R(\mathrm{i}))\,R(\mathrm{i}) = R_0(\mathrm{i})\,VR(\mathrm{i})^2 \in \mathscr{L}_\infty(\mathscr{H}_0) \subset \mathscr{L}_{H,P_H^c}(\mathscr{H}_0)_a \cap \mathscr{L}_{-H,P_H^c}(\mathscr{H}_0)_a.$$

Since $R_0(\mathrm{i}) - R(\mathrm{i})$ is bounded, this proves that

$$|\mathrm{A}|\text{-}\lim_{t\to\pm\infty} \mathrm{e}^{\mathrm{i}tH}(R_0(\mathrm{i}) - R(\mathrm{i}))\,\mathrm{e}^{-\mathrm{i}tH}P_H^c u = 0$$

on the dense set $\mathscr{D} = R(\mathrm{i})\,\mathscr{H}_0$, and also for all of $\mathscr{H}_0$. That is, $R_0(\mathrm{i}) - R(\mathrm{i}) \in \mathscr{L}_{H,P_H^c}(\mathscr{H}_0)_a \cap \mathscr{L}_{-H,P_H^c}(\mathscr{H}_0)_a$. Thus assumption (IV) of Theorem 15 is fulfilled.

It remains to prove assumption (III) of Theorem 15. We prove it only for the + sign. The other case is similar. We have

$$\begin{aligned}\|(R_0(\mathrm{i}) - R(\mathrm{i}))\,\mathrm{e}^{-\mathrm{i}tH_0}F_+\| &\leq \|VR_0(\mathrm{i})\,\mathrm{e}^{-\mathrm{i}tH_0}F_+\| \\ &\leq \|A_1\,\mathrm{e}^{-\mathrm{i}tH_0}F_+\,\mathrm{e}^{\mathrm{i}tH_0}\| + \|B_2R_0(\mathrm{i})\,\mathrm{e}^{-\mathrm{i}tH_0}F_+\,\mathrm{e}^{\mathrm{i}tH_0}\| \qquad (17)\end{aligned}$$

where $A_1$ and $B_2$ are the multiplication operators by $a_1(x)$ and $b_2(x)$ respectively. The operator $\mathrm{e}^{-\mathrm{i}tH_0}F_+\,\mathrm{e}^{\mathrm{i}tH_0}$ is the multiplication operator by the function $\chi_{[0,\infty)}(x - t) = \chi_{[t,\infty)}(x)$ (recall $(\mathrm{e}^{-\mathrm{i}tH_0}u)(x) = u(x - t)$). Thus $A_1\,\mathrm{e}^{-\mathrm{i}tH_0}F_+\,\mathrm{e}^{\mathrm{i}tH_0}$ is the multiplication operator by the function $a_1(x)\,\chi_{[t,\infty]}(x)$ which is zero, for $t > R$ because of $\operatorname{supp} a_1(\cdot) \subseteq [-R, R]$. From (16) we obtain that

$$(B_2R_0(\mathrm{i})\,\mathrm{e}^{-\mathrm{i}tH_0}F_+\,\mathrm{e}^{\mathrm{i}tH_0})\,(x, y) = -\mathrm{i}b_2(x)\,\mathrm{e}^{y-x}\chi_{(-\infty,x]}(y)\,\chi_{[t,\infty)}(y),$$

which is zero if $t > x$. Since $\operatorname{supp} b_2(\cdot) \subseteq [-R, R]$, this implies that $B_2\,\mathrm{e}^{-\mathrm{i}tH_0} = 0$ for $t > R$. Thus (17) is integrable in $t$ over the region $[0, \infty)$ and this proves that (III) of Theorem 15 is fulfilled. This finishes the proof that the strong wave operators $W_\pm(H, H_0)$ exist, are complete and $H$ has no singularly continuous spectrum.

### 15.2.4. Application: Potential scattering

Let $H_0 = -\Delta$ and $H = H_0 + V$ on $\mathscr{H} = \mathscr{H}_0 = L^2(\mathbb{R}^3, \mathrm{d}x)$ where $V$ denotes a multiplication operator by the function $q(x)$. Assume $q(x) = q_1(x) + q_2(x)$ with $q_1 \in L^\infty(\mathbb{R}^3, \mathrm{d}x)$ and $q_2 \in L^2(\mathbb{R}^3, \mathrm{d}x)$. Further, assume $|q(x)| \leq c\,|x|^{-1-\varepsilon}$ for $|x| > R_0 > 0$ and some $\varepsilon > 0$, $c > 0$. We want to prove that $H$, $H_0$ satisfy the assumptions of Corollary 19 and thus that the complete strong wave operators $W_+(H, H_0)$, $W_-(H, H_0)$ exist.

Let $F$ denote the unitary operator on $\mathscr{H}$ which transforms the function $u(\cdot) \in L^2(\mathbb{R}^3, \mathrm{d}x)$ into its Fourier transform $\hat{u}(\cdot) \in L^2(\mathbb{R}^3, \mathrm{d}k)$. We have $F\mathscr{H} = \hat{\mathscr{H}} = L^2(\mathbb{R}^3, \mathrm{d}k)$ and $FH_0F^* = \hat{H}_0$ where $\hat{H}_0$ is the multiplication operator by $k^2$ in $L^2(\mathbb{R}^3, \mathrm{d}k)$. Set $k = (\lambda, \omega)$ with $\lambda = |k|$ and $\omega = k/|k|$, and let $\Omega$ denote the unit sphere in $\mathbb{R}^3$. Set $\tilde{\mathscr{H}} := L^2(\mathbb{R}_+, \mathrm{d}\lambda; \mathscr{K})$ with $\mathscr{K} = L^2(\Omega, \mathrm{d}\omega)$, and let $U$ be the unitary operator from $\hat{\mathscr{H}}$ to $\tilde{\mathscr{H}}$ given by

$$(Uv)\,(\lambda, \omega) = 2^{1/2}\lambda^{-1/2}v(\lambda^2, \omega), \qquad v \in \hat{\mathscr{H}}.$$

Then it is easy to see that $U\hat{H}_0U^* = \tilde{H}_0$ where $\tilde{H}_0$ denotes the multiplication operator by $\lambda$ in $\tilde{\mathscr{H}}$. Thus $\tilde{\mathscr{H}}$ gives a spectral representation for $H_0$.

Next we come to the definition of the operators $F_\pm$. We can embed $L^2(\mathbb{R}_+, \mathrm{d}\lambda; \mathscr{K})$ into $L^2(\mathbb{R}, \mathrm{d}\lambda; \mathscr{K})$ and denote the projection onto $L^2(\mathbb{R}_+, \mathrm{d}\lambda; \mathscr{K})$ by $Z_+$. Let $F_0$ be the unitary operator on $L^2(\mathbb{R}, \mathrm{d}\lambda; \mathscr{K})$ defined by

$$(F_0u)\,(p) = \hat{u}(p), \qquad u \in L^2(\mathbb{R}, \mathrm{d}p; \mathscr{K}).$$

Further, let $K_0$ and $K$ be selfadjoint operators on $L^2(\mathbb{R}, \mathrm{d}p; \mathscr{K})$ given by $p$ and $-\mathrm{i}d/dp$ respectively. Then we have $F_0{}^*K_0 = KF_0{}^*$ and $Z_+\tilde{H}_0 = K_0Z_+$. Now we define

$$\tilde{F}_\pm := Z_+F_0\chi_\pm F_0{}^*Z_+$$

where $\chi_\pm$ are the multiplication operators by the characteristic functions $\chi_{\mathbb{R}_\pm}(p)$ on $L^2(\mathbb{R}, \mathrm{d}p; \mathscr{K})$. This gives

$$\tilde{F}_\pm\,\mathrm{e}^{-\mathrm{i}t\tilde{H}_0} = Z_+F_0(\chi_\pm\,\mathrm{e}^{-\mathrm{i}tK})\,F_0{}^*Z_+.$$

From the proof of Theorem 7.4 we obtain $\operatorname*{s-lim}_{t\to\pm\infty} \chi_\pm\,\mathrm{e}^{-\mathrm{i}tK} = 0$. Thus $\tilde{F}_+ \in \ker \mu_s{}^{-\tilde{H}_0}$ and $\tilde{F}_- \in \ker \mu_s{}^{\tilde{H}_0}$. Then the operators $F_\pm := F^*U\tilde{F}_\pm U^*F$ have the desired asymptotic properties. $F_\pm \in \ker \mu_s{}^{\mp H_0}$, and $F_+ + F_- = 1$ follows from $\chi_+ + \chi_- = 1$ since $F$, $U$ and $F_0$ are unitary operators and $Z_+$ is unitary on $L^2(\mathbb{R}_+, dp; \mathscr{K})$.

From the properties of the function $q$ it follows immediately (see KATO [1, p. 304]) that $V$ is $H_0$-compact and thus $V$ is also $H$-compact (see WEIDMANN [1, p. 258]). That means $R_0(i) - R(i) = R(i)\, VR_0(i)$ is compact and thus (III) of Corollary 19 is satisfied. Now we factor $V = B^*A$ where $A$ is the multiplication operator by the function $a(x) = (1 + |x|)^{-1-\varepsilon/2}$. Then $B^*$ corresponds to a function $b(x)$ that is again a sum of a function from $L^\infty(\mathbb{R}^3, dx)$ and of a function from $L^\infty(\mathbb{R}^3, dx)$ and satisfies $|b(x)| \leqq c(1 + |x|)^{-\varepsilon/2}$ for $|x| > R_0$. This shows that $R(i)\, B^* = (BR(-i))^*$ is compact because of the same statements which say that $V$ is $H$-compact. Thus we can use the estimate

$$\begin{aligned}\|(R(i) - R_0(i))\, \varphi(H_0)\, e^{-itH_0}F_\pm\| &= \|R(i)\, V\, R_0(i)\, \varphi(H_0)\, e^{-itH_0}F_\pm\| \\ &\leqq \|BR(-i)\|\, \|AR_0(i)\, \varphi(H_0)\, e^{-itH_0}F_\pm\|\end{aligned} \tag{18}$$

in proving the existence of the integrals in $t$ (Assumption (IV)).

Let $B_{a|t|}$ denote the multiplication operator in $L^2(\mathbb{R}^3, dx)$ by the characteristic function of the set $\{x \in \mathbb{R}^3;\, |x| < a\,|t|\}$ with $a > 0$. Clearly, $B_{a|t|}$ is a projection. Split $A$ into

$$A = AB_{a|t|} + A(B_{a|t|})^\perp. \tag{19}$$

Now $\|A(B_{a|t|})^\perp\| \leqq (1 + a\,|t|)^{-1-\varepsilon/2}$. This proves that

$$\|A(B_{a|t|})^\perp\, R_0(i)\, \varphi(H_0)\, e^{-itH_0}F_\pm\| \leqq \|A(B_{a|t|})^\perp\|\, \|\varphi(H_0)\|\, \|F_\pm\|$$

is integrable in $t$. Thus, by (18), (19), and $\|AB_{a|t|}D_t\| \leqq \|A\|\, \|B_{a|t|}D_t\|$, it is sufficient for the proof of assumption (IV) to show that

$$\left|\int_0^{\pm\infty} dt\, \|B_{a|t|}R_0(i)\, \varphi(H_0)\, e^{-itH_0}F_\pm\|\right| < \infty \tag{20}$$

for some $a > 0$ where $a$ may depend on $\varphi$. As set $\mathscr{A}_\Gamma$ we choose the set of functions $\varphi$ that belong to $C_0^\infty(\mathbb{R})$ and which vanish in a neighborhood of 0. Clearly, $\psi(H_0) = R_0(i)\, \varphi(H_0)$ is then such a function as well. It is easy to verify that (20) follows from the estimate

$$|(e^{\mp itH_0}\psi(H_0)\, F_\pm u)\, (x)|^2 \leqq c' t^{-2l+1}\, \|u\|^2 \tag{21}$$

for $|x| < at$, $t > 0$, and arbitrary $l \in \mathbb{N}$ where $c'$ depends on $l$ and $\psi$. Therefore the last step in the proof is to show (21). We shall prove (21) only for the upper sign. The other case can be treated similarly.

For the proof of (21) we need the following elementary

**Lemma 20.** *Let $\psi \in C_0^\infty(\mathbb{R})$ with $\operatorname{supp}\psi \subseteqq [a, b]$, $0 < a < b < \infty$. Let $|y| < 2^{-1}a|s|$. Then for every $l \in \mathbb{N}$ the estimate*

$$\left|\int_0^\infty e^{2i\lambda y + i\lambda^2 s}\psi(\lambda)\, d\lambda\right| \leqq c_1\, |s|^{-l} \tag{22}$$

*is valid where the constant $c_1$ depends only on $\psi$ and $l$.*

Proof. Set $ys^{-1} = x$, $\lambda + x = \varrho$. Then the left side of (22) is equal to

$$\left|\int_x^\infty e^{is\varrho^2}\psi(\varrho - x)\, d\varrho\right| = \left|\int_0^\infty e^{is\varrho^2}\psi(\varrho - x)\, d\varrho\right| = \left|2^{-1}\int_0^\infty e^{i\lambda s}\psi_x(\lambda)\, d\lambda\right| \tag{23}$$

where $\psi_x(\lambda) = \lambda^{-1/2}\psi(\lambda^{1/2} - x)$ and we have used the support property of $\psi$ and $|s|^{-1}\, y < 2^{-1}a$. From this equality, the fact that the support of the functions $\psi_x(\lambda)$ lies in a common interval because of $|x| < 2^{-1}a$ and the fact the derivatives of $\psi_x(\lambda)$ can be estimated by constants independent of $x$ we obtain the desired estimate (22) by integration by parts on the right side of (23). ■

Now we have

$$(e^{-itH_0}\psi(H_0)\, u)(x) = (2\pi)^{-3/2} \int_{\mathbb{R}^3} dk\, e^{-ixk-ik^2t}\psi(k^2)\, \hat{u}(k)$$

$$= (2\pi)^{-3/2} \int_{\Omega} d\omega \int_0^\infty d\lambda\lambda^2\, e^{-i\lambda x\omega-i\lambda^2 t}\psi(\lambda^2)\, \hat{u}(\lambda, \omega). \tag{24}$$

Set $u = F_+v$, $\hat{u} = U^*Fu$, $\hat{u} = Fu$, $\tilde{v} = U^*Fv$. Then by definition of $F_+$, we get $\hat{u} = UZ_+F_0w$ with $w = \chi_+F_0{}^*Z_+\tilde{v}$. Using the definition of $U$, $Z_+$ and $F_0$, we obtain

$$\hat{u}(\lambda, \omega) = \pi^{-1/2}\lambda^{-1/2} \int_0^\infty e^{-i\lambda^2 s} w(s, \omega)\, ds.$$

Inserting this expression in (24), we have

$$(e^{-itH_0}\psi(H_0)\, F_+v)\,(x)$$

$$= 2^{-3/2}\pi^{-2} \int_{\Omega} d\omega \left\{ \int_0^\infty ds w(s, \omega) \int_0^\infty e^{-i\lambda x\omega-i\lambda^2(t+s)}\lambda^{3/2}\psi(\lambda^2)\, d\lambda \right\}.$$

Now suppose $\operatorname{supp}\psi \subset [a, b]$ with $0 < a < b < \infty$ and $|x| < at$. Further, set $y = -2^{-1}x\omega$, $p = -(t + s)$ and write $\varphi(\lambda) = \psi(\lambda^2)\, \lambda^{3/2}$. With these notations the integral $\int_0^\infty \{\cdot\}\, d\lambda$ in the above expression satisfies the assumptions of Lemma 20 (replace $\psi$, $s$ by $\varphi$, $p$). Thus this integral does not exceed $c_{l,\psi}|t + s|^{-l}$ where $l \in \mathbb{N}$ is arbitrary. Using this and the Schwarz inequality in the integral over $\omega$ and $s$, we obtain

$$|(e^{-itH_0}\psi(H_0)\, F_+v)\,(x)|^2 \leq c'_{l,\psi} \int_{\Omega} d\omega \int_0^\infty ds\, |w(s, \omega)|^2 \int_0^\infty ds(s + t)^{-2l}. \tag{25}$$

By $w = \chi_+F_0{}^*Z_+\tilde{v} = \chi_+F_0{}^*Z_+U^*Fv$, we have that

$$\int_{\Omega} d\omega \int_0^\infty ds\, |w(s, \omega)|^2 \leq \|v\|^2.$$

This and an elementary calculation in (25) show that (21) is valid and completes the proof that the strong complete wave operators for $H$, $H_0$ exist. Moreover, by Corollary 19, $H$ has no singular continuous spectrum and the eigenvalues of $H$ can accumulate at most at the point zero.

## 15.3. An invariance principle

In Part III it was shown that the existence of the wave operators $W_\pm(H, H_0; J)$ implies the existence of $W_\pm(\alpha(H), \alpha(H_0); J)$ for a wide class of functions. Here we prove a slightly stronger result under additional assumptions, namely that the falloff condition

$$\|(R(i)\, J - JR_0(i))\, e^{-itH_0}u\| \leq C_u|t|^{-(1+\varepsilon)}, \qquad \varepsilon > 0,$$

on a special dense set $\mathcal{D} \ni u$ implies the existence of the wave operators $W_\pm(\alpha(H), \alpha(H_0); J)$ in the strong sense.

**Theorem 21.** *Let $\alpha$ be a real-valued function on $\mathbb{R}$. Assume there is a decomposition of $\mathbb{R}$, $\mathbb{R} = \bigcup_{n=1}^{N} I_n$ with $I_n = (a_n, a_{n+1}]$ such that $\alpha$ is strictly increasing and twice continuously differentiable on each $(a_n, a_{n+1})$. Let $\mathscr{D}$ be a generating set in $P_0^{\mathrm{ac}}\mathscr{H}_0$ such that $\mathscr{D} \subseteq \mathscr{M}_\infty(H_0)$ and for each $u \in \mathscr{D}$, $u \in E_0([c, d])\,\mathscr{H}_0$ with $[c, d]$ contained in any interval $(a_n, a_{n+1})$, $n \in \{1, 2, \dots, N\}$. Suppose $\big(R(\mathrm{i})\,J - JR_0(\mathrm{i})\big)\,R_0(\mathrm{i})^n \in \mathscr{L}_\infty\,(\mathscr{H}_0, \mathscr{H})$* for some $n \in \mathbb{N}$ and

$$\big\|\big(R(\mathrm{i})\,J - JR_0(\mathrm{i})\big)\,\mathrm{e}^{-\mathrm{i}tH_0}u\big\| \leqq c_u|t|^{-(1+\varepsilon)}$$

*for some positive constants $\varepsilon$ and $c_u$ where $c_u$ may depend on $u$. Then the strong wave operators $W_\pm\big(\alpha(H), \alpha(H_0); J\big)$ exist and are independent of $\alpha$.*

**Proof.** 1. First we note that $\{H, H_0, J\}$ satisfies the assumptions of Theorem 4. Thus the strong wave operators $W_\pm = W_\pm(H, H_0; J)$ exist. Further, set

$$A(t) := \mathrm{e}^{\mathrm{i}tH}J'\,\mathrm{e}^{-\mathrm{i}tH_0}, \quad J' := R(\mathrm{i})\,JR_0(\mathrm{i}). \tag{26}$$

Then $A(t)$ is strongly differentiable and we obtain

$$\begin{aligned}\frac{\mathrm{d}}{\mathrm{d}t}A(t) &= \mathrm{e}^{\mathrm{i}tH}(H\,J' - J'H_0)\,\mathrm{e}^{-\mathrm{i}tH_0} = \mathrm{i}\,\mathrm{e}^{\mathrm{i}tH}R(\mathrm{i})\,VR_0(\mathrm{i})\,\mathrm{e}^{-\mathrm{i}tH_0}\\ &= \mathrm{i}\,\mathrm{e}^{\mathrm{i}tH}\big(R(\mathrm{i})\,J - JR_0(\mathrm{i})\big)\,\mathrm{e}^{-\mathrm{i}tH_0},\end{aligned} \tag{27}$$

with $V := HJ - JH_0$.

2. In the following we prove the statement only for the $+$ sign. The other case is similar. Let us consider the vector function

$$\begin{aligned}I_u(t) &:= R(\mathrm{i})\,[\mathrm{e}^{\mathrm{i}t\alpha(H)}J\,\mathrm{e}^{-\mathrm{i}t\alpha(H_0)} - W_+]\,R_0(\mathrm{i})\,u\\ &= R(\mathrm{i})\,\mathrm{e}^{\mathrm{i}t\alpha(H)}(J - W_+)\,\mathrm{e}^{-\mathrm{i}t\alpha(H_0)}R_0(\mathrm{i})\,u = \mathrm{e}^{\mathrm{i}t\alpha(H)}(J' - W_+')\,\mathrm{e}^{-\mathrm{i}t\alpha(H_0)}u\end{aligned}$$

where $W_+' = R(\mathrm{i})\,W_+R_0(\mathrm{i})$ and $u \in \mathscr{D}$. Here we have used the intertwining relation $HW_+ = W_+H_0$. We claim that

$$\operatorname*{s-lim}_{t\to\infty} I_u(t) = 0, \qquad u \in \mathscr{D}. \tag{28}$$

This implies first that

$$\operatorname*{s-lim}_{t\to\infty} R(\mathrm{i})\,(\mathrm{e}^{\mathrm{i}t\alpha(H)}J\,\mathrm{e}^{-\mathrm{i}t\alpha(H_0)} - W_+)\,u = 0 \tag{29}$$

for all $u \in \mathscr{D}' = R_0(\mathrm{i})\,\mathscr{D}$, and since $\mathscr{D}'$ is generating in $P_0^{\mathrm{ac}}\mathscr{H}_0$, for all $u \in P_0^{\mathrm{ac}}\mathscr{H}_0$. Now we have

$$\operatorname*{s-lim}_{t\to\infty} \mathrm{e}^{\mathrm{i}t\alpha(H)}\big(R(\mathrm{i})\,J - JR_0(\mathrm{i})\big)\,\mathrm{e}^{-\mathrm{i}t\alpha(H_0)}P_0^{\mathrm{ac}} = 0 \tag{30}$$

because $\big(R(\mathrm{i})\,J - JR_0(\mathrm{i})\big)\,R_0(\mathrm{i})^n$ is compact. Using this and the intertwining relation $R(\mathrm{i})\,W_+ = W_+R_0(\mathrm{i})$, we obtain from (29) the existence of

$$\operatorname*{s-lim}_{t\to\infty} \mathrm{e}^{\mathrm{i}t\alpha(H)}J\,\mathrm{e}^{-\mathrm{i}t\alpha(H_0)}P_0^{\mathrm{ac}} = W_+,$$

that is, the desired result about the existence of the strong wave operators $W_\pm(\alpha(H), \alpha(H_0); J)$.

3. The last step is to prove (28), that is

$$\operatorname*{s-lim}_{t\to\infty} (J' - W_+') \, e^{-it\alpha(H_0)} u = 0, \qquad u \in \mathscr{D}.$$

For this we use the notations and the results of the proof of Prop. 8.28. Let supp $u \in [c, d] \subset (a_n, a_{n+1})$. Take a function $\beta$ that is twice continuously differentiable with compact support in $(a_n, a_{n+1})$ such that $\beta(x) = 1$ on $[c, d]$. Then $\beta(H_0) u = u$. This gives

$$(J' - W_+') \, e^{-it\alpha(H_0)} \beta(H_0) \, u = \int_{-\infty}^{\infty} ds g(t, s) \, (J' \, e^{-isH_0} - e^{-isH} W_+') \, u$$
$$= \int_{-\infty}^{\infty} ds g(t, s) \, e^{-isH} (A(s) - W_+') \, u. \tag{31}$$

Because $A(s)$ is strongly differentiable with a derivative which is uniformly bounded and can be estimated by $c_u |t|^{-(1+\varepsilon)}$, $\varepsilon > 0$, we can use the same argument as in the proof of Prop. 8.28 to prove that (31) tends strongly to zero. ■

To illustrate the use of this statement we consider the following simple example

**Example 22** (relativistic potential scattering). Let $\mathscr{H}_0 = \mathscr{H} = L^2(\mathbb{R}^3, dx)$ and let $J = 1$. Assume that $H$ and $H_0$ are selfadjoint operators given by the expressions $\sqrt{-\Delta + m^2}$ and $\sqrt{-\Delta + m^2} + q(x)$ respectively, where $m$ is a nonnegative number. Further assume that $q(x) \geqq 0$ and $q \in L^\infty(\mathbb{R}^3, dx) \cap L^2(\mathbb{R}^3, dx)$. This gives that $H_0$ and $H$ are nonnegative operators and thus $H = (H^2)^{1/2}$, $H_0 = (H_0^2)^{1/2}$. Set $K_0 = H_0^2$ and $K = H^2$. Then we shall prove the assumptions of Theorem 21 for $H_0, H$ replaced by $K_0, K$. Since the function $\alpha(x)$ defined by $\alpha(x) = \sqrt{x}$ on $\mathbb{R}_+$ and $\alpha(x) = -\sqrt{|x|}$ on $\mathbb{R}_-$ satisfies the assumptions in Theorem 21 and $\alpha(K_0) = H_0$, $\alpha(K) = H$, we then obtain the existence of the strong wave operators $W_\pm(H, H_0)$.

We have $K_0 = -\Delta + m^2$ and $K = K_0 + K^{1/2}Q + QK_0^{1/2} = K_0 + V$ where $Q$ denotes the bounded multiplication operator by $q(x)$ and where ima $Q \subseteqq$ dom $H$ is assumed. Thus

$$\|(R_K(i) - R_{K_0}(i)) \, e^{-itK_0} u\| = \|R_K(i) \, V R_{K_0}(i) \, e^{-itK_0} u\|$$
$$\leqq \|R_K(i) \, K^{1/2}\| \, \|Q \, e^{-itK_0} R_{K_0}(i) \, u\| + \|R_K(i)\| \, \|Q \, e^{-itK_0} K_0^{1/2} R_{K_0}(i) \, u\|. \tag{32}$$

Let $\mathscr{D}$ be the set of all $u(\cdot) \in L^2(\mathbb{R}^3, dx)$ so that $\hat{u}(\cdot) \in C_0^\infty(\mathbb{R}^3 \setminus \{0\})$. Thus $\mathscr{D}$ satisfies the conditions of Theorem 21. If $u \in \mathscr{D}$, then $u_1 = R_{K_0}(i) \, u$ and $u_2 = K_0^{1/2} R_{K_0}(i) \, u$ are also elements of $\mathscr{D}$. Furthermore, it is obvious that $R_K(i) \, K^{1/2} = (K^{1/2} R_K(-i))^*$ is bounded. (32) shows that it is therefore enough to prove that $\|Q \, e^{-itK_0} u\| \leqq c_u \, |t|^{-(1+\varepsilon)}$ for all $u \in \mathscr{D}$ to fulfil the estimate in Theorem 21. But we have already proved in Section 1.2 (see (3)) that this is satisfied for $\varepsilon = 1/2$ and all $u$ with $u(\cdot) \in L^1(\mathbb{R}^3, dx)$ and therefore also for all $u$ from the smaller set $\mathscr{D}$. This concludes the proof that the wave operators $W_\pm(H, H_0)$ exist in the strong sense; the compactness of $(R_K(i) - R_{K_0}(i)) \, R_{K_0}(i)^n$ follows easily from

$$R_K(i) - R_{K_0}(i) = R_K(i) \, (K^{1/2}Q + QK_0^{1/2}) \, R_{K_0}(i)$$

and $q \in L^2(\mathbb{R}^3, dx)$ (see Lemma 16.8).

# 16. Trace Class Methods

## 16.1. Wave operators for trace class perturbations

### 16.1.1. A general theorem about trace class perturbations

In this section we want to derive the existence of strong wave operators under the assumption that $HJ - JH_0$ is a trace operator. In the next section this result is used to obtain further useful statements on trace class perturbations of more general kind. The proof of the following theorem needs only the well-known Rosenblum lemma on Hilbert-Schmidt operators (see Prop. 3.24) and the fact that for compact operators $C$ the expression $C\,\mathrm{e}^{-\mathrm{i}tH_0}P_0^{\mathrm{ac}}$ converges strongly to zero as $t \to \pm\infty$. Further recall the concept of admissible functions (cf. Definition 8.16).

**Theorem 1.** *Suppose that $V := HJ - JH_0 \in \mathcal{L}_1(\mathcal{H}_0, \mathcal{H})$ (in the form sense). Then for each admissible function $\Phi$ the strong wave operators $W_+(\Phi(H), \Phi(H_0); J)$, $W_-(\Phi(H), \Phi(H_0); J)$ exist and are independent of $\Phi$. In particular, the wave operators $W_+(H; H_0; J)$, $W_-(H; H_0; J)$ exist in the strong sense.*

Proof. We give the proof only for the $+$ sign. Set $W_t := \mathrm{e}^{\mathrm{i}tH}J\,\mathrm{e}^{-\mathrm{i}tH_0}$. By the uniform boundedness of $W_t$ it suffices to prove that

$$\|W_tu - W_su\| \to 0 \quad \text{as} \quad t > s \to \infty \tag{1}$$

for all $u \in \mathcal{M}_\infty(H_0)$ in order to show that $W_+(H, H_0; J)$ exists in the strong sense. The idea of the proof is to write

$$\|(W_t - W_s)\,u\|^2 = \big(u, W_t^*(W_t - W_s)\,u\big) - \big(u, W_s^*(W_t - W_s)\,u\big), \tag{2}$$

that is to write the strong limit as a weak one and to transform the operators on the right side of (2) into integrals over terms of the form $\mathrm{e}^{\mathrm{i}pH_0}X_{t,s}\,\mathrm{e}^{-\mathrm{i}pH_0}$, where $X_{t,s}$ is uniformly bounded in $t$ and $s$ and of trace class. The last step is then to use Prop. 3.24.

1. For fixed $t$ and $s$ we have

$$W_t - W_s = \mathrm{i}\int_s^t \mathrm{e}^{\mathrm{i}pH}V\,\mathrm{e}^{-\mathrm{i}pH_0}\,\mathrm{d}p.$$

Since $V$ is compact, we get that $W_t - W_s$ and $W_r^*(W_t - W_s)$ are compact (note that the integral is norm convergent and the values of the integrand are members of $\mathcal{L}_\infty(\mathcal{H}_0, \mathcal{H})$). Thus, by Prop. 6.70,

$$\operatorname*{s-lim}_{a\to\infty} W_r^*(W_t - W_s)\,\mathrm{e}^{-\mathrm{i}aH_0}u = 0. \tag{3}$$

Inserting this into (2), we find

$$\begin{aligned}\|(W_t - W_s)\,u\|^2 &= \lim_{a\to\infty} \big\{\big(u, W_t^*(W_t - W_s)\,u\big) - \big(u, e^{iaH_0}W_t^*(W_t - W_s)\,e^{-iaH_0}u\big)\\ &\quad - \big(u, W_s^*(W_t - W_s)\,u\big) + \big(u, e^{iaH_0}W_s^*(W_t - W_s)\,e^{-iaH_0}u\big)\big\}\\ &= \lim_{a\to\infty} \big[\big(u, \{W_t^*(W_t - W_s) - e^{iaH_0}W_t^*(W_t - W_s)\,e^{-iaH_0}\}\,u\big)\\ &\quad - \big(u, \{W_s^*(W_t - W_s) - e^{iaH_0}W_s^*(W_t - W_s)\,e^{-iaH_0}\}\,u\big)\big].\end{aligned}$$

Now we have

$$W_t^*W_s - e^{iaH_0}W_t^*W_s\,e^{-iaH_0} = -i\int_0^a e^{ipH_0}(H_0W_t^*W_s - W_t^*W_sH_0)\,e^{-ipH_0}\,dp$$

and we can express $H_0W_t^*W_s - W_t^*W_sH_0$ by $V$ and $V^*$:

$$\begin{aligned}H_0W_t^*W_s - W_t^*W_sH_0 &= e^{itH_0}H_0J^*\,e^{-itH}\,e^{isH}J\,e^{-isH_0} - e^{itH_0}J^*\,e^{-itH}\,e^{isH}JH_0\,e^{-isH_0}\\ &= e^{itH_0}(H_0J^*\,e^{i(s-t)H}J - J^*\,e^{i(s-t)H}JH_0)\,e^{-isH_0}\\ &= e^{itH_0}\big((-V^* + J^*H)\,e^{i(s-t)H}J - J^*\,e^{i(s-t)H}(HJ - V)\big)\,e^{-isH_0}\\ &= e^{itH_0}(-V^*\,e^{i(s-t)H}J + J^*\,e^{i(s-t)H}V)\,e^{-isH_0}.\end{aligned}$$

For abbreviation we put

$$X_{t,s} := e^{itH_0}(J^*\,e^{i(s-t)H}V - V^*\,e^{i(s-t)H}J)\,e^{-isH_0} \tag{4}$$

and

$$F_a(X) := -i\int_0^a e^{ipH_0}X\,e^{-ipH_0}\,dp, \qquad X \in \mathcal{L}(\mathcal{H}_0). \tag{5}$$

Then we have

$$\|(W_t - W_s)\,u\|^2 = \lim_{a\to\infty} \big(u, F_a(X_{t,t} - X_{t,s} - X_{s,t} + X_{s,s})\,u\big). \tag{6}$$

2. Formula (6) shows that it is sufficient for the verification of (1) to estimate the terms $\big(u, F_a(X_{t,s})\,u\big)$ uniformly with respect to $a$ and to get the bounds small if $t$ and $s$ are sufficiently large. First, by Schwarz' inequality we obtain a simple estimate for $\big(u, F_a(e^{itH_0}X^*Y\,e^{-isH_0})\,u\big)$ with $X, Y \in \mathcal{L}(\mathcal{H}_0)$:

$$\begin{aligned}\big|\big(u, F_a(e^{itH_0}X^*Y\,e^{-isH_0})\,u\big)\big| &\le \int_0^a |(X\,e^{-i(t+p)H_0}u, Y\,e^{-i(s+p)H_0}u)|\,dp\\ &\le \left(\int_t^{a+t} \|X\,e^{-ipH_0}u\|^2\,dp\right)^{1/2} \left(\int_s^{a+s} \|Y\,e^{-ipH_0}u\|^2\,dp\right)^{1/2}.\end{aligned} \tag{7}$$

Obviously it is sufficient to consider one of the four terms of (6), for example $X_{t,s}$. Because the arguments are the same it is enough to consider one of the two terms of

$X_{t,s}$, for example $J^* e^{i(s-t)H} V$. In this case we put

$$Y = |V|^{1/2} \in \mathcal{L}_2(\mathcal{H}_0), \qquad X = |V|^{1/2} (\operatorname{sgn} V^*) e^{-i(s-t)H} J \in \mathcal{L}_2(\mathcal{H}_0).$$

Then, by (7) and Prop. 3.24, we obtain

$$\begin{aligned}
&|(u, F_s(e^{itH_0} J^* e^{i(s-t)H} V e^{-isH_0}) u)| \\
&\leq \left( \int_{-\infty}^{\infty} \| |V|^{1/2} (\operatorname{sgn} V)^* e^{-i(s-t)H} J e^{-ipH_0} u\|^2 dp \right)^{1/2} \left( \int_s^{\infty} \| |V|^{1/2} e^{-ipH_0} u\|^2 dp \right)^{1/2} \\
&\leq (2\pi)^{1/2} \| |V|^{1/2} (\operatorname{sgn} V)^* e^{-i(s-t)H} J\|_2 \, |||u|||_\infty \left( \int_s^{\infty} \| |V|^{1/2} e^{-ipH_0} u\|^2 dp \right)^{1/2} \\
&\leq (2\pi)^{1/2} \|J\| \cdot \| |V|^{1/2}\|_2 \cdot |||u|||_\infty \left( \int_s^{\infty} \| |V|^{1/2} e^{-ipH_0} u\|^2 dp \right)^{1/2}.
\end{aligned} \tag{8}$$

Once more using Prop. 3.24 we obtain the existence of $W_+(H, H_0; J)$ in the strong sense.

**3.** It remains to show the existence of $W_+(\Phi(H), \Phi(H_0); J)$ in the strong sense. For this it is enough to prove that

$$\|W_+ u - e^{iq\Phi(H)} J e^{-iq\Phi(H_0)} u\| = \|(W_+ - J) e^{-iq\Phi(H_0)} u\| \to 0 \text{ as } q \to \infty \text{ for } u \in \mathcal{M}_\infty(H_0).$$

Now we have

$$\|(W_+ - J) e^{-iq\Phi(H_0)} u\| = \lim_{t\to\infty} \|(W_t - J) e^{-iq\Phi(H_0)} u\|.$$

Using (6) with $s = 0$ and replacing $u$ by $e^{-iq\Phi(H_0)} u$, from (6) and (8) we obtain

$$\begin{aligned}
&\|(W_+ - J) e^{-iq\Phi(H_0)} u\|^2 \\
&\leq 8(2\pi)^{1/2} \|J\| \cdot \| |V|^{1/2}\|_2 \cdot |||u|||_\infty \left( \int_0^{\infty} dp \, \| |V|^{1/2} e^{-ipH_0} e^{-iq\Phi(H_0)} u\|^2 \right)^{1/2} \\
&= \text{const} \left( \sum_n \mu_n \int_0^{\infty} dp \, |(\varphi_n, e^{-ipH_0} e^{-iq\Phi(H_0)} u)|^2 \right)^{1/2},
\end{aligned} \tag{9}$$

where $|V| = \sum_n \mu_n \varphi_n(\varphi_n, \cdot)$, $\|\varphi_n\| = 1$, $\{\varphi_n\}$ an orthonormal system. Because $\Phi$ is admissible, we get from Lemma 8.19 (cf. formula (14.22)) that the integral converges to zero as $q \to \infty$ for each $n$. Since the integrals are uniformly bounded in $n$ and $\sum_n \mu_n < \infty$, we get that the right-hand side tends to zero as $q \to \infty$. This concludes the proof. ∎

**Remark 2.** Relation (8) also gives the speed of convergence of $W_t$ to $W_+$ for the vectors $u$ from $\mathcal{M}_\infty(H_0)$. Furthermore, from (9) we obtain the estimate

$$\|(W_+ - J) u\|^2 \leq 16\pi \|J\| \cdot |||u|||_\infty{}^2 \cdot \|V\|_1.$$

**Remark 3.** From $HJ - JH_0 \in \mathcal{L}_1(\mathcal{H}_0, \mathcal{H})$ it follows that $H_0 J^* - J^* H \in \mathcal{L}_1(\mathcal{H}, \mathcal{H}_0)$ and thus the triple $\{H_0, H, J^*\}$ in place of $\{H, H_0; J\}$ also satisfies the assumptions of Theorem 1. Hence the strong wave operators $W_\pm(\Phi(H_0), \Phi(H); J^*)$ exist as well, and this together with our usual assumption on $J$ and $J^*$ ($J$ is asymptotic left inverse to $J^*$ and $J^*$ is asymptotic left inverse to $J$, see Prop. 9.39 and 9.40) may be used to get the completeness of the wave operators $W_\pm(\Phi(H), \Phi(H_0); J)$.

### 16.1.2. Other trace class results

Often $HJ - JH_0$ is not a trace operator, but some function of $H$, $H_0$ and $J$ is. Then it is also possible to prove results as strong as those in Theorem 1. The next statement uses the semiboundedness of $H$ and $H_0$ as additional assumptions.

**Proposition 4.** *Suppose that the operators $H, H_0, J$ satisfy the following assumptions:*

(I) $H_0 \geqq -(c-1)$ *and* $H \geqq -(c-1)$ *for some* $c \geqq 0$.

(II) $(H+c)^{-n} J - J(H_0+c)^{-n} \in \mathscr{L}_1(\mathscr{H}_0, \mathscr{H})$ *for some* $n \in \mathbb{N}$.

*Then the strong wave operators $W_+(\Phi(H), \Phi(H_0); J)$, $W_-(\Phi(H), \Phi(H_0); J)$ exist for each admissible function $\Phi$.*

Proof. Define $\Psi(x)$ by $\Psi(x) = (-x)^{-1/n} - c$ on $(-\infty, 0)$ and $\Psi(x) = x - c$ on $[0, \infty)$. Let $\Psi_1(x) = -(x+c)^{-n}$ on $(-c, \infty)$ and $\Psi_1(x) = x + c$ on $(-\infty, -c]$. Then $\Psi(\Psi_1(x)) = x$ on $(-c, \infty)$. Now let $\Phi$ be an admissible function. Then the function $\tilde{\Phi}(x) := \Phi(\Psi(x))$ is welldefined, and it is easy to verify from $\Psi'(x) = (1/n)(-x)^{-(1+1/n)} > 0$ on $(-\infty, 0)$, $\Psi'(x) = 1$ on $[0, \infty)$ and from the fact that $\Psi$ is twice continuously differentiable on $(-\infty, 0) \cup (0, \infty)$ that $\tilde{\Phi}$ is admissible. Further, we have $\Psi_1(H_0) = -(H_0+c)^{-n}$, $\Psi_1(H) = -(H+c)^{-n}$. Thus we obtain

$$\Phi(H_0) = \tilde{\Phi}\big(-(H_0+c)^{-n}\big), \qquad \Phi(H) = \tilde{\Phi}\big(-(H+c)^{-n}\big). \tag{10}$$

Using Theorem 1 with $H$, $H_0$ replaced by $-(H_0+c)^{-n}$, $-(H+c)^{-n}$, we obtain that the strong wave operators $W_\pm\big(\tilde{\Phi}(-(H_0+c)^{-n}), \tilde{\Phi}(-(H+c)^{-n}); J\big)$ exist. This together with (10) gives the desired result. ∎

Sometimes one can factor the perturbation $V := HJ - JH_0$ in the form $V = B^*A$ such that $BR(i)^n$ and $AR_0(i)^n$ are Hilbert-Schmidt operators for some $n \in \mathbb{N}$. Then the following theorem is useful.

**Proposition 5.** *Suppose that $V = B^*A$ (in the form sense) such that:*

(I) *$B$ is $H$-bounded and $A$ is $H_0$-bounded.*

(II) $BR(z)^n \in \mathscr{L}_{r_1}(\mathscr{H}, \mathscr{H}_0)$ *and* $AR_0(z)^n \in \mathscr{L}_{r_2}(\mathscr{H}_0, \mathscr{H}_0)$ *for some* $z \in \operatorname{res} H \cap \operatorname{res} H_0$, $n \in \mathbb{N}$ *and* $r_1, r_2 \in \mathbb{N} \cup \{\infty\}$ *with* $r_1^{-1} + r_2^{-1} = 1$. *Then the strong wave operators $W_\pm(\Phi(H), \Phi(H_0); J)$ and $W_\pm(\Phi(H_0), \Phi(H); J^*)$ exist for each admissible function $\Phi$.*

Proof. Set $J' := R(\bar{z})^{2n} JR_0(z)^{2n}$. Then

$$\begin{aligned} HJ' - J'H_0 &= R(\bar{z})^{2n} HJR_0(z)^{2n} - R(\bar{z})^{2n} JH_0R_0(z)^{2n} \\ &= R(\bar{z})^{2n} (HJ - JH_0) R_0(z)^{2n} = R(\bar{z})^{2n} VR_0(z)^{2n} \\ &= R(\bar{z})^n \big(BR(z)^n\big)^* \big(AR_0(z)^n\big) R_0(z)^n. \end{aligned} \tag{11}$$

Since $BR(z)^n \in \mathscr{L}_{r_1}(\mathscr{H}, \mathscr{H}_0)$ implies $\left(BR(z)^n\right)^* \in \mathscr{L}_{r_1}(\mathscr{H}_0, \mathscr{H})$ and since $C \in \mathscr{L}_{r_1}(\mathscr{H}_0, \mathscr{H})$, $D \in \mathscr{L}_{r_2}(\mathscr{H}_0, \mathscr{H}_0)$ implies $CD \in \mathscr{L}_1(\mathscr{H}_0, \mathscr{H})$ (see Section 1.3.4), we obtain from (11) that $HJ' - J'H_0 \in \mathscr{L}_1(\mathscr{H}_0, \mathscr{H})$. From Theorem 1 we then get that the strong wave operators $W_\pm\left(\Phi(H), \Phi(H_0); J'\right)$ exist. Because of the definition of $J'$ this result and a density argument prove that

$$\operatorname*{s-lim}_{t\to\pm\infty} e^{it\Phi(H)} R(\bar{z})^{2n} J\, e^{-it\Phi(H_0)} P_0^{ac} \tag{12}$$

also exist. Now we claim that

$$\operatorname*{s-lim}_{t\to\pm\infty} \left(R(\bar{z})^{2n} J - JR_0(\bar{z})^{2n}\right) e^{-it\Phi(H_0)} P_0^{ac} = 0. \tag{13}$$

(12) and (13) then imply that

$$\operatorname*{s-lim}_{t\to\pm\infty} e^{it\Phi(H)} J\, e^{-it\Phi(H_0)} P_0^{ac} u$$

exist for all $u \in \mathscr{D} := R_0(\bar{z})^{2n} \mathscr{H}_0$. $\mathscr{D}$ is dense in $\mathscr{H}_0$ and thus this gives the existence of the strong wave operators $W_\pm\left(\Phi(H), \Phi(H_0); J\right)$. It remains to prove (13). Using succesively $R(\bar{z})\, J - JR_0(\bar{z}) = R(\bar{z})\, VR_0(\bar{z})$ we find

$$\begin{aligned} R(\bar{z})^{2n} J &= JR_0(\bar{z})^{2n} + \sum_{l=1}^{2n} R(\bar{z})^l\, VR_0(\bar{z})^{2n+1-l} \\ &= JR_0(\bar{z})^{2n} + \sum_{l=1}^{2n} \left(BR(z)^l\right)^* AR_0(\bar{z})^{2n+1-l}. \end{aligned}$$

Clearly either $l$ or $2n + 1 - l$ is greater or equal to $n$. Thus each term in the finite sum is an operator from $\mathscr{L}_r(\mathscr{H}_0, \mathscr{H})$ with $r = \max\{r_1, r_2\}$ because of our assumptions on $AR_0(z)^n$, $BR(z)^n$ and the fact that $AR_0(z)^n \in \mathscr{L}_{r_2}(\mathscr{H}_0, \mathscr{H}_0)$ implies $AR_0(\bar{z})^n \in \mathscr{L}_{r_2}(\mathscr{H}_0, \mathscr{H}_0)$ as a simple calculation with resolvents shows. Hence $R(z)^{2n}J - JR_0(z)^{2n}$ is compact and this proves, by Prop. 6.70, that (13) is true. Since the assumptions are symmetric with respect to $H$ and $H_0$, we also get the existence of the strong wave operators $W_\pm\left(\Phi(H_0), \Phi(H); J^*\right)$. ■

**Remark 6.** From the proof of this statement we see that the assumptions can be replaced by $R(z)^n\, VR_0(z)^n \in \mathscr{L}_1(\mathscr{H}_0, \mathscr{H})$ and $R(z)\, VR_0(z) \in \mathscr{L}_\infty(\mathscr{H}_0, \mathscr{H})$.

## 16.2. A stationary proof of a trace class theorem

Because of the importance of the trace class methods on the one hand, and in order to illustrate that the assumptions of the general stationary criterion for the existence of wave operators can be verified without great difficulties on the other hand, we present here a stationary proof of a statement like Theorem 1. The essential elements of the proof are some results from Part I for operators $A\, e^{itH}$, $AR(\lambda + i\varepsilon)$, $AE(\Delta)$ and $\int_\Delta E(d\lambda)\, A\{\cdot\}$ where $A$ is in the Hilbert-Schmidt class.

**Theorem 7.** *Suppose* $V := HJ - JH_0 \in \mathcal{L}_1(\mathcal{H}_0, \mathcal{H})$ *(in the form sense). Then the strong wave operators* $W_+(H, H_0; J)$, $W_-(H, H_0; J)$ *exist.*

**Proof.** We give the proof only for the $+$ sign. It requires several steps. First we deduce with the help of Theorem 13.22 that $\operatorname*{s-lim}_{\varepsilon\to+0} P^{ac}\Omega_\varepsilon P_0^{ac}$ exists. Second we show that $\operatorname*{s-lim}_{\varepsilon\to+0} E(\Delta)\,\Omega_\varepsilon P_0^{ac} = 0$ for null sets $\Delta$, which implies that we may omit in $\operatorname*{s-lim}_{\varepsilon\to+0} P^{ac}\Omega_\varepsilon P_0^{ac}$ the projection $P^{ac}$ ($(P^{ac})^\perp = E(\Delta)$ for some null set $\Delta$), i.e. the second step yields the existence of $\operatorname*{s-lim}_{\varepsilon\to+0} \Omega_\varepsilon P_0^{ac} = \Omega_+$. The last step is to verify that $\operatorname*{s-lim}_{t\to\infty} (\Omega_+ - J)\,e^{-itH_0}P_0^{ac} = 0$, where we use the fact that

$$\Omega_+ - J = -\operatorname*{s-lim}_{\varepsilon\to+0} \int_{-\infty}^{\infty} P^{ac}E(d\lambda)\, VR_0(\lambda + i\varepsilon)\, P_0^{ac}$$

(see Prop. 13.2). By Prop. 9.42 this proves that the strong wave operator $W_+(H, H_0; J)$ exists.

1. First we claim that $\|E(\Delta)\, P^{ac}\, e^{itH} J\, e^{-itH_0}u\|^2$ is equiabsolutely continuous, that is

$$\lim_{|\Delta|\to 0} \sup_{t\geq 0} \|E(\Delta)\, P^{ac}\, e^{itH} J\, e^{-itH_0}u\|^2 = 0 \tag{14}$$

for all $u \in \mathcal{M}_\infty(H_0)$. This then implies that $\|E(\Delta)\, P^{ac}\Omega_\varepsilon P_0^{ac}u\|$ is equiabsolutely continuous in $\varepsilon \in (0, 1)$ for all $u \in \mathcal{M}_\infty(H_0)$ because of the estimate

$$\begin{aligned}\sup_{\varepsilon\in(0,1)} \|E(\Delta)\, P^{ac}\Omega_\varepsilon P_0^{ac}u\| &\leq \sup_{\varepsilon\in(0,1)} \int_0^\infty dt\, \varepsilon\, e^{-\varepsilon t}\, \|E(\Delta)\, P^{ac}\, e^{itH} J\, e^{-itH_0}P_0^{ac}u\| \\ &\leq \sup_{t\geq 0} \|E(\Delta) P^{ac}\, e^{itH} J\, e^{-itH_0}P_0^{ac}u\|.\end{aligned} \tag{15}$$

Since $\mathcal{M}_\infty(H_0)$ is dense in $P_0^{ac}\mathcal{H}_0$ and $\Omega_\varepsilon$ is uniformly bounded in $\varepsilon$, this result can be extended to all vectors $u \in P_0^{ac}\mathcal{H}_0$.

Now we verify (14). We do this by deriving of an estimate for the function

$$g_\Delta(t) := \|E(\Delta)\, J\, e^{-itH_0}u\|^2 = (u, e^{itH_0}J^*E(\Delta)\, J\, e^{-itH_0}u).$$

Using $g_\Delta'(t) = i\big(u, e^{itH_0}(H_0J^*E(\Delta)\, J - J^*E(\Delta)\, JH_0)\, e^{-itH_0}u\big)$, we get

$$\begin{aligned} g_\Delta(t) &= \big(u, J^*E(\Delta)\, Ju\big) \\ &\quad + i\int_0^t \big(u, e^{i\tau H_0}(H_0J^*E(\Delta)\, J - J^*E(\Delta)\, JH_0)\, e^{-i\tau H_0}u\big)\, d\tau. \end{aligned} \tag{16}$$

An elementary calculation using $V = HJ - JH_0$ and $V^* = J^*H - H_0J^*$ gives

$$H_0J^*\big(E(\Delta)\, J\big) - \big(J^*E(\Delta)\big)\, JH_0 = -V^*E(\Delta)\, J + J^*E(\Delta)\, V.$$

Inserting this into (16), we obtain

$$\begin{aligned} g_\Delta(t) &= \|E(\Delta)\, Ju\|^2 - i\int_0^t d\tau\big(u; e^{i\tau H_0}(V^*E(\Delta)\, J - J^*E(\Delta)\, V)\, e^{-i\tau H_0}u\big) \\ &= \|E(\Delta)\, Ju\|^2 - 2\,\mathrm{Im}\int_0^t d\tau\big(u, e^{i\tau H_0}J^*E(\Delta)\, V\, e^{-i\tau H_0}u\big) \end{aligned}$$

where we have used that $V^*E(\Delta)\, J = \big(J^*E(\Delta)\, V\big)^*$. Factoring $V = B^*A$ with $A = |V|^{1/2} \in \mathscr{L}_2(\mathscr{H}_0)$ and $B^* = (\operatorname{sgn} V)\, |V|^{1/2} \in \mathscr{L}_2(\mathscr{H}_0, \mathscr{H})$, using the Schwarz inequality and Prop. 3.24, this equality gives

$$\begin{aligned} g_\Delta(t) &\leqq \|E(\Delta)\, Ju\|^2 + 2 \int_0^t \big|\big(BE(\Delta)\, J\, \mathrm{e}^{-\mathrm{i}\tau H_0}u,\, A\, \mathrm{e}^{-\mathrm{i}\tau H_0}u\big)\big|\, \mathrm{d}\tau \\ &\leqq \|E(\Delta)\, Ju\|^2 + 2 \int_0^t \|A\, \mathrm{e}^{-\mathrm{i}\tau H_0}u\| \cdot \|BE(\Delta)\, J\, \mathrm{e}^{-\mathrm{i}\tau H_0}u\|\, \mathrm{d}\tau \\ &\leqq \|E(\Delta)\, Ju\|^2 + 2 \left(\int_0^\infty \|A\, \mathrm{e}^{-\mathrm{i}\tau H_0}u\|^2\, \mathrm{d}\tau \int_0^\infty \|BE(\Delta)\, J\, \mathrm{e}^{-\mathrm{i}\tau H_0}u\|^2\, \mathrm{d}\tau\right)^{1/2} \\ &\leqq \|E(\Delta)\, Ju\|^2 + 4\pi\, \|A\|_2\, |||u|||^2\, \|BE(\Delta)\, J\|_2. \end{aligned} \tag{17}$$

In particular we obtain from (17)

$$\|E(\Delta)\, P^{\mathrm{ac}}\, \mathrm{e}^{\mathrm{i}tH} J\, \mathrm{e}^{-\mathrm{i}tH_0}u\|^2 \leqq \|E(\Delta)\, P^{\mathrm{ac}} Ju\|^2 + 4\pi\, \|A\|_2\, |||u|||^2\, \|J\| \cdot \|BP^{\mathrm{ac}}E(\Delta)\|_2.$$

By Prop. 3.28, $\|BP^{\mathrm{ac}}E(\Delta)\|_2$ is absolutely continuous. Thus (15) is proved.

To prove the existence of $\operatorname*{s-lim}_{\varepsilon\to+0} P^{\mathrm{ac}}\Omega_\varepsilon P_0^{\mathrm{ac}}$ it remains to verify that

$$d_{\varepsilon_1,\varepsilon_2}(\lambda) := |G_{\varepsilon_1}(\lambda)\, u - G_{\varepsilon_2}(\lambda)\, u|_{E,\lambda}^2 = |VR_0(\lambda + \mathrm{i}\varepsilon_1)\, u - VR_0(\lambda + \mathrm{i}\varepsilon_2)\, u|_{E,\lambda}^2$$

is convergent in measure to zero as $\varepsilon_1, \varepsilon_2 \to +0$ with respect to $\lambda \in \mathbb{R}$ (recall the definition of $|\cdot|_{E,\lambda}$ in Section 3.5.5, and recall Prop. 13.3). We have, setting $R_{\varepsilon_1,\varepsilon_2}(\lambda) := R_0(\lambda + \mathrm{i}\varepsilon_1) - R_0(\lambda + \mathrm{i}\varepsilon_2)$,

$$\begin{aligned} d_{\varepsilon_1,\varepsilon_2}(\lambda) &= \big(VR_{\varepsilon_1,\varepsilon_2}(\lambda)\, u,\, E(\mathrm{d}\mu)\, VR_{\varepsilon_1,\varepsilon_2}(\lambda)u\big)/\mathrm{d}\mu|_{\mu=\lambda} \\ &= \big(AR_{\varepsilon_1,\varepsilon_2}(\lambda)\, u,\, \{(BE(\mathrm{d}\mu)\, B^*)/\mathrm{d}\mu\}\, AR_{\varepsilon_1,\varepsilon_2}(\lambda)\, u\big)|_{\mu=\lambda} \\ &\leqq \|(BE(\mathrm{d}\lambda)\, B^*)/\mathrm{d}\lambda\| \cdot \|AR_{\varepsilon_1,\varepsilon_2}(\lambda)\, u\|^2. \end{aligned}$$

The first factor on the right-hand side exists a.e. (mod $|\cdot|$) (see Prop. 3.13) and the second is a Cauchy net as $\varepsilon_1, \varepsilon_2 \to +0$ for almost all $\lambda$ (mod $|\cdot|$) (see Prop. 3.15). This implies the desired result and, applying Theorem 13.22, we obtain the existence of $\operatorname*{s-lim}_{\varepsilon\to+0} P^{\mathrm{ac}}\Omega_\varepsilon P_0^{\mathrm{ac}}$.

2. The second step requires to prove that $\operatorname*{s-lim}_{\varepsilon\to+0} E(\Delta)\, \Omega_\varepsilon P_0^{\mathrm{ac}} = 0$ for null sets $\Delta$. For this it is sufficient to show that

$$\operatorname*{s-lim}_{t\to\infty} E(\Delta)\, J\, \mathrm{e}^{-\mathrm{i}tH_0} P_0^{\mathrm{ac}} = 0, \qquad |\Delta| = 0. \tag{18}$$

For the proof of (18) it is enough to show that

$$\operatorname*{s-lim}_{t\to\infty} E(\Gamma)\, J\, \mathrm{e}^{-\mathrm{i}tH_0} P_0^{\mathrm{ac}} E_0(\mathbb{R} \setminus \Gamma) = 0 \tag{19}$$

for each open set $\Gamma$, since we can associate to a null set $\Delta$ a sequence of open sets $\Gamma_n \supset \Delta$ such that $|\Gamma_n| \to 0$ as $n \to \infty$ and the set of vectors $u \in \bigcup_{n=1}^\infty E(\mathbb{R} \setminus \Gamma_n)\, P_0^{\mathrm{ac}}\mathscr{H}_0$ is dense in $P_0^{\mathrm{ac}}\mathscr{H}_0$. First we prove (19) for open intervals $(a, b)$. It is sufficient to prove this relation for all vectors of the form $u = E_0(M)\, u$ with some closed set $M \subset \mathbb{R} \setminus [a, b]$, because this set of vectors $u$ is dense in $P_0^{\mathrm{ac}} E_0(\mathbb{R} \setminus (a, b))\, \mathscr{H}_0$. Then $\operatorname{dis}(M, (a, b)) > 0$. Now an easy calculation yields

$$E((a, b))\, J E_0(M) = \int_a^b E(\mathrm{d}\lambda) \int_M (\lambda - \mu)^{-1}\, V E_0(\mathrm{d}\mu).$$

Since $V \in \mathscr{L}_1(\mathscr{H}_0, \mathscr{H}) \subset \mathscr{L}_\infty(\mathscr{H}_0, \mathscr{H})$ and the operator spectral integrals are convergent with respect to the operator norm (cf. Sections 5.1.1 and 5.1.2), we obtain $E((a, b))\, J E_0(M) \in \mathscr{L}_\infty(\mathscr{H}_0, \mathscr{H})$, i.e. (19) is valid for open intervals. Then it easily follows that (19) is also true for open sets $\Gamma$ being a union of finitely many disjoint open intervals. Now let $\Gamma$ be arbitrarily open. Then there exists a decomposition $\Gamma = \bigcup_{\varrho=1}^{\infty} (a_\varrho, b_\varrho)$, where $(a_\varrho, b_\varrho)$ are disjoint intervals. We put $\Gamma_n = \Gamma \setminus \bigcup_{\varrho=1}^{n} (a_\varrho, b_\varrho)$. For $\Gamma$ replaced by $\bigcup_{\varrho=1}^{n} (a_\varrho, b_\varrho)$, (19) is true. Thus the proof of (19) for $\Gamma$ can be reduced to the question wether

$$\lim_{n\to\infty} \sup_{t \geq 0} \|E(\Gamma_n)\, J\, \mathrm{e}^{-\mathrm{i}tH_0} P_0^{\mathrm{ac}} E_0(\mathbb{R} \setminus \Gamma)\, u\| = 0 \tag{20}$$

for a dense set of vectors $u$ in $P_0^{\mathrm{ac}}\mathscr{H}_0$. For this set we take $\mathscr{M}_\infty(H_0)$. From the estimate (17) we obtain

$$\sup_{t \geq 0} \|E(\Gamma_n)\, J\, \mathrm{e}^{-\mathrm{i}tH_0} P_0^{\mathrm{ac}} E_0(\mathbb{R} \setminus \Gamma)\, u\|^2 \leq \|E(\Gamma_n)\, J P_0^{\mathrm{ac}} E_0(\mathbb{R} \setminus \Gamma)\, u\|^2$$
$$+ 4\pi\, \|A\|_2\, |||u|||^2\, \|J\| \cdot \|B E(\Gamma_n)\|_2.$$

Since $\operatorname*{s-lim}_{n\to\infty} E(\Gamma_n) = 0$ this implies (20) (note that $\|B E(\Gamma_n)\|_2^2 = \|E(\Gamma_n)\, B^*\|_2^2 = \sum_\varrho \|E(\Gamma_n)\, B^* e_\varrho\|^2 \leq \sum_\varrho \|B^* e_\varrho\|^2$ with some orthonormal basis). Thus (19), hence also (18), is proved.

3. The last step is to prove $\operatorname*{s-lim}_{t\to\infty} (\Omega_+ - J)\, \mathrm{e}^{-\mathrm{i}tH_0} P_0^{\mathrm{ac}} = 0$. We put $\Omega_+ - J = Y_+$. Because of (18) $\operatorname*{s-lim}_{t\to\infty} Y_+\, \mathrm{e}^{-\mathrm{i}tH_0} P_0^{\mathrm{ac}} = 0$ is equivalent with $\operatorname*{s-lim}_{t\to\infty} P^{\mathrm{ac}} Y_+\, \mathrm{e}^{-\mathrm{i}tH_0} P_0^{\mathrm{ac}} = 0$. We prove the latter relation. Let $\Delta_1 \subset \Delta_2 \subset \cdots$ be a sequence of Borel sets with $|\mathbb{R} \setminus \Delta_N| \to 0$ as $N \to \infty$. Further properties of this sequence are fixed later. Then we can write

$$P^{\mathrm{ac}} Y_+\, \mathrm{e}^{-\mathrm{i}tH_0} u = P^{\mathrm{ac}} E(\Delta_N)\, Y_+\, \mathrm{e}^{-\mathrm{i}tH_0} u + P^{\mathrm{ac}} E(\mathbb{R} \setminus \Delta_N)\, Y_+\, \mathrm{e}^{-\mathrm{i}tH_0} u. \tag{21}$$

Now it suffices to show that the first term on the right side tends strongly to zero for each fixed $N$ and the second term goes to zero uniformly in $t$ as $N \to \infty$. For the second term we use again the estimate (17) for $u \in \mathscr{M}_\infty(H_0)$, leading to

$$\|P^{\mathrm{ac}} E(\mathbb{R} \setminus \Delta_N)\, Y_+\, \mathrm{e}^{-\mathrm{i}tH_0} u\| \leq \|P^{\mathrm{ac}} E(\mathbb{R} \setminus \Delta_N)\, \Omega_+\, \mathrm{e}^{-\mathrm{i}tH_0} u\| + \|P^{\mathrm{ac}} E(\mathbb{R} \setminus \Delta_N)\, J\, \mathrm{e}^{-\mathrm{i}tH_0} u\|$$
$$\leq \|P^{\mathrm{ac}} E(\mathbb{R} \setminus \Delta_N)\, \Omega_+ u\| + \big(\|E(\mathbb{R} \setminus \Delta_N)\, P^{\mathrm{ac}} J u\|^2 + 4\pi\, \|A\|_2\, |||u|||^2\, \|J\| \cdot \|B P^{\mathrm{ac}} E(\mathbb{R} \setminus \Delta_N)\|\big)^{1/2}$$

Since $|\mathbb{R} \setminus \Delta_N| \to 0$ as $N \to \infty$ the right side tends to zero as $N \to \infty$. For the first term we specify the sequence $\Delta_N$: Let

$$\Delta_N = \{\lambda \in \mathbb{R};\ \|(BP^{ac}E(\mathrm{d}\lambda)\, B^*)/\mathrm{d}\lambda\| \leq N\}.$$

Because $\|(BP^{ac}E(\mathrm{d}\lambda)\, B^*)/\mathrm{d}\lambda\| < \infty$ a.e. (mod $|\cdot|$) we have $|\mathbb{R} \setminus \Delta_N| \to 0$ as $N \to \infty$. For $u \in \mathcal{M}_\infty(H_0)$ we get

$$P^{ac}E(\Delta_N)\, Y_+\, \mathrm{e}^{-\mathrm{i}tH_0}u = -\lim_{\varepsilon \to +0} \int_{\Delta_N} P^{ac}E(\mathrm{d}\lambda)\, B^*AR_0(\lambda + \mathrm{i}\varepsilon)\, \mathrm{e}^{-\mathrm{i}tH_0}u.$$

By Prop. 5.16, $\int_{\Delta_N} P^{ac}E(\mathrm{d}\lambda)\, B^*\{\cdot\}$ is a bounded operator from $L^2(\mathbb{R}, \mathrm{d}\lambda; \mathcal{H}_0)$ into $\mathcal{H}$. On the other hand, according to Remark 3.26, $AR_0(\lambda + \mathrm{i}\varepsilon)\, \mathrm{e}^{-\mathrm{i}tH_0}u$ is a member of the Hardy-class $\mathcal{H}_+^2(\mathbb{R}, \mathrm{d}\lambda; \mathcal{H}_0)$ and (3.39) yields

$$\begin{aligned}\lim_{\varepsilon \to +0} \int_{-\infty}^{\infty} \mathrm{d}\lambda\, \|AR_0(\lambda + \mathrm{i}\varepsilon)\, \mathrm{e}^{-\mathrm{i}tH_0}u\|^2 &= \int_{-\infty}^{\infty} \mathrm{d}\lambda\, \|AR_0(\lambda + \mathrm{i}0)\, \mathrm{e}^{-\mathrm{i}tH_0}u\|^2 \\ &= 2\pi \int_0^{\infty} \|A\, \mathrm{e}^{-\mathrm{i}(t+s)H_0}u\|^2\, \mathrm{d}s.\end{aligned}$$

Clearly, this tends to zero as $t \to \infty$. Thus, this and the boundedness of $\int_{\Delta_N} P^{ac}E(\mathrm{d}\lambda)\, B^*\{\cdot\}$ imply that the first term on the right side of (21) tends to zero as $t \to \infty$ in the strong sense. This concludes the proof of the last step. ■

## 16.3. Applications

### 16.3.1. Wave operators for $H = H_1 + H_2$, $H_0 = H_1 \oplus H_2$

Let $H_1$, $H_2$ be two bounded selfadjoint operators on the Hilbert space $\mathcal{H}$. Suppose that $H_1H_2$ is a trace operator. Define $H := H_1 + H_2$ on $\mathcal{H}$ and $H_0 := H_1 \times H_2$ on $\mathcal{H}_0 := \mathcal{H} \times \mathcal{H}$. Clearly, $H$ and $H_0$ are again selfadjoint operators. From $H_1H_2 \in \mathcal{L}_1(\mathcal{H})$ we easily get that

$$H_1E_{H_2}(\Delta) \in \mathcal{L}_1(\mathcal{H}), \quad H_2E_{H_1}(\Delta) \in \mathcal{L}_1(\mathcal{H}), \quad E_{H_1}(\Delta)\, E_{H_2}(\Delta) \in \mathcal{L}_1(\mathcal{H}) \tag{22}$$

for each Borel set $\Delta \subseteq (-\infty, -\varepsilon) \cup (\varepsilon, \infty)$, $\varepsilon > 0$.

We want to show that the absolutely continuous parts of $H_0$ and $H$ are unitarily equivalent. For this we shall prove that the strong wave operators $W_\pm(H, H_0; J)$ exist, are partially isometric and are complete where the identification operator $J: \mathcal{H}_0 \mapsto \mathcal{H}$ is defined by

$$Ju = u_1 + u_2, \qquad u = \{u_1, u_2\} \in \mathcal{H} \times \mathcal{H} = \mathcal{H}_0. \tag{23}$$

First we define $\hat{J}: \mathcal{H}_0 \mapsto \mathcal{H}_0$ by $\hat{J}u = \{u_1 + u_2, 0\}$ with $u = \{u_1, u_2\} \in \mathcal{H}_0$. Then the existence, completeness and partial isometry of $W_\pm(H, H_0; J)$ follow from the corresponding properties of the wave operators $W_\pm(H \times 0, H_0; \hat{J})$, where $H \times 0$ is a selfadjoint operator on $\mathcal{H}_0$ given by $(H \times 0)\, u = \{Hu_1, 0\}$ with $u = \{u_1, u_2\}$. Note that for the proof of these properties of $W_\pm(H \times 0, H_0; \hat{J})$ it suffices to show that

$$W_\pm(\Delta) := \operatorname*{s-lim}_{t \to \pm\infty} \mathrm{e}^{\mathrm{i}t(H \times 0)}\, \hat{J}\mathrm{e}^{-\mathrm{i}tH_0}P_0^{ac}E_0(\Delta) \tag{24}$$

exist and $W_\pm(\Delta)^* W_\pm(\Delta) = P_0^{ac} E_0(\Delta)$, $W_\pm(\Delta) W_\pm(\Delta)^* = P_{H\times 0}^{ac} E_{H\times 0}(\Delta)$ for every set $\Delta = \Delta(\varepsilon) = \mathbb{R}\setminus(-2\varepsilon, 2\varepsilon)$ with $\varepsilon > 0$ arbitrary but fixed. Further, let $J(\Delta) = \mathscr{H}_0 \to \mathscr{H}_0$ be defined by

$$J(\Delta) := \begin{pmatrix} E_{H_1}(\Delta_0) & E_{H_1}(\mathbb{R}\setminus\Delta_0) \\ E_{H_1}(\mathbb{R}\setminus\Delta_0) & E_{H_1}(\Delta_0) \end{pmatrix}$$

with $\Delta_0 = \mathbb{R}\setminus(-\varepsilon, \varepsilon)$ (the same $\varepsilon$ as in the definition of $\Delta$). Then

$$\begin{aligned}(J(\Delta) - \hat{J})\, E_0(\Delta) &= (J(\Delta) - \hat{J})\,(E_{H_1}(\Delta) \times E_{H_2}(\Delta)) \\ &= \begin{pmatrix} 0 & -E_{H_1}(\Delta_0)\, E_{H_2}(\Delta) \\ 0 & E_{H_1}(\Delta_0)\, E_{H_2}(\Delta) \end{pmatrix} \in \mathscr{L}_1(\mathscr{H}_0)\end{aligned}$$

because of (22). This shows that $\operatorname*{s-lim}_{t\to\pm\infty} (J(\Delta) - \hat{J})\, e^{-itH_0} P_0^{ac} E_0(\Delta) = 0$ and therefore we can replace $\hat{J}$ in (24) by $J(\Delta)$. Next we introduce the operators $H_{1\Delta} = H_1 E_{H_1}(\Delta_0)$, $H_{2\Delta} = H_2 E_{H_2}(\Delta_0)$, $H_{0\Delta} = H_{1\Delta} \times H_{2\Delta} = H_0 E_0(\Delta_0)$, $T_\Delta = H_2 E_{H_2}(\mathbb{R}\setminus\Delta_0) + E_{H_2}(\mathbb{R}\setminus\Delta_0)\, H_1 E_{H_1}(\mathbb{R}\setminus\Delta_0)\, E_{H_2}(\mathbb{R}\setminus\Delta_0)$ and $\tilde{H}_{0\Delta} = H_{1\Delta} \times (H_{2\Delta} + T_\Delta)$. Clearly all these operators are selfadjoint and we have that

$$H_{0\Delta} \restriction E_0(\Delta)\, \mathscr{H}_0 = \tilde{H}_{0\Delta} \restriction E_0(\Delta)\, \mathscr{H}_0, \qquad E_{H_{0\Delta}}(\Delta) = E_{\tilde{H}_{0\Delta}}(\Delta).$$

This follows from the fact that $H_{0\Delta} = H_{1\Delta} \times H_{2\Delta}$ and $0 \times T_\Delta$ are different from zero only in the mutually orthogonal subspaces $E_{H_1}(\Delta_0)\mathscr{H} \times E_{H_2}(\Delta_0)\, \mathscr{H}$ and $\{0\} \times E_{H_2}(\mathbb{R}\setminus\Delta_0)\, \mathscr{H}$ and from $\|T_\Delta\| \leqq 2\varepsilon$. Namely, if $P = E_{H_0}(\Delta_0)$ and $Q$ denote the projections onto these subspaces, we have

$$E_{\tilde{H}_{0\Delta}}(\Sigma) = E_{H_{0\Delta}}(\Sigma)\, P + E_{0\times T_\Delta}(\Sigma)\, Q, \qquad \Sigma \text{ Borel set},$$

where $E_{0\times T_\Delta}((-2\varepsilon, 2\varepsilon))\, Q = Q$. Hence $E_{\tilde{H}_{0\Delta}}(\Delta) = E_{H_{0\Delta}}(\Delta)\, P = E_{H_{0\Delta}}(\Delta) = E_{H_0}(\Delta_0)\, E_{H_0}(\Delta \cap \Delta_0) \times E_{H_0}(\Delta_0) = E_{H_{0\Delta}}(\Delta)$. Summing up these remarks, we find that in (24) one can replace $H_0$ by $\tilde{H}_{0\Delta}$ and $E_{H_0}(\Delta)$ by $E_{\tilde{H}_{0\Delta}}(\Delta)$. Thus, for the verification of the properties of $W_\pm(\Delta)$ it is sufficient to show that the complete strong wave operators $W_\pm(H \times 0, \tilde{H}_{0\Delta}; J(\Delta))$ exist and are partial isometries for each $\Delta$. A simple calculation shows, that $J(\Delta)$ is unitary on $\mathscr{H}_0$. Therefore these assertions follow from Theorem 1 combined with Remark 3 provided that

$$V := (H \times 0)\, J(\Delta) - J(\Delta)\, \tilde{H}_{0\Delta} \in \mathscr{L}_1(\mathscr{H}_0).$$

We have

$$V = \begin{pmatrix} H_2 E_{H_1}(\Delta_0) & (H_1 + H_2)\, E_{H_1}(\mathbb{R}\setminus\Delta_0) - E_{H_1}(\mathbb{R}\setminus\Delta_0)\, (H_{2\Delta} + T_\Delta) \\ 0 & -E_{H_1}(\Delta_0)\, (H_{2\Delta} + T_\Delta) \end{pmatrix}.$$

Then $V \in \mathscr{L}_1(\mathscr{H}_0)$ follows from (22) together with the relations

$$\begin{aligned}
&E_{H_1}(\Delta_0)\, (H_{2\Delta} + T_\Delta) = E_{H_1}(\Delta_0)\, H_2 - E_{H_1}(\Delta_0)\, E_{H_2}(\Delta_0)\, H_1 E_{H_1}(\mathbb{R}\setminus\Delta_0)\, E_{H_2}(\mathbb{R}\setminus\Delta_0),\\
&(H_1 + H_2)\, E_{H_1}(\mathbb{R}\setminus\Delta_0) - E_{H_1}(\mathbb{R}\setminus\Delta_0)\, (H_{2\Delta} + T_\Delta)\\
&= H_2 - H_2 E_{H_1}(\Delta_0) + E_{H_1}(\mathbb{R}\setminus\Delta_0)\, H_1 - H_2 + E_{H_1}(\Delta_0)\, H_2 - E_{H_1}(\mathbb{R}\setminus\Delta_0)\, H_1 E_{H_2}(\mathbb{R}\setminus\Delta_0)\\
&\qquad + E_{H_1}(\mathbb{R}\setminus\Delta_0)\, E_{H_2}(\Delta_0)\, H_1 E_{H_2}(\mathbb{R}\setminus\Delta_0) - E_1(\mathbb{R}\setminus\Delta_0)\, E_{H_2}(\Delta_0)\, H_1 E_{H_1}(\Delta_0)\, E_{H_2}(\mathbb{R}\setminus\Delta_0)\\
&= \Gamma_\Delta + E_{H_1}(\mathbb{R}\setminus\Delta_0)\, (H_1 - H_1 E_{H_2}(\mathbb{R}\setminus\Delta_0))\\
&= \Gamma_\Delta + E_{H_1}(\mathbb{R}\setminus\Delta_0)\, H_1 E_{H_2}(\Delta_0)
\end{aligned}$$

where $\Gamma_\Delta$ is a sum of terms which clearly belong to $\mathscr{L}_1(\mathscr{H})$. This concludes the proof of the existence of the strong complete wave operators $W_\pm(H, H_0; J)$.

### 16.3.2. Wave operators for Dirac operators

Let $\mathscr{H}_0 = \mathscr{H} = L^2(\mathbb{R}^3, dx)^4 = L^2(\mathbb{R}^3, dx) \times L^2(\mathbb{R}^3, dx) \times L^2(\mathbb{R}^3, dx) \times L^2(\mathbb{R}^3, dx)$. Elements $u$ of this space are 4-tuples $u = \{u_1, u_2, u_3, u_4\}$. Further, let $D_j$ be the selfadjoint operators given by

$$(D_j u)(x) = -i\{(\partial/\partial x_j)\, u_1(x), \dots, (\partial/\partial x_j)\, u_4(x)\}, \quad j = 1, 2, 3 \quad \text{and} \quad x = \{x_1, x_2, x_3\}.$$

The free Dirac operator $H_0$ is then defined by

$$(H_0 u)(x) = \sum_{j=1}^{3} \alpha_j (D_j u)(x) + \beta u(x) \tag{25}$$

where $\alpha_j$ and $\beta$ are the following symmetric matrices:

$$\beta = \begin{pmatrix} 1 & 0 & 0 & 0 \\ 0 & 1 & 0 & 0 \\ 0 & 0 & -1 & 0 \\ 0 & 0 & 0 & -1 \end{pmatrix}, \alpha_j = \begin{pmatrix} 0 & \sigma_j \\ \sigma_j & 0 \end{pmatrix}, \ \sigma_1 = \begin{pmatrix} 0 & 1 \\ 1 & 0 \end{pmatrix}, \ \sigma_2 = \begin{pmatrix} 0 & -i \\ i & 0 \end{pmatrix}, \ \sigma_3 = \begin{pmatrix} 1 & 0 \\ 0 & -1 \end{pmatrix}.$$

$H_0$ is essentially selfadjoint on $C_0^\infty(\mathbb{R}^3)^4$ (see WEIDMANN [1, p. 310]). A simple calculation shows that

$$H_0^2 u = \{(-\Delta + 1)\, u_1, \dots, (-\Delta + 1)\, u_4\}. \tag{26}$$

Now we consider perturbations of $H_0$ by symmetric matrix functions $V = \{q_{ij}(x)\}$. For simplicity we assume that the functions $q_{ij}(x)$ are once continuously differentiable and $q_{ij}(x)$ and $(\partial/\partial x_s)\, q_{ij}(x)$ are uniformly bounded. Further we assume that $q_{ij}(\cdot) \in L^1(\mathbb{R}^3, dx)$. This ensures that $V$ is bounded, dom $H$ = dom $H_0$ with $H = H_0 + V$ and also that $H^2 = H_0^2 + Q$ where $Q$ has the bound zero; therefore dom $H^2$ = dom $H_0^2$. This proves that $(z - H_0^2)(z - H^2)^{-1}$ is bounded for each $z$ with $\operatorname{Im} z \neq 0$. Further, it is obvious that $(z - H_0^2)(z - H_0)^{-2}$ is bounded. These two facts together show that

$$(z - H)^{-2} = (z - H)^{-2} (z - H^2)(z - H^2)^{-1} (z - H_0^2)(z - H_0^2)^{-1} = G_z (z - H_0^2)^{-1}$$

where $G_z$ is a bounded operator. Using this, we get

$$\begin{aligned}(i - H)^{-2}\, V (i - H_0)^{-2} &= G_i (i - H_0^2)^{-1}\, V (i - H_0)^{-2} \\ &= G_i [(i - H_0^2)^{-1}\, V_1] \, [V_2 (i - H_0^2)^{-1}] \, (i - H_0^2)(i - H_0)^{-2}\end{aligned}$$

where $V = V_1 V_2$ is such that $V_1$ is a multiplication operator by a function $q_1(\cdot) \in L^2(\mathbb{R}^3, dx)$ and the matrix operator $V_2 = \{\tilde{q}_{ij}(x)\}$ contains functions $\tilde{q}_{ij}(\cdot) \in L^2(\mathbb{R}^3, dx)$. Such a factorization is possible because $q_{ij}(\cdot) \in L^1(\mathbb{R}^3, dx)$. For the proof of the existence of the strong wave operators $W_\pm(H, H_0)$ and their completeness it suffices, by Prop. 5, to show that $(i - H)^{-2}\, V_1$, $V_2 (i - H_0)^{-2}$ are Hilbert Schmidt operators, and this is equivalent to $(i - H_0^2)^{-1}\, V_1$, $V_2 (i - H_0^2)^{-1} \in \mathscr{L}_2(\mathscr{H})$. Let $F$ denote the Fourier operator on $L^2(\mathbb{R}^3, dx)$. Then $(i - H_0^2)^{-1} = F_0 f F_0^{-1}$ where $F_0 u = \{F u_1, \dots, F u_4\}$ and $(f u)(x) = \{f(x)\, u_1(x), \dots, f(x)\, u_4(x)\}$. Clearly, $f(\cdot) \in L^2(\mathbb{R}^3, dx)$. Now we have

**Lemma 8.** *Let $f, g \in L^2(\mathbb{R}^n, dx)$. Then the operator $A := f F g F^{-1}$, where $f, g$ are the multiplication operators by the functions $f(x), g(x)$ and $F$ is the Fourier operator on $L^2(\mathbb{R}^n, dx)$, belongs to $\mathscr{L}_2(L^2(\mathbb{R}^n, dx))$.*

From this lemma, the properties of the functions $q_1$ and $\tilde{q}_{ij}$, and $f(\cdot) \in L^2(\mathbb{R}^3, dx)$ we easily get that $(i - H_0^2)^{-1}\, V_1$, $V_2 (i - H_0^2)^{-1} \in \mathscr{L}_2(\mathscr{H})$ which proves the desired result, namely the existence and completeness of the strong wave operators $W_\pm(H, H_0)$.

Proof of Lemma 8. We have

$$(Au)(x) = f(x)\, (2\pi)^{-n/2} \int dk\, e^{-ikx}\, g(k)\, \hat{u}(-k) = (2\pi)^{-n} \int dx'\, f(x)\, \check{g}(x - x')\, u(x')$$

that is, $A$ is an integral operator with the kernel $a(x, x') = f(x)\,\check{g}(x - x')\,(2\pi)^{-n}$. Now $g(\cdot) \in L^2(\mathbb{R}^n, dx)$ implies $\check{g}(\cdot) \in L^2(\mathbb{R}^n, dx)$. Thus

$$(2\pi)^{2n} \int |a(x, x')|^2 \, dx\, dx' = \int dx\, |f(x)|^2 \int dx'\, |\check{g}(x - x')|^2$$
$$\leq \|\check{g}\|^2 \int dx\, |f(x)|^2 \leq \|f\|^2 \cdot \|g\|^2 < \infty.$$

This proves that $A \in \mathscr{L}_2(L^2(\mathbb{R}^n, dx))$. ■

## 16.3.3. Wave operators for a weakly uniformly propagative system

In this section we present an existence result within the framework of weakly uniformly propagative systems (see Section 9.7.2). Here we use the weighted $L^2$-space $L_\delta{}^2(\mathbb{R}^n, dx)$ consisting of all $f$ such that $\|\{(1 + |x|^2)^{\delta/2} f(x)\}\|_{L^2(\mathbb{R}^n, dx)} < \infty$.

**Theorem 9.** *Let $H_0$ be weakly uniformly propagative, and assume that $H_0$ has no zero energy mode (that is, $H_0$ has no eigenvalue 0). Further, let $H$ be defined by (9.67) and suppose that $E(x)$ satisfies (9.70) and $(E(x) - 1)_{i,j} \in L^2_{(n+1)/2}(\mathbb{R}^n, dx)$ for each index pair $i, j \in \{1, \ldots, p\}$. Then the strong wave operators $W_\pm(H, H_0; J)$ exist, are partially isometric and complete.*

First, recall that $\mathscr{H}_0 = L^2(\mathbb{R}^n, dx; \mathbb{C}^p)$, $u \in \mathscr{H}$ if $u \in \mathscr{H}_0$ and $\|u\|$ is defined by $\|u\|^2 = \int dx(u(x), E(x)\,u(x))$ where $E(x)$ is a symmetric strictly positive matrix-valued function satisfying:

$$0 < E_1 \leq E(x) \leq E_2 < \infty, \quad \|E(x) - 1\|_{\mathbb{C}^p} \to 0 \quad \text{as} \quad |x| \to \infty,$$
$$(E(x) - 1)_{i,j} \in L_\delta{}^2(\mathbb{R}^n, dx) \quad \text{for some} \quad \delta > n/2.$$

Further, recall that $H_0 = -\mathrm{i} \sum_{j=1}^{n} A_j\, \partial/\partial x_j$ where $A_j$ are symmetric matrices in $\mathbb{C}^p$, independent of $x$, and that $H = E^{-1}JH_0J^{-1}$ on $\mathscr{H}$ where $J\colon \mathscr{H}_0 \to \mathscr{H}$ is given by $Ju = u$ and $E^{-1}$ is the operator in $\mathscr{H}$ corresponding to the matrix-valued function $E(x)^{-1}$. Under the assumptions of Theorem 9, $P_0^{ac} = 1$. Further note that $J$ is asymptotically equivalent (with respect to $H_0$) to the isometric operator $J_2$, $J \underset{\pm H_0}{\approx} J_2$ (see Section 9.7.2). This ensures that the wave operators $W_\pm(H, H_0; J)$ provided they exist, are partial isometries with initial subspaces $\mathscr{H}_0$. From the definition of $J$ it follows that $J^*$ is given by $(J^*u)(x) = E(x)\,u(x)$. Thus $J^*$ is bounded invertible and this shows that the initial subspace of the wave operators $W_\pm(H_0, H; J^*)$ is $P^{ac}\mathscr{H}$, provided they exist. All these facts together imply that the existence of the wave operators $W_\pm(H, H_0; J)$, $W_\pm(H_0, H; J^*)$ already leads to their partial isometry and completeness.

Thus it remains to prove the existence of the strong wave operators $W_\pm(H, H_0; J)$, $W_\pm(H_0, H; J^*)$, and this will be done with the help of Prop. 5 and Remark 6. That is, we shall show that $R(\mathrm{i})\,VR_0(\mathrm{i}) \in \mathscr{L}_\infty(\mathscr{H}_0, \mathscr{H})$ and $VR_0(\mathrm{i})^m \in \mathscr{L}_1(\mathscr{H}_0, \mathscr{H})$ with $m = n + 2$.

We have $V = HJ - JH_0 = E^{-1}JH_0J^{-1}J - JH_0 = (E^{-1}J - J)\,H_0$. Now $Ju = u$ and $(E^{-1}u)(x) = E^{-1}(x)\,u(x)$ imply that $E^{-1}J = J\tilde{E}^{-1}$, where $\tilde{E}$ is given by the same matrix-valued function as $E$, i.e. $(\tilde{E}u)(x) = E(x)\,u(x)$, $u \in \mathscr{H}_0$. This gives

$$V = J(\tilde{E}^{-1} - 1)\,H_0 = J\tilde{E}^{-1}(1 - \tilde{E})\,H_0. \tag{27}$$

Thus, to prove $VR_0(\mathrm{i})^m \in \mathscr{L}_1(\mathscr{H}_0, \mathscr{H})$ it suffices to prove that $(1 - \tilde{E})\,H_0R_0(\mathrm{i})^m \in \mathscr{L}_1(\mathscr{H}_0)$, or equivalently that $(1 - \tilde{E})\,R_0(\mathrm{i})^{m-1} \in \mathscr{L}_1(\mathscr{H}_0)$. For this we use (9.69) which says that

$$(R_0(\mathrm{i})^j\, u)^\wedge(k) = ((\mathrm{i} - H_0)^{-j}\, u)^\wedge(k) = \sum_{q=1}^{l} (\mathrm{i} - \lambda_q(k))^{-j}\, P_q(k)\,\hat{u}(k)$$
$$= \sum_{q=1}^{l} (\mathrm{i} - |k|\,\lambda_q(\hat{k}))^{-j}\, \hat{u}_q(k) \tag{28}$$

with $\|u_q\| = \|P_q u\| \leq \|u\|$ and $\hat{k} = k/|k|$. Because of our assumptions in the theorem, all roots $\lambda_q(k)$ satisfy the estimate $0 < a_1 \leq \lambda_q(\hat{k}) \leq a_2 < \infty$ for all $|\hat{k}| = 1$. This shows that the functions

$\{f_q(k)\} := \{(\mathrm{i} - |k|\, \lambda_q(\hat{k})^{(-m-1)})\} \in L^2_{(n+1)/2}(\mathbb{R}^n, \mathrm{d}k)$. By assumption, $(1 - E(\cdot))_{i,j} \in L^2_{(n+1)/2}(\mathbb{R}^n, \mathrm{d}x)$. Thus, from (27) and these relations we see that the inclusion $(1 - \tilde{E})\, R_0(\mathrm{i})^{m-1} \in \mathscr{L}_1(\mathscr{H}_0)$ is a simple consequence of the following lemma.

**Lemma 10.** *Let* $f(\cdot), g(\cdot) \in L_\delta^2(\mathbb{R}^n, \mathrm{d}x)$ *with some* $\delta > n/2$. *Further, let* $F$ *denote the Fourier operator on* $L^2(\mathbb{R}^n, \mathrm{d}x)$ *and let* $f$, $g$ *denote the multiplication operators by* $f(x)$, $g(x)$ *on* $L^2(\mathbb{R}^n, \mathrm{d}x)$. *Then* $fFgF^{-1} \in \mathscr{L}_1(L^2(\mathbb{R}^n, \mathrm{d}x))$.

**Proof.** See the Notes and Remarks. ■

The last step is to show that $R(\mathrm{i})\, VR_0(\mathrm{i}) \in \mathscr{L}_\infty(\mathscr{H}_0, \mathscr{H})$. A simple calculation leads to

$$(\mathrm{i} - H)^{-1} = (\mathrm{i}E - JH_0J^{-1})^{-1}\, E = J(\mathrm{i}\tilde{E} - H_0)^{-1}\, J^{-1}E.$$

This yields

$$\begin{aligned} R(\mathrm{i})\, VR_0(\mathrm{i}) &= J(\mathrm{i}\tilde{E} - H_0)^{-1}\, J^{-1}EJ\tilde{E}^{-1}(1 - \tilde{E})\, H_0(\mathrm{i} - H_0)^{-1} \\ &= J(\mathrm{i}\tilde{E} - H_0)^{-1}\, (1 - \tilde{E})\, H_0(\mathrm{i} - H_0)^{-1}. \end{aligned}$$

Since $\tilde{E}$ is a bounded operator, $(\mathrm{i}\tilde{E} - H_0)^{-1}\, (\mathrm{i} - H_0)$ is bounded. Using this and the equality above, we get

$$R(\mathrm{i})\, VR_0(\mathrm{i}) = (J(\mathrm{i}\tilde{E} - H_0)^{-1}\, (i - H_0))\, (\mathrm{i} - H_0)^{-1}\, (1 - \tilde{E})\, (H_0(\mathrm{i} - H_0)^{-1}).$$

Thus, it suffices to prove $(\mathrm{i} - H_0)^{-1}\, (1 - \tilde{E}) \in \mathscr{L}_\infty(\mathscr{H}_0)$. The same reasoning as above for proving that $VR_0(\mathrm{i})^m \in \mathscr{L}_1(\mathscr{H}_0, \mathscr{H})$ with the help of (28) can now be used to show that it is enough to prove $fFgF^{-1} \in \mathscr{L}_\infty(L^2(\mathbb{R}^n, \mathrm{d}x))$, where $f$, $g$ are multiplication operators by bounded functions $f(x)$, $g(x)$ vanishing at infinity and where $F$ is the Fourier operator on $L^2(\mathbb{R}^n, \mathrm{d}x)$. Since $f$ and $g$ can be approximated in norm by functions $f_m, g_m \in L^2(\mathbb{R}^n, \mathrm{d}x)$, since $f_mFg_mF^{-1} \in \mathscr{L}_2(L^2(\mathbb{R}^n, \mathrm{d}x))$ by Lemma 8, and since $f_mFg_mF^{-1} \to fFgF^{-1}$ in norm and the norm limit of Hilbert Schmidt operators is compact, $fFgF^{-1} \in \mathscr{L}_\infty(L^2(\mathbb{R}^n, \mathrm{d}x))$. This concludes the proof of the existence of the strong, partially isometric and complete wave operators $W_\pm(H, H_0; J)$.

# 17. Smooth Perturbations

## 17.1. Smooth operators

### 17.1.1. Basic notions

First we introduce the basic definition for this chapter.

**Definition 1.** Let $X$ be a closed operator from $\mathscr{H}$ to another Hilbert space $\mathscr{H}'$ with dom $X$ dense in $\mathscr{H}$. $X$ is called *$H$-smooth* if and only if for each $u \in \mathscr{H}$ and each $\varepsilon \neq 0$, $R(\lambda + i\varepsilon)\, u \in \operatorname{dom} X$ for almost all $\lambda \in \mathbb{R}$ (mod $|\cdot|$) and

$$\|X\|_H^2 := \sup_{\|u\|=1,\varepsilon>0} (2\pi)^{-2} \int_{-\infty}^{\infty} \{\|XR(\lambda + i\varepsilon)\, u\|^2 + \|XR(\lambda - i\varepsilon)\, u\|^2\}\, d\lambda < \infty. \tag{1}$$

Note that this definition is symmetric with respect to $H$ and $-H$, i.e. $\|X\|_H < \infty$ implies $\|X\|_{-H} < \infty$ and the equality of these two numbers. Later we shall consider some examples showing that the class of $H$-smooth operators is not empty if $P^{ac} > 0$.

Next we prove that one can weaken the assumptions in Definition 1.

**Proposition 2.** *Let $X$ have the properties given in Definition 1 except (1). Suppose that for each vector $u \in \mathscr{H}$*

$$\sup_{\varepsilon>0} \int_{-\infty}^{\infty} \{\|XR(\lambda + i\varepsilon)\, u\|^2 + \|XR(\lambda - i\varepsilon)\, u\|^2\}\, d\lambda \leq c_u < \infty. \tag{2}$$

*Then relation (1) is also true.*

**Proof.** 1. For fixed $\varepsilon > 0$, $D_\varepsilon^\pm := XR(\lambda \pm i\varepsilon)$ considered as operators from $\mathscr{H}$ into $L^2(\mathbb{R}\, d\lambda; \mathscr{H}')$ are defined on the whole Hilbert space $\mathscr{H}$ because of (2). Now we show that $D_\varepsilon^+$ is closed, and this implies then that $D_\varepsilon^+$ is bounded because of the closed graph theorem (see Weidmann [1, p. 94]).

Let $u_n \in \mathscr{H}$ be strongly convergent to $u \in \mathscr{H}$. Assume further that $D_\varepsilon^+ u_n$ converges strongly to $v \in L^2(\mathbb{R}, d\lambda; \mathscr{H}')$. This implies that there exists a subsequence $u_{n_k}$ such that $(D_\varepsilon^+ u_{n_k})(\lambda)$ converges strongly in $\mathscr{H}'$ to $v(\lambda)$ a.e. (mod $|\cdot|$) (see Natanson [1, p. 185]). On the other hand, there is a null set $\Delta_0$, $|\Delta_0| = 0$, so that $R(\lambda + i\varepsilon)\, u_{n_k} \in \operatorname{dom} X$ and $R(\lambda + i\varepsilon)\, u \in \operatorname{dom} X$ for all $\lambda \in \mathbb{R} \setminus \Delta_0$ and $n_k$. Therefore

$$XR(\lambda + i\varepsilon)\, u_{n_k} = (D_\varepsilon^+ u_{n_k})(\lambda), \quad XR(\lambda + i\varepsilon)\, u = (D_\varepsilon^+ u)(\lambda) \quad \text{are defined for}$$

$\lambda \in \mathbb{R} \setminus \Delta_0$, and for these values of $\lambda$ we have $XR(\lambda + i\varepsilon)\, u_{n_k} \to XR(\lambda + i\varepsilon)\, u$ strongly; i.e. $(D_\varepsilon^+ u_{n_k})(\lambda)$ converges a.e. (mod $|\cdot|$) to $(D_\varepsilon^+ u)(\lambda)$. Thus $(D_\varepsilon^+ u)(\lambda) = v(\lambda)$ a.e. (mod $|\cdot|$); and this implies $D_\varepsilon^+ u = v = \operatorname{s-lim}_{n\to\infty} D_\varepsilon^+ u_n$. But this means precisely that $D_\varepsilon^+$ is closed. Similarly we can prove that $D_\varepsilon^-$ is closed and thus bounded.

2. By (2) we have for each vector $u$ that $\|D_\varepsilon^\pm u\| \leqq c_u < \infty$ for all $\varepsilon > 0$. Applying the principle of uniform boundedness to the family of bounded operators $D_\varepsilon^\pm$, we get that $\|D_\varepsilon^\pm\| \leqq c < \infty$ for all $\varepsilon > 0$, and this is relation (1). ∎

Note that (2) implies that $XR(\lambda \pm \mathrm{i}\varepsilon)\, u$ are elements of the Hardy class $\mathscr{H}_\pm^2(\mathbb{R}, \mathrm{d}\lambda; \mathscr{H}')$. In particular, this ensures that the limits $\operatorname*{s-lim}_{\varepsilon\to+0} XR(\lambda \pm \mathrm{i}\varepsilon)\, u$ exist a.e. (mod $|\cdot|$), $u \in \mathscr{H}$, and these limits are elements of $L^2(\mathbb{R}, \mathrm{d}\lambda; \mathscr{H}')$ (see Section 1.2.3). The next proposition gives a definition of $H$-smoothness in terms of the unitary group $\mathrm{e}^{-\mathrm{i}tH}$, i.e. a so-called time-dependent definition.

**Proposition 3.** *$X$ is $H$-smooth if and only if for all $u \in \mathscr{H}$, $\mathrm{e}^{-\mathrm{i}tH} u \in \operatorname{dom} X$ for almost all $t$(mod $|\cdot|$) and*

$$c_0 := \sup_{\|u\|=1} (2\pi)^{-1} \int_{-\infty}^{\infty} \|X\, \mathrm{e}^{-\mathrm{i}tH} u\|^2 \,\mathrm{d}t < \infty. \tag{3}$$

*Then* $c_0 = \|X\|_H^2$.

**Proof.** The statement is a simple consequence of the equations

$$\int_{-\infty}^{\infty} \|XR(\lambda \pm \mathrm{i}\varepsilon)\, u\|^2 \,\mathrm{d}\lambda = \pm 2\pi \int_0^{\pm\infty} \mathrm{e}^{-2\varepsilon|t|}\, \|X\, \mathrm{e}^{-\mathrm{i}tH} u\|^2 \,\mathrm{d}t$$

or $\varepsilon > 0$ and all $u \in \mathscr{H}$ (see (3.39)) and (3.40)) and Definition 1. ∎

**Remark 4.** Looking at the proof above, we see that

$$\int_{-\infty}^{\infty} \left(\|XR(\lambda + \mathrm{i}\varepsilon)\, u\|^2 + \|XR(\lambda - \mathrm{i}\varepsilon)\, u\|^2\right) \mathrm{d}\lambda$$

is a monotone decreasing function in $\varepsilon$. Consequently, it suffices to consider the supremum in Definition 1 over $\varepsilon \in (0, \varepsilon_0)$ with $\varepsilon_0 > 0$ arbitrary.

Now we are able to prove some simple properties of $H$-smooth operators.

**Theorem 5.** *If $X$ is $H$-smooth, then*

*(i) $X$ is $H$-bounded with relative bound zero,*

*(ii) ima $X^* \subseteq P^{\mathrm{ac}}\mathscr{H}$, i.e. $Xu = 0$ if $u \in (P^{\mathrm{ac}})^\perp \mathscr{H}$.*

**Proof.** (i) From (3.8) we obtain, for $u' \in \operatorname{dom} X^*$

$$\mathrm{i}\left(X^*u', R(\lambda + \mathrm{i}\varepsilon)\, u\right) = \int_0^{\infty} (X^*u', \mathrm{e}^{-\mathrm{i}tH} u)\, \mathrm{e}^{\mathrm{i}t\lambda}\, \mathrm{e}^{-\varepsilon t} \,\mathrm{d}t.$$

Using the Schwarz inequality and Prop. 3, we find

$$\begin{aligned} \left|\left(X^*u', R(\lambda + \mathrm{i}\varepsilon)\, u\right)\right|^2 &\leqq (2\varepsilon)^{-1} \int_0^{\infty} |(X^*u', \mathrm{e}^{-\mathrm{i}tH} u)|^2 \,\mathrm{d}t \\ &\leqq (2\varepsilon)^{-1} \|u'\|^2 \int_0^{\infty} \|X\, \mathrm{e}^{-\mathrm{i}tH} u\|^2 \,\mathrm{d}t \leqq (\pi/\varepsilon)\, \|u'\|^2\, \|X_H\|^2\, \|u\|^2. \end{aligned} \tag{4}$$

(4) shows that $R(\lambda + i\varepsilon) u \in \operatorname{dom} X^{**} = \operatorname{dom} X$ and therefore that $\operatorname{dom} X \supseteq \operatorname{dom} H$. Furthermore, the estimate (4) yields

$$\|XR(\lambda + i\varepsilon)\|^2 \leqq (\pi/\varepsilon) \|X\|_H^2. \tag{5}$$

Setting in (4), $\lambda = 0$ and $u = (H + i\varepsilon) v$, we get

$$\|Xv\|^2 \leqq (\pi/\varepsilon) \|X\|_H^2 \|(H + i\varepsilon) v\|^2$$

leading to $\|Xv\| \leqq \pi^{1/2} \|X\|_H (\varepsilon^{-1/2}\|Hv\| + \varepsilon^{1/2} \|v\|)$. Since the set of admissible vectors $v$ (in this estimate) equals dom $H$ and $\varepsilon > 0$ is arbitrary, $X$ is $H$-bounded with $H$-bound zero.

(ii) Let $u \in \operatorname{dom} X^*$. Then, using the first resolvent identity,

$$\int_a^b \left|\operatorname{Im}\left(X^*u, R(\lambda + i\varepsilon) X^*u\right)\right|^2 d\lambda$$

$$= 1/4 \int_a^b \left|\left(u, XR(\lambda + i\varepsilon) X^*u\right) - \left(u, XR(\lambda - i\varepsilon) X^*u\right)\right|^2 d\lambda$$

$$= \varepsilon^2 \int_a^b \|R(\lambda + i\varepsilon) X^*u\|^4 d\lambda \leqq \varepsilon^2 \int_a^b \|R(\lambda + i\varepsilon) X^*\|^4 \|u\|^4 d\lambda.$$

By (5), this gives the estimate

$$\int_a^b \left|\operatorname{Im}\left(X^*u, R(\lambda + i\varepsilon) X^*u\right)\right|^2 d\lambda \leqq |b - a| \pi^2 \|X\|_H^4 \|u\|^4,$$

proving, similarly as in the proof of Prop. 3.21, that for each bounded intervall $(a, b)$, $E\big((a, b)\big) X^*u \in P^{ac}\mathscr{H}$. Since this is true for each vector $u \in \operatorname{dom} X^*$ and interval $(a, b)$, we obtain the desired result ima $X^* \subseteq P^{ac}\mathscr{H}$. ∎

This theorem states that the set of $H$-smooth operators is defined on dom $H$ and that only the absolutely continuous part of $H$ is important. If $H$ is bounded, then it is obvious that all $H$-smooth operators are bounded.

It is also easily verified that the set of $H$-smooth operators with the same final space $\mathscr{H}'$ forms a Banach space with the norm $\|\cdot\|_H$.

The following statement says that there are many equivalent definitions of $H$-smoothness.

**Theorem 6.** *Let $X$ be a closed operator from $\mathscr{H}$ to another Hilbert space $\mathscr{H}'$. Then the following statements are equivalent:*

(i) *$X$ is $H$-smooth,*

(ii) dom $H \subseteq$ dom $X$ *and* $c_1 := \sup\limits_{\|u\|=1, -\infty<a<b<\infty} \dfrac{\left\|XE\big((a, b)\big) u\right\|^2}{b - a} < \infty,$

(iii) $c_2 := \sup\limits_{\substack{u \in \operatorname{dom} X^*, \|u\|=1 \\ -\infty<a<b<\infty}} \dfrac{\left\|E\big((a, b)\big) X^*u\right\|^2}{b - a} < \infty,$

(iv) $c_3 := (2\pi)^{-1} \sup\limits_{\substack{u \in \operatorname{dom} X^*, \\ \|u\|=1, z \notin \mathbb{R}}} \left|\big(X^*u, (R(z) - R(\bar z))\, X^*u\big)\right| < \infty,$

(v) $c_4 := \pi^{-1} \sup\limits_{\substack{u \in \operatorname{dom} X^*, \\ \|u\|=1, z \notin \mathbb{R}}} \|R(z)\, X^*u\|^2 \, |\operatorname{Im} z| < \infty,$

(vi) $\operatorname{dom} H \subseteq \operatorname{dom} X$ *and* $c_5 := \pi^{-1} \sup\limits_{\|u\|=1, z \notin \mathbb{R}} \|XR(z)\, u\|^2 \, |\operatorname{Im} z| < \infty,$

(vii) $\operatorname{dom} H \subseteq \operatorname{dom} X$ *and* $c_6 := \pi^{-1} \sup\limits_{\|u\|=1, z \notin \mathbb{R}} \{\|XR(z)\, u\|^2 + \|XR(\bar z)\, u\|^2\} \, |\operatorname{Im} z| < \infty.$

*Moreover, if any (and thus all) of the numbers $c_1, c_2, \dots, c_6$ is finite, then all $c_i$ are equal to $\|X\|_H{}^2$.*

Proof. 1. First we show that $\sup\limits_{u \in \mathscr{H}} \|XBu\|/\|u\| = \alpha_1$ is equal to $\alpha_2 := \sup\limits_{u' \in \operatorname{dom} X^*} \|B^*X^*u'\|/\|u'\|$ for $B \in \mathscr{L}(\mathscr{H})$, $\operatorname{ima} B \subseteq \operatorname{dom} X$, and $X$ closed. If $\alpha_1 < \infty$ then $|(u, B^*X^*u')| = |(XBu, u')| \le \alpha_1 \|u\| \cdot \|u'\|$ for any $u \in \mathscr{H}$ and $u' \in \operatorname{dom} X^*$. Hence $\|B^*X^*u'\| \le \alpha_1 \|u'\|$ and $\alpha_1 \ge \alpha_2$. If $\alpha_2 < \infty$ then $|(Bu, X^*u')| = |(u, B^*X^*u')| \le \alpha_2 \|u\| \cdot \|u'\|$. Thus $Bu \in \operatorname{dom} X^{**} = \operatorname{dom} X$ with $\|XBu\| \le \alpha_2 \|u\|$ so that $\alpha_1 \le \alpha_2$. This gives the desired relation $\alpha_1 = \alpha_2$.

2. From $\alpha_1 = \alpha_2$ we get that $c_1 = c_2$ and $c_4 = c_5$. Further, we have that

$$\big(X^*u, (R(z) - R(\bar z))\, X^*u\big) = -2\,\mathrm{i} \operatorname{Im} z \, \|R(z)\, X^*u\|^2.$$

This proves $c_3 = c_4$. Thus it is sufficient to verify the relations $\|X\|_H^2 \le c_3 \le c_2 \le \|X\|_H^2$ and $c_3 \le c_6 \le \|X\|_H{}^2$.

3. $\|X\|_H^2 \le c_3$. We may assume $c_3 < \infty$. If $\operatorname{Im} z < 0$, then the operator

$$(2\pi\mathrm{i})^{-1} \big(R(z) - R(\bar z)\big) = -\pi^{-1} \operatorname{Im} z \, R(z)\, R(\bar z)$$

is symmetric and nonnegative. Let $K(z)$ be the nonnegative square root of this operator. Then

$$\|K(z)\, X^*u'\|^2 = (2\pi\mathrm{i})^{-1} (X^*u', [R(z) - R(\bar z)]\, X^*u') \le c_3 \|u'\|^2.$$

By the relation $\alpha_1 = \alpha_2$ in 1., we have that for every $u \in \mathscr{H}$, $XK(z)\, u \in \mathscr{H}'$ exists and

$$\|XK(z)\, u\|^2 \le c_3 \|u\|^2. \tag{6}$$

Further,

$$\begin{aligned} &\int_{-\infty}^{\infty} \|X\big(R(\lambda - \mathrm{i}\varepsilon) - R(\lambda + \mathrm{i}\varepsilon)\big)\, u\|^2 \, d\lambda \\ &= 4\pi^2 \int_{-\infty}^{\infty} \|XK(\lambda - \mathrm{i}\varepsilon)^2\, u\|^2 \, d\lambda \le 4\pi^2 c_3 \int_{-\infty}^{\infty} \|K(\lambda - \mathrm{i}\varepsilon)\, u\|^2 \, d\lambda \\ &\le 4\pi^2 c_3 (2\pi\mathrm{i})^{-1} \int_{-\infty}^{\infty} \big(u, (R(\lambda - \mathrm{i}\varepsilon) - R(\lambda + \mathrm{i}\varepsilon))\, u\big) \, d\lambda = 4\pi^2 c_3 \|u\|^2 \end{aligned} \tag{7}$$

because of the first resolvent identity and (3.10). Now by (3.8),

$$X\big(R(\lambda - \mathrm{i}\varepsilon) - R(\lambda + \mathrm{i}\varepsilon)\big)\, u = \mathrm{i} \int_{-\infty}^{\infty} \mathrm{e}^{-\varepsilon|t|}\, \mathrm{e}^{\mathrm{i}\lambda t} X\, \mathrm{e}^{-\mathrm{i}tH} u \, dt.$$

Then Parseval's equation says that

$$\int_{-\infty}^{\infty} \|X(R(\lambda - i\varepsilon) - R(\lambda + i\varepsilon))\, u\|^2 \, d\lambda = 2\pi \int_{-\infty}^{\infty} e^{-2\varepsilon|t|} \|X\, e^{-itH} u\|^2 \, dt. \tag{8}$$

By Prop. 3 and (7), we get $\|X\|_H^2 \leq c_3$.

4. $c_3 \leq c_2$. We have

$$R(\lambda - i\varepsilon) - R(\lambda + i\varepsilon) = \int_{-\infty}^{\infty} 2i\varepsilon\left(\varepsilon^2 + (\lambda - \lambda')^2\right)^{-1} E(d\lambda').$$

This gives for $u' \in \operatorname{dom} X^*$

$$\left(X^*u', [R(\lambda - i\varepsilon) - R(\lambda + i\varepsilon)]\, X^*u'\right) = 2i \int_{-\infty}^{\infty} \varepsilon\left(\varepsilon^2 + (\lambda - \lambda')^2\right)^{-1} \|E(d\lambda')\, X^*u'\|^2.$$

We may assume $c_2 < \infty$. Then $\|E(\Delta)\, X^*u'\|^2 \leq c_2\, |\Delta|\, \|u'\|^2$ and so $\|E(d\lambda)\, X^*u'\|^2 \leq c_2 \|u'\|^2 \, d\lambda$. Thus the expression above taken in absolute value is smaller then or equal to $2\pi c_2 \|u'\|^2$. This proves the relation $c_3 \leq c_2$.

5. $c_3 \leq c_6$. Since $c_3 = c_4 = c_5$ and $c_4 \leq c_6$ is obvious, this relation is verified.

6. $c_2 \leq \|X\|_H^2$. Let $u \in \operatorname{dom} X^*$. Suppose that $a$ and $b$ are not eigenvalues of $H$. Then, by (3.16),

$$\begin{aligned}
\left|\left(X^*u, E((a, b))\, u'\right)\right|^2 &= (4\pi^2)^{-1} \lim_{\varepsilon \to +0} \left| \int_a^b \left(X^*u, [R(\lambda - i\varepsilon) - R(\lambda + i\varepsilon)]\, u'\right) d\lambda \right|^2 \\
&\leq (4\pi^2)^{-1} \|u\|^2\, |b - a| \lim_{\varepsilon \to +0} \int_{-\infty}^{\infty} \left\|X\left(R(\lambda - i\varepsilon) - R(\lambda + i\varepsilon)\right) u'\right\|^2 d\lambda \\
&\leq |b - a|\, \|X\|_H^2\, \|u\|^2\, \|u'\|^2
\end{aligned}$$

where we have used (8) and Prop. 3 in the last step. By a limit process we can extend this relation to the case where $a$ and/or $b$ are eigenvalues. This gives $c_2 \leq \|X\|_H^2$.

7. $c_6 \leq \|X\|_H^2$. We have (see (3.8))

$$-i\left(X^*u', R(\lambda \pm i\varepsilon)\, u\right) = \int_0^{\pm\infty} (X^*u', e^{-itH}u)\, e^{it\lambda}\, e^{-\varepsilon|t|} \, dt.$$

Using the Schwarz inequality and integrating the first factor, we obtain

$$\left|\left(X^*u', R(\lambda \pm i\varepsilon)\, u\right)\right|^2 \leq (2\varepsilon)^{-1} \|u'\|^2 \left| \int_0^{\pm\infty} \|X\, e^{-itH}u\|^2 \, dt \right|.$$

Summing up the two terms, for the + and the − sign, we obtain

$$\|XR(\lambda + i\varepsilon)\, u\|^2 + \|XR(\lambda - i\varepsilon)\, u\|^2 \leq (2\varepsilon)^{-1} \int_{-\infty}^{\infty} \|X\, e^{-itH}u\|^2 \, dt.$$

From Prop. 3 and the definition of $c_6$ the desired relation follows. ∎

### 17.1.2. Representation as integral operators

This section is devoted to the proof that for each $H$-smooth bounded operator $X$ the operator $X^*X$ is an integral operator with $L^\infty$-kernel with respect to a spectral representation of $H$. The proof will be done in several steps.

**Lemma 7.** *Let $\mathscr{H} = L^2(\Delta, \mathrm{d}x; \mathscr{K}_0)$, $H$ be the multiplication operator by $x$ in $\mathscr{H}$, and let $X \in \mathscr{L}(\mathscr{H}, \mathscr{H}')$ be $H$-smooth. Then there is an $\mathscr{L}(\mathscr{H}', \mathscr{K}_0)$-valued strongly $|\cdot|$-measurable function $F(x)$ defined on $\Delta$ such that $\|F(x)\| \leqq \|X\|_H$ and $(X^*u')(x) = F(x)\, u'$ for every $u' \in \mathscr{H}'$.*

**Proof.** Let $u' \in \mathscr{H}'$ and set $v := X^*u' \in \mathscr{H}$ and

$$w(I) := \int_{I \cap \Delta} v(x)\, \mathrm{d}x \in \mathscr{K}_0, \tag{9}$$

where $v(x)$ is the representer of $v$ with respect to $\mathscr{H} = L^2(\Delta, \mathrm{d}x; \mathscr{K}_0)$ and $I$ runs through all finite intervals. Then

$$\begin{aligned} \|w(I)\| &\leqq \int_{I \cap \Delta} \|v(x)\|\, \mathrm{d}x \leqq |I|^{1/2} \Big( \int_{I \cap \Delta} \|v(x)\|^2\, \mathrm{d}x \Big)^{1/2} \\ &= |I|^{1/2}\, \|E_H(I \cap \Delta)\, v\| = |I|^{1/2}\, \|E_H(I)\, X^*u'\| \\ &\leqq |I| \cdot \|X\|_H\, \|u'\|, \end{aligned} \tag{10}$$

by Theorem 6. Now the derivative $w'(x) = \operatorname*{s-lim}_{\varepsilon \to +0} w(I_{x,\varepsilon})/\, |I_{x,\varepsilon}|$, $I_{x,\varepsilon} = (x - \varepsilon, x + \varepsilon)$ exists a.e. (mod $|\cdot|$) (set $v(x) = 0$ on $\mathbb{R} \setminus \Delta$ and use the results of Section 1.2.4). Choose a countable dense set of vectors $u_n$ in $\mathscr{H}'$. Such a set exists because $\mathscr{H}'$ is separable. Furthermore, define the corresponding quantities $v_n$, $w_n(I)$ and $w_n'(x)$ as above. Obviously, there is a null set $\Gamma_0$ such that $w_n'(x)$ exists for all $n$ if $x \in \mathbb{R} \setminus \Gamma_0$. Let $u' \in \mathscr{H}'$ and $x \in \mathbb{R} \setminus \Gamma_0$. For any $\varepsilon > 0$ we can find a vector $u_n$ so that $\|u' - u_n\| < \varepsilon$. Then we have, by (10)

$$\|w(I) - w_n(I)\| \leqq \|X\|_H\, |I|\, \varepsilon$$

and therefore

$$\limsup_{\delta \to +0} \|\, |I_{x,\delta}|^{-1}\, w(I_{x,\delta}) - w_n'(x)\| \leqq \varepsilon\, \|X\|_H, \qquad x \in \mathbb{R} \setminus \Gamma_0.$$

Thus

$$\limsup_{\delta_1, \delta_2 \to +0} \|\, |I_{x,\delta_1}|^{-1}\, w(I_{x,\delta_1}) - |I_{x,\delta_2}|^{-1}\, w(I_{x,\delta_2})\| \leqq 2\varepsilon\, \|X\|_H.$$

Since $\varepsilon > 0$ was arbitrary, the left-hand side must vanish. This means that $w'(x)$ exists for all $x \in \mathbb{R} \setminus \Gamma_0$ and $u' \in \mathscr{H}'$. Clearly, $w'(x)$ depends linearly on $u'$ and $\|w'(x)\| \leqq \|X\|_H \times \|u'\|$, by (10). Thus setting $F(x)\, u := w'(x)$ for $x \in \mathbb{R} \setminus \Gamma_0$ and $F(x)\, u = 0$ for $x \in \Gamma_0$, we get an operator $F(x)\colon \mathscr{H}' \mapsto \mathscr{K}_0$ for $x \in \mathbb{R}$ with $\|F(x)\| \leqq \|X\|_H$. Since $F(x)\, u = w'(x) = v(x) = (X^*u')\,(x)$ a.e. (mod $|\cdot|$) and $v(x)$ is $|\cdot|$-measurable, $F(x)$ is also strongly $|\cdot|$-measurable. This concludes the proof. ∎

**Lemma 8.** *Under the assumptions of Lemma 7 one has*

$$Xu = \int_\Delta F(x)^* u(x) \, dx \quad \text{for} \quad u \in \operatorname{dom} H. \tag{11}$$

Proof. Note that $F(x)^* \in \mathscr{L}(\mathscr{K}_0, \mathscr{H}')$ so that $F(x)^* u(x) \in \mathscr{H}'$ for each $x \in \Delta$ and $u \in \mathscr{H} = L^2(\Delta, dx; \mathscr{K}_0)$. $F(x)^* u(x)$ is $|\cdot|$-measurable because $F(x)$ is strongly $|\cdot|$-measurable and $u(x)$ is $|\cdot|$-measurable. Since

$$\|F(x)^* u(x)\|_{\mathscr{K}_0} \leq \|X\|_H \|u(x)\|_{\mathscr{K}_0},$$

by Lemma 7, $F(x)^* u(x)$ is Bochner integrable for $u \in \operatorname{dom} H$. This follows from $u(x) = (x + i\varepsilon)^{-1} \tilde{u}(x)$, $\tilde{u} \in \mathscr{H}$ because of $u \in \operatorname{dom} H$ and the fact that $\tilde{u}(\cdot) \in L^2(\Delta, dx; \mathscr{K}_0)$ and $(\cdot + i\varepsilon)^{-1} \in L^2(\mathbb{R}, dx)$. Now (11) follows from

$$\int \big(F(x)^* u(x), v\big) \, dx = \int \big(u(x), F(x) v\big) \, dx$$
$$= (u, X^*v) = (Xu, v) \quad \text{for} \quad u \in \operatorname{dom} H, \qquad v \in \mathscr{H}'. \blacksquare$$

**Lemma 9.** *Under the assumptions of Lemma 7, $D := X^*X$ is an integral operator with kernel $D(x, x')$ such that*

$$(Du)(x) = \int_\Delta D(x, x') \, u(x') \, dx', \qquad u \in \operatorname{dom} H,$$

*and* $\|D(x, x')\| \leq \|X\|_H^2$ a.e. (mod $|\cdot|$).

Proof. It follows from Lemma 7 and Lemma 8 that if $v = X^*Xu$, $u \in \operatorname{dom} H$, then

$$v(x) = F(x) \, Xu = F(x) \int_\Delta F(x')^* u(x') \, dx' = \int_\Delta D(x, x') \, u(x') \, dx'$$

a.e. (mod $|\cdot|$) with $D(x, x') := F(x) \, F(x')^*$. Further

$$\|D(x, x')\| = \|F(x) \, F(x')^*\| \leq \|F(x)\| \cdot \|F(x')\| \leq \|X\|_H^2$$

a.e. (mod $|\cdot|$) because of Lemma 7. This proves the lemma. $\blacksquare$

**Lemma 10.** *Let $H$ and $\mathscr{H}$ be as in Lemma 7. Further, let $D$ be a bounded nonnegative operator on $\mathscr{H} = L^2(\Delta, dx; \mathscr{K}_0)$ of the form*

$$(Du)(x) = \int_\Delta D(x, x') \, u(x') \, dx'$$

*for $u \in \operatorname{dom} H$, with* $\operatorname{ess\,sup}_{x, x' \in \Delta} \|D(x, x')\| = b < \infty$. *Then $D^{1/2}$ is $H$-smooth with* $\|D^{1/2}\|_H = b^{1/2}$.

Proof. Let $u \in E(I) \mathscr{H}$ so that $u(x)$ has support in $\Delta \cap I$ where $I$ is a finite interval. Then

$$|(Du, u)| = \int_\Delta \int_\Delta \big(D(x, x') \, u(x'), u(x)\big) \, dx \, dx'$$
$$\leq b \int_\Delta \|u(x')\| \, dx' \int_\Delta \|u(x)\| \, dx \leq b \left(\int_I \|u(x)\| \, dx\right)^2 \leq b \, |I| \, \|u\|^2.$$

Thus $\|D^{1/2}u\|^2 = |(Du, u)| \leq b\,|I|\,\|u\|^2$ for each $u \in E(I)\,\mathscr{H}$. This gives $\|D^{1/2}E(I)\| \leq b\,|I|$ and therefore that $D^{1/2}$ is $H$-smooth by Theorem 6. On the other hand, by Lemma 9, we have

$$\operatorname*{ess\,sup}_{x,x' \in \Delta} \|D(x, x')\| = b \leq \|D^{1/2}\|_H^2.$$

Thus

$$\|D^{1/2}\|_H^2 = b = \operatorname*{ess\,sup}_{x,x' \in \Delta} \|D(x, x')\|. \blacksquare$$

**Remark 11.** From Lemma 10 we get that in Lemma 9 one can replace $\|D(x, x')\| \leq \|X\|_H^2$ by $\operatorname*{ess\,sup}_{x,x' \in \Delta} \|D(x, x')\| = \|X\|_H^2$.

Because of Theorem 5 we can restrict ourselves to the case where $H$ is absolutely continuous. Let $x$, $L^2(\Delta, \mathrm{d}x; \mathscr{H}_x; \mathscr{S})$ be a spectral representation of $H$, $\mathscr{H}$ ($\mathscr{S}$ separable). Then one can assume without loss of generality that all $\mathscr{H}_x \subseteq \mathscr{H}_0$ where $\mathscr{H}_0$ is any separable Hilbert space. Thus $\mathscr{H} = L^2(\Delta, \mathrm{d}x; \mathscr{H}_x; \mathscr{S})$ may be regarded as a subspace of $\tilde{\mathscr{H}} = L^2(\Delta, \mathrm{d}x; \mathscr{H}_0)$ and $H$ as a restriction of the multiplication operator $\tilde{H}$ by $x$ in this space, i.e. $H = \tilde{H} \restriction \mathscr{H}$. Further, let $X$ be an integral operator in $\mathscr{H} = L^2(\Delta, \mathrm{d}x; \mathscr{H}_x; \mathscr{S})$ with the kernel $X(x, x')$. Then $X$ can be regarded as a restriction to $\mathscr{H}$ of an integral operator $\tilde{X}$ in $\tilde{\mathscr{H}}$ with $\tilde{X} = X$ on $\mathscr{H}$ and $\tilde{X} = 0$ on $\tilde{\mathscr{H}} \ominus \mathscr{H}$. Thus $\operatorname*{ess\,sup}_{x,x' \in \Delta} \|\tilde{X}(x, x')\| = \operatorname*{ess\,sup}_{x,x' \in \Delta} \|X(x, x')\|$. Furthermore, $X$ is $H$-smooth iff $\tilde{X}$ is $\tilde{H}$-smooth, and $\|X\|_H = \|\tilde{X}\|_{\tilde{H}}$. These considerations reduce the general case to the special one treated in the lemmas. Then the following theorem is an easy consequence of the lemmas.

**Theorem 12.** *Let $H$ be absolutely continuous and $X \in \mathscr{L}(\mathscr{H}, \mathscr{H}')$. Then $X$ is $H$-smooth if and only if $D := X^*X$ is an integral operator given by*

$$(Du)(x) = \int_\Delta D(x, x')\, u(x')\, \mathrm{d}x', \qquad u \in \operatorname{dom} H,$$

*with respect to a spectral representation $x$, $L^2(\Delta, \mathrm{d}x; \mathscr{H}_x; \mathscr{S})$ of $H$, $\mathscr{H}$ ($\mathscr{S}$ separable) such that $\operatorname*{ess\,sup}_{x,x' \in \Delta} \|D(x, x')\| = b < \infty$. Furthermore, one then has $b = \|X\|_H^2$.*

**Proposition 13.** *If $H$ is bounded and $HP^{\mathrm{ac}}$ has no part with infinite multiplicity, then each $H$-smooth operator is a Hilbert-Schmidt operator.*

**Proof.** It is sufficient to treat the case in which $H$ is simple and is absolutely continuous. Then, if $X$ is $H$-smooth, $X^*X$ is an integral operator with kernel $D(x, x')$ with respect to the spectral representation $x$, $L^2(\Delta, \mathrm{d}x)$ of $H$, $\mathscr{H}$. Thus

$$\int_\Delta \int_\Delta |D(x, x')|^2\, \mathrm{d}x\, \mathrm{d}x' \leq \operatorname*{ess\,sup}_{x,x' \in \Delta} |D(x, x')|^2\, |\Delta|^2.$$

Since $|\Delta| < \infty$ ($H$ is bounded) and $\operatorname*{ess\,sup}_{x,x' \in \Delta} |D(x, x')| < \infty$ ($X$ is $H$-smooth), this proves that the right side is finite and therefore $X^*X \in \mathscr{L}_2(L^2(\Delta, \mathrm{d}x))$. $\blacksquare$

**Remark 14.** For each selfadjoint operator $H$ there is a sequence of projections $P_n$ such that $P_n \in \operatorname{com} H$, $\sum_n P_n = 1$ and $HP_n$ are bounded and simple. This shows that every $H$-smooth operator $X$ is "locally" a Hilbert-Schmidt operator, where "locally" means that $XP_n$ are Hilbert-Schmidt operators.

### 17.1.3. Local smoothness

In this section we want to generalize the notion of $H$-smoothness. The reason is that in applications the requirement that $X$ is $H$-smooth is often too strong.

**Definition 15.** A closed operator $X$ from $\mathscr{H}$ to another Hilbert space $\mathscr{H}'$ is called *$H$-smooth in $\Delta$, $\Delta$* a Borel set, if $XE(\Delta)$ is $H$-smooth.

**Definition 16.** Let $X$ be a closed operator from $\mathscr{H}$ to another Hilbert space $\mathscr{H}'$. Assume there is a decomposition $-\infty = a_1 < a_2 < \cdots < a_n = \infty$ such that $X$ is $H$-smooth on each closed interval $\Delta$ contained in any set $(a_j, a_{j+1})$, $j = \{1, 2, ,.., n - 1\}$. Then $X$ is called *locally $H$-smooth.*

**Remark 17.** From this definition it immediately follows that the set of locally $H$-smooth operators with the same final space forms a linear space.

The next theorem states that local $H$-smoothness is invariant with respect to changing $H$ into $\alpha(H)$ where $\alpha$ are certain real-valued functions.

**Theorem 18.** *Let $X$ be a locally $H$-smooth operator. Let $\alpha(\cdot)$ be a real-valued differentiable function such that there is a decomposition $-\infty = c_1 < c_2 < \cdots < c_n = \infty$, $n \in \mathbb{N}$, with the property that $\alpha'(x) > 0$ (or $\alpha'(x) < 0$) on each open interval $(c_j, c_{j+1})$. Then $X$ is also locally $\alpha(H)$-smooth.*

Proof. 1. It is obvious that $XE(\Delta)$ is locally $H$-smooth for each Borel set $\Delta$ if $X$ is locally $H$-smooth. Now

$$X = \sum_{j=1}^{n-1} XE\big((c_j, c_{j+1})\big) + XE\left(\bigcup_{j=2}^{n-1} \{c_j\}\right).$$

Thus each summand is also locally $H$-smooth. It holds $E\left(\bigcup_{s=2}^{n-1} \{c_j\}\right) \leqq E_{\alpha(H)}\left(\bigcup_{s=1}^{n-1} \{d_s\}\right)$ with $n_1 \in \mathbb{N}$ and $d_s \in \mathbb{R}$. Thus $XE\left(\bigcup_{j=2}^{n-1} \{c_j\}\right)$ is also locally $\alpha(H)$-smooth. By Remark 17, it therefore suffices to show that each operator $XE_H\big((c_j, c_{j+1})\big)$ is locally $\alpha(H)$-smooth, $j \in \{1, 2, \ldots, n - 1\}$.

Now we have $E_H\big((c_j, c_{j+1})\big) \leqq E_{\alpha(H)}\big(\alpha((c_j, c_{j+1})\big)$. Thus $Y_j := XE_H\big((c_j, c_{j+1})\big)$ is $\alpha(H)$-smooth on $\mathbb{R} \setminus \alpha\big(c_j, c_{j+1})\big)$. It remains to show that there exists a decomposition $\alpha(c_j) = a_1 < a_2 < \cdots < a_s = \alpha(c_{j+1})$ such that $XE_{\alpha(H)}([a, b])$ is $\alpha(H)$-smooth for each interval $[a, b] \subset (a_{j'}, a_{j'+1})$, $j' \in \{1, 2, \ldots, s - 1\}$. Since $X$ is locally $H$-smooth, there is a decomposition $c_j = e_1 < \cdots < e_p = c_{j+1}$ such that $XE\ ([a', b'])$ is $H$-smooth for each interval $[a', b'] \subset (e_{j'}, e_{j'+1})$, $j' \in \{1, 2, \ldots, p - 1\}$. Now we set $a_s := \alpha(e_s)$. Then $E([a', b']) \leqq E_{\alpha(H)}\big(\alpha([a', b'])\big) = E_{\alpha(H)}([a, b])$ with $\alpha(a') = a$, $\alpha(b') = b$ and $[a, b]$ $(a_{j'.}, a_{j'+1}) \subseteq \big(\alpha(c_j), \alpha(c_{j+1})\big)$. Set $X \restriction E([a', b'])\,\mathscr{H} = \tilde{X}$, $E([a', b'])\,\mathscr{H} = \tilde{\mathscr{H}}$ and $\alpha(\tilde{H}) = \alpha(H) \restriction E([a'; b'])\,\mathscr{H}$. Then we see that the proof of local $\alpha(H)$-smoothness of $X$ is reduced to the question:

Suppose that the selfadjoint operator $\tilde{H}$ on $\mathscr{H}$ is bounded and absolutely continuous, that $\tilde{X}$ is $\tilde{H}$-smooth, and that $\alpha'(x) > 0$ on the spectrum of $\tilde{H}$. Does it follow that $\tilde{X}$ is $\alpha(\tilde{H})$-smooth? The case $\alpha'(x) < 0$ is similar since $H$-smoothness is invariant with respect to changing $H$ into $-H$.

2. Now we prove that the answer to this question is yes. By using the spectral representation of $\tilde{H}$, $\mathscr{H}$ and Theorem 5 it is easy to see that it is enough to consider the special case in which $\tilde{H}$ is the multiplication operator by $x$ on the Hilbert space $\mathscr{H} = L^2([a', b'], \mathrm{d}x; \mathscr{H}_0)$. Then we know from Theorem 12 that $\tilde{X}^*\tilde{X}$ is an integral operator with a kernel $D(x, x')$ such that $\|D(x, x')\| \leqq \|\tilde{X}\|_H^2 < \infty$ a.e. (mod $|\cdot|$). For the proof of the $\alpha(\tilde{H})$-smoothness of $\tilde{X}$ it is sufficient, by the same theorem, to show that $\tilde{X}^*\tilde{X}$ is an integral operator with respect to the spectral representation of $\alpha(\tilde{H})$, $\mathscr{H}$ such that its kernel $D_\alpha(x, x')$ is essentially bounded. Now the transformation from $H$, $\mathscr{H}$ to the spectral representation of $\alpha(\tilde{H})$, $\mathscr{H}$ is given by

$$u(x) \to \tilde{u}(x) = u\big(\alpha^{-1}(x)\big) \big(\alpha'(\alpha^{-1}(x))\big)^{-1/2}.$$

Then the integral operator $\tilde{X}^*\tilde{X}$ with the kernel $D(x, x')$ goes over into an integral operator with the kernel

$$D_\alpha(x, x') = \big(\alpha'(\alpha^{-1}(x))\, \alpha'(\alpha^{-1}(x'))\big)^{-1/2} D\big(\alpha^{-1}(x), \alpha^{-1}(x')\big),$$

under this transformation. Thus, note that $\mathscr{H} \to L^2(\alpha([a', b'])\,\mathrm{d}x; \mathscr{H}_0)$, $\operatorname{ess\,sup}_{x,x' \in \alpha([a',b'])} \|D_\alpha(x, x')\|$ $\leqq \operatorname{ess\,sup}_{x,x' \in [a',b']} \|D(x, x')\| \operatorname{ess\,sup}_{x,x' \in [a',b']} \big(\alpha'(x)\, \alpha'(x')\big)^{-1/2}$. The right side is finite because $\alpha'(x) > 0$ on $[a', b']$ and $\|D(x, x')\| \leqq \|\tilde{X}\|_H^2 < \infty$ a.e. (mod $|\cdot|$). This proves the theorem. ∎

It is possible to generalize the class of functions in this statement. One can also formulate such a theorem for $H$-smooth operators, but then the class of admissible functions is much smaller. For example, if $X$ is $H$-smooth, then $X$ is in general not $H^2$-smooth.

**Proposition 19.** *Let $X$ be a closed operator from $\mathscr{H}$ to another Hilbert space $\mathscr{H}'$. Suppose that* $\operatorname{dom} X \supseteq \operatorname{dom} H$ *and let $\Delta \subseteq \mathbb{R}$. Further suppose that*

$$\text{(I)} \qquad \sup_{0<|\varepsilon|<\varepsilon_0, \lambda \in \Delta} |\varepsilon|\, \|XR(\lambda + \mathrm{i}\varepsilon)\|^2 < \infty, \qquad \varepsilon_0 > 0,$$

*or*

$$\text{(II)} \qquad \sup_{0<\varepsilon<\varepsilon_0, \lambda \in \Delta} \|XR(\lambda + \mathrm{i}\varepsilon)\, X^*\| < \infty, \qquad \varepsilon_0 > 0.$$

*Then $X$ is $H$-smooth on* clo $\Delta$.

Proof. (1) For each $\varepsilon \neq 0$, $\|XR(\lambda + \mathrm{i}\varepsilon)\|$ is continuous in $\lambda$, so the bound $c := \sup_{\lambda \in \Delta, 0<|\varepsilon|<\varepsilon_0} |\varepsilon|\, \|XR(\lambda + \mathrm{i}\varepsilon)\|^2$ extends to all $\lambda \in \operatorname{clo} \Delta$. Now assume that $\lambda \in \mathbb{R} \setminus \operatorname{clo} \Delta$. Choose $\lambda_0 \in \operatorname{clo} \Delta$ such that $|\lambda - \lambda_0| = \operatorname{dis}(\lambda, \operatorname{clo} \Delta)$. Then

$$\begin{aligned}
&|\varepsilon|\, \|XR(\lambda + \mathrm{i}\varepsilon)\, E(\operatorname{clo} \Delta)\, u\|^2 \\
&= |\varepsilon|\, \|XR(\lambda_0 + \mathrm{i}\varepsilon) \big(1 - (\lambda - \lambda_0)\, R(\lambda + \mathrm{i}\varepsilon)\big) E(\operatorname{clo} \Delta)\, u\|^2 \\
&\leqq |\varepsilon|\, \|XR(\lambda_0 + \mathrm{i}\varepsilon)\|^2\, \|\big(1 - (\lambda - \lambda_0)\, R(\lambda + \mathrm{i}\varepsilon)\big) E(\operatorname{clo} \Delta)\, u\|^2 \\
&\leqq 4\, |\varepsilon|\, \|XR(\lambda_0 + \mathrm{i}\varepsilon)\|^2\, \|u\|^2
\end{aligned}$$

where we have used the first resolvent identity and $|\lambda - \lambda_0| \cdot |\lambda + i\varepsilon - x|^{-1} < 1$ for any $x \in \operatorname{clo} \Delta$. Thus

$$\sup_{\lambda \in \mathbb{R}, 0<|\varepsilon|<\varepsilon_0} |\varepsilon| \, \|XR(\lambda + i\varepsilon) \, E(\operatorname{clo} \Delta)\|^2 < \infty$$

and this proves, by Theorem 6 and Remark 4, that $XE(\operatorname{clo} \Delta)$ is $H$-smooth.

(2) Since $\big(XR(\lambda + i\varepsilon) \, X^*\big)^* = XR(\lambda - i\varepsilon) \, X^*$, we get from (II) that

$$c := \sup_{\lambda \in \Delta, 0<|\varepsilon|<\varepsilon_0} \|XR(\lambda + i\varepsilon) \, X^*\| < \infty.$$

Thus

$$\sup_{\lambda \in \Delta, 0<|\varepsilon|<\varepsilon_0} |\varepsilon| \, \|R(\lambda + i\varepsilon) \, X^*\|^2 = \sup_{\lambda \in \Delta, 0<|\varepsilon|<\varepsilon_0} |\varepsilon| \, \|XR(\lambda + i\varepsilon) \, R(\lambda - i\varepsilon) \, X^*\|$$

$$= \sup_{\lambda \in \Delta, 0<|\varepsilon|<\varepsilon_0} (1/2) \, \|X[R(\lambda + i\varepsilon) - R(\lambda - i\varepsilon)] \, X^*\| \leqq c < \infty.$$

By $\|R(\lambda + i\varepsilon) \, X^*\| = \|XR(\lambda - i\varepsilon)\|$, this implies that assumption (I) holds, and therefore $XE(\operatorname{clo} \Delta)$ is $H$-smooth. ■

The next proposition relates positive commutators to $H$-smooth operators and absolute continuity of $H$.

**Proposition 20.** *Let $A$ be a bounded selfadjoint operator on $\mathscr{H}$. Suppose that there is a closed operator $C$ such that $C = i(HA - AH) \geqq 0$ in the form sense. Then $C^{1/2}$ is $H$-smooth. In particular, if* $\ker C = \{0\}$, *then $H$ is absolutely continuous.*

**Proof.** From the definition of $C$, $(u, Cu') = i[(Hu, Au') - (Au, Hu')]$ for $u, u' \in \operatorname{dom} H$, we find that $\operatorname{dom} C^{1/2} \supseteq \operatorname{dom} H$. Now, for $u \in \operatorname{dom} H$,

$$\frac{d}{dt} (u, e^{itH} A \, e^{-itH} u) = i(H \, e^{-itH} u, A \, e^{-itH} u) - i(e^{-itH} u, AH \, e^{-itH} u) = (u, e^{itH} C \, e^{-itH} u).$$

Thus

$$\int_s^t (u, e^{i\tau H} C \, e^{-i\tau H} u) \, d\tau = (u, e^{itH} A \, e^{-itH} u) - (u, e^{isH} A \, e^{-isH} u).$$

This gives

$$\int_s^t \|C^{1/2} \, e^{-i\tau H} u\|^2 \, d\tau \leqq 2 \, \|A\| \cdot \|u\|^2.$$

Since $s$ and $t$ are arbitrary, we obtain for $u \in \operatorname{dom} H$

$$2 \, \|A\| \cdot \|u\|^2 \geqq \int_{-\infty}^{\infty} e^{-2\varepsilon|t|} \, \|C^{1/2} \, e^{-itH} u\|^2 \, dt$$

$$= \frac{1}{2\pi} \int_{-\infty}^{\infty} d\lambda \, \{\|C^{1/2} R(\lambda + i\varepsilon) \, u\|^2 + \|C^{1/2} R(\lambda - i\varepsilon) \, u\|^2\}.$$

By taking the supremum of the right side over $\varepsilon > 0$ and $\|u\| = 1$, we find that $C^{1/2}$ satisfies the assumptions of Definition 1, that is, $C^{1/2}$ is $H$-smooth. Here we have used that the right side is defined for all $u$ because $\operatorname{dom} C^{1/2} \supseteq \operatorname{dom} H$. The other statement of this proposition follows from Theorem 5. ∎

To illustrate that the theory of $H$-smoothness is not empty, we present some simple examples.

**Example 21.** Let $H$ be the multiplication operator by $x$ on $\mathscr{H} = L^2([a, b], dx)$. Assume that $X$ is an integral operator with a uniformly bounded kernel $X(x, x')$, i.e.

$$(Xu)(x) = \int_a^b X(x, x')\, u(x')\, dx' \quad \text{and} \quad \sup_{x,x' \in [a,b]} |X(x, x')| = c < \infty.$$

Then $X$ is $H$-smooth. This follows from the fact that $X^*X$ then also has a uniformly bounded kernel and from Theorem 12. When $X(x, x')$ is not uniformly bounded but $\sup_{x \in [a,b], x' \in \Delta} |X(x, x')| < \infty$ for all closed intervals $\Delta \subset (a, b)$, then $X$ is locally $H$-smooth. This is a consequence of the fact that $XE(\Delta)$ is $H$-smooth for each closed $\Delta \subset (a, b)$ since the kernel of the integral operator $XE(\Delta)$ is given by $X(x, x')\, \chi_\Delta(x')$.

**Example 22.** Let $H = -i\, d/dx$ on $\mathscr{H} = L^2(\mathbb{R}, dx)$. Let $X$ be the multiplication operator by the function $a(x)$ with $a(\cdot) \in L^2(\mathbb{R}, dx)$. Then we have

$$\int_{-\infty}^{\infty} \|X\, e^{-itH} u\|^2\, dt = \int_{-\infty}^{\infty} dt \int_{-\infty}^{\infty} dx\, |a(x)\, u(x - t)|^2 \tag{12}$$

because of $(e^{-itH} u)(x) = u(x - t)$. Now

$$\int_{-\infty}^{\infty} dx \int_{-\infty}^{\infty} dt\, |a(x)|^2\, |u(x - t)|^2 = \|a\|^2\, \|u\|^2.$$

By Fubini's theorem, we get that the integral (12) also exists and equals $\|a\|^2\, \|u\|^2 < \infty$. By Prop. 3, this proves that $X$ is $H$-smooth. Furthermore, by Theorem 18, we get that $X$ is also locally $H^2$-smooth. By a more careful consideration one can see that $X$ is $H^2$-smooth on each set $\Delta \subset \mathbb{R}$ with $\Delta \cap (-\varepsilon, \varepsilon) = \emptyset$ for some $\varepsilon > 0$.

## 17.2. Existence of wave operators

### 17.2.1. Symmetrical case

In this and the following section we apply the smoothness theory to the questions of existence and completeness of wave operators for the triple $\{H, H_0; J\}$. Moreover, the assumption that the perturbation $V = HJ - JH_0$ is in a certain sense smooth with respect to $H_0$ and $H$ will provide us with additional information about the singular parts of $H$ and $H_0$.

**Theorem 23.** *Suppose that:*

(I) $V := HJ - JH_0 = B^*A$ *in the form sense.*

(II) $A$ *is* $H_0$*-smooth and* $B$ *is* $H$*-smooth.*

*Then the limits*

$$\underset{t\to\pm\infty}{\text{s-lim}}\, e^{itH} J\, e^{-itH_0}, \qquad \underset{t\to\pm\infty}{\text{s-lim}}\, e^{itH_0} J^*\, e^{-itH}$$

*exist for both signs, + and −. In particular, the strong wave operators $W_\pm(H, H_0; J)$, $W_\pm(H_0, H; J^*)$ exist for both signs.*

**Proof.** By (II) and Theorem 5, dom $B \supseteq$ dom $H$ and dom $A \supseteq$ dom $H_0$. Thus the expression

$$w(t) := \left(v, e^{itH}(HJ - JH_0)\, e^{-itH_0}u\right) = (B\, e^{-itH}v, A\, e^{-itH_0}u)$$

is well-defined for all $u \in \text{dom}\, H_0 \cap P_0^{ac}\mathscr{H}_0$ and $v \in \text{dom}\, H$. This gives, using the Schwarz inequality and Prop. 3

$$\sup_{v\in \text{dom} H, \|v\|=1} \int_s^t |w(\tau)|\, d\tau \leq \sup_{\substack{v\in \text{dom} H\\ \|v\|=1}} \left\{\int_s^t \|B\, e^{-i\tau H}v\|^2\, d\tau \int_s^t \|A\, e^{-i\tau H_0}u\|^2\, d\tau\right\}^{1/2}$$

$$\leq \left\{2\pi\, \|B\|_H{}^2 \int_s^t \|A\, e^{-i\tau H_0}u\|^2\, d\tau\right\}^{1/2}.$$

Thus, the system $\{H, H_0, J\}$ satisfies the assumptions of Theorem 15.2 with $\mathscr{D}_0 = \text{dom}\, H_0 \cap P_0^{ac}\mathscr{H}_0$ and $\mathscr{D} = \text{dom}\, H$, and this implies that the limit $\underset{t\to\infty}{\text{s-lim}}\, e^{itH}J\, e^{-itH_0}P_0^{ac}$ exists. By Theorem 5 and (II), $A(P_0^{ac})^\perp = V(P_0^{ac})^\perp = 0$. This implies according to Lemma 14.15 the existence of $\underset{t\to\infty}{\text{s-lim}}\, e^{itH}J\, e^{-itH_0}(P_0^{ac})^\perp = J(P_0^{ac})^\perp$. Therefore the strong limit of $e^{itH}J\, e^{-itH_0}$ exists. Similarly, we can prove the existence of the other limits. This concludes the proof. ■

**Remark 24.** Assume the homogeneous parts of $H_0 \mid P_0^{ac}\mathscr{H}_0$ and $H \mid P^{ac}\mathscr{H}$ have finite multiplicity. Then we see from Prop. 13 that (II) implies $E(\Delta)\,(HJ - JH_0)\, E_0(\Delta') \in \mathscr{L}_1(\mathscr{H}_0, \mathscr{H})$ for all finite intervals $\Delta$, $\Delta'$. Thus, as far as the existence of wave operators is concerned, smooth perturbations can also be treated by trace class methods.

**Corollary 25.** *In addition to the assumptions of Theorem 23 assume that $J$ is bounded invertible and* $\underset{t\to\pm\infty}{\text{A-lim}}\, \|J\, e^{-itH_0}u\|^2 = \|u\|^2$ *for all $u \in \mathscr{H}_0$. Then $H$ and $H_0$ are unitarily equivalent and the wave operators $W_+(H, H_0; J)$ and $W_-(H, H_0; J)$ are partial isometries and complete.*

**Proof.** From Theorem 23 we obtain the existence of the limits

$$J_\pm := \underset{t\to\pm\infty}{\text{s-lim}}\, e^{itH}J\, e^{-itH_0}, \qquad \tilde{J}_\pm := \underset{t\to\pm\infty}{\text{s-lim}}\, e^{itH_0}J^*\, e^{-itH}.$$

Further we have $J_\pm{}^* = \tilde{J}_\pm$ because of $(e^{itH}J\, e^{-itH_0})^* = e^{itH_0}J^*\, e^{-itH}$. Since $J$, and thus also $J^*$, is bounded invertible. Prop. 9.39 implies that the operators $J_\pm$ and $\tilde{J}_\pm$ have support $\mathscr{H}_0$ and $\mathscr{H}$ respectively. Further,

$$\operatorname*{A-lim}_{t\to\pm\infty} \|J\, \mathrm{e}^{-\mathrm{i}tH_0}u\|^2 = \operatorname*{A-lim}_{t\to\pm\infty} \|\mathrm{e}^{\mathrm{i}tH}J\, \mathrm{e}^{-\mathrm{i}tH_0}u\|^2 = \|J_\pm u\|^2 = \|u\|^2$$

yields that $J_\pm$ are isometric and thus even unitary operators (supp $J_\pm{}^* =$ supp $\tilde{J}_\pm = \mathscr{H}$). By Prop. 9.10 ($P^{\mathrm{ex}} = 1$ here) we obtain that $J_\pm$ are intertwining operators for $H, H_0$; i.e. $H$ and $H_0$ are unitarily equivalent. Now $W_\pm(H, H_0; J) = J_\pm P_0^{\mathrm{ac}}$ and $W_\pm(H_0, H; J^*) = \tilde{J}_\pm P^{\mathrm{ac}}$. Then the unitarity of $J_\pm$ leads to the assertions about the wave operators. ∎

**Theorem 26.** *Let $\Delta$ be a bounded open interval. Suppose that:*

*(I) $V := HJ - JH_0 = B^*A$ in the form sense.*
*(II) $A$ is $H_0$-bounded and $B$ is $H$-bounded.*
*(III) $A$ is $H_0$-smooth on $\Delta$ and $B$ is $H$-smooth on $\Delta$.*

*Then the limits*

$$W_\pm{}^\Delta := \operatorname*{s-lim}_{t\to\pm\infty} \mathrm{e}^{\mathrm{i}tH}J\, \mathrm{e}^{-\mathrm{i}tH_0}E_0(\Delta), \quad \tilde{W}_\pm{}^\Delta := \operatorname*{s-lim}_{t\to\pm\infty} \mathrm{e}^{\mathrm{i}tH_0}J^*\, \mathrm{e}^{-\mathrm{i}tH}E(\Delta)$$

*exist for both signs, + and −, and $(W_\pm{}^\Delta)^* = \tilde{W}_\pm{}^\Delta$. In particular, the strong wave operators $W_\pm(H, H_0; JE_0(\Delta))$, $W_\pm(H_0, H; J^*E(\Delta))$ exist for both signs and $W_\pm(H, H_0; JE_0(\Delta))^* = W_\pm(H_0, H; J^*E(\Delta))$.*

**Proof.** By Theorem 23, the limits

$$\operatorname*{s-lim}_{t\to\pm\infty} \mathrm{e}^{\mathrm{i}tH}E(\Delta)\, JE_0(\Delta)\, \mathrm{e}^{-\mathrm{i}tH_0}, \qquad \operatorname*{s-lim}_{t\to\pm\infty} \mathrm{e}^{\mathrm{i}tH_0}E_0(\Delta)\, J^*E(\Delta)\, \mathrm{e}^{-\mathrm{i}tH}$$

exist. Now we claim that

$$\operatorname*{s-lim}_{t\to\pm\infty} E(\mathbb{R}\setminus\Delta)\, \mathrm{e}^{\mathrm{i}tH}J\, \mathrm{e}^{-\mathrm{i}tH_0}E_0(\Delta) = 0. \tag{13}$$

Clearly, this implies that $W_\pm{}^\Delta$ exist. Because of the symmetry in our assumptions we then get the existence of $\tilde{W}_\pm{}^\Delta$. Furthermore, $W_\pm{}^\Delta\mathscr{H}_0 \subseteq E(\Delta)\,\mathscr{H}$ and $\tilde{W}_\pm{}^\Delta\mathscr{H} \subseteq E_0(\Delta)\,\mathscr{H}_0$ (see Prop. 9.20). This and

$$\operatorname*{w-lim}_{t\to\pm\infty} E_0(\Delta)\, \mathrm{e}^{\mathrm{i}tH_0}J^*\, \mathrm{e}^{-\mathrm{i}tH} = \operatorname*{w-lim}_{t\to\pm\infty} E_0(\Delta)\, \mathrm{e}^{\mathrm{i}tH_0}J^*\, \mathrm{e}^{-\mathrm{i}tH}E(\Delta) = (W_\pm{}^\Delta)^*$$

(following from the existence of $W_\pm{}^\Delta$) imply $(W_\pm{}^\Delta)^* = \tilde{W}_\pm{}^\Delta$. The assertions about the wave operators follow from these statements and $W_\pm(H, H_0; JE_0(\Delta)) = W_\pm{}^\Delta P_0^{\mathrm{ac}}$, $W_\pm(H_0, H; J^*E(\Delta)) = \tilde{W}_\pm{}^\Delta P^{\mathrm{ac}}$.

It remains to prove (13). First, we note that it suffices to prove (13) with $E_0(\Delta)$ replaced by $E_0(\sigma)$ where $\sigma$ is any arbitrary closed subinterval of the open interval $\Delta$. Let $\mathscr{C}$ be a contour enclosing $\sigma$ and have a nonvanishing distance to $\sigma$ and $\mathbb{R}\setminus\Delta$. Then

$$\begin{aligned}
&E(\mathbb{R}\setminus\Delta)\, \mathrm{e}^{\mathrm{i}tH}J\, \mathrm{e}^{-\mathrm{i}tH_0}E_0(\sigma)\, u \\
&= -(2\pi\mathrm{i})^{-1} \oint_{\mathscr{C}} E(\mathbb{R}\setminus\Delta)\, (z - H)^{-1}\, \mathrm{e}^{\mathrm{i}tH}J\, \mathrm{e}^{-\mathrm{i}tH_0}E_0(\sigma)\, u\, \mathrm{d}z \\
&\quad + (2\pi\mathrm{i})^{-1} \oint_{\mathscr{C}} E\,(\mathbb{R}\setminus\Delta)\, \mathrm{e}^{\mathrm{i}tH}J\, \mathrm{e}^{-\mathrm{i}tH_0}(z - H_0)^{-1}\, E_0(\sigma)\, u\, \mathrm{d}z \\
&= -(2\pi\mathrm{i})^{-1} \oint_{\mathscr{C}} E(\mathbb{R}\setminus\Delta)\, \mathrm{e}^{\mathrm{i}tH}\big(R(z)\, J - JR_0(z)\big)\, \mathrm{e}^{-\mathrm{i}tH_0}E_0(\sigma)\, u\, \mathrm{d}z
\end{aligned} \tag{14}$$

where we have used the operator version of the Cauchy integral formula

$$(2\pi i)^{-1} \oint_{\mathscr{C}} (z - z')^{-1} \, dz = \begin{cases} 1 & \text{if } z' \text{ is in the interior of } \mathscr{C}, \\ 0 & \text{if } z' \text{ is outside of } \mathscr{C}, \end{cases}$$

and the fact that the spectrum of $H_0 E_{H_0}(\sigma)$ is in the interior of $\mathscr{C}$ and that of $HE_H(\mathbb{R} \setminus \Delta)$ is outside of $\mathscr{C}$.

Because the integrand on the last line of (14) is uniformly bounded on $\mathscr{C}$ it suffices to prove

$$\underset{t \to \pm\infty}{\text{s-lim}} \left(R(z) J - JR_0(z)\right) e^{-itH_0} E_0(\sigma) = 0$$

for all nonreal $z$. But this follows from the estimate

$$\begin{aligned} \left|\left(v, (R(z) J - JR_0(z)) e^{-itH_0} E_0(\sigma) u\right)\right| &= \left|\left(BR(\bar{z}) v, AR_0(z) e^{-itH_0} E_0(\sigma) u\right)\right| \\ &\leq \|v\| \, \|BR(\bar{z})\| \, \|AE_0(\sigma) e^{-itH_0} R_0(z) u\| \end{aligned}$$

if we use that $BR(z)$ is bounded and $AE_0(\sigma)$ is $H_0$-smooth. Then by Prop. 3, $\|AE_0(\sigma) e^{-itH_0} R_0(z) u\|^2$ is integrable in $t$ and furthermore, it is uniformly continuous in $t$ because $AR_0(z)$ is bounded and $e^{-itH_0}u$ is uniformly continuous. Thus, by Lemma 15.1, $\|AE_0(\sigma) e^{-itH_0} R_0(z) u\|^2 \to 0$ as $t \to \infty$. This concludes the proof. ∎

**Proposition 27.** *Suppose that assumptions* (I) *and* (II) *of Theorem* 26 *are satisfied. Further suppose that*

(III′) *$A$ is locally $H_0$-smooth and $B$ is locally $H$-smooth.*
*Then the limits*

$$\underset{t \to \pm\infty}{\text{s-lim}} \, e^{itH} J \, e^{-itH_0} P^c_{H_0}, \qquad \underset{t \to \pm\infty}{\text{s-lim}} \, e^{itH_0} J^* \, e^{-itH} P^c_H$$

*exist for both signs. In particular, the strong wave operators $W_\pm(H, H_0; J)$, $W_\pm(H_0, H; J^*)$ exist for both signs.*

**Proof.** By definition of local-$H$-smoothness, there exist two decompositions $-\infty = a_1 < a_2 < \cdots < a_n = \infty$ and $-\infty = b_1 < \cdots < b_n = \infty$ such that $A$ is $H_0$-smooth on each closed interval $\Delta \subset (a_j, a_{j+1})$ and $B$ is $H$-smooth on each closed interval $\Delta' \subset (b_j, b_{j+1})$. Clearly, there is a common decomposition, $-\infty = c_1 < \cdots < c_m = \infty$ such that $A$ is $H_0$-smooth and $B$ is $H$-smooth on each closed interval from $(c_j, c_{j+1})$. By Theorem 26, the strong limits of

$$e^{itH} J \, e^{-itH_0} u, \qquad e^{itH_0} J^* \, e^{-itH} v$$

exist for all $u \in \mathscr{D}_0 = \{u;\, u \in E_0(\Delta) P^c_{H_0} \mathscr{H}_0,\ \Delta$ closed interval in $(c_j, c_{j+1}),\ j \in \{1, 2, \ldots, m-1\}\}$ and $v \in \mathscr{D} = \{v;\, v \in E(\Delta) P^c_H \mathscr{H},\ \Delta$ closed interval in $(c_j, c_{j+1})$ with $j \in \{1, 2, \ldots, m-1\}\}$.

These sets are generating in $P^c_{H_0} \mathscr{H}_0$ and $P^c_H \mathscr{H}$ respectively. Thus by the uniform buondedness of the operators $e^{itH} J \, e^{-itH_0}$ and $e^{itH_0} J^* \, e^{-itH}$ the strong limits exist also for all vectors from $P_{H_0}^c\mathscr{H}_0$ and $P_H^c\mathscr{H}$ respectively, proving the assertions of this propositon. ■

From this statement we easily get the following

**Corollary 28.** *In addition to the assumptions of Prop. 27 assume that $J$ is bounded invertible and* $\operatorname{A-lim}_{t\to\pm\infty} \|J\,e^{-itH_0}P_{H_0}^c u\|^2 = \|u\|^2$. *Then $HP_H^c$ and $H_0P_{H_0}^c$ are unitarily equivalent and the wave operators $W_+(H, H_0; J)$, $W_-(H, H_0; J)$ are partially isometric and complete.*

**Proof.** From Prop. 27 we get that the limits

$$W_\pm := \operatorname*{s-lim}_{t\to\pm\infty} e^{itH}J\,e^{-itH_0}P_{H_0}^c, \qquad \widetilde{W}_\pm := \operatorname*{s-lim}_{t\to\pm\infty} e^{itH_0}J^*\,e^{-itH}P_H^c$$

exist. Since $J$, and thus also $J^*$, is bounded invertible, the supports of $W_\pm$ and $\widetilde{W}_\pm$ are $P_{H_0}^c\mathscr{H}_0$ and $P_H^c\mathscr{H}$ respectively. Furthermore, from $\operatorname{A-lim}_{t\to\pm\infty} \|e^{itH}J^*\,e^{-itH_0}P_{H_0}^c u\|^2 = \|W_\pm u\|^2 = P_{H_0}^c u\|^2$ it follows that $W_\pm$ are partially isometric with initial subspaces $P_{H_0}^c\mathscr{H}$. Since $W_\pm$ and $\widetilde{W}_\pm$ exist, this gives

$$\begin{aligned}\operatorname*{w-lim}_{t\to\pm\infty} P_{H_0}^c\,e^{itH_0}J^*\,e^{-itH} &= \operatorname*{w-lim}_{t\to\pm\infty} P_{H_0}^c\,e^{itH_0}J^*\,e^{-itH}P_H^c = W_\pm{}^*\\ &= \operatorname*{s-lim}_{t\to\pm\infty} P_{H_0}^c\,e^{itH_0}J^*\,e^{-itH}P_H^c = P_{H_0}^c\widetilde{W}_\pm = \widetilde{W}_\pm.\end{aligned}$$

Thus the final subspaces of $W_\pm$ are equal to the initial subspaces of $\widetilde{W}_\pm$, which coincide with $P_H^c\mathscr{H}$. This proves that the intertwining operators $W_\pm$ for $H$ and $H_0$ are partial isometries from $P_{H_0}^c\mathscr{H}_0$ onto $P_H^c\mathscr{H}$. Therefore $HP_H{}^c$ and $H_0P_{H_0}^c$ are unitarily equivalent. The assertions about the wave operators follow from the properties of $W_\pm$. ■

### 17.2.2. Smallness condition

Since it is often difficult to verify the $H$-smoothness of the operator $B$ from the factorization $V = B^*A$, we want to avoid this assumption by stronger requirements on $A$, $B$ with respect to $H_0$. The way to do this is similar to that in the general factorization method (see Section 14.2.3).

**Theorem 29.** *Suppose that:*

(I) *$J$ is bounded invertible,*

(II) *$V := HJ - JH_0 = B^*A$ in the form sense with $A$*: $\operatorname{dom} A \mapsto \mathscr{H}_0$,

(III) *$A$ is $H_0$-bounded and $B$ is bounded,*

(IV) *$A$ and $BJ$ satisfy assumption* (I) *of Prop. 19 with respect to $H_0$ and for an open interval $\Delta$. Further,* $\sup_{\lambda\in\Delta, 0<|\varepsilon|<\varepsilon_0} \|AR_0(\lambda + i\varepsilon)\,J^{-1}B^*\| < 1$ *for some* $\varepsilon_0 > 0$.

*Then the limits*

$$W_\pm{}^\Delta := \operatorname*{s-lim}_{t\to\pm\infty} e^{itH}J\,e^{-itH_0}E_0(\Delta), \qquad \widetilde{W}_\pm{}^\Delta := \operatorname*{s-lim}_{t\to\pm\infty} e^{itH_0}J^*\,e^{-itH}E(\Delta)$$

*exist for both signs. Further,* $(W_\pm{}^\Delta)^* = \widetilde{W}_\pm{}^\Delta$, *the support of $W_\pm{}^\Delta$ is $E_0(\Delta)\,\mathscr{H}_0$ and* clo ima $W_\pm{}^\Delta = E(\Delta)\,\mathscr{H}$. *Im particular the wave operators* $W_\pm(H, H_0; JE_0(\Delta))$, $W_\pm(H_0, H; J^*E(\Delta))$

*exist in the strong sense, their supports are* $E_0(\Delta)\, P_0^{ac}\mathscr{H}_0$, $E(\Delta)\, P^{ac}\mathscr{H}$ *respectively, and we have* $W_\pm(H, H_0; JE_0(\Delta))^* = W_\pm(H_0, H; J^*E(\Delta))$.

**Proof.** For the proof we use Theorem 26. Thus, we have to verify that $B$ is $H$-smooth on $\Delta$. Using the second resolvent identity, we obtain (formally)

$$R(z)\, J = JR_0(z) + JR_0(z)\, J^{-1}B^* \sum_{n=0}^{\infty} \left(AR_0(z)\, J^{-1}B^*\right)^n AR_0(z). \tag{15}$$

By (IV), this sum is convergent in norm and it is easy to show that the right side equals the left side. Now we multiply from the left with $B$. This gives

$$BR(z)\, J = BJR_0(z) \left(1 + X(z)\right) AR_0(z) \tag{16}$$

where $X(z)$ is an operator-valued function satisfying $\|X(\lambda + i\varepsilon)\| < c < \infty$ for all $\lambda \in \Delta$ and $0 < |\varepsilon| < \varepsilon_0$. Using (IV) and Prop. 19 we easily obtain from this equation that $B$ is $H$-smooth on $\Delta$. Thus, from Theorem 26 we get that $W_\pm{}^\Delta$ and $\tilde{W}_\pm{}^\Delta$ exist and $(W_\pm{}^\Delta)^* = \tilde{W}_\pm{}^\Delta$. The assertion on the supports of these operators follow immediately from (I). Then the wave operators clearly exist and

$$W_\pm{}^\Delta P_0^{ac} = W_\pm(H, H_0; JE_0(\Delta)), \qquad \tilde{W}_\pm{}^\Delta P^{ac} = W_\pm(H_0, H; J^*E(\Delta)).$$

This gives the other assertions about the wave operators. ■

This statements says, among other things, that $H_0E_0(\Delta)$ and $HE(\Delta)$ are unitarily equivalent because of the intertwining property $W_\pm{}^\Delta H_0 = HW_\pm{}^\Delta$ and the assertion on the supports of $W_\pm{}^\Delta$ and $(W_\pm{}^\Delta)^*$.

When $J$ is unitary, the theorem is considerably simpler. Then $J^{-1}V = J^*V = J^*HJ - H_0$ is selfadjoint. Thus the polar decomposition of $J^{-1}V$ is $W\,|V| = |V|\, W$ with $W = W^*$, as a simple calculation and the properties of the polar decomposition of self-adjoint operators (see Kato [1, p. 335]) show. Now take the factorization $V = B^*A = B^*\,|V|^{1/2}$. Therefore $J^{-1}B^* = J^*B^* = W\,|V|^{1/2} = |V|^{1/2}\, W = BJ$. This gives

$$\sup_{\lambda\in\Delta, 0<|\varepsilon|<\varepsilon_0} \|AR_0(\lambda + i\varepsilon)\, J^{-1}B^*\| = \sup_{\lambda\in\Delta, 0<|\varepsilon|<\varepsilon_0} \|\,|V|^{1/2}\, R_0(\lambda + i\varepsilon)\,|V|^{1/2}\|.$$

Thus, if this expression is finite, $A = |V|^{1/2}$ and also $BJ = W\,|V|^{1/2}$ satisfy assumption (II) (hence also assumption (I)) of Prop. 19 (i.e. this implies that $A$, $BJ$ are $H_0$-smooth on $\Delta$). That is, it suffices to replace (IV) in Theorem 29 by the condition $\sup_{\lambda\in\Delta, 0<|\varepsilon|<\varepsilon_0} \|\,|V|^{1/2} \times R_0(\lambda + i\varepsilon)\,|V|^{1/2}\| < 1$. Moreover, in this case the condition that $B$ is bounded in (III) of Theorem 29 can be omitted, as one can see from (15), (16) and $BJ = J^{-1}B^* = W\,|V|^{1/2}$. Naturally, the wave operators $W_\pm(H, H_0; JE_0(\Delta))$ are partially isometric and complete in this case.

**Corollary 30.** *Suppose that assumptions* (I)–(III) *of Theorem 29 are satisfied. Further suppose that*

(IV) *$A$ and $BJ$ satisfy assumption* (I) *of Prop.* 19 *on $S$ and*

$$\sup_{\lambda\in S, 0<|\varepsilon|<\varepsilon_0} \|AR_0(\lambda + i\varepsilon)\, J^{-1}B^*\| < 1 \text{ for some } \varepsilon_0 > 0$$

*where* $S = \bigcup_{n=1}^{N} (a_n, a_{n+1})$ *with* $-\infty = a_1 < a_2 < \cdots < a_N = \infty$.

*Then the limits*

$$J_\pm := \text{s-}\lim_{t\to\pm\infty} e^{itH} J\, e^{-itH_0} P^c_{H_0}, \quad \tilde{J}_\pm := \text{s-}\lim_{t\to\pm\infty} e^{itH_0} J^*\, e^{-itH} P^c_H$$

*exist. Further, $J_\pm^* = \tilde{J}_\pm$, the support of $J_\pm$ is $P^c_{H_0}\mathscr{H}_0$ and* clo ima $J_\pm = P^c_H\mathscr{H}$. *In particular, the strong wave operators $W_\pm(H, H_0; J)$ exist and are complete.*

**Proof.** The proof is a simple consequence of Theorem 29 and the fact that $P^c_H = E_H(S)\, P^c_H$, $P^c_{H_0} = E_{H_0}(S)\, P^c_{H_0}$. ∎

Sometimes it is useful to replace the estimate of a sandwiched resolvent $AR_0(z)\, A^*$ by the corresponding estimate for the sandwiched evolution $A\, e^{-itH_0}A^*$.

**Proposition 31.** *Let $C \in \mathscr{C}(\mathscr{H}_0, \mathscr{H})$. Suppose that* dom $C \supseteq$ dom $H$ *and*

$$\int_{-\infty}^{\infty} \|C\, e^{-itH_0}C^*\|\, dt \leq a < \infty.$$

*Then* $\sup_{\lambda\in\mathbb{R}, 0<|\varepsilon|<\varepsilon_0} \|CR_0(\lambda + i\varepsilon)\, C^*\| < a$, $\varepsilon_0 > 0$, *and $C$ is $H_0$-smooth.*

**Proof.** The proof is a simple consequence of the equation (use (3.8))

$$\left(CR_0(\lambda + i\varepsilon)\, C^*u, u'\right) = i\int_0^{\pm\infty} e^{-\varepsilon|t|}\, e^{-i\lambda t}(C\, e^{-itH_0}\, C^*u, u')\, dt$$

and Prop. 19. ∎

## 17.3. Applications

### 17.3.1. Smooth perturbations of $-i(d/dx)$

Let $H_0 = -i(d/dx)$ on $\mathscr{H}_0 = L_2(\mathbb{R}, dx)$. Further, let $V$ be a symmetric integral operator with kernel $v(x, x')$ satisfying

$$\int_{-\infty}^{\infty} dx \int_{-\infty}^{\infty} dx'\, |f(x)^{-1}|^2\, |f(x')^{-1}|^2\, |v(x, x')|^2 \leq c < \infty \tag{17}$$

where $f(x) = (1 + |x|)^{-(1+\varepsilon)}$ with some $\varepsilon > 0$. From this assumption we conclude that $v(x, x')$ is square integrable, and thus $V$ is a Hilbert-Schmidt operator. Therefore $H_\alpha = H_0 + \alpha V$ is selfadjoint on the same domain as $H_0$ for all $\alpha \in \mathbb{R}$. We want to prove that the complete wave operators $W_\pm(H_\alpha, H_0)$ exist and are unitary for all sufficiently small $\alpha$ (here we have $P_0^{ac} = 1$).

First we factor the perturbation $V = B^*A$ where $B^* = B$ is the multiplication operator by $f(x)$ on $L^2(\mathbb{R}, dx)$ and we verify that $\sup_{\lambda\in\mathbb{R}, 0<|\varepsilon|<\varepsilon_0} \|AR_0(\lambda + i\varepsilon)\, B\| < c_1 < \infty$. This will be done with the help of Prop. 31. We have

$$\begin{aligned}
\int dt\, \|A\, e^{-itH_0}Bu\| &= \int dt \Big(\int dx\, \Big|\int dx'\, f(x)^{-1}\, v(x, x')\, f(x' - t)\, u(x' - t)\Big|^2\Big)^{1/2} \\
&\leq \int dt \Big(\int dx \Big|\int dx'\, \{|f(x)^{-1}| \cdot |f(x')^{-1}| \cdot |v(x,x')|\}\, \{|f(x')\, f(x' - t)\, u(x' - t)|\}\Big)^{1/2} \\
&\leq \int dt \Big(\int dx \int dx'\, |f(x)^{-1}|^2\, |f(x')^{-1}|^2\, |v(x, x')|^2 \int dx'\, |f(x')\, f(x' - t)\, u(x' - t)|^2\Big)^{1/2} \\
&\leq \sqrt{c} \int dt \Big(\int dx'\, |f(x')\, f(x' - t)\, u(x' - t)|^2\Big)^{1/2}
\end{aligned}$$

where we have used the Schwarz inequality and (17). Furthermore,

$$f(x')\, f(x' - t) \leq \big|1 + |x'| + |x' - t|\big|^{-(1+\varepsilon)} \leq (1 + |t|)^{-(1+\varepsilon)}.$$

Inserting this in the right side of the inequality above, we obtain

$$\int dt\, \|A\, e^{-itH_0}Bu\| \leq \sqrt{c} \int dt\, |1 + |t||^{-(1+\varepsilon)}\, \|u\| \leq b < \infty$$

proving that, for sufficiently small $\alpha$, $\sup_{\lambda \in \mathbb{R}, 0<|\varepsilon|<\varepsilon_0} \alpha \|AR_0(\lambda + i\varepsilon)\, B\| < 1$. Thus, according to the remarks after the proof of Theorem 29, the assumptions of Theorem 29 for $\Delta = \mathbb{R}$ and $H$, $V$ replaced by $H_\alpha$, $\alpha V$ are satisfied for all $\alpha \in [0, \alpha_0)$ with some $\alpha_0 > 0$. This proves that the strong wave operators $W_\pm(H_\alpha, H_0)$ exist for $\alpha \in (0, \alpha_0)$, are partially isometric and have final subspace $\mathscr{H}$. Thus, the wave operators are also unitary. This in turn implies that $H_\alpha$ is absolutely continuous for all $\alpha \in (0, \alpha_0)$.

### 17.3.2. Potential scattering

Let $H_0 = -\Delta$ on $\mathscr{H}_0 = \mathscr{H} = L^2(\mathbb{R}^3, dx)$. Further let $V$ be the multiplication operator by the real-valued function $q(\cdot) \in L^2(\mathbb{R}^3, dx) \cap L^1(\mathbb{R}^3, dx)$. Then $H_\alpha := H_0 + \alpha V$, $\alpha \in \mathbb{R}$, is selfadjoint with $\operatorname{dom} H_0 = \operatorname{dom} H_\alpha$. We want to prove that the strong wave operators $(W_\pm H_\alpha, H_0)$ exist and are unitary for all sufficiently small $\alpha$.

Let $V = B^*A$ where $A$ is multiplication by the function $|q(x)|^{1/2}$ which belongs to $L^2(\mathbb{R}^3, dx)$. Then $A = A^*$ and $B = B^*$. For the proof that the unitary strong wave operators $W_\pm(H_\alpha, H_0)$ exist, we use Theorem 29 and the remarks after it. In our case $\Delta = \mathbb{R}$; $H = H_\alpha$ and $V$ is to replace by $\alpha V$. First we note that (IV) implies that $A$ is $H_0$-bounded; namely (IV) implies that $A$ is $H_0$-smooth in case $\Delta = \mathbb{R}$ and $H_0$-smoothness provides $H_0$-boundedness (see Theorem 5). Naturally $B = (\operatorname{sgn} V)\, A = A(\operatorname{sgn} V)$ is also $H_0$-bounded. It remains to show that $\sup_{\lambda \in \mathbb{R}, 0<|\varepsilon|<\varepsilon_0} \|AR_0(\lambda + i\varepsilon)\, A^*\| < \infty$ for some $\varepsilon_0 > 0$. By Prop. 31 we can replace this estimate by

$$\int_{-\infty}^{\infty} \|A\, e^{-itH_0}A^*\|\, dt \leq a < \infty \tag{18}$$

and $\operatorname{dom} A \supseteq \operatorname{dom} H_0$. $\operatorname{dom} A \supseteq \operatorname{dom} H_0$ is fulfilled since $A$ is multiplication by a square integrable function (see KATO [1, p. 302]). It remains to verify (18). Now formula (15.3) says that $\|A\, e^{-itH_0}u\|^2 \leq |4\pi t|^{-3}\, \|\, |q(\cdot)|^{1/2}\|_{L^2}^2\, \|u\|_{L^1}^2$ where $u \in L^1(\mathbb{R}^3, dx) \cap L^2(\mathbb{R}^3, dx)$. Set $u = A^*v$ with $v \in L^2(\mathbb{R}^3, dx)$. Then $u \in L^1(\mathbb{R}^3, dx)$ and

$$\begin{aligned} \|u\|_{L^1}^2 = \|Av\|_{L^1}^2 &= \Big(\int_{\mathbb{R}^3} dx\, |q(x)|^{1/2}\, |v(x)|\Big)^2 \\ &\leq \int_{\mathbb{R}^3} dx\, |v(x)|^2 \int_{\mathbb{R}^3} dx\, |q(x)| = \|v\|^2\, \|q\|_{L^1}. \end{aligned}$$

Inserting this into the estimate above, we obtain

$$\|A\, e^{-itH_0}A^*v\|^2 \leq |4\pi t|^{-3}\, \|\, |q(\cdot)|^{1/2}\|_{L^2}^4\, \|v\|^2.$$

This shows that (18) is satisfied and concludes the proof that the strong wave operators $W_\pm(H_\alpha, H_0)$ xist for all $|\alpha| < \alpha_0$, with some $\alpha_0 > 0$, and are unitary.

## Notes and Remarks to Part IV

The general historical development of the methods for proving existence and completeness of wave operators was already described in the Introduction. Here we present further details.

*Chapter 14.* Stationary methods date back to FRIEDRICHS [1938a], [1948a] who constructed intertwining operators for a pair $H_0$, $H = H_0 + V$ by studying solutions $\Omega_+$ of the equation

$$\Omega_+ = 1 + \Gamma_H(V\Omega_+) := 1 + i \int_0^\infty e^{itH_0}V\Omega_+\, e^{-itH_0}\, dt$$

(for $\Gamma_H$ see Section 7.4.1) and proving the existence of the limits of $VR_0(\lambda \pm i\varepsilon)$ as $\varepsilon \to +0$ in a certain sense. Here $H_0$ is multiplication by $x$ on $L^2(\Delta, dx; \mathscr{K}_0)$ and $V$ is an integral operator satisfying a smallness condition in a so-called *gentleness-norm* (an example of a gentleness-norm is the norm $\|\cdot\|_{\alpha,\beta}$, see Section 7.4.4). Since $\Omega_+$ coincides with the wave operator $W_+(H, H_0)$, FRIEDRICHS indirectly gave the first nontrivial example for the existence (and also completeness) of a wave operator (indirectly in the sense that the notion of the wave operator was not rigorously defined at the time and did not appear in his paper). Applications and generalizations of the theory of gentle perturbations were given by LADYŽENSKAJA/FADDEEV [1958a], SCHWARTZ [1960a], REJTO [1963a], [1964a], FADDEEV [1964a], FRIEDRICHS [1], USHIJIMA [1966a] and MOCHIZUKI [1967a]. An abstract version of FRIEDRICHS' method leading to intertwining operators for a pair $H_0$, $H$, but in general not to wave operators, can be found in DUNFORD/SCHWARTZ [3]. LIN [1969a], [1969b] showed that some types of gentle perturbations can be included in the class of smooth perturbations studied in Chapter 17.

Another type of stationary theory was proposed by REJTO [1967a], [1967b], [1969a] and independently, by HOWLAND [1967a]. The essential idea is the introduction of a Banach space $\mathfrak{X}$ with $\mathfrak{X} \cap \mathscr{H}_0$ dense in $\mathscr{H}_0$ and $\mathfrak{X}$ such that the resolvents $R(\lambda \pm i\varepsilon)$, $R_0(\lambda \pm i\varepsilon)$ have weak limits and $VR_0(\lambda \pm i\varepsilon)$, $VR(\lambda \pm i\varepsilon)$ have strong limits in the space $\mathfrak{X}$. See also HÖEGH-KROHN [1968a], [1968b].

A third concept of stationary methods has its origin in the papers of KATO [1957a] and KURODA [1963a], [1967a] who used certain spectral representations of $H_0$ and $H$ and culminated in two important works of KATO/KURODA [1970a], [1971a]. This stationary theory includes the former ones and is applicable to gentle, smooth and trace class perturbations (see KATO/KURODA [1971a]). It is based on the notion of a spectral form and uses auxiliary manifolds $\mathfrak{X}_0$, $\mathfrak{X}$ on which the operators $VR_0(\lambda + i\varepsilon)$, $VR(\lambda + i\varepsilon)$ have limits as $\varepsilon \to \pm 0$. Modifications (with special attention to multichannel systems and two-space theory), improvements and applications can be found in HOWLAND [1976a], KATO [1977a], [1977b] and YAJIMA [1978a]. In the text we essentially present a variant of KATO/KURODA [1971a] and KATO [1977b] using the abstract stationary theory developed in Chapter 13.

The factorization method as described in Paragraph 2 has its roots in the paper of SCHWARTZ [1960a]. There the perturbation $V$ was factored in the form $V = B^*A$ and it was assumed that $AR_0(\lambda + i\varepsilon) B^*$ has limits as $\varepsilon \to \pm 0$. Is was further developed and applied to some scattering problems by KURODA [1967a], KATO/KURODA [1970a], [1971a] and in the two-space framework by DEIČ [1973a] and SCHECHTER [1974b]. See also WEDER [1978a], SCHECHTER [1979c], [1979d], and DEIČ/JAFAEV [1980a]. In the papers of SCHECHTER [1974b], [1979c] one can find a little more general statements than ours in the text. The factorization of the perturbation $V$ is widely used as a technical tool in scattering theory also outside stationary methods. For example, see its use in the other chapters of this part. Lemma 19 is already to be found in KURODA [1967a]. The proof given in the text follows the one in KATO/KURODA [1971a].

Another variant of the stationary methods, more connected with the approach to scattering problems which one finds in more or less old quantum mechanics books (see e.g. LANDAU/LIFSCHITZ [1]) constructs the wave and scattering operators via generalized eigenfunctions for the continuous spectra of the perturbed and unperturbed operators (for eigenfunctions of the continuous spectra of differential operators see e.g. TITCHMARSH [1, 2]). This method gives detailed results for differential operators and was first used by POVZNER [1955a] and IKEBE [1960a], [1970a]. It was further developed and applied to some differential operators by SHENK [1966a], THOE [1967a], IKEBE [1967a], [1968a], GOLDSTEIN [1969a], [1969b], SHENK/THOE [1971a], ALSHOLM/SCHMIDT [1971a], SCHULENBERGER/WILCOX [1972a], THOMPSON [1972a], SAITO [1972a], AGMON [1975a], SCHECHTER [1976c], and others. This approach leads to additional results about the absence of the singular continuous spectrum and about the position and multiplicity of the eigenvalues. For details see in particular AGMON [1975a] and SCHECHTER [1976c]. It should be remarked that this method gave for a long time the best results for Schrödinger operators. Later the more abstract methods of the stationary theory and the new approach of ENSS (see the Notes and Remarks to Chapter 15) led to the same results. The eigenfunction method is connected with the so-called Limiting Absorption Method which studies the limiting behavior of solutions $u_{\lambda \pm i\varepsilon}$ of $(H - (\lambda \pm i\varepsilon))\, u_{\lambda \pm i\varepsilon} = f$ as $\varepsilon \to +0$ (the results are often called a Principle of Limiting Absorption). For details and further literature we refer the reader to VAINBERG [1966a] and GUSTAFSON [1978a].

In the framework of the stationary theory working with auxiliary manifolds $\mathscr{X}_0$ and $\mathscr{X}$ one can introduce perturbed generalized eigenfunctions and eigenfunction expansions in an abstract setting if generalized eigenfunctions are given for the unperturbed operator $H_0$. The generalized eigenfunctions can then be interpreted as elements of the dual spaces of $\mathscr{X}_0$ and $\mathscr{X}$. For details we refer to KURODA [1967b], HOWLAND [1968a], KATO/KURODA [1970a], [1971a] and KURODA [1973a], [1973b]. See also YAJIMA [1975a].

In Section 1.4 it was shown that the wave operators $W_\pm(H, H_0; J)$ even exist in the strong sense when the auxiliary manifold $\mathscr{X}$ is a Hilbert space. There are other criteria which ensure that the wave operators (whose existence is proved with the help of statements from Paragraph 1) exist in the strong sense. For instance, assume that there is a subset $\mathscr{D}_0$ of $\mathscr{X}_0$ such that $P_0^{ac}\mathscr{D}_0$ is generating with respect to $P_0^{ac}\mathscr{H}_0$ and $H_0$ and $f_+^{H_0}(\lambda, u; v)$, $u \in \mathscr{D}_0$ and $v \in \mathscr{X}_0$ is locally Hölder continuous on $\mathbb{R} \setminus \Gamma$ where $\Gamma$ is a closed null set (see Section 1.1 for $f_+^{H_0}$ and $\Gamma$). Then $W_+(H, H_0; J)$ exists even in the strong sense. This is a trivial two-space generalization of a result of HOWLAND [1967a]. For other criteria, see HOWLAND [1967a] and OBERMANN/WOLLENBERG [1978a].

Stationary methods were also used for trace class perturbations (see e.g. KATO [1957a], KURODA [1963a], BIRMAN/ÉNTINA [1964a], [1967a] and smooth perturbations (see KATO [1966a]). KURODA [1967a] showed that trace class perturbations and smooth perturbations (at least in the unsymmetrical case), in case $\mathscr{H} = \mathscr{H}_0$ and $J = 1$, can be treated with stationary methods. In Section 3.1 it was shown how one can treat gentle perturbations in the framework of the stationary theory.

*Nonselfadjoint perturbations.* The stationary methods can also applied to nonselfadjoint perturbations $V$. For gentle perturbations, see FRIEDRICHS [1948a], SCHWARTZ [1960a], MOCHIZUKI [1967b], and DUNFORD/SCHWARTZ [3]. For the stationary methods working with auxiliary spaces, see for example GOLDSTEIN [1970a], [1971a]. Eigenfunction methods were used by MOCHIZUKI [1967a], [1968a], SUZUKI [1973a], and SAITO [1974a].

*Chapter 15.* Corollary 3 appeared first in COOK's paper [1957a] in the concrete situation: $\mathscr{H} = \mathscr{H}_0 = L^2(\mathbb{R}^3, dx)$, $H_0 = -\Delta$, $H = H_0 + V(x)$. This statement was extended to a variant of Prop. 5 by SCHECHTER [1977a], [1979c], [1979d] and SIMON [1977a]. After that KATO [1979a] gave a simple and beautiful proof of Theorem 4 (and Prop. 5) that includes all former versions as special cases.

The applications of the abstract criteria presented in Sections 1.2, 1.3, and 1.4 are simplifications or modifications of results to be found in the literature: (a) The example in Section 1.2 in the special case $V = q(x)$ is already contained in KUPSCH/SANDHAS [1966a] who first showed that local singularities in potential scattering are unessential when one only considers the existence of wave operators. The generalization to more general $V$ as in the text is trivial. (b) The result of Section 1.3 can be found in numerous papers about hard core potentials (or perturbations of the boundary conditions of the Laplacian). See e.g. HUNZIKER [1967a]. (c) For some values of the parameters $\alpha$ and $\gamma$ the result of Section 1.4 is contained in MATVEEV/SKRIGANOV [1972a] whose idea of proof, namely the construction of a suitable identification operator, we have taken over. It is likely that our example for all values of $\alpha$ and $\gamma$ can be fitted in the more general assumptions on the potential in the papers about oscillating potentials but we have not checked it.

The general statements of Section 1.1 have been successfully applied to many scattering problems, in particular to problems in which $H_0$ is a constant coefficient differential operator and $V$ is multiplication by a function. In the last mentioned situation one can use stationary phase methods for estimating $(e^{-itH_0}u)(x)$ by using the Fourier transform. Besides the papers of COOK [1957a] and KURODA [1959a] we note the papers of HACK [1958a] (who extended COOK's result to potentials with $|x|^{-(1+\varepsilon)}$ falloff), BROWNELL [1962a] (who gave an extension to $n$ space dimension), HACK [1959a] (an extension to $n$-body systems), JAUCH/ZINNES [1959a], and LUNDQUIST [1967a]. More general perturbations of constant coefficient differential operators were treated, with the help of falloff methods, by IKEBE/TAYOSHI [1968a], JÖRGENS/WEIDMANN [1973a], VESELIC/WEIDMANN [1973a], [1974a], SCHECHTER [1977a], AVRON/HERBST [1977a], BERTHIER/COLLET [1977a], and others.

Prop. 8 is a simple consequence of the intertwining property of the wave operators and the spectral assumptions. The basic idea already appeared in KURODA [1959a] and is to be found (in slightly weaker form) in different papers and books on scattering theory. For example, see AMREIN/JAUCH/SINHA [1].

The approach to the completeness of wave operators via time-falloff conditions described in Section 2.2 was introduced in the concrete situation of potential scattering and in a less general formulation, by ENSS [1978a]. This approach allows to treat the completeness of wave operators in a direct time-dependent way without using stationary methods (included trace class and smooth methods). The new concept was further developed and applied to different scattering problems by ENSS [1979a], [1979b], MOURRE [1979a], JAFAEV [1979b], DAVIES [1980b], and PERRY [1980a]. In particular, see the important work of SIMON [1979a] which contains extensions and simplifications of ENSS' paper and numerous applications (e.g. Dirac operators and Schrödinger operators with absorption). In our presentation of this approach we essentially follow the abstract version developed in AMREIN/PEARSON/WOLLENBERG [1980a] which also contains applications to strongly singular potential scattering.

The proof in Section 2.4 is taken from JAFAEV [1979b].

The invariance principle of Paragraph 3 is due to CHANDLER/GIBSON [1976a], [1979a]. A slightly weaker result was proven earlier by DONALDSON/GIBSON/HERSH [1973a]. Further references to invariance principles appear in the Notes and Remarks to Chapters 10 and 11.

Strong continuity of the wave and scattering operators in the perturbation $H_\lambda - H_0 = V_\lambda$ under time-falloff conditions was proved in NARNHOFER [1975a].

*Nonselfadjoint perturbations and time-dependent perturbations.* It is not hard to see that if we replace $e^{itH}$ in $e^{itH}J\,e^{-itH_0}$ by a semigroup uniformly bounded in $t$ or by the propagator $U(t, s)$ for a time-dependent selfadjoint operator $H(t)$, then we can also use the time-falloff criteria for the existence of the strong limits of the corresponding expressions. The first paper using this idea is STANKEVIČ [1965a]. Other works are SCHMIDT [1975a], YAJIMA [1977a], KURODA/MORITA [1977a], SIMON [1979a], and DAVIES [1980a].

*Chapter 16.* The idea of using resolvents in proving the unitary equivalence of $H_0P_0^{ac}$, $HP^{ac}$ is due to PUTNAM [1957a]. The Rosenblum-Kato theorem on trace class perturbations (see the Introduction) was extended to cases where $R_0(z) - R(z)$ is a trace operator by BIRMAN [1962b] ($H_0$ and $H$ are bounded from below), BIRMAN/KREĬN [1962a], DE BRANGES [1962a], and BIRMAN [1963a]. Extensions of the trace class theorems to the two-space situation were given by BELOPOL'SKĬ/BIRMAN [1968a], BIRMAN [1970a], JAFAEV [1974a], LYFORD [1975a], and BELOPOL'SKIĬ [1975a]. The general two-space version of the ROSENBLUM-KATO theorem (Theorem 1) was proved by PEARSON [1978a] and NEIDHARDT [1978a]. See also DEIFT [1976a].

The proof of Theorem 1 is taken from PEARSON [1978a] who used suggestions by J. GINIBRE and T. KATO. The proof of Theorem 7 uses some ideas from the paper of NEIDHARDT [1978a]. Prop. 4 and 5 ($n = 1$) in case $\mathscr{H} = \mathscr{H}_0$ and $J = 1$ go back to KURODA [1959b]. See also STANKEVIČ [1962a].

Sometimes the following local trace class theorem using the notion of "subordination" is useful: Suppose that $H$ and $H_0$ are selfadjoint operators acting on the same Hilbert space $\mathscr{H}$. Assume that

(I) $E(\Delta)(H - H_0)E_0(\Delta) \in \mathscr{L}_1(\mathscr{H})$ for each bounded interval $\Delta$,

(II) $H$ and $H_0$ are mutually subordinate.

Then the strong wave operators $W_\pm(H, H_0)$ exist and are complete. Here "mutually subordinate" is defined as follows: $H$ is *subordinate* to $H_0$ if there are continuous functions $f$ and $g$ on $\mathbb{R}$ with $f(x) \geqq 1$, $g(x) \geqq 1$ and $\lim_{|x|\to\infty} f(x) = \infty$ such that $\operatorname{dom} g(H_0) \subseteqq \operatorname{dom} f(H)$, hence $f(H)\,g(H_0)^{-1}$ is bounded. If $H$ is subordinate to $H_0$ and $H_0$ is subordinate to $H$, then we say they are *mutually subordinate*. For instance $H$ and $H_0$ are mutually subordinate if $\operatorname{dom} H = \operatorname{dom} H_0$. The proof of this local trace class theorem and the notion of subordination go back to BIRMAN [1968a]. See also the paper of PEARSON [1972a] where the assumption of mutual subordination is replaced by an asymptotic condition.

One can show that the factorization method of Paragraph 14.2 includes the trace class results, at least in the case where $J$ is bijective and is asymptotically equivalent to an isometric operator (with respect to $H_0P_0^{ac}$), using the same ideas of proof as KURODA [1967a] in case $\mathscr{H} = \mathscr{H}_0$ and $J = 1$. The first proofs of trace class results were of stationary nature. A proof of Theorem 1 (in case $\mathscr{H} = \mathscr{H}_0$ and $J = 1$) which uses time-falloff methods were first given in KATO [1, first edition 1966].

Formulas for wave and scattering operators under trace class conditions can be found in KATO [1957a], KURODA [1963a], BIRMAN/ÈNTINA [1964a], [1967a], and ÈNTINA [1971a], [1972a]. See Sections 9.7.1 and 19.1.1.

In connection with trace class perturbations one should note the important result of KURODA [1958a], which extends a known result of v. NEUMANN [1935a], that Theorem 1 cannot be extended to trace classes $\mathscr{L}_p(\mathscr{H}_0, \mathscr{H})$ with $p > 1$, because for every selfadjoint $H_0$ and $p > 1$, one can find a selfadjoint operator $H$ on the same space with pure point spectrum such that $H - H_0 \in \mathscr{L}_p(\mathscr{H}_0)$. Note also the paper of CAREY/PINCUS [1976a] on unitary equivalence modulo trace operators.

The first application of trace class results to a physically interesting case is due to KURODA [1959]. He applied a version of Prop. 5 to the question of the completeness of wave operators in potential scattering (the potential $q$ satisfies $q \in L^1(\mathbb{R}^3, dx) \cap L^2(\mathbb{R}^3, dx)$. Other applications are given by PUTNAM [1957a], [1960a], ROSENBLUM [1960a], BIRMAN [1962a], [1969b], [1971a], IKEBE/TAYOSHI [1968a], COMBES [1969a], LAVINE [1970a], DEIČ [1971a], SCHULENBERGER [1972a], DEIFT [1976a], REED/SIMON [1977a], DAVIES [1979a], and SUSLOV [1979a]. From the last paper we have taken the proof in Section 3.1. It contains another proof of a theorem of ISMAGILOV [1963a].

The result on Dirac operators in Section 3.2 is not very strong. Better results with other methods were obtained by ECKHARDT [1974b]. Lemma 8 needed in the proof is a special version of a more general theorem proved in SEILER/SIMON [1975a]. See also REED/SIMON [3, Appendix 2 to XI.3]. The results and the method of proof in Section 3.3 are taken from DEIFT [1976a] whose paper contains considerably stronger results on scattering for classical systems. Lemma 10 is an easy consequence of a more general result of BIRMAN/SOLOMJAK [1969a]. See also REED/SIMON [3, Appendix to XI.3].

Further criteria for operators to be trace class can be found in GOHBERG/KREĬN [1] and SIMON [3]. We remark that there is an approach which uses the Wiener path integral and the Feynman-Kac formula to prove that some functions of $H$ and $H_0$ are trace operators and applies this to scattering problems. The first paper in this direction is ARSEN'EV [1968a]. Later and independently of this paper DEIFT/SIMON [1976a] and BAUMGÄRTEL/DEMUTH [1979a] used this approach for the proof of the existence of complete wave operators in potential scattering in the case of strongly local singularities of the potentials. Further applications can be found in DEMUTH [1980a]. See also SIMON [4].

That the wave operator $W_+(H_\lambda, H_0)$ is strongly continuous in $\lambda$ if $H_\lambda$ is continuous in trace norm can be found in KATO [1, p. 553]. A corresponding result for the two-space wave operator can be easily obtained from Remark 16.2.

*Nonselfadjoint and time-dependent perturbations.* Theorem 1 in the special case $J = 1$ was extended to the case where $H_0$ is selfadjoint and $-i(H - H_0) \geqq 0$, i.e. $H$ is dissipative, by STANKEVIČ [1965a] (under additional assumptions on $H$) and by SAHNOVIČ [1968a]. A special version of Theorem 1, namely the one in which $J$ is bijective, was first proved by SAHNOVIČ [1968b]. Recall that the two-space theory includes the scattering theory for nonselfadjoint operators which are similar to selfadjoint ones. NABOKO [1977a], [1980a] proved the existence of strong wave operators for $H$ and $H_0$ where $H_0$ is selfadjoint and $H - H_0 \in \mathscr{L}_1(\mathscr{H}_0)$ under small additional assumptions. Here $H$ is not necessarily dissipative. DAVIES [1978a] proved the existence of the wave operator $\text{s-lim}_{t\to\infty} e^{itH_0} J e^{-itH} P^{ac}$ under the assumption that $JH - H_0J \in \mathscr{L}_1(\mathscr{H}, \mathscr{H}_0)$ and $e^{-itH}$ is a contraction semigroup. See also DAVIES [1980a]. Note that the definition of the absolutely continuous subspace of $H$ in the papers of NABOKO [1977a] and SAHNOVIČ [1968a], [1968b] is different from the one in the works of DAVIES [1978a], [1980a] and SIMON [1979a], and it is not clear if they agree. DAVIES defines $P_H^{ac}$ as the projection onto the subspace clo spa $\{u \in \mathscr{H}; \int_0^\infty |(e^{-itH}u, v)|^2 \, dt \leqq c_u \|v\|^2 \text{ for all } v \in \mathscr{H}\}$. An extension of the results of SAHNOVIČ, NABOKO and DAVIES can be found in NEIDHARDT [1981a].

Time dependent perturbations with trace class assumptions were treated by SCHMIDT [1975a] and HOWLAND [1979a].

*Chapter 19.* The presentation of the smoothness theory except Theorem 18 is essentially taken from the basic papers of KATO [1966a], [1968a] and LAVINE [1971a], [1972a].

The assertion that $H$ is absolutely continuous in Prop. 20 ($C$ bounded) is due to PUTNAM [1]. Prop. 20 and its proof go back to KATO [1968a]. Especially LAVINE [1969a], [1971a], [1972a], [1973a] extended this statement and found useful applications to Schrödinger operators. See also ARAI [1972a].

The smoothness theory was applied to problems in potential scattering by LAVINE (see the papers cited above and [1973b)], KATO [1966a], GINIBRE/MOULIN [1974a], ROBINSON [1974a], FERRERO/PAZZIS/ROBINSON [1974a] and NAJMAN [1979a]. Smooth perturbations of $-\mathrm{i}d/dx$ were considered by LIN [1969a]. This paper contains stronger results than ours in Section 3.1. Certain smooth perturbations of the multiplication operator by $x$ on $L^2(\Delta, dx; \mathcal{K})$ (gentle perturbations) were also studied by LIN [1969b].

In a certain sense the papers of SCHWARTZ [1960a] and PROSSER [1964a] already contain some ideas of the smoothness theory. PROSSER [1964a] used the estimate

$$\sup_{\|u\|=1} \int_{-\infty}^{\infty} dt \, \| \, |V|^{1/2} \, e^{-itH_0} \, |V|^{1/2} u \| < \infty$$

for certain potentials $V$ (see Prop. 31).

Continuity of the wave operators $W_\pm(H_\lambda, H_0)$ in norm, under certain assumptions on the continuity of $H_\lambda - H_0$ in the smoothness norm, was already proved by KATO [1968a]. Such assertions can easily be extended to two-space wave operators.

*Nonselfadjoint and time-dependent perturbations.* From the beginning the smoothness theory was developed for not necessarily selfadjoint operators $H$ and $H_0$ (see KATO [1966a]) in Hilbert spaces. LIN [1969a] extended it to semigroups in Banach spaces. See also EVANS [1976a]. DAVIES [1974a] and EVANS [1976a] applied this theory to scattering theory for time-dependent perturbations. See also MOCHIZUKI [1967a] and KAKO/YAJIMA [1976a].

*Spectral properties of wave operators.* In case $\mathcal{H} = \mathcal{H}_0$ and $J = 1$ spectral properties of $W_\pm(H, H_0)$ were also treated in the literature. In particular, we refer to PUTNAM [1962a], [1963a], [1964a], [1]. Spectral properties of the scattering operator can be found in Part V. Some methods of this part can also be used to obtain spectral properties of wave operators.

*Scattering theory for differential operators.* There is an extensive literature on scattering problems for differential operators, in particular those related to potential scattering in quantum mechanics. Partly, we have cited some works in connection with the methods presented in the text. In what follows we give some further references to this subject.

(a) General differential operators: BIRMAN [1962a], [1969a] [1971a], KURODA [1973b], SCHECHTER [1974a], [1975a], [1976a], HÖRMANDER [1976a], WEIDMANN [1977a], BERTHIER/COLLET [1977a], AGMON [1979a], and MURATA [1979a].

(b) Potential scattering: AGUILAR/COMBES [1971a], PEARSON/WHOULD [1973a], PEARSON [1975b], DEIFT/SIMON [1976a], BALSLEV [1976a], ENSS [1978a], SIMON [1979a], and JENSEN/KATO [1979a]. See also SIMON [1] and REED/SIMON [3].

(c) Strongly singular potentials: BERTOCCHI/FUBINI/FURLAN [1965a], KUPSCH/SANDHAS [1966a], HUNZIKER [1967a], ARSEN'EV [1968a], [1], AMREIN/GEORGESCU [1974a], PEARSON [1974]a, [1975b], SEMENOV [1977a], COMBESCURE/GINIBRE [1978a], AMREIN/PEARSON/WOLLENBERG [1980a].

(d) Oscillating potentials: MATVEEV/SKRIGANOV [1972a], SCHECHTER [1977b], COMBESCURE/GINIBRE [1976a], BAETEMAN/CHADAM [1976a], SCHECHTER [1977b], DOLLARD/FRIEDMAN [1978a], BOURGEOIS [1979a], PEARSON [1979a], AMREIN/PEARSON [1980a].

(e) Anisotropic potentials: DEIČ/KOROTJAEF/JAFAEV [1977a], RUIJSENAARS/BONGAARTS [1977a], DAVIES [1977b], DAVIES/SIMON [1978a].

(f) Potential scattering with electric and/or magnetic fields: HERBST [1977a], AVRON/HERBST [1977a], VESELIC/WEIDMANN [1977a], AVRON/HERBST/SIMON [1978a], [1978b], HERBST [1979a], YAJIMA [1979a], HUNZIKER [1980a] (and papers cited therein).

(g) Dirac operator: PROSSER [1963a], MOCHIZUKI [1964a], YAMADA [1973a], [1976a], GUILLOT/SCHMIDT [1974a], ECKHARDT [1974a], [1974b], NAJMAN [1976a], PEARSON [1977a].

(h) Klein-Gordon equation: STRAUSS [1963a], CHADAM [1969a], [1970a], VESELIC [1970a], LUNDBERG [1973a], SCHECHTER [1976b], [1980a], WEDER [1978a], DEIČ/JAFAEV [1980a]. See also DEIFT [1976a] and the references in the Notes and Remarks to Chapter 12.

(i) Scattering in crystals: KURODA [1969a], THOMAS [1973a], TROIANIELLO [1974a], BENTOSELA [1976a].

(j) Boltzmann equation: HEJTMANEK [1974a] first considered wave operators for the BOLTZMANN equation. Further papers are SIMON [1974a], PROTOPOPESCU [1976a], VOIGT [1977a], and HEJTMANEK [1979a].

# Part V. Some Properties of the Scattering Operator, the Scattering Matrix, and the Scattering Amplitude

In this part we derive representations of the scattering operator for a scattering system $\{H, H_0; J_0\}$ within the framework of the stationary theory developed in Chapter 13. Further, under additional assumptions on the scattering system (e.g. smoothness conditions) explicit formulas for the scattering matrix $\hat{S}(\lambda)$ are established when a representation of $P_0^{ac}\mathscr{H}_0$ by a direct integral with respect to $H_0$ is given. Starting with such formulas, the behavior of the so-called *scattering amplitude* $\hat{T}(\lambda) := \hat{S}(\lambda) - 1_{\mathscr{H}_\lambda}$ is investigated, for example conditions for the analytic continuation are discussed. Also the resonance problem is pointed out, in particular for the Friedrichs model. Further, we present properties of $\hat{T}(\lambda)$ within the framework of trace class perturbations (see Chapter 16), for example in this case we have $\hat{T}(\lambda) \in \mathscr{L}_1(\mathscr{H}_\lambda)$. Moreover, the property $\hat{T}(\lambda) \in \mathscr{L}_p(\mathscr{H}_\lambda)$, $p > 1$, is proved in other cases, for instance within the framework of time-falloff methods (see Chapter 15). In particular, the property $\hat{T}(\lambda) \in \mathscr{L}_2(\mathscr{H}_\lambda)$ is of interest in physical applications because in this case the physical important quantity $\sigma(\lambda)$, the *total scattering cross section*, is finite.

As in Parts III and IV, $H_0$ and $H$ always denote selfadjoint operators acting on separable Hilbert spaces $\mathscr{H}_0$ and $\mathscr{H}$, and $J$ denotes a bounded operator from $\mathscr{H}_0$ to $\mathscr{H}$. Further, we use the notations $E_0(\cdot) := E_{H_0}(\cdot)$, $E(\cdot) := E_H(\cdot)$, $P_0^{ac} := P_{H_0}^{ac}$, $P^{ac} := P_H^{ac}$, $R_0(z) := R_{H_0}(z)$ and $R(z) := R_H(z)$. In the present part we use direct integral representations of $P_0^{ac}\mathscr{H}_0$ with respect to $H_0$ of the form $L^2(\Delta_0, d\lambda; \mathscr{H}_\lambda; \mathscr{S}) = \bigtimes_{l=1}^{N} L^2(\Delta_l, d\lambda; \mathscr{K}_l)$ or $L^2(\Delta_0, d\lambda; \mathscr{H}_\lambda; \mathscr{S}) = L^2(\Delta_0, d\lambda; \hat{\mathscr{K}}_\lambda)$, with $\mathscr{S}$ separable. Sometimes we also use the notation $\mathscr{H}_\lambda$ (as value space) for $\hat{\mathscr{K}}_\lambda$, e.g. in relations as $\hat{T}(\lambda) \in \mathscr{L}_p(\mathscr{H}_\lambda)$. Throughout this part we assume that the wave operators $W_+ = W_+(H, H_0; J)$, $W_- = W_-(H, H_0; J)$ exist, are partially isometric and complete (some or all of these assumptions often follow from the other assumptions of the corresponding statements in this part), unless otherwise stated. The scattering operator, the scattering matrix and the scattering amplitude for the scattering system $\{H, H_0; J\}$ are denoted by $S$, $\hat{S}(\lambda)$ and $\hat{T}(\lambda)$.

# 18. Representations of the Scattering Operator and the Scattering Amplitude, and Analyticity Properties of the Scattering Amplitude

## 18.1. General formulas and properties for $S$ and $\hat{T}(\lambda)$ within the framework of the stationary theory

### 18.1.1. Scattering amplitude and total scattering cross section

Let $\hat{S}(\lambda)$ be the scattering matrix of a scattering system $\{H, H_0; J\}$. Then the operator function $\hat{T}(\lambda) := \hat{S}(\lambda) - 1_{\mathscr{H}_\lambda}$ is called the *scattering amplitude*. Its definition, like that of $\hat{S}(\lambda)$, depends on the special direct integral representation of $P_0^{ac}\mathscr{H}_0$ with respect to $H_0$. A scattering system is called *trivial* if $\hat{S}(\lambda) = 1_{\mathscr{H}_\lambda}$ for all $\lambda \in \Delta_0 = \operatorname{spec}^c(H_0 \mid P_0^{ac}\mathscr{H}_0)$. The scattering amplitude $\hat{T}(\lambda)$ describes the deviation of the scattering system $\{H, H_0; J\}$ from a trivial one.

A physical important quantity is the *total scattering cross section* $\sigma(\lambda)$ that is proportional to $\|\hat{T}(\lambda)\|_2^2$. In potential scattering $\sigma(\lambda) = (\pi/\lambda)\, \|\hat{T}(\lambda)\|_2^2$ and in other scattering situations $\sigma(\lambda) = \big(\pi/\alpha(\lambda)\big)\, \|\hat{T}(\lambda)\|_2^2$ where $\alpha$ is a nonnegative increasing function on $\Delta_0$ (for details see the literature cited in the Notes and Remarks). Therefore, to know that $\sigma(\lambda)$ is finite, it is important to verify that $\hat{T}(\lambda) \in \mathscr{L}_2(\mathscr{H}_\lambda)$. In particular, this is true if $\hat{T}(\lambda) \in \mathscr{L}_1(\mathscr{H}_\lambda)$. In this case we have

$$\|\hat{T}(\lambda)\|_2^2 = \operatorname{tr}\big(\hat{T}(\lambda)^*\, \hat{T}(\lambda)\big) = -\operatorname{tr}\big(\hat{T}(\lambda)^* + \hat{T}(\lambda)\big) = -2\operatorname{Re}\big(\operatorname{tr}\hat{T}(\lambda)\big) \tag{1}$$

which is a consequence of the unitarity of $\hat{S}(\lambda)$, namely of

$$\begin{aligned} 1_{\mathscr{H}_\lambda} &= \hat{S}(\lambda)^*\, \hat{S}(\lambda) = \big(1_{\mathscr{H}_\lambda} + \hat{T}(\lambda)\big)^* \big(1_{\mathscr{H}_\lambda} + \hat{T}(\lambda)\big) \\ &= 1_{\mathscr{H}_\lambda} + \hat{T}(\lambda)^*\, \hat{T}(\lambda) + \hat{T}(\lambda)^* + \hat{T}(\lambda). \end{aligned}$$

(1) shows that the behavior of the total scattering cross section is also determined by that of $\operatorname{tr}\hat{T}(\lambda)$. Obviously, $\sigma(\lambda)$ is independent of the special direct integral representation of $P_0^{ac}\mathscr{H}_0$.

### 18.1.2. A representation of the scattering operator by a spectral integral

Let a scattering system $\{H, H_0; J\}$ be given. In order to get a representation of the scattering operator $S$ (as a double limit of an operator spectral integral) we prove

**Lemma 1.** *Let $g(\cdot)\colon \mathbb{R} \to \mathscr{H}$ be a function such that $g(\cdot)$, $g'(\cdot)$ are strongly continuous on $\mathbb{R}$ and $\int_{-\infty}^{\infty} E(\mathrm{d}x)\, g(x)$ exists. Then for each finite interval $[a, b]$ the vector spectral integral*

$\int_a^b E_0(\mathrm{d}\lambda)\, J^* \varepsilon R(\lambda + \mathrm{i}\varepsilon)\, g(\lambda)$ *exists for all* $\varepsilon > 0$ *and the relation*

$$\operatorname*{w-lim}_{\varepsilon\to+0} \left\{ P_0^{\mathrm{ac}} \int_a^b E_0(\mathrm{d}\lambda)\, J^* \varepsilon R(\lambda + \mathrm{i}\varepsilon)\, g(\lambda) \right.$$

$$\left. - P_0^{\mathrm{ac}} \int_a^b E_0(\mathrm{d}\lambda)\, J^* \varepsilon R(\lambda + \mathrm{i}\varepsilon) \int_{-\infty}^{\infty} E(\mathrm{d}x)\, g(x) \right\} = 0$$

*is valid.*

**Proof.** Note that for operator (vector-) spectral integrals $\int_a^b := \int_{[a,b)}$ (cf. Section 5.1.1). The existence of the vector spectral integrals which appear above is obvious (see 5.1.2). Further, we have

$$\operatorname*{s-lim}_{\varepsilon\to+0} \int_a^b E_0(\mathrm{d}\lambda)\, J^* \varepsilon R(\lambda + \mathrm{i}\varepsilon) \int_\Delta E(\mathrm{d}x)\, g(x) = 0$$

if $\Delta$ is a finite interval with $\operatorname{dis}(\Delta, [a, b]) > 0$. Recall that $\mathrm{i} \int_a^b E_0(\mathrm{d}\lambda)\, J^* \varepsilon R(\lambda + \mathrm{i}\varepsilon) = E_0([a, b))\, \Omega_\varepsilon^*$ and thus $\left\| \int_a^b E_0(\mathrm{d}\lambda)\, J^* \varepsilon R(\lambda + \mathrm{i}\varepsilon) \right\| \leqq \|J\|$ (see 13.3). Therefore, applying also Lemma 13.6 for $\Lambda = \{b\}$, it is sufficient to prove.

$$\operatorname*{w-lim}_{\varepsilon\to+0} \left\{ P_0^{\mathrm{ac}} \int_a^b E_0(\mathrm{d}\lambda)\, J^* \varepsilon R(\lambda + \mathrm{i}\varepsilon)\, g(\lambda) \right.$$

$$\left. - P_0^{\mathrm{ac}} \int_a^b E_0(\mathrm{d}\lambda)\, J^* \varepsilon R(\lambda + \mathrm{i}\varepsilon) \int_a^b E(\mathrm{d}x)\, g(x) \right\} = 0.$$

Using the definition of vector spectral integrals and the formula for $E_0([a, b))\, \Omega_\varepsilon^*$ cited above, we easily get

$$P_0^{\mathrm{ac}} \int_a^b E_0(\mathrm{d}\lambda)\, J^* \varepsilon R(\lambda + \mathrm{i}\varepsilon)\, g(\lambda) = P_0^{\mathrm{ac}} \int_a^b E_0(\mathrm{d}\lambda) \left\{ \int_a^b E_0(\mathrm{d}x)\, J^* \varepsilon R(x + \mathrm{i}\varepsilon) \right\} g(\lambda)$$

$$= -\mathrm{i} P_0^{\mathrm{ac}} \int_a^b E_0(\mathrm{d}\lambda)\, \Omega_\varepsilon^* g(\lambda). \tag{2}$$

Further we have

$$P_0^{\mathrm{ac}} \int_a^b E_0(\mathrm{d}\lambda)\, J^* \varepsilon R(\lambda + \mathrm{i}\varepsilon) \int_a^b E(\mathrm{d}x)\, g(x)$$

$$= -\mathrm{i} P_0^{\mathrm{ac}}\, E_0([a, b))\, \Omega_\varepsilon^* \int_a^b E(\mathrm{d}x)\, g(x). \tag{3}$$

Subtracting the right side of (3) from the right side of (2) and integrating by parts (see (5.5)), we get

$$-\mathrm{i}P_0^{\mathrm{ac}}E_0([a,b))\left\{\int_a^b E_0(\mathrm{d}\lambda)\,\Omega_\varepsilon^* g(\lambda)-\Omega_\varepsilon^*\int_a^b E(\mathrm{d}x)\,g(x)\right\}$$

$$=-\mathrm{i}P_0^{\mathrm{ac}}E_0([a,b))\left\{[E_0(b)\,\Omega_\varepsilon^*-\Omega_\varepsilon^*E(b)]\,g(b)-[E_0(a)\,\Omega_\varepsilon^*-\Omega_\varepsilon^*E(a)]\,g(a)\right.$$

$$\left.-\int_a^b[E_0(\lambda)\,\Omega_\varepsilon^*-\Omega_\varepsilon^*E(\lambda)]\,g'(\lambda)\,\mathrm{d}\lambda\right\}. \tag{4}$$

Since $\operatorname*{s-lim}_{\varepsilon\to+0}\Omega_\varepsilon P_0^{\mathrm{ac}}=W_+(H,H_0;J)$ exists, we have that $\operatorname*{w-lim}_{\varepsilon\to+0}P_0^{\mathrm{ac}}\Omega_\varepsilon^*$ exists and is intertwining. Applying this and the estimate $\|(E_0(\lambda)\,\Omega_\varepsilon^*-\Omega_\varepsilon^*E(\lambda))\,g'(\lambda)\|\leqq 2\,\|J\|\cdot\|g'(\lambda)\|$ to the right side of (4), we obtain, with the help of Lebesgue's dominated convergence theorem, that the left side vanishes in weak sense as $\varepsilon\to+0$, and this implies the desired assertion. ■

From the proof of Lemma 1 we see that one can replace the weak limit in the assertion by the strong one if either $\operatorname*{s-lim}_{\varepsilon\to+0}P_0^{\mathrm{ac}}\Omega_\varepsilon^*$ exists or $\operatorname*{s-lim}_{\varepsilon\to+0}\Omega_\varepsilon^*P^{\mathrm{ac}}$ exists and $g(\lambda)=P^{\mathrm{ac}}g(\lambda)$. The latter case is given if not only the wave operators $W_\pm(H,H_0;J)$ but also $W_\pm(H_0,H;J^*)$ exist. Thus we obtain

**Corollary 2.** *In addition to the assumptions of Lemma* 1 *assume that the wave operators* $W_\pm(H_0,H;J^*)$ *exist. Then*

$$\lim_{\varepsilon\to+0}\left\|\int_a^b E_0(\mathrm{d}\lambda)\,J^*\varepsilon R(\lambda+\mathrm{i}\varepsilon)\,P^{\mathrm{ac}}g(\lambda)\right.$$

$$\left.-\int_a^b E_0(\mathrm{d}\lambda)\,J^*\varepsilon R(\lambda+\mathrm{i}\varepsilon)\,P^{\mathrm{ac}}\int_{-\infty}^{\infty}E(\mathrm{d}x)\,g(x)\right\|=0.$$

Note that if $\mathscr{H}_0=\mathscr{H}$ and $J=1$, then the existence of $W_\pm(H_0,H)$ follows from the existence and completeness of $W_\pm(H,H_0)$ (see Section 9.3.2 for other criteria).

Further from the proof of Lemma 1 we can easily see that the following corollary is valid.

**Corollary 3.** *In addition to the assumption of Lemma* 1 *assume that* $\int\|g'(\lambda)\|\,\mathrm{d}\lambda<\infty$. *Then the vector spectral integral* $\int E_0(\mathrm{d}\lambda)\,J^*\varepsilon R(\lambda+\mathrm{i}\varepsilon)\,g(\lambda)$ *exists and we have*

$$\operatorname*{w-lim}_{\varepsilon\to+0}\left\{P_0^{\mathrm{ac}}\int_{-\infty}^{\infty}E_0(\mathrm{d}\lambda)\,J^*\varepsilon R(\lambda+\mathrm{i}\varepsilon)\,g(\lambda)\right.$$

$$\left.-P_0^{\mathrm{ac}}\int_{-\infty}^{\infty}E_0(\mathrm{d}\lambda)\,J^*\varepsilon R(\lambda+\mathrm{i}\varepsilon)\int_{-\infty}^{\infty}E(\mathrm{d}x)\,g(x)\right\}=0.$$

With the help of Lemma 1 (Corollary 3) we are able to derive an important representation of the scattering operator.

**Theorem 4.** *Assume* $J \operatorname{dom} H_0 \subseteq \operatorname{dom} H$ *and put* $V := HJ - JH_0$. *Then*

$$S - P_0^{ac} = \operatorname*{s-lim}_{\delta\to+0} \operatorname*{w-lim}_{\varepsilon\to+0} \Big\{ -\int_{-\infty}^{\infty} P_0^{ac} E_0(d\lambda) \big(J^*V + V^*R(\lambda+i\varepsilon) V\big) \big(R_0(\lambda - i\delta) - R_0(\lambda + i\delta)\big) P_0^{ac} \Big\}. \tag{5}$$

**Proof.** We have $S - P_0^{ac} = W_+^* W_- - W_+^* W_+ = W_+^*(W_- - W_+)$. Using the integral representations for $W_\pm$ and $W_\pm^*$ (see (13.2) and (13.3)), we get

$$S - P_0^{ac} = \operatorname*{w-lim}_{\varepsilon\to+0} \Big\{ iP_0^{ac} \int_{-\infty}^{\infty} E_0(d\lambda)\, J^*\varepsilon R(\lambda + i\varepsilon) \Big\} \operatorname*{s-lim}_{\delta\to+0} \Big\{ -i \int_{-\infty}^{\infty} E(dx)\, J\delta\big(R_0(x - i\delta) + R_0(x + i\delta)\big) P_0^{ac} \Big\}$$

$$= \operatorname*{s-lim}_{\delta\to+0} \operatorname*{w-lim}_{\varepsilon\to+0} \Big\{ P_0^{ac} \int_{-\infty}^{\infty} E_0(d\lambda)\, J^*\varepsilon R(\lambda + i\varepsilon) \int_{-\infty}^{\infty} E(dx)\, J\delta\big(R_0(x - i\delta) + R_0(x + i\delta)\big) P_0^{ac} \Big\}.$$

Applying now Corollary 3, we obtain

$$S - P_0^{ac} = \operatorname*{s-lim}_{\delta\to+0} \operatorname*{w-lim}_{\varepsilon\to+0} \Big\{ P_0^{ac} \int_{-\infty}^{\infty} E_0(d\lambda)\, J^*\varepsilon R(\lambda + i\varepsilon)\, J\delta\big(R_0(\lambda - i\delta) + R_0(\lambda + i\delta)\big) P_0^{ac} \Big\}.$$

Because of $J^*\varepsilon R(\lambda + i\varepsilon) = -i\big(J^* + (J^*H - \lambda J^*)\, R(\lambda + i\varepsilon)\big)$ and the fact that in the spectral integral we can replace $\lambda$ by $H_0$ (see Prop. 5.11),

$$S - P_0^{ac} = \operatorname*{s-lim}_{\delta\to+0} \operatorname*{w-lim}_{\varepsilon\to+0} \Big\{ -iP_0^{ac} \int_{-\infty}^{\infty} E_0(d\lambda) \big(J^* + V^*R(\lambda + i\varepsilon)\big) J\big(\delta R_0(\lambda - i\delta) + \delta R_0(\lambda + i\delta)\big) P_0^{ac} \Big\}.$$

Now without changing the value of $\operatorname*{w-lim}_{\varepsilon\to+0} \{\cdot\}$ we can replace $\delta R_0(\lambda - i\delta) + \delta R_0(\lambda + i\delta)$ by $(\delta + \varepsilon)\, R_0(\lambda - i\delta) + (\delta - \varepsilon)\, R_0(\lambda + i\delta) = -i(\lambda + i\varepsilon - H_0)\big(R_0(\lambda - i\delta) - R_0(\lambda + i\delta)\big)$. Thus

$$S - P_0^{ac} = \operatorname*{s-lim}_{\delta\to+0} \operatorname*{w-lim}_{\varepsilon\to+0} \Big\{ -P_0^{ac} \int_{-\infty}^{\infty} E_0(d\lambda) \big(J^* + V^*R(\lambda + i\varepsilon)\big) J(\lambda + i\varepsilon - H_0) \times \big(R_0(\lambda - i\delta) - R_0(\lambda + i\delta)\big) P_0^{ac} \Big\}$$

Therefore, in order to prove the assertion, we have to manipulate the expression $(J^* + V^*R(\lambda + i\varepsilon)) J(\lambda + i\varepsilon - H_0) = J^*J(\lambda + i\varepsilon - H_0) + V^*R(\lambda + i\varepsilon) J(\lambda + i\varepsilon - H_0)$. For the second term on the right side we get

$$V^*R(\lambda + i\varepsilon) J(\lambda + i\varepsilon - H_0) = V^*R(\lambda + i\varepsilon) J(\lambda + i\varepsilon) - V^*R(\lambda + i\varepsilon) (HJ - V)$$
$$= V^*R(\lambda + i\varepsilon) (\lambda + i\varepsilon - H) J + V^*R(\lambda + i\varepsilon) V = V^*J + V^*R(\lambda + i\varepsilon) V. \tag{6}$$

Concerning the first term, without changing the value of $\operatorname{w-lim}_{\varepsilon\to+0} \{\cdot\}$, the summand $i\varepsilon J^*J$ can be omitted. Furthermore, the summand $\lambda J^*J$ can be replaced by $H_0J^*J$ in the spectral integral. Thus $J^*J(\lambda + i\varepsilon - H_0)$ can be replaced by $H_0J^*J - J^*JH_0 = (J^*H - V^*) J - J^*(HJ - V) = -V^*J + J^*V$ in $\operatorname{w-lim}_{\varepsilon\to+0} \{\cdot\}$. This together with (6) implies the desired assertion (5). ■

If not only the wave operators $W_\pm(H, H_0; J)$ exist but also $W_\pm(H_0, H; J^*)$, then Corollary 2 is applicable. Thus in (5) we can replace $\operatorname{w-lim}_{\varepsilon\to+0}$ by $\operatorname{s-lim}_{\varepsilon\to+0}$ when we also replace $J$ by $P^{ac}J$. For instance, if $\mathscr{H}_0 = \mathscr{H}$ and $J = 1$, then the wave operators $W_\pm(H_0, H)$ exist and we obtain the following result for the one-space scattering system.

**Corollary 5.** *Let $\mathscr{H}_0 = \mathscr{H}$ and $J = 1$. Assume that $W_\pm(H_0, H)$ exist,* $\operatorname{dom} H_0 \subseteq \operatorname{dom} H$ *and put $V := H - H_0$. Then*

$$S - P_0^{ac} = \operatorname{s-lim}_{\delta\to+0} \operatorname{s-lim}_{\varepsilon\to+0} \left\{ -P_0^{ac} \int_{-\infty}^{\infty} E_0(d\lambda) \left(P^{ac}V + VP^{ac}R(\lambda + i\varepsilon)\right) V\left(R_0(\lambda - i\delta) - R_0(\lambda + i\delta)\right) P_0^{ac} \right\}. \tag{7}$$

It is also possible to derive representations of $S$ with the help of the other type of an operator spectral integral, namely $\int B(\lambda) E_0(d\lambda)$. For example the analog of Corollary 5 is given by

$$S - P_0^{ac} = \operatorname{s-lim}_{\delta\to+0} \operatorname{s-lim}_{\varepsilon\to+0} \left\{ -P_0^{ac} \int_{-\infty}^{\infty} \left(R_0(\lambda - i\delta) - R_0(\lambda + i\delta)\right) \left(VP^{ac} + VP^{ac}R(\lambda + i\varepsilon) V\right) E_0(d\lambda) P_0^{ac} \right\}.$$

The representations of the scattering operator given by expression (5) and (7) are not yet useful enough for the purpose to obtain formulas for $\hat{T}(\lambda)$. The main defect of these expressions is that $S$ is not represented by an operator spectral integral but by a (double) limit of such an integral. Therefore we have to look for conditions which allow to interchange limit and spectral integration.

### 18.1.3. A general formula for $\hat{T}(\lambda)$

In spite of the foregoing remark in the preceding section we can derive from (5) or (7) a formula for the scattering amplitude which is, however, rather abstract. The derivation proceeds similarly as the one for the wave matrix (see Section 13.4.3). For simplicity we give this formula only in the special case $J = 1$, that is, the starting point is (7). Note that in the case of (5) we need the formula with strong limit in $\delta$ as well as in $\varepsilon$. We put

$$T_{\delta,\varepsilon} = \int_{-\infty}^{\infty} E_0(d\lambda)\, T_{\delta,\varepsilon}(\lambda)$$

where

$$T_{\delta,\varepsilon}(\lambda) = -P_0^{\mathrm{ac}}\big(P^{\mathrm{ac}}V + VP^{\mathrm{ac}}R(\lambda + \mathrm{i}\varepsilon)\, V\big)\big(R_0(\lambda - \mathrm{i}\delta) - R_0(\lambda + \mathrm{i}\delta)\big)\, P_0^{\mathrm{ac}}.$$

Then we obtain, for $f \in \mathscr{H}_0$ and with $T = S - P_0^{\mathrm{ac}}$,

$$\|T_{\delta,\varepsilon}f - Tf\| = \int_{-\infty}^{\infty} |T_{\delta,\varepsilon}(\lambda)\, f - Tf|^2_{E_0,\lambda}\, \mathrm{d}\lambda,$$

hence, by Corollary 5,

$$\operatorname*{m-lim}_{\delta\to+0}\ \operatorname*{m-lim}_{\varepsilon\to+0} |T_{\delta,\varepsilon}(\lambda)\, f - Tf|_{E_0,\lambda} = 0.$$

Similarly as in Section 13.4.3 we get

$$\operatorname*{m-lim}_{\delta\to+0}\ \operatorname*{m-lim}_{\varepsilon\to+0} |D_\lambda\{T_{\delta,\varepsilon}(\lambda)\, f\} - \hat{T}(\lambda)\, \hat{f}(\lambda)|_{E_0,\lambda} = 0$$

where $\hat{T}(\lambda)$ denotes the scattering amplitude with respect to a representation of $P_0^{\mathrm{ac}}\mathscr{H}_0$ by $L^2(\Delta_0, \mathrm{d}\lambda; \hat{\mathscr{K}}_\lambda)$, $\hat{f}(\lambda)$ denotes the representer of $f$ in this representation and $D_\lambda$ denotes the value-operator. Hence, if $f$ is fixed, then there are sequences $\varepsilon_n \to +0$, $\delta_n \to +0$ such that

$$\operatorname*{s-lim}_{\delta_n\to+0}\ \operatorname*{s-lim}_{\varepsilon_n\to+0} D_\lambda\{T_{\delta_n,\varepsilon_n}(\lambda)\, f\} = \hat{T}(\lambda)\, \hat{f}(\lambda) \ \text{ a.e. mod } |\cdot|$$

(strong convergence in the value-space $\hat{\mathscr{K}}_\lambda$).

### 18.1.4. An explicit formula for $\hat{T}(\lambda)$

The starting point in this section is a factorization of the perturbation $V$, $V = B^*A$ where $A \in \mathscr{L}(\mathscr{H}_0, \mathscr{K})$, $B \in \mathscr{L}(\mathscr{H}, \mathscr{K})$ with $\mathscr{K}$ some Hilbert space, and certain convergence conditions with respect to the resolvents $R$, $R_0$. We begin with a few preparatory remarks.

Let $\mathscr{M} \subset P_0^{\mathrm{ac}}\mathscr{H}_0$ be a spectral manifold with respect to $E_0(\cdot)$, that is, $(k, k')_{E_0,\lambda} = \big(k, E_0(\mathrm{d}\lambda)\, k'\big)/\mathrm{d}\lambda$ exists for all $k, k' \in \mathscr{M}$ and all $\lambda \in \Lambda \subseteq \Delta_0$, where $\Lambda$ denotes the associated spectral core. Recall the definitions of $\hat{\mathscr{M}}_\lambda = \mathrm{clo}_{|\cdot|_{E_0,\lambda}}\big(\mathscr{M} \bmod (\ker |\cdot|_{E_0,\lambda})\big)$, $\mathscr{D}_\lambda$, $\hat{\mathscr{D}}_\lambda$, $D_\lambda$ and the relations $\hat{\mathscr{M}}_\lambda \subseteq \hat{\mathscr{D}}_\lambda$, $D_\lambda k \in \hat{\mathscr{D}}_\lambda$, $(k, k')_{E_0,\lambda} = (D_\lambda k, D_\lambda k')_{\hat{\mathscr{M}}_\lambda}$ for $k, k' \in \mathscr{M}$ (see 3.5.5 and 5.2.4). When $\mathscr{M}$ is also generating with respect to $P_0^{\mathrm{ac}}\mathscr{H}_0$ and $H_0$, then we obtain the direct integral representation $L^2(\Lambda, \mathrm{d}\lambda; \hat{\mathscr{M}}_\lambda)$ of $P_0^{\mathrm{ac}}\mathscr{H}_0$ with respect to $H_0$. Moreover, $D_\lambda k$, $k \in \mathscr{M}$ can be regarded as element of $\hat{\mathscr{M}}_\lambda$ and can be identified with the "value" $\hat{k}(\lambda)$ of the representer $\hat{k}(\cdot)$ of $k$ at $\lambda$ with respect to this direct integral representation (cf. Prop. 5.15).

In general, if we have a direct integral representation $L^2(\Delta_0, \mathrm{d}\lambda; \mathscr{H}_\lambda; \mathscr{S})$, $\mathscr{S}$ separable, of $P_0^{ac}\mathscr{H}_0$ with respect to $H_0$ and an arbitrary spectral manifold $\mathscr{M} \subset P_0^{ac}\mathscr{H}_0$ (with corresponding spectral core $\Lambda$) then, without loss of generality, we may assume $\mathscr{H}_\lambda \subset \hat{\mathscr{D}}_\lambda$ and, moreover, we have

$$D_\lambda f = \hat{f}(\lambda), \qquad f \in \mathscr{M},$$

a.e. (mod $|\cdot|$), where $\hat{f}(\lambda)$ denotes the representer of $f$ with respect to $L^2(\Delta_0, \mathrm{d}\lambda; \mathscr{H}_\lambda; \mathscr{S})$; the exceptional set may vary with $f$ (but note that the left side $D_\lambda f$ is well-defined for all $\lambda \in \Lambda$ and all $f \in \mathscr{M}$). This means that $D_\lambda f$ is nothing else than the value of the representer of $f$ at $\lambda$.

We omit details of the proof; note that it is sufficient to consider the case $L^2(\Delta_0, \mathrm{d}\lambda; \mathscr{K})$, where $\mathscr{K}$ is a fixed Hilbert space and where $\Delta_0$ is compact. In this case the injective embedding $\mathscr{K} \subset \hat{\mathscr{D}}_\lambda$ for all $\lambda \in \Delta_0$ is obvious: the element $k \in \mathscr{K}$ corresponds to the $\mathscr{K}$-valued function $k(\lambda) \equiv k$ for all $\lambda \in \Delta_0$, which belongs to $\mathscr{D}_\lambda$ mod (ker $[\cdot]_{E_0,\lambda}$) and equals 0 only if $k = 0$. The equality $D_\lambda f = \hat{f}(\lambda)$ a.e. mod $|\cdot|$ (as an equation in $\hat{\mathscr{D}}_\lambda$) can be proved similarly as Prop. 5.15.

Furthermore, if $\mathscr{M}$ is generating with respect to $P_0^{ac}\mathscr{H}_0$ and $H_0$, without loss of generality we may assume that $\{D_\lambda f; f \in \mathscr{M}\}$ is a submanifold of $\mathscr{H}_\lambda$ and dense in $\mathscr{H}_\lambda$ for almost all $\lambda \in \Delta_0$ (mod $|\cdot|$).

Let $C \in \mathscr{L}(\mathscr{H}_0, \mathscr{K})$ and assume ima $C^* \subseteq \mathscr{M}$. Consider the map $F_C(\lambda): \mathscr{K} \to \mathscr{H}_\lambda$ given by

$$F_C(\lambda) f := D_\lambda C^* f; \qquad f \in \mathscr{K} \quad \text{and} \quad \lambda \in \Lambda. \tag{8}$$

Now we obtain

**Lemma 6.** *Let* $C \in \mathscr{L}(\mathscr{H}_0, \mathscr{K})$, ima $C^* \subseteq \mathscr{M}$ *and* $F_C(\lambda) \in \mathscr{L}(\mathscr{K}, \mathscr{H}_\lambda)$ *for all* $\lambda \in \Lambda$. *Then:*

(i) *For each* $f \in \mathscr{M}$ *and all* $\lambda \in \Lambda$, *there is a vector from* $\mathscr{K}$, *formally denoted by* $(CE_0(\mathrm{d}\lambda) f)/\mathrm{d}\lambda$ *such that* $\big(g, (CE_0(\mathrm{d}\lambda) f)/\mathrm{d}\lambda\big) = \big(g, CE_0(\mathrm{d}\lambda) f\big)/\mathrm{d}\lambda$, $g \in \mathscr{K}$.

(ii) *For fixed* $\lambda \in \Lambda$ *the vector* $(CE_0(\mathrm{d}\lambda) f)/\mathrm{d}\lambda$ *depends only on* $D_\lambda f$ *(when* $f$ *runs through* $\mathscr{M}$*).*

(iii) *The operator* $G(\lambda): \mathscr{H}_\lambda \to \mathscr{K}$ *given by* $G(\lambda) D_\lambda f = (CE_0(\mathrm{d}\lambda) f)/\mathrm{d}\lambda$ *coincides with* $F_C(\lambda)^*$ *on the set* $\{D_\lambda f; f \in \mathscr{M}\}$, $\lambda \in \Lambda$.

Proof. First note that $\big(g, CE_0(\mathrm{d}\lambda) f\big)/\mathrm{d}\lambda = \big(C^*g, E_0(\mathrm{d}\lambda) f\big)/\mathrm{d}\lambda$ exists for all $g \in \mathscr{K}$, $f \in \mathscr{M}$ and $\lambda \in \Lambda$ because $C^*g \in \mathscr{M}$. Furthermore, we immediately obtain

$$\big(C^*g, E_0(\mathrm{d}\lambda) f\big)/\mathrm{d}\lambda = (C^*g, f)_{E_0,\lambda} = (D_\lambda C^*g, D_\lambda f)_{\mathscr{H}_\lambda}, \qquad \lambda \in \Lambda.$$

This equation says that the left side depends only on $D_\lambda f$ for fixed $\lambda$ and $C^*g$. Moreover, since $D_\lambda C^*g = F_C(\lambda)\, g$ and $F_C(\lambda) \in \mathscr{L}(\mathscr{K}, \mathscr{H}_\lambda)$, the left side defines a continuous linear functional with respect to $g$. Hence, according to Riesz' theorem (see e.g. Weidmann [1, p. 64]), we get a vector from $\mathscr{K}$, denoted by $(CE_0(\mathrm{d}\lambda) f)/\mathrm{d}\lambda$, such that $\big(g, (CE_0(\mathrm{d}\lambda) f)/\mathrm{d}\lambda\big) = \big(g, CE_0(\mathrm{d}\lambda) f\big)/\mathrm{d}\lambda$.

Finally, we have $\big(g, (CE_0(\mathrm{d}\lambda) f)/\mathrm{d}\lambda\big) = \big(F_C(\lambda)\, g, D_\lambda f\big)_{\mathscr{H}_\lambda}$. This means that $G(\lambda) = F_C(\lambda)^*$ on the set of vectors $D_\lambda f$, $f \in \mathscr{M}$. ■

Since $F_C(\lambda)^*$ is bounded and the set $\{D_\lambda f; f \in \mathscr{M}\}$ is dense in $\mathscr{H}_\lambda$ for $\lambda \in \Lambda$, $G(\lambda)$ can be extended to a bounded operator from $\mathscr{H}_\lambda$ to $\mathscr{K}$ again denoted by $G(\lambda)$.

Now we are able to prove an explicit formula for the scattering amplitude.

**Theorem 7.** *Assume that* $J(\operatorname{dom} H_0) \subseteq \operatorname{dom} H$ *and put* $V = HJ - JH_0$. *Let* $V$ *be given in the form* $V = B^*A$ *with* $A \in \mathscr{L}(\mathscr{H}_0, \mathscr{K})$, $B \in \mathscr{L}(\mathscr{H}, \mathscr{K})$ *and assume that* $A$ *and* $BJ$ *(in place of* $C$*) satisfy the conditions of Lemma 6. Further, assume that*

(I) $\sup\limits_{\varepsilon>0} \|BR(\lambda + i\varepsilon) B^*\| \leqq \alpha(\lambda)$ *where* $\alpha(\lambda) < \infty$ *a.e.* mod $|\cdot|$ *on* $\Delta_0$,

(II) $\operatorname*{s-lim}\limits_{\delta\to+0} BR(\lambda + i\delta) B^* =: BR(\lambda + i0) B^*$ *exists a.e.* mod $|\cdot|$ *on* $\Delta_0$,

(III) $\operatorname*{s-lim}\limits_{\delta\to+0} A\big(R_0(\lambda - i\delta) - R_0(\lambda + i\delta)\big) f$ *exists a.e.* mod $|\cdot|$ *on* $\Delta_0$, $f \in \mathscr{M}$.

*Then there is a sequence* $\Delta_j$, $\Delta_1 \subseteq \Delta_2 \subseteq \cdots$, *with* $\bigcup\limits_{j=1}^{\infty} \Delta_j = \Delta_0(\text{mod } |\cdot|)$ *such that for each compact set* $\Delta \subset \mathbb{R}$ *we have*

$$\big(u, E_0(\Delta)\,(S - P_0^{ac})\,v\big) = -2i\pi \lim_{j\to\infty} \int\limits_{\Delta\cap\Delta_j} \Big\{\big(F_{BJ}(\lambda)^* D_\lambda u, F_A(\lambda)^* D_\lambda v\big)_{\mathscr{K}}$$

$$+ \big(F_A(\lambda)^* D_\lambda u, BR(\lambda + i0)\, B^* F_A(\lambda)^* D_\lambda v\big)_{\mathscr{K}}\Big\}\, d\lambda, \quad u, v \in \mathscr{M}\,. \tag{9}$$

**Proof.** Inserting the factorization $V = B^*A$ into formula (5), we obtain

$$\big(u, E_0(\Delta)\,(S - P_0^{ac})\,v\big) = -\lim_{\delta\to+0} \int\limits_{\Delta} \big(BJE_0(d\lambda)\,u, A\big(R_0(\lambda - i\delta) - R_0(\lambda + i\delta)\big)\,v\big)$$

$$- \lim_{\delta\to+0} \lim_{\varepsilon\to+0} \int\limits_{\Delta} \big(AE_0(d\lambda)\,u, BR(\lambda + i\varepsilon)\, B^*A\big(R_0(\lambda - i\delta) - R_0(\lambda + i\delta)\big)\,v\big)$$

for $u, v \in \mathscr{M} \subseteq P_0^{ac}\mathscr{H}_0$. Since Lemma 6 is applicable to $A$ and $BJ$, this equation can be transformed in the following manner:

$$\big(u, E_0(\Delta)\,(S - P_0^{ac})\,v\big) = -\lim_{\delta\to+0} \int\limits_{\Delta} \Big\{\big(BJE_0(d\lambda)\,u, A\big(R_0(x - i\delta)$$

$$- R_0(x + i\delta)\big)\,v\big)/d\lambda|_{x=\lambda}\Big\}\, d\lambda - \lim_{\delta\to+0} \lim_{\varepsilon\to+0} \int\limits_{\Delta} \Big\{\big(AE_0(d\lambda)\,u, BR(x + i\varepsilon)$$

$$\times B^*A\big(R_0(x - i\delta) - R_0(x + i\delta)\big)\,v\big)/d\lambda|_{x=\lambda}\Big\}\, d\lambda$$

$$= -\lim_{\delta\to+0} \int\limits_{\Delta} \big(\big(BJE_0(d\lambda)\,u\big)/d\lambda, A\big(R_0(\lambda - i\delta) - R_0(\lambda + i\delta)\big)\,v\big)\, d\lambda$$

$$- \lim_{\delta\to+0} \lim_{\varepsilon\to+0} \int\limits_{\Delta} \big(\big(AE_0(d\lambda)\,u\big)/d\lambda, BR(\lambda + i\varepsilon)\, B^*A\big(R_0(\lambda - i\delta) - R_0(\lambda + i\delta)\big)\,v\big)\, d\lambda$$

$$= -\lim_{\delta\to+0} \int\limits_{\Delta} \big(F_{BJ}(\lambda)^* D_\lambda u, A\big(R_0(\lambda - i\delta) - R_0(\lambda + i\delta)\big)\,v\big)\, d\lambda$$

$$- \lim_{\delta\to+0} \lim_{\varepsilon\to+0} \int\limits_{\Delta} \big(F_A(\lambda)^* D_\lambda u, BR(\lambda + i\varepsilon)\, B^*A\big(R_0(\lambda - i\delta) - R_0(\lambda + i\delta)\big)\,v\big)\, d\lambda. \tag{10}$$

From (III) it follows that $\big(f, 2i\pi AE_0(d\lambda)\,v\big)/d\lambda = \lim\limits_{\delta\to+0} \big(f, A\big(R_0(\lambda - i\delta) - R_0(\lambda + i\delta)\big)\,v\big)$ exists, a.e. mod $|\cdot|$ on $\Delta_0$, $f \in \mathscr{K}$. This relation, (III) and Lemma 6 imply

$$2i\pi\big(AE_0(d\lambda)\,v\big)/d\lambda = \operatorname*{s-lim}_{\delta\to+0} A\big(R_0(\lambda - i\delta) - R_0(\lambda + i\delta)\big)\,v \quad \text{a.e. mod } |\cdot| \text{ on } \Delta_0.$$

Put $\beta(\lambda) = \sup_{\varepsilon>0} \left\|A\big(R_0(\lambda - i\varepsilon) - R_0(\lambda + i\varepsilon)\big) v\right\|$. Then $\beta(\lambda) < \infty$ a.e. mod $|\cdot|$ on $\Delta_0$. Define Borel sets $\Lambda_n$, $n \in \mathbb{N}$, by $\Lambda_n = \{\lambda \in \Delta_0;\ \max\{\beta(\lambda), \alpha(\lambda), \|F_{BJ}(\lambda)^*\|, \|F_A(\lambda)^*\|\} < n\}$. Then we claim

$$\big(u, E_0(\Delta \cap \Lambda_n)\,(S - P_0^{ac})\,v\big) = -2i\pi \int_{\Delta\cap\Lambda_n} \big(F_{BJ}(\lambda)^* D_\lambda u, \big(AE_0(d\lambda)\,v\big)/d\lambda\big)\,d\lambda$$

$$-2i\pi \int_{\Delta\cap\Lambda_n} \big(F_A(\lambda)^* D_\lambda u, BR(\lambda + i0)\,B^*\big((AE_0(d\lambda)\,v)/d\lambda\big)\big)\,d\lambda. \tag{11}$$

But this follows from the estimates

$$\left|\big(F_{BJ}(\lambda)^* D_\lambda u, A\big(R_0(\lambda - i\delta) - R_0(\lambda + i\delta)\big) v\big)\right| \leqq \beta(\lambda)\,\|F_{BJ}(\lambda)^* D_\lambda u\| \leqq n^2\,\|D_\lambda u\|_{\mathscr{H}_\lambda},$$

$$\left|\big(F_A(\lambda)^* D_\lambda u, BR(\lambda + i\varepsilon)\,B^*A\big(R_0(\lambda - i\delta) - R_0(\lambda + i\delta)\big) v\big)\right|$$

$$\leqq \beta(\lambda)\,\alpha(\lambda)\,\|F_A(\lambda)^* D_\lambda u\| \leqq n^3\,\|D_\lambda u\|_{\mathscr{H}_\lambda},$$

$$\int_\Delta \|D_\lambda u\|_{\mathscr{H}_\lambda}\,d\lambda \leqq |\Delta|^{1/2} \Big(\int_\Delta \|D_\lambda u\|^2_{\mathscr{H}_\lambda}\,d\lambda\Big)^{1/2} \leqq |\Delta|^{1/2}\,\|u\|,$$

the pointwise convergence of the integrand in (10) (a.e. mod $|\cdot|$ on $\Delta_0$) and the Lebesgue dominated convergence theorem. Replacing $\big(AE_0(d\lambda)\,v\big)/d\lambda$ by $F_A(\lambda)^* D_\lambda$ in (11) and observing that the sequence $\Lambda_n$ satisfies the other assertions, we obtain (9). ■

**Remark 8.** The conditions (I) and (II) in Theorem 7 are obviously satisfied if $\operatorname{n-lim}_{\varepsilon\to+0} BR(\lambda + i\varepsilon)\,B^*$ exists a.e. mod $|\cdot|$ on $\Delta_0$.

Note that a representation of the scattering amplitude $\hat{T}(\lambda)$ with respect to $L^2(\Delta_0, d\lambda; \mathscr{H}_\lambda; \mathscr{S})$ easily follows from (9). First, formula (9) may be extended to $u, v \in P_0^{ac}\mathscr{H}_0$ (note the remark after the proof of Lemma 6 and the proof of (9) itself). Second, $\mathscr{S}$ is separable, that is, there is a generating sequence $u_n \in P_0^{ac}\mathscr{H}_0$ (this means $\{\hat{u}_n(\lambda), n = 1, 2, \ldots\}$ is generating in $\mathscr{H}_\lambda$ for almost all $\lambda$ (mod $|\cdot|$), see Section 4.3.1). Third, $\big(u, E_0(\Delta)\,(S - P_0^{ac})\,v\big) = \int_{\Delta\cap\Delta_0} \big(\hat{u}(\lambda), \hat{T}(\lambda); \hat{v}(\lambda)\big)_{\mathscr{H}_\lambda}\,d\lambda$ for each compact set $\Delta$ and all $u, v \in P_0^{ac}\mathscr{H}_0$. Therefore we obtain

**Corollary 9.** *Let the assumptions of Theorem 7 be satisfied. Then*

$$\hat{T}(\lambda) = -2i\pi\{F_{BJ}(\lambda)\,F_A(\lambda)^* + F_A(\lambda)\,BR(\lambda + i0)\,B^*F_A(\lambda)^*\} \bmod |\cdot| \text{ on } \Delta_0 \tag{12}$$

*is valid for the representation of the scattering amplitude with respect to* $L^2(\Delta_0, d\lambda; \mathscr{H}_\lambda; \mathscr{S})$. *If* $\mathscr{H} = \mathscr{K} = \mathscr{H}_0$, $J = 1$ *and* $V = BA = AB$ *where* $A$, $B$ *are selfadjoint, then* (12) *can be written in the form*

$$\hat{T}(\lambda) = -2i\pi F_B(\lambda)\,\big(1 + AR(\lambda + i0)\,B\big)\,F_A(\lambda)^* \bmod |\cdot| \text{ on } \Delta_0 \tag{13}$$

*provided that we replace* $BR(z)\,B^*$ *by* $AR(z)\,B$ *in* (I) *and* (II) *of Theorem* (7).

We conclude this section with the remark that Theorem 7 and Corollary 9 can be extended to unbounded perturbations $V$. For example, if we assume that $AE_0(\Delta)$ and $BJE_0(\Delta)$ are bounded for each finite interval $\Delta$ and satisfy the assumptions of Lemma 6 (at least for a set of intervals whose union covers $\Delta_0$ mod $|\cdot|$), then the assertions of Theorem 7 and Corollary 9 remain true.

### 18.4.5. An application to potential scattering

Let $\mathscr{H} = \mathscr{H}_0 = L^2(\mathbb{R}^3, dx)$, $J = 1$ and $H_0 = -\Delta$. Further, let $V = BA = AB$ where $V$ is self-adjoint on $\mathscr{H}_0$ and $A = |V|^{1/2}$. We assume that $V$ is multiplication by a real-valued function $v(\cdot) \in L^\infty(\mathbb{R}^3, dx) \cap L^1(\mathbb{R}^3, dx)$. Then $A$ and $B$ are multiplication operators by real-valued functions $a(\cdot)$ and $b(\cdot)$ belonging to $L^\infty(\mathbb{R}^3, dx) \cap L^2(\mathbb{R}^3, dx)$. Naturally, $v(\cdot)$ is also an element of $L^2(\mathbb{R}^3, dx)$.

Now we take over the results of Sections 17.2.2 and 17.3.2 for the scattering system $\{H = H_0 + \varkappa V, H_0\}$ where $\varkappa \in \mathbb{R}$ obeys the estimate $\sup_{\lambda \in \mathbb{R}, \varepsilon > 0} |\varkappa| \cdot \|AR_0(\lambda + i\varepsilon) B\| < 1$. Then we obtain:

(I) $A$ and $B$ are $H_0$-smooth. Hence, in particular $\operatorname{s-lim}_{\varepsilon \to +0} AR_0(\lambda \pm i\varepsilon) f$ exists a.e. mod $|\cdot|$ for each $f \in \mathscr{H}_0$.

(II) $A$ is $H$-smooth. Hence $\operatorname{s-lim}_{\varepsilon \to +0} AR(\lambda \pm i\varepsilon) Bf$ exists a.e. mod $|\cdot|$ for each $f \in \mathscr{H}_0$.

(III) $\sup_{\lambda \in \mathbb{R}, \varepsilon > 0} \|BR(\lambda + i\varepsilon) B^*\| = \sup_{\lambda \in \mathbb{R}, \varepsilon > 0} \|AR(\lambda + i\varepsilon) B\| < \infty$ because of the equality $1 + \varkappa AR(\lambda + i\varepsilon) B = (1 - \varkappa AR_0(\lambda + i\varepsilon) B)^{-1}$ and the estimate $\sup_{\lambda \in \mathbb{R}, \varepsilon > 0} |\varkappa| \, \|AR_0(\lambda + i\varepsilon) B\| < 1$ (the last estimate is valid for a $\varkappa > 0$).

(IV) The strong wave operators $W_\pm(H, H_0)$ exist and are unitary. The next step is to introduce a direct integral representation of $\mathscr{H}_0$ with respect to $H_0$, a manifold $\mathscr{M} \subset \mathscr{H}_0$ and the operator $D_\lambda$. Let $\Omega$ denote the unit sphere in $\mathbb{R}^3$, and let $\lambda = |p|^2$ and $\omega = p/|p|$ for $p \in \mathbb{R}^3$. Then the map

$$L^2(\mathbb{R}^3, dx) \ni f(\cdot) \to g(\cdot, \cdot) \in L^2(\mathbb{R}_+, d\lambda; L^2(\Omega, d\omega))$$

with $g(\lambda, \omega) = 2^{-1/2} |p|^{1/2} \hat{f}(p)$ leads to the direct integral representation $L^2(\mathbb{R}_+, d\lambda; L^2(\Omega, d\omega))$ of $\mathscr{H}_0$ with respect to $H_0$ (see also Section 15.2.4). Set $\mathscr{M} = \{f \in \mathscr{H}_0; f \in L^1(\mathbb{R}^3, dx)\}$. Then the operator $D_\lambda : \mathscr{M} \to \mathscr{H}_\lambda = L^2(\Omega, d\omega)$ is given by

$$(D_\lambda f)(\omega) = 2^{-1/2} \lambda^{1/4} (2\pi)^{-3/2} \int_{\mathbb{R}^3} e^{-i\lambda^{1/2}(\omega, x)} f(x) \, dx. \tag{14}$$

To conclude the proof that the operators $H$, $H_0$, $A$ and $B$ satisfy the assumptions of Corollary 9, we must establish that $A$ (and thus $B = (\operatorname{sgn} V) A = A(\operatorname{sgn} V)$ obeys the conditions of Lemma 6. Clearly, $\operatorname{ima} A^* \subseteq \mathscr{M}$ since $a(\cdot) \in L^\infty(\mathbb{R}^3, dx) \cap L^2(\mathbb{R}^3, dx)$. Further, for $f \in \mathscr{H}_0$ we have

$$(F_A(\lambda) f)(\omega) = (D_\lambda A f)(\omega) = (1/4) \lambda^{1/4} \pi^{-3/2} \int_{\mathbb{R}^3} e^{-i\lambda^{1/2}(\omega, x)} a(x) f(x) \, dx.$$

Since $a(\cdot) f(\cdot) \in L^1(\mathbb{R}^3, dx)$, $(D_\lambda A f)(\omega)$ is jointly continuous in $\lambda$ and $\omega$. Further, we easily obtain that $F_A(\lambda) \in \mathscr{L}_2(\mathscr{H}_0, L^2(\Omega, d\omega))$ and that $F_A(\lambda)$ is continuous in $\lambda \in \mathbb{R}_+$ with respect to $\|\cdot\|_2$. Thus we get the formula (13) for the scattering amplitude $\hat{T}(\lambda)$ of the scattering system $\{-\Delta + \varkappa V, -\Delta\}$ and moreover, we obtain that $\hat{T}(\lambda) \in \mathscr{L}_1(L^2(\Omega, d\omega))$ for all $\lambda \in \mathbb{R}_+$ and is continuous in $\lambda \in \mathbb{R}_+$ with respect to $\|\cdot\|_1$.

## 18.2. Smoothness conditions and the scattering amplitude

### 18.2.1. Some formulas for the scattering amplitude

The aim of this section is to derive formulas for $S - P_0^{ac}$ and $\hat{T}(\lambda)$ under smoothness conditions. The considerations proceed similarly as the ones for Theorem 7 and Corollary 9. We start with

**Lemma 10.** *Let $A \in \mathscr{C}(\mathscr{H}_0, \mathscr{H})$. If $A$ is $H_0$-smooth, then*

$$\int_{-\infty}^{\infty} E_0(d\lambda) A^* \{\cdot\} \in \mathscr{L}\big(L^2(\mathbb{R}, d\lambda; \mathscr{H}), \mathscr{H}_0\big). \tag{15}$$

**Proof.** Let $f \in L^2(\mathbb{R}, d\lambda; \mathcal{K})$ be piecewise absolutely continuous and let $\Delta$ be compact. Then, by Prop. 5.10, Prop. 5.14 and the fact that ima $A^* \subseteq P_0^{ac}\mathcal{H}_0$ ($A$ is $H_0$-smooth) and thus in (15) we can replace $E_0(d\lambda)$ by $P_0^{ac}E_0(d\lambda)$, we obtain

$$\left\| \int_\Delta E_0(d\lambda) A^*f(\lambda) \right\|^2 = \int_\Delta \big(f(\lambda), AE_0(dx) A^*f(\lambda)\big)/dx|_{x=\lambda} \, d\lambda.$$

Note that $\sup_{-\infty<a<b<\infty} |b-a|^{-1} \|AE_0([a,b]) A^*\| = c < \infty$ because of

$$\sup_{\|u\|=1} \|E_0([a;b]) A^*u\|^2 = \sup_{\|u\|=1} \big(E_0([a,b]) A^*u, E_0([a,b]) A^*u\big)$$

$$= \sup_{\|u\|=1} \big(u, AE_0([a,b]) A^*u\big) = \|AE_0([a,b]) A^*\|$$

and Theorem 17.6. Thus

$$\left\| \int_\Delta E_0(d\lambda) A^*f(\lambda) \right\|^2 \leqq c \int_\Delta \big(f(\lambda), f(\lambda)\big) \, d\lambda \leqq c \, \|f\|^2_{L^2(\mathbb{R}, d\lambda; \mathcal{K})}. \tag{16}$$

This implies that $\int E_0(d\lambda) A^*f(\lambda)$ exists and (16) is true for $\Delta = (-\infty, \infty)$. ∎

Now we are able to prove an analog of Theorem 7 under smoothness conditions.

**Theorem 11.** *Assume that* $J(\text{dom } H_0) \subseteq \text{dom } H$ *and put* $V = HJ - JH_0$. *Assume that* $V = B^*A$ *(in the form sense) where* $A \in \mathcal{C}(\mathcal{H}_0, \mathcal{K})$, $B \in \mathcal{C}(\mathcal{H}, \mathcal{K})$ *and* $\mathcal{K}$ *is a separable Hilbert space. Further assume that for some Borel sets* $\sigma \subseteq \mathbb{R}$, $A$ *and* $BJ$ *satisfy the condition* (I) *of Prop.* 17.19 *with respect to* $H_0$ *and* $\sigma$ *and that* $\sup_{\varepsilon<0, \lambda\in\sigma} \|BR(\lambda + i\varepsilon) B^*\| < \infty$ *(then* $BE(\sigma)$ *is* $H$*-smooth and* $\text{s-lim}_{\varepsilon\to+0} BR(\lambda + i\varepsilon) B^* =: BR(\lambda + i0) B^*$ *exists for all* $\lambda \in \sigma$*). Then*

$$E_0(\sigma)\,(S - P_0^{ac})\, f = -2i\pi \int_\sigma E_0(d\lambda)\, E_0(\sigma)\, \{J^*B^* + A^*BR(\lambda + i0)\, B^*\}$$

$$\times \big((AE_0(\sigma)\, E_0(d\lambda)\, f)/d\lambda\big), \quad f \in P_0^{ac}\mathcal{H}_0. \tag{17}$$

**Proof.** The starting point is formula (5). But in our case, in (5), the left side is to replace by $E_0(\sigma)\,(S - P_0^{ac})$ and the right side by

$$\text{s-lim}_{\delta\to+0} \text{w-lim}_{\varepsilon\to+0} \Big\{ -\int_\sigma E_0(d\lambda)\, E_0(\sigma)\, [J^*B^* + A^*BR(\lambda + i\varepsilon)\, B^*]\, AE_0(\sigma)$$

$$\times \big(R_0(\lambda - i\delta) - R_0(\lambda + i\delta)\big) \Big\}. \tag{18}$$

Here we have used that ima $E_0(\sigma) A^* \subseteq P_0^{ac}\mathcal{H}_0$, ima $E_0(\sigma) J^*B^* \subseteq P_0^{ac}\mathcal{H}_0$ and $AE_0(\sigma) = AE_0(\sigma) P_0^{ac}$. We shall prove that in this expression we can replace w-lim by s-lim and that (18) agrees with the right side of (17). From the $H_0$-smoothness of $AE_0(\sigma)$ and the $H$-smoothness of $BE(\sigma)$ we obtain the existence of

$$2i\pi\big(AE_0(\sigma)\, E_0(d\lambda)\, f\big)/d\lambda := \text{s-lim}_{\delta\to+0} AE_0(\sigma) \big(R_0(\lambda - i\delta) - R_0(\lambda + i\delta)\big) f,$$

$$BR(\lambda + i0)\, B^*g := \text{s-lim}_{\delta\to+0} BR(\lambda + i\delta)\, B^*g, \qquad f \in \mathcal{H}_0, \quad g \in \mathcal{K},$$

for almost all $\lambda \in \sigma$ (mod $|\cdot|$). Furthermore, note that this is also a $L^2(\mathbb{R}, d\lambda; \mathscr{K})$-convergence (this fact follows from the smoothness definition). Then it follows from Lemma 10 and the uniform boundedness of $\|BR(\lambda + i\varepsilon) B^*\|$ for $\varepsilon > 0$ and $\lambda \in \sigma$ that (18), with s-lim in place of w-lim, exists and gives (17). ∎

From (17) the following formula for the matrix elements of $S - P_0^{ac}$ is obtained: Let $\Delta \subset \sigma$ and $u, v \in \mathscr{H}_0$. Then

$$\left(u, E_0(\Delta)(S - P_0^{ac}) v\right) = -2i\pi \left\{ \int_\Delta \left( (BJE_0(\sigma) E_0(d\lambda) u)/d\lambda, (AE_0(\sigma) E_0(d\lambda) v)/d\lambda \right) d\lambda \right.$$
$$\left. + \int_\Delta \left( (AE_0(\sigma) E_0(d\lambda) u)/d\lambda, BR(\lambda + i0) B^* \left( (AE_0(\sigma) E_0(d\lambda) v)/d\lambda \right) \right) d\lambda \right\}.$$

Now recall the definitions and properties of $\mathscr{M}$, $D_\lambda$, $F_C(\lambda)$ and the direct integral representation $L^2(\Delta_0, d\lambda; \mathscr{H}_\lambda; \mathscr{S})$ of $P_0^{ac}\mathscr{H}_0$ with respect to $H_0$ (see Section 1.4). If we assume that ima $E_0(\sigma) A^* \subseteq \mathscr{M}$ and ima $E_0(\sigma) J^*B^* \subseteq \mathscr{M}$, then it is not difficult to prove that $F_{AE_0(\sigma)}(\lambda) \in \mathscr{L}(\mathscr{K}, \mathscr{H}_\lambda)$ and $F_{BJE_0(\sigma)}(\lambda) \in \mathscr{L}(\mathscr{K}, \mathscr{H}_\lambda)$ for almost all $\lambda \in \sigma$ (mod $|\cdot|$) (for the proof proceed similar as in the proof of Lemma 6 and use the smoothness properties of $AE_0(\sigma)$, $BJE_0(\sigma)$; in case $L^2(\Delta_0, d\lambda; \mathscr{H}_\lambda; \mathscr{S}) = L^2(\Delta_0, d\lambda; \mathscr{K}_0)$ Lemma 17.7 provides a proof). Thus $F_{AE_0(\sigma)}(\lambda)^* D_\lambda u = (AE_0(\sigma) E_0(d\lambda) u)/d\lambda$ and $F_{BJE_0(\sigma)}(\lambda)^* D_\lambda u = (BJE_0(\sigma) E_0(d\lambda) u)/d\lambda$. Using these properties and the separability of $\mathscr{S}$, we can reproduce (12) from the formula above for $(u, E_0(\Delta)(S - P_0^{ac}) v)$:

$$\hat{T}(\lambda) = -2i\pi\{F_{BJE_0(\sigma)}(\lambda) F_{AE_0(\sigma)}(\lambda)^* + F_{AE_0(\sigma)}(\lambda) BR(\lambda + i0) B^* F_{AE_0(\sigma)}(\lambda)^*\}. \quad (19)$$

### 18.2.2. A special case in the one-space theory

In this section we assume that $\mathscr{H}_0 = \mathscr{H}$, $J = 1$ and $V = A^*C A$ where $A \in \mathscr{L}(\mathscr{H}_0, \mathscr{K})$ and $C \in \mathscr{L}(\mathscr{K})$. Further, we assume that $A$ is $H_0$-smooth and moreover, $\sup_{\lambda \in \mathbb{R}, \varepsilon > 0} \|AR_0(\lambda + i\varepsilon) A^*\| < \infty$. Finally, we assume that $\text{n-lim}_{\varepsilon \to +0} AR_0(\lambda + i\varepsilon) A^*$ exists a.e. mod $|\cdot|$. Because of

$$CAR(\lambda + i\varepsilon) A^* = (1 - CAR_0(\lambda + i\varepsilon) A^*)^{-1} - 1,$$

for the existence of $\text{n-lim}_{\varepsilon \to +0} CAR(\lambda + i\varepsilon) A^* =: CAR(\lambda + i0) A^*$ it is sufficient to assume that $\|C\|$ is sufficiently small. Under these assumptions Theorem 11 is applicable.

Further, we suppose that we have the special direct integral representation $L^2(\Delta_0, d\lambda; \hat{\mathscr{M}}_\lambda)$ of $P_0^{ac}\mathscr{H}_0$ with respect to $H_0$ where $\mathscr{M} = $ ima $A^*$ (the assumption that ima $A^*$ is generating with respect to $P_0^{ac}\mathscr{H}_0$ and $H_0$ is made for simplicity, in general it suffices to assume $\mathscr{M} \supseteq$ ima $A^*$ for the following assertions). Then $F_A(\lambda) = D_\lambda A^*$ is well-defined and belongs to $\mathscr{L}(\mathscr{K}, \hat{\mathscr{M}}_\lambda)$ a.e. mod $|\cdot|$ (see the remarks at the end of the preceding section). Thus we obtain formula (19) which can be written in the form

$$\hat{T}(\lambda) = -2i\pi F_A(\lambda) \left(C + CAR(\lambda + i0) A^*C\right) F_A(\lambda)^*. \quad (20)$$

Now we define $M_0(\lambda) := AE_0(d\lambda) A^*/d\lambda := (1/2i\pi) \left(AR_0(\lambda - i0) A^* - AR_0(\lambda + i0) A^*\right)$. Clearly, $M_0(\lambda)$ is selfadjoint and nonnegative for all $\lambda$ where it exists. $M_0(\lambda)$ is connected with $F_A(\lambda)$ by the following relation.

**Lemma 12.** *Suppose that* n-$\lim_{\delta\to+0} (1/2i\pi) [AR_0(\lambda - i\delta) A^* - AR_0(\lambda + i\delta) A^* = M_0(\lambda)$ *exists for all* $\lambda \in \Delta_1 \subseteqq \mathbb{R}$ *and* $F_A(\lambda) \in \mathcal{L}(\mathcal{K}, \mathcal{H}_\lambda)$ *for all* $\lambda \in \Lambda \subseteqq \mathbb{R}$. *Then* $F_A(\lambda)^* F_A(\lambda) = M_0(\lambda)$ *for all* $\lambda \in \Delta_1 \cap \Lambda$.

Proof. $F_A(\lambda)^* F_A(\lambda) u = F_A(\lambda)^* D_\lambda A^* u = (AE_0(\mathrm{d}\lambda) A^* u)/\mathrm{d}\lambda$ (see Lemma 6). Further, $(v, AE_0(\mathrm{d}\lambda) A^*u)/\mathrm{d}\lambda = (v, (AE_0(\mathrm{d}\lambda) A^*u)/\mathrm{d}\lambda)$ for all $v \in \mathcal{K}$ and $\lambda \in \Lambda$ ($\Lambda$ independent of $v$ and $u$). Now

$$(v, (AE_0(\mathrm{d}\lambda) A^*u)/\mathrm{d}\lambda = \lim_{\delta\to+0} (v, (1/2i\pi) A(R_0(\lambda - i\delta) - R_0(\lambda + i\delta)) A^*u)$$

for all $\lambda$ where the left side exists (see Prop. 3.16). This implies the desired relation. ∎

Now we can use the polar decomposition of the operators $F_A(\lambda)$ to get a representation of the scattering amplitude $\hat{T}(\lambda)$ in the Hilbert space $\mathcal{H}_0$. From Lemma 12 it follows that $F_A(\lambda) = (\operatorname{sgn} F_A(\lambda)) |F_A(\lambda)| = (\operatorname{sgn} F_A(\lambda)) M_0(\lambda)^{1/2}$ a.e. mod $|\cdot|$. $F_A(\lambda)$ is an operator from $\mathcal{K}$ onto $\hat{\mathcal{M}}_\lambda$ since $F_A(\lambda) = D_\lambda A^*$ and ima $A^* = \mathcal{M}$ is, by definition, dense in $\hat{\mathcal{M}}_\lambda$. Thus $U_\lambda := \operatorname{sgn} F_A(\lambda)$ maps clo (ima $M_0(\lambda)^{1/2}$) onto $\hat{\mathcal{M}}_\lambda$ a.e. mod $|\cdot|$. Then we obtain from (20) that

$$\hat{T}(\lambda) = -2i\pi U_\lambda M_0(\lambda)^{1/2} (C + CAR(\lambda + i0) A^*C) M_0(\lambda)^{1/2} U_\lambda^*$$

a.e. mod $|\cdot|$ and that $\hat{T}(\lambda)$ is unitarily equivalent to the operator $\tilde{T}(\lambda)$ on $\mathcal{H}_0$ given by

$$\tilde{T}(\lambda) = -2i\pi M_0(\lambda)^{1/2} (C + CAR(\lambda + i0) A^*C) M_0(\lambda)^{1/2}. \tag{21}$$

The advantage of this formula is that it is independent of any special direct integral representation of $P_0^{ac}\mathcal{H}_0$ with respect to $H_0$. (21) contains only limits of "sandwiched" resolvents. Thus, if we are interested in assertions on $\hat{T}(\lambda)$ which are unitarily invariant, then it is senseful to use formula (21). $\tilde{T}(\lambda)$ can be regarded as the scattering amplitude pulled back into $\mathcal{H}_0$.

## 18.3. Analyticity properties of the scattering amplitude and resonances

### 18.3.1. General remarks

The interest of physicists in analyticity properties of the scattering amplitude came from the more or less tight connection between analyticity and causality ("more or less tight" refers to the different physical theories). For instance, in the Lax-Phillips theory (Chapter 12) analyticity of the scattering amplitude arises naturally from the basic assumptions reflecting the causality of the corresponding system. In this paragraph we present two abstract results on the analyticity of $\hat{T}(\lambda)$; in both cases analyticity properties of sandwiched resolvents $C_1R_0(z) C_2$ are used.

Another interesting point is the appearance of nonreal poles in the analytic continuation of the scattering amplitude $\hat{T}(\lambda)$ of the scattering system $\{H, H_0; J\}$. These poles are called *resonances* and one can sometimes prove that they agree with the poles of a certain sandwiched resolvent (called *virtual poles*). These virtual poles appear when $H_0$ has eigenvalues embedded in its absolutely continuous part and they are unstable with respect to the perturbation $V = HJ - JH_0$. The motivation for the study of such scattering systems comes from the following facts:

The measurement of cross sections, for example for nuclear reactions, shows that sometimes cross sections have sharp maxima near certain "resonance" energies $\lambda_0$ and that they can be represented in the simple form $\sigma(\lambda) \approx c\big((\lambda - \lambda_0)^2 + \Gamma^2\big)^{-1}$ near $\lambda_0$. The usual physical interpretation of these maxima is that during the scattering process the particles form a weakly unstable system, a so-called compound system with the bound energy $\lambda_0$, which subsequently decays. Now the idea is to introduce a suitable selfadjoint operator $H_\infty$, which has the resonance energies as eigenvalues, and to study the scattering system $\{H, H_\infty; J\}$ to get a better understanding of the behavior of the scattering amplitude for the original scattering system $\{H, H_0; J\}$. The results in this direction are very meager and we restrict ourselves to the discussion of a special model.

### 18.3.2. Analytic continuations of $\hat{T}(\lambda)$

For simplicity we consider the one-space case: $\mathscr{H} = \mathscr{H}_0$, $J = 1$. We assume that $P_0^{ac} H_0$ has constant multiplicity, and thus there is a direct integral representation $L^2(\Delta_0, \mathrm{d}\lambda; \mathscr{K}_0)$ of $P_0^{ac}\mathscr{H}_0$ with respect to $H_0$. Further we assume that $H - H_0 = V = A^*CA$ where $A \in \mathscr{C}(\mathscr{H}_0, \mathscr{F})$, $C \in \mathscr{L}(\mathscr{F})$ and $\mathscr{F}$ is some separable Hilbert space. With respect to $L^2(\Delta_0, \mathrm{d}\lambda; \mathscr{K}_0)$ we choose as in Section 1.4 a spectral manifold $\mathscr{M} \subseteq P_0^{ac}\mathscr{H}_0$ and assume without loss of generality that $D_\lambda \colon \mathscr{M} \to \mathscr{K}_0$ for all $\lambda \in \Delta_0$ such that $D_\lambda k = \hat{k}(\lambda)$, $k \in \mathscr{M}$, where $\hat{k}(\cdot)$ is the representer of $k$ with respect to $L^2(\Delta_0, \mathrm{d}\lambda; \mathscr{K}_0)$. In the following we fix an interval $(a, b) \subset \mathbb{R}$ and set $Q := E_0\big((a, b)\big)\,(P_0^{\,p})^\perp$. Then the main assumptions of this section are:

(I) $AQ$ is $H_0$-smooth and $AQR_0(z)\,A^*$, $\operatorname{Im} z > 0$, is analytic continuable into a region $\mathscr{G} \supset (a, b)$ of the complex plane.

(II) ima $QA^* \subseteq \mathscr{M}$ (thus $F_{AQ}(\lambda) \in \mathscr{L}(\mathscr{F}, \mathscr{K}_0)$ for almost all $\lambda \in (a, b)$; see the remarks at the end of Section 2.1) and $F_{AQ}(\lambda) = D_\lambda QA^*$ is holomorphic on $(a, b)$ and has an analytic continuation into $\mathscr{G} \cup \overline{\mathscr{G}}$ ($\overline{\mathscr{G}} = \{z; \bar{z} \in \mathscr{G}\}$).

(III) For simplicity, we assume that $\operatorname{eig} H_0 \cap (a, b) = \{\lambda_1, \lambda_2, \ldots, \lambda_r\}$ with $r < \infty$ and that $P_0 := \sum_{s=1}^{r} E_0(\{\lambda_s\})$ is finite-dimensional. (Note that this is an essential difference to Section 2.2 where eigenvalues are excluded).

(IV) The operator-valued function

$$\Gamma_+(z) := \big(1 - CAP_0^{\perp} R_0(z)\, A^*\big)^{-1}, \qquad \operatorname{Im} z > 0; \tag{22}$$

is meromorphic continuable into $\mathscr{G} \setminus \big(\mathbb{R} \setminus (a, b)\big)$ (e.g. if $\|CP_0^{\perp} R_0(z)\, A^*\| < 1$).

Note that the assumption of the analytic continuability in (II) is much stronger than the requirement that $\big(QA^*g, E_0(\mathrm{d}\lambda)\, QA^*f\big)/\mathrm{d}\lambda = (D_\lambda QA^*g, D_\lambda QA^*f)$ is analytic continuable which is implied by (I).

Now we have

$$\Phi_\pm(z) := CAR_0(z)\, A^* = \sum_{s=1}^{r} (z - \lambda_s)^{-1}\, CAE_0(\{\lambda_s\})\, A^* + CAP_0^{\perp}(z - H_0)^{-1} A^* \tag{23}$$

for $\operatorname{Im} z \gtrless 0$. From (I) it follows that $H_0 \restriction \operatorname{do} \{E_0(\Delta) f; f \in \operatorname{ima} QA^*, \Delta \text{ Borel set}\}$ has no singular spectrum and moreover that $\Phi_+(z)$ is analytic continuable into the region $\mathscr{G}_0 := \mathscr{G} \setminus \{(\mathbb{R} \setminus (a, b)) \cup \{\lambda_1, \lambda_2, \ldots, \lambda_r\}\}$.

An easy calculation shows that

$$\Psi_+(z) := CAR(z) A^* = \Gamma_+(z) + \Gamma_+(z) CAP_0\{H_+(z)\}^{-1} P_0 A^* \Gamma_+(z) - 1, \quad \operatorname{Im} z > 0, \tag{24}$$

where $H_+(z) = (zP_0 - H_0P_0 - P_0A^*\Gamma_+(z) CAP_0) \restriction P_0\mathscr{H}_0$ is the so-called *Livšic matrix* and where

$$H_+(z)^{-1} = \{P_0(z - H)^{-1} P_0\} \restriction P_0\mathscr{H}_0 \tag{25}$$

is the so-called *partial resolvent*. From (IV), we easily get that $\Psi_+(z)$ is meromorphic continuable into the region $\mathscr{G}$. The poles of $\Psi_+(z)$ are called *virtual poles*. If $\Gamma_+(z)$ is even holomorphic continuable into $\mathscr{G}$ (cf. assumption (IV)), then the virtual poles are precisely the poles of $H_+(z)^{-1}$ or, equivalently, the zeros of the determinant of the Livšic matrix. Under additional assumptions the real virtual poles form precisely the eigenvalue spectrum of $H$ in $(a, b)$. Nonreal virtual poles are not excluded, of course, they have necessarily a negative imaginary part.

**Proposition 13.** *Let the assumptions* (I)—(IV) *be satisfied. Then the scattering amplitude* $\hat{T}(\lambda)$ (*with respect to* $L^2(\Delta_0, \mathrm{d}\lambda; \mathscr{K}_0)$) *is given by*

$$\hat{T}(\lambda) = -2\mathrm{i}\pi F_{AQ}(\lambda) \left(C + CAR(\lambda + \mathrm{i}0) A^*C\right) F_{AQ}(\lambda)^*, \qquad \lambda \in (a, b),$$

*is holomorphic on* $(a, b)$ *and meromorphic continuable into* $\mathscr{G}_0$. *The poles of* $\hat{T}(\lambda)$ *in* $\mathscr{G}_0$ *are precisely the nonreal virtual poles.*

**Proof.** The representation of $\hat{T}(\lambda)$ by the formula above follows from the results of Section 2.1. The holomorphy and the meromorphic continuability are implied by the corresponding properties of $\Psi_+(\lambda)$ and $F_{AQ}(\lambda)$. It remains to show that the nonreal poles of $\Psi_+(z)$ in $\mathscr{G}$ are also poles of $\hat{T}(z)$. For this it suffices to verify that the poles of $\Psi_+(z)$ are poles of

$$CF_{AQ}(\lambda)^* \hat{T}(\lambda) F_{AQ}(\lambda) = -2\mathrm{i}\pi CF_{AQ}(\lambda)^* F_{AQ}(\lambda) \left(1 + \Psi_+(\lambda)\right) CF_{AQ}(\lambda)^* F_{AQ}(\lambda).$$

Recall that $CF_{AQ}(\lambda)^* F_{AQ}(\lambda) = \operatorname*{n-lim}_{\delta \to +0} (1/2\mathrm{i}\pi) CAQ\{R_0(\lambda - \mathrm{i}\delta) - R_0(\lambda + \mathrm{i}\delta)\} A^*$ for almost all $\lambda \in [a, b]$ (mod $|\cdot|$) under our assumptions (see Lemma 12). Thus, by $\operatorname*{n-lim}_{\delta \to +0} Q^\perp(R_0(\lambda - \mathrm{i}\delta) - R_0(\lambda + \mathrm{i}\delta)) = 0$ for all $\lambda \in (a, b) \setminus \{\lambda_1, \lambda_2, \ldots, \lambda_r\}$, we obtain that

$$CF_{AQ}(\lambda)^* F_{AQ}(\lambda) = \operatorname*{n-lim}_{\delta \to +0} CA\{R_0(\lambda - \mathrm{i}\delta) - R_0(\lambda + \mathrm{i}\delta)\} A^* = \Phi_-(\lambda) - \Phi_+(\lambda)$$

for $\lambda \in (a, b)$ (mod $|\cdot|$). Thus, for $z \in \mathscr{G}$ with $\operatorname{Im} z < 0$,

$$CF_{AQ}(z)^* T(z) F_{AQ}(z) = \left(\Phi_-(z) - \Phi_+(z)\right) \left(1 + \Psi_+(z)\right) \left(\Phi_-(z) - \Phi_+(z)\right).$$

From $1 + \Psi_+(z) = \left(1 - \Phi_+(z)\right)^{-1}$ we obtain

$$\Phi_+(z) \Psi_+(z) = \Psi_+(z) \Phi_+(z) = \Psi_+(z) - \Phi_+(z). \tag{26}$$

Assume $\xi$ is a nonreal pole of $\Psi_+(z)$. Then

$$\Psi_+(z) = (z - \xi)^{-m} D_\xi + C(z), \qquad D_\xi \neq 0 \quad \text{and} \quad m \geq 1,$$

and $(z-\xi)^m\, C(z) \underset{n}{\to} 0$ as $|z-\xi| \to 0$. Then we obtain

$$\operatorname*{n-lim}_{z\to\xi} (z-\xi)^m \big(\Phi_-(z) - \Phi_+(z)\big)\big(1 + \Psi_+(z)\big)\big(\Phi_-(z) - \Phi_+(z)\big)$$

$$= \big(\Phi_-(\xi) - \Phi_+(\xi)\big)\, D_\xi \big(\Phi_-(\xi) - \Phi_+(\xi)\big).$$

From (26) we have $D_\xi = \Phi_+(\xi)\, D_\xi = D_\xi \Phi_+(\xi) = \Phi_+(\xi)\, D_\xi \Phi_+(\xi)$. Inserting this into the relation above we obtain

$$\operatorname*{n-lim}_{z\to\xi} (z-\xi)^m \big(\Phi_-(z) - \Phi_+(z)\big(1 + \Psi_+(z)\big)\big(\Phi_-(z) - \Phi_+(z)\big)$$

$$= \big(1 - \Phi_-(\xi)\big)\, D_\xi \big(1 - \Phi_-(\xi)\big)\big). \tag{27}$$

Now $\big(1 - \Phi_-(\xi)\big) = 1 - CAR_0(\xi)\, A^*$ since $\operatorname{Im} \xi < 0$. This operator-valued function is clearly bounded invertible, namely $\big(1 - \Phi_-(\xi)\big)^{-1} = 1 + CAR(\xi)\, A^*$ for $\operatorname{Im} \xi < 0$. Thus the operator on the right side of (27) is not the zero operator. This implies that $\hat{T}(z)$ at the point $\xi$ has a pole of the order $m \geqq 1$. ∎

### 18.3.3. An application to potential scattering

Let $\mathscr{H} = \mathscr{H}_0 = L^2(\mathbb{R}^3, dx)$, $J = 1$ and $H_0 = -\Delta$. Further, let $V = A^*CA$ where $A$ is multiplication by $\alpha(x) = e^{-d|x|}$, $d > 0$, $C$ is multiplication by a real-valued function $c(\cdot) \in L^\infty(\mathbb{R}^3, dx)$. Then $V$ is multiplication by $v(x) = e^{-2d|x|}c(x)$ and is selfadjoint. We want to show that the system $H, H_0$ with the factorization $H - H_0 = V = A^*CA$ satisfies the assumptions of Prop. 13. First we note that the complete strong wave operators $W_\pm(H, H_0)$ exist (see e.g. Section 15.2.4). Further we take the same manifold $\mathscr{M}$, the direct integral representation $L^2(\mathbb{R}_+, d\lambda; L^2(\Omega, d\omega))$ and $D_\lambda$ as in Section 1.5. Then it is easy to see that $F_A(\lambda) = D_\lambda A^* = D_\lambda A$ is analytic continuable into the region $\mathscr{G} = \{z; |\operatorname{Im} z^{1/2}| < d\}$. Next we verify that $\Phi_+(z)$ is analytic continuable into $\mathscr{G}$.

**Proposition 14.** *For* $\operatorname{Im} z \notin \mathbb{R}_+$ *the operator-valued function* $AR_0(z)\, A^* \in \mathscr{L}_2(\mathscr{H}_0)$ *and it is analytic continuable into* $\mathscr{G}$.

Proof. For $\operatorname{Im} z^{1/2} > 0$ the kernel of $AR_0(z)\, A^*$ is given by

$$(AR_0(z)\, A^*)\,(x, y) = -\frac{e^{iz^{1/2}|x-y|}}{4\pi\, |x-y|}\, e^{-d|x|} e^{-d|y|}$$

(see e.g. KATO [1, p. 495]). For $x \neq y$ the kernel is holomorphic on the Riemannian surface of $z^{1/2}$, and for $\operatorname{Im} z^{1/2} > -a_0 > -d$ we have the estimate

$$\int_{\mathbb{R}^3}\int_{\mathbb{R}^3} \|(AR_0(z)\, A^*)\,(x, y)\|^2_{L^2(\Omega, d\omega)}\, dx\, dy$$

$$\leqq (1/16\pi^2) \int_{\mathbb{R}^3}\int_{\mathbb{R}^3} \left\{\frac{e^{-2\operatorname{Im} z^{1/2}|x-y|}}{|x-y|^2}\, e^{-2d(|x|+|y|)}\, dx\, dy\right.$$

$$\leqq (1/16\pi^2) \int_{\mathbb{R}^3}\int_{\mathbb{R}^3} \frac{e^{-2(d-a_0)|y|}\, e^{-2(d-a_0)|x|}}{|x-y|^2}\, dx\, dy < \infty.$$

Now the assertion is an easy consequence of this estimate. ∎

From Prop. 14 it follows that $\big(1 - CAR_0(z)\, A^*\big)^{-1}$ is meromorphic continuable into the region $\mathscr{G}$ since $CAR_0(z)\, A^*$ has this property and is even a $\mathscr{L}_2(\mathscr{H}_0)$-valued analytic function. Now we can apply Prop. 13 and obtain that $\hat{T}(\lambda)$ is meromorphic continuable into the region $\{z; |\operatorname{Im} z^{1/2}| < d\}$ and that the poles of $\hat{T}(\lambda)$ (if any exist) agree with the poles of $CAR(z)\, A^*$.

### 18.3.4. The finite-dimensional Friedrichs model

First we note the assumptions of this model:

(I) $\operatorname{spec} H_0 = \operatorname{spec}_{ac} H_0 = \operatorname{spec}^c H_0 = [a, b]$, $P^{sc}_{H_0} = 0$ and $H_0 \restriction P^{ac}_0 \mathscr{H}_0$ has constant multiplicity on $[a, b]$. There is precisely one eigenvalue $\lambda_0 \in (a, b)$ of $H_0$, which has finite multiplicity $n$. We set $P_0 = E_0(\{\lambda_0\})$.

(II) Let $\Gamma: P_0\mathscr{H}_0 \to P_0{}^\perp\mathscr{H}_0$ be an operator with $\dim(\operatorname{ima} \Gamma) = n$. The extension of $\Gamma$ onto $\mathscr{H}_0$ given by $\Gamma \restriction P_0{}^\perp\mathscr{H}_0 = 0$ is also denoted by $\Gamma$. We assume that ima $\Gamma$ is a generating manifold with respect to $P^{ac}_0\mathscr{H}_0$ and $H_0$, that is, $P^{ac}_0\mathscr{H}_0 = \operatorname{clo} \operatorname{spa} \{E_0(\Delta) f;$ $f \in \operatorname{ima} \Gamma, \Delta \subset \mathscr{B}_{\mathbb{R}}\}$. Obviously ima $\Gamma$ is spectral and $B(\lambda) := (\Gamma^* E_0(d\lambda) \Gamma)/d\lambda$ exists for almost all $\lambda \in [a, b]$ (mod $|\cdot|$).

(III) The selfadjoint operator $H$ is defined by $H = H_0 + \Gamma + \Gamma^*$ on $\mathscr{H}_0$.
The system $\{H, H_0\}$ is called a *(finite-dimensional) Friedrichs model.*

From (II) it follows that $L^2([a, b], d\lambda; \hat{\mathscr{K}}_\lambda)$ with $\mathscr{K} = \operatorname{ima} \Gamma$ is a direct integral representation of $P^{ac}_0\mathscr{H}_0$ with respect to $H_0$ (see Chapter 4). In this case $\hat{\mathscr{K}}_\lambda = \mathscr{K} \operatorname{mod} (\ker |\cdot|_{E_0,\lambda})$ because the multiplicity of $H_0 \restriction P^{ac}_0\mathscr{H}_0$ turns out to be $n$. Since $H - H_0 = \Gamma + \Gamma^* \in \mathscr{L}_1(\mathscr{H}_0)$, the complete strong wave operators $W_\pm(H, H_0)$ exist. Now we calculate the scattering amplitude $\hat{T}(\lambda)$ with respect to $L^2([a, b], d\lambda; \hat{\mathscr{K}}_\lambda)$. First, we note that $F_{\Gamma^*}(\lambda) = D_\lambda \Gamma \in \mathscr{L}(\mathscr{H}_0, \hat{\mathscr{K}}_\lambda)$ a.e. mod $|\cdot|$. This is a simple consequence of the fact that ima $\Gamma = \mathscr{K}$ and ima $\Gamma$ is finite dimensional. Further, we have $VP^{ac}_0 = \Gamma^* P^{ac}_0 = \Gamma^*$ and $P^{ac}_0 V = P^{ac}_0 \Gamma = \Gamma$. Thus $P^{ac}_0 V P^{ac}_0 = 0$. Then Theorem 4 provides

$$S - P^{ac}_0 = \operatorname*{s-lim}_{\delta\to+0} \operatorname*{w-lim}_{\varepsilon\to+0} \left\{ -\int_{-\infty}^{\infty} E_0(d\lambda)\, \Gamma R(\lambda + i\varepsilon)\, \Gamma^* \big(R_0(\lambda - i\delta) - R_0(\lambda + i\delta)\big) \right\}$$

$$= \operatorname*{s-lim}_{\delta\to+0} \operatorname*{w-lim}_{\varepsilon\to+0} \left\{ -\int_{-\infty}^{\infty} E_0(d\lambda)\, \Gamma P_0 R(\lambda + i\varepsilon) P_0 \Gamma^* \big(R_0(\lambda - i\delta) - R_0(\lambda + i\delta)\big) \right\}. \quad (28)$$

Using that $\operatorname*{n-lim}_{\varepsilon\to+0} P_0 R(\lambda + i\varepsilon) P_0 =: P_0 R(\lambda + i0) P_0$ exist a.e. mod $|\cdot|$ ($\dim P_0\mathscr{H}_0 < \infty$) and that $\operatorname*{s-lim}_{\delta\to+0} \Gamma^* R_0(\lambda \pm i\delta) f$ exists a.e. mod $|\cdot|$, we can make the same conclusions as in the proof the Theorem 7 and Corollary 9. This leads to the formula

$$\hat{T}(\lambda) = -2i\pi F_{\Gamma^*}(\lambda)\, P_0 R(\lambda + i0)\, P_0 F_{\Gamma^*}(\lambda)^* \quad (29)$$

for the scattering amplitude $\hat{T}(\lambda)$. Note that $P_0 R(\lambda + i0) P_0$ is the partial resolvent introduced in Section 3.2., that is,

$$P_0 R(z) P_0 \restriction P_0\mathscr{H}_0 = H(z)^{-1}$$

where the Livšic matrix $H(z)$ in this case is given by

$$H(z) = z - \lambda_0 - \int_a^b (z - x)^{-1} B(x)\, dx = z - \lambda_0 - \Gamma^*(z - H_0)^{-1} \Gamma. \quad (30)$$

Because $H_0 \upharpoonright P_0^{ac}\mathscr{H}_0$ is of constant multiplicity $n$, the direct integral representation of $P_0^{ac}\mathscr{H}_0$ with respect to $H_0$ can also be given in the form $L^2([a, b], d\lambda; \mathscr{N})$ with a fixed Hilbert space $\mathscr{N}$ of dimension $n$. According to the remarks at the beginning of Section 1.4 we may assume that $F_{\Gamma^*}(\lambda) = D_\lambda\Gamma \in \mathscr{L}(\mathscr{H}_0, \mathscr{N})$ a.e. mod $|\cdot|$. Now the decisive assumption for analyticity assertions is:

(IV) The operator-valued function $F_{\Gamma^*}(\lambda) = D_\lambda\Gamma: \mathscr{H}_0 \to \mathscr{N}$ is holomorphic continuable from $[a, b]$ into a region $\mathscr{G} \supset [a, b]$ of the complex plane which is symmetric to the real line.

**Proposition 15.** *If assumptions* (I)—(IV) *are satisfied then $H(z)$ is holomorphic continuable (from $\mathbb{C}_+$) into the region $\mathbb{C}_+ \cup \mathscr{G} \setminus (\mathbb{R} \setminus [a, b])$.*

Proof. It follows from

$$\left(u, \Gamma^*(z - H_0)^{-1} \Gamma v\right) = \left(\Gamma u, (z - H_0)^{-1} \Gamma v\right) = \int_a^b (z - x)^{-1} \left(F_{\Gamma^*}(x)\, u, F_{\Gamma^*}(x)\, v\right)_{\mathscr{N}} dx$$

and the well-known fact, that holomorphy of $\left(F_{\Gamma^*}(x)\, u, F_{\Gamma^*}(x)\, v\right)_{\mathscr{N}}$ implies the holomorphic continuation of the Cauchy integral above. ■

From this statement immediately follows that $H(z)^{-1}$ is meromorphic continuable into the region $\mathbb{C}_+ \cup \mathscr{G} \setminus (\mathbb{R} \setminus (a, b])$. In our case the virtual poles defined in Section 3.2 coincide with the zeros of $\det H(z)$. For sufficiently small $\|\Gamma\|$ one can prove that $\det \{H(z)\}$ has zeros because of $\det H(z) = (z - \lambda_0)^n$ in case $\Gamma = 0$ (the proof for small $\|\Gamma\|$ is a simple application of Rouche's theorem). Thus there exist virtual poles connected with the eigenvalue $\lambda_0$ of $H_0$. Either directly from (29) or with the help of Prop. 13 we can easily prove

**Proposition 16.** *If assumptions* (I)—(IV) *are satisfied, then $\hat{T}(\lambda)$ is meromorphic continuable into $\mathscr{G} \setminus \{\mathbb{R} \setminus [a, b]\}$ and*

$$\hat{T}(z) = -2i\pi F_{\Gamma^*}(z)\, H(z)^{-1}\, F_{\Gamma^*}(\bar{z})^*.$$

*$\xi$ is a pole of $\hat{T}(z)$ in $\mathscr{G} \setminus \{\mathbb{R} \setminus [a, b]\}$ iff $\xi$ is a virtual pole.*

# 19. Spectral Properties of the Scattering Amplitude

## 19.1. Trace class conditions and scattering amplitude

### 19.1.1. Calculation of the scattering amplitude for trace class perturbations

We begin with a representation of $S - P_0^{ac}$ by a vector spectral integral.

**Proposition 1.** *Let $V := HJ - JH_0 \in \mathscr{L}_1(\mathscr{H}_0, \mathscr{H})$ and let $V = B^*A$ with $A \in \mathscr{L}_2(\mathscr{H}_0, \mathscr{K})$, $B \in \mathscr{L}_2(\mathscr{H}, \mathscr{K})$ ($\mathscr{K}$ some separable Hilbert space). Then there is a sequence $\Lambda_n$, $\Lambda_1 \subseteqq \Lambda_2 \subseteqq \cdots$ and $\bigcup_{n=1}^{\infty} \Lambda_n = \mathbb{R}$ (mod $|\cdot|$), such that for each $f \in \mathscr{M}_\infty(H_0)$*

$$(S - P_0^{ac})\, f = (-2i\pi)\ \text{s-lim}_{n\to\infty} \int_{\Lambda_n} E_0(d\lambda)\, \{J^*B^* + A^*BR(\lambda + i0)\, B^*\} \big(AE_0(d\lambda)\, f/d\lambda\big). \quad (1)$$

Proof. Recall that $BR(\lambda + i\delta)\, B^*$ is $\|\cdot\|_2$-convergent as $\delta \to +0$ a.e. (mod $|\cdot|$) with respect to $\lambda$ (see Prop. 3.14). Further recall Prop. 5.16 and the fact that $g_\eta(\lambda) = A\big(R_0(\lambda - i\eta) - R_0(\lambda + i\eta)\big)\, f$ is a member of $L^2(\mathbb{R}, d\lambda; \mathscr{K})$ for all $\eta > 0$. Now we consider the expression $A^*BR(\lambda + i\delta)B^*\, g_\eta(\lambda)$.

Set $\beta(\lambda) = \sup_{0<\delta\leqq 1} \|BR(\lambda + i\delta)\, B^*\|$ and $\alpha(\lambda) = \|AE_0(d\lambda)\, A^*)/d\lambda\|$. Then $\beta(\lambda) < \infty$ and $\alpha(\lambda) < \infty$ a.e. mod $|\cdot|$ (see Prop. 3.14). Define Borel sets $\Lambda_n$ by $\max\{\alpha(\lambda), \beta(\lambda)\} \leqq n$ and $\Lambda_n \cap \operatorname{spec}^c \big(H_0 \restriction (P_0^{ac}\mathscr{H}_0)^\perp\big) = \emptyset$ for all $n \in \mathbb{N}$. Then $E_0(\Lambda_n)\, E_0(\cdot)$ is an absolutely continuous spectral measure and we can apply Prop. 5.16. This proves the existence of the spectral integrals $\int_{\Lambda_n} E_0(d\lambda)\, A^*BR(\lambda + i0)\, B^*g_\eta(\lambda)$ and the relation

$$\text{s-lim}_{\delta\to+0} \int_{\Lambda_n} E_0(d\lambda)\, A^*BR(\lambda + i\delta)\, B^*g_\eta(\lambda) = \int_{\Lambda_n} E_0(d\lambda)\, A^*BR(\lambda + i0)\, B^*g_\eta(\lambda)$$

for all $n \in \mathbb{N}$ and $\eta > 0$. Now recall that $g_\eta(\lambda)$ tends strongly to $2i\pi\big(AE_0(d\lambda)\, f\big)/d\lambda$ a.e mod $|\cdot|$ and also with respect to $L^2(\mathbb{R}, d\lambda; \mathscr{K})$ as $\eta \to +0$ (see Prop. 3.14 and Remark 3.26). Thus the spectral integrals $\int_{\Lambda_n} E_0(d\lambda)\, A^*BR(\lambda + i0)\, B^*\big(\big(AE_0(d\lambda)\, f\big)/d\lambda\big)$ exist for all $n \in \mathbb{N}$ and we have

$$\begin{aligned}
&\text{s-lim}_{\eta\to+0}\ \text{s-lim}_{\delta\to+0} \int_{\Lambda_n} E_0(d\lambda)\, A^*BR(\lambda + i\delta)\, B^*g_\eta(\lambda)\\
&= 2i\pi \int_{\Lambda_n} E_0(d\lambda)\, A^*\big(BR(\lambda + i0)\, B^*\big)\, \frac{AE_0(d\lambda)\, f}{d\lambda}.
\end{aligned} \quad (2)$$

Similarly, introducing a corresponding sequence $\Lambda_n'$; we obtain

$$\text{s-lim}_{\eta\to+0} \int_{\Lambda_n'} E_0(d\lambda)\, J^*B^*g_\eta(\lambda) = 2i\pi \int_{\Lambda_n'} E_0(d\lambda)\, J^*B^*\big(\big(AE_0(d\lambda)\, f\big)/d\lambda\big). \quad (3)$$

Denote the sequence $\Lambda_n \cap \Lambda_n'$ again by $\Lambda_n$. Then the relations above are true for this new sequence.

From Theorem 18.4 it follows that

$$(S - P_0^{ac}) f = \operatorname*{s-lim}_{n\to\infty} E_0(\Delta_n) (S - P_0^{ac}) f$$
$$= \operatorname*{s-lim}_{n\to\infty} \operatorname*{s-lim}_{\delta\to+0} \operatorname*{w-lim}_{\varepsilon\to+0} \left\{ -\int_{\Delta_n} E_0(d\lambda) [J^*B^* + A^*BR(\lambda + i\delta) B^* g_\eta(\lambda)] \right\}.$$

Inserting (2) and (3) into the right side, we get (1). ■

From Prop. 1 one can derive a formula for the scattering amplitude $\hat{T}(\lambda)$. In order to prepare this formula, let us recall some facts on Hilbert-Schmidt and trace operators.

Let $\mathscr{H}_0 = L^2(\Delta_0, d\lambda; \mathscr{K}_0)$ and $A \in \mathscr{L}_2(\mathscr{H}_0)$. Then $A$ acts as an integral operator on $\mathscr{H}_0$,

$$(Af)(x) = \int_{\Delta_0} A(x, y) f(y) \, dy, \qquad f \in \mathscr{H}_0,$$

where $A(x, y)$ is a.e. (mod $|\cdot|$ in $\mathbb{R}^2$) defined and belongs to $\mathscr{L}_2(\mathscr{K}_0)$. Further, $\|A\|_2^2 = \int_{\Delta_0} \int_{\Delta_0} \|A(x, y)\|_2^2 \, dx \, dy$. These facts follow easily from the Schmidt representation of $A$, $Af = \sum_n \alpha_n(g_n, f) e_n$, $\sum_n |\alpha_n|^2 < \infty$, $\{g_n\}$ and $\{e_n\}$ orthonormal systems, the formula $(Af)(x) = \int_{\Delta_0} \sum_n \alpha_n(g_n(y), f(y))_{\mathscr{K}_0} e_n \, dy$ and the definition $A(x, y) := \sum_n \alpha_n(g_n(y), \cdot)_{\mathscr{K}_0} e_n(x)$, which implies

$$\|A(x, y)\|_2^2 = \sum_n \alpha_n^2 \|g_n(y)\|_{\mathscr{K}_0}^2 \|e_n(x)\|_{\mathscr{K}_0}^2.$$

Further, recall that if $A, B \in \mathscr{L}_2(\mathscr{H}_0)$, then the so-called iterated kernel $T(x, y)$ corresponds to the product $T = AB \in \mathscr{L}_1(\mathscr{H}_0)$

$$T(x, y) := (AB)(x, y) = \int_{\Delta_0} A(x, \lambda) B(\lambda, y) \, d\lambda.$$

Note that the kernel $T(x, y)$ is well-defined for all $\{x, y\} \in \Delta_0 \setminus \Lambda \times \Delta_0 \setminus \Lambda$ where $\Lambda$ is some null set, $|\Lambda| = 0$. In particular, the "diagonal elements" $T(x, x)$ are well-defined by this procedure on $\Delta_0 \setminus \Lambda$. We can also introduce the kernel $T(x, y)$ via the Schmidt representation of $T$, namely if $T = \sum_n \beta_n(g_n, \cdot) e_n$ ($\{g_n\}$, $\{e_n\}$ orthonormal systems, $\sum_n |\beta_n| < \infty$); then we have $T(x, y) = \sum_n \beta_n(g_n(y), \cdot) e_n(x)$. From this definition the relation

$$\int_{\Delta_0} \|T(x, x)\|_1 \, dx \leq \|T\|_1 \tag{4}$$

immediately follows; use the estimate $\int_{\Delta_0} \|T(x, x)\|_1 \, dx \leq \sum_n |\beta_n| \int_{\Delta_0} \|g_n(x)\| \, \|e_n(x)\| \, dx \leq \sum_n |\beta_n| \left( \int_{\Delta_0} \|g_n(x)\|^2 \, dx \right)^{1/2} \left( \int_{\Delta_0} \|e_n(x)\|^2 \, dx \right)^{1/2}$. Another interesting relation is

$$\operatorname{tr} T = \int_{\Delta_0} \left( \operatorname{tr} T(x, x) \right) dx. \tag{5}$$

The proof follows from

$$\operatorname{tr} T = \sum_n (e_n, Te_n) = \sum_n \beta_n(g_n, e_n)$$
$$= \sum_n \beta_n \int_{\Delta_0} (g_n(x), e_n(x))_{\mathscr{K}_0} \, dx = \int_{\Delta_0} \sum_n \beta_n(g_n(x), e_n(x))_{\mathscr{K}_0} \, dx$$

and from the relation ($\{k_s\}$ denotes an orthonormal basis in $\mathscr{K}_0$)

$$\operatorname{tr} T(x, x) = \sum_s \left(k_s, T(x, x) k_s\right) = \sum_{s,n} \beta_n\left(g_n(x), k_s\right) \left(k_s, e_n(x)\right)$$
$$= \sum_n \beta_n \sum_s \left(g_n(x), k_s\right) \left(k_s, e_n(x)\right) = \sum_n \beta_n\left(g_n(x), e_n(x)\right)_{\mathscr{K}_0}.$$

**Proposition 2.** *Let $V = HJ - JH_0 \in \mathscr{L}_1(\mathscr{H}_0)$. Assume that $H_0$ has homogeneous multiplicity and that a spectral representation of $P_0^{ac}\mathscr{H}_0$ with respect to $H_0$ is given by $L^2(\Delta_0, d\lambda; \mathscr{K}_0)$. Then the scattering amplitude $\hat{T}(\lambda)$ (with respect to $L^2(\Delta_0, d\lambda; \mathscr{K}_0)$) belongs to $\mathscr{L}_1(\mathscr{K}_0)$ a.e.* mod $|\cdot|$ *and is given by*

$$\hat{T}(\lambda) = -2i\pi\left(P_0^{ac}[J^*V + V^*R(\lambda + i0) V] P_0^{ac}\right)(\lambda, \lambda) \quad \text{a.e. mod } |\cdot|. \tag{6}$$

Proof. First we introduce a special factorization $V = B^*A$ with $A \in \mathscr{L}_2(\mathscr{H}_0, P_0^{ac}\mathscr{H}_0)$, $B \in \mathscr{L}_2(\mathscr{H}, P_0^{ac}\mathscr{H}_0)$. Then the operators $AP_0^{ac}$, $P_0^{ac}J^*B^*$ and $P_0^{ac}A^*\left(BR(\lambda + i0) B^*\right)$ belong to $\mathscr{L}_2(P_0^{ac}\mathscr{H}_0)$ and can be represented by integral operators with certain kernels with respect to $L^2(\Delta_0, d\lambda; \mathscr{K}_0)$. Let $A(x, y)$ denote the kernel of $AP_0^{ac}$. Then $\left(AP_0^{ac}E_0(d\lambda) f\right)/d\lambda = \left(AE_0(d\lambda) f\right)/d\lambda$, $f \in P_0^{ac}\mathscr{H}_0$ is represented by $A(\cdot, \lambda) \hat{f}(\lambda)$ where $\hat{f}(\lambda)$ dentoes the representer of $f$ in $L^2(\Delta_0, d\lambda; \mathscr{K}_0)$. Therefore $P_0^{ac}\{J^*B^* + A^*BR(\lambda + i0) B^*\} \left(\left(AP_0^{ac}E_0(d\lambda) f\right)/d\lambda\right)$ is represented by

$$\int_\Delta \left(P_0^{ac}J^*B^* + P_0^{ac}A^*BR(\lambda + i0) B^*\right)(x, \varrho) A(\varrho, \lambda) \hat{f}(\lambda) d\varrho.$$

That is, we have to iterate the corresponding operators $P_0^{ac}J^*B^* + P_0^{ac}A^*BR(\lambda + i0) B^*$ and $AP_0^{ac}$ and obtain

$$\left(P_0^{ac}J^*VP_0^{ac} + P_0^{ac}V^*R(\lambda + i0) VP_0^{ac}\right)(x, \lambda) \hat{f}(\lambda)$$

for the representer of $P_0^{ac}\{J^*B^* + A^*BR(\lambda + i0) B^*\} \left(\left(AE_0(d\lambda) f\right)/d\lambda\right)$. Now recall that if $g = \int P_0^{ac}E_0(d\lambda) h(\lambda)$, then the representer $\hat{g}(\cdot)$ of $P_0^{ac}g$ with respect to the spectral representation $L^2(\Delta_0, d\lambda; \mathscr{K}_0)$ of $P_0^{ac}\mathscr{H}_0$ is given by $\hat{g}(\lambda) = \hat{h}(\lambda; \lambda)$, where $\hat{h}(\cdot; \lambda)$ denotes the representer of $P_0^{ac}h(\lambda)$ (see Section 5.2.4 and note that $L^2(\Delta_0, d\lambda; \mathscr{K}_0)$ is a special case of representations $L^2(\Delta, d\lambda; \hat{\mathscr{K}}_\lambda)$). Thus, using (1), we obtain that the representer of $Tf$, i.e. $\hat{T}(\lambda) \hat{f}(\lambda)$, is given by $-2i\pi \left(P_0^{ac}J^*VP_0^{ac} + P_0^{ac}V^*R(\lambda + i0) VP_0^{ac}\right)(\lambda, \lambda) \hat{f}(\lambda)$ a.e. mod $|\cdot|$. ■

### 19.1.2. A trace norm estimate of the scattering amplitude

Our aim in this section is to prove the simple estimate $\int_{\Delta_0} \|T(\lambda)\|_1 d\lambda \leq 2\pi \|V\|_1$. This estimate does not follow directly from (6). In order to derive it, we first prove an interesting lemma.

**Lemma 3.** *Let $A \in \mathscr{L}_1(P_0^{ac}\mathscr{H}_0)$ and let $L^2(\Delta_0, d\lambda; \mathscr{H}_\lambda; \mathscr{S})$ be a direct integral representation of $P_0^{ac}\mathscr{H}_0$ with respect to $H_0$ ($\mathscr{S}$ separable). Then*

$$-\infty < \int_{-\infty}^{\infty} (v, e^{itH_0}A\, e^{-itH_0}u) dt = 2\pi \int_{\Delta_0} \left(\hat{v}(\lambda), A(\lambda, \lambda) \hat{u}(\lambda)\right)_{\mathscr{H}_\lambda} d\lambda < \infty \tag{7}$$

*for all* $u, v \in \mathcal{M}_\infty(H_0)$, *where* $\hat{u}(\cdot)$, $\hat{v}(\cdot)$ *denote the corresponding representers in* $L^2(\Delta_0, d\lambda; \mathcal{H}_\lambda; \mathcal{S})$.

**Proof.** Let $u, v \in \mathcal{M}_\infty(H_0)$ and let $A = B^*C$ with $B, C \in \mathcal{L}_2(P_0^{ac}\mathcal{H}_0)$. Then (see Prop. 3.24)

$$\int_{-\infty}^{\infty} (v, e^{itH_0} A\, e^{-itH_0}u)\, dt = \int_{-\infty}^{\infty} (B\, e^{-itH_0}v, C\, e^{-itH_0}u)\, dt$$

$$= 2\pi \int_{-\infty}^{\infty} \big((BE_0(d\lambda)\, v)/d\lambda, (CE_0(d\lambda)\, u)/d\lambda\big)\, d\lambda,$$

and this integral is finite. Now we manipulate $\big((BE_0(d\lambda)\, v)/d\lambda, (CE_0(d\lambda)\, u)/d\lambda\big)$. Let $B(x, y)$, $C(x, y)$ denote the kernels of $B, C$ with respect to $L^2(\Delta_0, d\lambda; \mathcal{H}_\lambda; \mathcal{S})$.; the existence of these kernels is implied by the considerations on Hilbert-Schmidt operators on p. 401 and by the fact that $L^2_N(\Delta_0, d\lambda; \mathcal{H}_\lambda; \mathcal{S})$ is isomorphic to a direct sum $\bigtimes_{n=1} L^2(\Lambda_n, d\lambda; \mathcal{K}_n)$. Then the representers of $(BE_0(d\lambda)\, v)/d\lambda$, $(CE_0(d\lambda)\, u)/d\lambda$ are given by $B(\cdot, \lambda)\, \hat{v}(\lambda)$, $C(\cdot, \lambda)\, \hat{u}(\lambda)$. Hence

$$\big((BE_0(d\lambda)\, v)/d\lambda, (CE_0(d\lambda)\, u)/d\lambda\big) = \int_{\Delta_0} \big(B(\varrho, \lambda)\, \hat{v}(\lambda), C(\varrho, \lambda)\, \hat{u}(\lambda)\big)_{\mathcal{H}_\varrho}\, d\varrho$$

$$= \Big(\hat{v}(\lambda), \int_{\Delta_0} B(\lambda, \varrho)^*\, C(\varrho, \lambda)\, d\varrho\, \hat{u}(\lambda)\Big)_{\mathcal{H}_\lambda} = \big(\hat{v}(\lambda), A(\lambda, \lambda)\, \hat{u}(\lambda)\big)_{\mathcal{H}_\lambda}. \blacksquare$$

**Corollary 4.** *The operator-valued function* $A(\lambda) := A(\lambda, \lambda)$ *is uniquely determined* mod $|\cdot|$ *by* (7), *that is, if for all* $u, v \in \mathcal{M}_\infty(H_0)$ $\int_{\Delta_0} \big(\hat{v}(\lambda), A(\lambda)\, \hat{u}(\lambda)\big)_{\mathcal{H}_\lambda} d\lambda = 0$, *then* $A(\lambda) = 0$ mod $|\cdot|$.

**Proof.** $\|A(\lambda)\|_{\mathcal{H}_\lambda}$ is finite a.e. mod $|\cdot|$. Define Borel sets $\Lambda_n$ for each $n \in \mathbb{N}$ by $\|A(\lambda)\|_{\mathcal{H}_\lambda} \le n$. Then $\chi_{\Lambda_n}(\cdot)\, A(\cdot)\, \hat{u}(\cdot)$ is from $\mathcal{M}_\infty(H_0)$ if $u \in \mathcal{M}_\infty(H_0)$. Hence we obtain $\int_{\Lambda_n} \|A(\lambda)\, \hat{u}(\lambda)\|^2_{\mathcal{H}_\lambda}\, d\lambda = 0$, that is, $A(\lambda)\, \hat{u}(\lambda) = 0$ a.e. mod $|\cdot|$ on $\Lambda_n$ for all $u \in \mathcal{M}_\infty(H_0)$. Under our assumptions on $L^2(\Delta_0, d\lambda; \mathcal{H}_\lambda; \mathcal{S})$ (see the Introduction to this part) there is a countable set of vectors $u_n$ such that $\hat{u}_n(\lambda)$ is dense in $\mathcal{H}_\lambda$ a.e. mod $|\cdot|$. This implies that there also exists a countable set of vectors $v_n \in \mathcal{M}_\infty(H_0)$ with the same property. Thus $A(\lambda)\, \hat{u}(\lambda) = 0$ a.e. mod $|\cdot|$ implies $A(\lambda) = 0$ a.e. mod $|\cdot|$. ∎

Assume that the wave operators $W_\pm = W_\pm(H, H_0; J)$ exist even in the strong sense. Then from the definition of $W_\pm$ and $S - P_0^{ac} = W_+^*(W_- - W_+)$ we easily obtain, for all $u \in \operatorname{dom} H_0$ and $v \in \operatorname{dom} H_0 \cap P_0^{ac}\mathcal{H}_0$,

$$\big(u, (S - P_0^{ac})\, v\big) = -i \int_{-\infty}^{\infty} (u, e^{itH_0} W_+^* V\, e^{-itH_0}v)\, dt. \tag{8}$$

Clearly, this formula can be extended to all $u \in \mathcal{H}_0$, $v \in P_0^{ac}\mathcal{H}_0$ for which the right side exists.

Using formula (8), Lemma 3, and Corollary 4, we can prove

**Theorem 5.** *Let $V = HJ - JH_0 \in \mathscr{L}_1(\mathscr{H}_0, \mathscr{H})$ and let $L^2(\Delta_0, d\lambda; \mathscr{H}_\lambda; \mathscr{S})$ be a spectral representation of $P_0^{ac}\mathscr{H}_0$ with respect to $H_0$. Then the corresponding scattering amplitude $\hat{T}(\lambda)$ obeys the estimate*

$$\int_{\Delta_0} \|\hat{T}(\lambda)\|_1 \, d\lambda \leq 2\pi \|VP_0^{ac}\|. \tag{9}$$

Proof. Setting $A = W_+{}^*VP_0^{ac} \in \mathscr{L}_1(P_0^{ac}\mathscr{H}_0)$, we obtain from Lemma 3 and (8) that

$$\left(u, (S - P_0^{ac})\, v\right) = -2i\pi \int_{\Delta_0} \left(\hat{u}(\lambda), A(\lambda)\, \hat{v}(\lambda)\right)_{\mathscr{H}_\lambda} d\lambda.$$

On the other hand $\left(u, (S - P_0^{ac})\, v\right) = \int_{\Delta_0} \left(\hat{u}(\lambda), \hat{T}(\lambda)\, \hat{v}(\lambda)\right)_{\mathscr{H}_\lambda} d\lambda$. Hence, because of Corollary 4, we get $\hat{T}(\lambda) = -2i\pi A(\lambda, \lambda)$ a.e. mod $|\cdot|$. By (4), this implies (9). ∎

If $W_+$ is not necessarily partially isometric, then $\|W_+{}^*\| \leq \|J\|$. In this case (9) is to replace by $\int_{\Delta_0} \|\hat{T}(\lambda)\|_1 \, d\lambda \leq 2\pi \|J\| \, \|VP_0^{ac}\|_1$, where $T: = S - W_+^* W_+$.

Theorem 5 can be generalized to more general trace class perturbations.

**Proposition 6.** *Let $\alpha_1, \alpha_2$ be uniformly bounded from $\mathbb{R}$ to $\mathbb{C}$. Suppose $\alpha_1(H)\, V\alpha_2(H_0) \in \mathscr{L}_1(\mathscr{H}_0, \mathscr{H})$. Let $L^2(\Delta_0, d\lambda; \mathscr{H}_\lambda, \mathscr{S})$ be a spectral representation of $P_0^{ac}\mathscr{H}_0$ with respect to $H_0$. Then:*

(i) $\hat{T}(\lambda) \in \mathscr{L}_1(\mathscr{H}_\lambda)$ *for almost all $\lambda$ mod $|\cdot|$ for which $\alpha_1(\lambda)\, \alpha_2(\lambda) \neq 0$.*

(ii) $\int_{\Delta_0} |\alpha_1(\lambda)\, \alpha_2(\lambda)| \, \|\hat{T}(\lambda)\|_1 \, d\lambda \leq 2\pi \|\alpha_1(H)\, V\alpha_2(H_0)\, P_0^{ac}\|_1$.

Proof. It depends again on Lemma 3, Corollary 4 and (8), and it proceeds as the proof of Theorem 5. ∎

### 19.1.3. Example (potential scattering)

Let $\mathscr{H}_0 = L^2(\mathbb{R}^3, dx)$, $H_0 = -\Delta$ and $H = H_0 + V$ where $V$ is multiplication by the real-valued function $v(x)$ with $v \in L^1(\mathbb{R}^3, dx) \cap L^2(\mathbb{R}^3, dx)$. Then $V$ is $H_0$-bounded with bound zero and therefore dom $H_0$ = dom $H$ (see e.g. Kato [1, p. 302]). Further, let $v(x) = b(x)\, a(x)$ with $a(\cdot)$, $b(\cdot) \in L^2(\mathbb{R}^3, dx)$ and denote the corresponding operators by $B, A$. Thus the operators $A$ and $B$ satisfy the relation then $V = BA = AB$. The kernel of the integral operator $AR_0(\xi)$, Im $\xi^{1/2} > 0$, is given by (see Kato [1, p. 495])

$$(AR_0(\xi))\,(x, y) = -(4\pi\,|x - y|)^{-1} a(x) \exp\{i\xi^{1/2}\,|x - y|\}.$$

Hence

$$\iint dx\,dy\, |(AR_0(\xi))\,(x, y)|^2 = \iint |a(x)|^2\, (4\pi\,|x - y|)^{-2} \exp\{-2\,\mathrm{Im}\,\xi^{1/2}\,|x - y|\}\, dx\,dy < \infty,$$

that is, $AR_0(\xi) \in \mathscr{L}_2(\mathscr{H}_0)$ (one can easily prove this result also with the help of Lemma 16.8). Similarly, we get $BR_0(\xi)$, $R_0(\xi)\, B$, $R_0(\xi)\, A \in \mathscr{L}_2(\mathscr{H}_0)$. Since $R(\xi)\, B = R(\xi)\, (\xi - H_0)\, R_0(\xi)\, B$ and $R(\xi)\, (\xi - H_0) = ((\bar{\xi} - H_0)\, R(\bar{\xi}))^*$ is bounded (because of dom $H$ = dom $H_0$ = ima $R(\bar{\xi})$), this implies $R(\xi)\, B \in \mathscr{L}_2(\mathscr{H}_0)$ and therefore $R(\xi)\, BAR_0(\xi') = R(\xi)\, VR_0(\xi') \in \mathscr{L}_1(\mathscr{H}_0)$.

According to Prop. 16.5 and Remark 16.6, this ensures (take Im $\xi \neq 0$) that the wave operators $W_\pm(H, H_0)$, $W_\pm(H_0, H)$ exist and are complete. Furthermore, Prop. 6 is applicable, with $\alpha_1(\lambda) = (\xi - \lambda)^{-1}$ and $\alpha_2(\lambda) = (\bar{\xi} - \lambda)^{-1}$ for example. Thus, if we use the spectral representation $L^2(\mathbb{R}_+, d\lambda; L^2(\Omega, d\omega))$ of $\mathscr{H}_0$ with respect to $H_0$ where $\Omega$ is the unit sphere of $\mathbb{R}^3$ (see e.g. Section 18.1.5), then the corresponding scattering amplitude $\hat{T}(\lambda)$ is a trace operator from $\mathscr{L}_1(L^2(\Omega, d\omega))$

for almost all $\lambda \in \mathbb{R}_+$ (mod $|\cdot|$). Further, we have

$$\int_0^\infty ((\lambda - \operatorname{Re} \xi)^2 + (\operatorname{Im} \xi)^2)^{-1} \|T(\lambda)\|_1 \, d\lambda \leq 2\pi \|R(\xi) \, VR_0(\xi)\|_1 < \infty.$$

For example, let $\xi = i$. Because of $\sigma(\lambda) = (\pi/\lambda) \|\hat{T}(\lambda)\|_2^2 = -(2\pi/\lambda) \operatorname{Re} (\operatorname{tr} \hat{T}(\lambda)) \leq (2\pi/\lambda) \|\hat{T}(\lambda)\|_1$ the above inequality yields

$$\int_0^\infty \lambda(\lambda^2 + 1)^{-1} \sigma(\lambda) \, d\lambda < \infty,$$

which provides information on the behavior of $\sigma(\lambda)$ for $\lambda \to \infty$ (in particular, if $\sigma(\lambda)$ is uniformly continuous in $\lambda$, then $\sigma(\lambda) \to 0$ as $\lambda \to \infty$.

### 19.1.4. The spectral shift and the trace formula

In the present section we assume that $\mathscr{H} = \mathscr{H}_0$, $J = 1$ and $V = H - H_0 \in \mathscr{L}_1(\mathscr{H}_0)$. Our aim is to prove a theorem relating the notion of *perturbation determinant* and the notion of *spectral shift*. We begin with the definition of these notions.

The perturbation determinant $\Delta(z)$ (for $H_0$, $V$) is defined by

$$\Delta(z) := \det \big(1 - VR_0(z)\big), \qquad \operatorname{Im} z \neq 0. \tag{10}$$

From $\det (1 - AB) = \det (1 - BA)$ the relation $\Delta(z) = \overline{\Delta(\bar{z})}$ follows. Thus the formula

$$\begin{aligned}(-1/\pi) \arg \Delta(z) &= (-1/2i\pi) \big(\log \Delta(z) - \log \overline{\Delta(z)}\big) \\ &= (1/2i\pi) \big(\log \Delta(\bar{z}) - \log \Delta(z)\big) = (1/2i\pi) \log \big(\Delta(\bar{z})/\Delta(z)\big)\end{aligned}$$

is valid. For the expression $\Delta(\bar{z})/\Delta(z)$ we derive

**Lemma 7.** *The relation*

$$\Delta(\bar{z})/\Delta(z) = \det \Big(1 - \big(V + VR(z) \, V\big) \big(R_0(\bar{z}) - R_0(z)\big)\Big) \tag{11}$$

*is true for* $\operatorname{Im} z \neq 0$.

Proof. From the second resolvent equation $R(z) - R_0(z) = R_0(z) \, VR(z)$ it follows that $\big(1 - VR_0(z)\big) \big(1 + VR(z)\big) = 1 + V\big(R(z) - R_0(z) - R_0(z) \, VR(z)\big) = 1$. Thus $\det \big\{\big(1 - VR_0(z)\big) \big(1 + VR(z)\big)\big\} = \det \big(1 - VR_0(z)\big) \det \big(1 + VR(z)\big) = 1$ and this implies

$$\begin{aligned}\Delta(\bar{z})/\Delta(z) &= \det \big(1 - VR_0(\bar{z})\big) \det \big(1 + VR(z)\big) = \det \big\{\big(1 + VR(z)\big) \big(1 - VR_0(\bar{z})\big)\big\} \\ &= \det \{1 + VR(z) - VR_0(\bar{z}) - VR(z) \, VR_0(\bar{z})\}.\end{aligned} \tag{12}$$

Now we have

$$\big(V + VR(z) \, V\big) \big(R_0(z) - R_0(\bar{z})\big) = VR_0(z) + VR(z) \, VR_0(z) - VR_0(\bar{z}) - VR(z) \, VR_0(z)$$

$$= V[R_0(z) + R(z) \, VR_0(z)] - VR_0(\bar{z}) - VR(z) \, VR_0(\bar{z}) = VR(z) - VR_0(\bar{z}) - VR(z) \, VR_0(\bar{z}).$$

This equality and (12) lead to (11). ■

Now we define the spectral shift $\xi(\lambda)$ (for $H_0$, $V$) by

$$\xi(\lambda) := \lim_{\varepsilon\to+0} \{-(1/\pi) \arg \Delta(\lambda + \mathrm{i}\varepsilon)\}. \tag{13}$$

The following lemma states some properties of $\xi(\lambda)$.

**Lemma 8.** *The spectral shift $\xi(\lambda)$ exists a.e.* mod $|\cdot|$ *and can be calculated by*

$$\xi(\lambda) = (1/2\mathrm{i}\pi) \operatorname{tr} \log \{1 - 2\mathrm{i}\pi(1 + AR(\lambda + \mathrm{i}0) B^*) (AE_0(\mathrm{d}\lambda) B^*/\mathrm{d}\lambda)\} \tag{14}$$

*where* $V = B^*A$, $A, B \in \mathscr{L}_2(\mathscr{H}_0)$.

**Proof.** We have

$$(V + VR(z) V) (R_0(\bar{z}) - R_0(z)) = B^*(1 + AR(z) B^*) A(R_0(\bar{z}) - R_0(z))$$

and, by Lemma 7 and Prop. 1.8.,

$$\begin{aligned}(-1/\pi) \arg \Delta(z) &= (1/2\mathrm{i}\pi) \log \det \{1 - (V + VR(z) V) (R_0(\bar{z}) - R_0(z))\} \\ &= (1/2\mathrm{i}\pi) \operatorname{tr} \log \{1 - (V + VR(z) V) (R_0(\bar{z}) - R_0(z))\} \\ &= (1/2\mathrm{i}\pi) \operatorname{tr} \log \{1 - (1 + AR(z) B^*) (A(R_0(\bar{z}) - R_0(z))) B^*\}. \end{aligned} \tag{15}$$

Recall that $AR(\lambda + \mathrm{i}\varepsilon) B^*$ is $\mathscr{L}_2(\mathscr{H}_0)$-convergent as $\varepsilon \to +0$ a.e. mod $|\cdot|$, the limit is denoted by $AR(\lambda + \mathrm{i}0) B^*$, and that $A(R_0(\lambda - \mathrm{i}\varepsilon) - R_0(\lambda + \mathrm{i}\varepsilon)) B^*$ is $\mathscr{L}_1(\mathscr{H}_0)$-convergent to $2\mathrm{i}\pi(AE_0(\mathrm{d}\lambda) B^*)/\mathrm{d}\lambda$ a.e. mod $|\cdot|$ (see Prop. 3.14). Because of $\operatorname{tr} \log (1 - C) = \log \det (1 - C)$ and the continuity of $\det (1 - \cdot)$ with respect to $\|\cdot\|_1$ we obtain the existence of $\xi(\lambda)$ a.e. mod $|\cdot|$ and formula (14). ∎

The connection between $\xi(\lambda)$ and $\Delta(z)$ contained in definition (13) leads to the following statement.

**Proposition 9.** *The representation*

$$\log \Delta(z) = \int_{-\infty}^{\infty} (z - \lambda)^{-1} \xi(\lambda) \,\mathrm{d}\lambda, \qquad \operatorname{Im} z \neq 0, \tag{16}$$

*is valid.*

**Remark 10.** Note that $\Delta(z) \neq 0$ if $\operatorname{Im} z \neq 0$. Further, we have $\|R_0(z)\| \to 0$ as $|\operatorname{Im} z| \to \infty$ because of $|\operatorname{Im} z| \, \|R_0(z)\| \leq 1$. Hence for $\operatorname{Im} z > 0$ it follows that $\|VR_0(z)\|_1 \to 0$ as $\operatorname{Im} z \to \infty$ and therefore $\det (1 - VR_0(z)) \to 1$ as $\operatorname{Im} z \to \infty$, that is, $\Delta(z) \to 1$ as $\operatorname{Im} z \to \infty$. This shows that $\log \Delta(z)$ is uniquely determined by the convention $\lim_{\operatorname{Im} z\to\infty} \log \Delta(z) = 0$.

**Proof of Prop. 9.** Let $V = \sum_n V_n$ where $V_n = v_n(f_n, \cdot) f_n$ and $\{f_n\}$ is an orthonormal system. Clearly, $\sum_n |v_n| = \|V\|_1 < \infty$. Set $H_1 = H_0 + V_1$, $H_2 = H_1 + V_2, \ldots$, and $\Delta_n(z) = \det (1 - V_n R_{H_{n-1}}(z))$. Then, using the second resolvent equality and $\det \{(1 - A) (1 - B)\} = \det (1 - A) \det (1 - B)$, we obtain

$$\det \left(1 - \left(\sum_{n=1}^{N} V_n\right) R_0(z)\right) = \Delta_1(z) \, \Delta_2(z) \cdots \Delta_N(z).$$

Since $\lim_{N\to\infty} \det\left(1 - \left(\sum_{n=1}^{N} V_n\right) R_0(z)\right) = \det\left(1 - VR_0(z)\right)$ exists, the product $\prod_{n=1}^{\infty} \Delta_n(z)$ exists (for $\operatorname{Im} z \neq 0$) and

$$\log \det\left(1 - VR_0(z)\right) = \log \Delta(z) = \sum_{n=1}^{\infty} \log \Delta_n(z). \tag{17}$$

Now $\Delta_n(z) = 1 - v_n(f_n, R_{H_{n-1}}(z) f_n) = 1 - v_n \int (z - \lambda)^{-1} \sigma_n(d\lambda)$ where $\sigma_n(\Delta) = (f_n, E_{H_{n-1}}(\Delta) f_n)$. Therefore

$$\log \Delta_n(z) = \log \left\{1 - v_n \int (z - \lambda)^{-1} \sigma_n(d\lambda)\right\}.$$

Since $\operatorname{Im} \Delta_n(z) = \operatorname{Im} z)\, v_n \int \left((\lambda - \operatorname{Re} z)^2 + (\operatorname{Im} z)^2\right)^{-1} \sigma_n(d\lambda)$, we get, for $\operatorname{Im} z > 0$, either $\operatorname{Im} \Delta_n(z) > 0$ (for $v_n > 0$) or $\operatorname{Im} \Delta_n(z) < 0$ (for $v_n < 0$). Thus the imaginary part of $\log \Delta_n(z)$, which is $\arg \Delta_n(z)$, belongs either to the interval $(0, \pi)$ or to $(-\pi, 0)$. Furthermore, $|\operatorname{Im} z|\, |\Delta_n(z) - 1 + v_n/z| \to 0$ as $\operatorname{Im} z \to \infty$. Hence

$$|\operatorname{Im} z|\, |\log \Delta_n(z) + v_n/z| \to 0 \quad \text{as} \quad \operatorname{Im} z \to \infty.$$

For $z = iy$, $y \in \mathbb{R}$, this implies

$$|iy \log \Delta_n(iy) + v_n| \to 0 \quad \text{as} \quad y \to \infty,$$

and therefore $\sup_{y>0} |iy \log \Delta_n(iy)| < \infty$. Now we can apply a well-known theorem on the representation of holomorphic functions (in our case $\log \Delta_n(z)$ in the upper half plane (see e.g. Achieser/Glasmann [1, p. 192]). This provides

$$\log \Delta_n(z) = \int (z - \lambda)^{-1} \xi_n(\lambda)\, d\lambda$$

where $\xi_n(\lambda) = -(1/\pi) \arg \Delta_n(\lambda + i0)$, $\int |\xi_n(\lambda)|\, d\lambda = |v_n|$ and $\int \xi_n(\lambda)\, d\lambda = -v_n$. Thus, defining $\xi_0(\lambda)$ by

$$\xi_0(\lambda) = \sum_{n=1}^{\infty} \xi_n(\lambda), \tag{18}$$

where the convergence is a $L^1(\mathbb{R}, d\lambda)$-convergence, we obtain

$$\log \Delta(z) = \int (z - \lambda)^{-1} \xi_0(\lambda)\, d\lambda$$

and $\int |\xi_0(\lambda)|\, d\lambda = \|V\|_1$, $\int \xi_0(\lambda)\, d\lambda = -\operatorname{tr} V$. From (13), $\xi_0(\lambda) = \xi(\lambda)$ a.e. mod $|\cdot|$ follows. ■

In the course of the proof we have verified the relations

$$\int_{-\infty}^{\infty} |\xi(\lambda)|\, d\lambda = \|V\|_1, \quad \int_{-\infty}^{\infty} \xi(\lambda)\, d\lambda = -\operatorname{tr} V. \tag{19}$$

### 19.1.5. Phase shift and spectral shift

As in the preceding section we assume that $\mathscr{H}_0 = \mathscr{H}$, $J = 1$, and $V = H - H_0 \in \mathscr{L}_1(\mathscr{H}_0)$. Further, we assume that the multiplicity of $H_0$ in $P_0^{ac}\mathscr{H}_0$ is constant, and that $L^2(\Delta_0, d\lambda; \mathscr{K}_0)$ is a direct integral representation of $P_0^{ac}\mathscr{H}_0$ with respect to $H_0$. According to Prop. 2 the scattering matrix $\hat{S}(\lambda)$ with respect to $L^2(\Delta_0, d\lambda; \mathscr{K}_0)$ is given by

$$\hat{S}(\lambda) = 1_{\mathscr{K}_0} - 2i\pi\left(P_0^{ac}\left(V + VR(\lambda + i0)\, V\right) P_0^{ac}\right)(\lambda, \lambda),$$

where $\left(P_0^{ac}\left(V + VR(\lambda + i0)\, V\right) P_0^{ac}\right)(\lambda, \lambda')$ denotes the kernel of the trace operator $P_0^{ac}\left(V + VR(\lambda + i0)\, V\right) P_0^{ac} \in \mathscr{L}_1(P_0^{ac}\mathscr{H}_0)$ with respect to the spectral representation $L^2(\Delta_0, d\lambda; \mathscr{K}_0)$, taken on the diagonal line $\lambda = \lambda'$. Recall that this operator is well-defined a.e. mod $|\cdot|$. Thus det $\hat{S}(\lambda)$ is well-defined and $|\det \hat{S}(\lambda)| = 1$ a.e. mod $|\cdot|$ since $\hat{S}(\lambda)$ is unitary.

Now we define the *phase operator* $\Pi(\lambda)$ by

$$\Pi(\lambda) := (1/2i\pi) \log \left(1_{\mathscr{K}_0} - 2i\pi\left(P_0^{ac}(V + VR(\lambda + i0)\, V) P_0^{ac}\right)(\lambda, \lambda)\right) \tag{20}$$

(according to Dunford's calculus). Then

$$\hat{S}(\lambda) = \exp\{2i\pi \Pi(\lambda)\}$$

and

$$\det \hat{S}(\lambda) = \exp\{2i\pi \operatorname{tr} \Pi(\lambda)\}.$$

The *phase shift* $\delta(\lambda)$ is defined by

$$\begin{aligned}\delta(\lambda) &:= \operatorname{tr} \Pi(\lambda) = (1/2i\pi) \operatorname{tr} \log \left(1_{\mathscr{K}_0} + \hat{T}(\lambda)\right) \\ &= (1/2i\pi) \log \det \left(1_{\mathscr{K}_0} + \hat{T}(\lambda)\right) = (1/2i\pi) \log \det \hat{S}(\lambda).\end{aligned} \tag{21}$$

Thus $\det \hat{S}(\lambda) = \exp\{2i\pi\delta(\lambda)\}$.

Now it turns out that spectral shift and phase shift coincide. This is the content of the following famous theorem.

**Theorem 11.** *Let* $V = H - H_0 \in \mathscr{L}_1(\mathscr{H}_0)$. *Then*

$$\xi(\lambda) = \delta(\lambda) \quad \text{a.e. mod } |\cdot|. \tag{22}$$

Proof. Let $V = B^*A$ with $B, A \in \mathscr{L}_2(\mathscr{H}_0, P_0^{ac}\mathscr{H}_0)$. Set $M(\lambda) = AR(\lambda + i0)\, B^*$. Then from (21) and Lemma 8 we get that (22) is equivalent with

$$\begin{aligned}&\operatorname{tr} \log \left\{1_0 - 2i\pi(1 + M(\lambda)) \left((AE_0(d\lambda)\, B^*)/d\lambda\right)\right\} \\ &= \operatorname{tr} \log \left\{1_{\mathscr{K}_0} - 2i\pi\left(P_0^{ac}\left(V + VR(\lambda + i0)\, V\right) P_0^{ac}\right)(\lambda, \lambda)\right\}\end{aligned} \tag{23}$$

a.e. mod $|\cdot|$, where $1_0$ is the identity on $P_0^{ac}\mathscr{H}_0$. Note that $\left(AE_0(d\lambda)\, B^*\right)/d\lambda = \left(AE_0(d\lambda)\, P_0^{ac}B^*\right)/d\lambda$ a.e. mod $|\cdot|$. This follows from the fact that $\left(E_0(d\lambda)\,(P_0^{ac})^\perp f, f\right)/d\lambda$ exists and equals zero a.e. mod $|\cdot|$ (as derivative of a monotone function which contains no absolutely continuous part). Thus, in (23) we can replace $\left(AE_0(d\lambda)\, B^*\right)/d\lambda$ by $\left(AE_0(d\lambda)\, P_0^{ac}B^*\right)/d\lambda$.

In order to prove an equation of the form

$$\operatorname{tr} \log (1 - C) = \operatorname{tr} \log (1 - D); \qquad C \in \mathscr{L}_1(\mathscr{H}_0), \quad D \in \mathscr{L}_1(\mathscr{K}_0),$$

it is sufficient to prove (assume $1 \in \operatorname{res} C \cap \operatorname{res} D$)

$$\operatorname{tr} C^n = \operatorname{tr} D^n, \quad n \in \mathbb{N}. \tag{24}$$

Namely, $\operatorname{tr}\log(1-zC)$ and $\operatorname{tr}\log(1-zD)$ are holomorphic in a neighborhood of $z=0$ and the point 1 belongs to the region of holomorphy. If we can prove $\operatorname{tr}\log(1-zC) = \operatorname{tr}\log(1-zD)$ for some neighborhood of $z=0$, then this equality holds in the whole region of holomorphy. But $\operatorname{tr}\log(1-zC) = \operatorname{tr}\log(1-zD)$ follows from (24) (for small $z$) and $\log(1-zC) = \sum_{n\geq 1}(-z)^n C^n/n$.

Next we calculate

$$\operatorname{tr}\left(C_\lambda \frac{AP_0^{ac}E_0(\mathrm{d}\lambda)\,B^*}{\mathrm{d}\lambda}\right)^n = \operatorname{tr}\left\{\frac{\partial^n}{\partial x_1\ldots\partial x_n} C_\lambda AP_0^{ac}E_0(x_1)\,B^*\cdots C_\lambda AP_0^{ac}E_0(x_n)\,B^*\Big|_{x_j=\lambda}\right\}$$

$$= \frac{\partial^n}{\partial x_1\cdots\partial x_n}\operatorname{tr}\left\{\prod_{\varrho=1}^{n}\left(C_\lambda AP_0^{ac}E_0(x_\varrho)\,B^*\right)\right\}\Big|_{x_\varrho=\lambda} \tag{25}$$

where $C_\lambda = 2\mathrm{i}\pi\big(1+M(\lambda)\big)$. Note that the differentiation may be performed as $\mathscr{L}_1$-differentiation. Now we have

$$\operatorname{tr}\left\{\prod_{\varrho=1}^{n}\left(C_\lambda AE_0(x_\varrho)\,P_0^{ac}B^*\right)\right\} = \operatorname{tr}\left\{P_0^{ac}B^*\prod_{\varrho=1}^{n-1}\left(C_\lambda AE_0(x_\varrho)\,P_0^{ac}B^*\right)\left(C_\lambda AE_0(x_n)\,P_0^{ac}\right)\right\}$$

$$= \operatorname{tr}\left\{\prod_{\varrho=1}^{n}\left(P_0^{ac}B^*C_\lambda AP_0^{ac}E_0(x_\varrho)\right)\right\}.$$

Further, $P_0^{ac}B^*C_\lambda AP_0^{ac}\in\mathscr{L}_1(P_0^{ac}\mathscr{H}_0)$. The kernel of this operator (with respect to $L^2(\Delta_0, \mathrm{d}\lambda; \mathscr{K}_0)$ is denoted by $K_\lambda(\cdot,\cdot)$. Then the kernel of $P_0^{ac}B^*C_\lambda AP_0^{ac}E_0(\Delta_\varrho)$ is given by

$$K_{\lambda,\Delta_\varrho}(x,y) = K_\lambda(x,y)\,\chi_{\Delta_\varrho}(y),$$

and we obtain

$$\left(\prod_{\varrho=1}^{n}P_0^{ac}B^*C_\lambda AP_0^{ac}E_0(\Delta_\varrho)\right)(x,y)$$

$$= \int\int\cdots\int K_{\lambda,\Delta_1}(x,\eta_1)\,K_{\lambda,\Delta_2}(\eta_1,\eta_2)\cdots K_{\lambda,\Delta_n}(\eta_{n-1},y)\,\mathrm{d}\eta_1\,\mathrm{d}\eta_2\cdots\mathrm{d}\eta_{n-1}$$

$$= \int_{\Delta_1}\int_{\Delta_2}\cdots\int_{\Delta_{n-1}} K_\lambda(x,\eta_1)\,K_\lambda(\eta_1,\eta_2)\cdots K_\lambda(\eta_{n-1},y)\,\chi_{\Delta_n}(y)\,\mathrm{d}\eta_1\,\mathrm{d}\eta_2\cdots\mathrm{d}\eta_{n-1}.$$

Then

$$\operatorname{tr}\left\{\prod_{\varrho=1}^{n}\left(P_0^{ac}B^*C_\lambda AP_0^{ac}E_0(\Delta_\varrho)\right)\right\} = \int_{\Delta_n}\left\{\operatorname{tr}\left(\prod_{\varrho=1}^{n}\left(P_0^{ac}B^*C_\lambda AP_0^{ac}E_0(\Delta_\varrho)\right)\right)(x,x)\right\}\mathrm{d}x$$

$$= \int_{\Delta_1}\int_{\Delta_2}\cdots\int_{\Delta_n}\operatorname{tr}\left\{K_\lambda(x,\eta_1)\,K_\lambda(\eta_1,\eta_2)\cdots K_\lambda(\eta_{n-1},x)\right\}\mathrm{d}\eta_1\,\mathrm{d}\eta_2\cdots\mathrm{d}\eta_{n-1}\,\mathrm{d}x.$$

Therefore we obtain

$$\frac{\partial^n}{\partial x_1\,\partial x_2\cdots\partial x_n}\operatorname{tr}\left\{\prod_{\varrho=1}^{n}\left(P_0^{ac}B^*C_\lambda AP_0^{ac}E_0(x_\varrho)\right)\right\}\Big|_{x_\varrho=\lambda} = \operatorname{tr}K_\lambda(\lambda,\lambda)^n. \tag{26}$$

Now we have $P_0^{ac}B^*C_\lambda AP_0^{ac} = 2\mathrm{i}\pi P_0^{ac}\big(V+B^*M(\lambda)\,A\big)\,P_0^{ac} = P_0^{ac}\big(V+VR(\lambda+\mathrm{i}0)\,V\big)\,P_0^{ac}$. Thus (25), the definition of $K_\lambda(\lambda,\lambda)$ and (26) lead to

$$\operatorname{tr}\Big(\big(C_\lambda AP_0^{ac}E_0(\mathrm{d}\lambda)\,B^*\big)/\mathrm{d}\lambda\Big)^n = \operatorname{tr}\Big(2\mathrm{i}\pi\big(P_0^{ac}(V+VR(\lambda+\mathrm{i}0)\,V)\,P_0^{ac}\big)(\lambda,\lambda)\Big)^n$$

which concludes the proof of (22). ∎

**Remark 12.** One can easily extend this theorem to more general trace class perturbations. For example, if there is a function $\alpha(\cdot)$ such that $\alpha^{-1}(\cdot)$ exists and is admissible and $\alpha(H) - \alpha(H_0) \in \mathcal{L}_1(\mathcal{H}_0)$, then we can apply the invariance principle to the scattering systems $\{H, H_0\}$ and $\{\alpha(H), \alpha(H_0)\}$. In particular, it says that the scattering operators of these systems agree. Thus the corresponding scattering matrices $\hat{S}(\lambda)$ and $\hat{S}_\alpha(\lambda)$ are connected by $\hat{S}(\lambda) = \hat{S}_\alpha(\alpha(\lambda))$. Using this equation and applying Theorem 11 to $\{\alpha(H), \alpha(H_0)\}$, we get a relation between $\xi(\lambda)$ (for $\alpha(H_0), \alpha(H)$) and $\log \det \hat{S}(\lambda)$.

## 19.2. Time-falloff conditions and scattering amplitude

Throughout this paragraph we assume that the wave operators $W_\pm = W_\pm(H, H_0; J)$ exist in the strong sense.

### 19.2.1. Spectral properties of parts of $S - P_0^{ac}$

In the present section we use time-falloff assumptions on $V e^{-itH_0}$ (similarly as in Chapter 15) to prove that some parts of $S - P_0^{ac}$ are operators from $\mathcal{L}_p(\mathcal{H}_0)$. In particular, in what follows, we are interested in $p = 2$ (Hilbert-Schmidt operators) and $p = \infty$ (compact operators). First we need a preliminary lemma.

**Lemma 13.** *Let $G_t$, $t \in \Delta \subseteq \mathbb{R}$, be a Borel function from $\Delta$ into $\mathcal{L}_p(\mathcal{H}_0, \mathcal{H}')$, $1 \leq p \leq \infty$. Let $H'$ be a selfadjoint operator on the separable Hilbert space $\mathcal{H}'$. Then $G_1(t) = e^{itH'}G_t$ and $G_2(t) = G_t e^{-itH_0}$ are Borel functions from $\Delta$ into $\mathcal{L}_p(\mathcal{H}_0, \mathcal{H}')$.*

Proof. For $1 \leq p \leq \infty$, $\mathcal{L}_p(\mathcal{H}_0, \mathcal{H}')$ is a separable Banach space (for a proof, use that $\mathcal{L}_{fin}(\mathcal{H}_0, \mathcal{H}')$ is dense in $\mathcal{L}_p(\mathcal{H}_0, \mathcal{H}')$ with respect to $\|\cdot\|_p$). Thus (see Section 1.2.1) it is sufficient to prove that $e^{itH'}G_t$ and $G_t e^{-itH_0}$ are weakly $|\cdot|$-measurable. That is, we have to verify that $\operatorname{tr}\{X e^{itH'}G_t\}$ and $\operatorname{tr}\{XG_t e^{-itH_0}\}$ are Borel measurable where $X \in \mathcal{L}_q(\mathcal{H}', \mathcal{H}_0)$ and $q^{-1} + p^{-1} = 1$ (if $p = \infty$, then $q = 1$; if $p = 1$, then $X \in \mathcal{L}(\mathcal{H}', \mathcal{H}_0)$). Now let $\{f_n\}$ denote an orthonormal basis in $\mathcal{H}_0$. Then

$$\operatorname{tr}\{X e^{itH'}G_t\} = \sum_n (f_n, X e^{itH'}G_t f_n) = \sum_n (e^{-itH'}X^* f_n, G_t f_n).$$

Since $e^{-itH'}X^*f_n$ and $G_t f_n$ are Borel functions, their scalar product is Borel measurable and so is the sum $\sum_n \{\cdot\}$; i.e. $\operatorname{tr}\{X e^{itH'}G_t\}$ is Borel measurable. Similarly we can prove that $\operatorname{tr}\{XG_t e^{-itH_0}\}$ is Borel measurable. ■

Now we are in a position to prove some simple $\mathcal{L}_p$-properties of $S - P_0^{ac}$.

**Proposition 14.** *Let $\alpha_1, \alpha_2: \mathbb{R} \to \mathbb{C}$ be bounded Borel functions, and let $P \in \mathcal{L}(\mathcal{H}_0)$ with $P = P_0^{ac}P = PP_0^{ac}$ and $\alpha_2(H_0)\, P(\operatorname{dom} H_0) \subseteq \operatorname{dom} H_0 \cap P_0^{ac}\mathcal{H}_0$. Suppose that $G_t$ defined by $G_t := W_+^*\alpha_1(H)\, V\alpha_2(H_0)\, e^{-itH_0}P$ is a Borel function from $\mathbb{R}$ into $\mathcal{L}_p(\mathcal{H}_0)$, $1 \leq p \leq \infty$ such that*

$$\int_{-\infty}^{\infty} \|G_t\|_p \, dt < \infty. \tag{27}$$

*Then $\alpha_1(H_0)\,(S - P_0^{ac})\,\alpha_2(H_0)\, P \in \mathcal{L}_p(\mathcal{H}_0)$.*

**Proof.** If $u, v \in \operatorname{dom} H_0$, then $\alpha_1(H_0)^* u \in \operatorname{dom} H_0$ and $\alpha_2(H_0) Pv \in \operatorname{dom} H_0 \cap P_0^{ac}\mathscr{H}_0$. Thus in (8) we can replace $u$ and $v$ by $\alpha_1(H_0)^* u$ and $\alpha_2(H_0) Pv$. This gives

$$\left(u, \alpha_1(H_0)\,(S - P_0^{ac})\,\alpha_2(H_0)\,Pv\right) = -\mathrm{i}\int \left(u, \mathrm{e}^{\mathrm{i}tH_0}W_+^*\alpha_1(H)\,V\alpha_2(H_0)\,\mathrm{e}^{-\mathrm{i}tH_0}Pv\right)\mathrm{d}t$$

for all $u, v \in \operatorname{dom} H_0$. From Lemma 13 it follows that $\mathrm{e}^{\mathrm{i}tH_0}G_t$ is a Borel function from $\mathbb{R}$ into $\mathscr{L}_p(\mathscr{H}_0)$. By (27), $\mathrm{e}^{\mathrm{i}tH_0}G_t$ is Bochner integrable (with respect to $\|\cdot\|_p$) and the integral $\int \mathrm{e}^{\mathrm{i}tH_0}G_t\,\mathrm{d}t$ defines an operator $D \in \mathscr{L}_p(\mathscr{H}_0)$. Thus $\left(u, \alpha_1(H_0)\,(S - P_0^{ac})\,\alpha_2(H_0)\,Pv\right) = (u, -\mathrm{i}Dv)$ for all $u, v \in \operatorname{dom} H_0$. This implies the desired assertion. ∎

**Remark 15.** A natural condition which ensures that $G_t$ is a Borel function from $\mathbb{R}$ into $\mathscr{L}_p(\mathscr{H}_0)$ is given by $\alpha_1(H)\,V\alpha_2(H_0) \in \mathscr{L}_p(\mathscr{H}_0, \mathscr{H})$. The integrability of $\|G_t\|_p$ can often be obtained with the help of a suitable operator $P$.

**Proposition 16.** *Let $\alpha\colon \mathbb{R} \to \mathbb{C}$ be a bounded Borel function, and let $P_+, P_- \in \mathscr{L}(\mathscr{H}_0)$ such that $P_+ = P_0^{ac}P_+ = P_+P_0^{ac}$, $P_- = P_0^{ac}P_- = P_-P_0^{ac}$, $\alpha(H_0)\,P_+(\operatorname{dom} H_0) \subseteq \operatorname{dom} H_0 \cap P_0^{ac}\mathscr{H}_0$ and $\alpha(H_0)\,P_-(\operatorname{dom} H_0) \subseteq \operatorname{dom} H_0 \cap P_0^{ac}\mathscr{H}_0$. Suppose that $G_t^+$ and $G_t^-$, defined by $G_t^+ := V\alpha(H_0)\,\mathrm{e}^{-\mathrm{i}tH_0}P_+$ and $G_t^- := V\alpha(H_0)\,\mathrm{e}^{-\mathrm{i}tH_0}P_-$, are Borel functions from $\mathbb{R}_+$ into $\mathscr{L}_p(\mathscr{H}_0, \mathscr{H})$ and $\mathbb{R}_-$ into $\mathscr{L}_p(\mathscr{H}_0, \mathscr{H})$* for some $p$, $1 \leq p \leq \infty$, *such that*

$$\int_0^\infty \|G_t^+\|_p\,\mathrm{d}t < \infty, \qquad \int_{-\infty}^0 \|G_t^-\|_p\,\mathrm{d}t < \infty. \tag{28}$$

*Further, suppose that $P_+^*\alpha(H_0)^*\,(J^*J - P_0^{ac})\,\alpha(H_0)\,P_- \in \mathscr{L}_p(\mathscr{H}_0)$. Then $P_+^*\alpha(H_0)^* \times (S - P_0^{ac})\,\alpha(H_0)\,P^- \in \mathscr{L}_p(\mathscr{H}_0)$.*

**Proof. 1.** Our starting point are the formulas

$$\left(u, (W_+ - J)\,P_0^{ac}v\right) = \mathrm{i}\int_0^\infty (u, \mathrm{e}^{\mathrm{i}tH}V\,\mathrm{e}^{-\mathrm{i}tH_0}v)\,\mathrm{d}t,$$

$$\left(u, (W_- - J)\,P_0^{ac}v\right) = -\mathrm{i}\int_{-\infty}^0 (u, \mathrm{e}^{\mathrm{i}tH}V\,\mathrm{e}^{-\mathrm{i}tH_0}v)\,\mathrm{d}t$$

for all vectors $u \in \operatorname{dom} H$ and $v \in \operatorname{dom} H_0 \cap P_0^{ac}\mathscr{H}_0$, which can be easily verified. By our assumption on $\alpha(H_0)\,P_\pm$, we can replace $v$ by $\alpha(H_0)\,P_+v$ or $\alpha(H_0)\,P_-v$ in the above formulas. This gives

$$\left(u, (W_+ - J)\,\alpha(H_0)\,P_+v\right) = \mathrm{i}\int_0^\infty (u, \mathrm{e}^{\mathrm{i}tH}G_t^+v)\,\mathrm{d}t,$$

$$\left(u, (W_- - J)\,\alpha(H_0)\,P_-\,v\right) = -\mathrm{i}\int_{\infty}^0 (u, \mathrm{e}^{\mathrm{i}tH}G_t^-v)\,\mathrm{d}t.$$

From Lemma 13 it follows that $\mathrm{e}^{\mathrm{i}tH}G_t^\pm$ are Borel functions from $\mathbb{R}_\pm$ into $\mathscr{L}_p(\mathscr{H}_0, \mathscr{H})$. Moreover, they are Bochner integrable over $\mathbb{R}_\pm$. Thus $\int_0^\infty \mathrm{e}^{\mathrm{i}tH}G_t^+\,\mathrm{d}t$, $\int_{-\infty}^0 \mathrm{e}^{\mathrm{i}tH}G_t^-\,\mathrm{d}t$ define operators $D_+, D_- \in \mathscr{L}_p(\mathscr{H}_0, \mathscr{H})$. Because of $\left(u, (W_\pm - J)\,\alpha(H_0)\,P_\pm v\right) = (u, \pm\mathrm{i}D_\pm v)$ for all $v \in \operatorname{dom} H_0$, $u \in \operatorname{dom} H$, we obtain $(W_\pm - J)\,\alpha(H_0)\,P_\pm \in \mathscr{L}_p(\mathscr{H}_0, \mathscr{H})$.

2. The last relation implies that also $P_+{}^*\alpha(H_0)^*(W_+ - J)^* \in \mathscr{L}_p(\mathscr{H}, \mathscr{H}_0)$. This fact, the assumption $P_+{}^*\alpha(H_0)^*(J^*J - P_0^{\mathrm{ac}})\,\alpha(H_0)\,P_- \in \mathscr{L}_p(\mathscr{H}_0)$ and the equality

$$P_+{}^*\alpha(H_0)^*(S - P_0^{\mathrm{ac}})\,\alpha(H_0)\,P_- = P_+{}^*\alpha(H_0)^*[W_+{}^*W_- - P_0^{\mathrm{ac}}]\,\alpha(H_0)\,P_-$$
$$= P_+{}^*\alpha(H_0)^*\{[(W_+ - J)^* + J^*][(W_- - J) + J] - P_0^{\mathrm{ac}}\}\,\alpha(H_0)\,P_-$$
$$= P_+{}^*\alpha(H_0)^*(W_+ - J)^*\{W_-\alpha(H_0)\,P_-\}$$
$$+ \{P_+{}^*\alpha(H_0)^*J^*\}(W_- - J)\,\alpha(H_0)\,P_- + P_+{}^*\alpha(H_0)^*(J^*J - P_0^{\mathrm{ac}})\,\alpha(H_0)\,P_-$$

imply the desired result. ■

The statement just proved makes assumptions similar to those in Section 15.2. The assumption $P_+{}^*\alpha(H_0)^*(J^*J - P_0^{\mathrm{ac}})\,\alpha(H_0)\,P_- \in \mathscr{L}_p(\mathscr{H}_0)$ means that, roughly speaking, $J^*J$ is nearly a unitary operator on $P_0^{\mathrm{ac}}\mathscr{H}_0$.

To illustrate the use of Prop. 14, we present an easy example.

**Example 17.** Let $H_0 = -\mathrm{i}d/dx$ on $\mathscr{H}_0 = L^2(\mathbb{R}, dx; \mathscr{K}_0)$ and let $\mathscr{H} = \mathscr{H}_0$, $J = 1$. Suppose that $V$ is represented by a continuous operator-valued function $v(x)$ from $\mathbb{R}$ into $\mathscr{L}_2(\mathscr{K}_0)$ which has compact support. Further, suppose that $v(x) = v(x)^*$ for all $x \in \mathbb{R}$. Thus $V$ is selfadjoint and bounded, and we obtain $\operatorname{dom} H = \operatorname{dom} H_0$. Let $\alpha(x) = (\mathrm{i} - x)^{-1}$. Then $\alpha \in L^2(\mathbb{R}, dx) \cap C_\infty(\mathbb{R})$ and $\alpha(H_0)\,\mathscr{H}_0 = \operatorname{dom} H_0$. Further, let $P$ be a projection which is defined by $(Pu)(x) = \chi_{[a,b]}(x)\,u(x)$, $u \in \mathscr{H}_0$. Our aim is to show, with the help of Prop. 14, that the scattering operator $S$ for the system $\{H, H_0\}$ obeys the relation $\alpha(H_0)\,(S - 1)\,\alpha(H_0)\,P \in \mathscr{L}_2(\mathscr{H}_0)$.

First, we verify that the wave operators exist and are complete (in the strong sense) and that $V\alpha(H_0) \in \mathscr{L}_2(\mathscr{H}_0)$ (the proof of these facts is similar to the proof of the corresponding ones in Section 15.2.3 where $\mathscr{K}_0 = \mathbb{C}$. It uses that $V\alpha(H_0) = VA\alpha(H_0)$ where $A$ is multiplication by the characteristic function $\chi_\Delta(x)$, $\Delta$ is a finite interval). Recall that $\alpha(H_0) = (\mathrm{i} - H_0)^{-1}$ has the form $(\alpha(H_0)\,u)(x) = -\mathrm{i}\int_{-\infty}^{x} e^{y-x}u(y)\,dy$ (see Section 15.2.3). Thus $V\alpha(H_0)$ is an integral operator with the kernel $(V\alpha(H_0))(x, y) = -\mathrm{i}\chi_{(-\infty,x)}(y)\,e^{y-x}v(x)$ and

$$\int dx \int dy\, |(V\alpha(H_0))(x, y)|^2 = \int dx \|v(x)\|_2^2 \int dy\, |\chi_{(-\infty,x)}(y)\,e^{y-x}|^2$$
$$= \int_\Delta dx \{\|v(x)\|_2^2 \int_{-\infty}^{x} dy\, e^{2y-2x}] = (1/2)\int_\Delta dx\, \|v(x)\|_2^2.$$

It is easy to see that the right side is finite because $\Delta$ is a finite interval. Hence $V\alpha(H_0) \in \mathscr{L}_2(\mathscr{H}_0)$. The proof of the existence of complete wave operators (existing in the strong sense) can be done in the same way as in Section 15.2.3 and is omitted.

The next step is to prove that $V\alpha(H_0)\,e^{-\mathrm{i}tH_0}P$ is a Bochner integrable Borel function from $\mathbb{R}$ into $\mathscr{L}_2(\mathscr{H}_0)$. From Remark 15 it follows that $V\alpha(H_0)\,e^{-\mathrm{i}tH_0}$ is a Borel function from $\mathbb{R}$ into $\mathscr{L}_2(\mathscr{H}_0)$. Further, we have

$$\int dt\, \|V\alpha(H_0)\,e^{-\mathrm{i}tH_0}P\|_2 = \int dt\, \{\int dx \int dy\, \|v(x)\,\chi_{(-\infty,x)}(y)\,e^{y-x}\chi_{[a+t,b+t]}(y)\|_2^2\}^{1/2}$$
$$= \int dt \left\{\int_\Delta dx\, \|v(x)\|_2^2 \int_{-\infty}^{x} e^{2y-2x}\chi_{[a+t,b+t]}(y)\,dy\right\}^{1/2}$$
$$\le \left(\sup_{x\in\Delta} \|v(x)\|_2\right) \int dt \left\{\int_\Delta dx \int_{\min\{x,a+t\}}^{\min\{x,b+t\}} e^{2y-2x}\,dy\right\}^{1/2}.$$

The integrand under the integral $\int dt$ behaves like $e^t$ for large negative values of $t$ and is zero for all sufficiently large positive values of $t$. Thus the integral above exists and this shows that all assumptions of Prop. 14 are fulfilled for $H$, $H_0$, $\alpha = \alpha_1 = \alpha_2$ and $P$. Hence $R_0(\mathrm{i})\,(S - 1)\,R_0(\mathrm{i})\,P \in \mathscr{L}_2(\mathscr{H}_0)$.

### 19.2.2. The scattering amplitude for "locally" compact perturbations

For simplicity of notations we assume in this and the following section that $P_0^{ac}\mathscr{H}_0$ has constant multiplicity and that there is given a spectral représentation $L^2(\Delta_0, d\lambda; \mathscr{K}_0)$ of $P_0^{ac}\mathscr{H}_0$ with respect to $H_0$. According to this representation we consider operators $B$ on $P_0^{ac}\mathscr{H}_0$ having the form

$$B \cong B_0 \otimes 1_{\mathscr{K}_0} \tag{29}$$

if we use the representation $L^2(\Delta_0, d\lambda; \mathscr{K}_0) \cong L^2(\Delta_0, d\lambda) \otimes \mathscr{K}_0$ and where $B_0$ is a bounded operator from $\mathscr{L}(L^2(\Delta_0, d\lambda))$. That is, we have $(Bu)(\lambda) = (B_0 f)(\lambda)\, e$ if $u = f(\lambda)\, e$ with $f \in L^2(\Delta_0, d\lambda)$, $e \in \mathscr{K}_0$. Since clo spa $\{\chi_\Delta(\lambda)(B_0 f)(\lambda);\ f \in L^2(\Delta_0, d\lambda),\ \Delta \subset \mathscr{B}_{\mathbb{R}}\}$ $= L^2(\Delta_B, d\lambda)$ for some Borel set $\Delta_B \subseteqq \Delta_0$, we obtain

$$\text{clo spa}\,\{E_0(\Delta)\, Bu;\ u \in P_0^{ac}\mathscr{H}_0, \Delta \subset \mathscr{B}_{\mathbb{R}}\} \cong L^2(\Delta_B, d\lambda; \mathscr{K}_0) \tag{30}$$

with $\Delta_B \subseteqq \Delta_0$. In the following statements we shall use such operators $B$ and the notation $\Delta_B$.

**Theorem 18.** *Let the assumptions of Prop.* 14 *be satisfied for* $H, H_0, J, \alpha_1, \alpha_2, P$ and $p = \infty$. *Suppose further that* $P$ *(in place of* $B$*) has the form* (29). *Then the scattering amplitude* $\hat{T}(\lambda)$ *belongs to* $\mathscr{L}_\infty(\mathscr{K}_0)$ *for almost all* $\lambda \in \Delta_{\alpha_1(H_0)\alpha_2(H_0)P}$ (mod $|\cdot|$).

Proof. By Prop. 14, we have $\alpha_1(H_0)(S - P_0^{ac})\, \alpha_2(H_0)\, P \in \mathscr{L}_\infty(\mathscr{H}_0)$. Naturally, this implies $\beta(H_0)\, \alpha_1(H_0)(S - P_0^{ac})\, \alpha_2(H_0)\, P \in \mathscr{L}_\infty(\mathscr{H}_0)$ for all $\beta \in L^\infty(\mathbb{R}; d\lambda)$. Let $f \in L^2(\Delta_0, d\lambda)$. Then we define $A_f \in \mathscr{L}(\mathscr{K}_0, L^2(\Delta_0, d\lambda; \mathscr{K}_0))$ by

$$A_f e := f(\lambda)\, e, \qquad e \in \mathscr{K}_0.$$

This implies $A_f^* \in \mathscr{L}(L^2(\Delta_0, d\lambda; \mathscr{K}_0), \mathscr{K}_0)$ and

$$(A_f^*)\,(g(\cdot)) = \int_{\Delta_0} \overline{f(\lambda)}\, g(\lambda)\, d\lambda, \qquad g \in L^2(\Delta_0, d\lambda; \mathscr{K}_0).$$

Now we define $T_{f,g} \in \mathscr{L}(\mathscr{K}_0)$, $f, g \in L^2(\Delta_0, d\lambda)$ by

$$T_{f,g} := A_f^* \beta(H_0)\, \alpha_1(H_0)\, (S - P_0^{ac})\, \alpha_2(H_0)\, P A_g.$$

Then, obviously, $T_{f,g}$ is a compact operator on $\mathscr{K}_0$ and can be represented as a Bochner integral (with respect to $\|\cdot\|_{\mathscr{K}_0}$), $P \cong P_0 \otimes 1_{\mathscr{K}_0}$,

$$T_{f,g} = \int_{\Delta_0} \bar{f}(\lambda)\, \beta(\lambda)\, \alpha_1(\lambda)\, \alpha_2(\lambda)\, (P_0 g)(\lambda)\, \hat{T}(\lambda)\, d\lambda$$

since $\|\hat{T}(\lambda)\|_{\mathscr{K}_0} \leqq 2$ a.e. mod $|\cdot|$ and $\bar{f}(\cdot)\, \beta(\cdot)\, \alpha_1(\cdot)\, \alpha_2(\cdot)\, (P_0 g)(\cdot) \in L^1(\Delta_0, d\lambda)$. Now we set $\beta(\lambda) = \chi_{[a,a+h]}(\lambda)$ with $a \in \Delta_0$ and $h > 0$. Then for each $h > 0$, $h^{-1}T_{f,g}$ is compact and we obtain

$$h^{-1}T_{f,g} = h^{-1} \int_a^{a+h} \bar{f}(\lambda)\, \alpha_1(\lambda)\, \alpha_2(\lambda)\, (P_0 g)(\lambda)\, \hat{T}(\lambda)\, d\lambda$$

($f(\lambda)$, $(P_0 g)(\lambda)$ and $\hat{T}(\lambda)$ are zero outside of $\Delta_0$). The right side tends to $\bar{f}(a)\, \alpha_1(a)\, \alpha_2(a)$ $\times\, (P_0 g)(a)\, \hat{T}(a)$ for almost all $a \in \Delta_0$ (mod $|\cdot|$) in operator norm (see Section 1.2.4).

Thus, as limit of the sequence of compact operators $h^{-1}T_{f,g}$, $\bar{f}(a)\, \alpha_1(a)\, \alpha_2(a)\, (P_0 g)\,(a)\, \hat{T}(a) \in \mathscr{L}_\infty(\mathscr{H}_0)$ for almost all $a \in \Delta_0(\mathrm{mod}\, |\cdot|)$. Taking into account the support properties of $\alpha_1(\lambda)$, $\alpha_2(\lambda)$, $(P_0 g)\,(\lambda)$, we get the assertion. ■

**Theorem 19.** *Let $\mathscr{A}$ be a set of functions from $L^\infty(\mathbb{R}, \mathrm{d}x)$. Let the assumptions of Prop. 16 be satisfied for $H, H_0, J, P_+, P_-, p = \infty$ and each function $\alpha \in \mathscr{A}$. Suppose further that the operators $P_+, P_-$ (in place of $B$) have the form (29) and that $P_0^{\mathrm{ac}}\mathscr{H}_0 = \mathrm{clo\ spa}\,\{\alpha(H_0)\, P_\pm u;$ $u \in P_0^{\mathrm{ac}}\mathscr{H}, \alpha \in \mathscr{A}\}$ for both signs. Then the scattering amplitude $\hat{T}(\lambda)$ belongs to $\mathscr{L}_\infty(\mathscr{H}_0)$ for almost all $\lambda \in \Delta_0(\mathrm{mod}\, |\cdot|)$.*

**Proof.** The proof is similar to the corresponding one for Theorem 18. We again introduce operators $T_{f,g} \in \mathscr{L}_\infty(\mathscr{H}_0)$ by

$$T_{f,g} := A^*_{P_+ f}\alpha(H_0)^*\,(S - P_0^{\mathrm{ac}})\,\alpha(H_0)\,P_- A_g, \qquad f, g \in L^2(\Delta_0, \mathrm{d}\lambda),$$

($T_{f,g} \in \mathscr{L}_\infty(\mathscr{H}_0)$ follows easily from this definition and Prop. 16). Then we can represent $T_{f,g}$ as a Bochner integral (with respect to the operator norm in $\mathscr{H}_0$),

$$T_{f,g} = \int_{\Delta_0} g_1(\lambda)\, g_2(\lambda)\, \hat{T}(\lambda)\, \mathrm{d}\lambda \tag{31}$$

with $g_1(\lambda) = \overline{(P_{+0}f)\,(\lambda)}\; \overline{\alpha(\lambda)}$, $g_2(\lambda) = \alpha(\lambda)\,(P_{-0}g)\,(\lambda)$. The right side defines a compact operator for all $f, g \in L^2(\Delta_0, \mathrm{d}\lambda)$. Because of our assumptions on the functions $\alpha$ and operators $P_\pm$, the functions $g_1$ and $g_2$ run through a dense set of functions in $L^2(\Delta_0, \mathrm{d}\lambda)$. It is not difficult to verify that therefore the right side of (31) defines a compact operator for all $g_1, g_2 \in L^2(\Delta_0, \mathrm{d}\lambda)$. Taking a function $g_1(\lambda) = \chi_\Delta(\lambda)\, g_1(\lambda)$ and by reasoning as in the proof of Theorem 18, we obtain that $g_1(\lambda)\, \hat{T}(\lambda)\, g_2(\lambda) \in \mathscr{L}_\infty(\mathscr{H}_0)$ for almost all $\lambda \in \Delta_0(\mathrm{mod}\, |\cdot|)$. Since $g_1, g_2$ can run through $L^2(\Delta_0, \mathrm{d}\lambda)$, this implies $\hat{T}(\lambda) \in \mathscr{L}_\infty(\mathscr{H}_0)$ a.e. (mod $|\cdot|$). ■

To see how these theorems can be used in scattering theory, we consider the following two examples.

**Example 20.** Let $H$, $H_0$, $P$ and $\alpha$ as in Example 17, that is, $H_0 = -\mathrm{i}\,\mathrm{d}/\mathrm{d}x$ on $\mathscr{H}_0 = L^2(\mathbb{R}, \mathrm{d}x; \mathscr{K}_0)$, $V$ is multiplication by $v(x) \in \mathscr{L}_2(\mathscr{K}_0)$, $P$ is multiplication by $\chi_{[a,b]}(x)$ and $\alpha(x) = (\mathrm{i} - x)^{-1}$. It is easy to see that

$$\mathrm{clo\ spa}\,\{E_0(\Delta)\,\alpha(H_0)^2\,Pu;\, u \in \mathscr{H}_0, \Delta \subset \mathscr{B}_{\mathbb{R}}\} = \mathrm{clo\,spa}\,\{\mathrm{e}^{-\mathrm{i}tH_0}\alpha(H_0)^2\,Pu;\, t \in \mathbb{R}, u \in \mathscr{H}_0\} = \mathscr{H}_0$$

and thus $\Delta_{\alpha(H_0)\alpha(H_0)P} = \mathbb{R} = \mathrm{spec^c}\, H_0$. Since the system $\{H, H_0, \alpha = \alpha_1 = \alpha_2, P\}$ satisfies the assumptions of Prop. 14, we can apply Theorem 18 and obtain that the scattering amplitude $\hat{T}(\lambda) \in \mathscr{L}_\infty(\mathscr{H}_0)$ e.a. mod $|\cdot|$.

**Example 21** (potential scattering). Let $H_0$, $H$ and $P_\pm = F_\pm$ be as in Section 15.2.4. That is, $H_0 = -\Delta$ on $\mathscr{H}_0 = L^2(\mathbb{R}^3, \mathrm{d}x)$ and $V = H - H_0$ is multiplication by a real-valued function $v(x)$. In difference to the assumption in Section 15.2.4 we make the stronger assumption that $|v(x)| \leqq c(1 + |x|)^{-(1+\varepsilon)}$, $\varepsilon > 0$, for all $x \in \mathbb{R}^3$. This ensures that $VR_0(\mathrm{i})$ is compact (see Kato [1, p. 304]) and we can factor $V = B^*A$ where $A$ is multiplication by $\alpha(x) = (1 + |x|)^{-(1+\varepsilon)}$ and $B$ is bounded. Then the calculation in Section 15.2.4 shows that

$$\|VR_0(\mathrm{i})\,\beta(H_0)\,\mathrm{e}^{-\mathrm{i}tH_0}P_\pm\| \leqq \|B^*\|\,\|AR_0(\mathrm{i})\,\beta(H_0)\,\mathrm{e}^{-\mathrm{i}tH_0}P_\pm\|, \qquad \beta \in C_0^\infty(\mathbb{R}^3 \setminus \{0\}),$$

is integrable in $t$ over $\mathbb{R}_\pm$ respectively, and that the complete wave operators $W_\pm(H, H_0)$ exist in the strong sense. Thus the assumptions of Prop. 16 are satisfied for $H$, $H_0$, $P_+$, $P_-$ and the functions $\alpha(x) = (\mathrm{i} - x)^{-1} \beta(x)$, $\beta \in C_0^\infty(\mathbb{R} \setminus \{0\})$. Note that the operators $P_\pm = F_\pm$ constructed in Section 15.2.4 satisfy the other assumptions of Theorem 19. Then we obtain that the scattering amplitude $\hat{T}(\lambda) \in \mathcal{L}_\infty(L^2(\Omega, \mathrm{d}\omega))$ a.e. mod $|\cdot|$ on $\mathbb{R}_+$ where $L^2(\mathbb{R}_+, \mathrm{d}\lambda; L^2(\Omega, \mathrm{d}\omega))$ is the usual spectral representation of $\mathcal{H}_0$ with respect to $H_0 = -\Delta$.

### 19.2.3. The scattering amplitude for "locally" Hilbert-Schmidt perturbations

In the preceding section the main assumption was that $V\alpha(H_0) \, \mathrm{e}^{-\mathrm{i}tH_0} P_\pm$ or $\alpha_1(H) \, V\alpha(H_0) \, \mathrm{e}^{-\mathrm{i}tH_0} P$ is compact. Here we consider the case that $\alpha_1(H) \, V\alpha_2(H_0) \, \mathrm{e}^{-\mathrm{i}tH_0} P$ is a Hilbert-Schmidt operator. Usually this is a consequence of the assumption that $\alpha_1(H) \, V\alpha_2(H_0) \in \mathcal{L}_2(\mathcal{H}_0, \mathcal{H})$, this means that roughly speaking $V$ is "locally" (with respect to the spectral representation of $H_0$ and $H$) a Hilbert-Schmidt operator.

**Theorem 22.** *Let the assumptions of Prop. 14 be satisfied for $H, H_0, \alpha_1, \alpha_2, P$ and $p = 2$. Suppose further that $P$ (in place of $B$) has the form (29). Then the scattering amplitude $\hat{T}(\lambda)$ belongs to $\mathcal{L}_2(\mathcal{K}_0)$ for almost all $\lambda$(mod $|\cdot|$) in $\Delta_{\alpha_1(H_0)\alpha_2(H_0)P}$ (see (30) for $\Delta_B$) and*

$$\int_{\Delta_0} |g(\lambda)|^2 \, \|\hat{T}(\lambda)\|_2^2 \, \mathrm{d}\lambda < \infty \tag{32}$$

*for all functions $g \in L^2(\Delta_0, \mathrm{d}\lambda)$ given by $\big(\alpha_1(H_0) \, \alpha_2(H_0) \, Pu\big)(\lambda) = g(\lambda) \, e$ where $u$ runs through all vectors $u$ with $u(\lambda) = f(\lambda) \, e$, $e \in \mathcal{K}_0$ and $f \in L^2(\Delta_0, \mathrm{d}\lambda)$.*

Proof. According to Prop. 14, we obtain $(S - P_0^{\mathrm{ac}}) \, \alpha_1(H_0) \, \alpha_2(H_0) \, P \in \mathcal{L}_2(\mathcal{K}_0)$. Let $u \in \mathcal{H}_0$ with $u(\lambda) = f(\lambda) \, e$. Then

$$\big((S - P_0^{\mathrm{ac}}) \, \alpha_1(H_0) \, \alpha_2(H_0) \, Pu\big)(\lambda) = \hat{T}(\lambda) \, g(\lambda) \, e,$$
$$g(\lambda) \, e = \big(\alpha_1(H_0) \, \alpha_2(H_0) \, Pu\big)(\lambda).$$

Let $\|f\| = 1$ and $\{e_s\}$ be an orthonormal basis in $\mathcal{K}_0$. Then, denoting the vectors of the form $f(\lambda) \, e_s$ by $u_s$,

$$\|(S - P_0^{\mathrm{ac}}) \, \alpha_1(H_0) \, \alpha_2(H_0) \, P\|_2^2 \geq \sum_{s=1}^{\infty} \|(S - P_0^{\mathrm{ac}}) \, \alpha_1(H_0) \, \alpha_2(H_0) \, Pu_s\|^2$$
$$= \sum_{s=1}^{\infty} \int_{\Delta_0} \|\hat{T}(\lambda) \, g(\lambda) \, e_s\|^2 \, \mathrm{d}\lambda = \int_{\Delta_0} |g(\lambda)|^2 \, \|\hat{T}(\lambda)\|_2^2 \, \mathrm{d}\lambda.$$

The interchange of the integral with the sum is justified because the left side exists and the terms under the integral and sum are nonnegative. This estimate proves (32) and also $\hat{T}(\lambda) \in \mathcal{L}_2(\mathcal{K}_0)$ for almost all $\lambda$(mod $|\cdot|$) in $\Delta_{\alpha_1(H_0)\alpha_2(H_0)P}$ because of the definition of this set and the definition of $g(\lambda)$. ■

### 19.2.4. The scattering amplitude for a special scattering system

Let $H_0$ denote the multiplication operator by $\lambda$ on $\mathcal{H}_0 = \bigoplus_{j=1}^{N} L^2(\Delta_{0j}, \mathrm{d}\lambda; \mathcal{K}_{0j})$ where $\Delta_{0j}$ are mutually disjoint open intervals with $|\Delta_{0j}| < \infty$. Let $H$ be selfadjoint on $\mathcal{H} = \bigoplus_{l=1}^{M} L^2(\Delta_l, \mathrm{d}\lambda; \mathcal{K}_l)$ and let $J$ be a bounded operator from $\mathcal{H}_0$ into $\mathcal{H}$.

Suppose that $VE_0(\Delta_{0j}) := (HJ - JH_0)\, E_0(\Delta_{0j}) \in \mathscr{L}_2(\mathscr{H}_0, \mathscr{H})$ for each $j \in \{1, 2, \dots, N\}$. Then $VE_0(\Delta_{0j})$ is an integral operator (see Section 19.1.1), that is, there are Borel functions $V_{kj}(\lambda, \lambda'): \Delta_k \times \Delta_{0j} \to \mathscr{L}_2(\mathscr{K}_{0j}, \mathscr{K}_k)$ such that

$$\big(VE_0(\Delta_{0j})\, u\big)(\lambda) = \sum_{k=1}^{M} \int_{\Delta_{0j}} V_{kj}(\lambda, \lambda')\, u(\lambda')\, d\lambda' \tag{33}$$

and

$$\|VE_0(\Delta_{0j})\|_2^2 = \sum_{k=1}^{M} \int_{\Delta_k} d\lambda \int_{\Delta_{0j}} d\lambda'\, \|V_{kj}(\lambda, \lambda')\|_2^2. \tag{34}$$

Suppose further that each $V_{kj}(\lambda, \cdot)$ is an absolutely continuous $\mathscr{L}_2(\mathscr{K}_{0j}, \mathscr{K}_k)$-function; this means that there exists an operator $W$ from $\mathscr{H}_0$ into $\mathscr{H}$ such that $WE_0(\Delta_{0j}) \in \mathscr{L}_2(\mathscr{H}_0, \mathscr{H})$ for each $j \in \{1, 2, \dots, N\}$ and for almost all $\lambda \in \Delta_k$, $\lambda' \in \Delta_{0j} (\mathrm{mod}\, |\cdot|)$

$$V_{kj}(\lambda, \lambda') = \int_{\lambda_0}^{\lambda'} W_{kj}(\lambda, x)\, dx + V_{kj}^0(\lambda), \tag{35}$$

where $\lambda_0$ is a fixed point from $\Delta_{0j}$, the integral is a Bochner integral for the $\mathscr{L}_2(\mathscr{K}_{0j}, \mathscr{K}_k)$-valued function $W_{kj}(\lambda, x)$ and $V_{jk}^0(\lambda) \in \mathscr{L}_2(\mathscr{K}_{0j}, \mathscr{K}_k)$ for almost all $\lambda \in \Delta_k$ and is independent of $\lambda'$.

Now we introduce a family of projections $P_j(\varrho)$ in $\mathscr{H}_0$ by

$$\big(P_j(\varrho)\, u\big)(\lambda) := \varrho(\lambda) \int_{\Delta_{0j}} d\lambda' \varrho(\lambda')\, u(\lambda'), \qquad u \in \mathscr{H}_0, \tag{36}$$

where $\varrho(\cdot) \in C_0^\infty(\Delta_{0j})$, $\int d\lambda |\varrho(\lambda)|^2 = 1$ and $\varrho(\cdot)$ is real-valued. Clearly, $P_j(\varrho)\, u \in L^2(\Delta_{0j}, d\lambda; \mathscr{K}_{0j})$ for all vectors $u \in \mathscr{H}_0$. The vectors $u_{jl}$ with $u_{jl}(\lambda) = \varrho(\lambda)\, e_{jl}$, $\{e_{jl}\}$ an orthonormal basis in $\mathscr{K}_{0j}$, form an orthonormal basis of the image (range) of $P_j(\varrho)$. Thus, if we assume for a moment that $(S - 1)\, P_j(\varrho) \in \mathscr{L}_2(\mathscr{H}_0)$, we obtain

$$\|(S - 1)\, P_j(\varrho)\|_2^2 = \sum_l \|(S - 1)\, u_{jl}\|^2 = \int_{\Delta_{0j}} \varrho(\lambda)^2\, \|\hat{T}(\lambda)\|_2^2\, d\lambda < \infty. \tag{37}$$

The next step is to prove $(S - 1)\, P_j(\varrho) \in \mathscr{L}_2(\mathscr{H}_0)$. From (8) we obtain the estimate

$$\|(S - 1)\, P_j(\varrho)\|_2 \leq \int_{-\infty}^{\infty} dt\, \|V\, e^{-itH_0} P_j(\varrho)\|_2. \tag{38}$$

Thus, if we can estimate the right side of (38), we have also an estimate for the expressions in (37).

By the Schwartz inequality, we have

$$\left( \int_{-\infty}^{\infty} dt\, \|V e^{-itH_0} P_j(\varrho)\|_2 \right)^2 \leq \int_{-\infty}^{\infty} dt (1 + t^2)^{-1} \int_{-\infty}^{\infty} dt (1 + t^2)\, \|V\, e^{-itH_0} P_j(\varrho)\|_2^2$$

$$\leq c \left\{ \int_{-\infty}^{\infty} dt\, \|V\, e^{-itH_0} P_j(\varrho)\|_2^2 + \int_{-\infty}^{\infty} dt\, t^2\, \|V\, e^{-itH_0} P_j(\varrho)\|_2^2 \right\} \tag{39}$$

with a constant $c < \infty$. Now we estimate both terms on the right side of (39). For the first term we get

$$\int_{-\infty}^{\infty} dt \, \|V \, e^{-itH_0} P_j(\varrho)\|_2^2 = \int_{-\infty}^{\infty} dt \, \|P_j(\varrho) \, e^{itH_0} V^*\|_2^2$$

$$= \sum_{l=1}^{M} \int_{-\infty}^{\infty} dt \int_{\Delta_{0l}} d\lambda \int_{\Delta_l} d\lambda' \, |\varrho(\lambda)|^2 \left\| \int_{\Delta_{0l}} dx \; \varrho(x) \; V_{kj}(\lambda', x)^* \, e^{ixt} \right\|_2^2$$

$$= \sum_{l=1}^{M} 2\pi \int_{\Delta_l} d\lambda' \int_{\Delta_{0j}} \|\varrho(x) \, V_{kj}(\lambda', x)^*\|_2^2 \, dx \leqq 2\pi \, |\varrho(\cdot)|^2 \, \|VE_0(\Delta_{0j})\|_2^2 < \infty \tag{40}$$

where we have used the Parseval equation with respect to the Fourier transform $\int dx \, e^{ixt}$, $\int \varrho(\lambda)^2 \, d\lambda = 1$, and $\|VE_0(\Delta_{0j})\|_2^2 = \|E_0(\Delta_{0j}) \; V^*\|_2^2$. To estimate the second term on the right side of (39), we first write this term with the help of the kernel of $VE_0(\Delta_{0j})$ (as above) and then we integrate by parts with respect to $x$. Then we make the same transformations as above for the first term on the right side of (39). This gives

$$\int_{-\infty}^{\infty} dt \, t^2 \, \|V \, e^{-itH_0} P_j(\varrho)\|_2^2 = \int_{-\infty}^{\infty} dt \, t^2 \, \|P_j(\varrho) \, e^{itH_0} V^*\|_2^2$$

$$= 2\pi \sum_{l=1}^{M} \int_{\Delta_l} d\lambda' \int_{\Delta_{0j}} dx \left\| \frac{d}{dx} \, [\varrho(x) \; V_{kj}(\lambda', x)^*] \right\|_2^2$$

$$= 2\pi \sum_{l=1}^{M} \int_{\Delta_l} d\lambda' \int_{\Delta_{0j}} dx \, \|\varrho'(x) \; V_{kj}(\lambda', x)^* + \varrho(x) \; W_{kj}(\lambda', x)^*\|_2^2$$

$$\leqq 4\pi \{|\varrho'(\cdot)|^2 \, \|VE_0(\Delta_{0j})\|_2^2 + |\varrho(\cdot)|^2 \, \|WE_0(\Delta_{0j})\|_2^2\},$$

and proves, together with (38) to (40) that $(S - 1) \, P_j(\varrho) \in \mathscr{L}_2(\mathscr{H}_0)$. Further we get

$$\int_{\Delta'_j} d\lambda \, \varrho(\lambda)^2 \, \|T(\lambda)\|_2^2 \leqq c\{\|VE_0(\Delta_{0j})\|_2^2 \, (2\pi \, |\varrho(\cdot)|^2 + 4\pi \, |\varrho'(\cdot)|^2) + 4\pi \, |\varrho(\cdot)|^2 \, \|WE_0(\Delta_{0j})\|_2^2\} < \infty,$$

is valid for all $j$. For the proof we have assumed that the strong wave operators $W_\pm(H, H_0; J)$ exist, are partially isometric, and are complete. Note that the existence of the strong wave operators follows from the finiteness of the right side of (38) for all $\varrho$ (which we have proved) if we use Prop. 15.4. For the completeness and partial isometry of $W_\pm$ we need additional assumptions on $H, H_0; J$.

# Notes and Remarks to Part V

*Chapter 18.* Stationary formulas for the scattering operator like those given in Theorem 4 and Corollary 5 were developed first by Amrein/Georgescu/Jauch [1971a] and independently by Pearson [1971a]; see also Chandler/Gibson [1972a]. Explicit formulas for the scattering amplitude in case of potential scattering have been known for many years. Usually they were derived by means of the generalized eigenfunctions of $H = -\Delta + V(x)$. There are many papers on this subject. The following list includes some of them: Green/Lanford [1960a], Grossman/Wu [1961a], Hunziker [1961a], [1963a], Kuroda [1962a], Ikebe [1965a], Alsholm/Schmidt

[1971a], Kato [1971a], Agmon [1975a]. There are only few papers dealing with explicit formulas for the scattering amplitude of abstract scattering systems: Birman/Éntina [1964a], [1967a] derived such a formula under trace class assumptions on $H$, $H_0$ (see Section 19.1.1); under smoothness assumptions an expression for the scattering amplitude was given in the paper of Baumgärtel/Demuth/Wollenberg [1978a]; in the framework of the stationary theory Kato [1978a] developed explicit formulas for the scattering amplitude similar to those in Section 1.4.

The idea of using the operators $F_G(\lambda)$ to obtain formulas for the scattering amplitude can already be found in Amrein/Jauch/Sinha [1] in the concrete situation: $\mathscr{H}_0 = L^2(\mathbb{R}^3, dx)$, $H_0 = -\Delta$, $H = -\Delta + V(x)$.

For the definition of the total scattering cross section and its relation to the scattering amplitude $\hat{T}(\lambda)$ the reader may consult Amrein/Jauch/Sinha [1] and Amrein/Pearson [1979a].

*Analytic properties of* $\hat{T}(\lambda)$. A discussion of the physical aspects of analyticity can be found in the survey article of Brenig/Haag [1959a], in Goldberger/Watson [1], and in Newton [1]. There one can also find many references concerning the earliest proofs of analyticity of the scattering amplitude in potential scattering. In particular, the result that $\hat{T}(\lambda)$ is meromorphic continuable if the potential $V(x)$ falls off exponentially (see Section 3.3) was proved by Hunziker [1961a] (spherically symmetric potentials), Grossman/Wu [1961a], [1962a], Muldowney [1965a], Dolph/Mc Leod/Thoe [1966a], Howland [1969a] under more or less weak additional assumptions. Further results, concerning analytic properties of the scattering amplitude for one-channel and multichannel systems, were obtained by Federbush [1967a], Riahi [1969a], Jensen [1977a], Balslev [1978a], [1979a], Hagedorn [1979a], and Chadan/Martin [1979a].

Abstract results about analyticity of the scattering amplitude are meager. Howland's paper [1969a] contains some elements of an abstract approach to analyticity via the sandwiched resolvent. This idea was abstracted and further developed in Baumgärtel/Demuth/Wollenberg [1978a].

*Resonances.* Resonances are not uniquely defined in literature. Sometimes poles of certain matrix elements of the resolvent are called resonances (such poles we call virtual poles) whereas our definition is that resonances are nonreal poles of the scattering amplitude $\hat{T}(\lambda)$. For a discussion of resonances, maxima of the scattering cross section and virtual poles we refer to Brenig/Haag [1959a], Newton [1], Simon [1973a], Baumgärtel [1977b], [1978a], [1981a], Shamra/Ogunsulire [1978a], and Paravicini/Gorini/Sudarshan [1980a].

For virtual poles and their interpretation as "complex eigenvalues" of $H$, see Grossman [1964a], Howland [1969b], [1974b], [1975a], Aguilar/Combes [1971a], Horwitz/Marchand [1971a], Simon [1973a], Baumgärtel [1973b], [1974a], [1975a], [1976a], [1979a], Sudarshan/Chiu/Gorini [1978a], and Bohm [1979a]. See also Davies [1975a].

The equality of resonances and virtual poles was first established by Lax/Phillips [1] in the concrete situation: $\mathscr{H}_0 = L^2(\mathbb{R}^n, dx)$, $n$ odd, $H_0 = -\Delta$ and $H = -\Delta$ with a Dirichlet boundary condition. The extension of this result to the case that $H_0 = -\Delta$ and $H = -\Delta + V(x)$ (where $V(x)$ falls off exponentially or has compact support) was given by Shenk/Thoe [1972a], Simon [1973a], Babitt/Balslev [1976a], Jensen [1977a], and Balslev [1978a]. See also C. Goldstein [1973a], Rauch [1980a]. For corresponding assertions for multiparticle systems we refer to Balslev [1979a] and Hagedorn [1979a]. Abstract results about the equality of resonances and virtual poles can be found in Baumgärtel/Demuth/Wollenberg [1978a]. Prop. 13 is a slight generalization of some results of this paper.

The content of Section 3.4 is essentially taken from Baumgärtel [1979a].

*Chapter 19.* The first result on trace class properties of $\hat{T}(\lambda)$ appeared in Birman/Krein [1962a] ($\hat{T}(\lambda) \in \mathscr{L}_1(\mathscr{H}_0)$ if $R(z) - R_0(z) \in \mathscr{L}_1(\mathscr{H}_0)$). Prop. 2 is essentially due to Birman/Éntina [1964a], [1967a]. Lemma 3 and Corollary 4 were proved by Jauch/Sinha/Misra [1972a] with other methods. Theorem 5 and Prop. 6 (in a slightly other form) were first derived by Jauch/Sinha [1972a] and Martin/Misra [1973a]. Applications of these statements to potential scattering and other scattering problems (e.g. relativistic potential scattering) can be found in the cited papers and in Martin/Misra [1974a].

*Spectral shift and phase shift.* The notion of spectral shift first appeared in Lifšic [1952a]. Kreĭn [1953a], [1962a], proved and studied the relation $\operatorname{tr}(\alpha(H) - \alpha(H_0)) = \int \xi(\lambda)\, \alpha'(\lambda)\, d\lambda$ for a class of functions $\alpha$. The content of Section 1.4 is essentially taken from Kreĭn [1964a]. For other properties of $\xi(\lambda)$ see also Birman/Solomjak [1972a]. Theorem 11 first appeared in Birman/

KREĬN [1962a]. This theorem was extended to more general scattering systems by KOPLIENKO [1977a] (local trace class assumptions on $H$, $H_0$), [1979a] (two-space scattering systems). Relations between spectral shift and phase shift in potential scattering were proved by BUSLAEV/FADDEEV [1960a], BUSLAEV [1962a], BUSLAEV/MERKUR'EV [1969a], BUSLAEV [1972a], [1973a]. See also BUSLAEV [1970a], JAFAEV [1980a].

*Scattering cross section and $\mathscr{L}_p$-properties ($p > 2$) of $\hat{T}(\lambda)$.* The finiteness of the total scattering cross section in potential scattering has been the subject of many papers. Some relevant references are GREEN/LANFORD [1960a] (spherically symmetric potentials), HUNZIKER [1963a] (asymptotic behavior of $\sigma(\lambda)$ for smooth potentials), KURODA [1970a] (he gave sufficient conditions for $\hat{T}(\lambda) \in \mathscr{L}_p$) JAUCH/SINHA [1972a] (trace class methods), P. MARTIN [1972a], KURODA [1973b] ($\sigma(\lambda)$ is finite for potential $|V(x) \approx c\,|x|^{-(2+\varepsilon)}$), MARTIN/MISRA [1974a], A. MARTIN [1979a]. In a series a papers AMREIN/PEARSON ([1979a], [1979b], [1979c)] developed new approaches to the problem of the finiteness of the total scattering cross section and the $\mathscr{L}_2$-property of the scattering amplitude $\hat{T}(\lambda)$. In AMREIN/PEARSON [1980a] the methods were applied to scattering with oscillating potentials and in AMREIN/PEARSON/SINHA [1979a], to scattering in $N$-body systems. See also ENSS/SIMON [1980a].

*Sections 2.3 and 2.4* are essentially taken from AMREIN/PEARSON [1979c]. Theorem 19 is an abstraction of a result from JAFAEV [1979b]. The result that $\hat{T}(\lambda) \in \mathscr{L}_\infty$ for the scattering system $H_0 = -\Delta$, $H = -\Delta + V(x)$ with $|V(x)| \leqq c(1 + |x|)^{-(1+\varepsilon)}$ (see Example 21) is a special case of a result of AGMON [1975a]. He proved it under the assumptions: $V(\cdot)$ is locally $L^2$-integrable and $V(x) \approx c\,|x|^{(-1+\varepsilon)}$ for large $|x|$.

Spectral properties of the scattering matrix for abstract scattering systems were also considered in KOPLIENKO [1971a] and in ROTFEL'D [1972a], [1974a].

*Levinson's theorem and time delay.* We close the Notes and Remarks to this part with a brief description of two topics which we have not treated in the text.

LEVINSON's theorem connects the number of bound states of a Schrödinger operator with spherically symmetric potential with the phase shift $\delta(\lambda)$. More precisely, if $H_{0l} = -\mathrm{d}^2/\mathrm{d}x^2 + l(l+1)\,x^{-2}$ and $H_l = H_{0l} + V(x)$ on $\mathscr{H}_0 = L^2([0, \infty), \mathrm{d}x)$, if $\int_0^\infty |V(x)|\, x\, \mathrm{d}x < \infty$ and $\int_0^\infty |V(x)|\, x^2 \mathrm{d}x < \infty$ and if $\delta_l(x)$ denotes the corresponding phase shift for the scattering system $H_{0l}$, $H_l$; then

$$\delta_l(\infty) - \delta_l(0) = \pi(n + (1/2)\, y)$$

where $n = \dim E_{H_l}(\mathrm{eig}\, H_l)\, \mathscr{H}_0$ and $y$ takes the values 0 or 1 (1 is only possible for $l = 0$ and if there is a so-called zero energy resonance). This theorem was proved by LEVINSON [1949a]. A detailed discussion can be found in NEWTON [1960a]. A generalization for nonspherically symmetric potentials was given first by BUSLAEV [1962a] ($V(\cdot) \in \mathscr{S}(\mathbb{R}^3)$ and no eigenvalue at 0) and later by NEWTON [1977a], WOLLENBERG [1977c], OSBORN/BOLLE [1977a], and DREYFUS [1978a], [1978b] under weaker assumptions on the potential. There were some attempts to extend LEVINSON's theorem to abstract scattering systems. The following list includes papers of this nature: JAUCH [1957a], MARTIN [1958a], BUSLAEV [1970a], DREYFUS [1976a], WOLLENBERG [1977a], NEWTON [1977b].

In potential scattering a quantity can be introduced which measures the excess time that the particle spends in the scattering region compared to a free particle. This quantity is called the time delay. For the definition one needs the concept of a position operator which is not available in the abstract scattering theory. In the Lax-Phillips theory the scattering region is well-defined and therefore a concept of time delay can be introduced; for instance see LAX/PHILLIPS [1978b] and AMREIN/WOLLENBERG [1981a]. It turns out that the time delay operator $Q_0(\lambda)$ can be written in terms of $\hat{S}(\lambda)$, namely $Q_0(\lambda) = -\mathrm{i}\hat{S}(\lambda)^* \left(\frac{\mathrm{d}\hat{S}(\lambda)}{\mathrm{d}\lambda}\right)$. This is also true in potential scattering. For details we refer to JAUCH/SINHA/MISRA [1972a], AMREIN/JAUCH/SINHA [1], SINHA [1978a], and BOLLE/OSBORN [1979a].

# Bibliography

## Books and Monographs

ACHIESER, N. I., and I. M. GLASMANN
[1] Theorie der linearen Operatoren im Hilbert-Raum. 6th ed., Akademie-Verlag, Berlin 1975 (8th revised and extended ed. 1981).
AGRANOVIČ, Z. S., and V. A. MARČENKO (Агранович, З. С., и В. А. Марченко)
[1] Обратная задача теории рассеяния. Харьковск. университет, Харьков 1960 (English transl.: AGRANOVICH, Z. S., and V. A. MARCHENKO, The inverse problem of scattering. Gordon and Breach, New York 1963).
AMREIN, W. O., J. M. JAUCH, and K. B. SINHA
[1] Scattering theory in quantum mechanics. W. A. Benjamin, Reading, Massachusetts, 1977.
ARSEN'EV, A. A. (Арсеньев, А. А.)
[1] Сингулярные потенциалы и резонансы. Московск. университет, Москва 1974.
ARVESON, W.
[1] An invitation to $C^*$-algebra. Springer, New York—Heidelberg—Berlin 1976.
BAUMGÄRTEL, H., G. LASSNER, A. PIETSCH, and A. UHLMANN (eds.)
[1] Proceedings of the international conference on operator algebras, ideals, and their applications in theoretical physics (Leipzig, September 12—20, 1977). B. G. Teubner, Leipzig 1978.
BIRMAN, M. Š., and M. Z. SOLOMJAK (Бирман, М. Ш., и М. З. Соломяк)
[1] Спектральная теория самосопряженных операторов в гильбертовом пространстве. Ленингр. университет, Ленинград 1980.
BOGOLJUBOV, N. N., A. A. LOGUNOV, and I. T. TODOROV (Боголюбов, Н. Н., А. А. Логунов и И. Т. Тодоров)
[1] Основы аксиоматического подхода в квантовой теории поля. Наука, Москва 1969 (English transl.: BOGOLIUBOV, N. N., A. A. LOGUNOV, and I. T. TODOROV, Introduction to axiomatic quantum field theory. Benjamin, New York 1975).
CHADAN, K., and P. SABATIER
[1] Inverse problems in quantum scattering theory (with a foreword by R. G. NEWTON). Springer, New York—Heidelberg—Berlin 1977.
DIESTEL, J., and J. J. UHL, Jr.
[1] Vector measures. Amer. Math. Soc., Providence (R.I.) 1977.
DIEUDONNE, J.
[1] Grundzüge der modernen Analysis 2. VEB Deutscher Verlag der Wissenschaften, Berlin 1975 (transl. from French).
DIXMIER, J.
[1] Les algebres d'operateurs dans l'espace Hilbertien (Algebres de von Neumann). Gauthier-Villars, Paris 1969.
DUNFORD, N., and J. T. SCHWARTZ
[1] Linear operators, Part I: General theory. Wiley, New York—London 1958.
[2] Linear operators, Part II: Spectral theory, selfadjoint operators in Hilbert space. Wiley, New York—London 1963.
[3] Linear operators, Part III: Spectral operators. Wiley, New York—London—Sydney—Toronto 1971.
ENZ, C. P., and J. MEHRA (eds.)
[1] Physical reality and mathematical description. Reidel, Dordrecht 1974.
FRIEDRICHS, K. O.
[1] Perturbation of spectra in Hilbert space. Amer. Math. Soc., Providence (R.I.) 1965.

GELFAND, I. M., and G. E. SCHILOW
[1] Verallgemeinerte Funktionen (Distributionen) III. VEB Deutscher Verlag der Wissenschaften, Berlin 1964 (transl. from Russian).
GELFAND, I. M., and N. J. WILENKIN
[1] Verallgemeinerte Funktionen (Distributionen) IV. VEB Deutscher Verlag der Wissenschaften, Berlin 1964 (transl. from Russian).
GOHBERG, I. C., and M. G. KREĬN (Гохберг, И. Ц., и М. Г. Крейн)
[1] Введение в теории линейных несамосопряженных операторов. Москва 1965 (English transl.: GOHBERG, I. C., and M. G. KREĬN, Introduction to the theory of linear nonselfadjoint operators. Amer. Math. Soc., Providence (R.I.) 1969).
GOLDBERGER, M. L., and K. M. WATSON
[1] Collision theory. Wiley, New York 1964.
HALMOS, P.
[1] Measure theory. Springer, Berlin—Heidelberg—New York 1974 (reprint of edition published by Van Nostrand 1950).
GRADŠTEĬN, I. S., and I. M. RYŽIK (Градштайн, И. С., и И. М. Рыжик).
[1] Таблицы интегралов, сумм, рядов и произведений. изд. пямое, Наука, Москва 1971 (German transl.: RYSHIK, I. M., und I. S. GRADSTEIN, Summen-, Produkt- und Integraltafeln. 2nd, revised ed., VEB Deutscher Verlag der Wissenschaften, Berlin 1963).
HARDY, G. H.
[1] Divergent series. Oxford University Press, Oxford 1949.
JOST, R.
[1] The general theory of quantized fields. Amer. Math. Soc., Providence (R.I.) 1965.
KATO, T.
[1] Perturbation theory for linear operators. 2nd ed., Springer, Berlin—Heidelberg—New York 1976.
LANDAU, L. D., and E. M. LIFSCHITZ
[1] Lehrbuch der theoretischen Physik III: Quantenmechanik. 3rd ed., Akademie-Verlag, Berlin 1967 (transl. from Russian).
LA VITA, J. A., and J.-P. MARCHAND (eds.)
[1] Scattering theory in mathematical physics. Reidel, Dordrecht—Boston 1974.
LAX, P., and R. PHILLIPS
[1] Scattering theory. Academic Press, New York—London 1967.
[2] Scattering theory for automorphic functions. Princeton University Press, Princeton (N.J.) 1974.
MUSCHELISCHWILI, N. I.
[1] Singuläre Integraloperatoren. Akademie-Verlag, Berlin 1965 (transl. from Russian).
NAĬMARK, M. A. (Наймарк, М. А.)
[1] Нормированные кольца. Наука, Москва 1968.
NATANSON, I. P.
[1] Theorie der Funktionen einer reellen Veränderlichen. 4th ed., Akademie-Verlag, Berlin 1975.
NEUMANN, J. VON
[1] Mathematische Grundlagen der Quantenmechanik. Springer, Berlin 1932 (reprint 1968).
NEWTON, R. G.
[1] Scattering theory of waves and particles. Mc Graw Hill, New York 1966.
PUTNAM, C. R.
[1] Commutation properties of Hilbert space operators and related topics. Springer, Berlin—Heidelberg—New York 1967.
REED, M.
[1] Abstract non-linear wave equations. Lecture Notes in Mathematics, vol. 507, Springer, New York—Heidelberg—Berlin 1976.
REED, M., and B. SIMON
[1] Methods of modern mathematical physics I: Functional analysis. Academic Press, New York—London 1974.
[2] Methods of modern mathematical physics II: Fourier analysis, selfadjointness. Academic Press, New York—San Francisco—London 1975.

[3] Methods of modern mathematical physics III: Scattering theory. Academic Press, New York—San Francisco—London 1979.
[4] Methods of modern mathematical physics IV: Analysis of operators. Academic Press, New York—San Francisco—London 1978.
Rellich, F.
[1] Perturbation theory of eigenvalue problems. Gordon and Breach, New York 1969.
Saito, Y.
[1] Spectral representations for Schrödinger operators with long-range potentials. Lecture Notes in Mathematics, vol. 727. Springer, Berlin—Heidelberg—New York 1979.
Sakai, S.
[1] $C^*$-algebras and $W^*$-algebras. Springer, Berlin—Heidelberg—New York 1971.
Segal, I. E.
[1] Mathematical problems of relativistic physics. Amer. Math. Soc., Providence (R.I.) 1963.
Simon, B.
[1] Quantum mechanics for Hamiltonians defined as quadratic forms. Princeton University Press, Princeton (N.J.) 1971.
[2] The $P(\varphi)_2$ Euclidian (quantum) field theory. Princeton University Press, Princeton (N.J.) 1974.
[3] Trace ideals and their applications. London Math. Soc. Lecture Notes Series, vol. 35, Cambridge University Press, Cambridge—New York—Melbourne 1979.
[4] Functional integration and Quantum physics. Academic Press, New York 1979.
Stein, E. M., and G. Weiss
[1] Introduction to Fourier analysis on Euclidean spaces. Princeton University Press, Princeton (N.J.) 1971.
Takesaki, M.
[1] Theory of operator algebras. Springer, New York—Heidelberg—Berlin 1979.
Thirring, W.
[1] Lehrbuch der Mathematischen Physik 1: Klassische dynamische Systeme. Springer, Wien—New York 1977.
[2] Lehrbuch der Mathematischen Physik 3: Quantenmechanik von Atomen und Molekülen. Springer, Wien—New York 1979.
Titchmarsh, E. C.
[1] Eigenfunction expansions associated with second-order differential equations, Part I. Clarendon Press, Oxford 1962.
[2] Eigenfunction expansions associated with second-order differential equations, Part II. Clarendon Press, Oxford 1958.
Triebel, H.
[1] Höhere Analysis. VEB Deutscher Verlag der Wissenschaften, Berlin 1972.
Weidmann, J.
[1] Lineare Operatoren in Hilberträumen. B. G. Teubner, Stuttgart 1976.
Wilcox, C. H.
[1] Scattering theory for the d'Alembert equation in exterior domains. Lecture Notes in Mathematics, vol. 442, Springer, Berlin—Heidelberg—New York 1975.
Yosida, K.
[1] Functional analysis. Springer, Berlin—Göttingen—Heidelberg 1965 (4th ed. 1974).

## Articles

Adamjan, V. M., and D. Z. Arov (Адамян, В. М., и Д. З. Аров)
[1965a] Об одном классе операторов рассеяния и характеристических оператор-функций сжатий. Докл. Акад. Наук СССР 160, 9—12.
[1965b] Об операторах рассеяния и полугруппах сжатий в гильбертовом пространстве. Докл. Акад. Наук СССР 165, 9—12.

AGMON, S.
[1975a] Spectral properties of Schrödinger operators and scattering theory. Ann. Scuola Norm. Sup. Pisa, Ser. 4, 2, 151–218.
[1979a] Some new results in spectral and scattering theory of differential operators on $\mathbb{R}^n$. Seminaire Goulaouic-Schwartz Exp. No. 2 (1978/79), Centre Math. Ecole Polytech., Palaiseau.
AGUILAR, J., and J. M. COMBES
[1971a] A class of analytic perturbations for one-body Schrödinger Hamiltonians. Comm. Math. Phys. 22, 269–279.
ALSHOLM, P.
[1977a] Wave operators for long-range scattering. J. Math. Anál. and Appl. 59, 550–572.
ALSHOLM, P., and T. KATO
[1973a] Scattering with long range potentials. *In*: Partial Differential Equations. Amer. Math. Soc., Providence (R.I.), pp. 393–399.
ALSHOLM, P., and G. SCHMIDT
[1971a] Spectral and scattering theory for Schrödinger operators. Arch. Rational Mech. Anal. 40, 281–311.
AMREIN, W. O., and V. GEORGESCU
[1973a] On the characterization of bound states and scattering states in quantum mechanics. Helv. Phys. Acta 46, 635–658.
[1974a] Strong asymptotic completeness of wave operators for highly singular potentials. Helv. Phys. Acta 47, 517–533.
AMREIN, W. O., V. GEORGESCU, and J. M. JAUCH
[1971a] Stationary state scattering theory. Helv. Phys. Acta 44, 407–434.
AMREIN, W. O., V. GEORGESCU, and P. MARTIN
[1974a] Approche algebrique de la theorie non-relativiste de la diffusion a canaux multiples. *In*: ENZ/MEHRA (eds.) [1], pp. 255–276.
AMREIN, W. O., P. MARTIN, and B. MISRA
[1970a] On the asymptotic condition of scattering theory. Helv. Phys. Acta 43, 313–344.
AMREIN, W. O., and D. B. PEARSON
[1979a] Commutators of Hilbert-Schmidt type and the scattering cross section. Ann. Inst. H. Poincaré, Sect. A (NS) 30, 89–107.
[1979b] A time-dependent approach to the total scattering cross section. J. Phys. A 12, 1469 to 1492.
[1979c] The $S$-matrix for abstract scattering systems. Lett. Math. Phys. 3, 83–86.
[1980a] The scattering matrix for rapidly oscillating potentials. J. Phys. A 13, 1259–1264.
AMREIN, W. O., D. B. PEARSON, and K. B. SINHA
[1979a] Bounds on the total scattering cross section for $N$-body systems. Nuovo Cimento A (11) 52, 115–131.
AMREIN, W. O., D. B. PEARSON, and M. WOLLENBERG
[1980a] Evanescence of states and asymptotic completeness. Helv. Phys. Acta 53, 335–351.
AMREIN, W. O., and M. WOLLENBERG
[1981a] On the Lax-Phillips scattering theory. Proc. Royal Soc. Edinburgh, Sect. A 87, 219–230.
ANTONEC, M. A. (Антонец, М. А.)
[1978a] О задаче рассеяния в классической механики. Функц. анализ и приложен. 12, 68–69.
ARAI, M.
[1972a] Absolute continuity of Hamiltonian operators with repulsive potentials. Publ. Res. Inst. Math. Sci. 7, 621–635.
ARAKI, H., and R. HAAG
[1967a] Collision cross sections in terms of local observables. Comm. Math. Phys. 4, 77–91.
ARONSZAJN, N.
[1957a] On a problem of Weyl in the theory of singular Sturm Liouville equations. Amer. J. Math. 79, 597–610.
AROV, D. Z. (Аров, Д. З.)
[1974a] Об унитарных сцеплениях с потерями (теория рассеяния с потерями). Функц. анализ и приложен. 8, вып. 4, 5–22.

Arsen'ev, A. A. (Арсеньев, А. А.)
[1968a] Рассеяние плоской волны на сингулярном потенциале. Журн. вычислит. мат. и мат. Физ. 8, 544—572.
Asano, K.
[1967a] Notes on Hilbert transforms of vector valued functions in the complex plane and their boundary values. Proc. Japan Acad., Ser. A Math. Sci. 43, 572—577.
Avila, G. S. C., and D. G. Costa
[1980a] Asymptotic properties of general symmetric hyperbolic systems. J. Funct. Anal. 35, 49—63.
Avron, J. E., and I. W. Herbst
[1977a] Spectral and scattering theory of Schrödinger operators related in the Stark effect. Comm. Math. Phys. 52, 239—254.
Avron, J., J. Herbst, and B. Simon
[1978a] Schrödinger operators with magnetic fields, I. General interactions. Duke Math. J. 45, 847—883.
[1978b] Schrödinger operators with magnetic fields, II. Separation of the center of mass in homogeneous magnetic fields. Ann. Physics 114, 431—451.
Babitt, D., and E. Balslev
[1976a] Local distortion techniques and unitarity of the $S$-matrix for the 2-body problem. J. Math. Anal. and Appl. 54, 316—349.
Baeteman, M. L., and K. Chadan
[1976a] Scattering theory with highly singular oscillating potentials. Ann. Inst. H. Poincaré, Sect. A (NS) 24, 1—16.
Balslev, E.
[1978a] Analytic scattering theory of two-body Schrödinger operators. J. Funct. Anal. 29, 375—396.
[1980a] Analytic scattering theory of quantum mechanical three-body systems. Ann. Inst. H. Poincaré, Sect. A (NS) 32, 125—160.
Baumgärtel, H.
[1967a] Integraldarstellung der Wellenoperatoren von Streusystemen. Mber. Dt. Akad. Wiss. 9, 169—174.
[1969a] Eine Bemerkung zur Theorie der Wellenoperatoren. Math. Nachr. 42, 359—363.
[1973a] Zu einem Problem von M. G. Krein. Math. Nachr. 58, 279—294.
[1973b] Spectral concentration and factorization. Amer. J. Math. 95, 849—884.
[1974a] Kompakte Störungen nichtisolierter Eigenwerte endlicher Vielfachheit. Math. Nachr. 59, 265—273.
[1974b] Wellenoperatoren, Abelscher Limes und Vertauschungsrelation. Math. Nachr. 62, 104—109.
[1975a] Partial resolvent and spectral concentration. Math. Nachr. 69, 107—121.
[1976a] Resonances of perturbed selfadjoint operators and their eigenfunctionals. Math. Nachr. 75, 133—151.
[1977a] Remarks on multichannel scattering theory. ZIMM-Preprint, Akad. Wiss. DDR, Berlin.
[1977b] On resonances in multi-channel scattering theory. ZIMM-Preprint, Akad. Wiss. DDR, Berlin.
[1978a] A decoupling approach to resonances for $N$-particle quantum scattering systems. *In*: Baumgärtel/Lassner/Pietsch/Uhlmann (eds.) [1], pp. 350—363.
[1979a] Mathematical remarks to resonances and their eigenfunctionals in decay-scattering systems, demonstrated by means of Friedrichs' model. *In*: Beiglböck, W., A. Böhm, and E. Takasugi (eds.), Group theoretical methods in physics. Lecture Notes in Physics, vol. 94, Springer, Berlin—Heidelberg—New York, pp. 202—211.
[1981a] Asymptotic resonance properties of the finite-dimensional Friedrichs model. *In*: Gustafson, K., and W. P. Reinhardt (eds.), Quantum Mechanics in Mathematics, Chemistry and Physics. Plenum, New York.
[1982a] Stationary completeness criterion for wave operators. Math. Nachr. 105, 281—294.
Baumgärtel, H., and M. Demuth
[1976a] Perturbation of unstable eigenvalues of finite multiplicity. J. Funct. Anal. 22, 187—203.
[1979a] Decoupling by a projection. Rep. Math. Phys. 15, 173—185.

Baumgärtel, H., M. Demuth, and M. Wollenberg
[1978a] On the equality of resonances (poles of the scattering amplitude) and virtual poles. Math. Nachr. 86, 167–174.
[1978b] Weakly asymptotic properties of the operator evolution $e^{itHB} e^{-itH}$. ZIMM-Preprint P-12/78, Akad. Wiss. DDR, Berlin.
Baumgärtel, H., H. Neidhardt, and J. Rehberg
[1983a] On identification operators between free and interacting quantum fields. *To appear in*: Math. Nachr.
Baumgärtel, H., P. Obermann, and M. Wollenberg
[1979a] Abel wave operators. III. Stationary theory. ZIMM-Preprint P-32/79, Akad. Wiss. DDR, Berlin.
Baumgärtel, H., J. Rehberg, and M. Wollenberg
[1981a] The inverse problem of the abstract scattering theory within a relativistic framework. Colloquia Math. Soc. János Bolyai 27, 79–91.
Belopol'skiĭ, A. L. (Белопольский, А. Л.)
[1975a] Задачи рассеяния для пары пространств с неограниченным оператором отождествления. Изв. ВУЗ мат. 11 (162), 11–20.
Belopol'skiĭ, A. L., and M. Š. Birman (Белопольский, А. Л., и М. Ш. Бирман)
[1968a] Существование волновых операторов в теории рассеяния для пары пространств. Изв. Акад. Наук СССР 32, 1162–1175.
Bentosela, F.
[1976a] Scattering from impurities in a crystal. Comm. Math. Phys. 46, 153–166.
Berthier, A. M., and P. Collet
[1977a] Existence and completeness of the wave operators in scattering theory with momentum dependent potentials, J. Funct. Anal. 26, 1–15.
[1977b] Wave operators for momentum dependent long range potential. Ann. Inst. H. Poincaré, Sect. A (NS) 26, 279–294.
Bertocchi, L., S. Fubini, and G. Furlan
[1965a] On the theory of scattering by singular potentials. Nuovo Cimento 35, 633–643.
Birman, M. Š. (Бирман, М. Ш.)
[1962a] Возмущения непрерывного спектра сингулярного эллиптического оператора при изменении границы и граничных условий. Вестник ЛГУ, Серия мат. мех. астрон. №. 1, 22–55.
[1962b] Об условиях существования волновых операторов. Докл. Акад. Наук СССР 143, 506–509.
[1962c] Об одном признаке существования волновых операторов. Докл. Акад. Наук СССР 147, 1008–1009.
[1963a] Об условиях существования волновых операторов. Изв. Акад. Наук СССР 27, 883–906.
[1964a] Локальный признак существования волновых операторов. Докл. Акад. Наук СССР 159, 485–488.
[1968a] Локальный признак существования волновых операторов. Изв. Акад. Наук СССР 32, 914–942.
1969a] Некоторые приложения локального признака существования волновых операторов. Докл. Акад. Наук СССР 185, 735–738.
[1969b] Задачи рассеяния для дифференциальных операторов. Функц. анализ и приложен. 3, вып. 3, 1–16.
[1970a] Признак существования полных волновых операторов в теории рассеяния для пары пространств. Проблемы мат. физики, ЛГУ, вып. 4, 22–26.
[1971a] Задачи рассеяния для дифференциальных операторов при возмущении пространства. Изв. Акад. Наук СССР 35, 440–455.
Birman, M. Š., and S. B. Éntina (Бирман, М. Ш., и С. Б. Энтина)
[1964a] О стационарном подходе в абстрактной теории рассеяния. Докл. Акад. Наук СССР 155, 506–508.
[1967a] Стационарный подход в абстрактной теории рассеяния. Изв. Акад. Наук СССР 31, 401–430.

BIRMAN, M. Š., and M. G. KREĬN (Бирман, М. Ш., и М. Г. Крейн)
[1962a] К теории волновых операторов и операторов рассеяния. Докл. Акад. Наук СССР 144, 475—478.
BIRMAN, M. Š., and M. Z. SOLOMJAK (Бирман, М. Ш., и М. З. Соломяк)
[1966a] Двойные операторные интегралы Стилтьеса. Проблемы мат. физики, ЛГУ, вып. 1, 33—67.
[1967a] Двойные операторные интегралы Стилтьеса, II. Проблемы мат. физики, ЛГУ, вып. 2, 26—60.
[1969]a Об оценках сингулярных чисел интегральных операторов. III. Операторы в неограниченных областях. Вестник ЛГУ, Серия мат. мех. астрон. 24, №.1, 35—48.
[1972a] Замечание о функция спектрального сдвига. Записки научных семинаров ЛОМИ 27, 33—46.
[1973a] Двойные операторные интегралы Стилтьеса, III. Проблемы мат. физики, ЛГУ, вып. 6, 27—53.
BOHM, A.
[1979a] Gamow state vectors as functionals over subspaces of the nuclear space. Lett. Math. Phys. 3, 455—461.
BOLLÉ, D., and T. A. OSBORN
[1979a] Time delay in $N$-body scattering. J. Math. Phys. 20, 1121—1134.
BORCHERS, H. J.
[1962a] On the structure of the algebra of fields operators. Nuovo Cimento 24, 214—236.
BOURGEOIS, B.
[1979a] On scattering theory for oscillatory potentials of slow decay. Ann. Physics 121, 415—431.
BRANGES, L. DE
[1962a] Perturbation of self-adjoint transformations. Amer. J. Math. 84, 543—560.
BRENIG, W., and R. HAAG
[1959a] Allgemeine Quantentheorie der Stoßprozesse. Fortschr. Phys. 7, 183—242.
BROWNELL, F. H.
[1962a] A note on Cook's wave-matrix theorem. Pacific J. Math. 12, 47—52.
BUSLAEV, V. S. (Буслаев, В. С.)
[1962a] Формулы следов для оператора Шредингера в трехмерном пространстве. Докл. Акад. Наук СССР 143, 1067—1070.
[1970a] Спектралные тождества и формула следа в модели Фридрихса. Проблемы мат. физики, ЛГУ, вып. 4, 43—53.
[1972a] О формулах следа в многоканальных задачах. Записки научных семинаров ЛОМИ 27, 47—70.
[1973a] Формула следа и сингулярности матрицы рассеяния для системы трех одномерных частиц. Теорет. Мат. Физ. 10, 245—249.
BUSLAEV, V. S., and L. D. FADDEEV (Буслаев, В. С., и Л. Д. Фаддеев)
[1960a] О формулах следов для дуфференциального сингулярного оператора Штурма-Лиувилла. Докл. Акад. Наук СССР 132, 13—16.
BUSLAEV, V. S., and V. B. MATVEEV (Буслаев, В. С., и В. Б. Матвеев)
[1970a] Волновые операторы для уравнения Шрёдингера с медленно убывающим потенциалом. Теорет. Мат. Физ. 1, 367—376.
BUSLAEV, V. S., and S. P. MERKUR'EV (Буслаев, В. С., и С. П. Меркурьев)
[1969a] Формула следа для системы трех частиц. Докл. Акад. Наук СССР 189, 269 до 272.
CAREY, R. W., and J. D. PINCUS
[1976a] Unitary equivalence modulo the trace class for self-adjoint operators. Amer. J. Math. 98, 481—514.
CHADAM, J. M.
[1969a] The asymptotic behavior of the Klein-Gordon equation with external potential, II. Pacific J. Math. 31, 19—31.
[1970a] The asymptotic behavior of the Klein-Gordon equation with external potential I. J. Math. Anal. and Appl. 31, 334—348.
[1972a] Asymptotics for $\Box u = m^2u + G(x, t, u, u_x, u_t)$. I. Ann. Scuola Norm Sup. Pisa 26, 33—65.

[1972b] Asymptotics for $\square u = m^2u + G(x, t, u, u_x, u_t)$. II. Ann. Scuola Norm Sup. Pisa 26, 65–95.

CHADAN, K., and A. MARTIN

[1979a] Scattering theory and dispersion relations for a class of long-range oscillating potentials. Comm. Math. Phys. 70, 1–27.

CHANDLER, C.

[1978a] Nonrelativistic multichannel quantum scattering theory in a two Hilbert space formulation. *In*: BAUMGÄRTEL/LASSNER/PIETSCH/UHLMANN (eds.) [1], pp. 308–319.

CHANDLER, C., and A. G. GIBSON

[1972a] On a recent paper of Amrein, Georgescu and Jauch. Helv. Phys. Acta 45, 734–737.

[1973a] Transition from time-dependent to time-independent multichannel quantum scattering theory. J. Math. Phys. 14, 1328–1335.

[1974a] Time-dependent multichannel Coulomb scattering theory. J. Math. Phys. 15, 291–294.

[1976a] Invariance principle for scattering with long-range (and other) potentials. Indiana Univ. Math. J. 25, 443–460.

[1977a] *N*-body quantum scattering theory in two Hilbert spaces, I. The basic equations. J. Math. Phys. 18, 2336–2347.

[1978a] *N*-body quantum scattering theory in two Hilbert spaces, II. Some asymptotic limits. J. Math. Phys. 19, 1610–1616.

[1979a] Long-range invariance principle revisited. Indiana Univ. Math. J. 28, 389–403.

[1980a] A two-Hilbert space formulation of multichannel scattering theory. In: Mathematical methods and applications of scattering theory. Springer, Berlin—Heidelberg—New York, pp. 134–148.

CHEN, K.-H., and C. C. YANG

[1978a] Spectral theory for the acoustic wave equation with generalized Neumann boundary conditions in exterior domain. J. Math. Phys. 19, 924–929.

COESTER, F.

[1965a] Scattering theory for relativistic particles. Helv. Phys. Acta 38, 7–23.

COESTER, F., S. COHEN, B. DAY, and C. M. VINCENT

[1970a] Variation in nuclear-matter binding energies with phase-shift-equivalent two-body potentials. Phys. Rev. C 1, 769–776.

COMBES, J. M.

[1969a] An algebraic approach to scattering theory. Preprint C.N.R.S., Centre de Physique Theorique, Marseille.

[1969b] Time-dependent approach to nonrelativistic multichannel scattering. Nuovo Cimento A 64, 111–144.

COMBESCURE, M., and J. GINIBRE

[1976a] Spectral and scattering theory for the Schrödinger operator with strongly oscillating potentials. Ann. Inst. H. Poincaré, Sect. A (NS) 24, 17–30.

[1978a] Scattering and local absorption for the Schrödinger operator. J. Funct. Anal. 29, 54–73.

COOK, J. M.

[1957a] Convergence to the Møller wave matrix. J. Math. Phys. 36, 82–87.

[1967a] Banach algebras and asymptotic mechanics. *In*: LURCHIT, F. (ed.), Cargese Lectures in Theoretical Physics. Gordon and Breach, New York.

CORBETT, J. V.

[1970a] Galilean symmetry, measurement, and scattering as an isomorphism between two subalgebras of observables. Phys. Rev. D 1, 3331–3344.

[1975a] Particles and simple scattering theory. J. Math. Phys. 16, 271–274.

DALECKIĬ, JU. L., and S. G. KREĬN (Далецкий, Ю. Л., и С. Г. Крейн)

[1956a] Интегрирование и дифференцирование функций эрмитовых операторов и приложения к теории возмущений. Труды Воронежск. госуд. универс. вып. 1, 81–105.

DAVIES, E. B.

[1974a] Time-dependent scattering theory. Math. Ann. 210, 149–162.

[1975a] Resonances, spectral concentration and exponential decay. Lett. Math. Phys. 1, 31–35.

[1977a] Eigenfunction expansions for singular Schrödinger operators. Arch. Rational Mech. Anal. 63, 261–272.

[1977b] Scattering from infinite sheets. Math. Proc. Camb. Phil. Soc. 82, 327–334.

[1978a] Two-channel Hamiltonians and the optical model of nuclear scattering. Ann. Inst. H. Poincaré, Sect. A (NS) **29**, 395—413.
[1979a] Scattering from a spatially restricted atom, I. Adv. in Math. **34**, 28—45.
[1980a] Non-unitary scattering and capture. I. Hilbert space theory, Comm. Math. Phys. **71**, 277—288.
[1980a] On Enss' approach to scattering theory. Duke Math. J. **47**, 171—185.
DAVIES, E. B., and B. SIMON
[1978a] Scattering theory for systems with different spatial asymptotics on the left and right. Comm. Math. Phys. **63**, 277—301.
DEĬČ, V. G. (Дейч, В. Г.)
[1971a] Приложения метода ядерных возмущений в теории рассеяния для пары пространства. Изв. ВУЗ мат. №. **6**, 33—42.
[1973a] Стационарний локальный метод в теории рассеяния для пары пространства. Проблемы мат. физики, ЛГУ, вып. **6**, 76—90.
DEĬČ, V. G., and D. R. JAFAEV (Дейч, В. Г., и Д. Р. Яфаев)
[1980a] Спектральная теория и рассеяние для оператора Даламбера с векторным потенциалом. Труды Мат. инст. им. Стеклова **147**, 5—13.
DEĬČ, V. G., E. L. KOROTJAEV, and D. R. JAFAEV (Дейч, В. Г., Е. Л. Коротяев и Д. Р. Яфаев)
[1977a] Теория потенциального рассеяния при учете пространственной анизотропии. Записки научных семинаров ЛОМИ **78**, 35—51.
DEIFT, P. A.
[1976a] Classical scattering theory with a trace condition. Diss., Princeton University, Princeton (N.J.).
[1978a] Applications of a commutation formula. Duke Math. J. **45**, 267—310.
DEIFT, P. A., and B. SIMON
[1976a] On the decoupling of finite singularities from the question of asymptotic completeness in two body quantum systems. J. Funct. Anal. **23**, 218—238.
[1977a] A time-dependent approach to the completeness of multiparticle quantum systems. Comm. Pure Appl. Math. **30**, 573—583.
DEIFT, P. A., and E. TRUBOVITZ
[1979a] Inverse scattering on the line. Comm. Pure Appl. Math. **32**, 127—257.
DEMUTH, M.
[1980a] Scattering of diffusion process generators. Lett. Math. Phys. **4**, 417—424.
DEMUTH, M., H. NEIDHARDT, and M. WOLLENBERG
[1976a] On the identification operator in the two space theory. ZIMM-Preprint, Akad. Wiss. DDR, Berlin.
DOLLARD, J. D.
[1964a] Asymptotic convergence and the Coulomb interaction. J. Math. Phys. **5**, 729—738.
[1966a] Adiabating switching in the Schrödinger theory of scattering. J. Math. Phys. **7**, 802—810.
[1969a] Scattering into cones: I. Potential scattering. Comm. Math. Phys. **12**, 193—203.
[1973a] Scattering into cones: II. $N$-body problems. J. Math. Phys. **14**, 708—718.
[1976a] Interpretation of Kato's invariance principle in scattering theory. J. Math. Phys. **17**, 46—51.
DOLLARD, J. D., and C. N. FRIEDMAN
[1978a] Existence of the Møller wave operators for $r^{-\beta}\lambda \sin(\mu r^{\alpha})$. Ann. Physics **111**, 251—266.
[1979a] A new proof of the invariance principle in scattering theory. Ann. Physics **119**, 71—75.
DOLPH, C. L., J. B. MCLEOD, and D. THOE
[1966a] The analytic continuation to the unphysical sheet of the resolvent kernel and the scattering operator associated with the Schrödinger operator. J. Math. Anal. and Appl. **16**, 311—332.
DONALDSON, J. A., A. G. GIBSON, and R. HERSH
[1973a] On the invariance principle of scattering theory. J. Funct. Anal. **14**, 131—145.
DREYFUS, T.
[1976a] Levinson's theorem for non-local interactions. J. Phys. A **9**, L187—L191.
[1978a] The number of states bound by non-central potentials. Helv. Phys. Acta **51**, 321—329.
[1978b] The determinant of the scattering matrix and its relation to the number of eigenvalues. J. Math. Anal. and Appl. **64**, 114—134.

ECKARDT, K.-J.
[1974a] On the existence of wave operators for Dirac operators. Manuscripta Math. 11, 359–371.
[1974b] Scattering theory for Dirac operators. Math. Z. 139, 105–131.
EKSTEIN, H.
[1956a] Scattering in field theory. Nuovo Cimento 4, 1017–1058.
[1956b] Theory of time-dependent scattering for multichannel processes. Phys. Rev. 101, 880 to 890.
[1960a] Equivalent Hamiltonians in scattering theory. Phys. Rev. 117, 1590–1595.
ENSS, V.
[1978a] Asymptotic completeness for quantum mechanical potential scattering. Comm. Math. Phys. 61, 285–291.
[1979a] Asymptotic completeness for quantum mechanical potential scattering. II. Singular and long range potentials. Ann. Physics 119, 117–132.
[1979b] Two-cluster scattering of $N$ charged particles. Comm. Math. Phys. 65, 151 to 165.
ENSS, V., and B. SIMON
[1980a] Finite total cross sections in nonrelativistic quantum mechanics. Comm. Math. Phys. 76, 177–209.
ÈNTINA, S. B. (Энтина, С. Б.)
[1971a] Интегральные представления для волновых операторов в абстрактной теорий рассеяния. Проблемы мат. физики ЛГУ, вып. 5, 113–122.
[1972a] Интегральные представления для волновых операторов в абстрактной теорий рассеяния II. Проблемы мат. физики ЛГУ, вып. 6, 138–141.
EVANS, D. E.
[1976a] Smooth perturbations in non-reflexive Banach spaces. Math. Ann. 221, 183–194.
[1976b] Time-dependent perturbations and scattering of strongly continuous groups on Banach spaces. Math. Ann. 221, 275–290.
FADDEEV, L. D. (Фаддеев, Л. Д.)
[1963a] Математические вопросы квантовой теории рассеяния для системы трех частиц. Труды Мат. инст. им. Стеклова 69, 1–121.
[1964a] О модели Фридрихса в теории возмущений непрерывного спектра. Труды Мат. инст. им. Стеклова 73, 292–313.
[1974a] Обратная задача квантовой теории рассеяния. *В*: Итоги Науки и Техники, Серия: Современные проблемы математики, Т. 3, Москва, 93–180.
FEDERBUSH, P.
[1967a] Results on the analyticity of many-body scattering amplitudes in perturbation theory. J. Math. Phys. 8, 2415–2419.
FERRERO, P., O. DE PAZZIS, and D. ROBINSON
[1974a] Scattering theory with singular potentials, II. The $N$-body problem and hard cores. Ann. Inst. H. Poincaré, Sect. A (NS) 21, 217–232.
FOIAS, C.
[1975a] On the Lax-Phillips nonconservative scattering theory. J. Funct. Anal. 19, 272–301.
FRIEDRICHS, K. O.
[1938a] Über die Spektralzerlegung eines Integraloperators. Math. Ann. 115, 249–272.
[1948a] On the perturbation of continuous spectra. Comm. Pure Appl. Math. 1, 361–406.
GALINDO TIXAIRE, A.
[1959a] Scattering integral equations in Hilbert space. Helv. Phys. Acta 32, 412–422.
GELL-MANN, M., and M. L. GOLDBERGER
[1953a] The formal theory of scattering. Phys. Rev. 91, 398–408.
GEORGESCU, V.
[1974a] Méthodes stationaires pour des potentiels à longue portée à symmetrie spherique. These, Université de Geneve, Geneve.
[1977a] Sur l'existence des opérateurs d'onde dans la théorie algébrique de la diffusion. Ann. Inst. H. Poincaré, Sect. A (NS) 27, 9–29.
GIBSON, A. G., and C. CHANDLER
[1974a] Time-independent multichannel scattering theory for charged particles. J. Math. Phys. 15, 1366–1377.

GINIBRE, J.
[1977a] Spectral and scattering theory of the Schrödinger equation for three-body systems. *In*: The Schrödinger Equation, Acta Physica Austriaca Suppl. 17, 95–138.
GINIBRE, J., and M. MOULIN
[1974a] Hilbert space approach to the quantum mechanical three-body problem. Ann. Inst. H. Poincaré, Sect. A (NS) 21, 97–145.
GINIBRE, J., and G. VELO
[1978a] Existence of solutions and scattering theory for the nonlinear Schrödinger equation. *In*: BAUMGÄRTEL/LASSNER/PIETSCH/UHLMANN (eds.) [1], pp. 320–334.
[1978b] On a class of nonlinear Schrödinger equations. III. Special theories in dimensions 1, 2 and 3. Ann. Inst. H. Poincaré, Sect. A (NS) 28, 297–316.
[1979a] On a class of nonlinear Schrödinger equations. II. Scattering theory, general case. J. Funct. Anal. 32, 33–71.
[1980a] On a class of nonlinear Schrödinger equations with nonlocal interactions. Math. Z. 170, 109–136.
GOLDSTEIN, C. I.
[1969a] Eigenfunction expansions associated with the Laplacian for certain domains with infinite boundaries. I. Trans. Amer. Math. Soc. 135, 1–31.
[1969b] Eigenfunction expansions associated with the Laplacian for certain domains with infinite boundaries. II. Applications to scattering theory. Trans. Amer. Math. Soc. 135, 33–50.
[1970a] Perturbation of non-selfadjoint operators. I. Arch. Rational Mech. Anal. 37, 268–296.
[1971a] Perturbation of non-selfadjoint operators. II. Arch. Rational Mech. Anal. 42, 380–402.
[1973a] The singularities of the $S$-matrix and Green's function associated with perturbations of $-\Delta$ acting in a cylinder. Bull. Amer. Math. Soc. 79, 1303–1306.
[1974a] Scattering theory for elliptic differential operators in unbounded domains. I. J. Math. Anal. and Appl. 45, 723–745.
[1974b] Scattering theory for elliptic differential operators in unbounded domains. II. J. Math. Anal. and Appl. 46, 671–699.
GOLDSTEIN, J.
[1975a] Temporally inhomogenous scattering theory. *In*: NACHBIN, L. (ed.), Analyse fonctionelle et applications. Hermann, Paris, pp. 125–132.
GOLDSTEIN, J., and C. MONLEZUN
[1976a] Temporally inhomogenous scattering theory. II. Approximation theory and second order equations. SIAM J. Math. Anal. 7, 276–290.
GOODMANSON, D., and J. R. TAYLOR
[1980a] Coulomb scattering as the limit of scattering of smoothly screened Coulomb potentials. J. Math. Phys. 21, 2202–2207.
GREEN, T. A., and O. E. LANFORD III
[1960a] Rigorous derivation of the phase shift formula for the Hilbert space scattering operator of a single particle. J. Math. Phys. 1, 139–148.
GROSSE, H.
[1979a] On the construction of Møller operators for the nonlinear Schrödinger equation. Physics Letter B 86, 267–271.
GROSSE, H., H. R. GRÜMM, H. NARNHOFER, and W. THIRRING
[1974a] Algebraic theory of Coulomb scattering. Acta Phys. Austriaca 40, 97–103.
GROSSMAN, A.
[1964a] Nested Hilbert spaces in quantum mechanics. J. Math. Phys. 5, 1025–1037.
GROSSMAN, A., and T. T. WU
[1961a] Schrödinger scattering amplitude I. J. Math. Phys. 2, 710–713.
[1962a] Schrödinger scattering amplitude III. J. Math. Phys. 3, 684–689.
GUILLOT, J. C., and G. SCHMIDT
[1974a] Spectral and scattering theory for Dirac operators. Arch. Rational Mech. Anal. 55, 193–206.
GUSTAFSON, K.
[1974a] Fourier scattering subspaces. *In*: ENZ/MEHRA (eds.) [1], pp. 277–285.
[1978a] Scattering theory and the principle of limiting absorption. *In*: BAUMGÄRTEL/LASSNER/PIETSCH/UHLMANN (eds.) [1], pp. 335–349.

GUSTAFSON, K., and G. JOHNSON
[1974a] On the absolutely continuous subspace of a self-adjoint operator. Helv. Phys. Acta 47, 163—166.
HAAG, R.
[1958a] Quantum field theories with composite particles and asymptotic conditions. Phys. Rev. 112, 669—673.
HACK, M. N.
[1958a] On convergence to the Møller wave operators. Nuovo Cimento 9, 731—733.
[1959a] Wave operators in multichannel scattering. Nuovo Cimento 13, 231—236.
HAGEDORN, G. A.
[1978a] Asymptotic completeness for a class of four particle Schrödinger operators. Bull. Amer. Math. Soc. 84, 155—156.
[1979a] A link between scattering resonances and dilation analytic resonances in few body quantum mechanics. Comm. Math. Phys. 65, 181—188.
[1979b] Born series for (1 cluster) → (2 cluster) scattering of 2,3,4-particle Schrödinger operators. Comm. Math. Phys. 66, 77—94.
[1980a] Asymptotic completeness for classes of two, three, and four particles Schrödinger operators. Trans. Amer. Math. Soc. 258, 1—76.
HEISENBERG, W.
[1943a] Die „beobachtbaren Größen“ in der Theorie der Elementarteilchen. Z. Phys. 120, 513 bis 538.
HEJTMANEK, J.
[1974a] Scattering theory of the linear Boltzmann operator. Comm. Math. Phys. 43, 109—120.
[1979a] Dynamics and spectrum of the linear multiple scattering operator in the Banach lattice $\mathscr{L}^1(R^3 \times R^3)$. Transport Theory Statist. Phys. 8, 29—44.
HENDRICKSON, J.
[1975a] Temporally inhomogenous scattering theory for modified wave operators. J. Math. Phys. 16, 768—771.
[1976a] $N$-body scattering into cones with long-range time dependent potentials. J. Math. Phys. 17, 729—733.
HELSTON, J. W., and J. V. RALSTON
[1976a] The first variation of the scattering matrix. J. Differential Equations 21, 378—394.
HEPP, K.
[1966a] On the connection between Wightman and LSZ quantum field theory. *In*: CHRETIEN, M., and S. DESER (eds.), Axiomatic Field Theory. Gordon and Breach, New York, pp. 135 to 246.
[1969a] On the quantum mechanical $N$-body problem. Helv. Phys. Acta 42, 425—458.
HERBST, I.
[1977a] Unitary equivalence of Stark Hamiltonians. Math. Z. 155, 55—70.
[1979a] Dilation analyticity in constant electric field. I. The two body problem. Comm. Math. Phys. 64, 279—298.
HÖEGH-KROHN, R.
[1968a] Partly gentle perturbations with application to perturbations by annihilation-creation operators. Comm. Pure Appl. Math. 21, 313—342.
[1968b] Gentle perturbations by annihilation-creation operators. Comm. Pure Appl. Math. 21, 343—357.
HOFMAN, G., and G. LASSNER
[1975a] Existence proofs for Wightman-type functionals. Wiss. Z. Karl-Marx-Universität Leipzig, Math. Nat. Reihe 24, 535—542.
HÖRMANDER, L.
[1976a] The existence of the wave operators in scattering theory. Math. Z. 146, 69—91.
HORWITZ, L. P., and J.-P. MARCHAND
[1971a] The decay-scattering system. Rocky Mountain J. Math. 1, 225—253.
HOWLAND, J. S.
[1967a] Banach space techniques in the perturbation theory of self-adjoint operators with continuous spectra. J. Math. Anal. and Appl. 20, 22—47.
[1968a] A perturbation-theoretic approach to eigenfunction expansions. J. Funct. Anal. 2, 1—23.

[1969a] Analytic continuation of the $S$-matrix for potential scattering. Proc. Amer. Math. Soc. **21**, 381–385.
[1969b] Spectral concentration and virtual poles. Amer. J. Math. **91**, 1106–1126.
[1972a] Nonexistence of asymptotic observables. Proc. Amer. Math. Soc. **35**, 175–176.
[1974a] Stationary scattering theory for time-dependent Hamiltonians. Math. Ann. **207**, 315 to 335.
[1974b] Puiseux series for resonances at an embedded eigenvalue. Pacific J. Math. **55**, 157–176.
[1975a] The Livsic matrix in perturbation theory. J. Math. Anal. and Appl. **50**, 415–437.
[1976a] Abstract stationary theory of multichannel scattering. J. Funct. Anal. **22**, 250–282.
[1979a] Scattering theory for Hamiltonians periodic in time. Indiana Univ. Math. J. **28**, 471–494.

HUNZIKER, W.
[1961a] Regularitätseigenschaften der Streuamplitude im Fall der Potentialstreuung. Helv. Phys. Acta **34**, 593–620.
[1963a] Potential scattering at high energies. Helv. Phys. Acta **36**, 838–856.
[1967a] Time-dependent scattering theory for singular potentials. Helv. Phys. Acta **40**, 1052–1062
[1968a] The $S$-matrix in classical mechanics. Comm. Math. Phys. **8**, 282–299.
[1980a] Schrödinger operators with electric or magnetic fields. *In*: OSTERWALDER, K. (ed.), Mathematical problems in theoretical physics. Lecture Notes in Physics, vol. 116, Springer, Berlin–Heidelberg–New York, pp. 25–44.

IKEBE, T.
[1960a] Eigenfunction expansions associated with the Schrödinger operators and their applications to scattering theory. Arch. Rational Mech. Anal. **5**, 1–34.
[1965a] On the phase shift formula for the scattering operator. Pacific J. Math. **15**, 511–523.
[1967a] On the eigenfunction expansion connected with the exterior problem for the Schrödinger equation. Japan. J. Math. **36**, 33–55.
[1967b] Scattering for the Schrödinger operator in an exterior domain. J. Math. Kyoto Univ. **7**, 93–112.
[1968a] Wave operators for $-\Delta$ in a domain with non-finite boundary. Publ. Res. Inst. Math. Sci. **4**, 413–418.
[1970a] Remarks on the orthogonality of eigenfunctions for the Schrödinger operator on $\mathbb{R}^n$. J. Fac. Sci. Univ. Tokyo, Sect. IA Math. **17**, 355–361.
[1975a] Spectral representation for Schrödinger operators with long-range potentials. J. Funct. Anal. **20**, 158–177.

IKEBE, T., and H. ISOZAKI
[1979a] Completeness of modified wave operators for long range potentials. Publ. Res. Inst. Math. Sci. **15**, 679–718.

IKEBE, T., and T. TAYOSHI
[1968a] Wave and scattering operators for second order elliptic operators in $\mathbb{R}^3$. Publ. Res. Inst. Math. Sci. **4**, 483–496.

INOUE, A.
]1972a] L'operateur d'onde et de diffusion pour on systéme evolutif $(d/dt) + iA(t)$. C. R. Acad. Sci. Paris, Ser. A-B **275**, 1323–1325.
[1973a] An example of temporally inhomogenous scattering. Proc. Japan Acad. **49**, 407–410.
[1974a] Wave and scattering operators for an evolving system $d/dt - iA(t)$. J. Math. Soc. Japan **26**, 608–624.

IORIO, R. J., and M. O'CAROLL
[1972a] Asymptotic completeness for multiparticle Schrödinger operators with weak potentials. Comm. Math. Phys. **27**, 137–145.

ISMAGILOV, R. S. (Исмагилов, Р. С.)
[1963a] О спектре матриц Тёплица. Докл. Акад. Наук СССР **149**, 769–772.

JAFAEV, D. R. (Яфаев, Д. Р.)
[1971a] О спектре возмущенного полигармонического оператора. Проблемы мат. физики ЛГУ, вып. **5**, 122–128.
[1978a] К теории многоканального рассеяния в паре пространств. Теорет. Мат. Физ. **37**, 48–57.
[1979a] On the breakdown of completeness of wave operators in potential scattering. Comm. Math. Phys. **65**, 167–179.

[1979b] On the proof of Enss of asymptotic completeness in potential scattering. Preprint LOMI E-2-79, Leningrad.
[1980a] О формуле следа в многоканальный модели Фридрихса. Труды Мат. инст. им. Стеклова 147, 194—201.
JAKUBOVSKIĬ, O. A. (Якубовский, О. А.)
[1967a] Об интегральных уравнениях теории рассеяния для $N$-частиц. Ядерная физика 5, 1312—1320.
JAUCH, J. M.
[1957a] On the relation between scattering phase and bound states. Helv. Phys. Acta 30, 143 to 156.
[1958a] Theory of the scattering operator. Helv. Phys. Acta 31, 127—158.
[1958b] Theory of the scattering operator II. Multichannel scattering. Helv. Phys. Acta 31, 661—684.
JAUCH, J. M., B. MISRA, and A. G. GIBSON
[1968a] On the asymptotic condition of scattering theory. Helv. Phys. Acta 41, 513—527.
JAUCH, J. M., and K. SINHA
[1972a] Scattering systems with finite total cross-sections. Helv. Phys. Acta 45, 580—586.
JAUCH, J. M., K. SINHA, and B. N. MISRA
[1972a] Time-delay in scattering processes. Helv. Phys. Acta 45, 398—426.
JAUCH, J. M., and I. I. ZINNES
[1959a] The asymptotic condition for simple scattering systems. Nuovo Cimento 11, 553—567.
JENSEN, A.
[1977a] Local distortion techniques, resonances, and poles of the $S$-matrix. J. Math. Anal. and Appl. 59, 505—513.
JENSEN, A., and T. KATO
[1978a] Asymptotic behavior of the scattering phase for exterior domains. Comm. Partial Differential Equations 3 (12), 1165—1195.
[1979a] Spectral properties of Schrödinger operators and time decay of the wave functions. Duke Math. J. 46, 583—611.
JORDAN, T. F.
[1962a] The quantum-mechanical scattering problem. J. Math. Phys. 3, 414—428.
[1964a] Lorentz invariant multichannel scattering formalism. J. Math. Phys. 5, 1345 to 1360.
JÖRGENS, K., and J. WEIDMANN
[1973a] Zur Existenz der Wellenoperatoren. Math. Z. 131, 141—151.
KAKITA, T., K. NISHIHARA, and C. TAMAMURA
[1979a] Time decay and scattering for some nonlinear wave equations. Publ. Res. Inst. Math. Sci. 15, 749—766.
KAKO, T.
[1976a] Spectral and scattering theory for the $J$-selfadjoint operators associated with the perturbed Klein-Gordon type equations. J. Fac. Sci. Univ. Tokyo, Sect. I A Math. 23, 199 to 221.
[1978a] Existence and equivalence of two types of long-range modified wave operators. J. Fac. Sci. Univ. Tokyo, Sect. I A Math. 25, 133—147.
KAKO, T., and K. YAJIMA
[1976a] Spectral and scattering theory for a class of non-selfadjoint operators. Scientific Papers of the College of Gen. Ed. Univ. Tokyo 26, 73—89.
KATO, T.
[1957a] On finite-dimensional perturbations of selfadjoint operators. J. Math. Soc. Japan 9, 239—249.
[1957b] Perturbation of continuous spectra by trace class operators. Proc. Japan Acad. 33, 260—264.
[1965a] Wave operators and unitary equivalence. Pacific J. Math. 15, 171—180.
[1966a] Wave operators and similarity for some non-selfadjoint operators. Math. Ann. 162, 258—279.
[1967a] Scattering theory with two Hilbert spaces. J. Funct. Anal. 1, 342—369.
[1968a] Smooth operators and commutators. Studia Math. 31, 535—546.

[1970a] Some results on potential scattering. *In*: Proc. Intern. Conference on Functional Anal. and Related Topics (Tokyo 1970), Univ. of Tokyo Press, Tokyo, pp. 206–215.
[1971a] Scattering theory. *In*: Studies in Mathematics 7, Math. Assoc. Amer., pp. 90–115.
[1977a] Stationary theory of scattering. *In*: The Schrödinger Equation, Acta Physica Austriaca, Suppl. 17, pp. 73–94.
[1977b] Two-space scattering theory, with applications to many-body problems. J. Fac. Sci. Univ. Tokyo, Sect. I A Math. 24, 503–514.
[1978a] Monotonicity theorems in scattering theory. Hadronic J. 1, 134–154.
[1979a] On the Cook-Kuroda criterion in scattering theory. Comm. Math. Phys. 67, 85–90.

Kato, T., and S. T. Kuroda
[1959a] A remark on the unitarity property of the scattering operator. Nuovo Cimento 14, 1102 to 1107.
[1970a] Theory of simple scattering and eigenfunction expansions. *In*: Browder, F. E. (ed.), Functional Analysis and Related Fields. Springer, Berlin—Heidelberg—New York, pp. 99–131.
[1971a] The abstract theory of scattering. Rocky Mountain J. Math. 1, 127–171.

Kato, Y., and N. Mugibayashi
[1963a] Regular perturbation and asymptotic limits of operators in quantum field theory. Prog. Theoret. Phys. 30, 103–133.

Kitada, H.
[1976a] A stationary approach to long-range scattering. Osaka J. Math. 13, 311–333.
[1977a] Scattering theory for Schrödinger operators with long range potentials. I. Abstract theory. J. Math. Soc. Japan 29, 665–691.
[1978a] Scattering theory for Schrödinger operators with long range potentials II. Spectral and scattering theory. J. Math. Soc. Japan 30, 603–632.

Koplienko, L. S. (Коплиенко, Л. С.)
[1971a] О спектре матрицы рассеяния. Изв. ВУЗ мат. №. 5, 54–61.
[1977a] Локальные признаки существования функции спектрального сдвига. Записки научных семинаров ЛОМИ 73, 102–117.
[1979a] Формула следа в абстрактной теории рассеяния для пары пространств. Проблемы мат. физики 9, 41–53.

Kobotjaev, E. L. (Коротяев, Е. Л.)
[1979a] Задача рассеяния для медленно убовающего потенциала периодически зависяющего от времени. Записки научных семинаров ЛОМИ 84, 114–116.

Košmanenko, V. D. (Кошманенко, В. Д.)
[1974a] О волновых операторах в теории рассеяния Хаага-Рюэлла. Докл. Акад. Наук СССР 218, 753–756.
[1975a] Теория рассеяния в терминах билинейных функционалов. Докл. Акад. Наук. СССР 224, 277–280.
[1976a] Обратная задача теории рассеяния в терминах билинейных функционалов. Докл. Акад. Наук СССР 229, 30–32.
[1979a] Теория рассеяния Хаага-Рюэлла как теория рассеяния в различных пространствах состояний. Теор. Мат. Физ. 38, 163–178.

Kreĭn, M. G. (Крейн, М. Г.)
[1953a] О формуле следов в теории возмущений. Мат. сборник 33, 597–626.
[1962a] Об определителях возмущения и формуле следов для унитарных и самосопряженных операторов. Докл. Акад. Наук СССР 144, 268–271.
[1964a] О некоторых новых исследованиях по теории возмущений самосопряженных операторов. *B*: сборник Первая летная матем. школа 1, Киев, 103–187.

Kupsch, J., and W. Sandhas
[1966a] Møller operators for scattering on singular potentials. Comm. Math. Phys. 2, 147–154.

Kuroda, S. T.
[1958a] On a theorem of Weyl-von Neumann. Proc. Japan Acad. 34, 11–15.
[1959a] On the existence and the unitarity property of the scattering operator. Nuovo Cimento 12, 431–454.
[1959b] Perturbation of continuous spectra by unbounded operators I. J. Math. Soc. Japan 11, 247–262.

[1960a] Perturbation of continuous spectra by unbounded operators II. J. Math. Soc. Japan 12, 243–257.
[1962a] On a paper of Green and Lanford. J. Math. Phys. 3, 933–935.
[1963a] Finite-dimensional perturbation and a representation of scattering operator. Pacific J. Math. 13, 1305–1318.
[1967a] An abstract stationary approach to perturbation of continuous spectra and scattering theory. J. d'Analyse Math. 20, 57–117.
[1967b] Perturbation of eigenfunction expansions. Proc. Nat. Acad. Sci. 57, 1213–1217.
[1969a] A stationary method of scattering and some applications. *In*: Proc. Intern. Conference on Functional Anal. and Related Topics, Tokyo, pp. 231–239.
[1970a] Some remarks on scattering for Schrödinger operators. J. Fac. Sci. Univ., Sect. I A Math. 17, 315–329.
[1973a] Scattering theory for differential operators, I. Operator theory. J. Math. Soc. Japan 25, 75–104.
[1973b] Scattering theory for differential operators, II. Selfadjoint elliptic operators. J. Math. Soc. Japan 25, 222–234.

Kuroda, S. T., and H. Morita
[1977a] An estimate for solutions of Schrödinger equations with time-dependent potentials and associated scattering theory. J. Fac. Sci. Univ. Tokyo, Sect. I A Math. 24, 459–475.

Ladyženskaja, O. A., and L. D. Faddeev (Ладыженская, О. А., и Л. Д. Фаддеев)
[1958a] К теории возмущений непрерывного спектра. Докл. Акад. Наук СССР 120, 1187 до 1190.

Lavine, R.
[1969a] Absolute continuity of Hamiltonian operators with repulsive potentials. Proc. Amer. Math. Soc. 22, 55–60.
[1970a] Scattering theory for long range potentials. J. Funct. Anal. 5, 368–382.
[1971a] Commutators and scattering theory I. Repulsive interactions. Comm. Math. Phys. 20, 301–323.
[1972a] Commutators and scattering theory II. A class of one body problems. Indiana Univ. Math. J. 21, 643–656.
[1973a] Completeness of the wave operators in the repulsive $N$-body problem. J. Math. Phys. 14, 376–379.
[1973b] Absolute continuity of positive spectrum for Schrödinger operators with long-range potentials. J. Funct. Anal. 12, 30–54.

Lax, P., and R. Phillips
[1962a] The wave equation in exterior domains. Bull. Amer. Math. Soc. 68, 47–49.
[1964a] Scattering theory. Bull. Amer. Math. Soc. 70, 130–142.
[1973a] Scattering theory for dissipative hyperbolic systems. J. Funct. Anal. 14, 172–235.
[1977a] Scattering of sound waves from an obstacle. Comm. Pure Appl. Math. 30, 195–233.
[1978a] Scattering theory for domains with non-smooth boundaries. Arch. Rational Mech. Anal. 68, 93–98.
[1978b] The time delay operator and a related trace formula. *In*: Topics in functional analysis. Advances in Math., Suppl. Studies 3, New York, pp. 197–215.

Levinson, N.
[1949a] On the uniqueness of the potential in a Schrödinger equation for a given asymptotic phase. Danske Vid. Selsk. Mat.-Fys. Medd. 25, No. 9.

Lifšic, I M. (Лифшиц, И. М.)
[1952a] Об одной задаче теории возмущений связанной с квантовой статистикой. Успехи Мат. Наук 7, 171–180.

Limic, N.
[1962a] Theory of scattering on highly singular potentials. Nuovo Cimento 26, 581–596.
[1965a] On the wave operators for scattering on strongly singular and not spherically symmetric potentials. Nuovo Cimento 36, 100–108.

Lin, S.-C.
[1969a] Wave operators and similarity for generators of semigroups in Banach spaces. Trans. Amer. Math. Soc. 139, 469–494.
[1969b] On smoothness of gentle perturbations. Bull. Amer. Math. Soc. 75, 445–449.

LIN, S.-C., and W. A. STRAUSS
[1978a] Decay and scattering of solutions of a nonlinear Schrödinger equation. J. Funct. Anal. 30, 245–263.
LUNDBERG, L.-E.
[1973a] Spectral and scattering theory for the Klein-Gordon equation. Comm. Math. Phys. 31, 243–257.
LUNDQUIST, E.
[1967a] On the existence of the scattering operator. Ark. Mat. 7, 145–157.
LYFORD, W.
[1975a] A two Hilbert space scattering theorem. Math. Ann. 217, 257–261.
[1978a] Spectral analysis of the Laplacian in distorted periodic wave guides. Math. Ann. 236, 255–284.
MAJDA, A.
[1976a] High frequency asymptotics for the scattering matrix and the inverse problem of acoustical scattering. Comm. Pure Appl. Math. 29, 261–291.
[1977a] A representation formula for the scattering operator and the inverse problem for arbitrary bodies. Comm. Pure Appl. Math. 30, 165–194.
MARTIN, A.
[1958a] On the validity of Levinson's theorem for non-local interactions. Nuovo Cimento 7, 607–627.
[1979a] Boundedness of total cross-sections in potential scattering. Comm. Math. Phys. 69, 89–98.
MARTIN, P.
[1972a] From the asymptotic condition to the cross-section. Helv. Phys. Acta 45, 794.
[1975a] Scattering theory with dissipative interactions and time delay. Nuovo Cimento B 30, 217–238.
MARTIN, P., and B. MISRA
[1973a] On trace class operators of scattering theory and the asymptotic behavior of scattering cross section at high energy. J. Math. Phys. 14, 997–1005.
[1974a] A refinement of the Kato-Rosenblum lemma and its application in scattering theory: Relativistic potential scattering, high energy behavior of scattering cross sections and the time delay problem. *In*: LA VITA/MARCHAND (eds.) [1], pp. 173–191.
MATVEEV, V. B. (Матвеев, В. Б.)
[1971a] Принцип инвариантности для обобщенных волновых операторов. Теор. Мат. Физ. 8, 49–54.
MATVEEV, V. B., and M. M. SKRIGANOV (Матвеев, В. Б., и М. М. Скриганов)
[1972a] Задача рассеяния для радиального уравнения Шредингера с медленно убывающим потенциалом. Теор. Мат. Физ. 10, 238–248.
MOCHIZUKI, K.
[1964a] On the perturbation of the continuous spectrum of the Dirac operator. Proc. Japan Acad. 40, 707–712.
[1967a] On the large perturbation by a class of non-selfadjoint operators. J. Math. Soc. Japan 19, 124–158.
[1968a] Eigenfunction expansions associated with the Schrödinger operators with a complex potential and the scattering theory. Publ. Res. Inst. Math. Sci. 4, 419–466.
[1969a] Spectral and scattering theory for symmetric hyperbolic systems in an exterior domain. Publ. Res. Inst. Math. Sci. 5, 219–258.
MONLEZUN, C.
[1974a] Temporally inhomogenous scattering theory. J. Math. Anal. and Appl. 47, 133 to 152.
MØLLER, C.
1945a] General properties of the characteristic matrix in the theory of elementary particles. Danske Vid. Selsk. Mat.-Fys. Medd. 23, no. 1.
MORAWETZ, C. S., and W. A. STRAUSS
[1972a] Decay and scattering of solutions of a nonlinear relativistic wave equation. Comm. Pure Appl. Math. 25, 1–31.
[1973a] On a nonlinear scattering operator. Comm. Pure Appl. Math. 26, 47–54.

Mourre, E.
[1973a] Quelques resultats dans la theorie algèbrique de la diffusion. Ann. Inst. H. Poincaré, Sect. A (NS) 18, 215–226.
[1977a] Applications de la méthode de Lavine un probleme à trois corpe. Ann. Inst. H. Poincaré, Sect. A (NS) 26, 219–262.
[1979a] Link between the geometrical and the spectral transformation approaches in scattering theory. Comm. Math. Phys. 68, 91–94.
Muldowney, J. S.
[1965a] On the analyticity of the $S$-matrix in potential scattering. Nuovo Cimento 35, 1138 to 1152.
Mulherin, D., and I. I. Zinnes
[1970a] Coulomb scattering. I. Single channel. J. Math. Phys. 11, 1402–1408.
Mullin, M.
[1976a] On the invariance of wave operators. J. Funct. Anal. 22, 242–249.
Murata, M.
[1979a] Rate of decay of local energy and wave operators for symmetric systems. J. Math. Soc. Japan 31, 451–480.
Naboko, S. N. (Набоко, С. Н.)
[1977a] Волновые операторы для несамосопряженных операторов и функциональная модель. Записки научных семинаров ЛОМИ 69, 129–135.
[1980a] Функциональная модель теории возмущений и ее приложения к теории рассеяния. Труды Мат. инст. им. Стеклова 147, 88–114.
Najman, B.
[1976a] Scattering for the Dirac operator. Glas. Mat. III. Ser. 11 (31), 63–80.
[1979a] Wave operators and similarity for some matrix operators with applications. Glas. Mat. III. Ser. 14 (34), 289–307.
Narnhofer, H.
[1975a] Continuity of the $S$-matrix. Nuovo Cimento B 30, 254–266.
Neidhardt, H.
[1978a] Zwei-Raum-Verallgemeinerung des Theorems von Rosenblum und Kato. Math. Nachr. 84, 195–211.
[1981a] Scattering theory for contraction semigroups. Report R-Math-05/81, Akad. Wiss. DDR, Berlin.
[1981b] On abstract linear evolution equations I. Math. Nachr. 103, 283–298.
[1981c] On abstract linear evolution equations II. Preprint P-Math-07/81, Akad. Wiss. DDR, Berlin.
Neĭmark, M. A., and S. V. Fomin (Неймарк, М. А., и С. В. Фомин)
[1955a] Непрерывные прямые суммы гилбертовых пространств и некоторые их применения. Успехи Мат. Наук 10, вып. 2 (64), 111–142.
Neumann, J. von
[1935a] Charakterisierung des Spektrums eines Integral-Operators. Actualites Sci. Indust. 229, 38–55.
Neumann, J. von, and B. O. Koopman
[1932a] Dynamical systems of continuous spectra. Proc. Nat. Acad. Sci. USA 18, 255 to 264.
Newton, R. G.
[1960a] Analytic properties of radial wave functions. J. Math. Phys. 1, 319–347.
[1977a] Noncentral potentials: The generalized Levinson theorem and the structure of the spectrum. J. Math. Phys. 18, 1348–1357.
[1977b] Nonlocal interactions: The generalized Levinson theorem and the structure of the spectrum. J. Math. Phys. 18, 1582–1588.
Obermann, P.
[1977a] Limitierungsverfahren in der Streutheorie. Dissertation, Akad. Wiss. DDR, Berlin.
[1982a] Adiabatic switching in abstract scattering theory Math. Nachr. 105, 7–25.
Obermann, P., and M. Wollenberg
[1978a] Abel wave operators, Part I: General theory. Math. Nachr. 85, 141–159.
[1978b] Abel wave operators. II. Wave operators for functions of operators. J. Funct. Anal. 30, 48–59.

O'Caroll, M.
[1972a] On the existence and completeness of wave operators in non-relativistic scattering. Nuovo Cimento B 8, 27–34.
Osborn, T. A., and D. Bolle
[1977a] An extended Levinson's theorem. J. Math. Phys. 18, 432–440.
Paravicini, G., V. Gorini, and E. C. G. Sudarshan
[1980a] Resonances, scattering theory and rigged Hilbert spaces. J. Math. Phys. 21, 2208–2226.
Pavlov, B. S. (Павлов, Б. С.)
[1979a] Парциальная матрица рассеяния, ее факторизация и аналитичность. Докл. Акад. Наук СССР 244, 291–295.
Pavlov, B. S., and L. D. Faddeev (Павлов, Б. С., и Л. Д. Фаддеев)
[1972a] Теория рассеяния и автоморфные функции. Записки научных семинаров ЛОМИ 27, 161–193.
Pearson, D. B.
[1971a] The wave and scattering operators for single-channel systems. Nuovo Cimento A 2, 853–880.
[1972a] Conditions for the existence of the generalized wave operators. J. Math. Phys. 13, 1490 to 1499.
[1974a] Time-dependent scattering theory for highly singular potentials. Helv. Phys. Acta 47, 249–264.
[1975a] An example in potential scattering illustrating the breakdown of asymptotic completeness. Comm. Math. Phys. 40, 125–146.
[1975b] General theory of potential scattering with absorption at local singularities. Helv. Phys. Acta 48, 639–653.
[1977a] The Dirac operator with highly singular potentials. Comm. Math. Phys. 57, 117–133.
[1978a] A generalization of the Birman trace theorem. J. Funct. Anal. 28, 182–186.
[1978b] Singular continuous measures in scattering theory. Comm. Math. Phys. 60, 13–36.
[1979a] Scattering theory for a class of oscillating potentials. Helv. Phys. Acta 52, 541–554.
Pearson, D. B., and D. H. Whould
[1973a] On the existence and asymptotic completeness of wave operators for spherically symmetric potentials. Nuovo Cimento A 14, 765–780.
Perry, P. A.
[1980a] Mellin transforms and scattering theory. I. Short range potentials. Duke Math. J. 47, 187–193.
Picard, R.
[1977a] Zur Existenz des Wellenoperators bei Anfangsrandwertproblemen vom Maxwell-Typ. Math. Z. 156, 175–185.
Povzner, A. Ja. (Повзнер, А. Я.)
[1955a] О разложении по собственным функциям уравнения Шредингера. Докл. Акад. Наук СССР 104, 360–363.
Prosser, R.
[1963a] Relativistic potential scattering. J. Math. Phys. 4, 1048–1054.
[1964a] Convergent perturbation expansions for certain wave operators. J. Math. Phys. 5, 708–713.
[1972a] On the asymptotic behavior of certain dynamical systems. J. Math. Phys. 13, 186–196.
Protopopescu, V.
[1976a] On the scattering matrix for the linear Boltzmann equation. Rev. Roumaine Phys. 21, 991–994.
Prugovecki, E.
[1969a] Rigorous derivation of generalized Lippman-Schwinger equations from time-dependent scattering theory. Nuovo Cimento B 2, 569–592.
[1971a] On time-dependent scattering theory for long-range interactions. Nuovo Cimento B 4, 105–123.
[1971b] Integral representations of the wave and transition operators in nonrelativistic scattering theory. Nuovo Cimento B 4, 124–134.
[1973a] Multichannel stationary scattering theory in two-Hilbert space formulation. J. Math. Phys. 14, 957–962 (Erratum: J. Math. Phys. 16 (1975), 442).

PRUGOVECKI, E., and A. TIP
[1974a] Scattering theory in a time-dependent external field. I. General theory. J. Phys. A 7, 572–585.
[1974b] Scattering theory in a time-dependent external field. II. Appplications. J. Phys. A 7, 586–596.
PRUGOVECKI, E., and J. ZORBAS
[1973a] Modified Lippman-Schwinger equations for two-body scattering theory with long-range interactions. J. Math. Phys. 14, 1398–1409 (Erratum; J. Math. Phys. 15 (1974), 268).
PUTNAM, C. R.
[1954a] On the continuous spectra of singular boundary value problems. Canad. J. Math. 6, 420–426.
[1957a] Continuous spectra and unitary equivalence. Pacific J. Math. 7, 993–995.
[1960a] A note on Toeplitz matrices and unitary equivalence. Boll. Un. Mat. Ital. 15, 6–9.
[1962a] Absolute continuity of certain unitary and half-scattering operators. Proc. Amer. Math. Soc. 13, 844–846.
[1963a] On the spectra of unitary half-scattering operators. Quart. Appl. Math. 20, 85–88.
[1964a] Commutators, absolutely continuous spectra, and singular integral operators. Amer. J. Math. 86, 310–316.
RAUCH, J.
[1980a] Perturbation theory for eigenvalues and resonances of Schrödinger Hamiltonians. J. Funct. Anal. 35, 304–315.
REED, M.
[1976a] Construction of the scattering operator for abstract nonlinear wave equations. Indiana Univ. Math. J. 25, 1017–1027.
REED, M., and B. SIMON
[1977a] The scattering of classical waves from inhomogenous media. Math. Z. 155, 163–180.
REHBERG, J.
[1979a] Streutheorie unitärer Darstellungen des $\mathbb{R}^n$, Dissertation. Akad. Wiss. DDR, Berlin.
REJTO, P. A.
[1963a] On gentle perturbations I. Comm. Pure Appl. Math. 16, 279–303.
[1964a] On gentle perturbations II. Comm. Pure Appl. Math. 17, 257–292.
[1967a] On partly gentle perturbations I. J. Math. Anal. and Appl. 17, 435–462.
[1967b] On partly gentle perturbations II. J. Math. Anal. and Appl. 20, 145–187.
[1969a] On partly gentle perturbations III. J. Math. Anal. and Appl. 27, 21–67.
[1971a] Some potential perturbations of the Laplacian. Helv. Phys. Acta 44, 708–736.
[1972a] On a limiting case of a theorem of Kato and Kuroda. J. Math. Anal. and Appl. 39, 541–557.
RIAHI, F.
[1969a] On the analyticity properties of the $N$-body scattering amplitude in non-relativistic quantum mechanics. Helv. Phys. Acta 42, 299–329.
ROBINSON, D.
[1974a] Scattering theory with singular potentials, I. The two body problem. Ann. Inst. H. Poincaré, Sect. A (NS) 21, 185–216.
ROSENBLUM, M.
[1957a] Perturbation of the continuous spectrum and unitary equivalence. Pacific J. Math. 7, 997–1010.
[1960a] The absolute continuity of Toeplitz's matrices. Pacific J. Math. 10, 987–996.
ROTFEL'D, S. JU. (Ротфельд, С. Ю.)
[1972a] О спектральных свойствах матрицы рассеяния. Изв. ВУЗ мат. №. 3, 52–60.
[1974a] О спектральных свойствах матрицы рассеяния II. Проблемы мат. физики ЛГУ вып. 7, 135–149.
RUELLE, D.
[1962a] On the asymptotic condition in quantum field theory. Helv. Phys. Acta 35, 147–163.
[1969a] A remark on bound states in potential scattering theory. Nuovo Cimento A 61, 655–662.
RUIJSENAARS, S.
[1980a] A positive energy dynamic and scattering theory for directly interacting relativistic particles. Ann. Physics 126, 399–449.

RUIJSENAARS, S., and P. BONGAARTS
[1977a] Scattering theory for one-dimensional step potentials. Ann. Inst. H. Poincaré, Sect. A (NS) 26, 1–17.
SAENZ, A. W., and W. W. ZACHARY
[1976a] On the equality of $S$-operators corresponding to unitarily equivalent Hamiltonians in single channel scattering. J. Math. Phys. 17, 409–418 (Erratum: J. Math. Phys. 17 (1976), 1616).
SAHNOVIČ, L. A. (Сахнович, Л. А.)
[1968a] Операторы подобные унитарным, с абсолютно непрерывным спектром. Функц. анализ и приложен. 2, № 1, 51–63.
[1968b] Диссипативные операторы с абсолютно непрерывным спектром. Труды. Моск. Мат. Общ. 19, 211–270.
[1970a] Обобщенные волновые операторы и регуляризация ряда теории возмущений. Теорет. Мат. Физ. 2, 80–86.
[1970b] Обобщенные волновые операторы. Мат. сборник 81, 209–227.
[1971a] Принцип инвариантности для обобщенных волновых операторов. Функц. анализ и приложен. 5, 61–68.
SAITO, Y.
[1972a] Spectral and scattering theory for second-order differential operators with operator-valued coefficients. Osaka J. Math. 9, 463–498.
[1974a] The principle of limiting absorption for the nonselfadjoint Schrödinger operator in $\mathbb{R}^N$ ($N \neq 2$). Publ. Res. Inst. Math. Sci. 9, 397–428.
SANDHAS, W.
[1966a] Definition and existence of multichannel scattering states. Comm. Math. Phys. 3, 358 to 374.
SCADRON, M., S. WEINBERG, and J. WRIGHT
[1964a] Functional analysis and scattering theory. Phys. Rev. B 135, 202–207.
SCHECHTER, M.
[1974a] Spectral and scattering theory for elliptic operators of arbitrary orders. Comment. Math. Helv. 49, 84–113.
[1974b] A unified approach to scattering. J. Math. Pures Appl. (9) 57, 373–396.
[1975a] Scattering theory for second order elliptic operators. Ann. Mat. Pura Appl. 55, 313–331.
[1976a] Scattering theory for pseudodifferential operators. Quart. J. Math. Oxford, Ser. (2) 27, 111–121.
[1976b] The Klein-Gordon equation and scattering theory. Ann. Physics 101, 601–609.
[1976c] Eigenfunction expansions for the Schrödinger equation. Ark. Mat. 14, 277–287.
[1977a] A new criterion for scattering theory. Duke Math. J. 44, 863–872.
[1977b] Wave operators for oscillating potentials. Lett. Math. Phys. 2, 127–132.
[1978a] Another look at scattering theory. Intern. J. Theoret. Phys. 17, 33–41.
[1979a] Scattering in two Hilbert spaces. J. London Math. Soc., II. Ser. 19, 175–186.
[1979b] The invariance principle. Comment. Math. Helv. 54, 111–125.
[1979c] Two space scattering theory using the subspace of continuity. Lett. Math. Phys. 3, 521–531.
[1979d] Completeness of wave operators in two-Hilbert spaces. Ann. Inst. H. Poincaré, Sect. A (NS) 30, 109–127.
[1980a] Wave operators for pairs of spaces and the Klein-Gordon equation. Aequationes Math. 20, no. 1, 38–50.
SCHMIDT, G.
[1975a] On scattering by time-dependent perturbations. Indiana Univ. Math. J. 24, 925–935.
[1978a] Scattering theory for the Schrödinger equation with a potential almost periodic in time. J. d'Analyse Math. 33, 121–145.
SCHULENBERGER, J. R.
[1972a] A local compactness theorem for wave propagation problems of classical physics. Indiana Univ. Math. J. 22, 429–432.
SCHULENBERGER, J. R., and C. H. WILCOX
[1971a] Completeness of the wave operators for perturbations of uniformly propagative systems. J. Funct. Anal. 7, 447–474.

[1972a] Eigenfunction expansions and scattering theory for wave propagation problems of classical physics. Arch. Rational Mech. Anal. 46, 280–320.
SCHWARTZ, J.
[1960a] Some non-selfadjoint operators. Comm. Pure Appl. Math. 13, 609–639.
SEGAL, I.
[1968a] Dispersion for nonlinear relativistic wave equation. Ann. Sci. Ecole Norm. Sup. 1, 459–497.
SEILER, E., and B. SIMON
[1975a] Bounds in the Yukawa$_2$ quantum field theory: Upper bound on the pressure, Hamiltonian bound and linear lower bound. Comm. Math. Phys. 45, 99–114.
SEMENOV, YU.
[1977a] Wave operators for the Schrödinger equation with strongly singular short range potentials. Lett. Math. Phys. 1, 457–461.
SEMON, M. D., and J. R. TAYLOR
[1975a] Scattering by potentials with Coulomb tails. Nuovo Cimento A 26, 48–58.
[1976a] Cross sections for screened potentials. J. Math. Phys. 17, 1366–1370.
SHARMA, C. S., and E. O. OGUNSULIRE
[1978a] Axiomatic quantum theory and resonances. J. Phys. A 11, 2159–2177.
SHENK, N.
[1966a] Eigenfunction expansions and scattering theory for the wave equation in an exterior region. Arch. Rational Mech. Anal. 21, 120–150.
[1968a] The invariance of wave operators associated with perturbations of $-\Delta$. J. Math. Mech. 17, 1005–1022.
SHENK, N., and D. THOE
[1971a] Eigenfunction expansions and scattering theory for perturbations of $-\Delta$. Rocky Mountain J. Math. 1, 89–125.
[1972a] Resonant states and poles of the scattering matrix. J. Math. Anal. and Appl. 37, 467–491.
SIGAL, I. M.
[1978a] On quantum mechanics of many-body systems with dilation analytic potentials. Bull. Amer. Math. Soc. 84, 152–154.
[1978b] Mathematical foundations of quantum scattering theory for multiparticle systems. Mem. Amer. Math. Soc. 16, no. 209.
SIMON, B.
[1971a] Wave operators for classical particle scattering. Comm. Math. Phys. 23, 37–48.
[1973a] Resonances in n-body quantum systems with dilatation analytic potentials and the foundations of time-dependent perturbation theory. Ann. of Math. 97, 247–274.
[1974a] Existence of the scattering matrix for the linearized Boltzmann equation. Comm. Math. Phys. 41, 99–108.
[1977a] Scattering theory and quadratic forms: On a theorem of Schechter. Comm. Math. Phys. 53, 151–153.
[1977b] Geometric methods in multiparticle quantum systems. Comm. Math. Phys. 55, 259–274.
[1978a] $N$-body scattering in the two cluster region. Comm. Math. Phys. 58, 205–210.
[1979a] Phase space analysis of simple scattering systems: Extensions of some work of Enss. Duke Math. J. 46, 119–168.
SINAǏ, JA. G. (Синай, Я. Г.)
[1961a] Динамические системы со счетнократным лебеговым спектром. Изв. Акад. Наук СССР 25, 899–924.
SINHA, K. B.
[1977a] On the absolutely and singularly continuous subspaces in scattering theory. Ann. Inst. H. Poincaré, Sect. A (NS) 26, 263–277.
[1978a] Relative scattering and time delay in presence of a uniform electric field. Rep. Math. Phys. 14, 65–73.
ŠIROKOV, JU. M. (Широков, Ю. М.)
[1979a] Единый формализм для квантовой и классической теорий рассеяния. Теорет. Мат. Физ. 38, 313–320.
SOHR, H.
[1978a] Über die Existenz von Wellenoperatoren für zeitabhängige Störungen. Monatsh. Math. 86, 63–81.

SOKOLOV, S. N.
[1979a] Classical analogoues of the Moeller operators, of the Pearson example and of the Birman-Kato invariance principle. Nuovo Cimento A 52, 1–22.
STANKEVIČ, I. V. (Станкевич, И. В.)
[1962a] К теории возмущения непрерывного спектра. Докл. Акад. Наук СССР 144, 279–282.
[1965a] Асимптотическое поведение при $t \to \infty$ решения нестационарного уравнения Шредингера с несамосопряженным гамильтонианом. Докл. Акад. Наук СССР 160, 1271 до 1274.
STRAUSS, W.
[1963a] Scattering for hyperbolic equations. Trans. Amer. Math. Soc. 108, 13–37.
[1974a] Nonlinear scattering theory. *In*: LA VITA/MARCHAND (eds.) [1], pp. 53–78.
[1979a] The existence of the scattering operator for moving obstacles. J. Funct. Anal. 31, 255–262.
SUDARSHAN, E. C. G., C. B. CHIU, and V. GORINI
[1978a] Decaying states as complex energy eigenvectors in generalized quantum mechanics. Phys. Rev. D 18, 2914–2929.
SUSLOV, A. V. (Суслов, А. В.)
[1979a] Об одной теорема Исмагилова. Проблемы мат. физики ЛГУ, вып. 9, 142–145.
SUZUKI, T.
[1973a] The limiting absorption principle for a certain non-selfadjoint operator. J. Fac. Sci. Univ. Tokyo, Sect. I A Math. 20, 401–412.
TAYLOR, J. R.
[1974a] A new rigorous approach to Coulomb scattering. Nuovo Cimento B 23, 313 to 334.
THOE, D.
[1967a] Eigenfunctions expansions associated with Schrödinger operators in $R_n$, $n \geqq 4$. Arch. Rational Mech. Anal. 26, 335–356.
THOMAS, L. E.
[1973a] Time dependent approach to scattering from impurities in a crystal. Comm. Math. Phys. 33, 335–343.
[1974a] On the algebraic theory of scattering. J. Funct. Anal. 15, 364–377.
[1975a] Asymptotic completeness in two- and three-particle quantum mechanical scattering. Ann. Physics 90, 127–165.
THOMPSON, M.
[1972a] Eigenfunction expansions and the associated scattering theory for potential perturbations of the Dirac equation. Quart. J. Math. Oxford, Ser. (2) 23, 17–55.
TROIANIELLO, G. M.
[1974a] Scattering theory for Schrödinger operators with $L^\infty$ potentials and distorted Bloch waves. J. Math. Phys. 15, 2048–2052.
UHLMANN, A.
[1962a] Über die Definition der Quantenfelder nach Wightman und Haag. Wiss. Z. Karl-Marx-Univ. Leipzig 11, 213–217.
[1968a] On positive functionals on algebras of test functions. Comm. Math. Phys. 7, 132–159.
USHIJIMA, T.
[1966a] Spectral theory of the perturbed homogenous elliptic operator with real constant coefficients. Sci. Papers College Gen. Ed. Univ. Tokyo 16, 27–42.
VAĬNBERG, B. R. (Вайнберг, Б. Р.)
[1966a] Принципы излучения, предельного поглощения и предельной амплитуды в общей теории уравнений с частным производными. Успехи Мат. Наук 21, 115–194.
VESELIC, K.
[1970a] A spectral theory for the Klein-Gordon equation with an external electrostatic field. Nuclear Phys. A 147, 215–224.
VESELIC, K., and J. WEIDMANN
[1973a] Existenz der Wellenoperatoren für eine allgemeine Klasse von Operatoren. Math. Z. 134, 255–274.
[1974a] Asymptotic estimates of wave functions and the existence of wave operators. J. Funct. Anal. 17, 61–77.
[1977a] Potential scattering in a homogenous electrostatic field. Math. Z. 156, 93–104.

VOIGT, J.
[1977a] On the existence of the scattering operator for the linear Boltzmann equation. J. Math. Anal. and Appl. 58, 541—558.
WEDER, R.
[1978a] Scattering theory for the Klein-Gordon equation. J. Funct. Anal. 27, 100—117.
WEIDMANN, J.
[1967a] Zur Spektraltheorie von Sturm-Liouville Operatoren. Math. Z. 98, 268—302.
[1977a] Scattering theory for partial differential operators. *In*: Differential Equations. Proc. Intern. Conf. Uppsala, pp. 189—198.
WICHNOWSKI, B. J.
[1976a] Temporally inhomogenous nonlinear scattering theory. J. Math. Anal. and Appl. 56, 570—585.
WILCOX, C. H.
[1966a] Wave operators and asymptotic solutions of wave propagation problems of classical physics. Arch. Rational Mech. Anal. 22, 37—78.
[1973a] Scattering states and wave operators in the abstract theory of scattering. J. Funct. Anal. 12, 257—274.
WINTER, C. v.
[1965a] Theory of finite systems of particles. II. Scattering theory. Mat. Fys. Sker. Dan. Vid. Selsk. 2, no. 10.
[1979a] A weighted Hilbert space for multiparticle scattering. J. Math. Anal. and Appl. 68, 35—67.
WOLLENBERG, M.
[1975a] The invariance principle for wave operators. Pacific J. Math. 59, 303—315.
[1977a] Levinsontheorem, instabile Eigenwerte und Streuquerschnittmaxima. Math. Nachr. 78, 223—230.
[1977b] On the inverse problem in the abstract theory of scattering, ZIMM-Preprint, Akad. Wiss. DDR, Berlin.
[1977c] Levinsontheorem für Schrödingeroperatoren. Math. Nachr. 78, 369—383.
[1977d] On the invariance principle of scattering theory with two Hilbert spaces. Math. Nachr. 79, 265—269.
[1978a] On the inverse problem in the abstract scattering theory. *In*: BAUMGÄRTEL/LASSNER/PIETSCH/UHLMANN (eds.) [1], pp. 372—376.
[1979a] Wave algebras and scattering theory, I. ZIMM-Report R-07/79, Akad. Wiss. DDR, Berlin.
[1979b] Wave algebras and scattering theory, II. ZIMM-Report R-10/79, Akad. Wiss. DDR, Berlin.
WOLLENBERG, M., and V. D. KOŠMANENKO (Волленберг, М., и В. Д. Кошманенко)
[1979a] О структуры волновых операторов. Докл. Акад. Наук СССР 244, 265—269.
YAJIMA, K.
[1975a] Eigenfunction expansions associated with uniformly propagative systems and their applications to scattering theory. J. Fac. Sci. Univ. Tokyo, Sect. I A Math. 22, 121—151.
[1977a] Scattering theory for Schrödinger equations with potentials periodic in time. J. Math. Soc. Japan 29, 729—743.
[1978a] An abstract stationary approach to the three body scattering. J. Fac. Sci. Univ. Tokyo, Sect. I A Math. 25, 109—123.
[1979a] Spectral and scattering theory for Schrödinger operators with Stark-effect. J. Fac. Sci. Univ. Tokyo, Sect. I A Math. 26, 377—390.
YAMADA, O.
[1973a] On the principle of limiting absorption for the Dirac operator. Publ. Res. Inst. Math. Sci. 8, 557—577.
[1976a] Eigenfunction expansions and scattering theory for Dirac operators. Publ. Res. Inst. Math. Sci. 11, 651—689.
ZACHARY, W. W.
[1973a] Comment on a paper of Amrein, Martin, and Misra. Helv. Phys. Acta 46, 327—330.
[1976a] Wave operators for multichannel scattering by longrange potentials. J. Math. Phys. 17, 1056—1063.

ZORBAS, J.
[1974a] A space cut-off approach to scattering involving Coulomb-like potentials. J. Phys. A **7**, 1557–1567.
[1976a] Stationary scattering for $N$-body systems involving Coulomb potentials. J. Math. Phys. **17**, 498–502.
[1976b] The Gell-Mann/Goldberger formula for long-range potential scattering. Rep. Math. Phys. **9**, 309–320.

*Literature added in proof* (At the end of bibliographic items there is occasionally a hint at the corresponding part of the book to which it belongs.)

ABRAHAM, P. B., B. DE FACIO, and H. E. MOSES
[1982a] An explicit example of a local and a nonlocal potential whose Hamiltonians are unitarily equivalent and whose scattering operators are identical. Stud. Appl. Math. **66**, 45–57. (11.2)
ALBEVERIO, S., and R. HÖEGH-KROHN
[1982/83a] Perturbation of resonances in quantum mechanisc. *To appear in*: J. Math. Anal. and Appl. (18.3)
ALSHOLM, P.
[1980a] Inverse scattering theory for perturbation of rank one. Duke Math. J. **47**, 391–398. (11.2)
ANDREA, S. A.
[1980a] Scattering theory and pseudodifferential operators. Acta Cient. Venezolana **31**, No. 2, 79–97.
AVRON, J., and B. SIMON
[1981a] Transient and recurrent spectrum. Refinement of the absolutely continuous spectrum. J. Funct. Anal. **43**, 1–31. (3.3.4 and 3.5)
BALSLEV, E.
[1980b] Analytic scattering theory for many body systems below the smallest three-body threshold. Comm. Math. Phys. **77**, 173–210. (10.1.3)
[1981a] Comment on "Analytic scattering theory for many body systems below the smallest three-body threshold". Comm. Math. Phys. **82**, 257–260. (10.1.3)
BIRMAN, M. Š., and D. R. JAFAEV (Бирман, М. Ш., и Д. Р. Яфаев)
[1981a] Асимптотика спектра матрицы рассеяния. Записки научных семинаров ЛОМИ **110**, 3–29. (19.1)
BOHM, A.
[1981a] Resonance poles and Gamow vectors in the rigged $H$-space formulation of quantum mechanics. J. Math. Phys. **22**, 2813–2823. (18.3)
BOLLE, D.
[1981a] On classical time delay. Acta Physica Austriaca Suppl. XXIII, 587–598. (Notes and Remarks to Part V)
BOURGEOIS, B.
[1982a] Quantum mechanical scattering theory for potentials of the form $V(r) = (\sin r)/r^\beta$, $1/3 < \beta \leqq 1/2$. J. Math. Phys. **23**, 790–797.
COLGEN, R.
[1981a] Stability theory for absolutely continuous parts of commutative $m$-tuples of selfadjoint operators in complex Hilbert spaces. J. Funct. Anal. **41**, 307–314. (Notes and Remarks to Part III, (5))
COMBES, J. M., and R. WEDER
[1981a] Spectral theory and unbounded obstacle scattering. *In*: Spectral theory of differential operators, North-Holland Math. Studies, vol. 55, North-Holland, Amsterdam, pp. 93–98.
CONSTANTIN, P.
[1981a] Scattering for Schrödinger operators in a class of domains with non-compact boundaries. J. Funct. Anal. **44**, 87–119.
COTTA-RAMUSINI, P., W. KRÜGER, and R. SCHRADER
[1979a] Quantum scattering by external metrics and Yang-Mills potentials. Ann. Inst. H. Poincaré, Sect. A (NS) **31**, 43–71.

DAVIES, E. B.
[1980c] Energy dependence of the scattering operator. Adv. in Appl. Math. 1, 300–323. (Part V)
[1980d] Nonunitary scattering and capture. II. Quantum dynamical semigroup theory. Ann. Inst. H. Poincaré, Sect. A (NS) 32, 361–375. (Notes and Remarks to Part III, (3))
DEMUTH, M.
[1981a] On time-dependent scattering theory in Kreĭn spaces. Math. Nachr. 102, 107–125.
[1982/83a] Scattering by potentials infinite on unbounded regions. *To appear in:* Math. Nachr.
[1982/83b] On necessary and sufficient conditions for the Cook-criteria. *To appear in:* Math. Nachr.
[1982/83c] On resonances in mathematical scattering theory. Banach Center Publications VIII, 337–340. (18.3)
DEVINATZ, A.
[1980a] The existence of wave operators for oscillating potentials. J. Math. Phys. 21, 2406–2411.
DOLLARD, J. D.
[1977a] On the definition of scattering subspaces in nonrelativistic quantum mechanics. J. Math. Phys. 18, 229–232. (9.2.1)
ECKARDT, K.-J.
[1980a] Scattering theory for the Klein-Gordon equation. Funct. Approx. Comment. Math. 8, 13–42.
EIDUS, D.
[1982a] Completeness properties of scattering problem solutions. Comm. Partial Differential Equations 7, 55–75.
ENSS, V.
[1981a] Geometric methods in scattering theory. Acta Physica Austriaca, Suppl. XXIII, 29–63. (15.2.2 and 15.2.4)
[1982/83a] Completeness of three body quantum scattering. *In:* Proc. of the Bielefeld Encounters in Physics and Mathematics, Bielefeld Nov. 30–Dec. 4, 1981. (10.1.3)
GAZDY, B.
[1980a] Theory of resonances in indefinite metric spaces. Phys. Letters A 78, 309–311. (18.3)
GINIBRE, J.
[1981a] La méthode dépendant du temps dans le probléme de la complétude asymptotique. Publ. IRMA (Strasbourg) RCP 25, 29, 1–66. (15.2.2)
GRABOWSKI, M.
[1980a] The asymptotic scattering condition for certain dynamical semigroups. Bull. Acad. pol. sci. Ser. sci. phys. et astron. 28, No. 3–4, 133–139. (Notes and Remarks to Part III, (5))
HORWITZ, L. P., and A. SOFFER
[1980a] On the existence of the wave operator in relativistic quantum scattering theory. Helv. Phys. Acta 53, 112–118.
HOWLAND, J.
[1980a] Born series and scattering by time-dependent potentials. Rocky Mountain J. Math. 10, 521–531. (Notes and Remarks to Part III, (1))
HOWLAND, J., and T. KATO
[1981a] On a theorem of Ismagilov. J. Funct. Anal. 41, 37–39. (16.3.1)
IKEBE, T., and H. ISOZUKI
[1982a] A stationary approach to the existence and completeness of long-range wave operators. Integral Equations Operator Theory 5, 18–49. (14.1 and 10.2)
JAFAEV, D. R. (Яфаев, Д. Р.)
[1980b] Волновые операторы для уравнения Шредингера. Теор. Мат. Физ. 45, 224–234.
[1980c] Об асимптотическам поведении решений нестационарного уравнения Шредингера. Мат. Сборник 111 (153), 169–188. (Notes and Remarks to Part III, (1))
JENSEN, A.
[1980a] Resonances in an abstract analytic scattering theory. Ann. Inst. H. Poincaré, Sect. A (NS) 33, 209–223. (18.3)
[1981a] Time delay in potential scattering theory, Some geometrical results. Comm. Math. Phys. 82, 435–456. (Notes and Remarks to Part V)
KOROTJAEV, E. L. (Коротяев, Е. Л.)
[1980a] Теория рассеяния для трех частиц с парными потенциалами периодическими по времени. Докл. Акад. Наук СССР 255, 836–839. (10.1.3)

KREĬN, M. G., and V. A. JAVRJAN (Крейн, М. Г., и В. А. Яврян)
[1981a] О функциях спектрального свига, восникающих при возмущениях положительного оператора. J. Operator Theory 6, 155–191. (11.2 and 19.1.4)
LAX, P. D., and R. S. PHILLIPS
[1981a] The translation representation theorem. Integral Equations Operator Theory 4, 416 to 421. (12.1.3)
MARTIN, A.
[1980a] Boundedness of total cross sections in potential scattering. Comm. Math. Phys. 73, 79–81, (19.1 and 19.2)
[1981a] Finiteness of total cross sections. Acta Physica Austriaca, Suppl. XXIII, 209–233. (19.1 and 19.2)
MARTIN, P.
[1981a] Time delay of quantum scattering processes. Acta Physica Austriaca, Suppl. XXIII, 157–208. (Notes and Remarks to Part V)
MERKURIEV, S. P.
[1981a] Three-body Coulomb scattering. Acta Physica Austriaca, Suppl. XXIII, 65–110. (10.1.3)
MOURRE, E.
[1980a] Absence of singular continuous spectrum for certain selfadjoint operators. Comm. Math. Phys. 78, 391–408. (3.3)
MURATA, M.
[1981a] Characterizations of the range of wave operators for symmetric systems and an application. Tokyo J. Math. 4, 113–132.
NAJMAN, B.
[1980a] Trace class perturbations and scattering for the equations of Klein-Gordon type. Glas. Mat. III Ser. 15 (35), 79–86.
NARNHOFER, H.
[1979a] Scattering theory for quasi-free time automorphisms of $C^*$-algebras and von Neumann algebras. Rep. Math. Phys. 16, 1–8. (Notes and Remarks to Part III, (5))
NIELSEN, O. A.
[1980a] Direct integral theory. Lecture Notes in Pure and Applied Mathematics, vol. 61, Marcel Dekker Inc., New York. (Chapter 4)
PERELMUTER, M. A., and YU. A. SEMENOV
[1981a] On decoupling of finite singularities in the scattering theory for the Schrödinger operator with a magnetic field. J. Math. Phys. 22, 521–533.
PERRY, P. A.
[1981a] Propagation of states in dilation analytic potentials and asymptotic completeness. Comm. Math. Phys. 81, 243–259. (15.2.2 and 15.2.4)
PERRY, P. A., I. M. SIGAL, and B. SIMON
[1980a] Absence of singular continuous spectrum in $N$-body quantum systems. Bull. Amer. Math. Soc. (N.S.) 3, 1019–1023. (3.3)
PITT, L. D.
[1982a] An example of stability of singular spectrum under smooth perturbations. Integral Equations Operator Theory 5, 114–126. (3.3)
PHILIPPS, R. S.
[1982a] Scattering theory for the wave equation with a short range perturbation. *To appear in*: Indiana Univ. Math. J. (10.3 and 12)
REHBERG, J.
[1982/83a] Scattering theory for unitary representations of $\mathbb{R}^n$ over cones. *To appear in*: Math. Nachr. (Notes and Remarks to Part III, (5))
ROBINSON, D.
[1980a] Propagation properties in scattering theory. J. Austral. Math. Soc. (Series B) 21, 474 to 485. (9.2.1)
SAMOĬLENKO, V. G. (Самойленко, В. Г.)
[1979a] О волновых операторах для оператора типа Шредингера в пространстве функций бесконечного числа переменных. *В*: сборник Операторы математической физики и бесконечномерный анализ, Институт Математики А. Н. УССР Киев, стр. 103–109.

**Schechter, M.**
[1981 a] Spectral and scattering theory for propagative systems. *In*: Spectral theory of differential operators (Birmingham/Ala., 1981), pp. 361–368. (9.7.2)
**Schrader, R.**
[1980 a] High energy behavior for non-relativistic scattering by external metrics and Yang-Mills potentials. Z. Physik C 4, 27–36.
**Sigal, I. M.**
[1982 a] Mathematical theory of single channel systems. Analyticity of scattering matrix. Trans. Amer. Math. Soc. 270, 409–437. (18.3)
**Strauss, W.**
[1978 a] Everywhere defined wave operators. *In*: Nonlinear evolution equations. Proc. Symp., Madison/Wis. 1977, pp. 85–102. (Notes and Remarks to Part III, (3))
[1981 a] Nonlinear scattering theory at low energy. J. Funct. Anal. 41, 110–133. (Notes and Remarks to Part III, (3))
**Tayoshi, T.**
[1980 a] Wave operators in the scattering problem related to nonfinite obstacles. Publ. Res. Inst. Math. Sci. 16, 627–634.
**Thirring, W.**
[1981 a] Classical scattering theory. Acta Physica Austriaca, Suppl. XXIII, 3–28. (Notes and Remarks to Part III, (2))
**Umeda, T.**
[1981 a] Scattering theory for pseudo-differential operators. The existence of wave operators. Osaka J. Math. 18, 361–392.
**Voigt, J.**
[1977 b] Perturbation theory for commutative *m*-tuples of selfadjoint operators. J. Funct. Anal. 25, 317–334. (Notes and Remarks to Part III, (5))
**Yajima, K.**
[1980 a] A multichannel scattering theory for some time dependent Hamiltonians, charge transfer problem. Comm. Math. Phys. 75, 153–178. (10.1)

# Subject Index

Bei Fragen zur Produktsicherheit wenden Sie sich bitte an:
If you have any questions regarding product safety,
please contact:

Walter de Gruyter GmbH
Genthiner Straße 13
10785 Berlin
productsafety@degruyterbrill.com